T0296851

ANCESTRAL DNA, HUMAN ORIGINS, AND MIGRATIONS

ANCESTRAL DNA, HUMAN ORIGINS, AND MIGRATIONS

Rene J. Herrera

Ralph Garcia-Bertrand

Academic Press is an imprint of Elsevier
125 London Wall, London EC2Y 5AS, United Kingdom
525 B Street, Suite 1650, San Diego, CA 92101, United States
50 Hampshire Street, 5th Floor, Cambridge, MA 02139, United States
The Boulevard, Langford Lane, Kidlington, Oxford OX5 1GB, United Kingdom

Notices
Knowledge and best practice in this field are constantly changing. As new research and experience broaden our understanding, changes in research methods, professional practices, or medical treatment may become necessary.

Practitioners and researchers must always rely on their own experience and knowledge in evaluating and using any information, methods, compounds, or experiments described herein. In using such information or methods they should be mindful of their own safety and the safety of others, including parties for whom they have a professional responsibility.

To the fullest extent of the law, neither the Publisher nor the authors, contributors, or editors, assume any liability for any injury and/or damage to persons or property as a matter of products liability, negligence or otherwise, or from any use or operation of any methods, products, instructions, or ideas contained in the material herein.

Library of Congress Cataloging-in-Publication Data
A catalog record for this book is available from the Library of Congress

British Library Cataloguing-in-Publication Data
A catalogue record for this book is available from the British Library

ISBN: 978-0-12-804124-6

For information on all Academic Press publications visit our website at https://www.elsevier.com/books-and-journals

Publisher: John Fedor
Acquisition Editor: Peter B. Linsley
Editorial Project Manager: Timothy Bennett
Production Project Manager: Punithavathy Govindaradjane
Cover Designer: Miles Hitchen

Typeset by TNQ Technologies

This book is dedicated to the strength and beauty of human diversity.

Contents

7. Dispersals Into India

8. The Occupation of Southeast Asia, Indonesia, and Australia

9. The Austronesian Expansion

10. From Africa to the Americas

11. The Bantu Expansion

12. Modern Humans in Europe

Preface

In February 2015, one of us (RJH) was contacted by the Elsevier Acquisitions Editor for human genetics, genomics, cancer research, and oncology, with concerns and questions on the need for a book on human population genetics. During preliminary conversations with the editor, we were impressed with the sparse amount of current foundational content and outdated information for those looking to read about the field. Based on our preliminary discussions, we decided to undertake the writing of this work that provides reference material and guidance to the nonprofessionals and useful information for researchers and practitioners in human population genetics. The aim is to inform and promote curiosity and the desire to learn using a simple and clear language and sentence structure with pertinent illustrations and updated references. Although the book is designed to inform the inquisitive reader, we aim at providing detail and extensive information on every subject. This makes our work useful to everyone interested in human origins and migrations.

During the ensuing 3 years, we worked closely with editors, designers, and staff at Elsevier in the production of this book. Our intent from the start was to write the book in an integrative style that would incorporate information not only from population genetics and evolution but also from other fields of knowledge that impact and enrich the subject matter. We believe that this approach enhances synergistically the value of our work by incorporating pertinent information that usually is not presented together in more traditional literature. There is tendency, as disciplines become more specialized, for information to become capsulated and compartmentalized creating a condition of intellectual isolation that compromise holistic understanding.

In writing the book, we also wanted to inform the reader beyond the scholarly subject of human population genetics and evolution. The aim was also to acquaint readers with useful knowledge about their origins, specific human dispersals and migrations that resulted in the peopling of the globe. Although, humans possess an innate curiosity about our past, the topic of human origins has experienced an unprecedented interest among the general public in recent years. Part of this phenomenon results from increased awareness of our ancestry and realization that every human is related to a common origin, likely originating in Africa. Today these notions are routinely considered and discussed by individuals who want to learn about their past and their forefathers. People want to know who their ancestors were, and in their efforts, they are increasingly turning to companies that use historical and DNA signals to assess ancestry. Thus, to address this inquisitiveness, the general theme throughout the book is the story of how our planet was populated; how humans and our hominin predecessors dispersed. Our efforts are now provided for the consideration and appreciation of the readers who will decide whether we succeeded in our attempts to make this book enjoyable, informative, and useful. For us, it was a pleasure to write it, and hope our enthusiasm is reflected in this final product.

The book is divided into 14 chapters addressing specific migrations that have occurred throughout the world since the great ape and human lineages separated about 6 million years ago. These chapters include the following: Chapter 1, The Nature of Evolution; Chapter 2, Early Hominins; Chapter 3, Origin of Modern Humans; Chapter 4, The Exodus Out of Africa; Chapter 5, The Settlement of the Near East; Chapter 6, Neanderthals, Denisovans, and Hobbits; Chapter 7, Dispersals into India; Chapter 8, The Occupation of South East Asia, Indonesia, and Australia; Chapter 9, The Austronesian Expansion; Chapter 10, From Africa to the Americas; Chapter 11, The Bantu Expansion; Chapter 12, Modern Humans in Europe; Chapter 13, The Agricultural Revolutions; Chapter 14, The Silk Roads.

It is a pleasure to acknowledge the contributions in the preparation of the book, from Audrey Dervarics (Colorado College, Colorado Springs, CO), Drs. Jason Somarelli (Duke University, Durham, NC), Robert Lowery (Indian River College, Fort Pierce, FL), and Dr. Diane J. Rowold (Foundation for Applied Molecular Evolution, Gainesville, FL). RG-B is grateful for the support of his colleagues in the Molecular Biology Department at Colorado College. Our research was financed by the Colorado College Natural Science Division, Colorado Springs, CO, the Howard Hughes Medical Institute Undergraduate Biological Sciences Program, and a grant from the Freeman Foundation, Stowe, VT, United States.

In addition, we are grateful to all who have contributed to the study of evolution. We are obliged to the field, laboratory, and computational researchers that have given generously to the study of human origins and migrations. And to the donors of biological material that allowed many of the investigations described in this book.

We also like to thank the countless number of people that have contributed to this book and to the field of population genetics with an emphasis on our origins and migration routes.

We also extend our gratitude and love to our family members for their patience and support during the course of the project.

Rene J. Herrera
Miami, Florida

Ralph Garcia-Bertrand
Colorado Springs, Colorado

CHAPTER

1

The Nature of Evolution

We must, however, acknowledge, as it seems to me, that man with all his noble qualities... still bears in his bodily frame the indelible stamp of his lowly origin. ***Charles Darwin***[1]

SUMMARY

This chapter describes the nature of forces that bring about biological evolution, from the genesis of genetic diversity to the mechanisms involved in natural selection. Mutations, both spontaneous and induced are the sources of variability, and most are detrimental. Nature does not select a priori for mutations with positive outcomes. It is after DNA is altered that the environment and natural selection allows the fittest to reproduce passing the beneficial mutations to subsequent generations. Because the environment is never static, it is imperative for the gene pool to maintain a healthy amount of genetic diversity and flexibility to adapt. Yet, because mutations, for the most part, are deleterious there is a limit to the amount of diversity that the gene pool can accommodate and tolerate before it becomes too heavy of a burden.

Another interesting characteristic of evolutionary change is that tissues and organs are not generated from scratch but are usually redesigned into new structures and functions as the organism evolves. This is due to the mutations in genes responsible for the development of tissues and organs over time.

Since the theory of evolution by natural selection was formulated, a number of observations have challenged its core and foundations as well as the idea that evolution only works gradually in geological time. One of these instances is the development of complex traits, especially when the organs in question are derived from different tissues and at different times during development. Keeping this in mind, what mechanisms provide the directionality in time and space to generate highly complex organs such as eyes and kidneys? Over the years, this evolutionary conundrum has fueled creationist arguments that at times evolutionary theory has not been able to respond satisfactorily. Thus, given that evolutionary processes cannot generate DNA changes that would only benefit the creation of specific structure and functions, how did highly complex structures evolve? Other cases involve adaptive changes that correlate with dramatic genetic alterations. This phenomenon at times involves genomic movement of reiterated or transposable elements or TEs (segments of DNA capable of moving within the genetic material of cells) that carry within them DNA capable of controlling gene expression

Ancestral DNA, Human Origins, and Migrations
https://doi.org/10.1016/B978-0-12-804124-6.00001-X

and may be responsible for instances of *punctuated evolution* such as the rapid encephalization and intellectual development seen in hominins. These dramatic processes contrast with the classic and neo-Darwinian principles of gradual evolution during the course of geological time. These issues are considered in the context of the genetic alterations observed during the course of hominin evolution.

ON THE NATURE OF EVOLUTION

Mechanisms

For the most part, evolution is an imperfect gradual process. It works on a geological time scale. Generally, changes in biological systems require thousands and often millions of years to be realized. Evolution relies on random changes in the genetic material known as mutations. These alterations can bring about positive, negative, or no consequences to organisms and populations. Some mutations may be neutral (neither beneficial nor detrimental) in terms of their impact on the anatomy or physiology of individuals. Thus, evolution does not get to select, a priori, for mutations with positive outcomes. Mutations occur aimlessly, and selection takes place only after the mutations happen. Mutations are blind to the impact that they are going to have on organisms. This mechanism of evolution could be considered rather inefficient compared with an optimized creation of species that acts and functions in a designed manner. Yet, in a changing environment, where living conditions are constantly fluctuating, this system provides a survival advantage. Because there is no way to anticipate how the habitat is going to be modified, having a repertoire of genetic and functional options could facilitate survival of any given species by providing a large number of evolutionary alternatives. Needless to say, this type of evolutionary system requires considerable genetic diversity to function properly; otherwise, certain environmental conditions and demands would go unmatched by the limited genetic variability available, possibly resulting in extinction.

Another important characteristic of the evolutionary possesses is that tissues, organs, and organ systems are not generated spontaneously or de novo. Evolution, for the most part, builds on what is already available. It modifies on structures and functions already possessed by organisms through mutations and subsequent selection of traits that confer a selective advantage. For example, the flippers of whales, dolphins, seals, and manatees are the modified appendages of ancestral terrestrial mammals that took to the sea about 80 million years ago (mya). These mammals that returned to the water may have looked like an extinct deer-like ungulate.[2] Another example, more pertinent to the human condition, is the changes seen in the hominin hand. The primate hand is an organ primarily designed for arboreal locomotion. It is a structure that evolved to grasp branches and swing from trees. While adopting bipedalism, the hand has developed into an organ for manipulating objects such as tools and weapons. Thus, evolution does not work from scratch like an engineer building a new home. It is more like the renovation of an historical building, which preserves the original elements of the edifice.

A consequence of this type of process is that at times organs retain characteristics reminiscent of previous ancestral organisms. That is, new structures evolve by modifying and superimposing on structures already in place. Such a mechanism may be considered cumbersome and somewhat inefficient because the evolution needs to work under the constraints of structures already in place created for a different purpose. Yet, it may be an economical way to create novel systems based on already available genes that provide the anatomy and physiology. At times, this type of mechanism leaves behind vestiges in the form of nonfunctional tissues and organs. For example, in humans the appendix has no known function but can be a target for infections and bleeding that can lead to death.

For early students and investigators of evolution, this evolutionary baggage provided developmental signals that allowed them to follow ancestral relationships among organisms. If they looked hard enough at development after fertilization (embryogenesis) the signature of ancestral organs could be detected. An example of this are the gills seen in developing vertebrate embryos, or the remnants of a tail (that humans are sometimes born with), which normally degenerate in vertebrates later in prenatal life. For example, to the untrained eye, the similarities among early embryos of distant organisms are so overwhelming that it is difficult to discriminate among incipient embryos of remote vertebrate species (Fig. 1.1). Initially evolutionary biologists felt that by observing human prenatal development, the observations of these ancestral organs and patterns were relics from previous ancestral lineages and evolutionary times. In other words, they believed these developmental observations were a reiteration of the evolutionary story, presented in fast-forward and evolutionary chronological order. This lead early developmental biologists to propose that this recapitulation during development contributed strong evidence for evolution and indicated the organism's phylogeny. This theory known as Ontogeny Recapitulates Phylogeny literally means that by looking at ontogeny (the development of an organism from fertilization) it summarizes phylogeny (the evolutionary history of the organism). This argument was later used by creationists to try to discredit evolutionary theory since it literally suggested that humans were once fish, mice, or even tortoises (since at one point in human prenatal development the human embryo

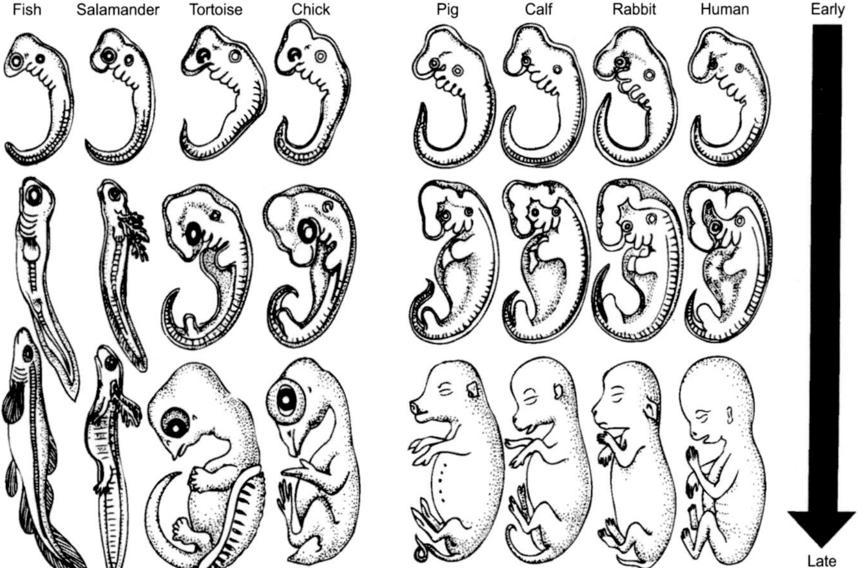

FIGURE 1.1 Similarities in the early development of vertebrate embryos. *Reproduced from Antranik.*

resembles a fish, a mouse, and a tortoise – the latter being a reptile not even in the same classification as humans). Of course, this is not the case as we will see throughout this book and the theory of ontogeny recapitulating phylogeny was discredited decades ago. Still the question remained why could you see similarities in the development of so many different organisms? As molecular developmental biology advanced the explanation for the similarities seen during embryogenesis began to emerge. As stated previously, evolution works on a random building process from genes and structures that already exist, and as genes involved in embryogenesis were discovered, it became obvious that evolution did not reinvent development for every organism but instead modified existing genes through random mutation. In other words, the genes that function during embryonic development are similar in a variety of organisms because they have been modified through mutation to differ in their expression by temporal, spatial, and quantitative regulation. So, in many instances we see similarities in embryonic development when comparing organisms, for example, the fruit fly and humans. During the earliest stages of fly development, (hours after fertilization) we see genes expressed that influence polarity of the organism (anterior and posterior), followed by genes that begin to start segmenting the embryo into regions that will contain the head, wings, legs, etc. When we look closely at the genes responsible for these polarity and segmentation phenotypes they show some similarity in both flies and human development. Obviously though flies and humans express some similar genes in setting up polarity and segmentation (both have an anterior and posterior and place organs and appendages in specific positions) humans were never flies as would be suggested by the theory of ontogeny recapitulates phylogeny. These early developmental genes that regulate the development of anatomical structures can be seen in all organisms (including plants) and are broadly known as homeotic genes. These genes strongly argue for evolution through the modification of existing genes, and their expression explains the phenotypic similarities we see during periods of embryogenesis. They represent modified ancestral genes that humans and other organisms have inherited, that provide some phenotypic similarities during embryo development but do not indicate that humans, for example, were ancestral species during their embryogenesis.

Mutations, the Raw Material of Evolution

Evolution is a slow, gradual, continuous process. Accidental changes in the genetic material provide the raw material for evolutionary change. And a number of evolutionary forces such as natural selection and random drift (aimless changes in the gene pool of populations) act on the unintentional genetic changes to drive evolution.

Alterations to the DNA of organisms can be characterized into two basic types, spontaneous or induced. Changes in which no extraneous agents induce the nucleic acid mutation are considered spontaneous, whereas induced alterations are caused by extrinsic factors such as radiation or chemicals acting as mutagens. Not all induced mutations are foreign agents from outside our bodies. Agents within our system can cause induced mutations as by-products of normal metabolism such as free radicals (highly reactive elements containing an unpaired electron) that attack the DNA. These reactive molecules are capable of modifying DNA or truncating chromosomes. Because organisms are constantly being bombarded by radiation and chemicals from different sources (internal metabolism and extraneous agents), it is difficult at times to assess if mutations are spontaneous or induced.

Independent of whether the mutations are spontaneous or induced, the changes need to be compatible with life and occur in the reproductive cells (sperm and eggs) to be transmitted into future generations. Some of the alterations in the DNA are so disrupting to life that when the resulting sperm and egg unite during fertilization, they produce an embryo that is not viable. It is thought that most spontaneous abortions stem from mutations incompatible with life. Spontaneous abortions when they occur early in pregnancy may go unnoticed by the parents. From an evolutionary perspective, this destruction (spontaneous abortion) by nature could be seen as an efficient mechanism that eliminates organisms that would represent a high genetic load (biological drain subject to negative selection) on the population. In humans, the less deleterious mutations that allow pregnancy to continue, or even allow birth to take place, could constitute greater energy and time expenditure for parents, family members, and/or society if they result in disease or physical disabilities. Thus, in the event that these detrimental DNA changes persist into postnatal life, they would impact the population increasing its genetic load.

The Random Nature of Mutations

To understand mutation, we must first understand the makeup of the genetic information system that comprises cells, tissues, and individuals. When we refer to mutations, we are referring to changes in DNA. DNA is the acronym of deoxyribonucleic acid, a large double helical molecule made up of units known as nucleotides. The sequence of four distinct nucleotides (adenine, cytosine, guanine, and thymidine) determines the nature of the gene. Although certain parts of the genome (the entire genetic makeup of an organism) such as repetitive DNA sequences and DNA regions rich in 3′guanine–cytosine (GpC) dinucleotides tend to experience more DNA mutations than others, mutations take place anywhere on the DNA. An organism can substitute any nucleotide (building blocks of the DNA) for a different one. In addition to substitutions, organisms can experience deletions and additions of nucleotides.

The environment plays no role in selecting specific nucleotides to be deleted, substituted, or replaced. Thus, there is no directionality to mutations. Yet, it is the environment, after the mutation occurs that dictates the kind of impact the mutation will have on the organism, although some detrimental mutations can be incompatible with life independent of the habitat. Thus, it is the nature of the environmental pressure that determines if a given DNA alteration is going to be selected for, against or be selectively neutral. This is an interesting point because depending on the environmental conditions the same mutation can be beneficial, deleterious, or neutral. Take for example, sickle cell anemia, a disease prevalent in several tropical regions worldwide. The classical example of sickle cell disease is caused by a single nucleotide change in a gene that codes for the beta-hemoglobin protein. Hemoglobin is the protein in red blood cells that carries oxygen and CO_2 to cells, tissues, and organs. The mutation is innately deleterious because affected individuals experience hemolysis (destruction of red blood cells) and limited capacity to transport oxygen and CO_2. This condition seriously compromises the well-being of individual carrying two copies of the mutation because it is deleterious and in many cases lethal. Thus, persons carrying this mutation would be selected against on the basis of these negative attributes. This negative selection takes the form of limited reproductive capacity due to premature death and/or incapacity to provide properly for their young. Both outcomes represent a genetic load on the population because they impact negatively on the genetic health of the group. Yet, in locations where malaria is endemic, carriers of this mutation enjoy some degree of immunity against malaria. It turns out that the malaria parasite does not survive in sickled cells and thus people with the mutation do not get infected with malaria as often. Therefore, in areas where malaria prevails, the mutation for sickle cell trait is selected for because of its benefits. The overall consequences of the sickle cell trait in these regions is determined when a balance between the negative selection pressures resulting from anemia, hemolysis (destruction of red blood cells) and poor oxygen and CO_2 carrying capacity, and the positive selections because of partial protection against malaria is reached. The two opposing forces exist in a state of dynamic equilibrium known to geneticists as *balanced polymorphism* in which benefits and disadvantages reach a balance, and the sickle cell mutation is kept in the gene pool at some level in spite of its health liabilities. The sickle cell case scenario clearly illustrates that environmental differences dictate whether a change in the DNA would be detrimental or beneficial.

The role of the environment on genetic diversity and evolution is further illustrated in instances in which habitat conditions change due to nature or human activity. In both cases, using again the example of the sickle cell trait, the eradication or reduction of the malaria parasite in a given region would shift the balance of positive and negative selection pressures. In human populations, diminution or elimination of a malady is a common occurrence, for example, when vaccines are developed providing immunity. Immunization against malaria, for example, would act to diminish the number of infections and bring about a decrease in positive selection for the sickle cell mutation because the parasite is being eradicated. Although malaria is a major international health problem, it is not common in certain parts of the world. In regions where malaria is not endemic, the positive selection pressure for the sickle cell trait is less or does not exist, and the mutation is entirely under negative selection. Thus, the nature of the selection pressure is environmentally dependent and it varies in time (i.e., past, present, and future) and space (i.e., in different parts of the world).

As stated before the environment cannot delete, substitute, or replace a nucleotide, however, recent discoveries have shown that the environment can modify nucleotides and gene expression. This modification of nucleotides without changing the genetic code is call epigenetics. One method of epigenetic modification is by adding or removing a methyl group (CH_3) on specific nucleotides. This nucleotide modification process can activate or silence genes where the modification has occurred. The evolutionary significance of this is that these modifications can be passed down to subsequent generations.

Beneficial or Deleterious Mutations

Humans and all organisms are the product of organic evolution that initiated life 4.1 billion years ago. We are all linked in the continuous chain of life made possible by evolution. Cells resembling present-day bacteria represent the most ancient type of life detected. Scholars today think that these organisms were chemoautotrophs capable of using carbon dioxide as a carbon source and oxidizing inorganic materials to extract energy.[3] All organisms, including us, are descendant from those original unicellular organisms, and we carry their heritage within our genetic material. As seen in the DNA molecule, we are all related and connected to those initial forms of life. Some species, such as our ancestral chemoautotrophs, represent very distant relatives whereas others, such as the great apes, are very close relatives. Alterations in the genetic material in combination with selection pressure and genetic drift, over time, have generated all the extant species that live today and the extinct organisms that have been detected as fossils.

As we previously stated, mutations rarely produce beneficial effects. The reason for this is simple. Organisms have been experiencing mutations at random since life began. This type of incidental process works under the principle of trial and error, and the possibility of a mutation improving on the existing condition of DNA and its function, is small since billion of years of mending and rectifying have generated relative efficient, fine-tuned metabolic systems. For this reason, most mutations are detrimental. Yet, there are still a small number of mutations that could improve the current physiological state of organisms. It is as if optimal conditions are never reached and there is always room for improvement if the environment remains unaltered. If the environment changes and given enough time then selection pressures most likely will deviate and the existing genetic constitution of organisms may not be appropriate for the new unique habitat. Thus, in a constantly changing environment, the genetic makeup is continuously being reevaluated by natural selection. This mechanism, in effect, permits the survival of populations over time. If the genome fails to generate the DNA diversity that would allow natural selection to act in the new environment, the populations may become extinct. In fact, most scholars believe that many species have ceased to exist not as the outcome of catastrophic events, such as the dramatic extinction of dinosaurs from a meteor impact, but as the result of organisms becoming overspecialized for a given niche and possessing a limited diversified gene pool. A good example of this phenomenon is the extinction of the Iris elk as a result of oversized antlers initially selected for as a secondary sex trait to attract females. When the elk's habitat changed to dense forest, it drove the species to extinction because it could not move fast enough within the forest to avoid predators. Maintaining a healthy and diversified gene pool is thought to provide for the survival of the population and a lack of genetic flexibility could lead to extinction in a changing environment.

When the Genetic Load Becomes Too Heavy?

Considering that most mutations are deleterious and environmental changes are not always abrupt, how much selectively negative or neutral diversity can a gene pool accommodate and tolerate before it becomes too heavy a burden conceivably causing the extinction of a population due to the energy expense and complications associated with maintaining a huge genome? In attempting to answer this question, it is important to consider that not all mutations are equally detrimental. As discussed previously, some genetic changes are so damaging that they are incompatible with life and cause death. Yet, some DNA modifications do not impact the phenotype (the way an organism looks or function) of individuals that much and are considered selectively neutral or only borderline negative by the process of natural selection. Because the adverse impact of these mutations is slight, this category of genetic changes can survive in the gene pool and remain undetected by natural selection. The disadvantageous effects of these mutations are so minimal that random genetic drift could allow these sequences to gradually become incorporated into the gene pool. Because they are undetected by selective evolutionary forces, these alterations drift slowly into the gene pool of populations. These slightly detrimental and neutral mutations represent part of the genetic diversity repertoire available for the creation of new genes with novel functions or the modification of existing sequences and corresponding functions. In either case, they provide the potential for genetic flexibility in a constantly changing environment.

Nevertheless, the maintenance of neutral or slightly deleterious genes in an organism or population is not without cost. Both cells and organisms need to replicate, repair, and distribute DNA to newly created cells at the expense of energy. Energy, in turn, requires the intake of nutrients. Thus, it is logical to expect that at a given point in time, after accumulating a certain level of slightly detrimental mutations, they could constitute a liability and an unbearable load on the gene pool of the population, and any further increment of *unutilized* diversity would be selected against by natural selection. So the benefits of possessing a well-diversified gene pool are outweighed by the energy expenditure involved in maintaining and keeping it. Yet, in instances of rapidly fluctuating environments, such as rapidly changing climatic conditions experienced during glacial fluctuations or dry–wet cycles, high levels of variability may be a determining factor for the survival of species.

The Evolution of Complex Genetic Traits

The development of complex traits represents somewhat of an evolutionary conundrum. Given that mutations are random and evolutionary processes cannot generate DNA changes that would only benefit the creation and evolution of specific structures and functions, how does highly complex structure evolve? Take, for example, an organ such as the eye. The eye derives from a number of unique tissues that develop at different times and places during development, with cellular migration occurring precisely to form the fine structure of the organ (Fig. 1.2).

Humans and vertebrates are made of three main types of tissues: ectoderm, mesoderm, and endoderm. All tissues and organs are derived from these three types of embryonic layers. Remarkably, the human eye is made up of two of these basic tissue layers, the mesodermal and the ectodermal tissues that come together in a precisely orchestrated manner in time and space to form the retina, ciliary body, iris, cornea, eyelid optic nerves, blood vessels, muscles, and vitreous. Specifically, the eye is generated from three types of tissues: neuroepithelium, surface ectoderm, and extracellular mesenchyme. The latter develops from both the neural crest and mesoderm (Fig. 1.2). The neuroepithelium forms the retina, ciliary body, iris, and optic nerves. Surface ectoderm creates the lens, corneal epithelium, and eyelid, whereas the extracellular mesenchyme generates the sclera, cornea, blood vessels, muscles, and vitreous.[4]

The fundamental question in the evolution of complex organs such as the vertebrate eye, kidneys, and the human brain: can mutations and natural selection, provided enough time to tinker and select, generate and fine-tune a very sophisticated biological system from different embryological origins? In other words, can time and random mutation provide the diversity and natural selection that would culminate in the proper organ structure and function? In terms of time, it has been theorized that it would take about 364,000 years for the vertebrate eye to evolve starting from a patch of photoreceptors seen in a number of invertebrates.[5] This time is an overestimation and likely a simplification based on the time required

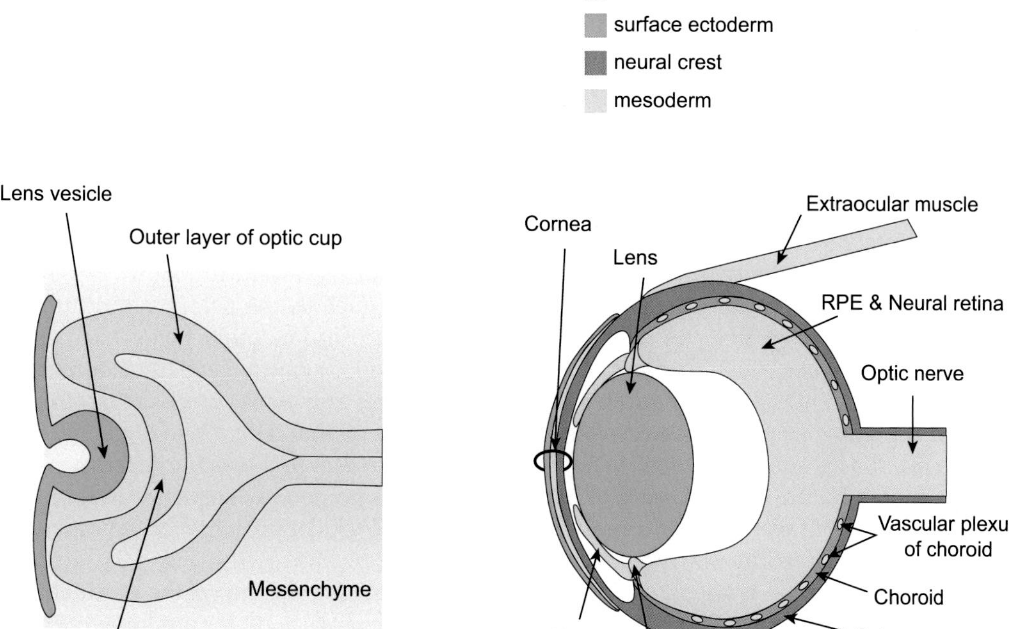

FIGURE 1.2 Organization and tissues of the eye. *From http://genesdev.cshlp.org/content/21/4/367/F1.large.jpg.*

for each key developmental stage given an estimated mutation rate and a generation time of 1 year exhibited by many small vertebrates. It is not clear how such a specific timeline could be realistically generated in the absence of facts such as specific rate of mutation of known and unknown genes and the nature and magnitude of the selection pressures involved. In thinking about these issues, it is necessary to realize that natural selection not only needs to keep beneficial mutations for specific functions, not yet realized, but also it requires to orchestrate, in time and space, the assembly of tissues from different origins into a very intricate and functional system. In other words, the contention that given enough time and variation very complex biological systems could evolve by natural selection to perform like a well-oiled machine is rather simplistic because it ignores the currently unknown mechanistic details and the serious problematic issues.

Currently, the molecular biology of genetic expression does not provide all the necessary details to explain how eye development takes place, especially given that our genome is made up of only about 20,000 protein-coding genes.[6] The limited amount of available functional protein-coding genetic material may restrict the amount of diversity necessary for natural selection to work with to develop highly complex functions. Moreover, knowledge of the developmental controlling signals and steps as well as the hierarchical matrixes of interactions and connections is minimal. It is clear that a large but currently undetermined number of genes need to be turned on and off precisely at very specific times during development. The formation of the eye, including the development of an optic vesicle and cup and the morphological and functional integration into neighboring tissues involve the coordinated action of a multitude of complex cellular and molecular events.

It is known that a master gene called *Pax-6* controls some of the early major genes for eye development. The name master gene is used to describe DNA that has the capacity to turn on or off other genes usually resulting in the stimulation or prevention of RNA production, respectively. RNA is then used to produce proteins for body structure and metabolic functions. This battery of subservient genes generally performs related functions. *Pax-6* is an ancient gene that is found in distantly related organisms such as the octopus and humans. In addition, it is known that the highly coordinated expression of other genes such as *Rx*, *Six-3*, *Lhx-2*, *Six-6/Optx-2*, ET, *tll*, *Hes1*, and *Otx-2* are needed for normal eye development. In biological systems, master genes allow for the simultaneous and consecutive developmental control of a number of genes in a cascade fashion, as a functional conga line of genes. *Pax-6* is also referred as a homeobox gene. The term homeobox refers to a specific repetitive DNA sequence that is found in this family of master genes. The homeobox sequence is about 180 nucleotides long that codes for 60 amino acids (building blocks of proteins) of the functional transcription (RNA production) factor proteins made by master genes (Fig. 1.3).[7] In other words, DNA-containing homeoboxes code for proteins that turn on and off the production of RNA in subordinate genes. These proteins that homeobox genes code for are known to bind to the DNA of other genes and in doing so turn their activity on or off. Yet, although the identification of the *Pax-6* gene and other genes have added to the body of knowledge concerning the control of eye development, what is known about its function falls short of explaining the evolution of vision.

Additional innovations that contribute to the complexity of eye development include color, polarization, focusing, and object location. Darwin in his book *On the Origin of Species* wrote about the evolution of the eye.

> ...the difficulty of believing that a perfect and complex eye could be formed by natural selection, though insuperable by our imagination, should not be considered as subversive of the theory. ***Charles Darwin***[8]

From the above passage, we can see that it was clear to Darwin that a complex system such as the eye provided a challenging case for natural selection to explain. Currently, a century and a half after Darwin's publication, scholars still wrestle with the issues of the evolution of the eye. Similar levels of structural and functional complexity are seen in other organs.

HOMININS: EVOLUTION ON STEROIDS

Since the mid-20th century a number of discoveries indicated the existence of evolutionary phenomena that seem to violate Darwinian evolutionary theory. Instead of the generally slow, smooth, and continuous evolutionary processes typically associated with natural selection, sudden jumps in the fossil record were clearly apparent. To many scholars, these observations were simply artifacts of an incomplete fossil record. Yet, as time went by, it became obvious that some of these episodes represented bona fide instances in which the

```
RRRKRTA-YTRYQLLE-LEKEFLF-NRYLTRRRRIELAHSL-
NLTERHIKIWFQNRRMKWKKEN
```

Amino acid key

One letter code	Three letter code	Name
A	ALA	Alanine
R	ARG	Arginine
D	ASP	Aspartic acid
N	ASN	Asparagine
C	CYS	Cysteine
E	GLU	Glutamic acid
Q	GLN	Glutamine
G	GLY	Glycine
H	HIS	Histidine
I	ILE	Isoleucine
L	LEU	Leucine
K	LYS	Lysine
M	MET	Methionine
F	PHE	Phenylalanine
P	PRO	Proline
S	SER	Serine
T	THR	Threonine
W	TRP	Tryptophan
Y	TYR	Tyrosine
V	VAL	Valine

FIGURE 1.3 Consensus homeodomain of 60 amino acids. Insertion sites known to bind to DNA turning genes on or off are indicated by *dashes*.

rate of change of traits was very fast (i.e., not on a geological time scale), whereas at other time periods there was no apparent change. The data, initially based on morphological traits, eventually were integrated into the concept of *punctuated evolution* (Fig. 1.4). This set of ideas contrasted sharply with the gradual nature of Darwinian evolutionary theory. Subsequently, a number of findings from the fields of genetics and, more recently, from molecular biology began to emerge providing mechanistic explanations for these dramatic spurts in novel characteristics that looked more like revolutions rather than evolution. A number of these cases involve unprecedented rapid rate of genetic change that have transformed the way we think about evolution.

Specifically, in the great ape and human lineages, the genomes have experienced approximately a twofold increment in the rate of DNA duplications (Fig. 1.5). Transpositions, core duplicons, and rapid amplification of repetitive elements are three important genetic mechanisms responsible for the dramatic alterations seen during hominin evolution and thus are described below.

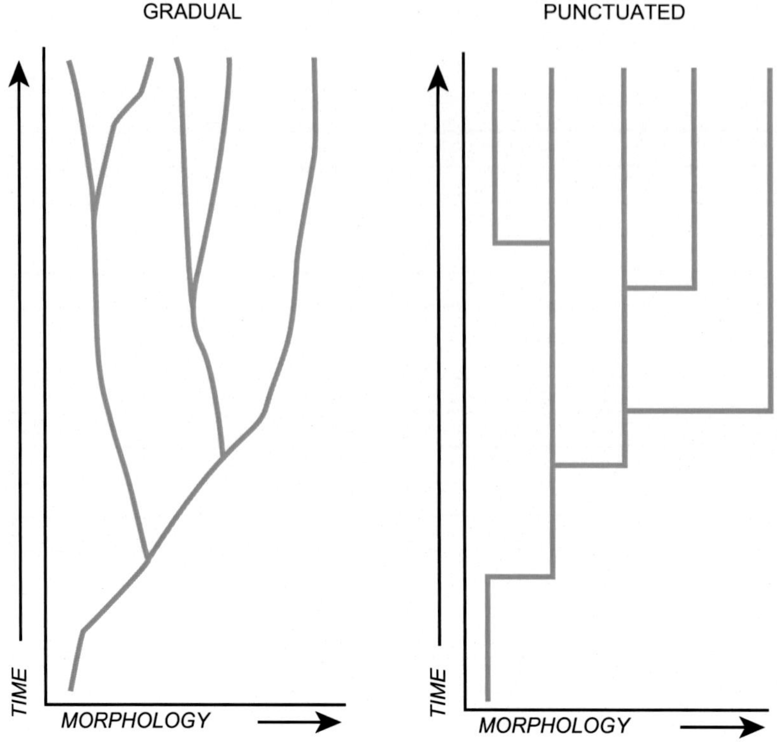

FIGURE 1.4 Gradual and punctuated evolution. *From http://eesc.columbia.edu/courses/v1001/gradpunct.html.*

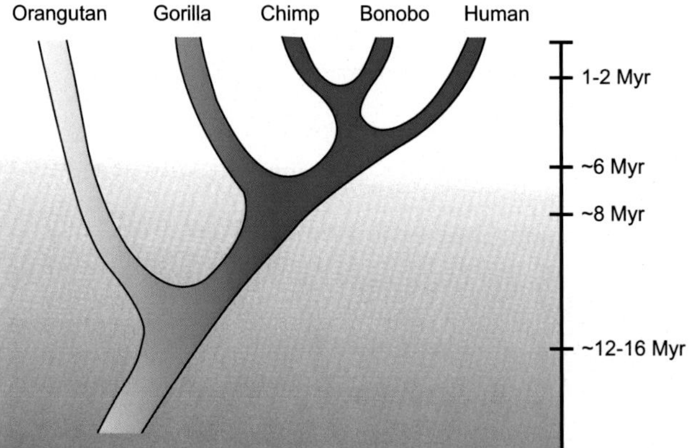

FIGURE 1.5 Increment in the rate of DNA duplications in the great ape and human lineages. The split between our great ape ancestors and other primates about 12 million years ago coincided with an outbreak of genetic duplication. Darker shape indicates higher levels of genetic duplication. *From https://www.quantamagazine.org/20140102-a-missing-genetic-link-in-human-evolution/.*

Transposons

Genetic diversity at the population level provides a reservoir of variability that fuels evolutionary change. It is thought that constant selection pressure by the environment and random genetic drift drives evolutionary change. Considering the random nature of mutations and genetic drift as well as the subsequent selection of beneficial DNA, evolution is expected to be a sluggish process, especially when it involves the creation of complex biological systems. Very slow rates of change have been one of the tenants of neo-Darwinism.

Yet, starting in the mid-20th century Barbara McClintock started noticing some unusual behavior involving the inheritance of certain traits. She discovered the existence of transposons (transposable elements or TEs) or pieces of DNA that, as the name implies, not only are able to move within the genome of cells but also change the activity of genes near or at their insertion site, turning the genes on or off and promoting high levels of mutations. Starting in 1944, McClintock began to study the action of two peculiar DNA sequences that seemed to move from location to location at will within the genome of corn.[9] McClintock named these TEs *dissociator* (*Ds*) and *activator* (*Ac*). The movement of these two DNA elements is actually responsible for the changes in color in maize kernels known as variegation or mosaicism (Fig. 1.6). McClintock's studies and results were initially misunderstood and disbelieved by the rather dogmatic, male-dominated, scientific orthodoxy that was not willing to accept that dramatic structural changes in the DNA of organisms could make certain cells, with identical genomes, function differently. One of the reasons for this reluctance to concede the validity of her results was that the data challenged the accepted core of evolutionary theory. McClintock's data were in defiance of the very core of traditional evolutionary theory that DNA was static and organisms inherited specific sequences of DNA that imparted certain traits that were not subject to random changes during the organism's lifetime due to DNA modifications. And the fact that she was a woman did not help. She demonstrated, for the first time, that a genome does not provide rigid set of instructions passed on only between generations. After decades of undermining her contributions, scientists in the 1970s

FIGURE 1.6 Multicolor corn showing the effect of transposon insertions. *From http://sceeephotography.com/img/s12/v171/p370769699-4.jpg.*

began to discover transposons in a variety of organisms and eventually isolated the genetic material responsible for transposition. McClintock received the Nobel Prize in Medicine and Physiology in 1983. She died in 1992.

Since McClintock's work, many TE types have been discovered in all organisms including humans.[10] In general, transposons in different organisms are characterized into two types, Class 1 transposons move from site to site within the DNA of cells via RNA intermediates, using the enzyme reverse transcriptase to make DNA during the insertion process. This process allows for the production of large numbers of these DNA sequences because a single transposon can serve as template to countless RNA intermediates, which then insert independently throughout the genome. This type of process known as RNA-mediated transposition or retrotransposition has a dramatic impact on the DNA of cells since the initial transposon sequence is amplified in the process leading to a potentially large number of insertion events and outcomes. Class 2 elements, on the other hand, change position within the genome by a cut-and-paste mechanism in which the DNA is not amplified by making large numbers of RNA copies before insertion but is rather simply excised from its original site, and it is then inserted into a new location (Fig. 1.7). Despite all that has been learned, it is still not clear what triggers transposition.

Transposons constitute 45% of the human genome, 37.5% of the mouse genome, and 41% of the dog genome.[11] Class 1 transposons in humans and great apes are, in turn, categorized into long and short interspersed nuclear elements (LINEs and SINEs, respectively). Long terminal repeat sequences (LTRs) and short stretches of DNA flank LINES and SINES, respectively.[12]

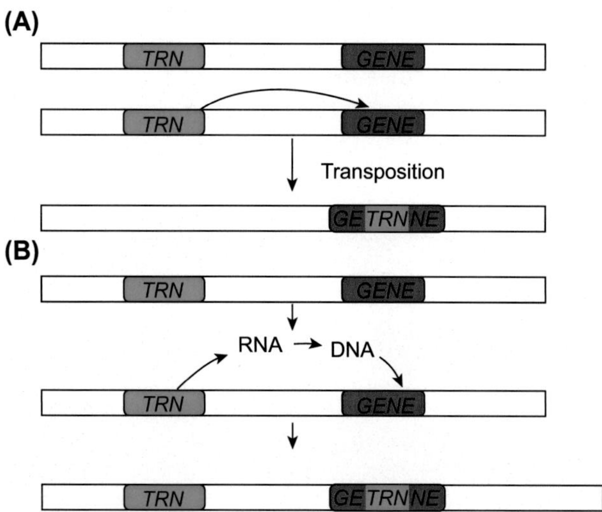

FIGURE 1.7 Transposons can move by cut and paste mechanisms or copy and paste mechanisms and disrupt gene activity. (A) Shows a cut and paste transposon (TRN) and a GENE on the same chromosome. The transposon is excised from its original position and inserts itself into the GENE thereby disrupting the activity of the gene. (B) Demonstrates the copy and paste mechanism of transposition. The transposon (TRN) is transcribed and then converted into a new double stranded DNA transposon. The new transposon is then inserted into the GENE, disrupting its function. Both of these examples show transposition along the chromosome that contains the transposon, however, cut and paste or copy and paste mechanisms can also occur across homologous or nonhomologous chromosomes. *From https://www.wiley.com/en-us/Genomes%2C+Evolution%2C+and+Culture%3A+Past%2C+Present%2C+and+Future+of+Humankind-p-9781118876404.*

LINEs are retrovirus-like elements that contain their own reverse transcriptase (enzyme or biological catalyst that makes DNA from an RNA intermediate before insertion into a new site) and can contain their own endonuclease (enzyme that cuts the DNA before insertion of a LINE). These are the copy-and-paste type of element that increases in number as they replicate and insert throughout the genome. The human genome contains more than 750,000 LINEs. The best studied of these are the L1 retrotransposons. L1 elements are considered the newest version of the LINEs and are the only ones capable of moving themselves. L1 insertions have accumulated at higher rates in bonobos and chimpanzees than in humans. They occupy about 30% of the human X chromosome and have been implicated in X-inactivation.[13,14]

SINEs are elements that are less than 500 base pairs (bp) in length and do not contain their own reverse transcriptase, thereby relying on LINE-encoded proteins for transposition. The human genome contains more than 1.5 million SINEs. The most common SINE family in primates, including humans, is the *Alu* sequences (named for their ability to be recognized by the *Alu* restriction enzyme that cuts the element into two). *Alus* are active nonautonomous retrotransposons as they require another type of elements, such as LINEs, to provide reverse transcriptase for retroposition into new sites on the DNA. *Alu* insertions are occasionally able to promote diseases in humans by eliciting insertions, deletions (Fig. 1.8), and inversions (Fig. 1.9) of DNA when different copies in distinct locations of the repetitive element family recombine.[15] Cell stress such as heat shock and viral infection can result in transcriptional activation of *Alus*.

SVAs (SINE-VNTR-*Alu*) are composite nonautonomous retroposons that include variable number tandem repeats (VNTRs) and SINE elements including *Alu* sequences. SVAs are seen in primate lineages, are transcribed by L1 reverse transcriptase, and have been associated with some genetic diseases.[16] Human endogenous retroviruses (HERVs) are more closely related to retroviruses such as HIV and have retained their ability to replicate themselves but

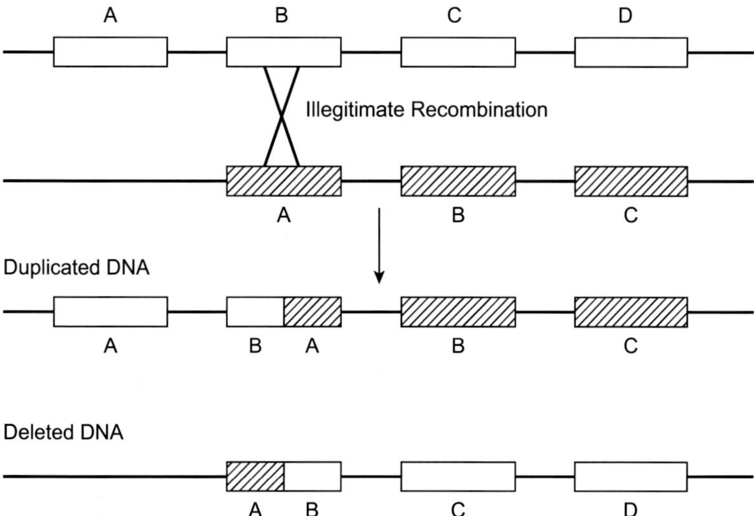

FIGURE 1.8 Deletion and additions involving duplicated transposable elements such as long and short interspersed nuclear elements.

have lost the ability to leave the cell. There are approximately 4000–5000 copies of HERVs per haploid genome. The SINE-R element was first reported as a retrotransposon derived from an HERV. Inactive transposons such as Mariner and MIR (microRNA) are also found in the human genome.

It is estimated that at about 40 human families (ancestrally related types) of transposons, totaling approximately 98,000 elements or 33 million nucleotides, have been moving around within the primate and hominin genome.[17] It has been ascertained that many of these families

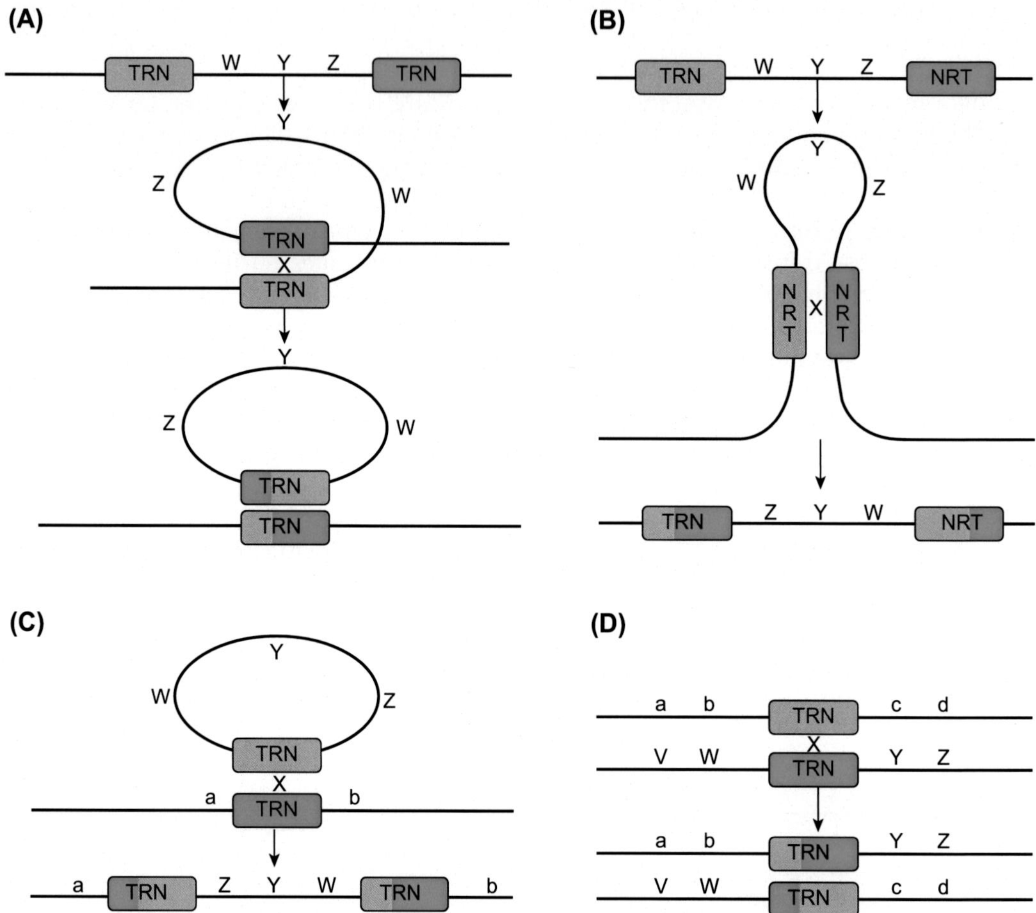

FIGURE 1.9 Figure represents the various types of recombination that can occur in a Non-allelic Homologous Recombination (NAHR) resulting in a deletion, inversion, insertion, and translocation. The figure shows a transposable element (TRN) that can act as a sight for NAHR. (A) Shows a chromosome loop forming as a result of two homologous TRN sequences existing as direct repeats on the chromosome. Recombination between the elements results in a deletion (ring chromosome) of the sequence between the two elements and a hybrid (red – green) transposon. (B) Shows the recombination between two inverted elements along the same chromosome resulting in an inversion of the intervening sequences. (C) Shows a ring chromosome with a TRN element and a non-homologous chromosome with a TRN element. Recombination between the two TRN elements results in an insertion of segment *ZYW* into the linear chromosome. (D) Shows two non-homologous chromosomes (*abcd* and *vwyz*) lining up as a result of two TRN sequences on the different chromosomes. Recombination between the TRN sequences results in a chromosome translocation, where *a* and *b* are now linked to *y* and *z*, and *v* and *w* are now linked to *c* and *d*. *From https://www.wiley.com/en-us/Genomes%2C+Evolution%2C+and+Culture%3A+Past%2C+Present%2C+and+Future+of+Humankind-p-9781118876404.*

were particularly very active in transposition during early primate evolution.[10] Recent studies involving other primate species such as the chimpanzee (*Pan troglodytes*) and the rhesus macaques (*Macaca mulatta*) demonstrate comparably high percentages of transposons.[18] Considering the proportion and diversity of jumping DNA in humans (Fig. 1.10), it is likely that they have played an important role in shaping the architecture of the primate genome and played a major role in hominin evolution.[19]

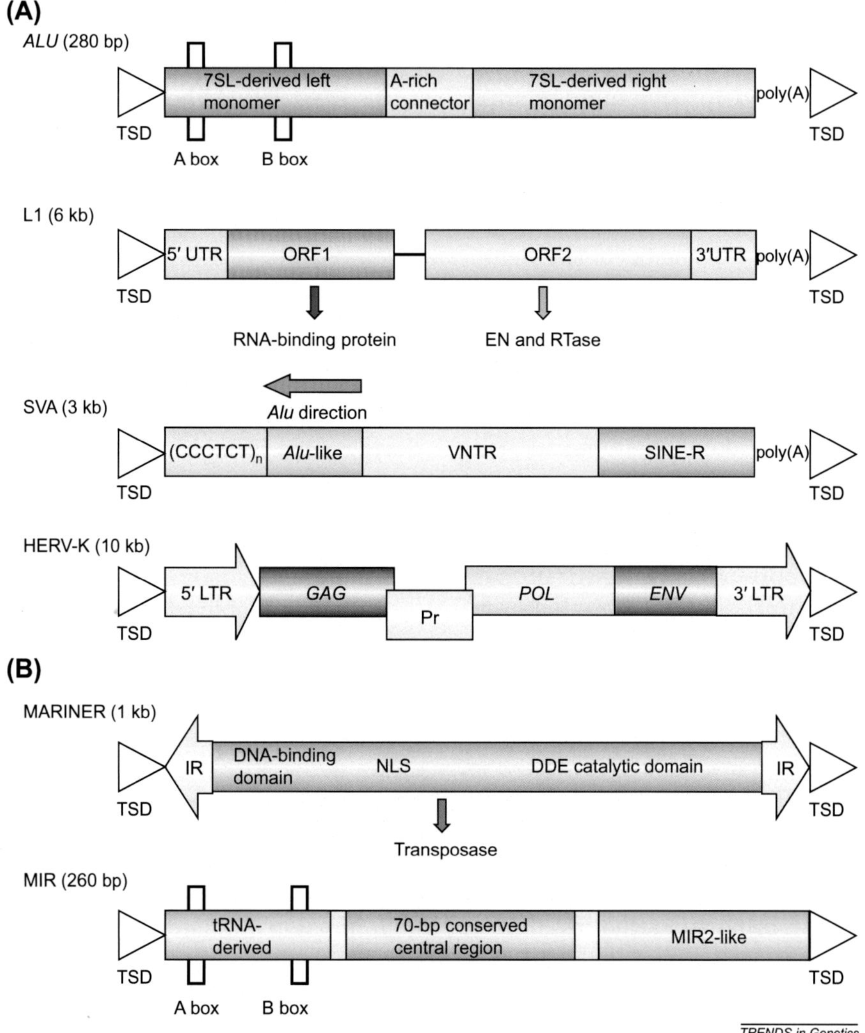

FIGURE 1.10 (A) The structures of human transposons that have been active over the past 6 million years. *TSD*, target site duplication; *ORF*, open reading frame; *UTR*, untranslated region; *VNTR*, variable number of tandem repeats; *SINE-R*, short interspersed repetitive element; *LTR*, long terminal repeat; *GAG*, gene for retroviral core protein; *POL*, gene for reverse transcriptase; *ENV*, gene for retroviral envelope protein; *Pr*, protease; (B) Structures of inactive transposons. *IR*, inverted repeats; *NLS*, nuclear localization structure; *DDE*, transposase for MARINER; *MIR*, mammalian-wide interspersed repeat. *Taken from Mills RE, et al. 2007. Which transposable elements are active in the human genome?* Trend in Genet **23**:*186–91. From https://www.wiley.com/en-us/Genomes%2C+Evolution%2C+and+Culture%3A+Past%2C+Present%2C+and+Future+of+Humankind-p-9781118876404.*

One characteristic of TEs that was not obvious from the seminal experiments of McClintock is that these jumping genes behave rather independently from the rest of the DNA and that some of the rearrangements caused in the genome may take place in germ line tissue and thus are passed on to the next generation. Their autonomous relocation behavior in relation to the rest of the genome has prompted some scholars to classify them within a group of DNA referred to as "selfish DNA" (nucleic acid that exists for the sake of their own existence). TEs have been implicated in disease, and genome rearrangements, including deletions, additions, inversions, and translocations (Fig. 1.9). They can be a source of novel genes, gene silencing mechanisms, creating transcription regulatory elements, and drivers of speciation.[20,21] Most TEs in humans have accumulated enough mutations to prevent them from replicating and/or moving and are sometimes referred to as fossilized TEs. Although many of these elements have lost their ability to replicate or relocate, many still remain transcriptionally active. However, of those that are still capable of making RNA, most are not translated nor undergo reverse transcription.

Germ line tissues are made up of cells that develop into ova and sperm that subsequent to conception generate the embryo. The significance of these observations is that drastic changes in the DNA with functional consequences can impact future generations and the path of evolution. Because TE transposition in humans is often associated with germinal tissues (as opposed to somatic transposition), these transposons in the germline will be passed on to the next generation. One possible mechanism associated with increased germline transposition is decreased DNA methylation in these tissues. Because methylation is proposed to down-regulate transposition, demethylation of DNA during meiosis could allow for a window of increased activity.[22] Thus, with new evidence that transpositional changes could occur with demethylation, scientists are contemplating evolutionary spurts of change during meiosis rather than gradual change.

Another important characteristic of many TEs is that they usually carry within them coding DNA for functional gene products and/or sequences capable of regulating the genetic expression of neighboring genes anywhere in the genome of the cell. This quality allows transposons affect the structure and function of genomes by simply inserting themselves and disturbing genes and their expression. Because they contain coding sequences and controlling DNA elements, they can change the nature of genes by adding new protein domains and dramatically altering gene expression by down- or upregulating gene expression. These characteristics of TEs have been proposed to strongly impact human evolution and to possibly have led to the rapid divergence between humans and other primates.

Transposons can interrupt the sequence coding for a necessary cellular function such as enzymes necessary for metabolism. The alteration of protein-coding DNA is usually detrimental and can bring about the cessation of protein function. Similarly, an insertion of these elements into a DNA region involved in transcription regulation could bring about inactivation of controlling sequences for proper gene expression. The effects of these events are usually detrimental because the transposons, for the most part, insert at random anywhere in the genome. The fact that these jumping genes carry within them protein-coding and regulatory DNA sequences provide another dramatic change to the genome's function because the protein-producing DNA may produce an abnormal protein in the new location (insertion site). Alternatively, the regulatory DNA that they ferry could trigger uncontrolled gene expression or transcription cessation of nearby genes. Although the consequences are usually deleterious, a minimal number of these insertion events elicit alterations that benefit the organism, thereby fueling evolutionary change.

The mode of dispersion of Class 1 TEs also helps to augment their evolutionary impact. As previously stated, the mechanism of movement of Class 1 TEs involves RNA intermediates. RNA is the immediate product of DNA transcription, and many copies of RNA are made from any given transposon. Transposon RNA is then reverse transcribed into DNA, which is then inserted all over the genome. This mass production of TE sequences allows for multiple insertions of transposon throughout the genome. This mode of movement, known as RNA-mediated transposition or retrotransposition, has the effect magnifying the number of subsequent potential insertion events and outcomes.

In humans, a number of TEs have been discovered and characterized. The *Alu* family of transposons, for example, is primate-specific and constitutes about 11% (by mass) of the human genome. This represents approximately 1 million *Alu* transposons per diploid human genome. *Alu*s are the most abundant group of transposons in the genome of great apes and humans (Fig. 1.11). Although *Alu* elements exhibit some insertion preference for GC-rich, (GC stands for guanine–cytosine) dinucleotide regions in the DNA, the *Alu* family is widely distributed throughout the genomes of these organisms (Fig. 1.12). The insertion preference of *Alu* transposons for DNA abundant in coding (functional) sequences, as opposed to noncoding spacer DNA, suggests that they may be particularly influential in impacting genes capable

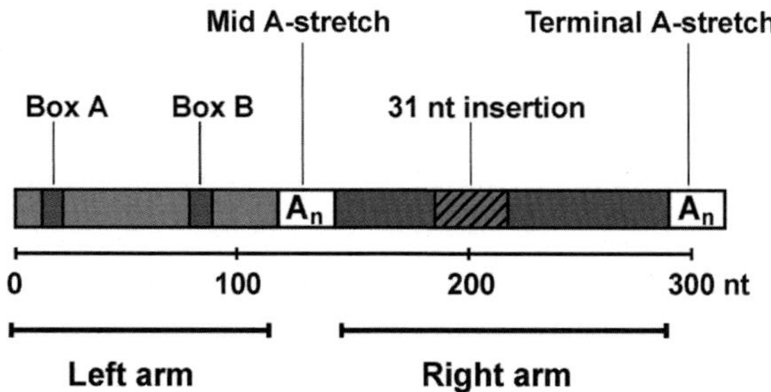

FIGURE 1.11 Anatomy of an *Alu* transposon. Box A and Box B, internal promoter sequences that control RNA transcription. Mid- and Terminal A-stretches, DNA sequences rich in Adenosine. Left and Right arms, duplicated DNA sequences of *Alus*. 31 nt insertion, 31 nucleotides inserted into the right arm after the duplication event occurred. *From https://insolemexumbra.files.wordpress.com/2014/10/screen-shot-2014-10-26-at-10-48-02-am.png.*

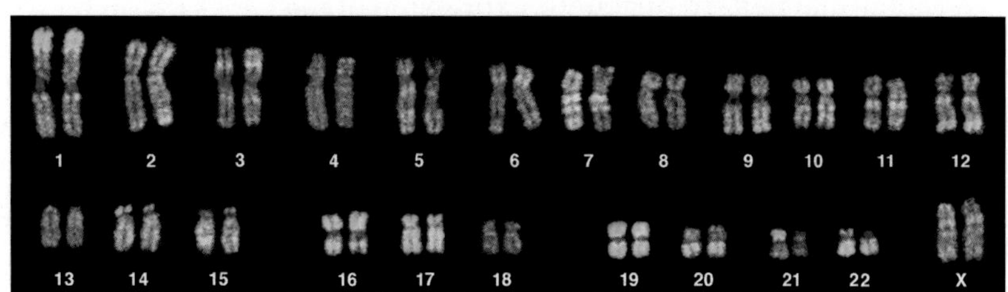

FIGURE 1.12 Distribution of *Alu* transposons within the human genome. Chromosomes were hybridized with a probe for Alu elements labeled to produce a green fluorescence. *From https://en.wikipedia.org/wiki/Alu_element#/media/File:PLoSBiol3.5.Fig7ChromosomesAluFish.jpg.*

of affecting evolution. On the other hand, for the same reason, because of their capacity to disrupt functional DNA, *Alu*s are particularly damaging as mutagens capable of inactivating genes by interrupting coding sequences. In addition, subsequent to the insertion event, in somatic tissue (body cells other than reproductive cells), *Alu*s are frequently involved in illegitimate recombination events with other *Alu*s throughout the genome leading to insertions, deletions (Fig. 1.8), and inversions (Fig. 1.9) of DNA. These recombination events are known as nonallelic homologous recombination because they involve different copies of the repetitive element family (e.g., *Alu*) in different locations of the genome. There are so many *Alu* elements distributed throughout our genome that their mutagenesis by illegitimate recombination is responsible for many of the known human genetic maladies. These secondary mutations are notoriously responsible for localized genetic diseases, such as cancer.[23]

The other major family of transposons, the L1s, is known to insert selectively at TT/AAAA sites, and therefore, is thought to show preference for AT-rich regions (adenine–thymidine) regions. Thus, given the low complexity of AT-rich regions and relative paucity of these sequences in protein-coding genes, contrary to *Alu*s, L1s may be less of a burden as a mutagen and less impacting as an evolutionary promoting agent.

Considering the speed in which transposons move and expand within genomes, it is logical to expect that they are capable of promoting dramatic evolutionary change in a short period of time, not geological time. For instance, the insertion of an *Alu* transposon into a coding sequence could lead to the creation of a truncated or inactive protein product for that gene or it may provide novel DNA (the *Alu* sequence) for the gene to evolve and modify its function. Alternatively, *Alu*s can insert in regions rich in transcription controlling elements, DNA involved in the activation and silencing of protein-coding genes. Such an event could bring about major changes in the mode in which genes are turn on and off, therefore, affecting dramatically organismal development. Profound developmental and evolutionary impact would be particularly expected when the DNA sequences affected are master genes known to work regulating large number of other genes downstream in the hierarchical functional cascades of developmental steps. In terms of evolutionary change, because transposons such as *Alu*s are particularly active in the germ line or tissue leading to ova and sperm, these insertions will be passed to future generations possibly modifying the evolutionary path of species.

We may envision a hypothetical scenario involving the incorporation of an *Alu* transposon within the region populated by transcription-controlling DNA that regulates the expression of a master gene that coordinates the activity of several genes responsible for cell movement during the development of fingers in the hominin hand. Such an event would likely trigger an abnormal pattern of genetic expression of the genes involved. This, in turn, may impact the size or number of fingers in the hominin hand and/or the type of muscle movement and dexterity possible. This kind of genetic change may be advantageous or detrimental to hominin living in a certain habitat. If the morphological alterations turn out to be beneficial, thus providing greater fitness (capacity to pass genes to the subsequent generations), the new trait will be selected for by natural selection, and the *Alu* insertion could increase its frequency in the gene pool because the carriers of the insertion would enjoy greater reproductive success.

The spread of repetitive elements, such as the members of the *Alu* family, occurs at the rate of one de novo insertion approximately every 20 births[24], whereas nucleotide substitution and deletion mutations take place at a frequency of about 1.1×10^{-8} per site per generation.[25] Considering the high frequency of insertions and the profound effects that transposons could have on neighboring sequences and the impact of illegitimate recombination among noncomplementary elements, *Alu* insertions are capable of promoting evolutionary change in thousands of years instead of millions of years. Of course as with all mutations many of these transposon insertions have no effect on phenotype and many are inserted into DNA with no apparent function. Those insertions that are detrimental are eliminated from the population.

Alu amplification rates vary in a lineage-specific manner (Fig. 1.13). For example, in the human lineage, two subfamilies of *Alu*s, Ya5, and Yb8 are responsible for most of the insertions subsequent to the split from the chimpanzee line. Since the separation of the human and chimpanzee lineages, approximately 6 mya, around 2400 and 5000 lineage-specific *Alu*s insertions, respectively, have occurred and are now fixed (all individuals of the species have them) in each species.[26] For comparison, the orangutan has experienced only 250 lineage-specific insertions in the last 12 million years.[26] It appears that the hominin line, for unknown

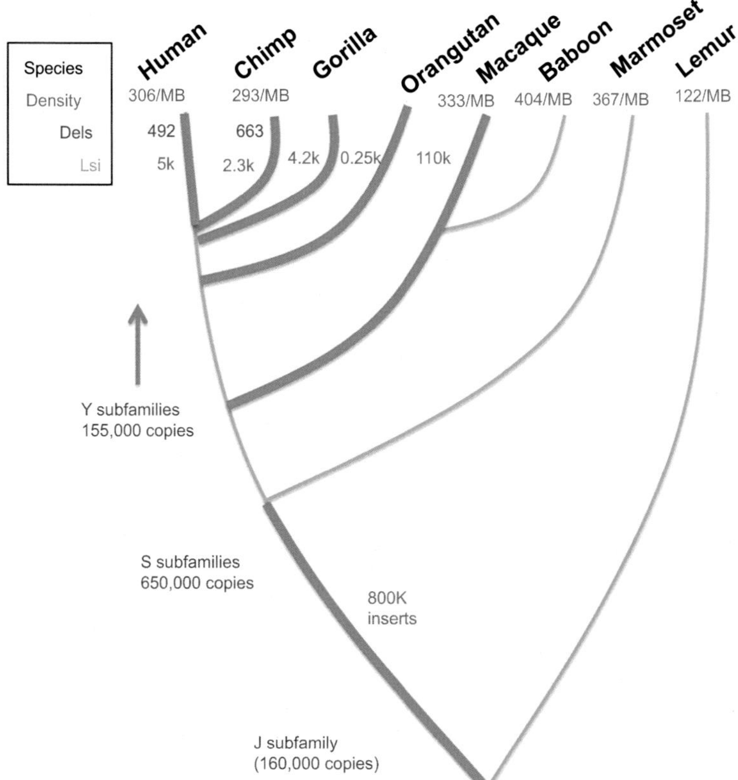

FIGURE 1.13 Primate evolutionary tree indicating number of *Alu*s per 1 million nucleotides of DNA. Thicker lines indicate the number of lineage-specific *Alu* insertions within a given evolutionary time periods. Numbers in red stand for number of deletions caused by illegitimate recombination involving *Alu* elements. *From* Genome Biol *2011;**12**(12):236.* Alu elements: know the SINEs. *Prescott Deininger. http://www.ncbi.nlm.nih.gov/pmc/articles/PMC3334610/.*

reasons, has experienced more *Alu* insertions compared with other primates. Interestingly, other transposon families such as L1 do not seem to have been differentially transposed among primate lineages.

In addition to differential rates of insertions, specific *Alu*s have been implicated with specific recent steps in hominin evolution, particularly with brain development in the human lineage and the separation between the chimpanzee and the hominin lines. It is known that an *Alu* insertion inactivated the of CMP-*N*-acetylneuraminic acid hydroxylase gene just before brain expansion during recent human evolution.[27] CMP-*N*-acetylneuraminic acid hydroxylase is an enzyme (protein catalyst of chemical reactions) responsible for the conversion of CMP-Neu5Ac into CMP-Neu5Gc. As a result, in humans, Neu5Gc is not made. CMP-Neu5Ac and CMP-Neu5Gc are important carbohydrates that are needed for the manufacturing of glycoprotein receptors (molecules located on the outside of cellular membranes that function in transmitting signals from the outside to the inside of cells) such as myelin-associated glycoprotein in brain tissue and other cellular receptors throughout the body. In addition, conversion of CMP-Neu5Ac into CMP-Neu5Gc may provide receptors for pathogenic microorganisms such as influenza A, rotaviruses, and the bacteria *Escherichia coli*. Also, Neu5Gc expression is developmentally regulated in a tissue-specific manner in a number of species suggesting an important role in ontogeny. Thus, these two molecules work by binding to a number of signaling molecules during brain development and pathogen–host cell interactions. The inactivation of the gene encoding for CMP-*N*-acetylneuraminic acid hydroxylase by an *Alu* insertion took place about 2.8 mya. Two Neanderthal specimens indicated that they possessed the inactive gene suggesting that the *Alu* insertion occurred before the common ancestor of humans and Neanderthals approximately 0.5 mya. Of particular note is that the timing of this inactivating *Alu* insertion coincides with a dramatic increase in brain size (from an average of 400–650 cm^3) observed in the fossil record as the hominin lineages transition from australopithecines to the genus *Homo*.

The microcephalin gene (*MCPH1*) represents another specific example of a gene involved in brain growth that has been affected by TEs.[28] *MCPH1* is a gene that is expressed during fetal brain development. The gene has 14 exons (protein-coding portion of the gene), and many of the introns (noncoding portion of the gene) include TE sequences. Overall, the *MCPH1* gene contains 57% TEs. Specifically, the 14th exon is made up of 88% of an *Alu*Y sequence. The TEs appear to have been part of the *MCPH1* introns for some time, considering the amount of mutation they have accumulated (determined by mismatches). Because the gene is functional, it appears that the insertion of TEs played a role in its evolution. Another example of TEs playing a role in development involves *Alu* RNA editing. In this mechanism of developmental control, *Alu* sequences function in changing or editing RNA sequences of genes. RNA editing comparisons by *Alu* sequences in the brain cells of humans, chimpanzees, and rhesus monkeys showed a higher level of adenosine-to-inosine nucleotide RNA editing in humans.[29] This suggests that increased *Alu* RNA editing may have been adapted by natural selection in hominins and may act as an alternative information mechanism for genes in the brain. Furthermore, a number of studies have examined TEs in the genomes of primates and other vertebrates to better understand their role in evolution. Mills and colleagues, for example, compared the human and chimpanzee genomes for unique TEs and found almost 11,000 transposons that were differentially present in humans and chimpanzees. *Alu*, SVA, and L1 insertions composed more than 95% of the total in both species, and about 34% of the insertions were located within genes. The data also indicated that there were more transpositions in humans and that the insertions represent species-specific variation that could have contributed to their divergence.[30] In a different study, Britten examined the sequences of transposons in vertebrates.[31] He found that of the 2732 TEs detected, around 1700 were found only in humans. Among all of the TEs only *Alus* made perfect matches, indicating recent insertion events. All of the examples with multiple copies in humans were *AluY* families and are considered to be young *Alus*. There were 655 perfect full-length matches in the human genome and only 283 in the chimpanzee genome that are considered recent events. When comparing humans and chimpanzees, 5530 new *Alus* were seen in humans, whereas only 1642 were seen in chimpanzees.

Darwin had no idea what cellular factors were involved in evolutionary change. Even after DNA was identified as the genetic material most investigators could not have envisioned the rate of change in the DNA afforded by these transposition events, and had no idea these types of DNA elements could exist. This lack of knowledge and the revolutionary nature of the implications to evolution were some of the reasons why Barbara McClintock's revolutionary results were ignored by the scientific community for so long.

Core Duplicons

Core duplicons are segments of DNA that are duplicated and dispersed *only* within the genome of hominins and great apes.[32] They are also referred to as segmental duplications. Core duplicons move throughout the genome and can randomly carry neighboring pieces of DNA with them (Fig. 1.14). In other words, every time core duplicons move, they can include additional genes or portions of genes with them. It is as if duplicons function as carriers of genetic material from one location in the genome to another. A dozen or so families of duplicons have been discovered. Several of these families are associated with cell proliferation and neuronal functions. As depicted in Fig. 1.14, it is thought that they move with their attached genes into novel genomic locations.

The genes that have been amplified by duplicons in the hominin- and great ape-specific lineages tend to be involved in signal transduction (intra- and intercellular metabolic

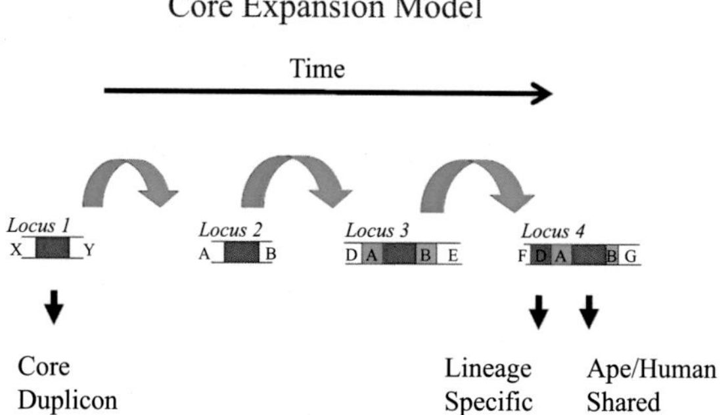

FIGURE 1.14 Hypothesis of core duplicon expansion. Initial core DNA is repeatedly duplicated, and over time the core duplicon is duplicated again bringing additional genes every time. *From https://evomics.org/2014/01/cnvs-and-humans-are-cool/.*

communication), neuronal activities such as neurotransmitter release and nerve impulse transmission as well as muscle contraction. In addition, a number of genes involved in human adaptations are found in high copy number associated with duplicons in hominins.[33] In comparison, in other lineages such as in the macaque line, duplicons contain genes for biological processes associated with amino acid metabolism or oncogenesis (the creation of cancerous cells). This type of data suggests a shift in the types of genes that have been amplified during hominin evolution.[33] Although duplicon-linked genes are expressed in a number of tissues, in humans they are particularly active in the brain, and more specifically in neurons. It is unknown how core duplicons originated and what provoked the spurt of duplication events early in hominin evolution. Also cloaked in mystery are the forces that fuel their expansion and the dispersion mechanisms that maintain their expansion.

Clearly genomic bursts in duplication activity have occurred at different times during hominin evolution. A significant spurt of 4- to 10-fold occurred just before the split between the human and the great ape lineages. It is interesting that this increment of duplicon insertions took place at a time when other mutational processes, such as single nucleotide changes and transposon amplification, were slowing down in the hominoid lineage. This increase in duplicon activity may have been the result of genomic destabilization at a period before and perhaps during hominin speciation.

Because duplicated pieces of DNA usually are not subject to intense natural selection, these duplicons and the genes they transport provide for rapid accumulation of mutations and the potential for evolutionary change. The duplicated sequences carried by duplicons tend to differentiate faster in comparison with other genes since they are expendable in nature because the original copies are still intact and functioning. They represent extra DNA that can afford to mutate without impacting the fitness of the organisms because the original copies are around to provide the essential functions. Also significant about these core duplicons is their speed of dispersal and their involvement with the dissemination of growth genes specifically expressed in neuronal tissue. Classical evolutionary theory or neo-Darwinism does not provide an explanation for the consequences of these observations.

Another noteworthy characteristic of duplicons include their involvement in dramatic episodes of chromosomal alteration. In hominins, core duplicon–mediated DNA alterations may represent one of the mechanisms unique to recent hominin evolution and an important driving force that shaped our species in a relatively short time period as opposed to slow geological time. Also significant is that the genes transported within these cassette-type duplicons are subject to strong selection pressure in comparison with other genes. In addition, these growth-promoting DNA sequences are disproportionally very active (transcriptionally), suggesting functional importance. In other words, the genes within core duplicons may represent the proverbial "fast-evolving" human genes long theorized to exist.

The GOLGA Duplicon

GOLGA is one of a growing number of core duplicons found to have been active during hominin evolution.[34] Genetic instability in the long arm (q) of chromosome 15, specifically in the 15q11-15q13 location, where *GOLGA* resides, compromises intellectual capacity. The name *GOLGA* derives from the golgin family of coiled-coil proteins known to facilitate transport to the Golgi apparatus (cellular organelle). *GOLGA* is responsible for the impressive expansion of a primate-specific golgin gene family during the last 20 million years of primate evolution. The majority of the sticking structural changes in this DNA region occurred during a very short evolutionary time period of 400,000 years (500,000–900,000 years ago). At this time, hominin species such as *Homo heidelbergensis* were evolving into archaic *Homo sapiens* in Africa and into Neanderthals in Europe and Denisovans in Asia. Currently in our species, the area exists in a polymorphic state with five different DNA configurations in different proportions among human populations.

In addition to the radical duplication and restructuring of the golgin genes in the *Homo* lineage, *GOLGA* is also implicated with rapid amplification of the RHGAP11B gene located in the same general region of the genome. RHGAP11B is a human-specific gene involved in proliferation of neurons and an increase in neocortex folding.[35] The duplication of RHGAP11B took place subsequent to the separation of the hominin and chimpanzee lineages but before the human–Neanderthal split. It is theorized that the duplication of this gene is implicated in the evolutionary expansion of the human neocortex. During that period of time the hominin brain increased in size from approximately 400 cm^3 to about 1200 cm^3, a dramatic threefold increase.

The SRGAP2 Gene

Another specific example of these core duplicon-based amplifications is provided by the formin-binding protein 2 (FNBP2).[36] FNBP2 is a protein that activates GTPases, a group of enzymes in nucleic acid metabolism. FNBP2 is coded by the *SRGAP2* gene (Fig. 1.15). The FNBP2 protein promotes neuronal movement and differentiation as well as nerve connections known as synapses. Remarkably, the *SRGAP2* gene has been duplicated several times in hominins to generate 23 paralogous loci (genes that are derived from the same ancestral gene by gene duplication) during hominin evolution (Fig. 1.15). Only one copy, the ancestral, is present in other mammals and primates. The original duplication occurred approximately 3.4 mya followed by a second duplication 2.4 mya and a third about 1 mya. The timing for these duplications are determined by the rate of mutation. Since all these duplications absorb different mutations they differ slightly in their nucleotide composition. Assuming a constant mutation rate scientists can estimate the time that has elapsed since the last duplication by following the number of mutations

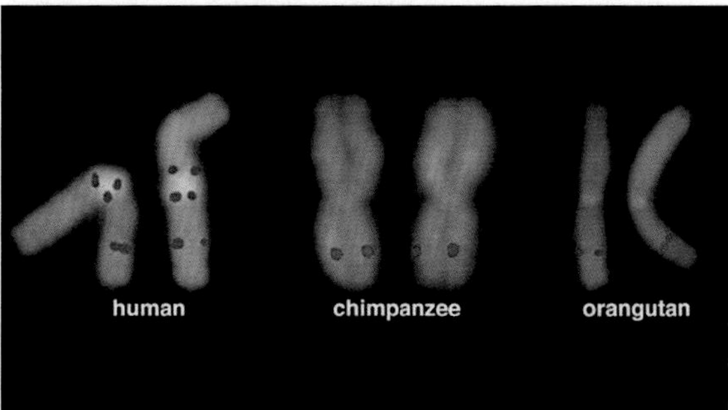

FIGURE 1.15 Duplicon-mediated amplification of the SRGAP2 gene in hominins. Chromosome pairs are aligned with the long arm (q) below the center (centromere) and the short arm (p) above the centromere. Red signals indicate the location of the duplicated SRGAP2 gene. *From https://www.quantamagazine.org/20140102-a-missing-genetic-link-in-human-evolution/.*

that have occurred since the duplication occurred. This so called molecular clock is not specific but provides a good estimate of time lines. These duplications gave rise to a number of new genes, *SRGAP2B*, *SRGAP2C*, and *SRGAP2D*. *SRGAP2C* is a truncated version of the original gene, *SRGAP2*, and it is known to inhibit its activity and promotes neuronal migration. In addition, *SRGAP2C* delays the aging process of neurons, increases neuronal density and the number of neuronal synapses for communication contacts between neurons. It is interesting that the duplication that created *SRGAP2C* took place about 2.4 mya at the time when cranial expansion dramatically started in the hominin lineage. During this period of time, species such as *Homo habilis* (a possible ancestor to modern humans) developed a brain size of 650 cm^3 compared with australopithecines' 400 cm^3 and *Homo erectus*' 930 cm^3, groups who lived before (3.0 mya) and after (1.8 mya) *Homo habilis*, respectively. Although these genetic alterations are congruent with increases in brain size, it is baffling why they happened (or at least were retained) so frequently and only in the hominin line. It is likely that these gene duplication events are more in line with punctuated evolutionary changes rather that with more gradual neo-Darwinism mechanisms because they provide a relatively quick source of new genetic material that can absorb mutations and possibly lead to new genes for evolution to act on.

Duplicon DUF1220

DUF1220 is another case of duplicon-type amplification that codes for a protein domain that exhibits a remarkable human lineage–specific increment in copy number.[37] The data from various disciplines strongly suggest that this duplicon is associated with nerve cell number, brain size, and possibly with hominin brain evolution (Fig. 1.16). This duplicon is responsible

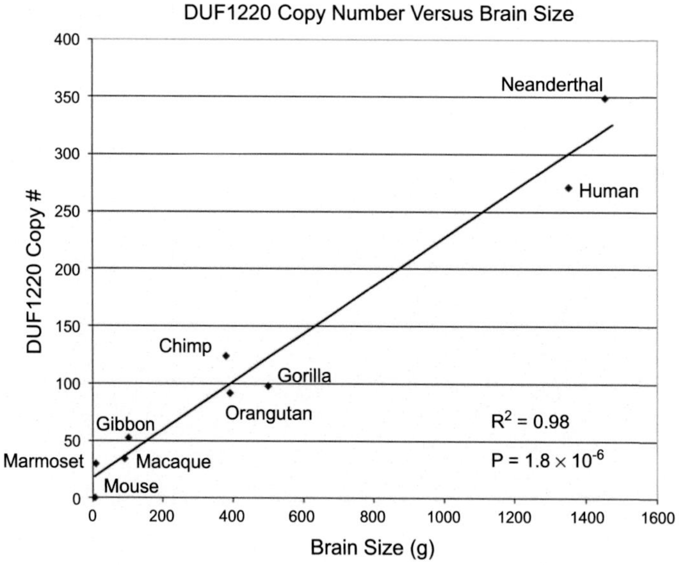

FIGURE 1.16 DUF1220 copy number and brain size. *From http://dienekes.blogspot.com/2012/08/our-big-human-brains-may-depend-on.html.*

for the largest increase in copy number of coding DNA in the human genome. The increase of *DUF1220* in primate and hominin species has been steady and dramatic. Over 270 copies are observed in our species in a state of polymorphin with some of us possessing different copy numbers, which are passed to the next generation. For instance, the chimpanzee genome has 125 copies, 145 less than contemporary humans, a twofold increment. These species are separated by only about 6 million years of evolution. Smaller copy numbers are seen in the orangutan (92) and Old World monkeys (35). Single or a few copies are present in nonprimate mammals and absent in nonmammals. It is theorized that our last common ancestor with the chimpanzee had about 102 *DUF1220* duplicons.[38]

DUF1220 regions are made up of DNA encoding proteins of 65 amino acids and constitute domains within proteins. In humans these duplicons are located mainly on the long arm (q) of chromosome 1, specifically in the region 1q21.1-q21.2, with additional duplicates in 1p36, 1p13.3, and 1p12. *DUF1220* duplicons are under strong positive selection pressure and are primarily expressed in the neurons of the brain. Of note, the ancestral DNA from which the original *DUF1220* derives is part of a gene known as NBPF (a neuroblastoma breakpoint family of proteins) that functions in regulating gene expression as a transcription factor that controls neuroblastoma formation. Remarkably, the NBPF family of proteins has been implicated in human evolution.[39]

Deletions and duplications of this duplicon have been associated with microcephaly and macrocephaly, respectively. Furthermore, a direct proportional relationship between number of *DUF1220* copies and amount of gray matter in the cerebral cortex of healthy humans has been reported.[40] Also interesting is that the number of *DUF1220* copies has been found to be directly proportional to scores in IQ (intelligence quotient) and mathematical aptitude tests.[37] In other words, *DUF1220* copy number linearly determines cognitive functions. In a recent study of European and North American males, particularly ranging in age from 6 to 11, an increase in *DUF1220* copy number was found to correspond with a statistically significant increase in the IQ test. In this group of subjects, each additional copy of the duplicon correlated with a 3.3-point increase in IQ. Similarly, in New Zealanders increments in *DUF1220* copy number correlated with an increase in math scores in aptitude tests.

The copy number of *DUF1220* in humans is also associated with a number of behavioral and anatomical disorders such as autism, schizophrenia, microcephaly, and macrocephaly. It seems that at least some of these core duplicons not only correlate with a revolutionary increase in brain size and intellectual ability but with genetic diseases as well. Thus, these fast-expanding duplicons may represent an evolutionary double-edge sword. On one hand, they may promote rapid evolutionary change, potentially advantageous for the survival of organisms, and on the other hand, the same fast-paced mechanism may promote chromosomal aberrations leading to pathological conditions related to neuronal tissue.

Rapidly Expanding Short Tandem Repeats

Short tandem repeats (STRs) are segments of DNA that consist of 2–100 nucleotides repeated in tandem and in the same orientation, that can be repeated 5 to over 200 times (Fig. 1.17).[41] Most STRs are located in no-coding regions (DNA regions not involved in RNA production) of the genome. STRs make up about 3% of the human genome and are notorious for their high duplicative mutation rates ranging from 1×10^{-3} to 1×10^{-4} per generation.[42] In other words,

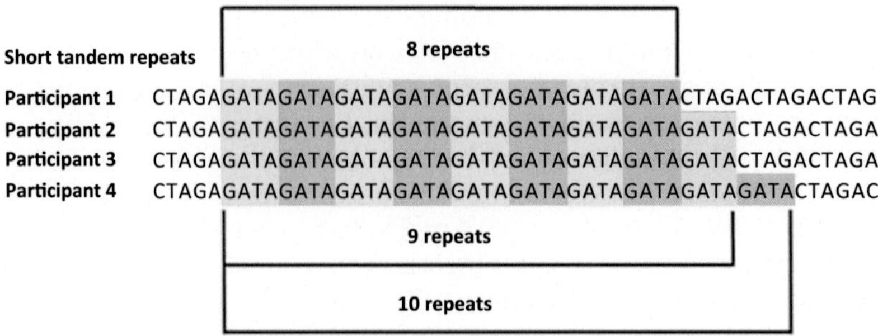

FIGURE 1.17 Diversity in numbers of short tandem repeats. The repeating unit is GATA. *From http://www.stewartsociety.org/images/bannockburn-tandem-repeats.jpg.*

Disease	Mode of Inheritance	Trinucleotide Repeat
Huntington disease	AD	CAG
Myotonic dystrophy 1	AD	CTG
Spinocerebellar ataxia 1 (SCA1)	AD	CAG
Dentatorubral-Pallidoluysian atrophy (DRPLA)	AD	CAG
Fragile X syndrome	XL	CGG
Oculopharyngeal muscular dystrophy	AD and AR	GCG
Friedreich ataxia	AR	GAA

FIGURE 1.18 Diseases caused by trinucleotide repeat expansion. *AD*, autosomal dominant; *AR*, autosomal recessive; *XL*, X linked. *From http://www.ncbi.nlm.nih.gov/bookshelf/br.fcgi?book=gene.*

any given STR site experiences changes in its number of repeat units (usually increasing) once in every 1000–10,000 generations. For comparison, other types of mutations such as nucleotide deletions or substitutions occur at an estimated rate of about 1.1×10^{-8} per site per generation. Thus, the rate of mutations of STRs is four to five orders of magnitude higher than that of nucleotide mutations. This elevated capacity to change makes them powerful agents of genetic diversity potentially affecting functions such as chromatin (DNA and chromosomal proteins) folding, transcription initiation, transcription termination, RNA processing, DNA recombination, and at times even creating additional protein domains.

In the early 1990s, a peculiar category of these repeated sequences was discovered in humans. This type of STRs experience unusually high mutation rates, compared with regular STRs, on the order of several duplications in one generation.[43] This type of STRs is referred to as *triplet repeat expansion sequences* and, as the name indicates, is made up of units of three nucleotides located in tandem and reiterated sometimes 100 of times. The trinucleotide repeat of CAG (cytosine-adenine-guanine) is an example that exists in humans at the Huntington's disease (HD) locus. Normal individuals have a CAG triplet repeat less than 35, where as individuals with 40 or more repeats will show the dominant disease phenotype. These triplet repeat expansion sequences are often found within protein-coding DNA. Medically, these STRs have been implicated in dozens of conditions. Fig. 1.18 illustrates some of the best-characterized

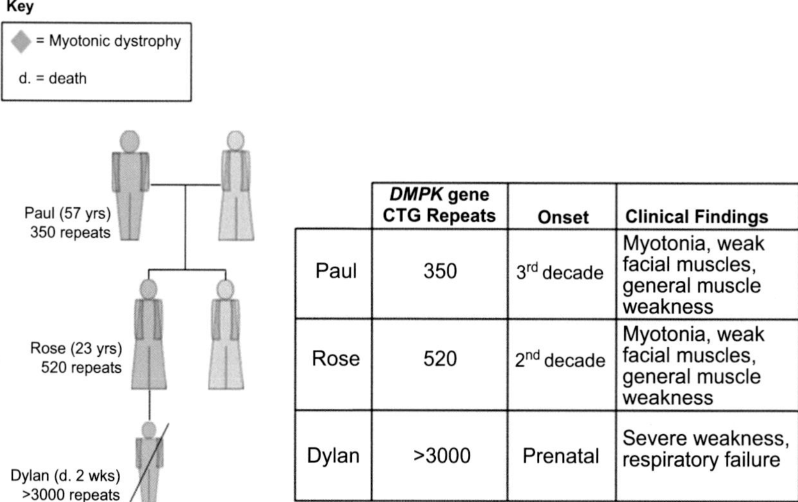

	DMPK gene CTG Repeats	Onset	Clinical Findings
Paul	350	3rd decade	Myotonia, weak facial muscles, general muscle weakness
Rose	520	2nd decade	Myotonia, weak facial muscles, general muscle weakness
Dylan	>3000	Prenatal	Severe weakness, respiratory failure

FIGURE 1.19 The phenomenon of anticipation in trinucleotide repeat expansion. *From http://www.ncbi.nlm.nih.gov/bookshelf/br.fcgi?book=gene.*

human maladies associated with these repeats. In general, reiteration on the order of a dozen repeats is the normal state. However, further expansions are known to promote a number of genetic disorders. In medical conditions, the abnormal phenotype deteriorates as the number of repeats increases from a few hundred to several hundred in one generation (Fig. 1.19). In addition, a phenomenon known as anticipation is observed in which the age of disease onset is advanced as the number of triplets increase. In the case of myotonic dystrophy, expansions of several thousand repeats are lethal prenatally (Fig. 1.19). In general, these STRs are also more unstable and prone to increase in size the longer they get. Interestingly, it has been observed that higher primates possess longer tracts of these repeats.

From an evolutionary perspective, the dynamic expansion mode that these STRs exhibit, as well as their mutator properties, make them powerful generators of genetic variability for potential evolutionary change. In other words, they may provide selective advantages in short periods of time as opposed to geological time. A number of documented examples hint at this possibility. For example, in the gene for spinocerebellar ataxia, hyperexpansion of the CAG repeat is known to cause the pathological condition (Fig. 1.18). Yet, the *long nonpathological* premutation state of the gene, exhibiting a limited number of repeats, seems to be under positive selection in a European population.[44] Similarly, data suggest that positive selection in northern European populations may have augmented the frequency of the shorter of two polythymidine repeat track at a transcription factor DNA-binding site in the human matrix metalloproteinase gene (*MMP3*).[45] Furthermore, carriers of HD have been shown to possess higher reproductive capacity probably resulting from elevated immune activity.[46] The data suggest that because health and reproductive output are positively related, the health benefits could positively select for the Huntington's disease expanded STR state. It is likely that these expansion sequences provide abundant genetic diversity for extremely fast (in generation time) adaptive changes.

References

1. Darwin C. *The descent of man, and selection in relation to sex*. London: John Murray; 1871.
2. Hyung-Soon Y, Sung Cho Y, Guang X, Gyun Kang S, et al. Minke whale genome and aquatic adaptation in cetaceans. *Nat Genet* 2014;**46**:88–92.
3. Bell EA, Boehnike P, Patrick T, Harrison M, et al. Potentially biogenic carbon preserved in a 4.1 billion-year-old zircon. *Proc Natl Acad Sci USA* 2015;**112**:14518–21.
4. Adler R, Canto-Soler MV. Molecular mechanisms of optic vesicle development: complexities, ambiguities and controversies. *Dev Biol* 2007;**305**(1):1–13.
5. Nilsson DE, Pelger S. A pessimistic estimate of the time required for an eye to evolve. *Proc Royal Society B* 1994;**256**(1345):53–8.
6. Pennisi E. Genomics. ENCODE project writes eulogy for junk DNA. *Science* 2012;**337**(6099):1159. 1161.
7. Scott MP, Weiner AJ. Structural relationships among genes that control development: sequence homology between the Antennapedia, Ultrabithorax, and fushi tarazu loci of Drosophila. *Proc Natl Acad Sci USA* 1984;**81**(13):4115–9.
8. Darwin C. *On the origin of species*. London: John Murray; 1859.
9. McClintock B. The origin and behavior of mutable loci in maize. *Proc Natl Acad Sci USA* 1950;**36**(6):344–55.
10. Rowold DJ, Herrera RJ. . Alu elements and the human genome. *Genetica* 2000;**108**:57–72.
11. Pace II JK, Feschotte C. The evolutionary history of human DNA transposons: evidence for intense activity in the primate lineage. *Genome Res* 2007;**17**(4):422–32.
12. Mills RE, et al. Which transposable elements are active in the human genome? *Trend Genet* 2007;**23**:186–91.
13. Marchetto MC, et al. Differential L1 regulation in pluri- potent stem cells of humans and apes. *Nature* 2013;**503**:525–9.
14. Bundo M, et al. Increased L1 retrotransposition in the neuronal genome in schizophrenia. *Neuron* 2014;**81**:306–13.
15. Sen SK, et al. Human genomic deletions mediated by recombination between Alu elements. *Am J Hum Genet* 2006;**79**:41–53.
16. Raiz J, et al. The non-autonomous retrotransposon SVA is trans-mobilized by the human LINE-1 protein machinery. *Nucleic Acids Res* 2012;**40**:1666–83.
17. Konkel MK, Walker JA, Batzer MA. LINEs and SINEs of primate evolution. *Evol Anthropol* 2010;**19**(6):236–49.
18. Cordaux R, et al. Estimating the retrotransposition rate of human Alu elements. *Gene* 2006;**373**:134–7.
19. Xing J, et al. Emergence of primate genes by retro-transposon-mediated sequence transduction. *Proc Natl Acad Sci USA* 2006;**103**:17608–13.
20. Hedges DJ, Batzer MA. From the margins of the genome mobile elements shape primate evolution. *Bioessays* 2005;**27**:785–94.
21. Cordaux R, Batzer MA. The impact of retrotransposons on human genome evolution. *Nat Rev Genet* 2009;**10**:691–703.
22. Suzuki S, et al. Retrotransposon silencing by DNA methylation can drive mammalian genomic imprinting. *PLoS Genet* 2007;**3**:e55.
23. Parks MM, Lawrence CE, Raphael BJ. Detecting non-allelic homologous recombination from high-throughput sequencing data. *Genome Biol* 2015;**16**:72.
24. Hedges DJ, Callinan PA, Cordaux R, Xing J, Barnes E, Batzer MA. Differential Alu mobilization and polymorphism among the human and chimpanzee lineages. *Genome Res* 2004;**14**:1068–75.
25. Hodgkinson A, Eyre-Walker A. Variation in the mutation rate across mammalian genomes. *Nat Rev Genet* 2011;**12**:756–66.
26. Locke DP, Hillier LW, Warren WC, Worley KC, Nazareth LV, Muzny DM, Yang SP, Wang Z, Chinwalla AT, Minx P, Mitreva M, Cook L, Delehaunty KD, Fronick C, Schmidt H, Fulton LA, Fulton RS, Nelson JO, Magrini V, Pohl C, Graves TA, Markovic C, Cree A, Dinh HH, Hume J, Kovar CL, Fowler GR, Lunter G, Meader S, Heger A, et al. Comparative and demographic analysis of orangutan genomes. *Nature* 2011;**469**:529–33.
27. Hayakawa T, Satta Y, Gagneux P, Varki A, Takahata N. Alu-mediated inactivation of the human CMP-N-acetylneuraminic acid hydroxylase gene. *PNAS USA* 2001;**98**:11399–404.
28. Shi L, et al. Functional divergence of the brain size regulatory gene MCPH1 during primate evolution and the origin of humans. *BMC Biol* 2013;**11**:1–6.
29. Bazak L, et al. A-to-I editing occurs at over a hundred million genomic sites, located in a majority of human genes. *Genome Res* 2014;**24**:365–76.

30. Mills RE, et al. Recently mobilized transposons in the human and chimpanzee genomes. *Am J Hum Genet* 2006;**78**:671–9.
31. Britten RJ. Transposable element insertions have strongly affected human evolution. *Proc Natl Acad Sci USA* 2010;**107**:19945–8.
32. Muehlenbein MP. *Human evolutionary biology*. London, UK: Cambridge University; 2010.
33. Marques-Bonet T, Kidd JM, Ventura M, Graves TA, Cheng Z, Hillier LW, Jiang Z, Baker C, Malfavon-Borja R, Fulton LA, Alkan C, Aksay G, Girirajan S, Siswara P, Chen L, Cardone MF, Navarro A, Mardis ER, Wilson RK, Eichler EE. A burst of segmental duplications in the African great ape ancestor. *Nature* 2009;**457**(7231):877–81.
34. Antonacci F, Dennis MY, Huddleston J, Sudmant PH, Meltz Steinberg K, Rosenfeld JA, Miroballo M, Graves TA, Vives L, Malig M, Denman L, Raja A, Stuart A, Tang J, Munson B, Shaffer LG, Amemiya CT, Wilson RK, Eichler EE. Palindromic GOLGA8 core duplicons promote chromosome 15q13.3 microdeletion and evolutionary instability. *Nat Genet* 2014;**46**(12):1293–302.
35. Florio M, Albert M, Taverna E, Namba T, Brandl H, Lewitus E, Haffner C, Sykes A, Wong FK, Peters J, Guhr E, Klemroth S, Prüfer K, Kelso J, Naumann R, Nüsslein I, Dahl A, Lachmann R, Pääbo S, Huttner WB. Human-specific gene ARHGAP11B promotes basal progenitor amplification and neocortex expansion. *Science* 2015;**347**(6229):1465–70.
36. Dennis MY, Nuttle X, Sudmant PH, Antonacci F, Graves TA, Nefedov M, Rosenfeld JA, Sajjadian S, Malig M, Kotkiewicz H, Curry CJ, Shafer S, Shaffer LG, de Jong PJ, Wilson RK, Eichler EE. Evolution of human-specific neural SRGAP2 genes by incomplete segmental duplication. *Cell* 2012;**149**(4):912–22.
37. Davis JM, Searles VB, Anderson N, Keeney J, Raznahan A, Horwood LJ, Fergusson DM, Kennedy MA, Giedd J, Sikela JM. DUF1220 copy number is linearly associated with increased cognitive function as measured by total IQ and mathematical aptitude scores. *Hum Genet* 2015;**134**(1):67–75.
38. O'Bleness MS, Dickens CM, Dumas LJ, Kehrer-Sawatzki H, Wyckoff GJ, Sikela JM. Evolutionary history and genome organization of DUF1220 protein domains. *G3 (Bethesda)* 2012;**9**:977–86.
39. Zhou F, Xing Y, Xu X, Yang Y, Zhang J, Ma Z, Wang J. NBPF is a potential DNA-binding transcription factor that is directly regulated by NF-κB. *Int J Biochem Cell Biol* 2013;**45**(11):2479–90.
40. Marques-Bonet T, Eichler EE. The evolution of human segmental duplications and the core duplicon,. *Cold Spring Harbor Symp Quant Biol* 2009Vol. 74:355–62.
41. Turnpenny P, Ellard S. *Emery's elements of medical genetics*. London: Elsevier; 2005.
42. Brinkmann B, Klintschar M, Neuhuber F, Huhne J, Rolf B. Mutation rate in human microsatellites: Influence of the structure and length of the tandem repeat. *Am J Hum Genet* 1998;**62**(6):1408–15.
43. Richards RI, Sutherland GR. Dynamic mutation: possible mechanisms and significance in human disease. *Trends Biochem Sci* 1997;**22**(11):432–6.
44. Yu F, et al. Positive selection of a preexpansion CAG repeat of the human SCA2 gene. *PLoS Genet* 2005;**1**(3):e41.
45. Rockman MV, et al. Positive selection on MMP3 regulation has shaped heart disease risk. *Curr Biol* 2004;**14**:1531–9.
46. Eskenazi BR, Wilson-Rich NS, Starks PT. A Darwinian approach to Huntington's disease: subtle health benefits of a neurological disorder. *Med Hypotheses* 2007;**69**:1183–9.

CHAPTER

2

Early Hominins

The proposition that humans have mental characteristics wholly absent in non-humans is inconsistent with the theory of evolution. ***Gary L. Francione***[1]

SUMMARY

This chapter deals with the habitats and environmental conditions that facilitated the adaptive changes seen in hominin evolution, such as bipedalism and enlargement of the brain. Specifically, we examine the role of the East African Ridge in development of bipedalism and encephalization in early hominin evolution. This ridge extends in a north-to-south direction, and its formation climatically partitioned the region in half. To the west, the habitat remained forested and humid, suitable for an arboreal existence. To the east, the landscape changed abruptly to a dry savanna in which tree dwelling was not an option. These differing climates may have driven early hominin adaptation toward a bipedal mode of locomotion in response to predation or other selective forces in the savanna.

This chapter also explores a number of peculiarities and themes observed in hominin evolution including the persistence of various proportions of ancestral and derived traits within organs. For instance, when the hominin fossil record is examined, different parts of the anatomy of the hand reflect the ancient arboreal way of life, whereas others exhibit the derived structures reflecting bipedal locomotion. Another theme within hominin evolution is the high degree of diversity in terms of numbers of closely related coexisting species and their cultural and reproductive interactions. Traditionally, evolutionary relationships among organisms have been illustrated in the form of tree branches, with a single line connecting, one to one, the species derived from a single ancestral group. Scholars are beginning to realize that this type of depiction represent an oversimplification in hominin evolution. The observed cohabitation of multiple groups and the mosaic of ancestral and derived traits seen in single organs of hominins are better depicted by a road that branches out into secondary paths in different directions only to reunite again further down the way. This chapter also examines the various groups of hominins that have been discovered in the fossil record with an effort to establish a continuity of the adaptive changes observed.

Ancestral DNA, Human Origins, and Migrations
https://doi.org/10.1016/B978-0-12-804124-6.00002-1

SUB-SAHARAN NORTHEAST AFRICA: VARIABLE HABITAT AND SELECTION PRESSURES

The Hominin Lineage Was Born at a Time of Dramatic Environmental Changes.

In the previous chapter, we explored a number of unique genetic mechanisms that have been active during hominin evolution. The resulting dramatic genetic changes were capable of generating genetic diversity in a short period of time and in doing so provided adaptive changes in a matter of generations as opposed to geological time. As discussed, a number of these rapid-paced alterations have been linked to neuronal development and increased brain size. Yet, to bring about adaptive evolutionary change, a second element is required. That is, in addition to the genetic diversity generated by mutations, the environment needs to exert selection pressure to allow the fittest individuals to reproduce and pass their DNA to the next generation while negatively affecting the reproduction of the less fit. In this respect, hominins' environmental conditions were also unusual. It turns out that simultaneously to the rapid genetic alterations experienced by great apes and hominins, Northeast sub-Saharan Africa provided for dramatic geological and climatic changes. Sub-Saharan Northeast Africa is the geographical region in which early hominins likely evolved from the common ancestor of the great ape and human lineages. In combination, the rapid genetic mechanisms and the dramatic habitat changes may have created a unique combination of circumstances and conditions that provided for the dynamic evolution experienced by hominins. In this section, the environmental conditions that impacted hominin evolution are described.

The earliest hominins discovered date to about 7 million years ago (mya). The dig sites are mainly concentrated in sub-Saharan Northeast Africa in what is today Ethiopia, Eritrea, and northern Kenya (Fig. 2.1). Although not all scholars agree, fossils classified under the genus *Sahelanthropus* could have been the first hominins. These early fossils represent preaustralopithecine individuals. Starting 4 mya, distribution of *Sahelanthropus* widens within Africa as the fossils become more abundant. The oldest *Sahelanthropus* fossil was discovered in Toros-Menalla, in current-day northern Chad, Central Africa, in a region that borders the arid Saharan desert. This specimen is exclusively comprised of a cranium. In Northeast Africa, the most ancient hominins are from the Aramis site (see the Middle Awash location in Fig. 2.1). Other sites in Northeast Africa

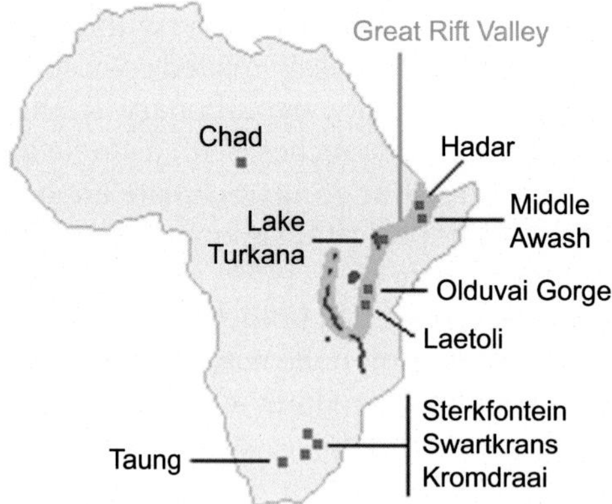

FIGURE 2.1 Major early hominin sites. *From http://anthro.palomar.edu/hominid/australo_1.htm.*

include Tugen Hills and Awash. All of the individuals of these ancient groups seem to have moved by some sort of primitive bipedalism, probably in a facultative (not all the time) mode. It is interesting that these very early hominins lived in locations characterized by a humid environment made up of forests instead of the grassland savannas, typical of the environment of later hominins. At this critical stage in hominin evolution, sparse woodlands, dry grasslands, and savannas were progressively replacing forests. In other words, the earliest locales, such as Toros-Menalla and Aramis, represented the ancient habitat sites where hominin and great ape lineages diverged as drier grassland savannas replaced the forest, and later hominins became progressively more bipedal. It is thought that in open grassland environments bipedalism would be advantageous.

The East African Rift

Sub-Saharan Northeast Africa is thought, by many, to be the origin of the hominin lineage,[2] although other locations such as Central and South Africa have been proposed.[3] Geologically, the region of Northeast Africa is extremely active as three tectonic plates are currently splitting from each other at a rate of about 7 mm per year.[4] These three plates, the Arabian, Nubian, and Somali, are separating from each other along the East African Rift Zone (Fig. 2.2). Eventually,

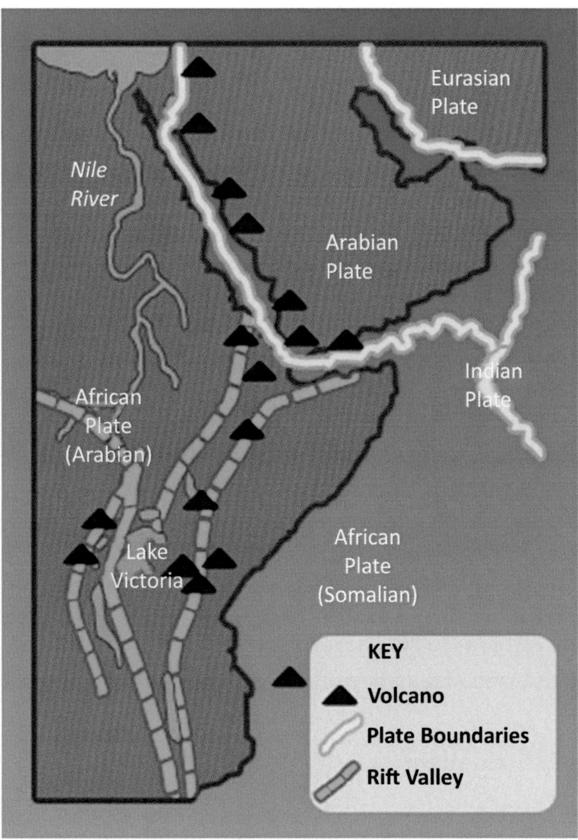

FIGURE 2.2 The East African Rift Zone. *From http://www.eclp.com.na/demo/geography/mod1page18.html.*

this process will tear sub-Saharan Northeast Africa from the rest of the continent. The forces driving the rift started about 25 mya, and it is expected that by 10 million years from now the Somali plate will separate completely forming a new island continent as it moves in a north-easterly direction. In the process, a new ocean between the African continent and the Horn of Africa east of the ridge will be created. It has been postulated that upwelling magna pressing up against the East African crust is creating the rift. The ridge that is being formed is so apparent that it can be seen from the space station at an altitude of 205–270 miles in the form of a depression and a progression of lakes starting with Lake Victoria at the northern end to Lake Malawi at the southern fridge of the rift. The valley created as part of the ridge reaches a maximum of 120 km in width in some areas. In addition, the rift valley is flanked by steep mountain ranges created by the tension and faulting of the tectonic plates.

It turns out that early hominin evolution and the dramatic geological activity splitting Northeast Africa and creating a new continent seem to be linked. It is not coincidental that a number of major events in hominin evolution likely transpired in Northeast Africa in the midst of rapid environmental and climatic changes. The geological episodes described in the previous paragraph most likely initiated an epoch of environmental changes that modified the habitat of the organisms in the region, including the common ancestors of the great apes and ours. This time period encompasses 10 million years of shifting habitats, from a high humidity forest to arid savanna.

The East African Rift Valley has altered the landscape to a fault basin with narrow, long, deep lakes. In addition, the same pressure from the magna pushing against the earth's crust created mountain chains, the East African Ridge. This ridge extends in a north-to-south direction, and its formation climatically partitioned the region in half. To the west, the habitat remained forest and humid. Yet, to the east the conditions progressively changed becoming increasingly arid, eventually extinguishing the forest. This drying trend eventually transformed the forest to the expanding savanna habitat that exists today. These dramatic changes in climate and habitats included extreme oscillations of dry and wet weather conditions in short periods of time. It has been proposed that these dramatic fluctuations in the environment played an important role in driving hominin evolution.[5] Specifically, it is thought that the transformation of forests to savannas provided for different and unique selection pressures that led to adaptive changes in the early hominins. Bipedalism was one of the initial developments in early hominin evolution as an adaptation to terrestrial locomotion in the treeless landscape. In addition, it has been theorized that drastic climatic variability east of the ridge that resulted in shifting selection pressures may have contributed to encephalization, technological innovations, culture, and the eventual migrations out of Africa. For instance, the extremely rapid brain enlargement experienced by *Homo erectus* around 1.8 mya coincided with the geological events and habitat changes east of the ridge. These sudden alterations in habitat in combination with the genetic changes previously discussed may have provided the setting for punctuated evolution episodes, which could have contributed to the revolutionary increase in brain size experienced within the hominin lineage.[6]

Upright Posture and Bipedalism

Various theories have been advanced to explain what triggered the split that created the road to humanity. From the fossil record, adoption of bipedalism was the major event early in hominin evolution. It seems that after the initial change from arboreal to terrestrial

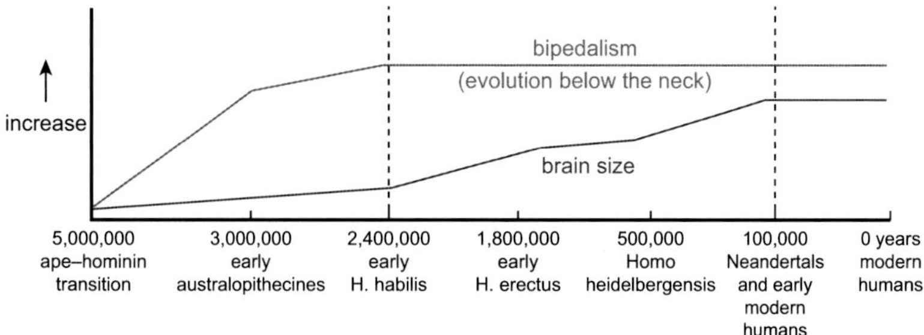

FIGURE 2.3 Timelines of bipedalism and brain size in hominins. *From http://anthro.palomar.edu/homo2/mod_homo_2.htm.*

locomotion as an adaptation to survive in the deforested savannas of Northeast Africa, early hominins began to evolve a number of additional traits that characterize humanity today.[7] In other words, bipedalism is considered the initial, critical step that initiated hominin evolution.

Upright posture and bipedal locomotion are seen early in the hominin lineage, evolving very rapidly soon after the ape–hominin separation (Fig. 2.3). During the first 2 million years of hominin evolution (5–3 mya), traits associated with bipedalism evolved at a high rate (Fig. 2.3). Then, amid the transition between early australopithecines and early *Homo habilis*, the rate of change involving characters related to bipedalism decelerated and eventually reached a plateau about 2.5 mya with *H. habilis*.

Yet, there are many scholars who believe that anatomically modern humans are not ready for bipedal locomotion. They are of the opinion that there has not been enough time for our backs to evolutionarily adapt to being upright. As attested by our numerous back problems, especially during our later years, bipedal locomotion is not working optimally in modern humans. It seems as if specific parts of our anatomy, especially the lumbar section of the spinal column, are still in need of further morphological change to achieve a less painful bipedal existence. Some anatomists also go to the extent of describing our backs as an engineering nightmare. For example, we are the only mammal capable of spontaneously fracturing vertebra. Evidence in support of this contention is seen in the anatomical changes experienced in the *H. erectus* to *Homo sapiens* stretch of hominin evolution; specifically the one less lumbar vertebra present in *sapiens* compared with *erectus*. This reduction could be interpreted as evolutionary forces still in effect selecting for stronger vertebrate columns in modern humans. In addition, a number of normal (nonpathological) anatomical variations in number of vertebrate, ranging from 32 to 35, and various morphological polymorphisms including the entire incorporation of the L5 vertebrate into the sacrum and the presence of an intervertebral disc between the S1 and S2 vertebrates are observed in contemporary humans. These various polymorphisms in the anatomy of the spinal column may be interpreted as a system in the midst of evolutionary change, in the process of experimenting with various morphological alternatives "in an effort to achieve" a less problematic upright posture.

Bipelalism predates other major developments in hominin evolution such as the rapid increase in brain size (Fig. 2.3), and the development of advanced stone tool traditions and culture. Although *Sahelanthropus tchadensis* may have been the first hominin capable of some form of facultative bipedalism for short distances, it is not until the australopithecines, about 4 mya, that hominins practiced sustained habitual erect locomotion.

The hallmark of bipedal locomotion is the position of the *foramen magnum* (an opening located at the base of the skull). This orifice connects the brain to the spinal cord. In hominin evolution, the *foramen magnum* gradually shifted from a posterior position seen in primates and other vertebrates to a more anterior location at the base of the skull. In addition, other morphological traits are associated with habitual (permanent) bipedalism. These include bicondylar angle, reduced and nonopposable big toe, higher foot arches, a posterior orientation of the anterior section of the iliac bone, a larger diameter of the femoral head, a larger femoral neck length and elongated femoral condyles.[8] Each of the abovementioned morphological modifications developed at different times and address specific and unique requirements for the various stages of bipedal locomotion, from facultative to habitual.

A number of ideas have been formulated to explain the development of bipedalism in hominins. Some involve the climatic changes and habitat alterations taking place at the time of the hominin separation from great ape lineage (see previous section). The savanna thesis, for example, suggests that a number of geological events including the creation of the East African Rift and Ridge prompted progressive ecological changes in the regional habitat transforming wet forest to arid grasslands and savannas. It is postulated that the mountains acted (as they still do) as barriers keeping rain precipitation on the west side of the East African Ridge. The lack of rain in the east of the range transformed the land into the relatively dry savanna seen today. As a result, it is envisioned that early arboreal hominin populations gradually became under strong selection pressure for traits that would allow them to survive in the new flat, rather desolate habitat. In other words, hominins needed to respond to the novel challenges to survive as species. They needed to evolve physiological and anatomical adaptations that would allow them to survive in the new dry plains of Northeast Africa. The transition from an arboreal existence to upright posture and bipedalism is observed in the fossilized skeletons of early hominins such as the australopithecines. Four million years ago, australopithecines display a medley of traits reflecting both arboreal and bipedal locomotion. In particular, members of the different species of the genus *Australopithecus* exhibited a number of lingering ancestral traits such as curved fingers designed for grasping to branches in a forest habitat while at the same time experiencing an anterior shift of the *foramen magnum* to allow for an upright posture.

Most of the models that have been put forward in an attempt to explain the evolution of bipedalism are based on the geological and climatic episodes that changed forest to savannas east of the Northeast African Range. One of the explanations for the evolution of bipedalism, for example, is that early hominins found it advantageous to adopt an upright posture to detect and frighten potential predators. This theory relies on aposematicism, and it is based on schemes designed to scare attackers away by appearing larger and as menacing as possible. Also, an upright posture may have been advantageous in the savanna as a method of surveillance against intruders and attracting mates during sexual advancements. In addition, bipedalism allows for the use of the arms and hands for various tasks other than *and during* locomotion. For instance, activities that are associated with hominin evolution such as toolmaking, hunting, and gathering/processing food as well as caring for the young, disabled or old were facilitated because of having free hands. In the rather structureless, flat landscape of grasslands, morphological and behavioral characteristics that increased the probabilities of survival were advantageous and were selected for because of the greater fitness provided to individuals.

Encephalization and the Hominin Brain

Although the hominin brain has the same overall morphology as the brains of other mammals, its cerebral cortex is far more developed. A number of other animals possess larger brains, but when the body size is taken into consideration, humans exhibit a relative brain size two times and three times larger than the bottlenose dolphin and chimpanzee, respectively. Much of the larger size of he human brain, which encompasses about 86 billion neurons, is part of the cortex and frontal lobes, areas known for a number of higher-level functions such as reasoning, abstract ideas, and consciousness.[9] The cortex is responsible for neuronal activities associated not only with self-awareness but also with mindfulness, creative thoughts, and cosmic awareness. This remarkable organ also provides for emotions, empathy, and altruism. The cortex itself is highly furrowed and made up of about 10 billion nerve cells, each interconnected to other neurons by 100 trillion synapses or junctions. These neurons communicate using about 100 neurotransmitters firing at a rate of about 10 billion times per second. Although research is continuously making advances in the neurobiology of humans and other species, it is likely that the human brain will remain as an intellectual singularity in biological evolution on this planet.

During the early stages in the evolution of traits associated with bipedalism in the hominin lineage, the brain started enlarging rather slowly (Fig. 2.3). Since the partitioning of the great ape–hominin lineages, approximately 5 mya, to the time of early *H. habilis* 2.5 mya, the brain experienced a modest increase in size from about 400 to 650 cm^3. Then, starting at about 2.5 mya, the hominin brain experienced a faster rate of growth (Fig. 2.3). Subsequently, after a growth plateau lasting about 800,000 years, during the period between *H. erectus* (1.5 mya) and *Homo heidelbergensis* (500,000 years ago [ya]), hominins experienced a second spurt that culminated with anatomically modern humans reaching an average cranial capacity of 1250–1400 cm^3 and the Neanderthals' 1600 cm^3. The larger Neanderthal brain derives from significantly bigger visual cortex and occipital lobes. It is interesting that the Neanderthal's and modern human's brains are the same size at birth, but that by adulthood, the Neanderthal's cranial capacity surpasses that of anatomically modern humans.

The hominin brain more than tripled in size (from approximately 400 to 1400 cm^3) from the early Austrolopithecines, 3 mya, to modern humans. It has been calculated that during this period of time the number of neurons have increased at an average rate of about 100,000 neurons per generation.[10] This impressive surge in the number of neurons may translate to a substantial increase in intellectual capacity. In addition to the increment in the number of neurons, the brain experienced a number of significant morphological changes including expansion and convolutions of the cerebral cortex and greater myelination of neurons, the latter changes generally implicated with faster neuronal transmission.

WHO ARE THE HOMININS?

By definition, hominins are all organisms, extinct and extant, within the human lineage after the split from the ancestors of the great apes. The divergence between these two evolutionary branches is thought to have occurred about 6–7 mya. The fossil record provides us with very limited information from that time period in hominin evolution. In essence, only one cranium from an organism, known by the name of *Sahelanthropus* has been discovered, and it is not clear whether it represented a common ancestor to humans and great apes, or alternatively, a member of the great ape or the hominin lineage.

Theme and Variations in Hominin Evolution

A reoccurring theme throughout hominin evolution is the various proportions of ancestral and derived traits seen together in any given extinct species or group. In this context, ancestral is defined as characteristics of more ancient primates and derived indicates more closely related with modern hominins. When the hominin fossil record is examined, different parts of the anatomy of individuals reflect ancient ways of life, for example arboreal, whereas other morphological characters are adaptations for bipedal locomotion. An example of this phenomenon is seen in the early hominin *Ardipithecus* who lived around 4.4 mya. *Ardipithecus'* hands retained certain ancestral characteristics of the primate hand such as long curved finger for grasping branches in a forest habitat, whereas the short human-like thumb and the presence of a third metacarpal styloid process represent derive traits for greater dexterity. In this process of anatomical adaptations to a rapidly changing environment, it is likely that basic morphological features are incorporated first followed by a series of more complex alterations that refine the structure and function of the organ. It is possible that the sudden geological, climatic, and habitat changes that transformed sub-Saharan Northeast Africa from forest to dry savanna provided intense selective pressure for upright posture, bipedalism, and dexterity. Such an extreme natural selection during a relatively short period of geological time encompassing less than 10 million years may have expedited specific physiological and anatomical alterations at different rates.

Another recently realized theme within hominin evolution is its high degree of complexity in terms of numbers of coexisting species and their cultural and reproductive interactions. Traditionally, phylogenetic (evolutionary) relationships among organisms have been illustrated in the form of trees connecting, with single lines, species that are thought to be derived from a most recent common ancestral group. An example of this type of depiction is illustrated in Fig. 2.4. Scholars are beginning to realize that this type of representation, at times, may represent an oversimplification of what actually happened millions of years in the past. Hominin evolution is a case in point. A number of recent discoveries from several fields have uncovered two previously unrecognized facts about hominin evolution. For example, from the field of molecular biology it is now known that introgression (interbreeding) had taken place at different times and places during hominin evolution. Along these lines, it is now known that gene flow occurred from Denisovan and Neanderthals into the modern humans' gene pool. Thus, these groups not only mated but also were able to bridge their genome (produce fertile offspring capable of transmitting their DNA to subsequent generations). In addition, from archeology, it is known that information transfer likely occurred among hominins. Consider, for example, the transmission of toolmaking technology between ancient *H. sapiens* and Neanderthals. These interactions were facilitated by the cohabitation of multiple types of hominin groups in close quarters. For instance, it is known that multiple species of australopithecines and other early *Homo* species coexisted in the same region of sub-Saharan Northeast Africa about 2 mya. Specifically, *H. erectus* cohabitated with *Homo rudolfensis, H. habilis,* and *Paranthropus boisei* in the area of the East African Rift, at times within the same cave, and at the end of its time span it coexisted with *H. sapiens* and *Homo floresiensis* in Southeast Asia. More recently, archaic modern humans lived side by side with Neanderthals

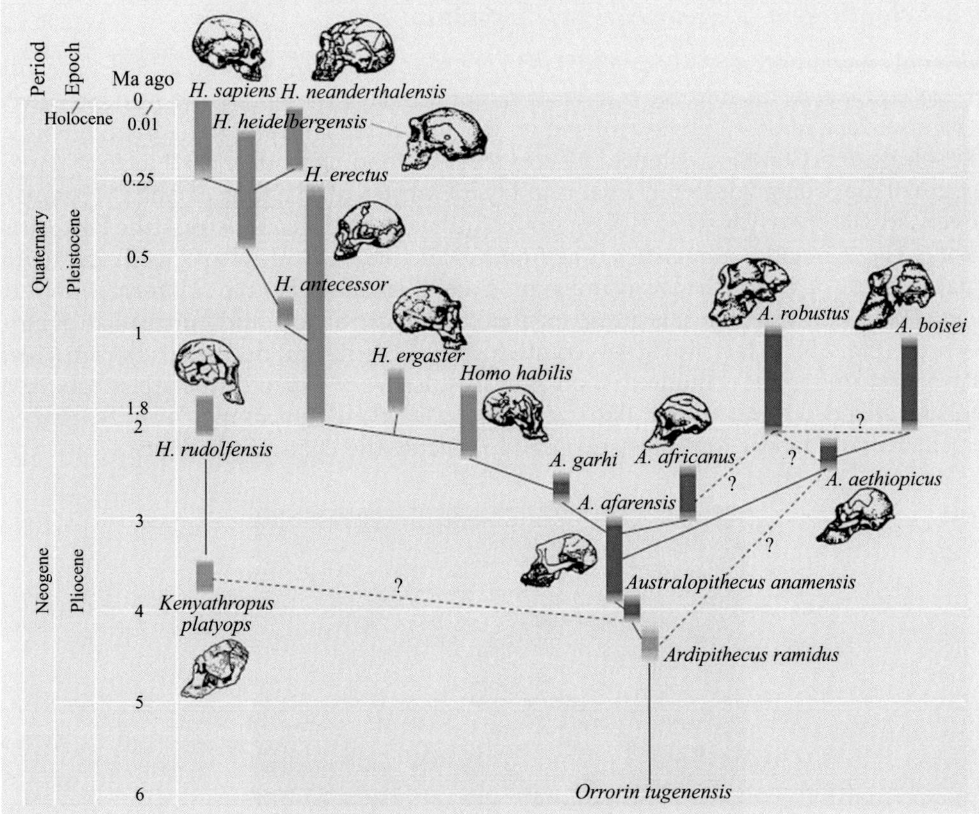

FIGURE 2.4 Phylogenetic tree illustrating the evolution of hominins. *From http://imgur.com/gallery/UFc94.*

about 120,000 ya in the Levant and subsequently from about 45,000 to 25,000 ya in Eurasia. These facts have obliterated the traditional paradigm of hominin groups gradually evolving and leading into new species in a linear one-to-one fashion.

It is likely that the two trends in hominin evolution discussed above are related. In other words, interbreeding may be responsible for the presence of both ancestral and derived traits in organs such as the hand or the pelvis. Just as there is evidence of introgression among several of the *Homo* species including *sapiens, neandethalensis, denisova,* and possibly *erectus* and *florensis,* it is possible that the collage of ancestral and derived characters observed in specific organs in all hominins is the result of genetic flow between groups with various combinations of ancestral and derived characteristics. Until recently, ancient DNA analysis was limited to selective remains no older than about 40,000 ya. Within that time period, ancient DNA sequencing was limited to recent hominins, species such as *sapiens, neandethalensis, and denisova.* Yet, technology is rapidly advancing, and currently, other specimens that are 10 times older, dating to 430,000 ya, are being sequenced. A good example of this new methodology was used in hominin fossils from Sima de los Huesos in Atapuerca, Northeast Spain. Thus, considering the various hominin species cohabitating during early hominin evolution throughout the world, recent advances in DNA analysis may reveal additional cases of interbreeding involving different permutations of species.

Was *Sahelanthropus* a Great Ape or a Hominin?

With only a cranium representing this extinct group, many questions still exist regarding where *Sahelanthropus* fits into the branch of recent human evolution. For a number of years after its discovery, intense debate centered on whether *Sahelanthropus* was a hominin or not.

The sole evidence for the existence of *S. tchadensis* is a finding made in 2001 in an arid region just south of the Saharan desert in what is today the west-central nation of Chad.[11] Before this discovery, fossils of early hominins were from Northeast sub-Saharan Africa (the East African Rift zone, Fig. 2.2) and South Africa. The finding consisted of a skull and teeth (nicknamed Toumaï or "hope of life" in the local language). From this limited source of information, scholars were able to surmise that this group exhibited a mosaic of ape- and human-like anatomical features (Fig. 2.5). In fact, the observed mixture of ancestral and derived characters created consternation among the orthodoxy soon after the discovery. Currently, scholars believe that the collage of ape-like and human-like traits as seen in *Sahelanthropus* may be indicative of an organism in evolutionary transition, exposed to diverse and changing habitats.

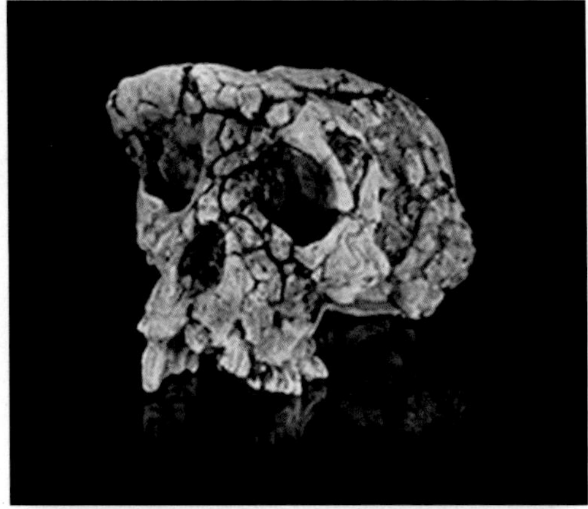

FIGURE 2.5 Cast of the *Sahelanthropus tchadensis'* skull and facial reconstruction. *From https://en.wikipedia.org/wiki/Sahelanthropus and http://humanorigins.si.edu/evidence/human-fossils/species/sahelanthropus-tchadensis.*

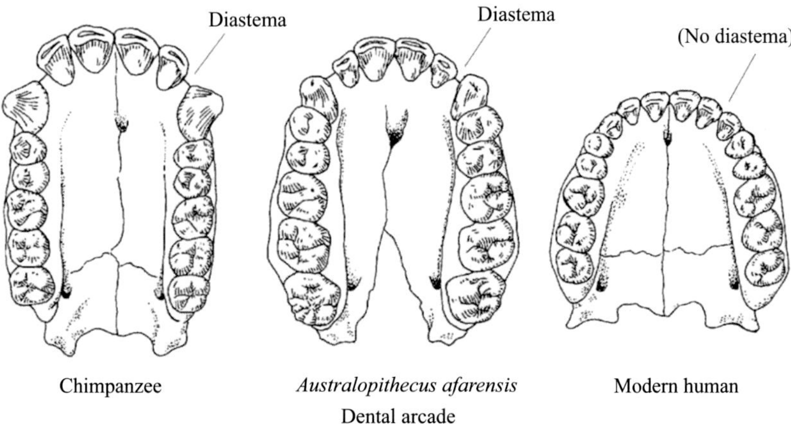

FIGURE 2.6 U-shaped dental arcade and diastema. *From* ***https://www.google.com/search?q=u-shape+dental+arcade&tbm=isch&tbo=u&source=univ&sa=X&ved=0ahUKEwifqI-h_7fZAhUGC6wKHacvCh4QsAQIYw&biw=1440&bih=756#imgdii=QrZRA3ugXMY10M:&imgrc=oCChOHyZ_LuAOM.***

The braincase of *Sahelanthropus* is about 380 cm^3 in volume, the same average size of a chimpanzee brain. Other primitive features include massive brow bridges, sloping face, u-shaped dental arcade, diastema (space between incisors and canines) (Fig. 2.6), and elongated skull. Among the derived traits reflecting hominin affinities are the small canines and the location of the *foramen magnum* beneath the skull. The position of the *foramen magnum* at the base of the braincase is a diagnostic sign of upright posture. Evolutionarily speaking, a shift to the base of the skull of the opening to the spinal cord is necessary to allow the vertebrate column to support the cranium in an erect position. The significance of a vertical posture is the potential for bipedal locomotion. The basal position of the *foramen magnum* underneath the skull in a rather ancient specimen such as *Sahelanthropus* dating back to the time of origin of the hominin lineage was initially surprising since bipedalism was thought to have evolved more recently, with the australopithecine. Also, considering the many ancestral characteristics exhibited by *Sahelanthropus* that indicate arboreal locomotion, it is logical to envision this group as facultatively bipedal as opposed to habitually bipedal.

A further consideration is that upright locomotion evolved in stages and that the various hominin groups practiced different forms of bipedalism. In other words, it is possible that bipedal locomotion evolved independently in the various groups, from primitive ape-like terrestrial motion, at times involving all four limbs, to fully upright, sustained erect locomotion.

In spite of the various human-like characteristics seen in Toumaï, not all scholars are convinced that *Sahelanthropus* was a hominin. Some experts believe that *Sahelanthropus* is not part of the human-specific branch generated after the separation from the ape lineage. A number of experts are of the opinion that the skull belonged to a common ancestor to both the great ape and hominin lineages. And still others are of the opinion that the specimen, although it is related to both humans and chimpanzees, was not ancestral to either lineage. It is likely that some of these questions will be answered with additional findings of the group, especially if they include parts of the postcranial skeleton.

Orrorin tugenensis

Although it is established that *Orrorin* ("original man" in the local language) was an early hominin that lived about 6.0 mya,[12] it is not clear how it fits into the modern human line of descent (Fig. 2.7). In other words, it is uncertain if *Orrorin* represents a side branch or is part of the ancestral line to contemporary humans. The first remains of this genus were discovered in Central Kenya in 2000. Since then, about two dozen bone fragments have been recovered including part of a mandible, teeth, digital bones as well as part of a humerus and fragments of femora. The orthodoxy believes that it had a small brain, about the size of a chimpanzee.

FIGURE 2.7 Facial reconstruction of *Orrorin tugenensis From https://upload.wikimedia.org/wikiversity/en/thumb/b/bf/Orrorin_2.jpg/200px-Orrorin_2.jpg and https://i.ytimg.com/vi/tyapyFFLYR0/hqdefault.jpg.*

Like many early hominin fossils, *Orrorin* exhibits a collage of ancestral and derived characters that, since their discovery, have generated intense debate among scholars. *Orrorin's* unique combination of ape-like and human-like traits has created a challenge for experts in assessing whether australopithecines are in fact the ancestors to modern humans. For example, relative to australopithecines, *Orrorin* dentition was smaller but similar in shape to the contemporary apes. Its thinner enamel is reminiscent of modern humans and not of ancient hominins such as *Ardipithecus* (see the next section). But probably the most significant diagnostic attribute of *Orrorin* is the morphology of its thighbone or femur with its interiorly oriented, ball-shaped head and the presence of a prominent, centrally located lesser trochanter (conical projection) (Fig. 2.8). The femur head articulates with the pelvis and in this frontal orientation, as seen in *Orrorin*, it allows for upright posture and is considered an indication of bipedal locomotion.[13] Considering the internal morphology of the thighbone, it is likely that *Orrorin* practiced some version of facultative bipedalism but was not yet committed to full habitual erect locomotion. In general, the postcranial skeleton indicates a mosaic of arboreal and terrestrial traits. For example, the hands exhibit long curved fingers, a condition seen repeatedly in early hominins and associated with grasping to branches, whereas thumbs are proportionally smaller, a feature linked to manipulation of objects, dexterity, and toolmaking.

Soon after the discovery of *Orrorin*, it became obvious that some reevaluation of australopithecines' position in the hominin lineage was required. *Orrorin* lived about 6 mya compared with australopithecines that first appeared in the fossil record around 4 mya, 2 million years of temporal differential. Yet, *Orrorin* possessed more traits in line with modern

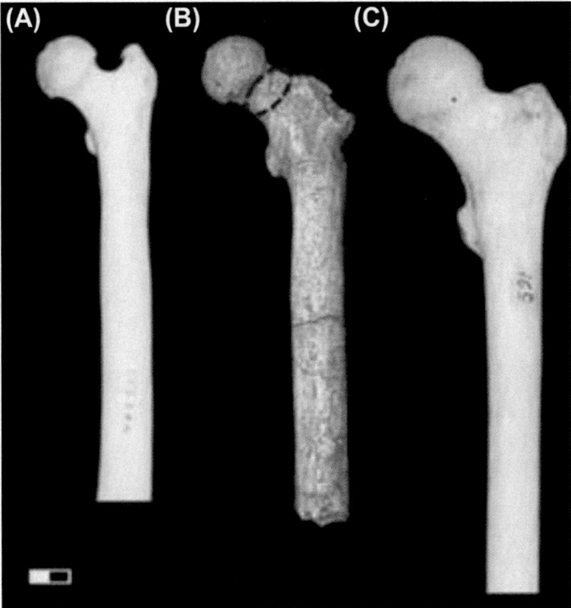

FIGURE 2.8 Femura of (A) chimpanzee; (B) *Orrorin tugenensis*; (C) modern human. *From https://whatmissinglink.wordpress.com/2014/05/04/upright-walking-a-long-standing-debate-pt-vi-the-fossils/.*

humans than *Australopithecus*. A possible explanation for this apparent conundrum is that the long-standing contention that australopithecines were direct ancestors to modern humans was not correct. In other words, it is possible that australopithecines represent a side branch in hominin evolution that eventually died out.

Ardipithecines: In-Between Worlds

The genus *Ardipithecus* is represented by two species, *Ardipithecus ramidus and Ardipithecus kadabba*, and they have been dated to approximately 4.4 and 5.6 mya, respectively (Fig. 2.9). It has been theorized that *A. kadabba* was a direct ancestor of *A. ramidus*. The ardipithecines remains were recovered from Middle Awash in the Ethiopia's Afar Depression at the northern ridge of the East African Rift.[14] Although arguments still remain concerning whether this genus belongs within the hominin or the great ape line, based on the numerous human-like attributes, most experts today are of the opinion that it was a hominin. During a period of

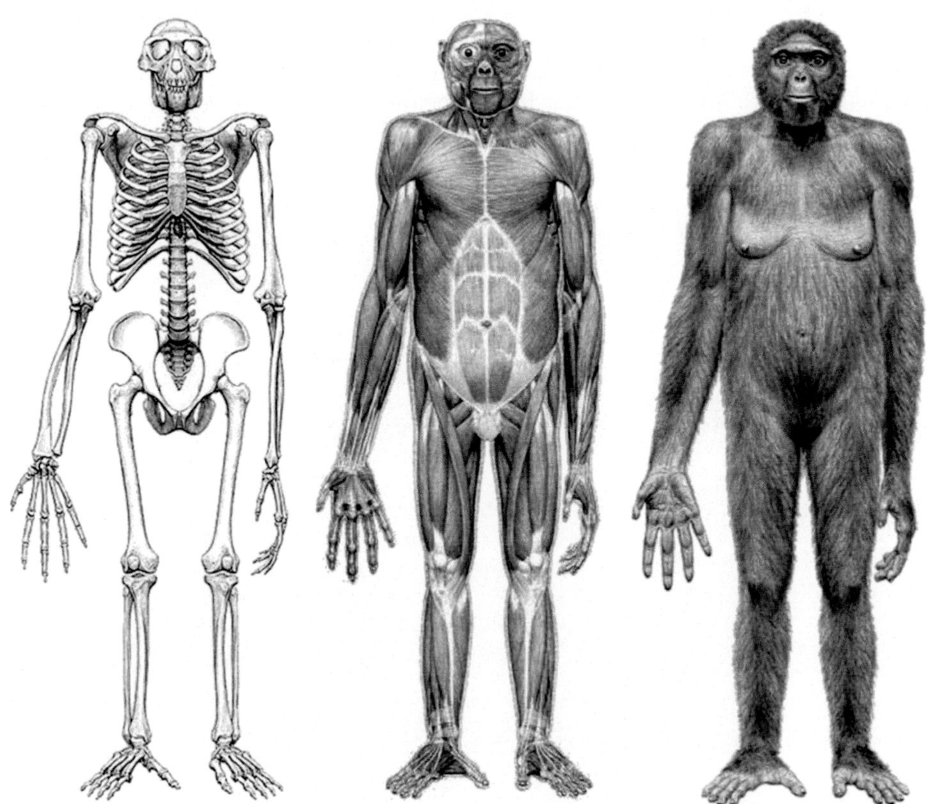

FIGURE 2.9 Full body representation of *Ardipithecus ramidus*. *Reproduced from DESCOPERA (https://www.pinterest.com/pin/505529126900941208/).*

9 years (1994–2003) about 110 fossilized bones from a minimum of 36 individuals have been recovered from these species, including a skull, mandible, teeth, and arm bones. An exceptionally well-preserved skeleton of one of the recovered individual is 45% complete. This particular individual was nicknamed Ardi.

Similarly to most early hominin groups, *Ardipithecus* displays a mosaic of ancestral and derived characteristics. Among the ape-like traits, *Ardipithecus* possessed a small brain similar to the size of a chimpanzee's brain, about 350 cm^3, opposable long toes for grasping to branches, protruding mandible, and a relative molar surface area reminiscent of apes. Just the pelvis itself is a collage of ancestral and derived traits. Specifically, the ilium (uppermost bone of the pelvis) is broad implying bipedalism, whereas the ischium (curved bone at the base of the pelvis) is longer suggesting arboreal locomotion. It is if this group enjoyed both terrestrial and arboreal existence.

Fundamentally, the human and ape feet play different roles. The former is an organ for propulsion and the latter performs a prehensile function. In terms of locomotion, Ardi possessed a *foramen magnum* underneath the skull indicative of bipedalism. In addition, the lengths of *Ardipithecus'* cuboid and navicular bones of the foot are greater than in apes but smaller than in humans. Yet, like apes, *Ardipithecus* has toes located away from the foot axis, long limbs, and an ape-like pelvis. Altogether, these characteristics suggest that *Ardipithecus* practiced facultative bipedal locomotion while on the ground and moved as a quadruped in the trees. In other words, this mosaicism of traits probably allowed these organisms to survive and exploit an arboreal and a terrestrial habitat, as a quadruped and biped, respectively. At a time of transition and dramatic environmental changes in sub-Sahara Northeast Africa, possessing such a unique combination of characters may have been advantageous.

Other derived characteristics include a bony projection in the metacarpal (bone between the wrist and the fingers) and a number of singularities of the skull such as a lateral relocation of the carotid canal (opening in the temporal bone through which the internal carotid artery enters the skull) and a diminution of the lateral tympanic cavity.

In terms of its diet, measurements of carbon and oxygen isotopes as well as the thickness of the enamel suggest that *Ardipithecus* consumed an arboreal and terrestrial diet in open fields and woodlands. In addition, the lack of specialization of its dentition for any specific type of food signals a diverse diet similar to that of an omnivorous species.

Particularly noteworthy of *Ardipithecus* are the small size of its canines. Canines are used as weapons during intra- and intertribal confrontations. They are also a symbol of strength and status within the group. Males routinely display their canines as a sign of social standing within the tribe and to warm intruders away. Unusual in ardipithecines is that the canines of males were not much larger than those of females, reflecting a lack of sexual dimorphism regarding the trait. The small size of the canines and the absence of size differences between the sexes have been interpreted by some scholars as an indication of key behavioral changes at this stage in hominin evolution. Specifically, it has been proposed that reduced canines in males reflect more docile and less belligerent interactions as well as bonding among males of the species and the beginning of social organization.[15] A diminution in the size of canines may have shifted the importance of aggression in male procreation and fitness to other traits such as provider of food and caretaker for the young. Significantly, if these assessments are correct, these behavioral changes occurred before the enlargement of the brain.

Australopithecines

Undoubtedly, the most well-known remains of an *Australopithecus* were discovered in 1974 in the Awash Valley of Ethiopia at the extreme north end of the East African Rift by Donald Johanson.[16] It belonged to a young adult *Australopithecus afarensis* female, and it includes about 40% of the skeleton. Inspection of the remains provided no clues as to the cause of her death. It was nicknamed Lucy after the Beatles' song *Lucy in the Sky with Diamonds*. Over the years over 300 *A. afarensis* individuals of this species have been uncovered.

Members of the genus *Australopithecus* constitute a highly successful and diverse group of hominins that appeared in the fossil record approximately 4 mya (Fig. 2.10). Some species of the genus survived to as recently as 1 mya (*Australopithecus robustus*). Hence, as a genus, *Australopithecus* were in existence for longer than *Homo* (thus far) by at least 1 million years. The general consensus among scholars is that one of the australopithecine groups gave rise to genus *Homo* approximately 2 mya. Thus, australopithecines are considered direct ancestors of the genus *Homo*. It is thought that it originated in sub-Saharan Northeast Africa in the region of the East African Ridge. Eventually, the genus speciated into a number of species distributed throughout subequatorial Africa, yet it never migrated out of the continent. A number of distinct groups exhibiting diverse characteristics have been recognized within the genus

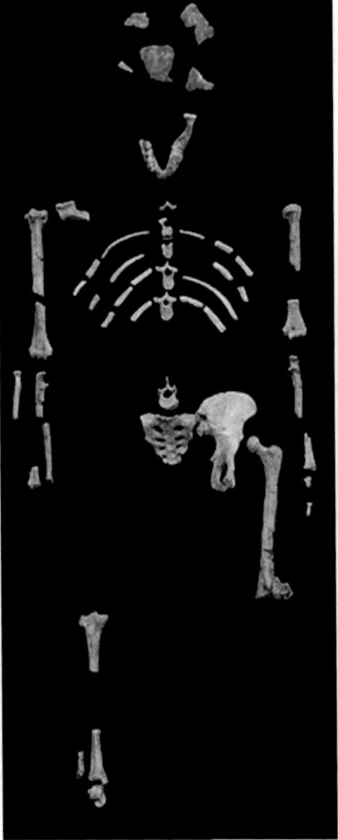

FIGURE 2.10 Skeleton and reconstruction of *Australopithecus aferencis. From http://blog.hmns.org/wp-content/uploads/2015/05/Lucy.jpg and http://humanorigins.si.edu/sites/default/files/styles/full_width/public/images/square/afarensis_JG_Recon_Head_CC_3qtr_lt_sq.jpg?itok=V748WLH_.*

that include *Australopithecus africanus, A. robustus, A. boisei, A. afarensis, A. sediba, A. anamensis, A. bahrelghazali, A. deyiremeda, A. garhi,* and *A. netiopicus.* All of the abovementioned groups except *boisei* and *robustus* are considered "gracile" because of their less massive bone structure. In addition, these two species are often described using the genus *Paranthapus* because of their diverse characteristics. *Boisei* and *robustus*, are known as "robust" forms that lived from about 1.8 mya to 1.0 mya and represent the most recent members of the genus. The other species existed from around 4 mya to 2 mya.

At least one group of australopithecines, *A. garhi*, that lived about 2.6 mya, has been found along with stone tool implements. It is likely that *A. garhi* was not a hunter and used the tools to scavenge meat from carcasses left behind by predators such as carnivores. If this finding is corroborated by additional discoveries, *A. garhi* was the first hominin group to possess a tool tradition. Another hominin group dated from 2.8 to 1.5 mya known to have used tools was *H. habilis* or *Australopithecus habilis*. Although *habilis* differs from most australopithecines in having a larger brain size (640 cm^3) and being capable of fabricating tools, most of its anatomical metrics resemble members of the genus *Australopithecus*. It is likely that *habilis*, being contemporary and cohabitating with other hominin species, contributed to the complexity and diversity of the australopithecine group by interbreeding with related groups.

Australopithecines, in general, had faces with forward-projecting jaws and conical rib cages as observed in great apes. The relatively long extremities with curved fingers and toes suggest the persistence of some degree of arboreal existence in spite the absence of opposable big toes. Yet, it is possible that some of the ancestral arboreal characters were retained longer because of their selective neutrality and/or other beneficial secondary functions to the individuals. Derived characters include smaller canines and molars compared with apes and a curved lumbar spinal region (Fig. 2.11). Also, the *foramen magnum* was at the base of the skull.

FIGURE 2.11 Skull and reconstruction of *Homo naledi. From http://www.news24.com/Multimedia/South-Africa/PICS-New-species-of-human-ancestor-revealled-20150910.*

The pelvis was more human-like compared with *Ardipithecus* with a short and broad ilium and a wide sacrum located in the back of the hip joint, both indicative of upright posture. Also, as in *Orrorin*, the femur of australopithecines points frontally toward the knee suggesting bipedal locomotion.

Members of one of the species of australopithecines are responsible for a striking set of footprints on volcanic ash created about 3.7 mya in Laetoli, Tanzania, 45 km south of Olduvai Gorge. The depressions represent the oldest unequivocal evidence of habitual bipedal locomotion by a hominin. No knuckle impressions indicative of quadruped locomotion are evident. Noteworthy in these imprints are the absence of an opposable big toe and the presence of a very human-like arch. In addition, the topology of the impressions reveal that the heels touched the ash first, and then the weight of the body was moved to the front of the foot before pushing off the toes. This pattern is very reminiscent of the human gait. In fact the tracks are so human-like that some experts refuse to accept that *Australopithecus* made them.

Considering the relatively small brain size of australopithecines, in the range of 420–500 cm^3, the indentations indicate that upright posture and bipedal locomotion evolved before the increase in brain size. In relation to brain size, it is interesting that SRGAP2 (see section on SRGAP2 in Chapter 1), a gene linked to neuronal movement and differentiation as well as nerve connections, started duplicating itself about 3.5 mya during australopithecine speciation and replicated a second time 2.5 mya coinciding with the appearance of the first *Homo* species from an australopithecines ancestor and the first burst in brain growth in hominins (Fig. 2.3).

The remarkable footprints were found by Mary Leakey in 1976. The imprints are seen in a stretch of 24 m and were made by three adult individuals walking in the same direction at the unhurried speed of 1 m per second, the regular pace of a modern human. There is no indication that the impressions were made by family members or at the same time. Yet, it has been observed that the impressions made by the smaller individual exhibit signs of uneven burden on one side. It has been suggested that a female carrying a child on her hip as she walked was responsible for the imprints. Although, the species that made the tracks has not been established, most investigators attribute the footprints to *A. afarensis*.

Homo naledi

The story of the recent discovery of *Homo naledi* (Fig. 2.11), a previously undescribed group of hominin, is remarkable in many ways. Although no direct dating of the fossils is currently available, age estimates from morphological traits suggests that the remains are about 2 million years old.[17] This time in hominin evolution corresponds to a pivotal period when some form of australopithecine evolved into the genus *Homo*. It is also an epoch that little was known about our evolution. The discovery of *H. naledi* took place on October 2013 near the city of Johannesburg in South Africa. Specifically, the remains were found in an extremely difficult place to reach known as the Dinaledi chamber in the Star cave system. Before the finding of the Dinaledi remains only a handful of fossils have been discovered from this critical transition period in hominin evolution.

To the scientific community, the fact that *H. naledi* lived in South Africa was a bit of a surprise because most previous early hominins were found in sub-Sahara Northeast Africa. It is possible that after the dispersal of australopithecines throughout the continent, early forms of the genus *Homo* originated in South Africa and subsequently migrated to the Northeast

African Ridge region. Alternatively, an ancient undiscovered form of *Homo* might have evolved in Northeast Africa, which then migrated and populated South Africa. Hopefully, future discoveries will clarify these issues.

The discovery was not totally accidental since American paleoanthropologist Lee Berger working at the University of the Witwatersrand had requested friends and recreational cavers such as Rick Hunter and Steven Tucker, two small and slender individuals, to be on the lookout for fossils during their explorations of cave systems in the region. The fact that the cavers were petite and nimble was critical in finding the fossils. Most human beings of average size would not have been able to squeeze through a 12 m passage at times only 8 inches wide to reach the small chamber where the bones were located. The cavers were quite surprised to see so many human-like bones in a single tiny place. The remains were found in strata indicating that they were not laid there at once but over a period of time. It was as if the bones were deposited in the minuscule atrium by some unknown force and now they lay randomly arranged in six inches of sediment. From the photographs the cavers took from the site, Lee Berger knew they had come across a unique finding. Since its discovery, the site has yielded a total of 1550 hominin specimens belonging to about 15–18 individuals. The quantity and quality of the fossils collected were exceptional. The fossils included every body part of old and young adults as well as infants. Of particular importance were the precious findings of an almost complete hand and foot. Together with the Sima de los Huesos site in Atapuerca, Spain, the Dinaledi remains represent two of the most significant archeological discoveries of the past 50 years.

Homo naledi, like many other early hominin groups, possessed a collage of ancestral and derived characteristics. Although some of the traits are reminiscent of more ancient groups such as australopithecines, other morphologies are seen in more recent groups such as *H. erectus* while other characteristics are unique to *H. naledi*. In general, the anatomy of these fossils above the hip reflects an ancestral body plan while below the hip, it resembles advanced human-like characters. Yet, overall, the characters exhibited by *H. naledi* align this group closer to the genus *Homo* than to *Austalopithecus*. In fact, some scholars consider *H. naledi* an early form of *H. erectus*.

In the Dinaledi assemblage, a total of four skulls were recovered, two males and two females. The skulls average 560 cm^3 for the males and 465 cm^3 for the females, suggesting some degree of sexual dimorphism. Considering that the brain size of Australopethicus ranged from 420 to 500 cm^3 and *H. erectus'* averaged about 900 cm^3, *H. naledi* cranial capacity resembled the former with dimensions that overlap. Yet, in spite of its diminutive brain, *H. naledi* possessed a less massive skull and temporal and occipital bulges, characteristics associated with *H. erectus*. In addition, a skull feature known as the postorbital constriction, a narrowing of the skull behind the eye sockets is considerably reduced in *H. naledi*. The postorbital constriction becomes less prominent as a function of hominin evolution, hence it is used as a reliable parameter to assess how derived a hominin is. In early *H. sapiens*, this feature is barely visible, and in contemporary humans it has disappeared altogether. The teeth are smaller than in australopithecines with crowns and five cusps like in modern humans. The muscles attached to the mandible are also smaller indicating a softer diet requiring less mastication compared with the ancestral condition.

The rib cage is conical in shape as in *Australopithecus* and great apes. The hands, on the other hand, like in earlier hominins, are a combination of ape-like and human-like traits.

Specifically, the thumb, wrist, and palm bones are human-like, whereas the fingers are still curved, an ancestral morphology designed for holding to branches during arboreal locomotion. For unknown reasons, curved fingers persistent longer than other ancestral conditions. The pelvis represents a collage of ancestral and derived characteristics. The base of the pelvis looks like that in very modern humans, whereas the blades spread outward as in australopithecines. Remarkably, the feet look like modern humans, whereas the shoulders are ape-like.

As in the previously described hominin groups, *H. naledi* was a mosaic of ancestral and derived features, again even within the same organ. The complexity and combinations of ape-like and human-like traits that are seen in the anatomy of this group may be the result of polyphyletic or hybrid origin derived from introgression. The observed theme of collages of ancestral and derived traits seen in hominin evolution is not properly illustrated in the one-to-one linear relationship of species typically seen in phylogenetic trees with a single root and unconnected branches (Fig. 2.4). What is seen in the fossil record is more reminiscent of closely related groups of organisms initially separating, creating different gene pools only to start interbreeding, after some time, allowing genetic flow to occur. This type of scenario would generate groups of hybrid organisms with a plethora of combinations of traits from different source populations. Groups of hominins interbreeding may not have been the exception to the rule. In fact, it may have been quite common considering that the different groups cohabitated.

The Dinaledi fossils are also remarkable because of the inaccessible location of the chamber where the remains were found and the exclusive presence of hominin bones. Except for an owl, all of the remains found belong to *H. naledi* individuals. So, how did the hominin bodies got to their final resting place? Could clan members have deposited the bodies there? Were the remains deposited in the chamber as part of a burial ceremony? The cave where the fossils were discovered is no older than 3 million years. Thus, the bodies were deposited in the chamber only 1 million years after the cave was formed. Although the cave system is quite popular among amateur explorers, the challenge of getting to the chamber makes it difficult to explain how the remains got there. It is possible that 2 mya the cave looked different and specifically the Dinaledi chamber was more easily accessible providing a bigger diameter chute and/or opening at the top that allowed tribal members to simply drop the bodies into the cavity. Since then, geological activity could have transformed the cave system to the present state. In addition, considering that the male *H. naledi* specimens found in the chamber were small by modern human standards (about 5 feet tall and 100 pounds), their entrance may have been facilitated, just as with the cavers' case, by their slender build. The likelihood of predators dragging the bodies into the enclosure is not very plausible considering that other that the dead owl, only *H. naledi* remains were found. Carnivores prey on various animals, not just hominins. In connection with this issue, it is interesting that most of the bones were disarticulated or separated at the joints, possibly the result of forcibly pulling the inanimate bodies throw a narrow opening and/or intentionally done to facilitate dragging the dead into the chamber. Based on these facts, it is likely that clan members somehow placed or threw the bodies into the chamber from an opening on top that no longer exist. Yet, the lack of illumination, as is the condition today, argues against intentionally and expressly laying the bodies by members of the clan.

The other interesting question regarding the Dinaledi site is whether *H. naledi* practiced ceremonial interments. Although the relative distribution of the bones pointed to previous visitors disturbing the bones in their resting area, possibly clan members or cavers, there is no indication that the bodies were deposited in any particular order or orientation. Also, no artifacts such as stone tools and personal effects that are part of more modern hominins' ceremonial assemblies were found. Considering that it is not known if *H. naledi* actually made tools or possessed cultural/spiritual awareness, the lack of ritualistic items would be expected.

Homo erectus, the First Out of Africa Migrants

The first *H. erectus* specimen was discovered in 1891 by Eugene Dubois, a Dutch surgeon/amateur archeologist doing fieldwork in Java, Indonesia, at the time a Dutch colony.[18] Dubois was heavily influenced by Charles Darwin and his theory of natural selection and mounted an expedition to find the ancestors of contemporary humans in Java. At the time, it was thought that human origins resided in Asia rather than in Africa. The initial discovery consisted of a skullcap and a femur and was named Java man by the popular press. Since then, hundreds of *H. erectus* sites have been found in Africa, Asia, Europe, and Oceania (Fig. 2.12). Today, this group is thought to have been the direct ancestors of modern humans and a number of findings since the 1950s, exhibiting archaic features in East Africa, solidified the premise that the *H. erectus* evolved there.

Anatomical differences are apparent within the geographical range of *H. erectus*, and pronounced variability has been detected even among residents of the same location such as Dmanisi. Considering that the Dmanisi skulls were collected in a single locale, it is remarkable that the degree of morphological variability falls within the range of modern human or chimpanzee variation *throughout* their entire geographical range. If *floresensis* is indeed a form of *H. erectus* that experienced island dwarfism, as many scholars contest, this pygmy population would add to the impressive anatomical diversity of *erectus*. The high anatomical

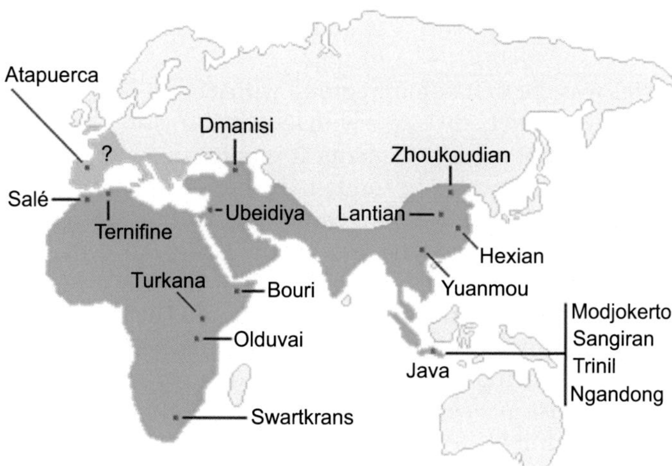

FIGURE 2.12 Geographical range of *Homo erectus*. *From http://anthro.palomar.edu/homo/homo_2.htm.*

diversity observed within the *H. erectus* conglomerate have prompted some scholars to suggest that the species should be divided into an African group with the name *Homo ergaster* and an Eurasian assembly that eventually evolved into more recent hominins such as *H. heidelbergensis*, *Homo neanderthalensis*, and *H. sapiens*.

As additional *erectus* remains were discovered and studied, it became obvious that this group was not just another group in the phylogenetic tree. *H. erectus* was a groundbreaker in terms of the number of important human-like characteristics not seen before in earlier hominins. For example, it is likely that in addition to being habitually upright, most of the time *erectus* was fully bipedal in a modern human sort of way. Tall and slender in physique, this group migrated out of Africa, exploiting various habitats and climatic zones, and possibly even using clothing for protection. Considering that it reached Eastern Indonesia (e.g., the island of Flores), it is likely that somehow it navigated through one or several bodies of water to reach it. During its 2-million-year tenure, the *erectus* brain increased dramatically in size and number of neurons. In addition, they made stone tools and were able to initiate and control fire for cooking and probably for protection against predators during the night. Anatomically, they would have been able to utter words, which facilitated social interactions such as communication during communal hunting, which they practiced. *H. erectus* was also altruistic in some of its behavior by caring for the old or disabled.

H. erectus evolved in sub-Saharan Africa from earlier hominin groups such as *H. naledi* or *H. habilis* (see above). As its name indicates, *H. erectus* was habitually bipedal and lived from two mya to as recently as 60,000 ya. In general, the Asian varieties of *erectus* became extinct more recently. If the remains of *H. floresiensis* (see Chapter 6) discovered in the island of Flores in Eastern Indonesia are in fact a dwarf form of *H. erectus*, the time range of the group could have extended even to more modern times, as recent as 12,000 ya. Remarkably, this would represent a period of overlap of about 45,000 years of *H. sapiens* and *H. erectus* cohabitating on the island. It is likely that being bipedal facilitated the dispersal of the group into diverse habitats because these hominins were not confined to forest and were capable of rapid movement on the ground. Considering an existence of almost 2 million years, compared with only 200,000 years for *H. sapiens*, *erectus* was quite successful in adapting to various diverse habitats.

In general, *H. erectus* was the first hominin group with a body plan that resembled modern humans. It possessed a very slim body frame with lengthy arms and legs. Its postcranial anatomy was very similar to that of modern humans. It is noteworthy that its limbs and extremities were basically the same in morphology and proportions compared with modern humans. They were taller than australopithecines and earlier member of the genus *Homo* such as *H. naledi*. In fact, the average African *H. erectus* was about 5.7 feet tall, a size that falls within the tallest 17% of the contemporary human male population. The discovery the Turkana Boy in 1984 provided a clear indication of the dimensions of *erectus*. The specimen belonged to an 8- to 12-year-old boy that was 5 feet 3 inches tall and would have been about 6 feet in adulthood. The Turkana Boy is the most complete *erectus* skeleton discovered thus far. Considering that the size of an adult male *H. naledi* was only 5 feet, the stature of *erectus* represents a substantial increase from previous hominins.

Heads of *H. erectus* were strikingly different in shape compared with those of modern humans. They had relatively strong muscles on the back of their necks. Their skulls were elongated with shallow foreheads sloping back from very prominent brow ridges (Fig. 2.13).

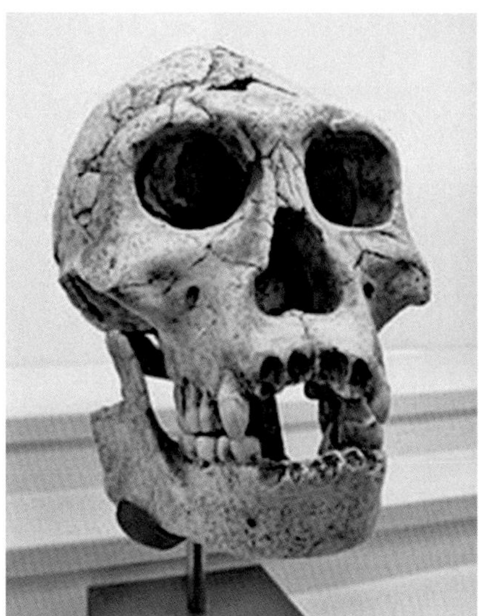

FIGURE 2.13 Skull and reconstruction of *Homo erectus. From https://en.wikipedia.org/wiki/Homo_erectus#/media/File:Homo_Georgicus_IMG_2921.JPG and http://humanorigins.si.edu/sites/default/files/styles/full_width/public/images/square/erectus_JC_Recon_Head_CC_f_sq.jpg?itok=USi9mPdW.*

During its tenure, the hominin brain experienced a dramatic spike in brain size possibly linked to a number of genetic alterations previously discussed in Chapter 1. The cranial capacity of *erectus* varied in time and space. For instance, the earliest *erectus* fossils exhibited brains about 850 cm^3 in volume, whereas more recent specimens from Java were considerable larger, around 1100 cm^3, within the range of modern humans. Regional differences are clearly evident in the remains discovered in Dmanisi, Republic of Georgia that exhibit considerably smaller crania, 546–600 cm^3 in volume. The frontal bone in *erectus* is less sloped, and the face is less protrusive than in earlier hominins. *H. erectus* had large brow ridges and prominent cheekbones (Fig. 2.13). The teeth were small in size in comparison with those of australopithecines, and the incisors were shovel shaped. It is significant that contemporary Asian and Native Americans possess the same type and degree of shovel-shaped incisors seen in *erectus*.

Possibly resulting from the increment in brain size, a number of technologies had their beginnings with *erectus*. *H. erectus* began to dramatically control its environment. The group started developing a culture and became more aware of their existence. Among these is the initiation and control of fire, the practice of cooking food, including meat, and toolmaking (e.g., hand axes) (Fig. 2.14). They probably were the first hominins that practiced a hunter-gatherer existence running after small game with their long legs and slender physique contrasting with the australopithecines that survived as slower-moving scavengers. It is possible that they formed band societies based on extended family groups, and as such, cared for the ill or disabled. Considering that certain remains such as the Turkana Boy possessed human-like neck structures and the brain region involved in speech, it is possible that *erectus* was able to articulate words using some sort of protolanguage. Some sort of word communication among clan members would have been critical during hunting of megafauna. All of these

FIGURE 2.14 Stone axe manufactured by *Homo erectus*. *From https://www.reference.com/history/did-homo-erectus-wear-clothes-2f140311aa756747.*

cultural–technological advances freed them from many climatic and environmental restrictions allowing worldwide dispersion into diverse habitats.

Although some investigators believed that *H. erectus* originated in Eurasian, most authorities are of the opinion that the group evolved in Africa. *H. erectus* was the first group of hominin that migrated out of Africa (Fig. 2.12) while all previous groups of hominins were confined to Africa. The fossil record indicates that shortly after *H. erectus* appeared in Africa; it started dispersing into North Africa and Eurasia reaching Dmanisi in the current Republic of Georgia, Southwest Asia[19] about 1.8 mya and moving toward Indonesia (Sangiran, Central Java and Trinil, East Java), Vietnam, China (Zhoukoudian and Shaanxi), and India by approximately 1 mya. It is highly likely that the upright posture and bipedal locomotion as well as technological advances were paramount in the remarkably fast dispersal of this group worldwide. Considering that *H. erectus* originated in sub-Sahara Africa about 2 mya and was then detected in Dmanisi West Asia 1.8 mya, it only took this early hominin, with limited technology, about 200,000 years to reach Southwest Asia. Stone tools found in Anatolia, Turkey indicates an upper limit of 1.2 mya to the incursion into Europe.[20]

In addition to the anatomical, intellectual, and technological changes experienced by *H. erectus* during their early evolution, environmental pressure likely played a role in their dispersal out of Africa. Starting about 3 mya, every 20,000 years or so earth completes a full circle of wobbling around its axis. The Earth's wobbling causes a weakening of the transport of warm upper waters to the north and cold deep water to the south of the planet triggering dry climate in the Sahara and Arabian Peninsula. This phenomenon is known as the Saharan pump, and it has the strength to change the Saharan desert into a jungle and vice versa. Currently, the Earth is 7000 years into a dry cycle. These pulses of extreme dry and wet weather likely played a pivotal role in the dispersal of *erectus* out of Africa in more than one occasion because these climatic fluctuations created contraction and expansion episodes of population movements and ranges in gateway regions of migration out of Africa such as the Levant and the Horn of Africa. During dry conditions populations contracted or migrated to other more livable terrains while in humid weather groups expanded. It is likely that these environmental forces motivated *erectus* to leave Africa.

Considering that *erectus* originated in the tropical and permissive environment of sub-Saharan Africa and then rapidly migrated to areas with diverse and fluctuating climatic conditions, it is logical to think that *erectus* was capable of manufacturing some type of apparel for covering and protection against adverse weather soon after its origin. For instance, the Dmanisi site in a mountainous region of the Caucasus demonstrates that 200,000 years after its origin in Africa, *erectus* was able to survive the frigid winter conditions of the region. The weather at the time in the Caucasus was much colder and drier than now. Although no articles of clothing have been recovered to date, possibly because of the disintegration of the organic matter that they were made of, it is likely that *erectus* used skin and fur from the game they killed to make clothing, covers, bedding material, and hide for protection of some sort of dwelling.

It is also possible that *erectus* built huts. Primitive shelters such as the ones in Terra Amata near Nice, France have been attributed to *erectus*. These shelters probably served as windbreakers and were fabricated out of hide extended over a wooden frame with foundations made up of stones and animal skins for floor. The Terra Amata huts were constructed around 400,000 ya as oval-shaped structures 26–50 feet long and 13–20 feet wide and reinforced with branches 3 inches in diameter and rocks at the bases. Other contemporary digs in Europe such as the one in Bilzingsleben, Germany exhibit similar circular 9–13 feet in diameter structures with a base made up of bones and stones. These rudimentary shelters attributed to *erectus* would have been of survival importance especially in regions experiencing extreme weather conditions.

H. erectus' ability to create and control fire probably meant that they used it to keep predators away and keep warm at night. In addition, they may have used the fire to cook. Cooking softens meat, release nutrients better than raw meat, and provides flavor. In relation to diet, it is noteworthy that cut marks created by tools have been detected in *erectus'* bones. This has been interpreted as evidence of some form of cannibalism, ritualistic, or as a mean of survival during lean times.

Another perplexing part of the trek to Oceania was the water crossing that *erectus* undertook to reach Eastern Indonesia (e.g., the island of Flores). It is likely that *erectus* was the first hominin that used rafts (deliberately or unintentionally) to cross open ocean.[21] During the Pleistocene from 2,588,000 to 11,700 ya, repeated glaciations lowered the sea to about 100 m compared with the current level. This means that at the glacial maxima, most of Western Indonesia was a continuous landmass. Yet, in two critical areas deep treacherous bodies of water persisted during the putative time of the *erectus* crossing. One was a 35 km body of water located just east of the island of Bali (the Wallace Line, a faunal boundary line) and the other, a 100 km stretch of ocean (the Makassar Strait) in between the islands of Borneo and Sulawesi. *H. erectus* had to cross one of these two bodies of water to reach the island of Flores. Yet, it is not clear how this feat was accomplished.

Chapter 5 specifically deals with introgression or interbreeding between hominins that evolved subsequent to *H. erectus* and modern humans. The possibility that *erectus* interbred with modern humans has been proposed.[22] A number of models have proposed that, as modern humans dispersed out of Africa and penetrated territory occupied by previous residents, they interbred with them.[23] Several lines of investigations suggest that introgression occurred between *erectus* and *H. sapiens*. Definitely, the opportunities presented themselves in time

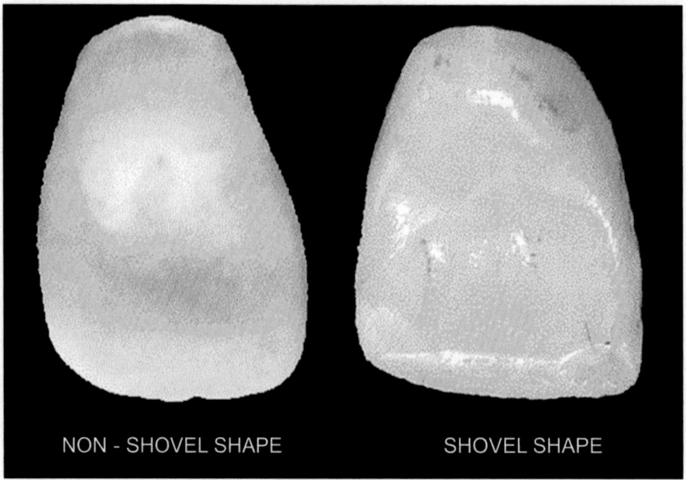

FIGURE 2.15 Shovel-shaped and non–shovel-shaped incisors. *From http://3.bp.blogspot.com/-A9ipYvA1R_U/UDr-Ftr1enI/AAAAAAAAACE/xDdo_MkIJsQ/s1600/sinodont+line+up.gif.*

and space. For example, in locations such as Southeast Asia and Oceania, *sapiens* and *erectus* temporarily overlapped for at least 10,000 years. In locations such as the island of Flores the cohabitation could have been much longer. And in Africa, the birthplace of both groups, the time of cohabitation would have been more extensive in the order of 50,000 years.

Signals of these interbreeding events involving *erectus* and *sapiens* derive from various disciplines. Let us take the specific pattern of marked shovel-shaped incisors seen in *erectus* and in most Asian and Native American populations (Fig. 2.15). Although shovel-shaped incisors are seen among non-Asian populations, the morphological patterns are different and the degrees are less. Albeit, it is possible that convergent evolution is responsible for independently generating identical teeth morphology, the absence of the trait in non–Asian-derived modern human populations implies that shovel-shaped incisors needed to evolve very rapidly in Asia de novo. Thus, it is possible that introgression was responsible for the parallelism. Other characters that exhibit continuity with *erectus* and may stem from introgression in Asia include flatness at the upper and middle part of the face resulting from blunt nasofrontal and zygomaxillary angles, respectively.

Evidence of introgression is also provided by the field of genetics. All the uniparental mitochondrial (mt) and Y chromosomal DNA of modern humans are of *recent African* origin. Thus, they do not corroborate admixture with *erectus* as they migrated into Asia or Europe. Considering the propensity of mtDNA and Y chromosome types to drop out of populations by random chance, it is possible that the reason for not seeing *erectus* uniparental DNAs in modern humans is because they were fortuitously left out since hybridization. Yet, fortunately, the rest of the DNA (transmitted by both parents) is subject to less of these incidental omissions. And when investigators proved the biparental DNA (autosomal DNA) of modern humans, they detected large pieces of DNA introduced into the modern human gene pool about the time *erectus* and *H. sapiens* confronted each other about 2 mya. Specifically, the *RRM2P4* site exhibits a large stretch (a region of 2400 nucleotides) of relatively undisturbed (conserved) DNA.[24] Moreover, this gene seems to have originated in Asia. By comparing the DNA sequence differences among contemporary human populations from around the world, investigators were able to extrapolate the sequences' time of divergence from each other and from that they assessed that the original *RRM2P4* gene was shared by a common hominin ancestor around 2 mya in Asia. This *RRM2P4* gene type is common in Asia and almost fixed (100% present) in certain Asian populations where *erectus* remains are most common. Other genes exhibiting ancient sequences suggesting introgression involving *erectus* include the sequences for 2′-5′-oligoadenylate synthetase 1 (*OAS1*), a protein required for the innate immune response to viral infection, and certain portions of the gene for hemoglobin. In addition to the putative gene flow events between *erectus* and *H. sapiens* in Eurasia, more recent data point to numerous archaic interbreeding events in the African continent.[25] Three identified pieces of DNA exhibit conserved sequences that were introduced into the modern human genome about 35,000 ya from a different hominin group that separated from the modern human lineage around 700,000 ya. In one instance, a 31,000-nucleotide segment of DNA suggests that hybridization occurred in Central Africa. These studies indicate that approximately 2% of *erectus* DNA was introduced into contemporary African populations. All of these findings are suggestive of introgression and reinforces the contention that recent human evolution is best illustrated as branches of a river that split and separate only to reunite later downstream.

References

1. Francione GL. *Animals as persons: essays on the abolition of animal exploitation*. New York: Columbia University Press; 2009.
2. Clark JD, et al. African *Homo erectus*: old radiometric ages and young Oldowan assemblages in the middle Awash Valley, Ethiopia. *Science* 1994;**264**:1907–9.
3. Balter M. Was North Africa the launch pad for modern human migrations? *Science* 2011;**331**:20–3.
4. Fernandes RMS, Ambrosius BAC, Noomen R, Bastos L, Combrinck L, Miranda JM, Spakman W. Angular velocities of Nubia and Somalia from continuous GPS data: implications on present-day relative kinematics. *Earth Planet Sci Lett* 2004;**222**:197–208.
5. Gibbons Profile A, Brunet M. One scientist's quest for the origin of our species. *Science* 2002;**298**:1708–11.
6. Conroy G. *Reconstructing human origins: a modern synthesis*. New York: W.W. Norton & Company; 2000.
7. Maslin MA, et al. East African climate pulses and early human evolution. *Quat Sci Rev* 2014;**101**:1e17.
8. Lovejoy CO. The natural history of human gait and posture. Part 1. Spine and pelvis. *Gait Posture* 2005;**21**:95–112.
9. Sylvester AD. Locomotor coupling and the origin of hominin bipedalism. *J Theor Biol* 2006;**242**:581–90.
10. Edelman GM. *Second nature: brain science and human knowledge*. New Haven, CT, USA: Yale University Press; 2006.
11. Brunet M, Guy F, Pilbeam D, Mackaye HT, Likius A, Ahounta D, Beauvilain A, Blondel C, Bocherens H, Boisserie JR, de Bonis L, Coppens Y, Dejax J, Denys C, Duringer P, Eisenmann V, Gongdibé F, Fronty P, Geraads D, Lehmann T, Lihoreau F, Louchart A, Mahamat A, Merceron G, Mouchelin G, Otero O, Pelaez Campomanes P, Ponce de León M, Rage JC, Sapanet M, Schuster M, Sudre J, Tassy P, Valentin X, Vignaud P, Viriot L, Zazzo A, Zollikofer C. A new hominid from the Upper Miocene of Chad, Central Africa. *Nature* 2002;**418**(6894):145–51.
12. Senut B, Pickford M, Gommery D, Mein P, Cheboi K, Coppens Y. First hominid from the Miocene (Lukeino Formation, Kenya). *Comptes Rendus de l'Académie de Sciences* 2001;**332**(2):137–44.
13. Richmond BG, Jungers WL. *Orrorin tugenensis* femoral morphology and the evolution of hominin bipedalism. *Science* 2008;**319**:1662–5.
14. White TD, Asfaw B, Beyene Y, Haile-Selassie Y, Lovejoy CO, Suwa G, WoldeGabriel G. *Ardipithecus ramidus* and the paleobiology of early hominids. *Science* 2009;**326**(5949):75–86.
15. Stanford CB. Chimpanzees and the behavior of *Ardipithecus ramidus*. *Annu Rev Anthropol* 2012;**41**:139.
16. Reardon S. The humanity switch. *N Sci* 2012;(2864):10–1.

17. Dembo M, Radovčić D, Garvin HM, Laird MF, Schroeder L, Scott JE, Brophy J, Ackermann RR, Musiba CM. The evolutionary relationships and age of *Homo naledi*: an assessment using dated Bayesian phylogenetic methods. *J Hum Evol* 2016;**97**:17–26.
18. Antón SC. Natural history of *Homo erectus*. *Am J Phys Anthropol* 2003;**122**:126–70.
19. Lordkipanidze D, et al. *Science* 2013;**342**:326–31.
20. Maddy D, Schreve D, Demir T, Veldkamp A, Wijbrans JR, van Gorp W, van Hinsbergen DJJ, Dekkers MJ, Scaife R, Schoorl JM, Stemerdink C, van der Schriek T. The earliest securely-dated hominin artefact in Anatolia? *Quat Sci Rev* 2015;**109**:68.
21. Gibbons A. Paleoanthropology: ancient island tools suggest *Homo erectus* was a seafarer. *Science* 1998;**279**(5357): 1635–7.
22. Whitfield J. Lovers not fighters. *Sci Am* 2008;**298**:20–1.
23. Stringer C. Evolution: what makes a modern human. *Nature* 2012;**485**:33–5.
24. Cox MP, Mendez FL, Karafet TM, Pilkington MM, Kingan SB, Destro-Bisol G, Strassmann BI, Hammer MF. Testing for archaic hominin admixture on the X chromosome: model likelihoods for the modern human RRM2P4 region from summaries of genealogical topology under the structured coalescent. *Genetics* 2008;**178**(1):427–37.
25. Hammera MF, Woernera AE, Mendezb FL, Watkinsc JC, Walld JD. Genetic evidence for archaic admixture in Africa. *Proc Natl Acad Sci USA* 2011;**108**(37):15123–8.

Further Reading

Jungers WL. Lucy's length: stature reconstruction in *Australopithecus afarensis* (A.L.288-1) with implications for other small-bodied hominids. *Am J Phys Anthropol* 1988;**76**(2):227–31.

CHAPTER

3

Origin of Modern Humans

A people without the knowledge of their past history, origin and culture is like a tree without roots. ***Marcus Garvey***[1]

SUMMARY

Three decades ago, archeological evidence and genetic studies suggested that the origins of modern humans and their evolution had been settled. The evidence pointed to our origin in East Africa evolving from *Homo erectus*, our single migration out of Africa into Europe and Eurasia, and the extinction of the Neanderthals most likely through our superior intellect and ability to outcompete them. Today we have come to understand that the picture of our evolution is much more complex than previously thought. Millions of years ago the first hominid species arose in Africa, and members of this archaic species evolved into the genus *Homo* approximately 2.8 million years ago (mya). Today we recognize several distinct archaic species of human, which are hominins that fall under the genus *Homo*, they include: *Homo habilis*, *Homo ergaster*, *H. erectus*, *Homo antecessor*, *Homo heildelbergensis*, *Homo naledi*, *Homo floresiensis*, and *Homo neanderthalensis*. The list and the number of human species vary depending on the group of paleoanthropologists that are consulted. The two paleoanthropologist groups fall under the titles lumpers and splitters, and, as their names imply, the lumpers tend to recognize fewer species than the splitters. Two distinguishing physical features that led to the genus *Homo* are increasing cranial capacity and a tendency toward bipedalism. Numerous species of the genus *Homo* evolved in Africa, and recent evidence points to different hominin species inhabiting Africa, Asia, and Europe at different times. African specimens include *H. habilis*, *H. ergaster*, *H. erectus*, *H. antecessor*, *H. heildelbergensis*, and *H. naledi*. *H. erectus* and *H. heildelbergensis* were presumably the first to leave Africa, with *H. heildelbergensis* later evolving into *H. neanderthalensis* and Denisovans in Europe and Northern Asia, respectively. All the recognized archaic *Homo* species that preceded *Homo sapiens* eventually went extinct, leaving *H. sapiens* as the only *Homo* species to survive. *H. sapiens* evolved in Africa around 200–400 kya, from the remaining groups of *H. heildelbergensis* (sometimes referred to as *Homo rhodesiensis*). *H. sapiens*, in their anatomically modern form, are believed to have moved throughout Africa, as supported by fossil findings. Eventually several

Ancestral DNA, Human Origins, and Migrations
https://doi.org/10.1016/B978-0-12-804124-6.00003-3

H. sapiens populations left Africa on several different occasions. *H. sapiens* that left Africa around 40–50 kya replaced the Neanderthals in Europe and the Denisovans in Northern Asia but not before inbreeding with Neanderthals, Denisovans, and possibly another *Homo* species. While there is no debate that *H. sapiens* developed to become the anatomically modern humans (AMHs), recent fossil finds have made the story of human evolution more complex, and the estimates of arrival of *H. sapiens* in various parts of the world continues to be debated. The remains of *H. sapiens* fossils, artifacts, and DNA from extinct humans that preceded us have provided an important contribution to the story of *H. sapiens* origin and evolution. In addition to archeological discoveries, scientists have turned to genetic evidence, looking at mitochondrial DNA, Y chromosomes, primate genomes, and the geographical distribution of genetic markers to better understand the origin and migratory patterns of humans. Owing to advances in technology, there is much more information that can be gained from archeological sites, allowing archeologists, geneticists, and paleoanthropologists to learn more about the migratory patterns and evolution of *H. sapiens*. This chapter explores what it means to be human, some of the hallmarks of human development, the possible migratory routes out of Africa and addresses the question, "Are *H. sapiens* still evolving?"

TOOLS, ENVIRONMENTS, AND ASSUMPTIONS USED TO STUDY THE ORIGINS OF *HOMO SAPIENS*

Although a plethora of new tools have been developed over the last 2 decades, and numerous new archeological sites and fossil remains have been uncovered, complications and limitations to data collection and analysis continue to arise. All the techniques used by anthropologists and geneticists require the use of fossils, ancient remains, and contemporary human and primate sources. The number of archeological sites, the state of remains, and the ability to date fossils varies from location to location. In many regions, fragmented fossils (teeth, skull, and other bone fragments) are found in caves, and in many cases the placement of the fossils have resulted from the fragments being washed into the cave or brought there by animals. These fossil remains are usually formed from magnesium carbonite (known as dolomite), which is in some cases hard to date because it is found in regions where the surrounding soil is formed from talus (the angular accumulation of soil and debris that does not form smooth stratigraphic layers) that cannot be geologically dated for the time of deposition. In other regions, fossils may be found in sedimentary rock formed near ancient lakes or river beds that provide open air sites where strata formed by volcanic activity and soil deposition provide more accurate methods of determining the geological time of deposition, and these dates can be used for comparisons in various locations across the region. Climatic conditions can also influence the ability to find fossils, and low oxygen levels can help stabilize and preserve fossil material. Although the fossil record continues to grow, most scientists believe that because fossil finds are rare, there is an underestimate of the number of species identified by the fossil evidence, and that it is difficult to distinguish different species based on the normally fragmented material that is recovered. The increase in the number of hominin fossils has complicated the development of phylogenetic trees because of the difficulty associated with the identification of different hominin species.[2]

Extant human and primate sources of genetic material also have limitations. Aboriginal populations (especially isolated populations in Africa) can be used to infer the genetics of our ancient ancestors that never left Africa. Although these studies are useful, they assume that the population represents low migration, low admixture, and a consistent mutation rate. The use of primates is also useful when looking for genetic changes that may have been responsible for the variation between humans and other primates and may help answer questions about human origins. One complication when comparing AMH with extinct humans, and other primates, is the ability to collect and compare gene expression data from different tissue types, and age groups, for comparisons and accurate analysis of genetic variation. In addition, as genetic data between humans and primates have accumulated, scientists have found that chimpanzees and humans are more closely related to each other than they are to other great apes. This has resulted in the human–great ape phylogenetic tree becoming more and more complex.[3]

Anthropologists and geneticists use an array of tools and methods to formulate and test hypotheses regarding the origin and migratory patterns of *H. sapiens*. The advent of molecular biology and recent archaeological advances has significantly changed the way that scientists study human origins. Today, molecular biologists use various molecular markers and techniques to study human evolution (see Chapter 1). The data from the use of several archeological and genetic tools and tests are mentioned throughout this chapter, thus a brief explanation of some of those tools follows.

Some Tools Mentioned in This Chapter

Optically stimulated luminescence (OSL) is a technique used to date fossils in geological sediments through ionized radiation to determine the last time a mineral was exposed to sunlight. OSL assumes that minerals were both constantly and sufficiently exposed to oxygen before being buried, noticeably bleaching the observed mineral or fossil. OSL is limited to dating substances within 100–200 kya.

Oxygen and carbon isotope analysis is used to study ancient tooth enamel to determine (1) diet, (2) the environment where an individual's teeth were formed, (3) changes in the climate based on different diets, and (4) occasionally where individuals traveled (based on where their teeth were formed and where their remains were found). The technology uses the natural absorption of oxygen by plants and animals and the mineral components fixed in teeth. In addition, carbon analysis is also used for plant identification. Teeth decay slowly relative to other human tissues and so they are frequently used as a source of information, including DNA analysis.

DNA analysis involves the isolation and sequencing of DNA from ancient samples. Today, genome sequencing uses a technology known as next-generation sequencing (NGS). NGS is used to determine ancient genomic sequences, the sequence of mitochondrial DNA, or the entire genome of ancient samples. Recent technology has allowed scientists to isolate mtDNA from Pleistocene sediment (dirt) surrounding archeological sites where no skeletal remains have been found.[4,5] The data collected from these sequences can be used to determine introgression from other species, physical traits (hair color, eye color, etc.), unique mutations or single-nucleotide polymorphisms (SNPs), genetic selection for specific alleles, and relationships between different populations. To date, the earliest hominin for which NGS has been used is for sequencing the genome of a 430 kyo *Homo* fossil found in Spain's Sima de los Huesos.[6]

DNA sequences can be used to compare different species and to date the divergence of species using molecular clocks. Molecular clocks depend on estimated mutation rates and assume that there is a constant rate of mutation. For example, the DRB1 gene, which is involved in immune system function, is often used to date primate species and has been estimated to have a mutation rate of 1.06×10^{-9} nucleotides per locus per year. This rate can be used to look at DRB1 sequences in ancient humans, and other primates to establish a time since divergence and to make coalescence maps.[7] Because different regions of the genome undergo different rates of mutation, this method of dating is controversial and is not as reliable as carbon and oxygen isotope dating and other methods recently developed.[8]

Genetic testing relies heavily on paternal and maternal DNA to create a picture of archaic male and female population structure. This type of research makes use of mitochondrial DNA (mtDNA) and Y chromosome DNA. mtDNA is passed on through a maternal lineage, has lower mutation rates than nuclear DNA, and is abundant in ancient DNA samples. Because mtDNA is passed on only from the female to all her offspring, the genetic material is regarded as a haplotype (i.e., a haplotype is an allelic variant which can be inherited intact and which defines all descendants carrying it). These characteristics make it an important tool for tracking maternal ancestry, and when coupled with other genetic tools can help create an ancestral tree or road map (see Chapter 1).

Y chromosome data are paternally transmitted and contain nonrecombining regions; within these nonrecombining regions, there are SNPs that can be used to track paternal ancestry. The unique display of SNPs on a Y chromosome makes up a haplotype. A haplogroup consists of a group of similar haplotypes (within a population) that all share the same SNP mutation within the genome (see Chapter 1). Haplogroups (a group of similar haplotypes) are used to differentiate between populations. Different haplogroups are caused primarily by genetic drift within a population. Genetic drift is the movement of alleles in a population caused by random circumstances, including the random chance that a parent will pass on their specific DNA to an offspring. While natural selection for a trait or marker will cause it to increase or persist within a population when it is favorable, genetic drift will cause it to fluctuate randomly and ultimately rise to fixation within the population or fall out of it completely if enough time passes. Because of this there is a chance that the haplotype will disappear from a population and be lost leading to a "dead end" when trying to trace its ancestry.

Giemsa staining and fluorescent in situ hybridization (FISH) can be used to examine and identify chromosome variation between humans and other primates. For example, although chimpanzees share 98% of our genes, many of those genes are on different chromosomes or arranged differently on the chromosomes, and chromosome rearrangements can affect gene regulation. Giemsa stain adheres to chromosome regions where the DNA is highly condensed. These regions are known as heterochromatin and indicate regions of the chromosome that are transcriptionally inactive. FISH is a technique that uses fluorescent probes to bind and illuminate regions of a chromosome that show sequence complementation to the probe. FISH can be used to identify specific DNA sequences on any chromosome that contains the sequence. The technique can be used to see how many copies of a DNA sequence are present and where they occur on the chromosome. The technique can also be used to examine mRNA inside of tissues and/or cells to indicate gene expression. This is done by making a probe that recognizes a specific mRNA molecule and then using it to determine the location and amount of mRNA in the sample.

Microarray analysis is a technique that can be used to look at the level of gene expression. The technique uses mRNA (which results from gene transcription) and converts it into cDNA (a more stable nucleic acid). The cDNA is fluorescently labeled and then quantitated using a small chip that contains the genes of interest. The intensity of the label determines the amount of mRNA present, reflecting the level of transcription.

Electron microscopy can be used to examine subcellular structure at very high resolution. The technique uses electrons instead of light to pass through the specimen. This not only gives higher resolution but can also detect organelle substructure. The technique can be used to detect chipping of tools and bones. It can also be used to examine tooth morphology and enamel.

WHAT DOES IT MEAN TO BE HUMAN?

Different Perspectives on Being Human, Self-Awareness, Degrees of Consciousness

When discussing the origin of modern humans, it is essential to describe what is meant by being human. Initially this may seem obvious and that it is easy to define the unique evolutionary characteristics that make us human; however, in the past people have tried to quantify human traits (specific brain sizes, average heights, dental characteristics, intellect, etc.) that would define an organism as human but have failed or only superficially quantified these traits.[9] This is because humans vary dramatically in there morphological, physiological, and mental characteristics. (For example, we might determine the average size range of the human skull and use this as a method for determining what is human but would not eliminate someone with microcephaly on this basis. We could use similar examples for all traits, including intelligence, height, physiology, etc.).

Being human can come down to the obvious and perceived differences between ourselves and other animals. In general, we perceive ourselves to be intellectually above all other animals, and this is emphasized in the genus and species name that we have given ourselves. The name *H. sapiens* is Latin for "wise person," not only attesting to our superior intellect but also to our early human ability to withstand the intense climate changes, displace our predecessors, and migrate to form sophisticated communities that occupy the world today.

Academicians look at humanity from different perspectives and define what it means to be human according to their specific discipline. Collectively their observations provide an idea of how we became human, what constitutes a human, and the basis for humanity.

From a paleoanthropological perspective, being human means that millions of years ago the first hominid species arose in Africa, and members of this archaic species evolved into the genus *Homo* approximately 2.8 mya.[10] They have pointed out how we became human through obvious adaptations that occurred during our evolution, including bipedalism, production of tools, brain enlargement, increased longevity, prolonged maturation, and advanced technology. Paleoanthropologists recognize several distinct archaic species of human that fall under the genus *Homo*, they include: *H. habilis*, *H. ergaster*, *H. erectus*, *H. antecessor*, *H. heildelbergensis*, *H. naledi*, *H. floresiensis*, and *H. neanderthalensis*. Today the consensus among paleoanthropologists is that modern-day humans (*H. sapiens*) evolved from *H. heildelbergensis* around 200–400 kya and that other archaic hominins went extinct for various reasons.

From a biological perspective, being human involves modern genetic analysis, which has shown that our essential physiological and physical components (heart, lungs, bones, brains, etc.) evolved from the genes of animals that preceded us and that much of human genetic information is similar to worms, flies, mice, and other primates. Biologists point out that mutations, variable gene expression, genetic recombination, genetic drift, and natural selection have resulted in traits that distinguish us from other animals. However, we hear, smell, feel, and see like many animals, and in fact many of our senses are not as acute as some animals. So why are we still classified as animals by biologists even though we distinguish ourselves as unique?

Answers for our uniqueness come from social and philosophical perspectives; our most obvious differences from other animals are our intellect, speech, self-awareness, emotions, distinctive cultures, and our behaviors. All these differences are thought to arise from our unique brains. Modern human brains are unique both in structure and in their ability to process information from our environment. Brain size, structure, and function have been a focus of human evolution, and studies have shown that brains have changed as humans evolved.

The Evolution of the Human Brain

In the last 2 million years, the human brain has nearly tripled in size (Figs. 3.1 and 3.2 and Table 3.1). The brain of *H. habilis*, the first species of the genus *Homo* discovered in the Olduvai Gorge in Tanzania, in 1960, was about 600 cubic cm. *H. habilis* fossils, considered by many as belonging to the first human, have been dated between 2.3 and 1.4 mya. Hominid brain size, in general, increased slowly until the arrival of *H. habilis*, leading to a trend of tremendous brain growth. Not coincidentally, the fossil record shows evidence of widespread stone tool use following the species' emergence. Appropriately named "handyman," *H. habilis* was the first known member of the genus *Homo* to make stone tools and is believed to have habitually used hand axes and sharp stone implements to carve meat off carcasses. Thousands of these tools were found in the Olduvai Gorge and are attributed to *H. habilis*.[19] The next million years following *H. habilis* saw continuing encephalization, with the brain of *H. erectus* measuring about 900 cubic cm.

Geological ages (millions of years ago), brain size (cm^3), estimated male and female body weights (kg), and posterior tooth surface areas (mm^2) for selected fossil hominid species

Species	Geological age (mya)	Brain size (cm^3)	Body weight		Posterior tooth surface area (mm^2)
			Male (kg)	Female (kg)	
Australopithecus afarensis	3.9–3.0	438	45	29	460
A. africanus	3.0–2.4	452	41	30	516
A. boisei	2.3–1.4	521	49	34	756
A. robustus	1.9–1.4	530	40	32	588
Homo habilis (*sensu strictu*)	1.9–1.6	612	37	32	478
H. erectus (early)	1.8–1.5	863	66	54	377
H.erectus (late)	0.5–0.3	980	60	55	390
H. sapiens	0.4–0.0	1350	58	49	334

FIGURE 3.1 Changes in hominid brain size and molar tooth surface areas suggest that dietary changes may have preceded rapid brain growth that began 2 million years ago. *Reyes-Centeno H, et al. Testing modern human out-of Africa dispersal models and implications for modern human origins.* J Hum Evol *2015;87:95–106.*

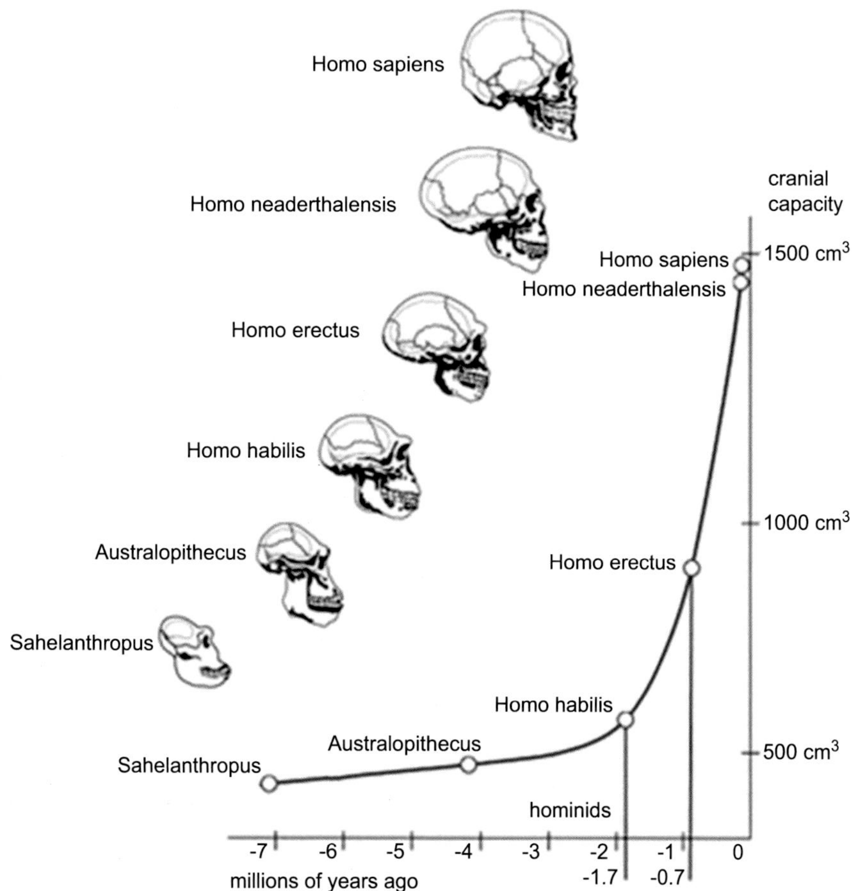

FIGURE 3.2 The growth rate of hominid cranial capacity began to rapidly increase, starting around 2 million years ago. *From Evolution du volume cerebral des Hominides. Le Journal du Net 2010. http://www.linternaute.com/science/biologie/dossiers/06/0608-memoire/8.shtml. Copyright 2011 Benchmark Group - 69-71 avenue Pierre Grenier 92517 Boulogne Billancourt Cedex, France*

TABLE 3.1 Differences in Brain Size of Several Hominins Over Time

Hominin	Age (mya)	Estimated Brain Weight (gr) [lbs]	Average Capacity of Braincase (cc)
Australopithecus afarensis	3.9–3.0	(435) [0.96]	440
Homo habilis	2.3–1.4	(600) [1.32]	640
Homo erectus	1.9–0.14	(850) [1.87]	930–1029
Homo neanderthalensis	0.6–0.04	(1497) [3.3]	1500–1600
Homo sapiens	0.02–0.0	(1370–1200) [2.7–3.1]	1250–1350

Brain size continually increased over time as hominins evolved; however, the brains of *H. sapiens* decreased when compared with *H. neanderthalensis*. The variation seen in human brain weights and braincase reflect the differences in male and female brain sizes.

The largest hominid brain belonged to *H. neanderthalensis*, with a brain of 1600 cm^3. Human brain size did not stabilize until the appearance of AMHs 200 kya, and then slightly shrank 100 kya to the current volume of approximately 1200 cubic cm.[11]

Shultz et al.[12] reviewed the arguments for the pressures driving brain expansion, quantitatively evaluated the time changes in brain size, and compared these to the environmental-based hypotheses. They concluded that using both absolute and residual brain size estimates showed that brain evolution resulted from a mixture of both gradualism and punctuated equilibrium and that punctuated changes in brain size were not temporally associated with changes in paleoclimate instability. In addition, their brain size estimates indicated that punctuated changes were observed at approximately 100 kya, 1 mya, and 1.8 mya in addition to gradual intralineage changes in *H. erectus* and *H. sapiens*.

The Anatomy and the Physiology of the Human Brain

The distinctive feature of the brain in modern humans compared with other animals is a very large brain relative to body mass. Compared with that of other species, the human brain is not only bigger but also has higher energy demands. The brain consumes 22 times the amount of energy as muscle tissue, consumes approximately 25% of the glucose metabolized by the body and 15% of the oxygen intake.[11]

The human brain is about 1200 cubic cm on average, contains approximately 100 billion neurons, over 100,000 km of interconnections, and has an estimated storage capacity of 1.25 trillion bytes. The distinctive feature of the human brain is extensive corticalization (wrinkling of the brain's cortex) with a six-layered cerebral cortex. The surface area of the modern brain is between 1500 and 2000 cm^2 (about the size of two full pages of newspaper). To fit such a large surface area inside the cranium, the cortex is highly folded into regions called gyri and grooves referred to as sulci.[13]

At the most basic level, the brain is an organ of soft tissue encased by the vertebrate skull that functions as the coordinating center of sensory information and of intellectual and nervous activity. The human brain is made up of three major parts: the forebrain, midbrain, and hindbrain (see Fig. 3.3).

The forebrain is mainly comprised of the cerebrum, which consists of right and left hemispheres connected by a bridge of nerve fibers. The cerebrum integrates sensory and neural functions, as well as initiates and coordinates voluntary activity in the body. The outer layer of the cerebrum is made up of gray matter, whereas the inner part is composed of white matter. Gray matter contains the cell bodies of neurons, whereas white matter contains the dendrites and axons that spread out from the cell bodies to connect to other neurons.[13] The two cerebral hemispheres are each divided into four lobes: the frontal lobe, responsible for thinking, making judgments, planning, decision-making, and conscious emotions; the parietal lobe, associated with spatial computation, body orientation, and attention; the temporal lobe, associated with hearing, language, and memory; and the occipital lobe, associated with visual processing.[11,13] The midbrain, or the upper part of the brain stem, contains nuclei that link the various parts of the brain responsible for motor functions (Fig. 3.3), eye movement, and auditory control.[14]

The hindbrain, or the lower part of the brain stem, is involved in the autonomic nervous system and facilitates unconscious actions and processes such as breathing and regulating heart rate. The medulla is the part of the brain stem that controls breathing and heartbeat. The cerebellum is also part of the hindbrain and coordinates and regulates timing of action with available sensory information. Located at the front of the cerebellum, the pons facilitates the timing coordination established by the cerebellum. The pons is primarily the white matter sending signals between the cerebrum and spinal cord to the cerebellum, which then routes output for action to these two areas (Fig. 3.4).

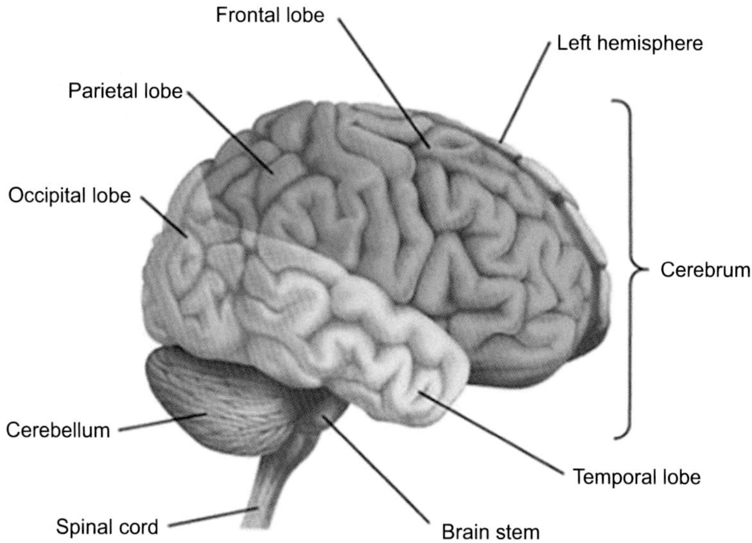

FIGURE 3.3 General anatomy of the brain. *From http://humandiagram.info/brain-anatomy-that-controls-our-body/anatomy-diagram-of-brain/#main.*

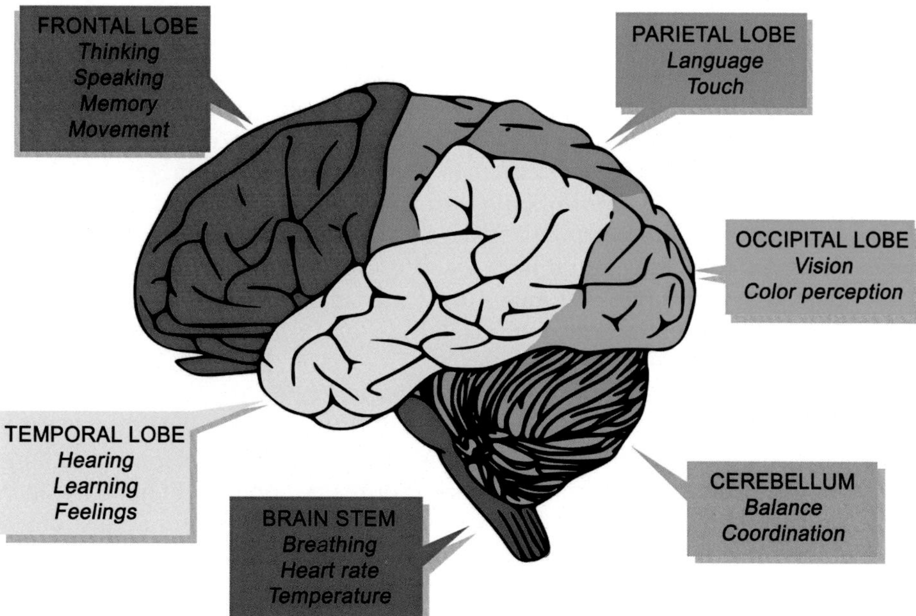

FIGURE 3.4 Brain anatomy and physiology.

The spinal cord can be viewed as a downward extension of the brain stem. It contains sensory and motor pathways from the body, as well as ascending and descending pathways from the brain. It also has reflex pathways like the knee-jerk reflex that function independently of the brain.[11,13]

The brain is hardwired with connections created by neurons gathering and transmitting electrochemical signals. Neurons are connected by synapses and can transmit signals such as nerve impulses and action potentials over long distances and can send messages to each other. This allows them to link sensory inputs and motor outputs with the different lobes of the cerebral cortex.

Some of the Characteristics That Make Us Human

Overall, when we consider what makes us human, the distinction is made based on several characteristics which are unique and ubiquitous to humans. These include intelligence, speech and language, self-awareness, emotion, culture, and the technology to overcome environmental and physical barriers. These will be discussed individually along with their evolutionary relevance.

Human Intellect

One outcome of the brain processing complex sensory information is biological intelligence. Intelligence is the ability for learning, problem-solving, creativity, and reasoning, resulting from the gathered information and interaction with the environment. Some neurologists have assigned certain aspects of intelligence to the prefrontal cortex, left temporal cortex, and parietal cortex. The thalamus relays sensory and motor signals to the cerebral cortex and regulates consciousness and sleep; however, the neural basis of all consciousness cannot be localized to a specific region of the brain. Consciousness involves all regions of the brain through whose activity an organism can construct an adequate model of its external world.[14]

Initially, increased intellect and creativity were considered a result of humans moving out of Africa around 40 kya. But earlier evidence for creativity has come from several studies showing that humans 70–100 kya demonstrated signs of invention. One study looking at excavations in the Sibudu Cave in South Africa found signs of bedding material made from a woody plant called *Cryptocarya woodi* that contains natural insecticides, indicating that the inhabitants understood the natural flora and some of its benefits. In addition, the use of other plant material for adhesives and snares to capture small animals were discovered at the site.[15,16] In other studies from the same period (100–200 kya), scientists have found bone awls, intricate jewelry from abalone shells, and tools.[15]

The rationale for the evolution of human intelligence is a heavily debated topic. Most scientists believe that the development of human intellect was closely linked with the evolution of brain size, the need for increased social interactions, the development of language, and some favorable genetic mutations. Brain size is only one component linked to increased intelligence, and we know that other animals have larger brains than humans but do not seem to have the same level of intelligence. One explanation for increased brain size and increased intelligence comes from studies of the prefrontal cortex in bonobos, chimpanzees, and humans. It appears that a region of the brain known as the Brodmann's area, a region in

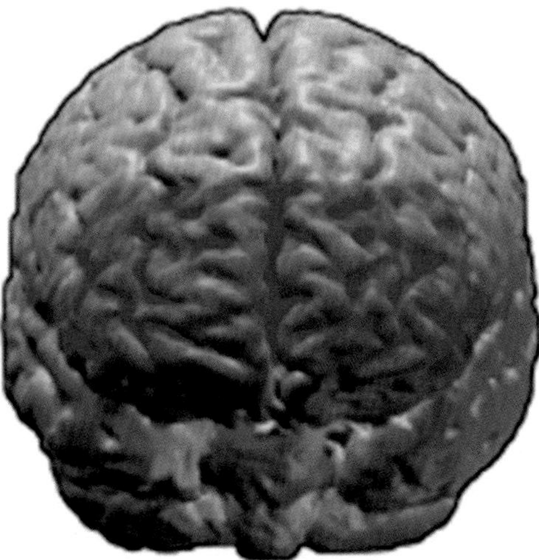

FIGURE 3.5 Brodmann's area 10 (in red) is thought to be the region involved in organization of sensory input and planning future actions. Measurements of Brodmann's area 10 show it nearly doubled in humans after the human and great ape lineages separated. *From https://upload.wikimedia.org/wikipedia/commons/1/14/Brodmann_area_10.png.*

the cerebral cortex divided into 52 areas from 11 tissue types, has been implicated in cognitive function and language. Specifically, Brodmann's area 10 (Fig. 3.5) is thought to be the region involved in organization of sensory input and planning future actions. Measurements of Brodmann's area 10 show it nearly doubled in humans after the human and great ape lineages separated. This increase in area also resulted in increased space between neurons and more complicated connections, which may lead to increased communication between neurons.[17]

Fundamental changes in the hominid diet are regarded as possible stimuli for the evolution of human intellect. Changes in diet and food availability, due to environmental conditions or migration patterns, were selective pressures that acted on both biological processes and anatomical features and may have had a dramatic and rapid impact in human evolution. Genome-wide and single-gene studies have shown that diet is an important evolutionary force that can have effects on gene regulation and expression. This type of environmental change (e.g., animal domestication, agriculture, lack of food sources, or appearance of new food) would have resulted in a rapid biological response to metabolism and nutrition.

More specifically, the advent of the omnivorous diet and the utilization of fire are believed to be driving factors behind the neurological boom (increased brain size) that gave rise to behaviorally modern humans. We know that cooking food increases the caloric intake and that it can also turn otherwise inedible food into good sources of energy. Recent studies comparing the daily energy expenditure between hunters and gathers and more sedentary individuals have shown that each consume the same number of calories per day (average 2600 for males and 1900 for females). Although this seems counterintuitive, many studies have confirmed this inflexible metabolic rate that humans maintain regardless of lifestyle. Studies comparing great ape metabolism with humans' have shown that humans burn more calories than any of our great ape relatives, when adjusting for body size. Scientists speculate that the extra calories are utilized to support our larger brains, increased reproductive abilities, and longevity. This shift in energy to support human-specific traits could explain some of the evolutionary acceleration of brain development.[18]

H. habilis is typically regarded as the first human ancestor to introduce meat to the hominid diet. The fossil record indicates the appearance of meat consumption 2–3 mya, which predates the use of stone tools.[19] Brain measurement studies reveal marked transverse expansion of the cerebrum, particularly the frontal and parieto-occipital parts, in *H. habilis* and increased bulk of the frontal and parietal lobes. This indicates that the hominid brain had attained a new evolutionary level of organization, which can be ascribed to the advent of meat consumption.[20]

Speech and Language

Although the origin of speech is still debated, it is one of the unique characteristics of humans and most likely arose in humans around 100 kya.[21] Other primates can communicate by producing sound, and in fact baboons can make vowel-like sounds similar to humans.[22] Some evidence for Neanderthal speech comes from their ability to make complex tools, which would have been difficult to teach and pass on to future generations without speech. In addition, Neanderthals had the same version as humans of the FOXP2 gene that controls oral-facial movements and has been loosely associated with language (see Chapter 6). But it is the complexity of language intertwined with vocalization that makes human's ability to communicate unique. Language does not have to be spoken (can be signed or written), but our spoken language makes sophisticated use of the lungs, lips, tongue, glottis, and larynx in a way that produces a unique variety of distinguishable languages. Studies of the brain and speech have concluded that Broca's area and Wernicke's area are the most prominent regions underlying speech. The ability of humans to learn and speak various languages is a key to forming logical and analytical skills and was instrumental in complex organization, migration, and the development of societies.

Self-Awareness

Self-awareness is our ability to perceive beyond physical and mental sensation to recognize ourselves as existing. Our ability for introspection and metacognition (being able to think about thinking, i.e., our higher-order thinking skills) and the ability to recognize oneself as an individual separate from the environment is a key factor in what sets *H. sapiens* apart from other organisms. While there are other organisms that are also self-aware, *H. sapiens* exceed all other species in their recognition of the universe and of their place within it. This stems from the evolution of higher degrees of consciousness within the human brain. Consciousness emerges from the brain's ability to construct a complex representation of the world and self. The human brain continuously creates the conscious experience of being in the world, with the self and the world representing models through which information is processed. These world representations are highly adaptive and have allowed *H. sapiens* to survive using this knowledge of the universe around them.[23]

Human consciousness has also given rise to questioning our existence; questions addressing why are we here, what is the purpose of life, is there an afterlife? These questions appear to be unique to humans, and in our own mind we use these questions as a distinguishing factor between ourselves and other animals. This level of consciousness and the questions it raises has given rise to religious beliefs, which we see very early on in Neanderthal practices and AMH evolutionary history, and that give us further reason to separate ourselves from all other organisms.

Emotions

Darwin speculated that emotions were part of human evolutionary heritage and research has shown that emotions are ubiquitous among humans. Many emotions can be seen through involuntary facial expressions, suggesting that they are a mechanism of evolution. Both basic (fear, sorrow, anger, happiness) and social (guilt, pride, shame) emotions arise as innate reactions to our environment. Although the degree of emotion varies between individuals and how we think, feel, and react to environmental cues, not all environmental cues elicit the same emotion in every person. This ability to react and feel differently is an aspect of emotion that is a part of being human. Paul Ekman, a noted psychologist and supporter of Darwin's beliefs on emotion, has postulated that human emotions are so powerful that they can override our drive for hunger and sex.[24,25]

Other animals also demonstrate some aspects of emotion. Elephants, for instance, are known to mourn their dead, demonstrating empathy, whereas many dogs demonstrate feelings of allegiance. However, the range of emotions shown by humans seems to exceed those of other animals. Some evidence for emotion in Neanderthals and early AMH comes from fossil remains of individuals that lost their teeth. These individuals would have had to be cared for and fed regularly by relatives or other members of a group. Feeding must have occurred using soft foods such as bone marrow or premasticated foods. Although other primates have demonstrated emotions like mourning the dead, and taking care of the weak within a population, many of the attributes ascribed to humans indicates an ancient social structure and caring that goes beyond the social activities of other primates.[26,27]

Culture

Another aspect of humanity is culture. Culture is the combination of beliefs, behavioral values, knowledge, and traditions shared by a group. Although some animals show some signs of culture especially within groups, humans are unique because they can transmit culture through actions and symbols (teaching from one generation to the next), and this ability allowed us to adapt to various environments. Culture has been implicated in human evolution, especially the acquisition of symbols, language, and our ability to cooperate. One idea on how culture may have steered evolution is that the creation of small groups, due to climate change, may have created both genetically and culturally unique populations that accelerated speciation. These small groups would have made it possible to create cooperative communities that would work together to outcompete other hominin species. In addition, these small groups would eventually construct social norms, work toward creating a more complex society, and develop societal goals unlike our great ape cousins. Robert Boyd, an evolutionary psychologist, feels that culture is an essential part of human adaptation and that cultural mechanisms interact with population dynamics to shape cultural variation, and this variation ultimately influences human evolution.[28]

Phenotypic variation has played a significant role in influencing human cultural practices, e.g., acceptance of different skin colors and facial features. All human phenotypic variations have resulted from adaptation to different environmental and climatic conditions. Climate has influenced the way we look, from our skin color to the shape of our nose.[29] In turn, different phenotypes within and between populations have resulted in the acceptance or rejection of certain individuals based on their appearance. The acceptance or rejection of individual phenotypes can lead to changes in sexual selection (positive or negative), which further influence evolution.

One aspect of sexual selection and culture that has shaped human evolution is the unusual practice of monogamy in many human societies. Less that 10% of mammals practice monogamy yet among humans it is the most common practice, even though there are obvious scenarios where the practice breaks down (divorce, affairs, polygamy, etc.). Data from the study of ancient hominins indicate that monogamy may have been the preferred practice for humans, especially as a result of bipedalism and concealing the external signals of female ovulation. There are several competing theories for the movement toward monogamy. One explanation for these "pair bonds" is that they helped solidify our complex social systems and create greater family ties. This movement from a promiscuous society to a monogamous one would not have occurred rapidly but may have occurred as societies became more complex and there was an increased need for a dependable mate to provide resources for offspring and to avoid violence from competing males.[30–32]

Although most scientists agree that culture played a role in evolution, it has been recognized that culture also limits humans by sometimes fixating them socially and making it difficult to adapt to extremes. For example, adapting from one extreme environment to another can pose difficulties with food, clothing, goals, and practices previously shared within a group.[33,34]

Debates on what it means to be human often turn into discussions of nature versus nurture, biology versus culture, and many scholars want to weigh each of these contributions separately; however, humans are influenced by how all these forces interact with one another. Humans are influenced by social conditioning and behave and respond in certain ways per the norms of society. Our biology dictates certain behaviors and emotions, but our culture influences choices. For example, our biology dictates that we need to eat, but our society influences what we eat, when we eat, and how we eat. We are conditioned by our social environment, desire for acceptance, and fear of rejection, which are all significant motivations for behavior.[28]

Most behavior has a genetic component, and all animals demonstrate various types of behavior, many of which we see in humans (hunger, self-defense, staying away from danger, etc.). In general, animal behavior can be innate, learned consciously or subconsciously, and voluntary or involuntary. Human behavior, however, is unique because of its complexity and often irrational nature, moved by subconscious impulses. Much of human behavior is dictated by our unique environment (our cultural biases, beliefs, social status, past experiences) and response to environmental ques. Because we live in a society, there are limits to most of our behaviors and an awareness to optimize benefit and reduce harm.

For many years, an explanation for the evolution of behavior and other complex phenotypes has challenged investigators. Recent studies have suggested that complex phenotypes such as behavior result from a combination of simpler functions, which required genomic instructions and were selectively favorable. These simple functions were then built upon through successive mutations that provided crucial new and more complex functions, i.e., built upon in a stepping-stone fashion.[35] This multistep mechanism is the current explanation for the genetic evolution and occurrence of most complex phenotypes, such as eye color, height, weight, intelligence, etc., that require many genes and can be influenced by the environment. In the case of behavior, recent genomic studies have identified "toolkit genes," which are highly conserved genes that regulate development. These ancient genes can be involved with common functions such as cell-to-cell communication, cell adhesion, and cell differentiation or can be associated with more complex phenotypes. Because they are ancient genes, they are similar across diverse species resulting in similar phenotypes. Using behavioral genomic studies, investigators have begun to identify genetic toolkits, regulated by many genes, which result in some behavioral phenotypes.[36]

Technology

Because of our intellect, cultural experiences, language, and self-awareness, we have been able to develop technology that has allowed us to overcome many natural obstacles and adapt to extreme environments. Technology has provided a major difference between ourselves and other animals and our ability to alter our environment while other animals must adapt to theirs. Our ability to make clothes, transport food and water, build shelters, develop energy sources, etc. has allowed humans to survive in extreme environments. For example, millions of people live in the deserts of Southern California, Nevada, and Arizona by pumping water hundreds of miles and cooling their living spaces with air conditioning.

Our ability to overcome some of the forces of nature through technology appears to be unique among all other animals and has led some scientists to assume we no longer live under the forces of natural selection. However, this type of logic fails to recognize that as our population grows so do the numbers of mutations, some of which may be advantageous. They also fail to recognize the evolution of other species that directly affect our daily lives, a good example being the microbiome that lives within and on us. And finally, these individuals fail to acknowledge the effects of social evolution created by our ability to move around the world and breed with other populations, our social breeding groups created from socioeconomic and educational forces, all leading to variable sources of recombination, mutation, genetic drift with the ultimate outcome of both natural and unnatural selection.[37]

ORIGINS OF MODERN HUMANS: HOW IT ALL BEGAN

Knowing what makes modern humans unique leads to the question of how did we get to be so different from other animals? It is obvious we came from primate ancestors and that other humans preceded us. Why did we survive when others perished? Some of these answers can come from looking at our closest living primate relatives and looking back at those hominins and humans that went extinct.

In the Beginning

Genetic and morphological evidence has determined that our closest living primate relatives are members of the great ape family. Evidence suggests that the great apes, gibbons, and humans emerged and diversified during the Miocene approximately 23–5 mya. A recent discovery of an ape cranium dated between 10 and 14 mya may be evidence of what the last common ancestor of humans and apes looked like.[38] Fossil evidence continues to indicate that at some point in our evolution we shared a common ancestor and at some point the lineages diverged. Estimates as to when the change took place are between 6 and 5.3 mya, and may have occurred due to dramatic climate changes during the Upper Miocene. Chaline and colleagues[39] have suggested that the divergence was a result of adaptations to ecological and geographical separation. They propose an allopatric speciation (speciation occurring in nonoverlapping geographical regions) model in which prechimpanzees were separated in the central and southwest region of Africa, pregorillas in northern Africa above the Zaire river, and preaustralopithecines in eastern and southeastern African Rift. They contend that the results of climatic changes during the Miocene across Africa would have led to adaptation pressures leading to speciation. Others have contended that the climatic changes were not significant enough to lead to the dramatic variations proposed.[40]

Because there is little known about the ancient common ancestor of gorillas, chimpanzees, and humans, nor about the regions they may have occupied, the theory of allopatric speciation remains a possibility but not one that is easily provable.[41]

Human Variation From Great Apes and Extinct Human Ancestors

It has been assumed that the divergence of chimpanzee from humans occurred around 6 mya, and DNA evidence suggests that they remain our closest living primate relatives. One method for examining the evolutionary forces that led to the origin of humans is to compare humans with the chimpanzee. Behavioral and genetic studies have revealed significant differences that may have been the earliest forces leading to the separation of humans from the great apes and humans from Neanderthals.

Genomic Variation Between Humans and Other Primates

Genomic comparisons between chimpanzees and humans became possible after the development of NGS and advances in bioinformatics. Initial comparisons disclosed that chimpanzees and human genomes were 98.7% identical, with the 1.3% of variation due to about 35 million single nucleotide variations and about 5 million insertions and deletions. In addition, more than 60 human-specific de novo genes have been identified.[42] Of the 35 million SNPs, about 40,000 resulted in amino acid substitutions, and most are thought to have little or no effect on protein function.[43] About 10%–20% of the amino acid differences between humans and chimpanzees have been fixed by positive selection and the remainder are neutral. Comparisons of genes from human and chimpanzees revealed that genes involved in sensory perception, reproduction, apoptosis, and immunity have the highest number of amino acid changes, and brain-expressed genes have the fewest amino acid substitutions.[44,45] The relatively small amount of variation (in comparison to our 3 billion base pairs that make

FIGURE 3.6 **Phenotypic similarities between humans and chimpanzees.** Chimpanzees and humans share many of the same phenotypic characteristics (i.e., similar structural polarity; placement of head, arms, hands, legs; five digits on each limb with nails not claws; similar ear anatomy; and similar mandible). The developmental similarity makes it easy to see why the genetic makeup of humans and chimpanzees are so similar. *From listverse.com/2012/02/14/10-comparisons-between-chimps-and-humans.*

up the human genome) makes sense as great apes and humans in general share many of the same phenotypic characteristics (i.e., similar structural polarity; placement of head, arms, hands, legs; five digits on each limb with nails not claws; similar ear anatomy; similar mandible; etc.) (Fig. 3.6); however, there are significant difference in behavior, metabolism, cognitive abilities, disease profiles, and morphology (Fig. 3.7 and Table 3.2).

The Neanderthals are considered today to be our closest extinct relatives, and we shared a common ancestor about 550–765 kya before diverging about 381–473 kya. Genetic evidence suggests that Neanderthal DNA is approximately 99.7% identical to modern humans. The 0.3% difference occurred after the split between humans and Neanderthals, resulting in two archaic human genomes, and although we share many of the same genes, new mutations gave rise to new derived alleles. NGS of archaic and extant human genomes indicate that some modern humans can have genetic material from Neanderthals, Denisovans, and/or another unidentified group.

Because humans mated with Neanderthals, modern-day humans whose ancestors originated outside of Africa contain on average of about 1%–4% of Neanderthal DNA. Some of the DNA inherited from Neanderthals has been shown to contain several genes with alleles that contribute to depression, addiction to tobacco, hypercoagulation, urinary tract infections, skin disorders, and some immune functions (see Chapter 6). Other DNA from Neanderthals appears to have no function, and these regions are referred to as DNA deserts.[46]

FIGURE 3.7 **Morphology of chimpanzee and human skulls.** Although chimpanzees and humans share some similar structures, they also vary in many phenotypic characteristics. In this example it is clear to see the variation in skull morphology. The top panel shows the skull of a chimpanzee and the lower panel, the skull of a modern-day human. *From listverse.com/2012/02/14/10-comparisons-between-chimps-and-humans/.*

Neanderthals also vary from modern-day humans in a variety of physical characteristics. Neanderthals had larger brains, thick and arched brow ridges, large incisors, larger hands with a longer thumb, a broad chest, their bones and teeth developed faster, and their young matured at a faster rate[41] (Fig. 3.8).

The phenotypic differences between humans, chimpanzees, and Neanderthals are thought to occur because of chromosome structural variation and the variability in the time and region of gene expression. Gene expression can be influenced by several factors, including (1) gene regulatory mutations that affect timing and level of expression, (2) posttranscriptional and epigenetic regulation, (3) the number of genes (copy number variation [CNV]), and (4) the arrangement of genes on the chromosomes. These variations have led to a variety of human-specific single gene and polygenic traits, and changes in gene expression have been hypothesized to have resulted in dramatic changes in cultural and demographic changes between humans and other primates.

TABLE 3.2 Differences Between Human and Chimpanzee Anatomy

Human Anatomy	Chimpanzee Anatomy
Fully terrestrial bipeds. Two arches in foot, relatively short toes, with large toes aligned with other digits. Relatively large ankle bones	Arboreal and terrestrial locomotion. Terrestrial quadrupedal locomotion, knuckle walking, occasionally bipedal, opposable large toe
Brain—Broca's area for processing speech. Overall brain size approximately 1400 cc. Spinal cord more centrally located at the base of the skull	Brain—Brodmann's area for noises, no speech. Overall brain size approximately 400 cc
High forehead, rounded skull, no brow ridge, no forward jutting jaw, chin, relatively small neck muscles	Sloping forehead, brow ridge, forward jutting jaw, heavy supraorbital torus (ridge bone above eye)
Teeth relatively small, small flattened or jagged canines resembling incisors. Relatively small incisors, relatively thick enamel. More rounded or arc-like dental structure with parallel back teeth	Large canine teeth with gaps in jaw to accommodate large canines. Relatively thin enamel and large incisors. Rectangular dental structure. Overall enlarged teeth, jaws, and chewing muscles
Five straight digits on the hand, relatively long opposable thumbs for fine grip	Five curved digits on the hand, relatively short thumb
Long legs and arms, twisted humerus, low shoulder, forwardly placed opening for spinal cord	Relatively short legs and arms
Vertebral column is S-shaped, pelvis with ilium is bowl-shaped	Vertebral column is bow-shaped, pelvis is long with flat ilium
Orbital margin (bony cavity in the skull containing the eye) set back in the skull, large orbital width (distance between eyelids).	Relatively small orbital margin and small opening between eyelids.
Size of gut relative to body mass is small compared with chimps	
Broad chest more cylindrical than chimps	Round, barrel-shaped rib cage that widens toward the base.
Slower development and prolonged childhood when compared with apes	

Many of these differences have resulted in the distinction between humans and other primates and are used in paleoarcheological studies to identify skeletal remains.
Adapted from Jobling, et al. Human evolutionary genetics. *2nd ed. New York NY: Garland Publishers; 2014. pp. 269–85.*

Changes in Gene Expression by Gene Regulatory Mutations

Looking at changes in gene expression is important because not all genes are turned on in all cells at the same time or at the same level. Specific genes are required for the development of the brain, skin, bones, etc., and since chimpanzees and other Apes carry many of our same genes, one mechanism for our differences may be the site and level of gene expression. The idea is that switches responsible for turning genes on or off in certain cell types at certain times may differ between humans and other apes.

Over the last decade, significant research into the molecular mechanisms of human evolution has focused upon the fastest evolving regions in the human genome known as human accelerated regions (HARs). HARs are a set of over 3000 segments of the human genome that

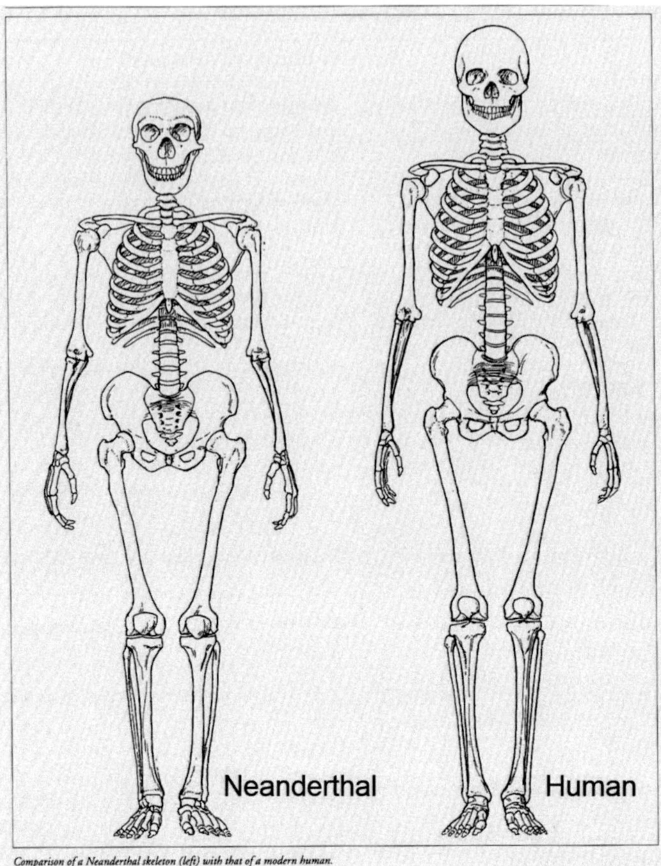

FIGURE 3.8 Human Neanderthal skeletal comparison. Although Neanderthals and anatomically modern humans are both representatives of the genus *Homo* and thus humans, there are distinct skeletal variations that are used by archeologists to distinguish human and Neanderthal skeletal remains. Variations in the scapula, wider hips, large outward rotated hip joint, thick walled femur shaft, large thick patella, large ankle joint and wide toe bones can all be used to distinguish the Neanderthal skeleton. *From http://ecodevoevo.blogspot.com/2014/01/walk-this-way-talk-this-way-roll-in-hay.html.*

are highly conserved elements in all mammals, except humans. These sequences are mostly outside of the protein coding regions of genes (in what has previous been referred to as junk DNA) and range in sizes from 100 to 400 bp and demonstrate human-specific changes when compared with other vertebrate genomes. Because these regions are variable in humans, they are considered potential regions for important functional elements. The goal behind looking at these elements is to determine if the data in a multiple sequence alignment are consistent with lineage-specific acceleration of selection, and if there are variations when compared with other lineages (i.e., are the sequences that show accelerated DNA substitutions consistent within the human lineage when compared with nonhumans). The search for these elements can be difficult, but the development of NGS and complex algorithms has identified several HARs to be potential candidates for driving important phenotypes during human evolution. The phenotypic impact of HARs is difficult to assess because of the limitations associated with human experimentation; however, the function of several HARs has been determined. Investigators have used creative nonhuman experiments to demonstrate the ability of several HARs to enhance the expression of specific genes, some of which appear to be involved in unique human phenotypes.[47]

To determine what gene activities HARs are responsible for, fertilized mouse embryos were injected with chimpanzee HAR sequences that were attached to a gene that expresses color (a so-called reporter gene). If the HAR sequences acted as enhancers to turn on genes in the embryo cells, those cells containing the active genes would turn blue. The investigators found that over 66% of the chimpanzee HARs turned on genes during mouse development. Human HARs were used in similar experiments, and eight HARs showed differences in enhancer activity when compared with the chimpanzee HARs. These included genes expressed in the eye (HAR25), developing limb (HAR2, HAR114), and central nervous system (HAR5 HAR142, HAR164, HAR170, HAR238).[48]

In the experiments using mouse embryos, human HAR5 (also referred to as HARE5) was shown to regulate expression of the Frizzled 8 gene, which affects the size and the development of the brain in mice. Experiments showed human HARE5 was active earlier in development, and affected a larger region of the brain neocortex, when compared with the chimpanzee version. Embryos carrying the human HARE5 sequences were 12% larger than those with the chimpanzee sequence. This was an important finding as the neocortex of humans differs dramatically in the rate of the cell cycle, prolonged cortex development, and overall increased size when compared with other great apes.[49]

Other HARs have also been identified inside genes and/or regulating genes that may have been involved in the split between humans and chimpanzees. HAR1 is part of a novel RNA gene that is expressed during neocortical development, whereas HAR2 is a noncoding sequence that functions as an enhancer in the developing limb bud, with the human-specific sequence enhancing expression in the anterior wrist and thumb.[50]

If HARs are responsible for characteristics that differentiate humans from other organisms, you would expect that mutations within these regions might affect social and cognitive function. Doan et al.[51] showed a significant number of biallelic point mutations in HARs in individuals with autism spectrum disorder. They identified disease-linked, biallelic HAR mutations in enhancers for CUX1, PTBP2, GPC4, CDKL5, and several other genes that are implicated in neural function. The studies indicate that several HARs show normal regulatory activity essential for neural development and that alterations in them may be responsible for altered social and cognitive behavior.

Using chimpanzee DNA for the identification of HARs can reveal ancient sequences that may have differentiated humans and their closest primate relatives 4 mya. However, what types of genetic changes may have occurred more recently (380–470 kya) that could have caused the differences between human and the Neanderthals and Denisovans? To address this question, comparisons of transcription factor binding sites (TFBSs) in humans, Denisovans, and Neanderthals revealed 25 sites that were human-specific yet remained in the ancestral state in Denisovans and Neanderthals.

Weyer and Paabo[52] analyzed 25 TFBSs in 50 ancestral and derived alleles in humans and Neanderthals to determine their ability to turn on reporter genes in three neuronal cell lines. They showed that 12 of the derived versions of the TFBSs differed in expression from the ancestral variants, suggesting that much of the variation between modern humans and Neanderthals may be due to differences in gene expression levels.

Several genes showing differences in amino acid composition and/or expression levels between humans and Neanderthals include SPAG17, which is involved in formation of the sperm tail; TTF1, a transcription factor; PCD16, which affects cell adhesion and expressed in the skin; CAN15, which is linked to visual development; DCHS-1, which codes for a protein involved in wound healing; RPTN, a multifunctional matrix protein expressed in the epidermis (sweat glands, hair, and tongue papillae); THADA, a gene expressed in the thyroid and involved in apoptosis; Runx2, which plays a part in cranial morphology; TRPM1, which influences melanin synthesis; and DYRK1A, NRG3, CADPS2, all three of which play a role in cognition. Many of the most significant changes in gene expression have occurred in genes associated with sperm production and skin morphology.[45]

Transposable Elements and Gene Regulation

Transposable element (TE) insertions have been shown to have a strong effect on gene regulation, changes in gene expression, and development (Fig. 3.9). Many of these changes have been traced back in the human lineage to about 4 mya.[53] TEs known as Alu are the most abundant TEs in the primate genome. Britten[54] maintains that Alu transposon insertions likely underlie rapid human evolution and may have possibly led to the rapid divergence between humans and other primates. TEs are known to have many effects, including increasing recombination frequency, unequal crossover, genome rearrangements, disease, a source of novel genes and exons, epigenetic silencing mechanisms, and cis-acting regulatory elements.

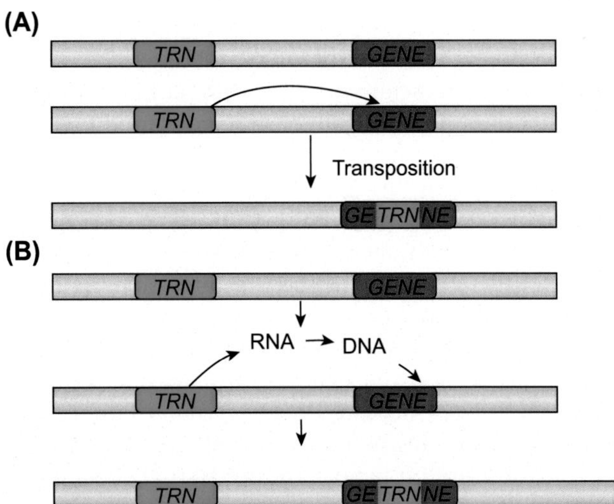

FIGURE 3.9 Action of transposable elements. Transposons (TRN) can move by a cut and paste mechanism, or copy and paste mechanism. (A) shows the cut and past TRN and a gene on the same chromosome. The TRN inserts itself into the gene thereby disrupting its function. (B) demonstrates the copy and paste mechanism whereby the TRN is transcribed and then converted to double-stranded DNA before insertion. Alu TRNs, mentioned in the text, move by the copy and paste mechanism. *Figure taken from Herrera RJ, et al.* Genomes evolution and culture: past present and future of humankind. *New York, NY: Wiley-Blackwell; 2016 [Chapter 11].*

One of the proposed functions of TEs in evolution is their ability to alter gene transcription either as enhancers, repressors, or insulators that block the interaction of enhancers and promoters. Because TEs have transcription regulatory sequences, they have been proposed to alter the rate of transcription or transcription factors that regulate gene activity. This occurs when a TE inserts near or inside of a gene and changes the TFBS, thereby altering the rate of gene transcription or the gene's tissue-specific activity. One example is the Alu-mediated loss of the cytidine monophosphate-*N*-acetylneuraminic acid hydroxylase gene (CMAH) that provides sialic acid distribution to cell surfaces. Various sialic acids cover the surface of cells and are involved in cell-to-cell communication. The result of the disruption of the CMAH gene is that human cells have very low levels of *N*-glycolylneuraminic acid (Neu5GC), which plays a role in pathogen adhesion. The lower levels of Neu5Gc could have been a result of positive selection and the altered gene form has been detected in the Neanderthal genome, indicating that it arose early during human evolution.[55] Altogether, there have been 10 human-specific changes in the sialic acid–binding immunoglobulin lectin genes, which include changes in enzyme-binding efficiency, gene expression pattern, and deletion or inactivation. Some of the deleted genes have functional relevance to immunity and nervous tissue and may have been a target for human evolution.[45] In another immune-related phenomenon, the GTPase M gene (IRGM) has been lost in ancestral primates but maintained in the human lineage.[56]

Lynch et al.[57] examined the gene regulatory mechanisms of endometrial cells and found that more than 1500 genes were expressed in endometrial tissue. Around 13% of these genes were within 200 kb of a eutherian-specific TE called MER20. The genes were regulated by the MER20 that carried progesterone (MER20 element was induced in the presence of progesterone). The genes in the MER20 region were involved in the differentiation of human endometrium. They concluded that MER20 contributed to the origin of the regulatory network that supports pregnancy and the development of the placenta. Years earlier, investigators found an envelope protein from the HERV-W virus, called syncytin, that had been sequestered to act in human placental morphogenesis.

Alus are not capable of autonomous movement but have been implicated in gene conversion, alternative splicing, changes in gene expression, and RNA editing. *Alus* are enriched in human genes related to neuronal functions and disease. One example of *Alu* regulating neuronal function is the deactivation of the CMP-N acetylneuraminic hydroxylase gene through the deletion of a 92-base pair exon. The deletion occurs in all humans but is not found in great apes, indicating that the deletion occurred early in the evolution of the hominid lineage. It is thought that the deactivation of the CMP-N acetylneuraminic hydroxylase gene caused significant changes in several lectin-mediated interactions. Sialic acids such as CMP-N acetylneuraminic hydroxylase are involved in intercellular cross talk involving specific vertebrate lectins, as well as in microbe–host recognition, involving a wide variety of pathogens. The loss of hydroxylase may have also been involved in the evolution of the human brain.[58] The microcephalin gene (*MCPH1*) is another example of a gene affected by TEs and involved in brain development. The *MCPH1* gene is expressed during fetal brain development and has been implicated in the functional changes that led to brain enlargement in humans during the evolution of primates. The gene has 14 exons and the 14th exon contains 88% of an *Alu*Y sequence. The *MCPH1* gene sequence contains 57% TEs, including those in the introns. The intron TEs appear to be an ancient part of *MCPH1*, considering the amount of mutation they have absorbed. Because the gene with its TE insertions is functional, it appears that the insertion of TEs played a role in its evolution.[59]

The chimpanzee genome has fewer *Alus* than the human genome, and comparisons show 12 new *Alus* in the human genome and only 5 in the chimpanzee genome, of which only 1 is active in each species. The implication is that the last common ancestor between chimpanzees and humans would have had the same number of *Alus* and that humans have generated more *Alus* since their divergence.[37]

RNA editing is a change in the mRNA nucleotide sequence that happens after transcription and before translation. It does not occur in all mRNAs but has been shown to occur in a variety of tissues leading to alternative forms of protein from the same gene. Comparisons of *Alu* sequences in the brain cells of chimpanzees, humans, and rhesus monkeys showed a higher level of adenosine-to-inosine nucleotide RNA editing in humans.[60] The finding indicates that increased *Alu* RNA editing may have been adapted by natural selection and may act as an alternative information mechanism for human genes in the brain.

Several studies have examined TEs in the genomes of primates and other vertebrates to better understand their role in evolution. Mills et al.[61] compared human and chimpanzee genomes for unique TEs and found that almost 11,000 TEs were differentially present in humans and chimpanzees and 34% of the insertions were located within genes. The data showed there were more transpositions in humans, and the insertions represented species-specific variation that could have contributed to human–chimpanzee divergence.[37]

Britten[54] compared the sequences of TEs in chimpanzees and humans and found 655 perfect full-length matches in the human genome and 283 in the chimpanzee genome considered to be recent events. When comparing humans and chimpanzees, 5530 new *Alus* were seen in humans, while 1642 were seen in chimpanzees.

CNV and TEs have played a role in the formation of the human leukocyte antigen (HLA) genes. The HLA gene complex is a set of genes on human chromosome 6 that are involved in the immune response. HLA encodes cell surface molecules that present antigen peptides to T cell receptors on T cells. CNV (explained below) and TEs have been shown to have been involved in antigen-D–related beta chain (HLA-DRB) genes. Nine different HLA-DRB genes have been described, some of which code for functional gene products, whereas others are pseudogenes with various insertions and deletions (indels). Several of the pseudogenes (e.g., *DRB2*, *DRB6*, and *DRB7*) appear to be the result of TE intron insertion and CNVs. Several of these pseudogenes are common in chimpanzees and humans but are not seen in other Old World monkeys.[62]

microRNA Regulation and Epigenetic Variation

microRNA (miRNA) has been implicated in dental differences between humans and Neanderthals. Genomic sequence comparisons in the human-derived allele compared with the Neanderthal allele for microRNA miR1304 shows a difference in humans. This miRNA in humans does not repress the enamelin and amelotin genes that are important in tooth formation.[45]

Many gene regulatory changes that occur in humans versus chimpanzees are in the expression of genes in the brain. Several mRNA sequences found in the prefrontal cortex of humans and chimpanzees have been found that show age-specific differences in expression. These mRNAs appear later in human brain development than in chimpanzees and cause significant changes in morphology. These temporal changes may be driven by several miRNA sequences, miR-92a, miR-320b, and miR454, that have been proposed to explain some of developmental alterations seen during human brain development.[63]

The effects of miRNA on gene expression in the brains of chimpanzees, macaques, and humans have indicated that there are significant changes in gene expression. The authors conclude that around 11% of the 325 expressed miRNAs varied between humans and chimpanzees, and that 31% varied between humans and macaques. Human-specific miRNA was localized in neurons and targeted genes involved in neural function, leading to variations in both mRNA and protein expression levels. In addition, they identified an upstream sequence in a miRNA (miR-34c-5p) with human-specific expression. The authors conclude that the miR-34c-5p expression took place after the split with Neanderthals and had an adaptive significance, indicating that changes in miRNA expression may have contributed to human cognitive functions.[64]

Microarray and RNA sequencing were used to examine human-specific gene expression in the cerebellum and prefrontal cortex of chimpanzees, macaques, and humans. Lui et al.[65] showed that most of the human-specific gene expression was found in genes for synaptic function in the prefrontal cortex of humans but not the cerebellum. The synaptic gene expression was not only higher in humans but also occurred later in development. The evidence was supported by both protein expression and increased synaptic density using transmission electron microscopy.

Cain and others[66] examined the histone epigenetic modifications in lymphoblast cell lines from humans, chimpanzees, and rhesus macaques. They focused on histone modification of H3K4me3 thought to be involved in transcriptional regulation. They observed interspecies differences in the histone methylation at the transcription start sites of genes that are differentially expressed in the different species. Although these results showed only a slight difference in interspecies gene expression, it indicated that epigenetic modifications could be an important mechanism for gene expression differences among primates.

Copy Number Variations in Primates

When comparing genomes between species or individuals, additional copies of DNA sequences or the absence of DNA sequences is known as Copy Number Variation (CNV). CNVs can be relatively short or long DNA sequences (50–1000's bp) and can involve genes or just repetitive DNA sequences. CNVs and their association with specific genes and/or gene regulatory regions may have had important roles in evolution and our divergence from other primates.[67,68] Dweep et al.[69] surmised that CNVs and miRNAs must have coevolved in the genome and interacted together to regulate dosage-sensitive gene expression. They demonstrated that miRNAs that targeted genes located in CNV regions in humans have special characteristics and regulate specific pathways that are absent in chimpanzees and seven other vertebrates. In addition, they showed that miRNAs that are within human CNV genes have a special evolutionary role and may be involved with some human diseases. This emphasizes the idea that the alteration of some genetic pathways may have very high evolutionary significance, leading to multiple human-specific changes.

Until the sequencing and analysis of entire genomes, geneticists assumed that the biological complexity of species would be correlated with the number of genes and the size of the genome. This led scientists to propose that the number of human genes would far exceed those of "lower" animals and plants. The duplication of genes was thought to be the most obvious reason for human morphological, physiological, and behavioral differences between humans and other primates. In addition, it was assumed that all genes (except for pseudogenes) were essential for life. So, once animal and plant genomes were sequenced, it was surprising to find nonfunctional alleles and determine that the average healthy individual has about 100 nonfunctional alleles, with about 20 in the homozygous state that show no apparent phenotype. This supports the idea that not all genes are functional and that many may be dispensable.[70]

In addition, genome sequencing revealed that the correlation between complexity and gene number was not true, and that in some cases the loss of certain genes or DNA segments were implicated in adaptation and the evolution of modern humans, whereas in other cases an increase in specific genes or DNA segments were associated with the differences between humans and their closest living relatives.[70–73]

The study of primate genomes and their comparisons may help us understand the genetic basis of phenotypes and emphasize the plasticity of the genome. Changes in the number and arrangement of base pairs in the genome are known as structural variations (SVs), and a significant proportion of the human genome has undergone SVs since the split of humans and chimpanzees. SVs include chromosome rearrangements and CNVs.

When comparing primate genomes, in some cases, humans have a greater number of gene copies and in other instances a lower number or complete absence of specific genes. Obvious examples are the primate genes for full body hair and increased muscle strength are missing in humans. Studies of DNA sequences, that have been lost in humans when compared with chimpanzees, have shown to be involved with the immune system, muscle development, metabolism, brain enlargement, upright walking, and mating behavior.

Many of the CNVs that exist in humans and nonhuman primates exist in regions with high numbers of segmental duplications (SDs). SDs are regions of DNA with identical or near-identical sequences. SDs are thought to be a source of new gene evolution and to have played a role in the divergence of humans from other primates. There are several CNVs that have persisted in all primates, and these are indicators of strong positive selection. CNV detection can help uncover ancient variation that has persisted or diminished in our ancestors.

CNV sequences that increase or decrease the number of nucleotides between different species are sometimes referred to as copy number differences. Prufer et al.[74] sequenced the genomes of bonobos, chimpanzees, and humans. They concluded that about 3% of the human genome is more closely related to either chimpanzees or bonobos than chimpanzees and bonobos are related to each other. They also speculated that, because of the similarities, the last common ancestor may have possessed an assortment of features, including those present in the bonobo, chimpanzee, and human.

Marques-Bonet et al.[75] examined CNVs in macaque, orangutan, chimpanzee, and humans. They discovered numerous examples of recurrent and independent gene-containing duplications within the gorilla and chimpanzee that are absent in humans. They concluded that CNVs differed significantly from other forms of mutations when comparing humans with other primates, concluding that a burst of duplication activity occurred during human evolution.

Jiang et al.[76] identified 4692 ancestral SD loci in the human genome and ordered them into 24 distinct groups of duplication blocks. They then compared SDs in the human genome with the genomes of chimpanzee and macaque. Their analysis showed SDs were often arranged in regions of transcriptional activity and primate-specific genes, indicating their involvement in positive selection.

Gokumen et al.[77] identified over 2000 human CNVs, with 170 that overlapped with chimpanzees and macaques. They placed these CNVs in 34 hotspot regions, many of which were functional regions of the genome. Several of these CNVs were in regions that included genes involved with the immune system and had significantly different expression levels between species, indicating species-specific positive selection.

CNV Gene and Sequence Deletions

McLean et al.[78] identified 510 deletions in noncoding regions that were human-specific, whereas these same regions were highly conserved (not deleted) in chimpanzees and other mammals. These deletions were enriched in sequences near genes involved in steroid signaling and neural function. One of the human-specific deletions was in the enhancer of the human androgen receptor (AR) gene. The enhancer, absent in humans, turns on the AR gene that is responsible for the production of ARs on cells. Once receptors have been formed, they respond to the presence of testosterone. The enhancer in chimpanzees initializes the development of sensory vibrissae (whiskers) and penile spines (keratinized surface projections on the penis). The deletion of the enhancer in humans removes these anatomical features. Studies have also found that the same *AR* enhancer deletion also existed in the Neanderthal and Denisovan genomes, demonstrating that the enhancer deletion is a characteristic of the human lineage and possibly have been involved with changes in our reproductive behavior due to less painful intercourse and increased intimacy among humans.[79,80]

A deletion was discovered in a noncoding region associated with a gene (*GADD45G*), involved in growth arrest and correlated with the expansion of specific brain regions in humans. The investigators reported finding that the intact DNA sequence in chimpanzees enhanced the expression of GADD45G, a neural gene involved in destroying surplus neurons produced in the embryo. The enhancer sequence that activated the GADD45G gene did so in very specific cells involved in the developing brain. This led the investigators to suggest that the loss of this enhancer sequence in humans may have resulted in the larger brain size in humans.[78,80,81]

Other studies have shown that the hydrocephalus-inducing homolog (HYDIN) found in mice has an additional human homolog on chromosome 1. This gene is thought to be associated with regulation of brain size. Deletions of the human HYDIN homolog lead to microcephaly, whereas duplications lead to macrocephaly.

CYP2D6, a gene involved in metabolism, is another that varies in copy number among primates. High copy numbers of the gene increase metabolism of a variety of drugs, whereas low copy numbers cause hypersensitivity to certain drugs. One hypothesis for the selection of low copy number of *CYP2D6* in humans is thought to be its ability to metabolize toxins and the idea that as food toxins became less prevalent in the human diet, *CYP2D6* copy numbers dropped, most likely due to genetic drift and a lack of selection pressure.[82]

Among primates, chimpanzees have lost 729 genes in comparison with humans who have lost 86 genes kept by chimpanzees over the same time period. The human Y chromosome is the primary example of lost genes in humans. A comparison of the male-specific sequences of the Y chromosome (MSY) in humans and chimpanzees showed that they differed dramatically in sequence structure and gene content. The chimpanzee sequences not only contained twice as many palindromes as the human sequence but also lost large segments of the MSY protein coding genes and gene families. The researcher's explanation for the differences ranged from the role of the MSY in sperm production, to genetic hitchhiking, frequent ectopic recombination, and differences in species mating behavior.[83] The Y chromosome has lost 640 genes that it once shared with the X chromosome. The Y chromosome has retained around 36 genes that are regulators of transcription and translation and that are dosage-sensitive and therefore required for both sexes. Several other genes are male-specific and are mostly associated with male reproduction.[70]

Although the Y chromosome provides the most drastic example of gene loss, there are other genes and DNA sequences that have been shown to be absent in normal humans. The CCR5 gene (a C–C chemokine type 5 receptor) and the ACKR1, also known as DUFFY (chemokine receptor 1) are both loss-of-function genes that impart some resistance to AIDS and malaria, respectively. These genes show a high frequency in populations exposed to these diseases.

The myosin heavy chain-16 gene (MYH16) has been lost in humans possibly because of dietary changes, and its loss correlates well with the anatomical alterations in the human lineage when compared with other primates. The gene used in the formation of strong jaw muscles was most likely dispensable and removed by purifying selection. It may have also been implicated in the expansion of cranial capacity and the increase in human brain size.[84]

The CASPASE12 gene, lost before human migration out of Africa, confers protection from sepsis, and as mentioned above (See section on TEs) the loss of the CMAH gene provides resistance to some pathogens.[55]

An investigation by Indjeian and others[85] found that a DNA sequence missing in humans was a switch that turned on a gene specifically in the toes and feet. They concluded that the loss of this DNA sequence (switch) in humans was responsible for the shorter toes in humans (toes 2–5) and responsible for alterations in foot morphology that improved the foot for walking.[81]

The variation in gene copy number (complete loss or gain) has not only evolutionary significance but also medical significance. Knowing the function of essential and nonessential genes in humans and other vertebrates could help in identifying model organisms for detecting disease resistance and identifying candidate genes that lead to disease.[86]

Chromosome Rearrangements

Chromosome rearrangements include inversions, reciprocal translocations, and occasionally other translocations that can have effects on gene expression through position effects (position of a gene on a chromosome). Chromosome rearrangements can be caused by exposure to radiation, and/or TEs have also been implicated in chromosome rearrangements (Fig. 3.10). Many of these rearrangements can be detected by chromosome painting, FISH, or Giemsa staining. Chromosomal rearrangements are a source for reproductive barriers (low hybrid fitness) and are suppressors of recombination. The most commonly referred to rearrangements involved in human evolution are the formation of the Y chromosome and the variation of karyotypes between primates, especially the formation of human chromosome 2.

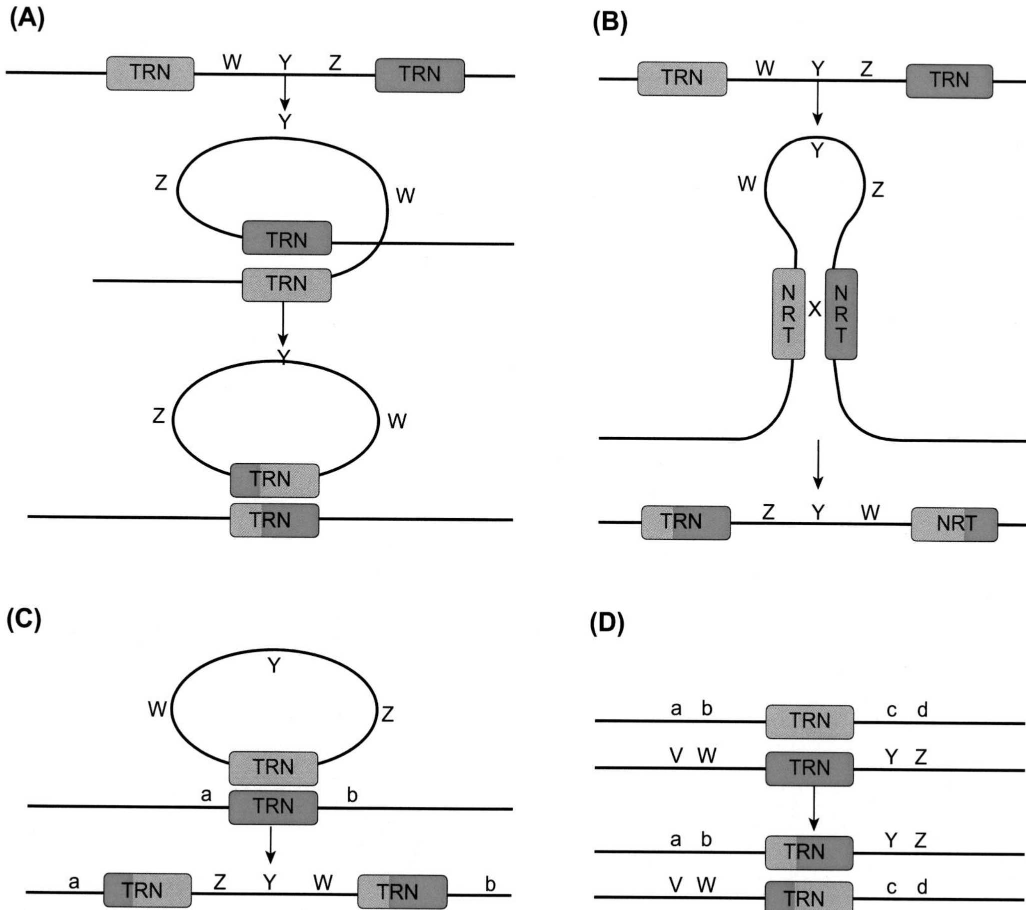

FIGURE 3.10 **Transposable elements and chromosome rearrangements.** The figures show a transposable element (TRN) that can act as a sight for recombination resulting in chromosome rearrangements. (A) shows a chromosome loop forming as a result of two TRNs coming together and subsequent recombination. The recombination results in a deletion. (B) shows the recombination between two inverted TRN sequences and subsequent recombination. The result is an inversion in the chromosome sequence. (C) shows a ring chromosome with a TRN and a linear chromosome with a TRN. Recombination between the TRN segments results in an insertion inside the linear chromosome. (D) shows two nonhomologous chromosomes (abcd and vwyz). Recombination between the TRN sequences results in a chromosome translocation, where a and b are now linked to y and z, and v and w are now linked to c and d. *Figure taken from Herrera RJ, et al.* Genomes evolution and culture: past present and future of humankind. *New York, NY: Wiley-Blackwell; 2016 [Chapter 11].*

Comparisons of primate karyotypes can reveal changes that may have led to phenotypic variability through chromosome rearrangements. An obvious difference between primate karyotypes is the number of chromosomes (humans with 46 and chimpanzees and gorillas with 48). The difference in chromosome number is due to the formation of human chromosome 2 by the terminal fusion of two chimpanzee chromosomes. Karyotype comparisons between chimpanzees and humans show very similar banding patterns between species, with the major changes involving inversions of various chromosomal segments and variations in the amount and placement of heterochromatin (highly condensed noncoding DNA). These rearrangements may be important in determining phenotypic differences between primates based on gene position effects. When genes are placed in different regions of the chromosome they can be regulated differently, i.e., turned down, turned up, or shut down. Although primates share a high percentage of genomic sequences, the arrangement of genes on the chromosomes is an important factor in gene regulation.[37]

Horovath et al.[87] found 14 ancestral loci that gave rise to duplicated DNA segments within the 700-kb region on human chromosome 2. They compared these to chimpanzee, gorilla, baboon, macaque, and orangutan. They concluded that the duplications occurred in a burst of activity during a very narrow time period between 10 and 20 mya (corresponding to the split between humans/ape and Old World monkey lineages as a result of punctuated duplicative transposition). Because duplications have been associated with novel gene development (as discussed in the CNV section), this is an import finding and provides evidence for punctuated events at the chromosomal level.

It is widely accepted that the structure of the human Y chromosome is the result of deletions, mutations, and rearrangements from an ancestral X chromosome. Comparisons of the X and Y chromosomes show divergence in structure and gene content. The Y chromosome has not only lost genetic material but has also acquired a series of repetitive sequences, including SINEs, endogenous retroviruses, and SDs (segments of DNA that are duplicated in tandem). This is most likely due to the lack of recombination (except at the telomeres) between the X and Y in male meiosis, whereas two X chromosomes can recombine in a female. The long nonrecombinant portion of the Y began to accumulate more mutations and the presence of mostly male sex-related genes. Rearrangements and mutation rates have resulted in a Y chromosome, with very little homology to the X chromosome (only 5% of the Y contains pseudoautosomal regions [recombining regions] confined to the ends of the chromosome). Comparisons of primate Y chromosomes reveal high sequence divergences between hominoid species; however, there is lower diversity within species. This low intraspecies diversity may be due to selection and genetic drift, and it may be related to hemizygosity.[37]

Primate Behavior

Young children and chimpanzees both can gesture but use it in very different ways. Chimpanzees use gesturing to give commands centered around an individual's needs (when they point at something they want). Pointing at something you want is a common gesture used by human toddlers, but as humans develop they use gesturing not only as commands but also to teach and pass along information for familiar activities. Humans can use gestures to convey abstract ideas (putting hand to mouth to indicate hunger) and to convey shared intentions (pointing to an athletic field to indicate play or competition). Gestures along with grunts or specific noises (moving hands to indicate specific objects and actions along with vocalization) could have been the initial sources of language. Collectively these types of activities may have led to the development of more complex communications. Furthermore, such complex communication has led to more elaborate social groups, an assemblage of social norms leading to ethic principles, and the eventual development of government, religious groups, and a more complex society.[88,89] In essence, it has led to what many would define as our humanity.

In 2007 Herrmann et al. tested spatial reasoning, ability to discriminate between quantities, and understanding of cause and effect relationships between 106 chimpanzees, 32 orangutans, and 105 human toddlers (2.5 years old). Chimpanzees and toddlers scored almost identically on these tests; however, toddlers far exceeded the apes on tests of social skills, including the ability to communicate and learn from others and evaluate the wishes of others. The group concluded that the general reasoning skills in young children are similar to chimpanzees but that children have a "cultural intelligence" that prepares them for learning later in life from the individuals they interact with on a daily basis.[90] In addition, young humans appear to have what psychologists refer to as a "theory of mind" (ability to discern other individual's desires, intentions, and perceptions), allowing them to share intentions and negotiate, a characteristic that is thought to be a unique evolutionary adaptation of humans. Chimpanzees have some ability to read peers' intentions but use it to compete with others and not necessarily as a way of cooperation.[91,92]

Cooperation between chimpanzees has been seen, especially when grooming and feeding, but is not as elaborate as humans. Studies looking at the relationship between oxytocin levels and cooperation in chimpanzees showed that increased levels occurred during food sharing and early mother–child interactions, suggesting it may be involved with cooperation. (Oxytocin is a peptide hormone released by the pituitary and plays a role in sexual and social bonding in both male and female humans. It also appears to help with mother–child bonding during breastfeeding.) Investigators found that levels of oxytocin also increased in chimpanzees just before conflict with rival groups. Initially this seemed counterintuitive but upon further consideration it seemed to be involved with galvanizing and increasing cooperation within the group, against a common enemy.[93]

A comparison of self-control and reactivity tests on 3- and 6-year-old humans and chimpanzees indicated they were very similar in abilities to resist immediate gratification, repeat successful actions, pay attention during distracting noises, and quit activities after repeated failures; however, 6-year-old children were more skilled at controlling their impulses. Overall, the data demonstrated that humans' self-control in decision-making is a part of general ape heritage beginning around age six in humans.[90] Shared intentionality, increased self-control, gestures, and language may not be the only behavioral mechanisms that separate chimpanzees from humans but were certainly instrumental in the evolution of humans.

Social behavior can vary tremendously among different chimpanzee populations and is dependent on some intergeneration transmission of behavior. However, there are some common social practices that differ from humans. Chimpanzee social structure is based around the group living mostly for protection from predators, feeding efficiency, and higher copulatory success. Groups can vary in size and individuals can leave and join different communities. The size of communities increases significantly when food is available and when females are in estrous. Chimpanzees are promiscuous, and sex parties include unisexual and bisexual intercourse, especially when the adult ratio is skewed.[94,95]

Sexual activity in chimpanzees is very different than that of humans. The comparisons with human matings are dramatic and include changes in time, intimacy, and childcare after birth. Chimpanzee copulation time is brief, approximately 10 s, and because of the size of the chimpanzee testicles they produce copious amount of sperm and copulate numerous times with numerous females. The presence of penile spines, made of keratin, on the chimpanzee penis result in a painful mating practice that does not involve intimacy, and females are reluctant to mate again with another male after initial intercourse. Once offspring are born, it is solely the females' responsibility to care for the young and females do not reproduce again until weaning is complete.[80]

Unlike humans, there is a distinct difference in the way males and females are treated after birth. Males remain in their communities, whereas females emigrate during adolescence, making it more likely that male chimpanzees in a community are related. After mating and childbirth, lactating females normally spend most of their time with their own progeny but can also be involved with nursing groups. Parental care is the responsibility of the female, and males have been known to practice infanticide when infants are unlikely to be their own.[96,97]

In addition to social behavior, chimpanzees and humans also vary in their susceptibility to disease. Because of the close neurological, physiological, and genetic similarities, one would assume that chimpanzees and humans are susceptible to the same diseases and in some cases this is true.[98] However, many human diseases cause only mild or no illness in chimpanzees, or diseases can result from different causes. Examples include HIV, hepatitis, and heart disease. In the case of hepatitis C, chimpanzees show a higher viral clearance, lack cirrhosis of the liver, rarely get liver cancer, and do not show mother-to-child transmission.[99] When diseases are found to be in common, studies involving comparative disease resistance and susceptibility can be useful for determining the genetic basis for susceptibility, development of vaccines, and creating disease models in chimpanzees to test treatments.

Sexual behavior between modern and archaic humans (Neanderthals and Denisovans) has been examined using DNA sequencing and new statistical methods to discover where and how often humans, Neanderthals, and Denisovans mated. Investigators have determined that East Asians have three Neanderthal ancestors in their lineage, Europeans two, and Melanesians one. In addition, it is estimated that Denisovans mated with Melanesians at least once. Collectively, Denisovans, Neanderthals, and modern humans all bred together at some point, but it is difficult to tell how often, and matings that did not produce offspring cannot be traced. Neanderthal and Denisovan DNA sequences are concentrated in certain regions of the human genome and missing from others[100] (see Chapter 6).

Archeological evidence suggests that Neanderthals did not congregate in large groups, populations were small, and individual families lived together. Genetic evidence supports these findings, showing that Neanderthals were much more inbred than modern humans of the same time period.[101]

Origin of Modern Humans

Three decades ago, archeological evidence and genetic studies suggested that the origins of modern humans and their evolution had been settled. The evidence pointed to our origin in East Africa around 4 mya evolving from *H. erectus*, our single migration out of Africa into Europe and Eurasia, and the extinction of the Neanderthals most likely through our superior intellectual ability to outcompete them.

Today we have come to understand that the picture of our evolution is much more complex than previously thought. New fossil discoveries and advances in genetic technology have opened our eyes to a variety of former misunderstandings. Seven-million-year-old (myo) hominin fossils (from a potential cousin, *Sahelanthropus tchadensis*) found in West Africa in northern Chad, two myo fossils from Malapa Nature Reserve in South Africa (*H. naledi*), and genetic evidence from extant individuals in South Africa suggests that we may not have originated in East Africa and in fact may be a conglomeration of *Homo* species from around Africa. The discovery of *H. heildelbergensis*, an intermediate between *H. erectus* and *H. sapiens* suggests that they were the species that ultimately evolved into modern-day humans. *H. heildelbergensis* is now known to have left Africa after *H. erectus* and evolved into Neanderthals and Denisovans. The discovery of Neanderthal's sophisticated tools, jewelry, and decorative body paint suggest that they were not the dullards we presumed them to be, and genetic evidence has shown that *H. sapiens* mated with them and other archaic species.

The Humans Before Us

H. habilis fossils found in Tanzania (East Africa) are some of the oldest ascribed to the genus *Homo* (2–1.5 myo) and therefore regarded by most anthropologists as the first humans. These individuals were also the first to leave Africa. Fossils resembling *H. habilis* have been found on the eastern shores of the Black Sea. In 1984, stone tools were found near the town of Dmanisi in the former Soviet Republic of Georgia. Fossils of *H. habilis* found there were dated at 1.77 myo and the stone tools at 1.85 myo.[102]

H. habilis is typically regarded as the first human ancestors to consume meat approximately 2–3 mya, which predates the use of stone tools. Around the same time, hominid molars began to shrink to make chewing meat easier. However, our ancestors' molars began to shrink before their brains began to grow, suggesting that the adage "food for thought" may be evolutionarily accurate.[17] Encephalometric studies revealed marked transverse expansion of the cerebrum, particularly the frontal and parieto-occipital parts, in *H. habilis*, and increased bulk of the frontal and parietal lobes. *H. habilis* was also the first hominid found to possess frontal and temporal brain regions that are structural markers of the neurological basis of spoken language. Collectively this indicates that the hominid brain had attained a new evolutionary level of organization, which can be ascribed to the preceding advent of meat consumption.[34]

H. habilis and *H. erectus* utilized fire 500–800 kya, most likely from opportunistic wildfires, because of the rare observations of hearths, wood ash, and coals from this time period, which would indicate controlled use of fire. Several anthropologists have proposed that the early use of fire allowed humans to cook and significantly impact their evolution. Cooked meat requires considerably less energy to digest than raw meat, leaving more energy left over from diet to be rerouted toward brain development. In addition, cooked food does not have to be heated to body temperature, is easier to chew, digests quicker, and is less likely to carry harmful bacteria. For these reasons, some anthropologists suggest cooking helped early humans survive and promoted further encephalization. Around 400 kya there is evidence of habitual use of fire and the presence of the larger-brained *H. heildelbergensis*. These findings of later use of controlled use of fire and cooking would suggest that cooking may have played a lesser role in larger brain development. However, cooking would have led to increased social interaction as hearths were places of social interaction used for warmth and for keeping away predators.[103]

Before cooking, there were no substantial changes in stone tool design for almost 2 million years. Then, around 200,000 years ago, hominids started to create new and better tools, which led to an increase in ingenuity. Around the same time, parts of the human genome responsible for metabolic functions of the brain began to undergo rapid changes. Because of these genetic changes, even more energy was directed to the increasingly asymmetric brains of early humans. Some view this as representing a different stage of cognitive development that eventually produced behaviorally modern humans.[34]

Recently discovered stone tools from Kenya that date to 3.3 mya have questioned the idea that the first sophisticated stone tool technology originated from the genus *Homo*. The original idea was that climate change had forced members of our genus to adapt and that the development of stone tools helped not only to adapt to the harsher climate but was also a force that led to expanded brain size. Because the earliest *Homo* fossils are dated to 2.8 mya, we may not have been the first to use stone tool technology.[104]

Although most investigators consider *H. habilis* as the first of the genus *Homo*, recent discoveries and classifications of fossils from South Africa have questioned these ideas. *Homo naledi* fossils from South Africa are considered by some to be in the same age range as *H. habilis* (2 myo), based on the morphology of the fossils.[105] However, more recent estimates have dated *H. naledi* at 200–300 kyo placing it in the same range of *H. sapiens*, suggesting that they may have been in Africa around the same time. If these latest dates are correct, and *H. sapiens* reached South Africa, they may have interacted with *naledi* and possibly interbred or led to their extinction.[106]

H. erectus (sometimes referred to as *H. ergaster*) were the first human species to emigrate out of Africa and populate Southeast Asia and Europe and were phenotypically different from *H. habilis* and *H. naledi*, with a larger brain (approximately 900 cm^3) and taller bodies (see Chapter 2).

The fossils of *H. heildelbergensis* have shared features of both *H. erectus* and *H. sapiens*. Because *H. heildelbergensis* had a greater cranial volume (1250 cm^3) than *erectus*, unique dental features, more advanced tools than *H. erectus*, and were found in Africa, Europe, and West Asia, it is considered a separate species from *H. erectus*. The African form of *H. heildelbergensis*, sometimes referred to as *H. rhodesiensis*, was first discovered in Zambia in 1921 by Arthur Smith Woodward. *H. heildelbergensis* lived 600–200 kya and are considered by most scientists to be the direct ancestor of Neanderthals, Denisovans, and modern humans. *H. heildelbergensis* left Africa between 400 and 300 kya, giving rise to Neanderthals in Europe and Denisovans in Northern Asia, while the remaining group (*H. rhodesiensis*) in Africa gave rise to modern humans.

The evidence for *H. heildelbergensis* giving rise to Neanderthals, Denisovans, and humans come from the many fossils found in Europe, Eurasia, and Africa. Over the years, the focus on human evolution and technological advances has come from European and North American institutions. However, recent interest from East Asia has revealed several unique transitional fossils found in modern-day China that are different from European and African *H. heildelbergensis*. The findings suggest that other intermediate forms of *Homo* may have existed, or that another group of hominins that descended from *H. erectus* gave rise to East Asian populations. This scenario would have involved a species like *H. erectus* interbreeding with groups from Africa and Eurasia, ultimately resulting in modern East Asians.[107]

Recent evidence from fossil skulls containing a mosaic of traits comes from Lingjing Xuchang in central China. The skulls were analyzed using OSL and are labeled Xuchang 1 and Xuchang 2 and dated between 100 and 130 kya. The skulls share several human traits; however, they have few traits indicative of more archaic humans from Eastern Eurasia. The skulls do indicate Old World trends in encephalization and in the suborbital, neurocranial vault. In addition, they show the distinctive Neanderthal occipital and temporal bony labyrinth of the inner ear, even though no Neanderthal remains have been discovered east of Siberia. Collectively the skulls show a mixture of traits not seen in prior samples. The findings question the traditional Out of Africa model that proposes most of our evolving took place in Africa before venturing out to other continents.[108]

H. sapiens, in their anatomically modern form, are believed to have arisen approximately 200–400 kya in sub-Saharan Africa and moved throughout Africa, as supported by fossil evidence. Initial fossil and skeletal remains were immediately recognized as being distinct from other hominids because of changes in facial and skeletal structure; the face and eyes were set under the globular braincase as opposed to an eye placement on a long and oval-shaped braincase. The oldest fossil remains for *H. sapiens* were discovered at two sites in Ethiopia and Morocco, suggesting origins in Africa and possible migratory routes out of Africa.

The Effects of Climate on Human Evolution

The geographical origin of humans in Africa is still debated, but the effects of climate change are regarded as a force that led to the survival and evolution of our species. Dramatic climate changes favor genes with certain advantageous traits (brain size, metabolic rates, ability to adapt to different diets, etc.), and dramatic changes in climate have been shown to involve mass extinctions.

Several theories linking the effects of the environment and climate change on adaptation and human evolution have been proposed. They include the savanna hypothesis, the woodland/forest hypothesis, the seasonality hypothesis, and the variability selection hypothesis. Each of these tries to explain the adaptation pressure and environmental scenarios that were involved in the origin of *H. sapiens* and their acquired traits.

The savanna hypothesis was one of the earliest and most popular that tried to explain the origin of bipedalism. This theory argued that life in the savanna (a region of grassy plains and sparingly placed trees and shrubs) would have favored bipedalism because it would allow individuals to see over tall grass, hunt more effectively, and avoid predators. In addition, the initial theory argued that the savanna environment would have encouraged the development of larger brains and toolmaking because of the fiercer competition (as compared with the woodlands) for resources; however, the paleo and climatological evidence has discredited the savanna theory, and the fossil record shows that early bipedal hominins were still climbing trees.

The woodland/forest hypothesis claims that hominins had evolved in and were primarily attracted to closed habitats during the Pleistocene. The wooded areas provided some camouflage from predators and a means of escape into the trees.

The seasonality hypothesis proposed that human evolution was cultivated by cold climate in higher latitudes or by seasonal variations in tropical and temperate zones; however, evidence has shown that the range of environmental change over time was more extensive and prolonged than the seasonality hypothesis predicted.

The variability selection hypothesis is supported by evidence for long successions of environmental changes and the persistence of humans and their flexibility to adapt to changes in climate and the wide range of habitat diversity.

Although climate certainly had an impact on traits, we did not acquire traits in a single instant but instead over long periods of time during bursts of climatic change. Scientists have determined that there was no onetime dramatic shift from forest to grassland, but instead wet–dry cycles of climate change took place in various regions of Africa and could have led to human migration and/or adaptation to survive. Evidence for these climatic changes come from geological studies near the African coasts, soil samples, and fossilized teeth. Core samples from the ocean indicate that wet and dry climes changed about every 23 ky, which coincide with the earth's orbital wobble. Soil sediment analysis of African landscapes and fossil evidence from teeth indicate these changes were associated with different fauna and flora and over millions of years the extinction and appearance of various hominin species, including our own (see Chapter 2). Carbon isotope analysis confirms that not only did the environments change but so did the diets of our predecessors. Plants can fix carbon in several ways. Woody species use a photosynthetic pathway called C3, whereas grasses and sedges use the C4 pathway to fix carbon. C4 grasses have a higher abundance of carbon 13 isotopes relative to carbon 12, whereas woody plants have a lower carbon 13 to carbon 12 ratio. This ratio can be used to study waxy plant deposits in soil and rock samples to determine the vegetation present and the regional landscape at the time. In addition, studies from fossilized teeth of the Turkana Basin indicate that 2 mya there was a split in the diets of our early *Homo* ancestors and those of the genus *Paranthropus*. The studies show that the diet of *Paranthropus* was very restrictive to C4 grasses, whereas the diets of the Homo group varied. This ability to ingest a variety of plants may explain the extinction of *Paranthropus* and the ability of the Homo group to adapt and survive.[109]

The idea that there was a transition species from an ancestor of the genus *Australopithecus* to the genus *Homo* is fairly well established; however, the relatively large number of australopithecine species has complicated the identification of the species most likely to have been the human ancestor. The discovery of the Bouri australopith (*Australopithecus garhi*) in 1999 has shed some light on this transition. The data collected from cranial and dental measurements, the discovery of tools, and the age of *A. garhi* (2.5 mya) have provided strong arguments for the possibility of *A. garhi* to be the direct ancestor of the genus *Homo*.[41]

Australopithecus sediba is also a strong candidate for the transitional species because of its craniodental measurements, its arched foot, and its apparent bipedalism and arboreal abilities. Other measurements that determined the precise age of *A. sediba* (1.97 mya) predates the appearance of the earliest *Homo* found in Africa and argues for a transitional form.[110,111]

Most paleoanthropologists agree that the earliest human was *H. habilis*, or the closely related *Homo rudolfensis*, with serious debates over the splitting of these two species. Many feel because of the similarities and sparse fossil remains that these are essentially the same species.[111] However, recently the more serious debate is over Richard Leakey's initial descriptions and conclusions about the two *H. habilis* fossil specimens. Recent studies suggest that the *H. habilis* specimens be removed from the *Homo* taxa. These arguments come from reexamination of the predicted body size and shape, relative brain size, dental measurements, and skeletal traits associated with locomotion. Initially, Wood and Collard argued that based on these traits, *H. habilis* is more closely associated with *Australopithecus*. Reexamination and reinterpretation of these data, and considering the limited availability of fossils from *H. habilis*, studies have concluded that *H. habilis* should be described as a partially arboreal hominin that should remain in the genus *Homo*.

In 2015 scientists revealed over 1500 fossils found in the Rising Star Cave near Johannesburg, South Africa (see Chapter 2). These fossils were eventually identified as a new human species *H. naledi*, and recent dating techniques were used to determine the samples were 236–335 kyo. The date of the specimens and new measurements of recently excavated specimens has led several investigators to suggest that *H. naledi's* primitive features link it to the earliest humans including *H. habilis* and that it may have emerged around the same time as *H. erectus*.[112] More controversial conclusions by Berger et al.[113] claim that *H. naledi* may have given rise to *H. erectus* or even *H. sapiens*. Although most consider the latter statements pure conjecture, it is clear from this and other studies that South Africa played a much larger role in the evolution of humans than previously thought.

While there is no debate that *H. sapiens* developed to become AMH, the estimates and arrival of *H. sapiens* in various parts of the world continue to be debated. There are two competing theories about *H. sapiens* exodus from Africa: (1) the recent single-origin (replacement or Out of Africa model) and (2) the multiregional (or polycentric) model.

The recent single-origin theory suggests that modern humans existed in a singular African geographical location and migrated around and out of Africa, replacing existing Hominid species, then left Africa 65,000 years ago through the northern Bab-el-Mandeb strait (Fig. 3.2). This suggests that a population bottleneck occurred in Africa and that there was an absence of any hybridization.

The single-origin theory is supported by cranial, linguistic, and skeletal evidence consistent with decreasing genetic differences with an increase in distances from sub-Saharan Africa.[1] A loss in genetic diversity with increases in genetic population creates a bottleneck effect, suggesting that modern humans may have mixed with other hominid populations whose genetic contributions were limited. Some genetic evidence insinuates that Neanderthals and *H. sapiens* coexisted in the Levant in attempts to escape undesirable conditions in Africa. As indicated by fossilized remains of temperature-sensitive animals, it is speculated that the Neanderthals occupied the Levant during cooler conditions and modern humans remained there in warmer conditions.[114] Recent genome sequencing and comparison of Neanderthals and *H. sapiens* indicates that interbreeding and genomic exchange between the two species occurred outside of Africa, most likely in Eurasia, where there is an estimated higher level (3.4%–7.3%) of admixture.[100]

Although believers of the recent single-origin hold the same principle that the modern human originated and eventually outlived archaic *Homo* species, the route taken by *H. sapiens* in the single origin is still highly debated. Archaeological, fossil, and genetic evidence provided supports both speculated routes, through the Horn of Africa or the Levantine corridor. Advances in technology have allowed for wide genome analysis and sequencing, leading to a better understanding of both gene diversity within populations and the origins of *H. sapiens*. Recent studies showed that the further away a population was from Africa, there was a greater chance of reductions both in genetic diversity and recombination rates.[12] This lack of genetic diversity is indicative of a more recent migration from Africa and gradual colonization process of Eurasia. Advances in archeological and genetic technology and evidence for a constantly changing climate over 100 kya suggest that neither theory is entirely correct; however, the Out of Africa theory is the one most espoused by scientists today.

Alternatively, the multiregional theory proposes that within Africa, there existed isolated populations of archaic modern humans, each possessing different characteristics. They then migrated and interbred with other hominid species, thereby diversifying the human genome until they possessed the genetic edge to outcompete other species and successfully spread both around and out of Africa. The multiregional theory implies a wide range of dispersal and that the most modern form of *H. sapiens* came into existence through a gradual geographically widespread manner rather than all at once. In effect, it suggests that multiple subtypes of *H. sapiens* evolved within distinct geographic regions, allowing for both gene flow between these populations and regionally specific gene flow from local archaic hominin populations. Over time, with growing support for the total replacement hypothesis, the multiregional hypothesis has grown from a model which proposed very distinct origins for each regional subtype of *H. sapiens* to one that suggests that subtypes developed with continual gene flow between them (a continuity model). Although opinions still vary on this subject, the current consensus seems to hold to an assimilation model that suggests that a replacement by AMHs occurred with some assimilation of archaic *Homo* DNA in the process.

Whatever theory holds true, it is understood that *H. sapiens* left Africa in waves under favorable and hospitable conditions. Additional evidence suggests two routes of exodus, one through the Levantine corridor through Egypt and a secondary route through the Horn of Africa, or Bab-el-Mandeb strait of northeastern Africa into Arabia (Fig. 3.1).

The remains of *H. sapiens* fossils and artifacts from extant aboriginal populations provide an important contribution to the *H. sapiens* origin story. Rather than relying solely on archeological discoveries, scientists have turned to genetic evidence, looking at mitochondrial DNA, Y chromosomes, and the geographical distribution of genetic markers to better understand the origin and migratory patterns of humans. Owing to advances in technology, there is now much more information that can be gained from ancient sites, allowing archeologists, geneticists, and paleoanthropologists to learn more about the migratory patterns and evolution of *H. sapiens,* as well as their contributions to the genomes and phenotypes of modern humans in Arabia, Asia, and Eurasia, suggesting a later back-to-Africa migration.

One of the flaws associated with the recent single-origin model is the Mitochondrial Eve theory. The theory claims that modern humans descended from a single woman, and the evidence from mtDNA is used to promote the theory. It is true that if you look at genes they gradually coalesce into fewer and fewer ancestors until arriving at a single ancestor; however, as you go back in time starting with individuals, more individuals appear as ancestors. The mtDNA that we received from Mitochondrial Eve was only a fraction of the DNA we inherited, and she had many ancestors that contributed to her whole genome. Likewise, other members of her population also had similar mtDNA. The same can be said for Y chromosomal DNA and the idea that we descended from a single male. Our DNA comes from many different male and female ancestors that existed in a population.[41]

The idea of a founder effect leading to *H. sapiens* diverging from their ancestors could have led to the distinctive features we see in modern humans compared with their extinct ancestors. One or more bottlenecks may have occurred, and theoretical and computer simulations suggest that these would have included 5–10 k individuals.[7,41]

The evidence from nuclear SNPs supports the idea that humans arose in sub-Saharan Africa and that individuals from the Khoe-San (Khoisan) from South Africa are the closest living group to the original *H. sapiens*. Schlebusch and others[115] examined 2.3 million SNPs in over 200 South Africans and determined that the divergence of the Khoe-San and other Africans took place over 100 kya, with the Khoe-San emerging about 35 kya.

Once humans began to spread across the earth they began to take on regional traits, including variations in skin color, hair color and texture, unique facial features, height, and body shape. These resulted from natural selection, genetic drift, sexual selection, and cultural practices. This has led to some geographical structure in the distribution of DNA in different populations. Rosenberg et al.[116] determined that the amount of geographically unique DNA is relatively small (3%–5%) when compared with differences between individuals within a population where the genetic diversity is around 93%–95%. They determined that when looking at populations in Africa, Europe, Central Asia, East Asia, and America they could distinguish groups based on unique DNA. However, Serre and Paabo[117] used a slightly different structural analysis and determined that most of their samples showed mixing so it was not possible to assign groups to continents. They did, however, show that there were DNA gradients that were associated with the geography in Africa, Eurasia, Oceania, and America.

CONCLUSION

The genus *Homo* has been transformed anatomically, behaviorally, and cognitively over the last several million years to become distinct from the rest of the hominin group. The development of stone tools, the adaptation to a terrestrial lifestyle, the changes in diet, adaptation to various climatic changes, and the concurrent changes in brain size resulted in the only species of the genus to avoid extinction.

Learning about the development and evolution of *H. sapiens* will allow scientists to understand the genetic changes and adaptations that our ancestors underwent to evolve into modern-day humans. This information will not only allow scientists to better understand genome differences and similarities between different human populations but also help us understand what the future of human evolution might bring and pave the way for new medical and genetic advances.

References

1. Garvey M. *The wise mind of Marcus Garvey*. Bensenville, IL: Lushena Books; 2014.
2. Wood B. Welcome to the family. *Sci Am* September 2014:43–7.
3. Klein RG. Issues in human evolution. *Proc Natl Acad Sci USA* 2016;**113**:6345–7.
4. Slon V, et al. Neanderthal and denisovan DNA from Pleistocene sediments. *Science* 2017;**356**. https://doi.org/10.1126/science.aam9695.
5. Wade L. DNA from cave soil reveals ancient human occupants. *Science* 2017;**356**:363.
6. Meyer M, et al. Nuclear DNA sequences from the Middle Pleistocene Sima de los Huesos hominins. *Nature* 2016;**531**:504–7.
7. Ayala FJ, Escalante A. The evolution of human population: a molecular perspective. *Mol Phylogenet Evol* 1996;**5**:188–201.
8. Callaway E. DNA mutation clock proves tough to set. *Nature* 2015;**519**:139–40.
9. Stringer C. The origin and evolution of *Homo sapiens*. *Phil Trans R Soc B* 2016;**371**:20150237.
10. Villmorare B, et al. Early *Homo* at 2.8 Ma from Ledi-Geraru, Afar, Ethiopia. *Science* 2015;**347**:1352–5.
11. Hofman M. Evolution of the human brain: when bigger is better. *Front Neuroanat* 2014;**8**. https://doi.org/10.3389/fnana.2014.00015.

12. Shultz S. Hominin cognitive evolution: identifying patterns and processes in the fossil and archeological record. *Phil Trans R Soc* 2012;**367**:2130–40.
13. Nowinski WL. In: Miller K, editor. *Biomechanics of the brain, biological and medical physics*. Berlin: Springer Science and Business Media; 2011. p. 5–40.
14. Phillips H. Introduction: the human brain. *N Sci* September 4, 2006; (2567). https://www.newscientist.com/article/dn9969-introduction-the-human-brain/.
15. Wadley AL. Were snares and traps used in middle stone age and does it matter? A review and a case study from Sibudu, South Africa. *J Hum Evol* 2010;**58**:179–92.
16. Wadley AL, et al. Middle stone age bedding construction and settlement patterns at Sibudu, South Africa. *Science* 2011;**334**:1388–91.
17. Shoenemann PT. Evolution of the size and functional areas of the brain. *Annu Rev Anthropol* 2006;**35**:379–406.
18. Pontzer H, et al. Metabolic acceleration and the evolution of human brain size and life history. *Nature* 2016;**533**:390–2.
19. Wells S. *The journey of man: a genetic odyssey*. Westminster, London: Penguin Books; 2003.
20. Tobias P. The brain of *Homo habilis*: a new level of organization in cerebral evolution. *J Hum Evol* 1987;**16**:741–61.
21. Perreault C, Mathew S. Dating the origin of language using phonemic diversity. *PLoS One* 2012;**7**:e35289.
22. Boe L-J, et al. Evidence of a Vocalic Proto-System in the Baboon (*Papio papio*) suggests pre-Hominin speech precursors. *PLoS One* 2017;**12**:e0169321.
23. Freudenrich C, Boyd R. How your brain works. *Science* June 2000. HowStuffWorks.com.
24. Ekman P. *Emotional awareness: overcoming the obstacles to psychological balance and compassion*. New York: Henry Holt and Co; 2008.
25. Kringelbach M, Phillips H. *Emotion: pain and pleasure in the brain*. Oxford University Press; 2014.
26. Lordkipanidze, et al. The earliest toothless hominin skull. *Nature* 2005;**434**:717–8.
27. Lorkipanidze, et al. Postcranial evidence from early *Homo* from Dmanisi. *Ga Nurs* 2007;**449**:305–10.
28. Boyd R. The cultural niche: why social learning is essential for human adaptation. *Proc Natl Acad Sci USA* 2011;**108**:10918–25.
29. Zaidi AA, et al. Investigating the case of human nose shape and climate adaptation. *PLoS Genet* 2017;**13**:e1006616. https://doi.org/10.1371/journal.pgen.1006616.
30. Henrich J, et al. The puzzle of monogamous marriage. *Philos Trans Royal Soc B* 2012;**367**:657–69.
31. Lukas D, Clutton-Brock TH. The evolution of social monogamy in mammals. *Science* 2013;**341**:526–30.
32. Opie C, et al. Male infanticide leads to social monogamy in primates. *Proc Natl Acad Sci USA* 2013;**110**:13328–32.
33. Herrmann E, et al. Humans have evolved specialized skills of social cognition: the cultural intelligence hypothesis. *Science* 2007;**317**:1360–6.
34. Tattersall I. New twist added to the role of culture in human evolution. *Sci Am* September 2014;**311**.
35. Lenski RE, et al. The evolutionary origin of complex features. *Nature* 2003;**423**:139–44.
36. Rittschof CC, Robinson GE. Chapter five-Behavioral genetic toolkits: toward the evolutionary origins of complex phenotypes. *Curr Top Dev Biol* 2016;**119**:157–204.
37. Herrera RJ, et al. *Genomes evolution and culture: past present and future of humankind*. New York, NY: Wiley-Blackwell; 2016. [Chapter 11].
38. Nengo I, et al. New infant cranium from the African Miocene sheds light on ape evolution. *Nature* 2017;**548**:169–74.
39. Chaline J, et al. Chromosomes and the origins of Apes and Australopithecines. *Hum Evol* 1996;**11**:43–60.
40. Kingston JD, et al. Isotopic evidence for neogene hominid paleoenvironments in the Kenya Rift valley. *Science* 1994;**264**:955–9.
41. Ayala FJ, Cela-Conde CJ. *Processes in human evolution: the journey from early hominins to Neanderthals and modern humans*. Oxford University Press; 2017. p. 220–1.
42. Wu DD, et al. De novo origin of human protein coding genes. *PLoS Genet* 2011;**7**:e1002379.
43. Ng PC, et al. Genetic variation in an individual human exome. *PLoS* 2008;**4**:e1000160.
44. Boyko AR, et al. Assessing the evolutionary impact of amino acid mutations in the human genome. *PLoS Genet* 2008;**4**:e1000083.
45. Jobling, et al. *Human evolutionary genetics*. 2nd ed. New York NY: Garland Publishers; 2014. p. 269–85.
46. Simonti CN, et al. The phenotypic legacy of admixture between modern humans and Neanderthals. *Science* 2016;**351**:737–41.
47. Hubisz MJ, Pollard KS. Exploring the genesis and functions of human accelerated regions sheds light on their role in human evolution. *Curr Opin Genet Dev* 2014;**29**:15–21.

48. Levchenko A, et al. Human accelerated regions and other human-specific sequence variations in the context of evolution and their relevance in brain development. *Genome Biol Evol* 2018;**10**:166–88.
49. Boyd JL, et al. Human-chimpanzee differences in a *FZD8* enhancer alter cell-cycle dynamics in the developing neocortex. *Curr Biol* 2015;**25**:772–9.
50. Katzman S, et al. G-C biased evolution near human accelerated regions. *PLoS* 2010;**6**:e1000960.
51. Doan RN, et al. Mutations in human accelerated regions disrupt cognitive and social behavior. *Cell* 2016;**167**:341–54.
52. Weyer S, Paabo S. Functional analysis of transcription factor binding sites that differ between present-day and archaic humans. *Mol Biol Evol* 2016;**33**:316–22.
53. Werren JH. Selfish genetic elements, genetic conflict, and evolutionary innovation. *Proc Natl Acad Sci USA* 2011;**108**:10863–70.
54. Britten RJ. Transposable element insertions have strongly affected human evolution. *Proc Natl Acad Sci USA* 2010;**107**:19945–8.
55. Chou HH, et al. Inactivation of CMP-N acetylneuraminic acid hydroxylase occurred prior to brain expansion during human evolution. *Proc Natl Acad Sci USA* 2002;**99**:11736–41.
56. Bekpen C, et al. Death and resurrection of the human IRGM gene. *PLoS Genet* 2009;**5**:e1000403.
57. Lynch VJ, et al. Transposon-mediated rewiring of gene regulatory networks contributed to the evolution of pregnancy in mammals. *Nat Genet* 2011;**43**:1154–9.
58. Hayakawa T, et al. Alu-mediated inactivation of the human N-acetylneuraminic acid hydroxylase gene. *Proc Natl Acad Sci USA* 2001;**98**:11399–404.
59. Shi L, et al. Functional divergence of the brain size regulatory gene *MCPH1* during primate evolution and the origin of humans. *BMC Biol* 2013;**11**:1–6.
60. Bazak L, et al. A-to-I editing occurs at over a hundred million genomic sites, located in a majority of human genes. *Genome Res* 2014;**24**:365–76.
61. Mills RE, et al. Recently mobilized transposons in the human and chimpanzee genomes. *Am J Hum Genet* 2006;**78**:671–9. 134.
62. Doxiadis GG, et al. Evolution of HLA-DRB genes. *Mol Biol Evol* 2012;**29**:3843–53.
63. Somel M, et al. MicroRNA-driven developmental remodeling in the brain distinguishes humans from other primates. *PLoS Biol* 2011;**9**:e1001214.
64. Hu HY, et al. MicroRNA expression and regulation in human, chimpanzee, and macaque brains. *PLoS* 2011;**7**:e1002327.
65. Lui X, et al. Extension of cortical synaptic development distinguishes humans from chimpanzees and macaques. *Genome Res* 2012;**22**:611–22.
66. Cain CE, et al. Gene expression differences among primates are associated with changes in histone epigenetic modification. *Genetics* 2011;**187**:1225–34.
67. James R, et al. tRNA gene copy number variation in humans. *Gene* 2014;**536**:376–84. 111.
68. Popesco MC, et al. Human lineage-specific amplification, selection and neuronal expression of DUF1220 domains. *Science* 2006;**313**:1304–7. 112.
69. Dweep H, et al. CNVs-microRNAs interactions demonstrate unique characteristics in the human genome. An interspecies in silico analysis. *PLoS One* 2013:e0081204.
70. Albalat R, Canestro C. Evolution by gene loss. *Nat Rev Genet* 2016;**17**:379–91.
71. Wang X, et al. Gene losses during human origins. *PLoS Biol* 2006;**4**:e52.
72. Dumas L, et al. Gene copy number variation spanning 60 million years of human and primate evolution. *Genome Res* 2007;**17**:1266–77.
73. Innan H, Kondrashov F. The evolution of gene duplications: classifying and distinguishing between models. *Nat Rev Genet* 2010;**11**:97–108.
74. Prufer K, et al. The bonobo genome compared with the chimpanzee and human genomes. *Nature* 2012;**486**: 527–31. 113.
75. Marques-Bonet T, et al. A burst of segmental duplications in the genome of the African great ape ancestor. *Nature* 2009;**457**:877–81.
76. Jiang Z, et al. Ancestral reconstruction of segmental duplications reveals punctuated cores of human genome evolution. *Nat Genet* 2007;**39**:1361–8.
77. Gokucumen O, et al. Refinement of primate copy number variation hotspots identifies candidate genomic regions evolving under positive selection. *Genome Biol* 2011. https://doi.org/10.1186/gb-2011-12-5-r52.

78. McLean CY, et al. Human-specific loss of regulatory DNA and the evolution of human-specific traits. *Nature* 2011;**471**:216–9.
79. Reno PL, et al. A penile spine/vibrissa enhancer sequence is missing in modern and extinct humans but is retained in multiple primates with penile spines and sensory vibrissae. *PLoS One* 2013;**8**:e84258.
80. Reno PL. Missing links. *Sci Am* May 2017:42–7.
81. McLean CY, et al. GREAT improves functional interpretation of cis-regulatory regions. *Nat Biotechnol* 2010;**28**:495–501.
82. Gaedigk A, et al. Detection of CYP2D6, SULT1A1 and UGT2B17 copy number variation using multiplex PCR amplification. *FASEB J Suppl* 2011;**25**:812.
83. Hughes JF, et al. Chimpanzee and human Y chromosomes are remarkably divergent in structure and gene content. *Nature* 2010;**463**:536–9.
84. Stedman HH. Myosin mutation correlates with anatomical changes in the human lineage. *Nature* 2004;**428**:415–8.
85. Indjeian VB, et al. Evolving new skeletal traits by cis-regulatory changes in bone morphogenetic proteins. *Cell* 2016;**164**:45–56.
86. Wang T, et al. Identification and characterization of essential genes in the human genome. *Science* 2015;**350**:1096–101.
87. Horovath J, et al. Punctuated duplication seeding events during the evolution of human chromosome 2p11. *Genome Res* 2005;**15**:914–27.
88. Stix G. The it factor. *Sci Am* 2014;**311**:72–9.
89. Tattersall I. If I had a hammer. *Sci Am* 2014;**311**:54–9.
90. Herrmann E, et al. Uniquely human self-control begins at school age. *Dev Sci* 2015;**18**:979–93.
91. Nanay B. An experimental account of creativity. In: Paul ES and Kaufman SB, editors. *The philosophy of creativity*, Oxford University Press, Oxford, England; 2014. 3–38.
92. de Wall F. One for all. *Sci Am* 2014;**311**:68–71.
93. Samuni L, et al. Oxytocin reactivity during intergroup conflict in wild chimpanzees. *PNAS USA* 2016;**114**:268–72.
94. Boesch C. Social grouping in Taï chimpanzees. In: McGrew WC, Marchant LF, Nishida T, editors. *Great ape societies*. Cambridge, England: Cambridge Univ; 1996. p. 101–13.
95. Mitani JC, et al. Ecological and social correlates of chimpanzee party size and composition. In: Boesch C, Hohmann G, Marchant LF, editors. *Behavioral diversity in chimpanzees and bonobos*. Cambridge, England: Cambridge University Press; 2002. p. 102–11.
96. Nishida T, et al. Demography, female life history and reproductive profiles among the chimpanzees of Mahale. *Am J Prim* 2003;**59**(3):99–121.
97. Pepper JW, et al. General gregariousness and specific social preferences among wild chimpanzees. *Int J Prim* 1999;**20**(5):613–32.
98. Yang S-H, et al. Towards a transgenic model of Huntington's disease in a non-human primate. *Nature* 2008;**453**:921–4.
99. Zhong J. Robust hepatitis C virus infection *in vitro*. *Proc Natl Acad Sci USA* 2005;**102**:9294–9.
100. Gibbons A. Rich sexual past between modern humans and Neanderthals revealed. *Science* 2016. https://doi.org/10.1126/science.aaf4207.
101. Harris K, Nielsen R. The genetic cost of Neanderthal introgression. *Genetics* 2016;**203**:881–91.
102. Gabunia I, et al. Earliest Pleistocene hominid cranial remains from Dmanisi republic of Georgia: taxonomy, geological setting, and age. *Science* 2000;**228**:1019–25.
103. Seddon C. *Humans: from the beginning*. San Bernardino, CA: Glanville Publications; 2015.
104. Harmand S, et al. 3.3 million-year-old tools from Lomekwi 3, West Turkana, Kenya. *Nature* 2015;**521**:310–5.
105. Berger LR, et al. *Homo naledi*, a new species of the genus *Homo* from the Dinaledi Chamber, South Africa. *Elife* 2015. https://doi.org/10.7554/eLife.09560.
106. Barras C. *Homo naledi* is only 250,000 years old – here's why that matters. *N Sci* April 2017. https://www.newscientist.com/article/2128834-homo-naledi-is-only-250000-years-old-heres-why-that-matters/.
107. Qiu J. The forgotten continent: fossil finds in China are changing ideas about the evolution of modern humans and our closest relatives. *Nature* 2016;**535**:218–20.
108. Li Z-Y, et al. Late Pleistocene archaic human crania from Xuchang, China. *Science* 2017;**355**:969–72.
109. de Menocal PB. New evidence shows how human evolution was shaped by climate change. *Sci Am* September 2014;**311**.
110. Berger LR, et al. *Australopithecus sediba*: a new species of *Homo*-like *australopith* from South Africa. *Science* 2010;**328**:195–204.

111. Pickering R, et al. *Australopithecus sediba* at 1.977 Ma and implications for the origins of Genus *Homo*. *Science* 2011;**333**:1421–3.
112. Wong K. New evidence of mysterious *Homo naledi* raises questions about how humans evolved. *Sci Am* May 2017.
113. Berger LR, et al. *Homo naledi* and Pleistocene hominin evolution in subequatorial Africa. *Elife Sci* 2017;**6**:e24234.
114. Adams JU. Human evolutionary tree. *Nat Educ* 2008;**1**:145.
115. Schlebusch CM, et al. Genomic variation in seven Khoe-San groups reveals adaptation and complex African history. *Science* 2012;**338**:374–9.
116. Rosenberg NA, et al. Genetic structure of human populations. *Science* 2002;**298**:2381–5.
117. Serre D, Paabo S. Evidence for gradients of human genetic diversity within and among continents. *Genome Res* 2004;**14**:1679–85.

CHAPTER

4

The Exodus Out of Africa

Of the gladdest moments in human life, methinks, is the departure upon a distant journey into unknown lands. Shaking off with one mighty effort the fetters of Habit, the leaden weight of Routine, the cloak of many Cares and the slavery of Civilization, man feels once more happy. ***Sir Richard Francis Burton***[1]

SUMMARY

The widespread rise of cities, advanced tools, agriculture, and other modern technology throughout the world would not be possible without the expansion and migration of humans thousands of years ago. Throughout history humans have shown a desire to travel. Even today, millions of humans travel within and across continents; however, travels today are not nearly as challenging as in the past. Over millions of years, various members of the genus *Homo* have evolved and gone extinct but not before migrating across the continent of Africa, and/or intercontinental travel. The first humans to migrate out of the African continent were *Homo erectus* that arose about 2 million years ago (mya) then left Africa soon after, populating Southeast Asia and Europe. They were followed by *Homo heidelbergensis* that appeared in Africa about 700 thousand years ago (kya) and then spread beyond Africa about 600 kya. *Homo sapiens* most likely evolved 200–300 kya from the members of *H. heidelbergensis* that remained in Africa. The earliest known dispersal out of Africa probably occurred 130–115 kya along the Nile Corridor and into the Levant as evidence by fossil remains in Israel, however many feel this resulted in a failed attempt to disperse into other regions. By the end of the Late Pleistocene, *H. sapiens* were distributed across every continent, except Antarctica, and were beginning to form complex interactions and cultures and establishing themselves as a dominant species. Although it is clear that anatomically modern humans (AMHs) originated in Africa, it is difficult to tell where they originated in Africa and when they left. Because it is not clear from archeological evidence which individuals initially migrated, survived, and established subsequent generations, it is difficult to determine which attempts to leave Africa were successful in establishing today's AMH. Fossil remains are sporadic and ancient DNA evidence is hard to access, so that the origin of migration and timelines concerning the departure of AMH from Africa vary by tens of thousands of years. There are two separate theories on the exodus of *H. sapiens*, often referred to as AMHs, from Africa. One theory suggests that there was one dispersal, whereas the other theory suggests there were multiple dispersals of AMH out of Africa, in at least two separate

Ancestral DNA, Human Origins, and Migrations
https://doi.org/10.1016/B978-0-12-804124-6.00004-5

waves, at separate times. All the species that left Africa left behind the comforts of home and eventually traveled great distances to explore the unknown. During their travels, they had to adapt to different climates, different food sources, different predators, and different diseases. These environmental forces caused AMHs to evolve, simultaneously diversifying the human genome and forming phenotypically and linguistically distinct populations in various locations around the world. The reason for the demise of most *Homo* species is unknown, but their fossil remains tell the story of their travels, and their artifacts tell the story of how they adapted. In addition, genetics and paleoanthropology tells a story of interbreeding and further phenotypic and genetic diversification. Studies of ancient travel help us to understand the origin of human diversity and our ancestral heritage. While the topic is still widely debated, this chapter explores migratory routes followed by *H. sapiens*, primarily through genetic and archaeological findings.

THE ARGUMENT FOR AN AFRICAN ORIGIN

To understand the exodus from the birthplace of humankind and the spread of humans across the globe, their site of origin needed to be determined. Initial theories on human origin were based on TH Huxley's comparisons of great ape and human anatomy, in the mid-17th century, and Darwin's proposal that the ancestors of humans arose in Africa. Since that time, investigators from a wide range of disciplines have shown that Darwin was correct. However, determining the African region where humans evolved and the number and locations of human migrations out of Africa has been more complex than initially thought. Recent evidence from archeological, genetic, and climatic data have provided important tools to study these events, but it is still unclear which theories are the most accurate. Were there multiple sites where humans evolved in Africa, were there multiple dispersals, what geographical route(s) were taken, how and when did our ancestors leave Africa? These are important questions whose answers can help us understand the phenotypic and genetic diversity seen in different geographical regions, and how they arose.

A more complete description of human origins can be found in Chapter 3; however, a short review is presented here, as the exodus from Africa and the origins of humans have been historically linked and are part of the migration models presented by investigators. Evidence from genetics and anthropology has confirmed that AMHs arose in Africa, although the specific site(s) of AMH evolution are still debated. The evidence for our African origin comes from fossil remains, genetic analysis, and linguistic analysis. Using these tools, it has been shown that genetic, phenotypic, and linguistic diversity decreases as you move further from sub-Saharan Africa. This conclusion can be explained by a process called a cascading bottleneck. When trying to determine the origin of a species you would expect that the individuals at the origin contained a high amount of phenotypic and genetic diversity. As a relatively small group of individuals moved away from the origin, you would expect that they brought with them a small sample of the genetic, phenotypic, and in the case of humans, linguistic diversity. As the original migrating group broke into smaller groups that ventured further, the smaller groups would again take a fraction of the diversity that was contained in the original group. Over time, environment, genetic drift, and selection would result in increased phenotypic differences between the groups, increased linkage disequilibrium (LD) (differences in recombination rates), and decreased heterozygosity. This type of analysis could also be used to infer the routes that individuals took as they dispersed, especially if there was a single dispersal.[2,3]

Initially you would expect little increase in genetic variation between the population that left Africa and the population that the migrants left behind, except for some adaptation to the different climates, which may have been mostly cultural (different clothes and accommodations) rather than genetic. However, changes in diet, contact with new animals and their contagious diseases, new parasites, and new regional diseases would invoke some natural selection. Some examples of changes that we see today due to natural selection in different parts of the world are pigmentation, lactose intolerance, resistance to malaria, and adaptation to high altitudes.[4]

During the latter part of the 20th century, discussions regarding the origin of modern humans and their migration out of Africa were focused on several theories: (1) the Out of Africa model (or recent single-origin), (2) the multiregional (or polycentric) model, and (3) the candelabra model.

The candelabra model is often combined with the multiregional model, but each are different and were proposed at different times by different investigators. The candelabra model was first proposed in the early 1960s and suggested that our early ancestors left Africa about 1 mya, and then as they spread out to other continents they independently evolved the features of AMH. Supporters of the candelabra theory surmised that this would explain the geographical diversity we see in extant populations and the intermediate forms of AMH we see in the fossil record. Genetic and anthropological studies have discredited this model, and the focus has since been on the Out of Africa and the multiregional models.[5,6]

The multiregional model was proposed in 1946 and was based on the idea that *H. erectus* was the precursor to *H. sapiens*. The theory states that through significant gene flow among subpopulations of *H. erectus* that lived in different regions of the world during the Pleistocene, various groups of *H. sapiens* evolved. This suggests that AMH ancestry can be attributed to multiple hominin groups, living in multiple regions and interacting, rather than independent evolution proposed by the candelabra model.[7]

The Out of Africa II model (Out of Africa I involved *H. erectus*) states that AMH evolved in Africa then moved out of Africa where they continued to evolve. The model recognizes that some interbreeding with other *Homo* species took place after leaving Africa. This model has been supported by mitochondrial DNA (mtDNA), Y chromosome DNA, and autosomal DNA studies, which indicated increased LD (mixing of genetic information) and decreased heterozygosity when moving away from Africa.[5]

Others have argued for an alternative evolutionary mechanism that asserts that rather than a single out-of-Africa dispersal and interbreeding with other *Homo* species, early modern humans were already divided into different populations in Pleistocene Africa. This argument infers a deep population substructure in Africa before members of those populations embarked on a complex set of migration patterns out of Africa. This scenario also suggests genetic drift followed migrations, leading to the geographic diversity that we see in extant human populations.[6,8–10]

Until recently, most anthropologists assumed that *H. sapiens* first evolved in sub-Saharan Africa about 200 kya, and Northeast Africa was considered by most experts to be the region where humans originated. Over the last several years human remains found in Southern and Northern Africa have caused researchers to question the idea that AMH originated only in northeastern Africa. Climate reconstructions now suggest that North Africa was much more hospitable than previously thought, and fossil remains dating to 300 kya that show a remarkable resemblance to AMH were recently uncovered in Jebel Irhoud, Morocco.[11] This new data pushed back the date of AMH origins initially ascribed to Eastern African remains commonly dated at 200 kya. Although the Jebel Irhoud remains are not quite modern humans, their faces resemble ours. Stone tools found at the site include Levallois flakes, scrapers, and unifacial points. Modified animal bones at the site and charcoal suggest the use of fire. Other studies in North Africa have uncovered fossil remains,[12] and studies by Fadhlaoui-Zid and others[13] analyzed Y chromosome and genome-wide data from North Africa, the Near East, and Europe that suggested there is evidence for a recent origin of human populations in North Africa.

Evidence for a Southern origin comes from the discovery of *Homo naledi* and is also supported by genetic evidence. Henn and coworkers[14] analyzed the genomes of extant tribes from sub-Saharan Africa: Central Africa, East Africa, and Southern Africa, and suggested that the LD and heterozygosity patterns of Southern hunter-gatherers were among the most genetically diverse of all human populations. These studies have been confirmed by other investigators, but because LD estimates vary among different extant tribes in regions of East and Southern Africa, it has been impossible to pinpoint an origin of AMH based solely on genetic evidence.[15,16]

The combination of the Jebel Irhoud discovery and the overlap in the dates ascribed to *H. naledi* from South Africa reinforce the idea that the evolution of modern humans was more dispersed than previously thought.[17]

Others have argued for a Central African origin before groups split and dispersed into Southern, Northern, and Eastern Africa.[18] While there is little to no mention of Western Africa, it cannot be ruled out. The focus on the East, North, and South is because these regions contain ancient fossil remains.

It is easy to see that the debate over the origin of AMH will continue to evolve as additional data are collected from ancient DNA, fossils, and climatic conditions. Although fossil remains, artifacts, and genetic data are useful in determining AMH origins and migration patterns, debate continues and data interpretations and conclusions require caution. This is because the data from fossils are fragmentary, not everyone agrees on the methods for data analysis, not all fossil remains have been found, nor can all ancient material be genetically sequenced. Caution in data interpretation is especially true when using extant aboriginal populations as a baseline for genetic analysis. Although these isolated populations are useful, the unknown extent of admixture and drift may skew genetic data that are used to represent the genetics of ancient populations. As additional data are collected, the story of human origins and migration is consistently changing and may end up concluding that AMH originated and migrated from various sites within Africa.

THE FIRST HUMAN TRAVELERS

The first individuals of the genus *Homo* to leave Africa around 2 mya were *H. erectus* (sometimes referred to as *Homo ergaster*, which is the African form of *H. erectus*). *H. erectus* remains have been found in East Africa, Morocco, Algeria, China, Georgia, and Indonesia. Researchers have also speculated that *H. erectus* occupied Europe, although there are limited numbers of fossilized remains.

H. erectus were the first human species to emigrate out of Africa and populate Southeast Asia and parts of Europe. Although there are slight phenotypic differences (shape of the brow

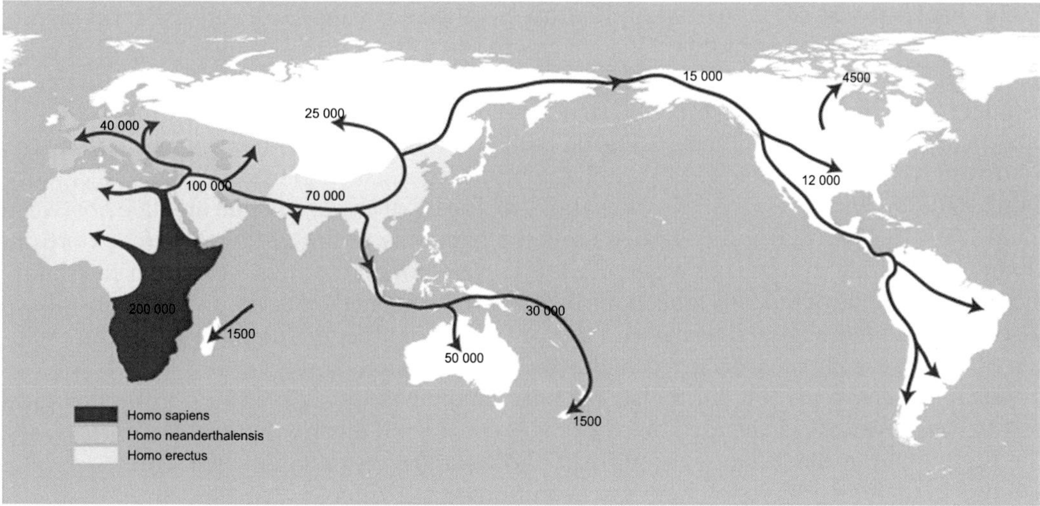

FIGURE 4.1 Map of early human migrations. *Homo erectus* is thought to have left Africa about 2 mya through a Northern route, eventually moving into the Middle East (approximately 1.9 mya), Asia, Indonesia (approximately 1.7 mya), and Southern Europe 1.2 mya. *Homo neanderthalensis* evolved in, and occupied, Europe around 300 kya. *Homo sapiens* moved initially into the Middle East, Southeast Asia, and Australia. Dates provided on the map are for *H. sapiens*. *Figure from https://en.wikipedia.org/wiki/Recent_African_origin_of_modern_humans#/media/File:Spreading_homo_sapiens_la.svg.*

ridges and smaller brain case) in the remains found in African and Asiatic *H. erectus*, they are considered by many archeologists to be the same species, with the slight differences due to environmental and founder effects[6,19] (see Chapter 2).

The migrations of *H. erectus* took place via the Levantine corridor and the Horn of Africa into Eurasia. Archeological evidence points to several migrations based on artifacts and fossil evidence (Fig. 4.1). Bar-Yosef and Belfer-Cohen[20] provided estimates for three migrations from tool artifacts. The first migration 1.8–1.6 mya based on Oldowan tools, a second migration between around 1.4 mya based on Acheulean tools, and the last migration based on leaf-shaped flakes about 0.8 mya. Fossil remains and/or stone tools of *H. erectus*, outside of Africa, have been found in Rawat, Pakistan (1.9 mya), Dmanisi in the Caucasus (1.8 mya), Java (1.7 mya), Nihewan Basin in China (1.6 mya), the Ubeidiya in the Levant (1.5 mya), and the Atapuerca Mountains of Western Europe (1.2 mya).

Stone tools found in Pakistan are considered the oldest remains of *H. erectus* and were dated to 1.9 mya, whereas the earliest fossils of *H. erectus*, found outside Africa, were within the town of Dmanisi in the Republic of Georgia, Southwest Asia.[21] The remains of five individuals dating to the same time strata were initially recovered in the 1990s, with the latest discovery of a well-preserved skull reported in 2013. The Dmanisi site provides the best evidence for a Levantine corridor dispersal from Africa. Beside the fossils, stone artifacts and animal bones with precise cut marks are evident at the site. Over 8000 choppers and scrapers and many flakes were recovered that were dated between 1.75 and 1.85 mya. Further evidence comes from Oldowan/Acheulean tools and tooth fragments found near Israel and the Jordan River Valley and ascribed to *H. erectus*.[22]

The dispersal of *H. erectus* to Asia and Europe, as attested by the fossil and artifact evidence, was relatively fast. Representatives of *H. erectus* reached Java approximately 1.7 million years ago (mya) and Atapuerca in Iberia, Spain about 1.2 mya. It is significant that *H. erectus* could populate the Caucasus despite its brutal winter weather conditions at the time.[23,24] Other sites include the island of Flores, Sumatra, Bali, Sulawesi, and the Philippines 88–80 kya. A controversial find were stone tools from Longgupo Cave located in Chongqing Municipality, China, and Renzidong (Renzi Cave) in Anhui Province, Eastern China that are dated between 2 and 2.5 mya.[25] Recent finds of early human stone tools from the Siwalik Hills, north of Chandigarh, India, in a village called Masol date to around 2.6 mya. Although the dating and interpretations of the tool collections have been questioned, collectively these finds have raised some doubts about *H. erectus* and the Out of Africa I theory, suggesting that the evolution of *H. erectus* may have occurred in the opposite direction.

Asia represents a pivotal middle stage in the hominin dispersals, and its anthropological history has repeatedly challenged many of the premises and theories accepted over the years. The findings from the Indian subcontinent allow for the reevaluation, refinement, or even rejection of theories addressing hominin evolution (see Chapter 7).

For example, the prevailing notion today is that hominins, other than Neanderthals, originated in Africa and then dispersed out of the continent into Eurasia. The accepted theory is that *H. erectus* was the first species to venture out of Africa around 2 mya and quickly dispersed into Asia and Europe. Yet, a growing number of discoveries do not fit into this view. For example, Masol has revitalized the interest in early hominin habitation and dispersal in Southeast Asia, a crucial region in the Out of Africa I dispersal toward the Far East. The Out of Africa I theory posits that the first hominins to disperse out of Africa were *erectus* groups shortly after their origin in Northeast Africa approximately 1.9 mya. It also postulates that an Indian coastal route was taken by these early presapiens on their way to East Asia.

Similarly, the presence of Acheulian implements (stone tools characterized by distinct oval- and pear-shaped hand axes) dating to 1.8 mya in the Dang Valley, Nepal, midway along the length of the Siwalik corridor, is not compatible with the Acheulian technology that was developing in Africa about the same time. There are several possible explanations for these findings: either (1) the Acheulian tradition originated earlier in Africa and then early hominin migrants transported it to South Asia or (2) the genesis of Acheulian tools was in South Asia or (3) the Acheulian technology originated independently in both regions, an unlikely scenario.

Although a coastal migration route to East Asia would have been longer, there are reasons that such a path would have been easier, especially for early hominin populations with limited technology. The route would have provided marine organisms and aquatic plants; in addition, the terrain close to the shoreline tends to be less mountainous and easier to navigate. Yet, no evidence has been found of early hominin habitation in coastal peninsular India. It could be argued that failure to detect such sites results not from their absence but from their present location under water as the sea level arose from the melting of glaciers during the current interglacial period.

Together, these archeological sites outline what is known as the Siwalik corridor. In spite of the relative ease of a coastal Indian route, archeological evidence suggests that the earliest hominins in South Asia employed a northern path along the Siwalik corridor just south of the Himalayas to get to East Asia (Fig. 4.2).

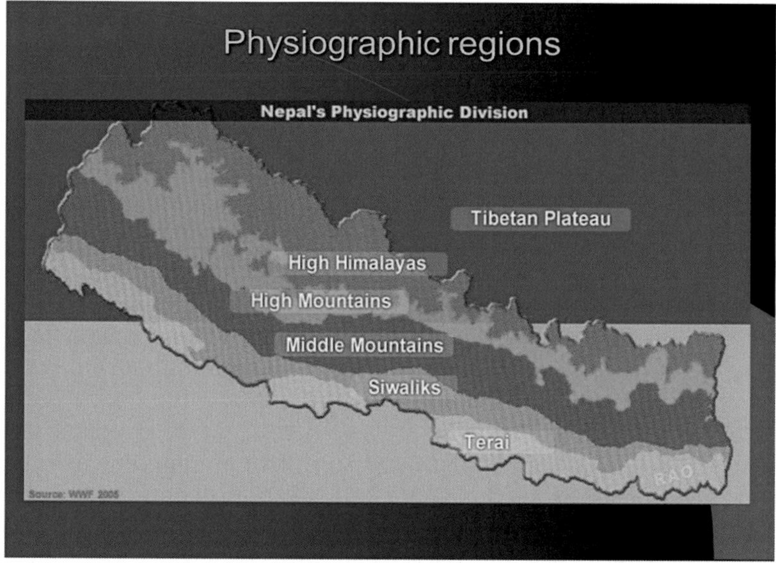

FIGURE 4.2 The Siwalik corridor runs south along the middle mountains of the Himalayas. *Figure from http://slideplayer.com/slide/6966190/24/images/7/Physiographic+regions.jpg.*

Archeological and fossil evidence found in the Siwalik range suggests its initial use as a passage through Asia during Out of Africa I, and that it was also used later during Out of Africa II (AMH dispersal), using more sophisticated Acheulean technology. Therefore, it is possible the earliest hominin dispersals to Southeast Asia were responsible for the settlement in Java.

Many authors have speculated that migration was based on climatic conditions; however, because the dates of fossils and tool discoveries do not necessarily coincide with the climatic cycles that have been used to explain migration, it appears that they were not necessarily a determinant of migration into Asia.[26] Instead, dispersals would have coincided with changes in landscape, climate, obstacles like high seas and high mountain ranges, culture, and food and water sources.

Other investigators have explained the impetus for migration into Asia to be due to ecological changes, declining resources in Africa, increased brain size, long distance endurance, technology, and social behavior.[27,28] A rise in aridity in Africa and the growing savannahs may have led to a decrease in primary productivity and less food for herbivores. Hunter-gatherers typically occupy large areas to gather food and other resources. As the population of *H. erectus* grew, it may have exceeded the carrying capacity of certain regions of Africa. Carrying capacity depends on climatic and biogeographic conditions which change over time. *H. erectus* had a larger brain and body mass than previous hominins, leading to increased meat consumption, which would have led to an extension of the home range, increased population size, and population distribution.[26,29] The dispersal of humans also coincided with the dispersal of large mammals such as horses, hippopotami, saber-toothed cats, wolves, and other primates about 1.8 mya. This could have been related to the human exit from Africa.[30]

The Saharan pump theory explains the migration of flora and fauna along the Levant land bridge. The theory suggests that climatic pulsations of wet and dry weather in the Sahara and Arabia created extended periods of rainfall, resulting in large lakes and rivers. These episodes of fluctuation of available water started about 3 mya. Since then, the Saharan region oscillates, transforming forest habitat to barren desert and back to forest approximately every 20 ky. These climatic fluctuations affected the expansion of plants and animals in the region and created emigrational gates that opened during wet expansion episodes and closed during dry contraction periods. The evidence for these pulses of migration is that the recurrent wet periods coincide with fossil evidence of migratory events by *H. erectus*, *H. heidelbergensis*, and early modern humans. In the case of *H. erectus* and AMH, there is some archeological evidence that migration may have taken place on more than one occasion. It has been suggested that these wet–dry cycles were particularly important in human evolution during the last 200 ky.[31]

H. heidelbergensis was the second human to leave Africa. *H. heidelbergensis* is distinct from *H. erectus* primarily due to its large brain size and body proportions, which are comparable to AMH. *H. heidelbergensis* inhabited Africa, Europe, and Asia from about 600–200 kya. It is thought that in Africa, *H. heidelbergensis* gave rise to *H. sapiens* around 200–300 kya.[32] Fossil evidence suggests that after *H. heidelbergensis* left Africa about 600 kya, they evolved into *Homo neanderthalensis* in Europe and Denisovans in Asia. Evidence from nuclear DNA extracted from fossil remains in the Sima de los Huesos cave site in the Atapuerca Mountains, Spain, dated to around 430 kya, indicates that Neanderthals had already evolved and had separated from Denisovans. mtDNA analysis of the fossils indicated they were Denisovan, whereas the nuclear DNA and phenotypic analysis clearly showed the fossils were of Neanderthals.[32] These individuals most likely represent a transition period in the evolution of Neanderthals and Denisovans.

WHICH WAY DID THEY GO?

While there is no debate that *H. sapiens* developed to become AMH that inhabit various parts of the world, the estimates of arrival times and routes taken continue to be debated. Traditionally, there are two competing theories describing how *H. sapiens* left Africa: (1) in either a single dispersal or (2) in multiple dispersals. Both theories are consistent with the evidence for decreasing genetic diversity as you move further from sub-Saharan Africa. In addition, two major routes have been proposed for both the single and multiple dispersion theories. A Northern route through the Levantine corridor through Egypt and the Sinai, and a Southern route through Ethiopia, the Bab-el-Mandeb strait of northeastern Africa into the southern Arabian Peninsula (Fig. 4.3). Genetic and archeological evidence has not been able to eliminate either route as the primary route.[6] Later migrations into Europe after reaching the Afro-Asian shore of the Mediterranean could have included the Levantine corridor, the Channel of Sicily, and the Strait of Gibraltar. The Levantine corridor is the easiest and most accepted route into Europe.[26]

In addition to debates over the routes used to exit Africa, there is continuing debate over the time of exodus. There are currently two major conflicting proposals, each differing by tens of thousands of years and not mutually exclusive. The first claims that the Eurasian dispersal took place around 50–60 kya, reaching Australia by 45–50 kya. The second claims a much

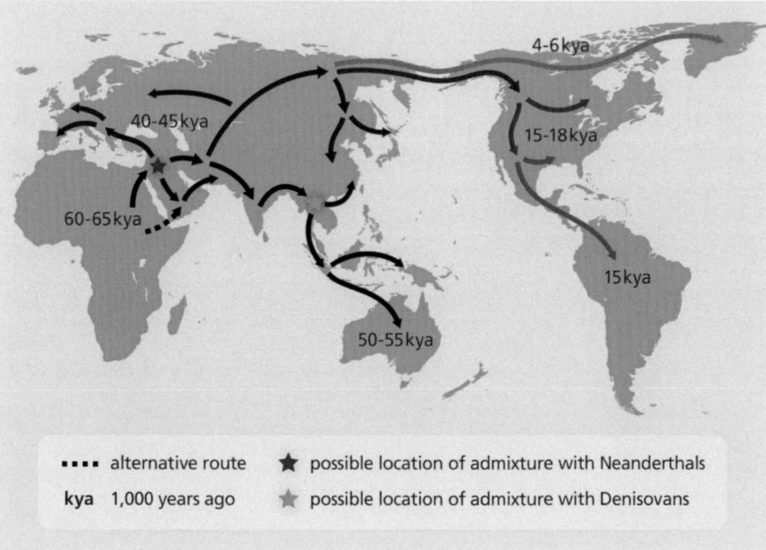

FIGURE 4.3 Depiction of possible migratory routes of *Homo sapiens* and admixture with Neanderthals and Denisovans. The figure shows the proposed Northern route out of Africa, with the Southern route indicated as an alternative (dotted lines). *Figure from http://www.yourgenome.org/stories/evolution-of-modern-humans.*

earlier dispersal around 100–130 kya based on tools found in volcanic ash supposedly from the eruption of Mount Toba in northern Sumatra, around 74 kya. The tools were uncovered in Jebel Faya (present-day United Arab Emirates) and the Nubian complex of Dhofar (present-day Oman) (Fig. 4.4) and have provided further support for an early migration via Arabia. Fossil remains found in the Levant in the Es Skhul and Qafzeh caves, of modern-day Israel, are dated at 120 kya and 100–90 kya, respectively. However, it is not clear that these early migrations resulted in individuals that survived and represent the ancestors of modern-day humans. The next AMH fossil remains in the Levant are dated to 55 kya, which represents a considerable gap in time. For this reason, archeological and genetic data analysis provide a wide range of dates for various out of Africa models.[6]

Both the single dispersal and multiple dispersal theories have been modified over the years. The single dispersal model has been modified to acknowledge a small genetic contribution due to introgression with other *Homo* species (Neanderthals and Denisovans) and suggests a single dispersion that bifurcated once outside Africa, most likely in Southwest Asia. The modified version acknowledges a Southern Asia route, occurring between 75 and 65 kya, which includes the coastal migration into Australia (45–50 kya). DNA evidence suggests there was a migration from Southeast Asia back into Africa about 4.5 kya. The evidence suggests that the likely source of back migration was early Neolithic farmers from modern-day Sardinia.[33]

In sum, the current multiple dispersion model, based on archeological and fossil evidence, suggests timelines for the first exodus, ranging from 131 to 114 kya (the estimated dates of African and Eurasian divergence) to between 75 and 62 kya, with a second migration occurring 38–25 kya.[2] The model includes Australians as an isolated population from the

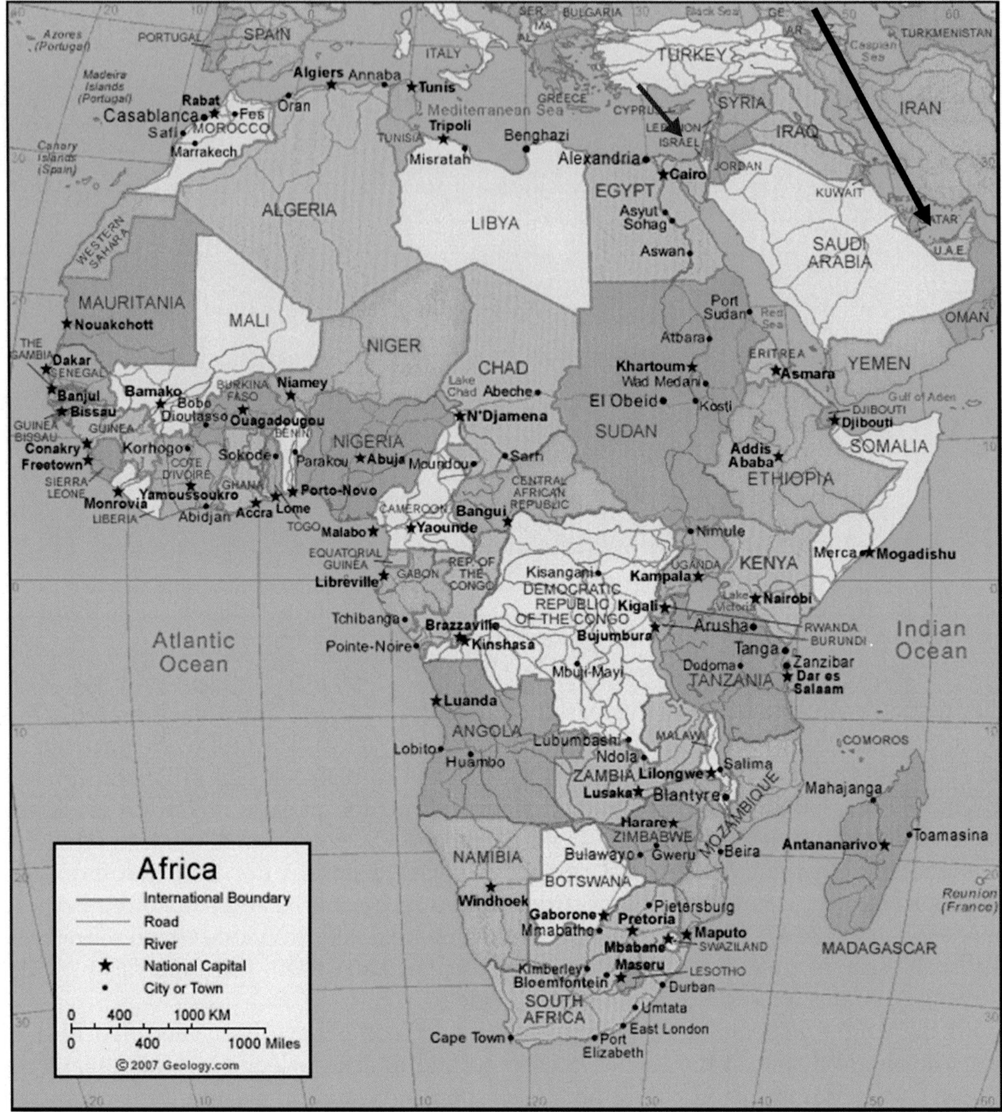

FIGURE 4.4 Map of Africa and the surrounding regions. The *black arrow* points to Oman and the United Arab Emirates. Ancient Nubian tools found in this region suggest that a Southern route out of Africa was taken 100–130 kya. The *red arrow* points to Israel where fossil remains found in the Levant in the Es Skhul and Qafzeh caves are dated at 120 kya and 100–90 kya, respectively. *Taken from http://geology.com/world/africa-map.gif.*

first southern coastal migration, with Papuans, Melanesians, and possibly the Aeta from the Philippines resulting from a second dispersal.

Advances in technology and data indicating a constantly changing climate over 100 kya suggest that neither theory is entirely correct; however, the strongest evidence suggests *H. sapiens* left Africa in waves under favorable and hospitable conditions. Additional evidence suggests two major routes of exodus, and archaeological, fossil, and genetic evidence supports both speculated routes, through the Horn of Africa or the Levantine corridor. However, genetic and archeological evidence do not always agree. Genetic studies point to a limited genetic diversity in Eurasia and are compatible with one or very few dispersals happening between 65 and 75 kya, whereas archeological evidence spans over 120–100 kya, indicating multiple dispersals. One explanation is that the first migrations resulted in failure and thus no genetic trace that can be seen today. Whatever the route taken, archeological evidence from tools and other artifacts indicates AMH met with Neanderthals, Denisovans, and *H. erectus* (local inhabitants of Asia and the Near East) over 100 kya. Probably the most important of these encounters would have been with Neanderthals and Denisovans, due to genetic evidence that indicates they contributed to the gene pool of AMH (see Chapter 6).

THE USE OF GENETICS TO DETERMINE WHEN AND WHERE HUMANS LEFT AFRICA

A genetic tool used to infer migration patterns and the origins of AMH comes from the study of small mutations. As discussed earlier in the chapter, the original migrants that left Africa would have taken a small sample of the population's genetic diversity with them. Over time, the accumulation of random mutations in the migrants would differ from the original African population, and as the migrating population split apart and moved (cascading bottleneck), the mutations would differ slightly between those migrant populations. These mutations can be used to trace back the source of the original genotype, to trace ancestry, and to infer migration routes.

Some of the most useful DNA sequences come from mtDNA and the nonrecombining region of the Y chromosome (NRY), which lead to the sequential accumulation of new mutations along radiating female or male lineages, respectively. The various mutations on these two types of DNA create monophyletic and evolutionary stable sequences known as haplotypes. Haplotypes are biallelic markers, most commonly single-nucleotide polymorphisms (SNPs), with a low mutation rate,[34] which can be arranged in a phylogenetic tree[35] (see Chapter 1).

To better understand the routes both accessible and preferred by ancestral AMH, researchers began studying mtDNA and NRY data from extant aboriginal populations and ancient DNA from fossil remains to determine the oldest established haplogroups (populations carrying haplotypes) by region and possible haplotype origins. mtDNA and NRY allows researchers to track ancestry and provides information regarding potential distances traveled by a specific population, geographical distributions, and discovery of possible interactions with other hominid species.

Using the phylogenetic information and the ethnogeographic distribution of haplogroups, geneticists can date some demographic processes behind human dispersal. In addition, by analyzing the worldwide distribution of the different biallelic markers, it is often possible to find one (or more) specific haplogroup confined to a restricted geographic region, allowing for the identification of its origin and the individuals that carried the marker.[4,36] The rate at which mtDNA mutates is known as the mitochondrial molecular clock. One study has reported that mtDNA encounters on average 1 mutation per 8000 years, making mtDNA less precise for genealogical dating than NRY DNA, which accumulates one mutation approximately every 10 years.[37]

Although geographical identification of haplogroups, and dates of origination, become discernible, offering researchers information relative to human migration events, it is worth noting that highly specific dating cannot be readily pinpointed but rather inferred and deducted from other clues and the use of geological, archeological, and linguistic data.

Several haplogroups are important in determining when and where AMH left Africa. mtDNA studies have suggested that almost all African lineages belong to the L clades. The mtDNA haplotype called L3 is a unique Eastern African haplotype that is thought to be involved in the out-of-Africa migration. Mutations in the L3 mtDNA haplotype then lead to all subsequent mtDNA haplotypes outside of Africa. Haplogroups N and M are considered descendants of L3 with estimates of M diverging from L3, 59–70 kya[38] and N diverging from L3, 57–65 kya.[39] These dates supply a reasonable estimate of when AMH left Africa. Haplogroup M spans all the continents, and all mtDNA haplogroups outside of Africa are considered descendants of haplogroup M or haplogroup N. The geographical distribution of mtDNA haplogroups M and N are most commonly found in Southeast Asia, Australia, and the Americas; however, haplogroup M is rare in western Eurasia (Fig. 4.5). The oldest clades of haplogroup N are found in Asia and Australia. A descendant of haplogroup N is haplogroup R found throughout Asia and Australasia.[39]

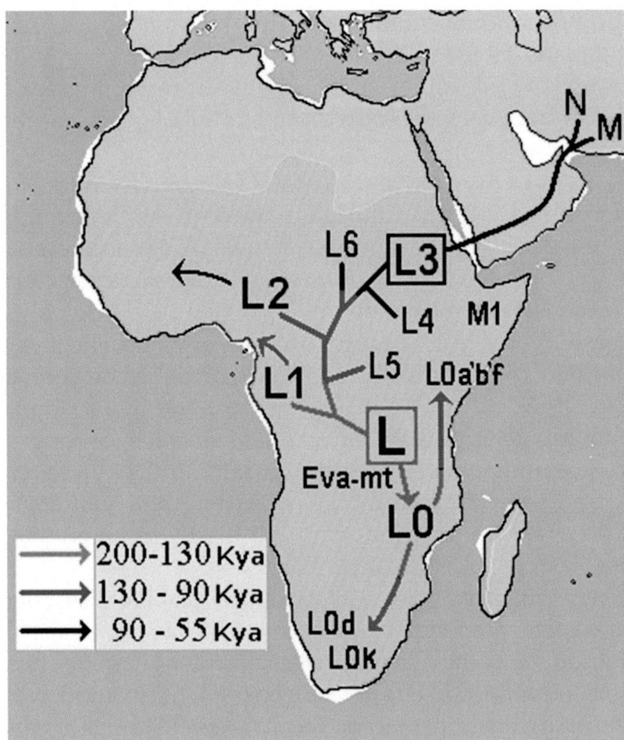

FIGURE 4.5 The figure shows the origin of the L mitochondrial haplotypes and their distribution in Africa. After crossing the Red Sea, the L haplotype evolved into the M and N haplotypes. *Taken from https://upload.wikimedia.org/wikipedia/commons/4/44/African_Mitochondrial_descent.PNG.*

Looking at the mtDNA haplogroup tree, the extensive splits in subhaplotypes in M, N, and R were so close together in time that few intermediate mutations accumulated. Many of these splits occurred in the southern Arabian regions 60–65 kya, where adaptation to inland conditions may have favored those of costal environments, leading to a rapid population expansion.[40,41]

Y chromosomal Adam (haplotype AOO) is considered to be the original NRY haplotype. AOO haplotype lead to the A haplotype, which is the one from which all modern paternal haplogroups descended. Other ancient African NRY haplotypes include BT, which arose approximately 140 kya and CT, which arose approximately 88 kya. Complete sequencing of 29 Y chromosomes provided estimates of the time that Y lineages gave rise to all lineages outside Africa (with a few inside Africa), with the data suggesting dates of 60–75 kya.[42] Other Y chromosome haplotypes from Northern Africa M173, and East Africa E3b*-M35, E1b1 and E1b1b-M35, have also been important in determining migration routes of AMH. Similar to mtDNA data, NRY data show a rapid expansion of male lineages after the out of Africa movement and dated to around 40–50 kya.[41,42]

Recent data on Y chromosomes by Mendez et al.[43] found evidence to support a model that placed the Neanderthal lineage as an outgroup to modern human Y chromosomes, including A00, the base haplogroup. They estimated that the time to the most recent common ancestor (MRCA) of Neanderthal and modern human Y chromosomes is 588 kya. The time estimate suggests that the Y chromosome divergence reflects the divergence of Neanderthals and modern humans, and it refutes alternative scenarios of a relatively recent or super-archaic origin of Neanderthal Y chromosomes. Although it is known that AMH and Neanderthals introgressed, the fact that the Neanderthal Y chromosome has never been observed in modern humans suggests that the lineage went extinct.

OUT OF AFRICA: A SOUTHERN ROUTE TO ARABIA?

A Southern exit leaving from the Horn of Africa, crossing the Red Sea into southern Arabia has previously been widely recognized as the lead theory due to skeletal remains, artifacts, and genetic evidence from mtDNA (Fig. 4.6). Evidence from geological surveys in southern Arabia reflects lake deposits estimated at times of 80, 100, and 125 kya. These climate changes would have made the Arabian Peninsula a hospitable savannah with shallow freshwater lakes and provided food for AMH expansion into Arabia.[44] In addition, the single Southern exit is consistent with the archaeological luminescence dates for colonization of the Philippines and Australia. This is consistent with the Last Glacial Maximum causing sea levels to drop and the idea that once the Red Sea was crossed by individuals, traveling the southern coast of the Arabian Peninsula toward India would have been possible with individuals, consuming shellfish and other marine animals (the so called "beachcomber" theory).[45]

Any migration through this area would have to have taken place before 75 kya when climatic conditions turned the region into an inhospitable desert environment, thus imposing alternative routes out of Africa for any populations after 75 kya. Many archaeological sites continue to be found in Arabia due to the constant rising and lowering of sea levels; although pinpointing dates and ages to these sites continues to pose problems.

The Southern route still poses the problem of crossing the Red Sea. Estimates of sea levels according to climatic conditions have estimated that at its lowest level the distance

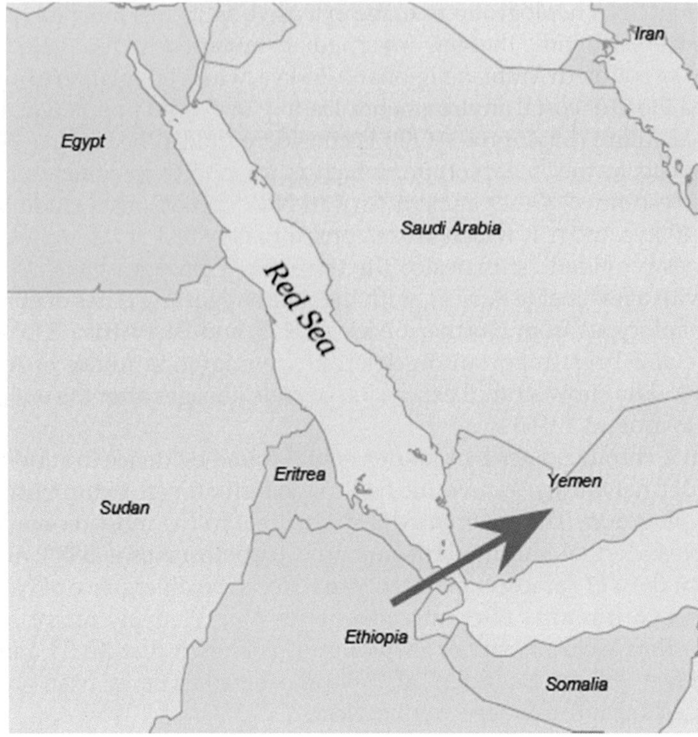

FIGURE 4.6 The proposed Southern route out of Africa crossing the Red Sea. The distance *Homo sapiens* would have had to travel into modern-day Yemen, calculated by climatic conditions at the time, would have been approximately 4 km and would have required some type of watercraft. *Figure taken from https://upload.wikimedia.org/wikipedia/commons/6/60/Red_Sea2.png.*

between Africa and southern Arabia would have been about 2.5 miles, requiring some type of simple watercraft to cross. These conditions occurred for sustained periods during the last two glacial cycles (during the Pleistocene) so that crossing could take place with a simple raft.[46]

Once AMH left Africa through the Southern route, they would have moved east into Southeast Asia and further into Oceania. mtDNA studies support this route. Studies have suggested that individuals with the L3 mtDNA haplotype migrated through the Southern route. Evidence from the geographical distribution of mtDNA haplogroups M and N (descendants of L3) show that movement through the Arabian route would seem to coincide with diversification of the mitochondrial genome. Haplogroups M and N are most commonly found in Southeast Asia, Australia, and the Americas (Fig. 4.5). The oldest clades of haplogroup N are found in Asia and Australia. The distribution of the oldest branches of haplogroups M, N, and R across Eurasia and Oceania suggest that there was a single Southern migration out of Africa. Further support for the Southern route comes from the findings that no L3 lineages, other than M and N, are found in India nor among non-African mitochondria. These mtDNA haplogroups and the geographic patterning and distribution (Fig. 4.7) essentially leaves a trail of genetic "breadcrumbs," pointing to a southern dispersal from Africa.[47]

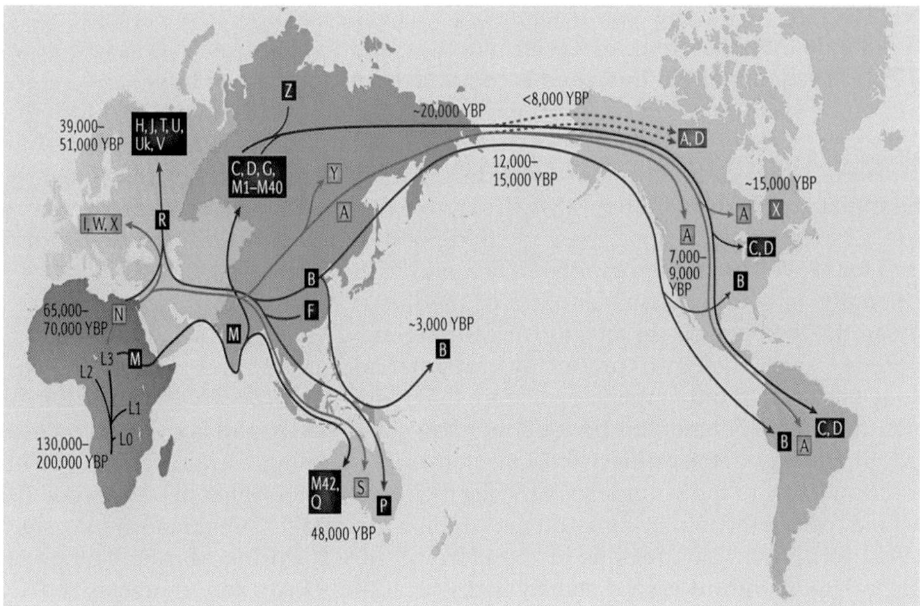

Nature Reviews | Genetics

FIGURE 4.7 The L mitochondrial haplogroups originated in sub-Saharan Africa 130–200 kya. The M and N haplogroups descended from L3 65–70 kya. When humans left Africa, haplogroup M moved into Asia later giving rise to haplogroups A, B, C, D, G, and F. Haplogroup N moved into Eurasia and later gave rise to haplogroup R, which is the root of haplogroups H, J, T, U, and V. Because mtDNA undergoes limited recombination, mutations leading to new haplogroups are useful in tracing maternal inherited ancestry. *From Stewart JB, Chinnery PF, The dynamics of mitochondrial DNA heteroplasmy: implications for human health and disease.* Nat Rev Genet *2015;**16**:530–42.*

In other mtDNA studies, supporting the Southern route, investigators looked at three minor west Eurasian haplogroups N1, N2, and X, which branch directly from the first non-African founder node, the root of haplogroup N, and coalesce with the movement of AMH 60 kya. The investigators sequenced 85 samples from Southwest Asians carrying the N1, N2, and X haplogroup, and compared them with 300 European samples. The results showed that these minor haplogroups (N1, N2, and X) have a relict distribution that suggests an ancient ancestry within the Arabian Peninsula. The authors suggest that AMH likely spread from the Gulf Oasis region toward the Near East and Europe 55–24 kya. The authors assume that this was the first successful attempt for AMH to leave Africa, resulting in Arabia as the first layover for AMH before their spread around the globe.[39]

In the above examples, mtDNA studies assume that the L3 marker is African-specific; however, Groucutt et al.[48] have argued that L3 could have arisen inside or outside of Africa based on gene flow from Africans and non-Africans after their initial divergence. This emphasizes the need to review archeological, environmental, fossil, as well as genetic data when trying to understand the expansion of AMH out of Africa. Also related to the origins of haplogroup N is whether ancestral haplogroups M, N, and R were part of the same migration out of Africa or whether haplogroup N left Africa via the Northern route through the Levant, and M left

Africa via Horn of Africa (Fig. 4.7). This theory was suggested because haplogroup N is by far the predominant haplogroup in western Eurasia, and haplogroup M is absent in western Eurasia but is predominant in India and is common in regions east of India.

Moreover, archeological findings complement the haplogroup distribution due to the similarities of artifacts found in both Arabia (Jebel Faya) and Africa. Hand tools found in Jebel Faya near the Persian Gulf have been dated to 125 kya and suggest they may have resulted from the initial southern migration of AMH out of Africa. The artifacts discovered ranged from side scrapers to small hand axes, speaking both to the types of jobs these stone tools were used for as well as the topography of the land at the time.

Additionally, in southern Arabia, relics of the late African Nubian Complex tools were found tying the Arabian inhabitants to African migrants. The Nubian Complex is a distinct Middle Stone Age technology of producing triangular and subtriangular points. In Africa, the technology has been found in northern Sudan, the middle and lower Nile Valley, the Red Sea Hills, and the eastern Sahara. It is divided into two stages (early and late) based on temporal and slight differences in the primary working surface. Using single-grain optically stimulated luminescence dating of the sediment layers, the tools were determined to be between 125 ± 16, 000 years old. Before the tool findings on the Arabian mountains, Nubian Complex sites were constrained to Africa, lithic regions in the Horn of Africa, during Marine Isotope Stage 5 (a geological temperature record which dates to 130–80 kya) and characterized by two subtypes, Type 1 and 2 (Fig. 4.8). While differences between Nubian Complex subtypes

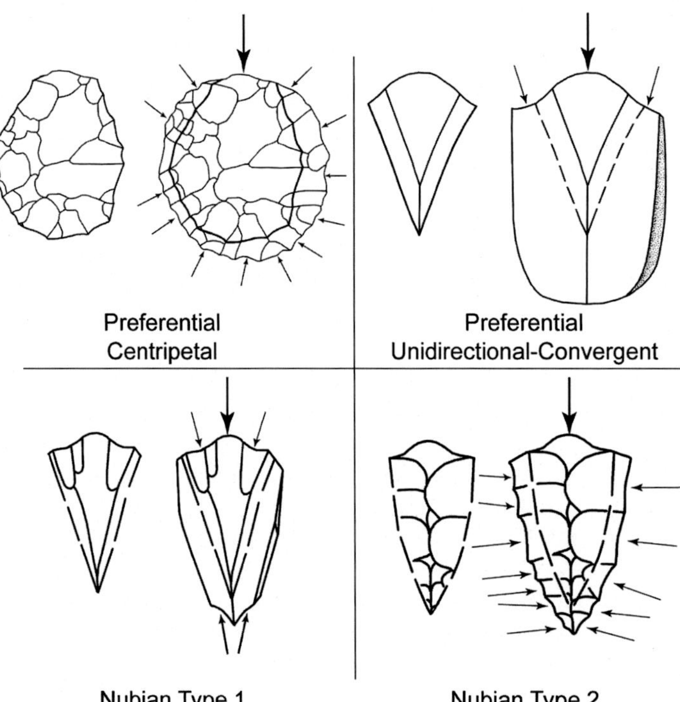

FIGURE 4.8 Type 1 and Type 2 Nubian Complex Depictions. *From http://journals.plos.org/plosone/article/figure/image?size=large&id=info:doi/10.1371/journal.pone.0028239.g002.*

are slight, Type 1 is more commonly associated with the later Nubian Complex of Northeast Africa. Type 1 Nubian Style Complex assemblages were primarily found on Arabian sites at Aybut al Auwal, Aybut Ath Thani, Mudayy As Sodh, and Jebel Sanoora (Fig. 4.9). The findings suggest a strong overlap between southern Arabian and Northeast African tools, indicating a very recent exchange of technology rather than a gradual technological evolution.[49,50] This finding is supported by the fact that the interglacial period commenced approximately 130–115 kya, causing sea levels to decrease and making the cross into Arabia fairly easy for *H. sapiens* at the time.[51] In other words, the two findings, i.e., similar tool types found in Africa and Arabia and the interglacial period causing decreased sea levels, strongly support the idea that AMH used a Southern route to cross the Red Sea approximately 125 kya.

The argument for the technological similarities between the artifacts found in Arabia and Africa are supported by the lowering of sea levels that occurred between Marine Isotope Stage 5 and 6 (130–191 kya) when crossing the Bab-el-Mandeb Strait would have allowed *H. sapiens* to escape the arid and uninhabitable conditions in Africa. During this interglacial period (130–115 kya) a Southern route would have been preferred both because of the ease of crossing the strait as well as the depletion of resources due to the infertile conditions

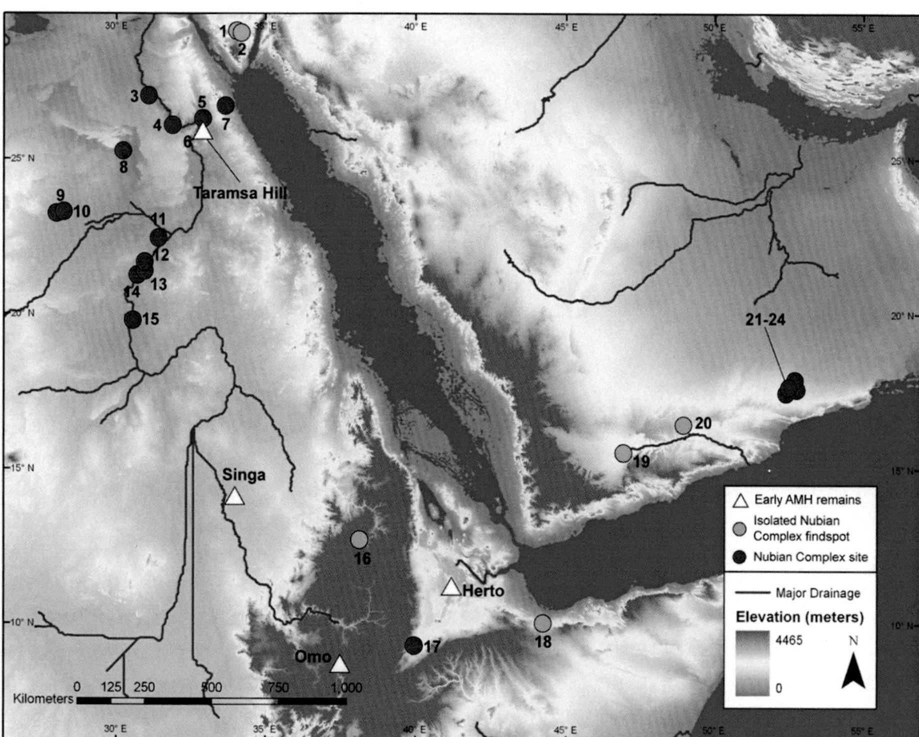

FIGURE 4.9 Map of Nubian Complex occurrences in Northeast Africa and Arabia. Nubian Complex sites include: Jebel Urayf (1), Jebel Naquah (2), Nazlet Khater (3), Abydos (4), Makhadma (5), Taramsa Hill (6), Sodmein Cave (7), Kharga Oasis (8), Bir Tarfawi (9), Bir Sahara (10), Abu Simbel (11), Jebel Brinikol (12), 1035 (13), 1038 (14), Sai Island (15), Gorgora Rockshelter (16), K'One (17), Hargeisa (18), Shabwa (19), Wadi Wa'shah (20), Aybut Al Auwal (21), Aybut Ath Thani (22), Mudayy As Sodh (23), and Jebel Sanoora (24). *From http://journals.plos.org/plosone/article/figure/image?size=large&id=info:doi/10.1371/journal.pone.0028239.g002.*

occurring in Africa. The wetter conditions in southern Arabia allowed for movement further into the Arabian interior of the country and less competition for resources, as was experienced in Africa, permitting more advanced hunter-gatherer styles. In addition, archeological evidence supporting the Southern route has shown that AMH were in southern Asia between 80–60 kya on their way to Australia and Oceania.

Liu W et al.[52] excavated 47 human teeth from the Fuyan Cave in Daoxian (southern China) dated to 80–120 kyo. The sample was unequivocally from AMH, resembling Middle-to-Late Pleistocene specimens and contemporary humans. The study confirmed that AMH were present in southern China 30–70 ky earlier than in the Levant and Europe. These data support a Southern route and infer that these were individuals that eventually made their way further East. It is important because archeological evidence for maritime resource exploitation (the beachcomber theory) is limited. Artifacts from the Red Sea and the Gulf Basin indicate habitation earlier than 100 kya; however, it is not clear whether these individuals survived and reproduced, thus representing modern-day humans. Instead, some investigators argue that these individuals went extinct and that movement along the coast would still have been difficult and slow.[6]

Microsatellite markers have also been used to establish a route out of Africa. Liu et al.[53] analyzed 783 autosomal microsatellites from 52 worldwide populations. Microsatellite data along with geographical and LD data were used to simulate the evolution of coalescence times during the colonization process. They specifically looked at the age of initial expansion, the number of individuals, and the growth rate and carrying capacity of colonized demes, and concluded that expansion started around 56 kya, starting from a source population of around 1000 individuals. Their data weakly supported a Southern route of expansion but could not rule out a Northern route.

AN EXIT THROUGH THE LEVANTINE CORRIDOR

The earliest known pathway out of Africa most likely occurred between 130–115 kya in what is described by geologists as Marine Isotope Stage 5. The Northern Route sometimes mistakenly referred to as "Out of Africa II" (because it was discovered relatively recently) proposes that *H. sapiens* left along the Nile corridor into the Levant, in Northeast Africa, to further colonize the rest of the surrounding continents and countries. This route is considered the easiest in terms of crossing, as it would not require the use of watercraft, capable of crossing the Red Sea. Support for the Northern route comes from some of the earliest *H. sapiens* remains found in Skhul and Qafzeh in present-day Israel dating to approximately 100–125 kya, alluding to AMH crossing the Sinai Peninsula into the Levant (Fig. 4.4). Many consider this early migration to be a failed attempt, as there are relatively few fossils this old from the region (although it may be too early to rule out this route as successful).[20]

Northeast Africa is also considered by many as the origin of the next major migration (after the southern exodus) out of Africa, (between 70 and 50 kya), but as with the Southern route, there is a considerable gap in the fossil evidence. Although Northeast Africa appears to be one of the sites of migration and expansion outside Africa, AMH appear to have left Africa in two or more waves, with researchers speculating that because of the glacial period 90 kya the first migrants did not survive in the Levant.

Data collected from mtDNA haplogroup frequencies of Near Eastern and sub-Saharan African populations along both the Levantine and Horn of Africa corridors allowed researchers to track the most recent gene flow of sub-Saharan Africans. The time since the most recent ancestor was calculated by using the mean transitional difference and human mtDNA mutation rates (in years) between lineages. These calculations produced the time since the most recent common ancestor (MRCA). This progression of sequential nucleotide changes illustrating the earliest mutations originating in Africa is one of the strongest arguments in support of AMH origins in Africa and the Out of Africa theory.

The sub-Saharan African populations were mostly of the traditionally African L haplogroups, whereas Near Eastern populations fit best into the West Eurasian haplogroups, indicating the migration patterns between Africa and the Near East. From this information, it can be inferred that the reported Near Eastern haplogroups did not evolve or originate in Africa. Haplogroups C, D (Eurasian origin), and M are geographically distributed in a manner that suggests an earlier dispersal from the Horn of Africa into Southern Asia approximately 45–50 kya.

Further genetic research observed M35 Y-chromosomes primarily in northern and Eastern African populations, with the E3b-M35 (an East African haplogroup) lineage present mostly in sub-Saharan African populations. M35 distribution in North Africa, the Near East, and Egypt is highly suggestive of a northbound migration through the Levantine corridor. This would suggest the initial North African migrations (assumingly through the Horn of Africa) involved populations of the E3b-M35 derivatives. M35 derivatives have also been found in Oman, Ethiopia, and Egypt, although at lower amounts than in Egypt. Despite similar geographical occurrences, the Omani M35 lineage was somewhat different from that found in Ethiopia.[36]

In addition, Pagani et al.[54] sequenced the genomes of 225 individuals from 6 modern Northeast African populations (Egyptian and Ethiopian populations). West Eurasian components were masked out of the analysis, and the remaining African haplotypes were compared with a panel of sub-Saharan African and non-African genomes. The data showed Northeast African haplotypes overall were more similar to non-African haplotypes and more frequently present outside Africa than were any sets of haplotypes derived from a West African population. In addition, Egyptian haplotypes showed these properties more markedly than the Ethiopian haplotypes, pointing to Egypt as the more likely route of exodus from Africa. Comparing several Ethiopian and Egyptian high-coverage genomes, the investigators estimated the genetic split times of Egyptians and Ethiopians from non-African populations at 55 and 65 kya, respectively, whereas that of West Africans were estimated to be 75 kya. This genomic evidence and Egypt's location on the outskirts of Africa suggest that a Northern route was the preferred migratory path for *H. sapiens*, with Egypt as the final stop before leaving (Fig. 4.10A).

Among Y chromosome lineages, the M173 haplogroup, characterized by the M173 mutation within haplogroup R1 (R*-M173), is predominantly found in Eurasia with the exception of Southeast Asia. However, Egypt and Oman were shown to be the only non-Eurasian populations containing all three M173 subtypes. Egypt's proximal location to Eurasia, the presence of the M173 subtypes, and the absence of the R1*-M173 mutation in Eastern, Central, and South African populations suggest that the Levantine corridor was the most likely passage. Together, the absence of E3b*-M35 in Oman, distribution of the M35 lineages, and the

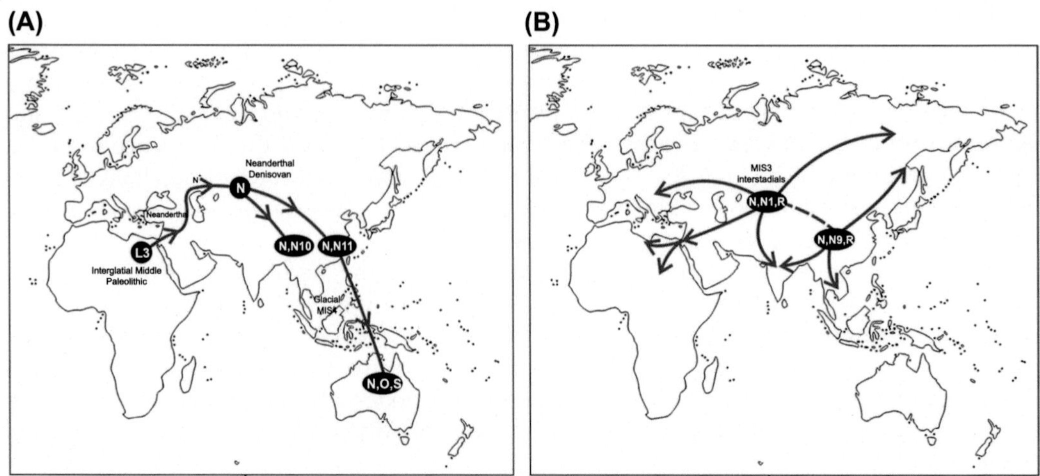

FIGURE 4.10 Possible movement of anatomically modern humans out and back into Africa through a Northern route and subsequent haplotypes. (A) Out of Africa migration, using the Northern route (B) secondary worldwide human expansions, deduced from the age and geographic distribution of L3 and N haplogroups. Dotted lines in B mean probable gene flow between populations from different dispersals. *(B) From http://journals.plos.org/plosone/article/figure/image?size=large&id=info:doi/10.1371/journal.pone.0129839.g001.*

presence of the R1*-M173 presence in non-Eurasian populations are suggestive of migratory movements between Africa and Eurasia by the Levantine corridor, insinuating genetic flow by way of the Horn of Africa as extremely limited.[55]

Luis et al.[56] analyzed 45 biallelic markers and 10 microsatellite loci on the nonrecombining region of the Y chromosome in extant males from Oman and northern Egypt. They concluded the Horn of Africa appeared to be of only minor importance in the human migratory movements between Africa and Eurasia based on the frequency distributions of E3b*-M35 and M173, both of which marked gene flow between Egypt and the Levant during the Upper Paleolithic and Mesolithic, supporting a Northern route for movements between Africa and Eurasia.

MULTIREGIONAL THEORIES

The multiregional theory proposes that the human species arose millions of years ago through interbreeding with other *Homo* species that existed in isolation in various parts of Africa and other regions of the world where *H. erectus* is found. Only through interactions and breeding within these populations did *H. sapiens* evolve until the modern human adapted to possess the necessary skills that would allow them to outlive their ancestors.

Within the multiregional theory, migratory routes are constantly debated. Genetic and fossil research show that *H. sapiens* may have left Africa either through the Levantine or the Horn of Africa corridors. Alternatively, the Horn of Africa would have consisted of a northeastern exodus, connecting Africa through the use of the Bab-el-Mandeb Strait to the Arabian Peninsula.

Additionally, climate changes, caused by the last glacial ice age resulted in the rising and lowering of sea levels during the last interglacial period. As a result, migrations through the Southern routes would have been highly dependent on favorable sea-faring conditions, leading to the belief that AMH may have been forced to leave the region in waves.

APPROACHING ASIA

It is highly probable that *H. sapiens* reached Asia through a southern dispersal route from Africa as determined by comparisons of both genomic and cranial phenotype data from modern Asian and African populations. As shown by past genetic research, Eurasian populations fall into the mtDNA M haplogroup, which is a subgroup of L3, the haplogroup from which all African populations originate (Fig. 4.5). Studies on Aboriginal Australian populations show a very strong relationship or trickle-down effect of genomes between the African, Eastern Asian, to present-day Aboriginal Australian populations. A southern dispersal suggests a possible timeline and initial group of *H. sapiens* dispersed by a coastal route approximately 62–75 kya, followed by a second group, through the Levant approximately 25–28 kya.[49]

Recent research suggests that *H. sapiens* arrived in East Asia through two separate North African expansions as determined by the geographical distribution of haplogroup D. It is believed that the first Southeast Asian expansion, which involved haplogroup D occurred around 60 kya, shown by geographical genetic distribution at higher frequencies in Tibet, Japan, and the Andaman Islands. Populations in the Andaman Islands are speculated to be part of a group of archaic African migrants from an initial wave of beachcombers because of the presence of modern Middle Paleolithic artifacts at Kara-Bom in Russia. A secondary expansion is believed to have occurred approximately 25–30 kya (after the first settlements in Australia) as suggested by geographical distribution of haplogroup O3 (an East Asian–dominated haplogroup). Genomic data of the 45 kyo, Ust'-Ishim (located in present-day Siberia) thighbone supports these two early migration theories by proposing that this modern individual belonged to a population that was not ancestral to either western or Eastern Eurasians.[57]

As observed in other population models, the greater the distance a population was from Africa, the more a decrease in genetic and biological diversity was observed, characteristic of a more recent, single dispersal, "Out of Africa" model. Alternatively, the decreased biological diversity may also be characteristic of the multiple dispersal model, suggesting that AMH left Africa sometime between 50 and 100 kya and traveled along the Arabian coast, and a secondary wave of *H. sapiens* left through the Levant around the same time and traveled into northern Eurasia.

Assessing biological and genetic distances using autosomal SNPs (to test spatial dispersal) between Asian and African populations showed that the multiple dispersal theory was the best supported model.[6] Among 10 populations, genetic difference was lowest between South African and Australian, Central Asian and North Indian populations. Similarly, a marginal phenotype difference was observed between South Africa and Australia.

Population divergence values revealed a large separation, 106,548 years between South African and Australian populations; however, a marginal difference of 77,815 and 75,592 calendar years was observed between South Africa, Central Asia, and Eastern Asia.[10] Together, the large genetic differences between South Africans and Australians, as compared with Southeast and South African and the Eastern Asian countries, would suggest a prolonged period of separation between Australians and South Africans, (see Chapter 8) as would be a consequence of the current model of a Southern Coastal route into Asia for AMH 62–75 kya.[54,58]

COASTAL MOVEMENT INTO AUSTRALIA

Although South Asia, as a territory, suffers from limited fieldwork, poor recording of data, and lack of systematic surveys, as well as undated sites, it is a region undoubtedly rich in archeological remains. In addition, it is also a strategic location in between the Near East and East which have been studied in more detail.

By about 250 kya, the subcontinent experienced a gradual technological transition from the large cutting lithics of the Late Acheulean to the prepared core and flake tools of the Middle Paleolithic. Evidence of this technological shift is seen in sites such as Bhimbetka in the Raisen District of Madhya Pradesh, Central India. This transition seems to represent an in situ process resulting from adaption to local conditions within the subcontinent. This capacity to adapt to habitats in the interior of the subcontinent by the migrants of the third phase of Out of Africa I signals the potential for transcontinental dispersals across India by hominins on their way to East Asia and Oceania.

The Indian subcontinent represents one of the first regions settled by AMHs and served as the main epicenter of dispersal to East and West Eurasia. Although the exact arrival time of AMHs to the subcontinent as part of the second movement (Out of Africa II) is still uncertain, evidence suggests that this dispersal originated in Northeast Africa and crossed the Bab-el-Mandeb Strait about 80 kya into the Near East. The range of departure time has been estimated between 50 and 100 kya. Once in the Near East, the migration continued to South Asia. mtDNA evidence provides time estimates of 75 to 50 kya for the arrival of AMHs to the subcontinent.[40]

There are a number of similarities between Y chromosome and mtDNA inheritance in AMHs with regard to the peopling of South Asia. For example, both sets of uniparental genetic systems indicate that the dispersals that led to the occupation of South Asia occurred soon after *H. sapiens* exited Africa into India and beyond. The absence of nucleotide differences in the gene-containing mtDNA among South Asian, Southeast Asian, and Oceania groups suggests a scenario of quick dissemination eastward during a time span of several thousand years. If the dispersal had been slow, the DNA would accumulate more mutations over the longer period of time.

Genetic evidence and the archaeological discoveries of human remains point to a coastal movement from southern Africa into Australia. The discovery of the "Australian Mungo Lady" estimated to be somewhere between 42 and 62 years old set the grounds for the initial explorations of human remains and burial sites in Australia. Using the human remains of a third Lake Mungo individual, researchers used radiocarbon dating to estimate the age of the individual to be somewhere between 62,000 ± 6000 years old. The Lake Mungo bodies had all been covered with red ocher, which was also used by archaic modern humans in burial sites in Africa. In addition to the red ocher used by both archaic African and Australian populations, comparable advanced tools, cave drawings, and sculptures were also found in various parts in Australia. Similarities in fossilized remains, in conjunction with the technological and cultural findings, indicate that the remains found in Australia were probable descendants of AMH.[59]

Comparing mitochondrial and nonrecombining Y chromosome DNA of fossilized *H. sapiens* remains and present-day Aboriginal Australians points to an early African dispersal sometime between 62 and 75 kya. Genome extractions and sequencing methods showed there was no evidence of recent genotype European admixture or contamination, eliminating the possibility of shared ancestry between Aboriginal Australians and Europeans, as was previously believed. Genome sequencing revealed more shared alleles between Aboriginal Australians and Asians, suggesting that *H. sapiens* arrived in Australia after traveling southward on the Asian coast to eventually form the present-day Aboriginal Australian populations.[60]

With knowledge that there was little to no relationship between European and Australian populations, further research was conducted to establish the claim that Aboriginal Australians are descendants from an originally African population who made their way up the coast from Asia (see Chapter 8).

BACK TO AFRICA MOVEMENT

Taking into the consideration the discrepancy of the migratory patterns of AMH as suggested by both mtDNA and Y chromosome data, there is some speculation of a back-to-Africa migration that would have contributed to the diversity of the human genome (Fig. 4.10B). The presence of mutations in African populations in countries such as Ethiopia and Yemen that are present only in Eurasian haplogroups seem to suggest that there was a backflow migration through the Horn of Africa causing discrepancies within the geographic haplogroup distribution. One example is the origin of mtDNA haplogroup M that has been related to its subclade haplogroup M1, which is the only variant of M found in Africa.[61] One possibility is that the presence of M1 in Africa is the result of a back migration from Asia which occurred sometime after the Out of Africa migration 40 kya.[62,63] It is also possible that M1 existed in very low frequency in Africa before AMH left Africa, but it is more likely the result of back migration.

A back-to-Africa migration from Asia to sub-Saharan Africa is supported by the analysis of Y chromosome haplotypes. Variations of 77 biallelic sites located in the nonrecombining portion of the Y chromosome were examined in 608 male subjects from 22 African populations. Correspondence analysis shows that three main clusters of populations can be identified: northern, eastern, and sub-Saharan Africans. The analysis suggests that a significant amount of the extant Khoisan gene pool from Southern Africa is East African in origin, and that Asia was the source of a back migration to sub-Saharan Africa. Y chromosomes with haplogroup IX appear to have been involved in the migration, the traces of which can now be seen mostly in northern Cameroon.[64]

Underhill et al.[35] created a Y chromosome phylogenetic tree of African subjects categorizing haplotypes into six distinct haplogroups based on the presence or absence of specific alleles within the phylogenetic tree. Researchers looked at haplogroup distribution to evaluate whether the genetic distribution could be attributed to more recent or archaic gene flow. Between Northern, Eastern, and Sub-Saharan Africa, sharing of haplotypes was limited; however, some mutations were found to be common in some areas but completely absent in other regions. For example, haplotypes with the M78 (present in northern and eastern Africa) and the M35 (predominantly eastern and southern Africa) mutations show genetic variation within distinct northern, eastern, and sub-Saharan African populations. Lineages carrying the M35, M34 (Eastern Africa), M78, and M81 mutations are only found in Northern Africa but completely absent or incredibly rare in Eastern Africa. These specific mutations, M34, M78, and M81, have also been found in European and Near Eastern populations, suggesting a high degree of geographical haplotype distribution within and out of Africa. A coalescence age for chromosomes carrying the M81 mutation was placed at approximately 2 kya, suggesting a more recent wide genetic expansion of M81 haplotypes in Northwestern Africa.[63,65]

In addition to the wide geographical distribution of E-M81, in Northwestern Africa but absent in other parts of Africa, the R-M207 and R-M173 mutations, (normally found in western Eurasia, Central, Northern, and Southern Asia), were found in high frequencies in sub-Saharan African populations such as Cameroon. Based on this Y chromosome data, the presence of a predominantly Asian mutation in sub-Saharan Africa seems to suggest a male-dominated backward migration from Asia to Africa. This is supported by the observation of K-M9 haplotypes in Asia and the reported discrepancies of haplotype O-M117. Haplotype O-M117 has been identified as an African haplotype with a coalescence age of 4.1 kya, possibly caused by an ancient Asian migration to Africa. Haplotype O-M117 is believed to be a part of the chromosomes present in Europe that derived from an ancient Eurasian migration. It is speculated that a prior Asian migration into both Europe and Cameroon may be responsible.[64]

Evidence of a back-to-Africa migration from Asia is further supported by the geographical distribution of haplogroup U5 (a branch of the North African haplogroup M), assumed to have originated in Europe sometime between 25 and 50 kya. This would seem to support the multiregional theory, suggesting that populations with various renditions of haplogroup U5 existed in various regions in Southeast Asia, and when climatic conditions allowed, these various populations migrated through the Levant reaching regions in Europe and North Africa.

Further evidence suggests an earlier backward migration from non-African countries into Africa through the Horn of Africa based on Y chromosomal and haplotype data. Identification of SNPs for African and non-African (Ethio-Somali) ancestries in the Horn of Africa revealed that those non-African ancestries are significantly different from non-African ancestries in North Africa, the Levant, and Arabia. The data showed that "typical" African populations in the Horn of Africa, namely Ethiopia, Eritrea, Djibouti, and Somalia had non-African ancestry based on Y chromosome and mtDNA data, hinting at a possibility of early migrations into the Horn of Africa. Taking into account mitochondrial, Y chromosome, paleoclimate, and archaeological data, researchers suggested that the time of the Ethio-Somali back-to-Africa migration was most likely prior to the onset of agriculture.[65]

Genetic data observations of Y chromosomes and haplogroups suggest the possibility of at least two non-African migrations into the Horn of Africa prior to 3 kya. First, the Horn of Africa populations contain characteristics of haplogroup E-M78, a subgroup of haplogroup E-V32, which is believed to have originated in North African populations almost 6 kya. Second, the presence of haplogroup T-M70 is present in moderate to high values in some Horn of Africa populations. Haplogroup T is found primarily in Northern Africa, the Near East, and Europe; however, it is believed to have originated in the Levant around 21 kya.[65,66]

Looking for shared gene identity and geographic distances between the Horn of Africa, Near Eastern, and North African samples revealed a decrease in genetic similarities with increases in distances between Arabian and Horn of African populations.[67] This would seem to indicate that there was long-term and steady gene flow between these populations. However, the most significant levels of gene flow were found between Horn of African,

Levantine Palestinian, and North African Algerian populations. It can be inferred from these data that secondary admixture of Arabian migrants with substantial non-African ancestry into the Horn of Africa may be responsible for the genetic-geographic distance gradient.[67]

WHAT DOES IT ALL MEAN

Clearly, at this time there is a lack of data to make any solid conclusions about the specific origins of AMH and their exodus out of Africa. New fossil evidence and comparisons of fossil remains from across Africa and beyond are beginning to form a new picture of when and where AMH first evolved and when they first left Africa. The past descriptions and distinctions of what constitutes archaic and modern humans is beginning to fade as new fossil discoveries emerge. As previously described, the fossil evidence from Jebel Irhoud, Morocco (dated around 300kya) has questioned previous dates of humans in Africa, and comparisons of the Irhoud fossils with the approximately 260 kya Florisbad fossil from South Africa show similar primitive ancestral features. In addition, similarities between the Jebel Irhoud fossils and those from Zuttiyeh and Tabun in Israel suggest that 300 kya corridors might have periodically linked northern Africa and western Asia. Additional fossil evidence and dating may begin to clarify the extent of the overlap between these individuals and the processes that may have led to AMH.[68]

Hershkovitz et al. recently discovered a fossil of a mouth part, a left hemimaxilla, with almost complete dentition at Mount Carmel, Israel. In addition to the jawbone, the sediments contained well-defined hearths, a rich stone-based industry, and animal remains. Analysis of the human remains and dating of the site and the fossil established an age of at least 177,000 years for the fossil. This finding represents the oldest member of Homo sapiens found outside Africa, and suggests that the northern route from Africa into the Levant, may have been the first route out of Africa.[69]

CONCLUSIONS

Although the genetic and archaeological evidence presented allow for inferences regarding *H. sapiens* migratory patterns and possibilities of secondary admixture leading to AMH, there is still much room for debate and the collection of more data. As discussed, the collections of fossils, artifacts, and genomic evidence from mtDNA, and Y chromosomes overwhelmingly support the "Out of Africa" II theory, but there is still debate over the recent single-origin theory, the multiple regional theory, and the various discrepancies associated with each model. Evidence of interbreeding with Denisovans and Neanderthals suggests that although we arose out-of-Africa, AMH have some multiregional contributions. It is still unclear if AMH left Africa in a single wave into Asia and eventually into Europe or multiple waves; one wave through Asia and into Oceania and another wave through a Northern route into the Levant. Evidence for both the Northern and Southern routes out of Africa indicate an early exodus around 100 kya; however, it is not clear if those early migrants lived and represent the AMH of populations today. Increasingly, geographical genomic patterns and genomic diversity supports the multiple wave theory, where AMH left Africa at various time points using both the Southern and Northern routes. New evidence also suggests that AMH may have also evolved in Northern Africa, near modern-day Morocco and may have migrated from there. Interest in these migration routes and the outcomes of dispersal are not only academic but also help explain the widespread rise of cities, advanced tools, agriculture, and modern technology throughout the world, which would not be possible without the expansion and migration of *H. sapiens* out of Africa. Learning more about the development of *H. sapiens* and their migration will not only allow scientists to better understand genome and phenotypic differences and similarities between extant aboriginal populations and urban dwellers but also pave the way for insight into medical and genetic advances. Understanding migratory routes and admixture between species creates a platform for understanding modern human's susceptibility to certain diseases and sickness. For example, what makes some populations more prone to ailments such as type 2 diabetes or lupus? What genomic differences enhance one population's ability to thrive in extreme conditions, while another is on the verge of extinction? Understanding migratory patterns gives clues to important cultural and genetic exchange information, to understand how AMHs came to populate the rest of the world and overcame physical and biological barriers that overpowered their other *Homo* ancestors.

References

1. Burton RF. 1821–1890. http://www.goodreads.com/quotes/498004-of-the-gladdest-moments-in-human-life-methinks-is-the.
2. Reyes-Centeno H, et al. Testing modern human out-of-Africa dispersal models and implications for modern human origin. *J Hum Evol* 2015;**87**:95–106.
3. Reyes-Centeno H, et al. Tracking modern human population history from linguistic and cranial phenotype. *Sci Reports* 2016;**6**:36645.
4. Chiaroni J, et al. Y chromosome diversity, human expansion, drift and cultural evolution. *Proc Natl Acad Sci USA* 2009;**106**:20174–9.
5. Templeton AR. Genetics and recent human evolution. *Evolution* 2007;**61**:1507–19.
6. Lopez S, et al. Human dispersal out of Africa: a lasting debate. *Evol Bioinf Online* 2016;**2**:57–68.
7. Weidenreich F. Some problems dealing with ancient man. *Am Anthropol* 1940;**42**:380–2.
8. Gunz P, et al. Early modern human diversity suggests subdivided population structure and a complex out-of-Africa scenario. *Proc Natl Acad Sci USA* 2009;**106**:6094–8.
9. Eriksson A, Manica A. Effect of ancient population structure on the degree of polymorphism shared between modern human populations and ancient hominins. *Proc Natl Acad Sci USA* 2012;**109**:13956–60.
10. Reyes-Centeno H, et al. Genomic and cranial phenotype data support multiple modern human dispersals from Africa and a southern route into Asia. *Proc Natl Acad Sci USA* 2014;**111**:7248–53.
11. Hublin J-J. New fossils from Jebel Irhoud, Morocco and the pan-African origin of *Homo sapiens*. *Nature* 2017;**546**:293–7.
12. Hublin J-J, Klien RG. Northern Africa could also have housed the source population for living humans. *Proc Natl Acad Sci USA* 2011;**108**:278–81.
13. Fadhlaoui-Zid K, et al. Genome-wide and paternal diversity reveal a recent origin of human populations in North Africa. *PLoS One* 2013;**8**:e80293.
14. Henn BM, et al. Hunter-gatherer genomic diversity suggests a southern African origin for modern humans. *Proc Natl Acad Sci USA* 2011;**108**:5154–62.
15. Schlebusch CM, et al. Genomic variation in seven Khoe-San groups reveals adaptation and complex African history. *Science* 2012;**338**:374–9.
16. Pickrell JK, et al. The genetic prehistory of southern Africa. *Nat Commun* 2012;**3**:1143.
17. Wong K. Meet Neo, a spectacular new fossil of *Homo naledi*. *Sci Am August* 2017:46–7.
18. Rito T, et al. 2013 the first modern human dispersals across Africa. *PLoS One* 2013;**8**:e80031.

19. Potts R, et al. Behavioral and environmental background to "out of Africa I" and the arrival of *Homo erectus* in East Asia. In: Fleagle JG, et al., editor. *Out of Africa I: the first hominin colonization of Eurasia*. Netherlands: Springer; 2010.
20. Bar-Yosef O, Belfer-Chohen A. From Africa to Eurasia –early dispersals. *Quart Int* 2001;**75**:19–28.
21. Hurcombe L. The lithic evidence from the Pabbi Hills. In: Dennell R, editor. *Early hominin landscapes in Northern Pakistan; investigations in the Pabbi hills*. New York, NY: Oxford University Press; 2004. p. 222–91.
22. Belmaker M, et al. In: Fleagle JG, et al., editor. *Out of Africa I: the first hominin colonization of Eurasia*. London, New York: Springer Dondrecht Heidelberg; 2010.
23. Carbonell E, et al. Out of Africa: the dispersal of the earliest technical systems reconsidered. *J Anthropol Archaeol* 1999;**18**:119–36.
24. Lordkipanidze D. A complete skull from Dmanisi, Georgia, and the evolutionary biology of early Homo. *Science* 2013;**342**:326–31.
25. Han F, et al. The earliest evidence of hominid settlement in China: combined electron spin resonance and uranium series (ESR/U-series) dating of mammalian fossil teeth from Longgupo cave. *Quat Int* 2017;**434**:75–83.
26. Ayala FJ, Cela-Conde CJ. *Processes in human evolution; the journey form early hominins to Neanderthals and modern humans*. New York, NY: Oxford University Press; 2017.
27. Bar-Yosef O, Belmaker J-G. Early and middle Pleistocene faunal and hominins dispersals through Southwestern Asia. *Quat Sci Rev* 2011;**30**:1318–37.
28. van der Made J. Biogeography and climate chnge as a context to human dispersal out of Africa and within Eurasia. *Quat Sci Rev* 2011;**223**:195–200.
29. Anton SC, et al. An ecomorphological model of the initial hominid dispersal from Africa. *J Hum Evol* 2002;**43**:773–85.
30. O'Regan HJ, et al. Hominins without fellow travelers? First appearance sand inferred dispersals of Afro-Eurasian large mammals in the Plio-Pleistocene. *Quat Sci Rev* 2011;**30**:1343–52.
31. Castañeda AS, et al. Wet phases in the Sahara/Sahel region and human migration patterns in North Africa. *Proc Natl Acad Sci USA* 2009;**106**:20159–63.
32. Mayer M, et al. Nuclear DNA sequences from the Middle Pleistocene Sima de los Huesos hominins. *Nature* 2016;**532**:504–7.
33. Llorente MG, et al. Ancient Ethiopian genome reveals extensive Eurasian admixture in eastern Africa. *Science* 2015;**350**:820–2.
34. Xue Y, et al. Human Y chromosome base-substitution mutation rate measured by direct sequencing in a deep-rooting pedigree. *Curr Biol* 2009;**19**:1453–7.
35. Underhill P, et al. Y chromosome sequence variation and the history of human populations. *Nature Genet* 2000:358–61.
36. Trombetta B, et al. Phylogeographic refinement and large-scale genotyping of human Y chromosome haplogroup E provide new insights into the dispersal of early Pastoralists in the African continent. *Genom Biol Evol* 2015;**7**:1940–50.
37. Loogvali E-L, et al. Explaining the imperfection of the molecular clock of hominid Mitochondria. *PLoS One* 2009;**4**(12):e8260.
38. Soares P, et al. The expansion of mtDNA haplogroup L3 within and out of Africa. *Mol Biol Evol* 2012;**29**:915–27.
39. Fernandes V, et al. The Arabian cradle: mitochondrial relicts of the first steps along the southern route out of Africa. *Am J Hum Genet* 2012;**90**:347–55.
40. Rajkumar R, et al. Phylogeny and antiquity of M macrohaplogroup inferred from complete mt DNA sequence of Indian specific lineages. *BMC Evol Biol* 2005;**5**:26–531.
41. Jobling, et al. *Human evolutionary genetics. Garland Science*. NY: Taylor and Francis group LLC; 2014.
42. Wei W, et al. A calibrated human Y chromosome phylogeny based on resequencing. *Genome Res* 2013;**23**:388–95.
43. Mendez FL, et al. The divergence of neandertal and modern human Y chromosomes. *Am J Hum Genet* 2016;**98**:728–34.
44. Rosenberg TM, et al. Humid periods in southern Arabia: windows of opportunity for modern human dispersal. *Geology* 2011;**39**:1115–8.
45. Oppenheimer S. Out-of-Africa, the peopling of continents and islands: tracing uniparental gene trees across the map. *Philos Trans R Soc Lond B Biol Sci* 2012;**367**:770–84.
46. Lambeck K, et al. Sea level and shoreline reconstructions for the Red Sea: isostatic and tectonic considerations and implications for hominin migration out of Africa. *Quat Sci Rev* 2011;**30**:3542–74.

47. Stewart JB, Chinnery PF. The dynamics of mitochondrial DNA heteroplasmy: implications for human health and disease. *Nat Rev Genet* 2015;**16**:530–42.
48. Groucutt HS, et al. Rethinking the dispersal of *Homo sapiens* out of Africa. *Evol Anthropol* 2015;**24**:149–64.
49. Seddon C. *Humans: from the beginning*. Glanville publications; 2015. p. 117–20. Open Library https://openlibrary.org/publishers.
50. Rose JI, et al. The Nubian complex of Dhofar, Oman: an African middle stone age industry in Southern Arabia. *PLoS One* 2011;**6**:e28239.
51. Crassard R, Hilbert YH. A Nubian complex site from Central Arabia: implications for Levallois taxonomy and human dispersals during the upper Pleistocene. *PLoS One* 2013;**8**:e69221.
52. Liu W, et al. The earliest unequivocally modern humans in southern China. *Nature* 2015;**526**:696–9.
53. Liu W, et al. A geographically explicit genetic model of worldwide human-settlement history. *Am J Hum Genet* 2006;**79**:230–7.
54. Pagani L, et al. Tracing the route of modern humans out of Africa by using 225 human genome sequences from Ethiopians and Egyptians. *Am J Hum Genet* 2015;**96**:986–91.
55. Rowold D, et al. Mitochondrial gene flow indicates preferred usage of the Levant corridor over the Horn of Africa passageway. *Japan Soc Hum Genet* 2007;**52**:436–47.
56. Luis JR, et al. The Levant versus the Horn of Africa: evidence for bidirectional corridors of human migrations. *Am J Hum Genet* 2004;**74**:532–44.
57. Fu Q, et al. Genome sequence of a 45,000-year-old modern human from western Siberia. *Nature* 2014;**514**:445–9.
58. Bolus M. Dispersals of early humans: adaptions, frontiers, and new territories. In: *Handbook of Paleoanthropology*. 2015. p. 2372–90.
59. Bowler JM, et al. Pleistocene human remains from Australia: a living site and human cremation from Lake Mungo, Western New South Wales. *World Archaeol* 1970;**2**(1):39–60.
60. Malaspina A-S, et al. A genomic history of Aboriginal Australia. *Nature* 2016;**538**:207–13.
61. Metspalu M, et al. Most of the extant mtDNA boundaries in South and Southwest Asia were likely shaped during the initial settlement of Eurasia by anatomically modern humans. *BMC Genet* 2004;**5**:26–9.
62. Gonzalez, et al. Mitochondrial lineage M1 traces and early human flow back to Africa. *BMC Genom* 2007;**8**:223–7.
63. Garcea E. Dispersals out of Africa and back to Africa: modern origins in North Africa. *Quat Int* 2016;**408**:79–89.
64. Francesca T, et al. Early modern human dispersal from Africa: genomic evidence for multiple waves of migration. *Invest Genet* 2015. https://doi.org/10.1186/s13323-015-0030-2.
65. Cruciani F, et al. A back migration for Asia to sub-Saharan Africa is supported by high resolution analysis of human Y-chromosome haplotypes. *Am J Hum Genet* 2002;**70**:1197–214.
66. Thorne A, et al. Australia's oldest human remains: age of the Lake Mungo 3 skeleton. *J Hum Evol* 1999;**36**:591–612.
67. Hodgson JA, et al. Early back-to-Africa migration into the Horn of Africa. *PLoS Genet* 2014. https://doi.org/10.1371/journal.pgen.1004393.
68. Stringer C, Galaway-Witham J. Paleoanthropology: On the origin of species. *Nature* 2017;**546**:212–14.
69. Hershkovitz et al. The earliest modern humans outside Africa. *Science* 2018;**359**:456–59.

CHAPTER

5

The Settlement of the Near East

To one who believes in the historical mission of Palestine, its archaeology possesses a value which raises it far above the level of the artifacts with which it must continually deal, into a region where history and theology share a common faith in the eternal realities of existence. ***William Foxwell Albright***[1]

SUMMARY

The region encompassing what is known today as the Near East is of special interest for the study of human evolution. The Arabian Peninsula and the Levant were the immediate points of entry for hominins traveling during the various Out of Africa episodes, and as such, these areas may have provided layover opportunities and incubation periods for the development of new technologies necessary for traveling and survival in the diverse habitats of Asia and Europe. Owing to its unique narrow topography, it has acted as a tapering funnel, canalizing hominin paths and restricting passageways to small areas, thus, facilitating interactions among various hominin populations. There is even evidence of cohabitation of specific shelters and caves by different hominin groups at various time periods. These close contacts undoubtedly brought about transfer of technology and culture and may even have served as a conduit for interbreeding that led to genetic flow among different populations. Specifically, the Near East is particularly unique in the numerous signals of *sapiens* and Neanderthal interactions in the Middle Paleolithic cultural framework. The Near East was, in fact, a sort of a melting pot of ideas, technology, and DNA. In more recent times, the Near East witnessed one of the most dramatic transitions experienced by our species: the transformation from a hunter-gathering lifestyle to agriculture and animal husbandry.

The Near East has been subjected to various incursions by hominins beginning at least with *Homo erectus* 2 million years ago (mya). *Homo sapiens* settled the region about 125,000 years ago (ya) and then suddenly disappeared from the Near East. More recent discoveries from a 300,000-year-old set of human fossils from Jebel Irhoud in Morocco and a 177,000–194,000 year-old *Homo sapiens* upper jaw from the Misliya Cave in Israel support the notion that *sapiens* probably began traveling out of Africa and settling the Near East at least 50,000 years earlier than previously thought. It is unclear whether these early *sapiens* vacated the Near East and moved on to East Asia, as recent findings from Daoxian, Zhirendong, and Xujiayao in current-day China dating to 120,000, 100,000, and 125,000 ya, respectively, suggest. Recent genetic evidence

Ancestral DNA, Human Origins, and Migrations
https://doi.org/10.1016/B978-0-12-804124-6.00005-7

also supports the notion that these early humans did not become extinct because their DNA persists in some contemporary human populations such as the Papua New Guinea aborigines of Oceania. Alternatively, they may have been forced to retreat back to Africa as a result of climatic changes and/or competition with other groups such as the Neanderthals. Although the extent to which and the nature of the interactions between *H. sapiens* populations and Neanderthals are not totally clear, the information gathered from the Near East sites indicates that the two groups knew of each other. The fact is that both groups of hominins occupied the same real state for at least 45,000 years. It is likely that the various hominins in the Near East interacted with each other depending on the specific circumstances at hand and the specific individuals involved, just as humans do today, at times collaborating, perhaps out of necessity, for example, during hunting, with the occasional belligerent confrontations. It is even possible that cooperation was driven at times by altruism and compassion.

Some of the hominin dispersals progressed from the Near East and traveled toward Europe, Asia, Oceania, the Indian Ocean, and America. Anatomically modern humans (AMHs) were not the only hominin group that left Africa for Eurasia. Today, we know that different members of our lineage moved into the Near East on a number of occasions. The various hominin dispersals out of Africa can be subdivided into two major epochs: (1) the ancient migrations involving *H. erectus* and earlier groups such as *Homo habilis* and possibly australopithecines and (2) the spreads that occurred subsequent to the *erectus'* occupation, which include the early *sapiens* and the AMHs.

In this chapter, in addition to the hominin's anatomical characteristics and the fauna present in the archeological remains, the cultural traditions, tools, and artifacts are compared among the different species and populations, as a function of time. Beginning with the potential settlements by *H. habilis* and then *erectus*, the major archeological sites are described including an account of the climatic conditions that allowed the migrations into the Near East. Subsequent to *erectus'* tenure in the region, putative *Homo heidelbergensis* incursions are described. Together, these pre–*sapiens* waves are known as Out of Africa I. The first wave of dispersals by *H. sapiens*, which resulted in an occupation from 125,000 to 80,000 ya is discussed as the first phase of Out of Africa II, and the *sapiens* involved in this first settlement are referred to as early *H. sapiens*. After an absence of about 20,000 years, *sapiens* returned to the Near East to stay as AMHs as part of the second phase of Out of Africa II dispersal.

THE FIRST HOMININ SETTLERS OF THE NEAR EAST

Climatic Conditions

Dispersal of organisms is strongly influenced by the environment. Although at times animals migrate due to harsh conditions in their habitat in search for subsistence, generally speaking, species tend to move into new territories where water and food are available. Poor decisions in the route or speed chosen may lead to death of individuals or even the extinction of the group. Thus, when examining patterns of hominin dispersal, knowledge of the local climate and ecology, in geological time, is paramount. This is especially the case in situations when the terrain is notoriously challenging as in the Near East, where prolonged periods of aridity predominate with short sporadic epochs of humidity.

Given that several species of hominins originated in Africa and subsequently traveled to Eurasia, these migrants had to travel through the Near East. Thus, the region has been a strategic location during hominin evolution and a bridge of dispersal from and to Africa. The region has served as a sort of geographical nexus or hop between the two continents for species of the genus *Homo* and possibly for earlier groups such as australopithecines. Currently, the outcome of the various hominin incursions into the Near East is not clear. In many instances, it is not known if the final outcome was extinction, return to Africa, migration to the east, or population continuity in the region, the latter implying adaptation to a harsh dry environment. What is clear is that the environment in the Near East has been a determining factor in hominin dispersal out and back to Africa.

What is known about the region today is that the Near East has experienced extreme environmental fluctuation during the Quaternary period, 2.6 mya to the present. And although the Near East, and especially Arabia, is generally thought as a perpetual arid desert land, the environment for the past 3 million years has been cyclical in terms of precipitation as well as the abundance of flora and fauna. That is, periods of rain have alternated with epochs of extreme dry weather. Precipitation allowed hominin settlements and transit to the east. Aridity prevented settlement and movement through it.

Fluctuations in climatic conditions in the Near East occur in two tiers. One is the broad level of dry glacial periods and interglacial wet phases. In the other one, oscillation in weather conditions takes place at a finer scale in shorter periods of time within glacial arid as well as interglacial humid periods.[2] Most of our knowledge regarding these cyclical climatic changes at both levels includes only the last 350,000 years, especially the most recent half of this time period (Fig. 5.1). Notice from Fig. 5.1 that the wet epochs occur periodically, but their duration

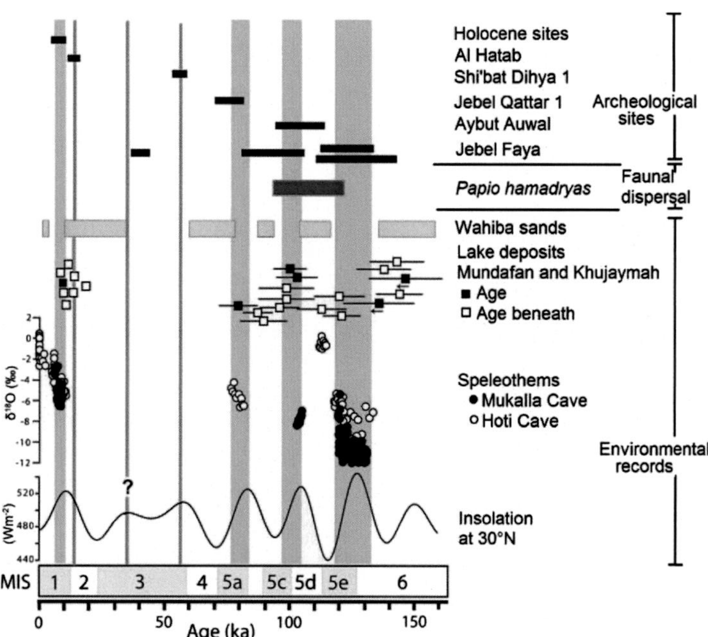

FIGURE 5.1 Arabian environmental, faunal, and archeological records over the last 150,000 years. Thick vertical blue lines represent key humid periods, thin gray lines represent short humid episodes. Note the correlation of archeological assemblages with periods of increased rainfall. *From https://www.researchgate.net/figure/227710326_fig5_Figure-2-Arabian-environmental-faunal-and-archeological-records-over-the-last-150-kyr.*

and the time intervals between them differ. For example, for the past 150,000 years there have been four humid periods about 5000 to 15,000 years in duration that correlated with the presence of archeological sites (Fig. 5.1).

The changes from dry to wet conditions and vice versa occurred fast in the Near East, within the span of a few centuries. So, any population inhabiting a lush, densely vegetated habitat will be caught dramatically, in a matter of generations, in extreme aridity risking extinction. It is known that these oscillations result from the earth's wobble around its axis, which takes about 26,000 years. In other words, our planet makes a full rotation around its axis during this period of time. Gravitational forces within our solar system cause this wobbling effect or tilt to occur. It seems that the earth wobble affects the movement of warm upper waters to the north (e.g., the Gulf Stream experienced in eastern United States) and the return of cold deep water toward the south. Interfering with this conveyor belt of water promotes dry conditions in the Near East as well as in Northern Africa creating the Saharan desert. Currently, the earth is about 7000 years into an arid phase of the cycle. Thus, at the end of this dry epoch, the Near East will start to experience a wet phase with frequent monsoons and the presence of abundant rivers and lakes (Fig. 5.2).

It has been postulated that these dramatic oscillations of extreme humid and arid weather conditions allowed for the expansion and contraction of organisms' ranges.[3] Populations expanded during wet periods and contracted during dry periods.

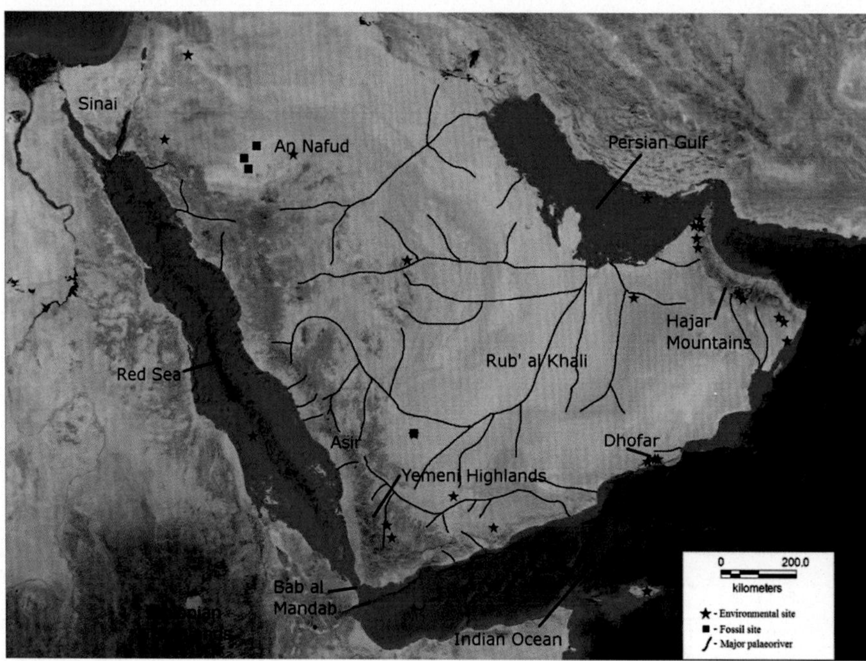

FIGURE 5.2 Geography of the Arabian Peninsula. The locations of Pleistocene fossil localities and key sites for paleoenvironmental reconstruction are shown. Note the southern and coastal concentration of the latter. Black lines depict major paleorivers. *From https://www.researchgate.net/figure/227710326_fig2_Figure-1-Geography-of-the-Arabian-Peninsula-The-locations-of-Pleistocene-fossil.*

In other words, this cyclical process dictated specific times of transit and settlement of hominins in the Near East. It is likely that the Near East acted as a door, opening and closing for the passage of populations dependent on the climatic conditions. Also, it is likely that the settlement and episodes of migrations of hominins across the Near East was restricted to the wet phase of the cycle as suggested by Fig. 5.1. Because a number of the important events in hominin evolution involved the Near East it is likely that these cyclical changes, generated by the earth's wobble, not only regulated the dispersal of hominins out and back to Africa to specific times but also affected the dynamics of human evolution and accelerated the speed of cultural, technological, and biological changes. In fact, it turns out that the recurrent wet periods coincide with known migratory events of various hominin forms including *H. erectus*, *H. heidelbergensis*, and early modern humans; in the case of *erectus* and modern humans on more than one occasion. It has been suggested that these wet–dry cycles have been particularly determinant in hominin migrations and recent human evolution during the last 200,000 years.[4]

Migration Signals

Early hominins migrated into the Near East from Africa on at least three occasions. During the first dispersal, migrants practicing Oldowan toolmaking techniques produced primitive artifacts such as choppers in the Near East about 1.8 mya. The second wave, approximately 1.4 mya, consisted of travelers practicing a more sophisticated lithic style, the early Acheulean industry. The third migration was made up of various cleaver-producing Acheulean groups about 0.8 mya. Together, these three waves are known as Out of Africa I.[5] The first and second waves of migrants could have been australopithecines, *H. habilis*, and/or *H. erectus*. Around 120,000 ya, hominins used the Near East corridor for a fourth time to disperse into Eurasia in what is known as the first Out of Africa II dispersal.

It is currently thought that the Strait of Gibraltar was not used significantly as a passageway for hominin migrations out of Africa. It seems that as the strait was only 10 km wide at the time of the potential crossings, the strong currents flowing into the Mediterranean may have prevented swimmers or rafters from reaching Europe. Another putative route out of Africa was across the Strait of Sicily. Currently, a 145-km body of water separates Sicily from Tunisia, North Africa, and undoubtedly during the early Pleistocene, this strait was even narrower due to the proximity of the earth's continents resulting from plate tectonic and lower sea levels. Furthermore, the Ain Hanech site dated at 1.8 mya in Algeria is just 400 km from the coast of North Africa and Oldowan tools have been uncovered in Sicily across the Mediterranean.[6]

Yet, currently, the Near East is the only region exhibiting extensive evidence for its use as an entranceway for early hominins into Eurasia. In spite of this unique distinction, information on the passage and settlement of the region is limited. This lacuna of knowledge is particularly evident when considering pre–*H. sapiens* hominin groups known to have migrated out of Africa. Prior to the occupation of modern humans, only two archeological sites stand out in Southwest Asia. One is Ubeidiya, in current-day Israel, and the second is Dmanisi in the Caucasus, what is presently the Republic of Georgia. Although a number of reasons are responsible for this deficit of information (see below), it is not totally clear why such a limited number of sites have been discovered and even why so few have been studied, considering the migrational monopoly that the region had for at least two million years.

In the case of *H. erectus* and modern humans, it is known that the migrants moved eastward rapidly subsequent to their arrival at the Near East. *Erectus* originated in Africa about 1.9 mya and started migrating northward into the Near East shortly after, reaching the Caucasus by about 1.8 mya and the island of Java in Indonesia by 1.6–1.8 mya. With about 10,000 km of distance separating Northeast Africa from the island of Java in Near Oceania by land, *erectus* moved at a speed of about 1 km every 10 years.

In spite of its importance in hominin evolution, archeological research in the Near East has been neglected. There is limited information on reliable dates, strata designations, systematic analysis of artifacts, description of faunal and flora, and hominin fossils. This archeological void can be attributed to a number of factors, including minimal fieldwork, lack of stratification, because the artifacts were never buried, or, conversely, the sites have been covered by extensive sand dunes. Some of the sites detected base their dating on correlations with similar technologies and styles of artifacts discovered outside the Near East such as Africa. It is possible that the small number of sites discovered in the Near East and the high speed of dispersal are related phenomena resulting from the arid and harsh environment that generally predominated in Southwest Asia at the time of the migration by *erectus*. In other words, the scarcity of resources including food, freshwater, and shelter made survival challenging in the Near East and hominins' stay in the region more of a transitory nature. In addition, the topology of the Levant and the Arabian Peninsula, lacking major geographical barriers, would have facilitated further northward and eastward dispersal. Arabia extends gently westward to the Mediterranean Sea providing a gateway to Europe, northward into what is today Turkey and Central Asia, and eastward to Persia. This type of terrain would have allowed steady and swift crossing of an arid landscape and a speedy arrival to Europe as well as Central and East Asia. Further causation for the limited archeological data from the Near East is the political unrest experienced in the region in recent decades, particularly in the Arabian Peninsula, Iran, Afghanistan, and Pakistan.

Yet, in spite of the limited amount of information on the first settlers of the Near East, a number of unique archeological and genetic data have emerged in recent years. The archeological information pertains specifically to the species that first populated the region and their high degree of anatomical diversity. The initial discovery of the Dmanisi site in 1991 in a location about 90 km from Tbilisi, the nation's capital, was followed by the finding of a number of hominin remains, notably five skulls, with the last, number 5, being the most controversial. With a diminutive brain size of 546 cc^3 and a height of only 146 cm (4.79 ft), this individual is characterized by many scholars as a member of a pre–*H. erectus* group.[7] The other four skulls exhibit brain capacity in the range of 601–730 cm^3, close to the range of an adult *H. erectus* of 750–1250 cm^3. The high degree of morphometric diversity exhibited by the Dmanisi individuals from the same location and archeological strata indicates to many that at least two species, and may be up to four, coexisted side by side in the region around 1.8 mya. Furthermore, the brain and stature differences observed among the individuals suggest that the smallest individual was a *H. habilis* or *habilis*-like specimen or even a more primitive australopithecine. If these assessments are in fact correct, it is possible that *H. erectus* was not the first group out of Africa. Alternatively, it is possible that *erectus* evolved from a *H. habilis*-like ancestor in the Near East, about 2.0 mya or earlier. In either case, the importance of the Near East in hominin evolution is paramount. The dating of the remains in the Ubeidiya and Erq el-Ahmar sites in Israel is congruent with a very early migration of *H. erectus* or of a pre–*H. erectus* group migrating out of Africa.

The possibility that *H. habilis* or even a form of australopithecine migrated out of Africa has been receiving support in recent years from genetic analyses. In a study published in 2005, data from 25 genomic markers indicate that an expansion out of Africa took place around 1.9 mya.[8] This dispersal event corresponds closely to the original expansion of *H. erectus* out of Africa into Eurasia as reflected in the archeological and fossil record. Continuous genetic flow starting 1.5 mya between Asian and African populations is also evident using the same 25 loci. A second expansion involving interbreeding with autochthonous populations is signaled approximately 700,000 ya. This migration out of Africa coincides with the spread of the Acheulean culture from Africa into much of Eurasia. A third migration that corresponds with the Out of Africa episode of early modern humans is also detected about 100,000 ya. In all of these dispersal events, total Eurasian replacement by the migrants is strongly rejected by the statistical analyses performed, implying that at least some of the indigenous populations interbreed with the incoming African migrants.

Altogether, the archeological and genetic data indicate that hominin evolution was a complex process involving a number of coexisting species that may have interbred instead of total replacement of groups by the incoming travelers and the disappearance of the original residents. Findings from a number of disciplines including molecular biology, archeology, and physical anthropology do not support the old notion of total extinctions resulting from belligerent confrontations with a technologically and culturally superior group. Currently, although the possibility of sporadic adversarial conformations among natives and newcomers cannot be ruled out, and likely occurred, it is clear that not all interactions were hateful and apparently there was time for love among the locals and the invaders.

The Controversial *Homo habilis*

H. habilis (handy man) (Fig. 5.3) exhibited a collage of ancestral and derived characteristics that makes it difficult to ascertain whether it belongs with the genus *Australopithecus* or *Homo*.[9] *Habilis* may represent the most ancestral group within the genus *Homo*, and thus, the least human-like group within the genus. Its larger brain size compared with australopithecines and its ability to make tools set *habilis* apart from the former, whereas its diminutive body dimensions and other ape-like traits indicate affinity to australopithecines. Still other paleoanthropologists consider the taxon invalid, made up of a mixed group of specimens, some being *Australopithecus* and some being part of the genus *Homo*.[9]

The original set of remains was discovered by Louis and Mary Leakey at Olduvai Gorge in Tanzania during their excavation during the period of 1960–63. *H. habilis* lived between 2.1 and 1.5 mya, and most scholars are of the opinion that it predates *H. erectus*[10]. An intermediate hominin between *Australopithecus* and *H. habilis* is primarily known by a portion of a jaw dated to 2.8 mya found in Afar, Ethiopia.[11]

H. habilis was small (1.3 m tall or 4 ft 3 in) with a brain less than half the size (550–687 cm^3) of a contemporary human (1350–1450 cm^3) and about twice the size of an early australopithecines specimen (380 cm^3). It possessed ape-like characteristics such as long arms and upper body proportions. Yet, it possessed certain attributes that set them apart from the australopithecines from which it descended. These included a less protruding face and the ability to make tools. Most scholars believe that *habilis* was the direct ancestor of *H. erectus*, and the archeological data indicate that the two species coexisted in Africa.[12]

FIGURE 5.3 Reconstruction of *Homo habilis*. *From https://en.wikipedia.org/wiki/Homo_habilis#/media/File:Homo_habilis.JPG.*

Habilis is also controversial in other ways. Although it is still a highly polemical proposition among scholars, it has been advanced that *H. habilis* may have been the first group of hominins that settled the Near East. In other words, *H. habilis* may have preceded *H. erectus* in an out of Africa dispersal. Yet, no fossil remains from this species have been found in the Near East or anywhere outside Africa. Thus, in the absence of direct evidence in the form of bones and artifacts, why was such a notion put forth in the first place? Well, it turns out that two early species within the genus *Homo* discovered in different parts of the world, outside Africa, exhibit greater anatomical similarities to the more ancient *H. habilis* than to *H. erectus*[13]. The two early groups were *Homo dmanisi* (Fig. 5.4) from the Republic of Georgia in Southwest Asia and *Homo floresiensis* from the island of Flores in Near Oceania. The hominins from Flores were so diminutive in stature (averaging 3.5 feet or 1.1 m) that they were nicknamed hobbits and initially were thought to represent cases of hypothyroidism resulting in a type of endemic cretinism in the island. Other early interpretations of the fossils from Flores included endemic microcephaly or dwarfism. Yet, as additional specimens were discovered in 2014 in Flores, it became clear that *H. floresiensis* was a bona fide population and not a pathological condition. As far as *H. dmanisi* is concerned, the initial publication reporting the finding suggested a population in the Caucasus exhibiting extensive morphological heterogeneity.[14] But subsequent evaluation of the five Dmanisi skulls recovered argued for the existence of as many as four different species, one represented by short individuals possessing small brains, reminiscent of *H. habilis*.[15]

The parallelisms among *H. dmanisi, H. floresiensis*, and *H. habilis* include overall smaller body size, a much smaller brain, diminutive mandible, and overall cranial morphology that

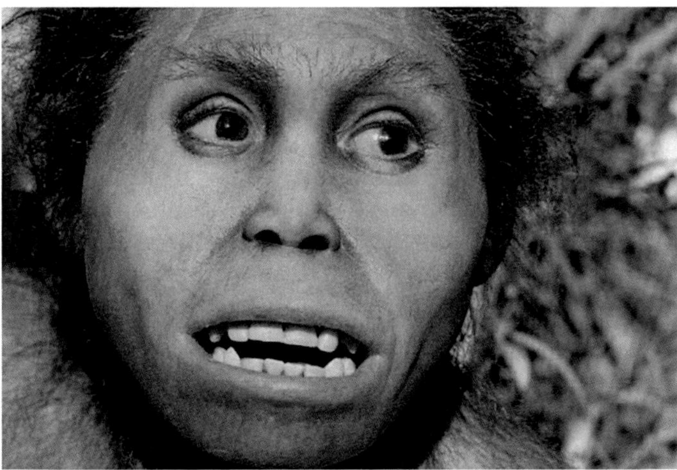

FIGURE 5.4 Reconstruction of *Homo dmanisi*. *From http://www.sci-news.com/othersciences/anthropology/science-dmanisi-human-skull-georgia-01474.html.*

contrasted *H. erectus*. In the case of *H. floresiensis*, its postcranial characteristics are more ape-like and similar to those of *H. habilis* or even australopiths than to those of *H. erectus*.[16] In other words, if *H. erectus* were the ancestor to *H. dmanisi* and *H. floresiensis*, it would require an evolutionary return to an ancestral state of small body and brain size that are rarely seen in the evolution of mammals. Thus, the geometric morphometric data allude to the possibility that *H. floresiensis* and *H. dmanisi* were direct descendants from a pre–*H. erectus* population of small-bodied hominins that dispersed out of Africa into Asia.[17,18] If *H. habilis* is indeed the earliest species within the genus *Homo* that preceded *erectus*, it is possible that *habilis* was the group that migrated out of Africa prior to *erectus* and then evolved into *dmanisi* and *floresiensis* in Asia. Yet, in spite of these arguments, *H. erectus* is still considered by the orthodoxy as the first hominin that migrated out of Africa about 1.9 mya.

Overall, fundamental to these considerations on the first settlers of the Near East is the need for an unbiased approach instead of thinking of just *Homo* species as candidates. All hominin groups should be consider as potential travelers into Southwest Asia. Even species of australopithecines could have migrated into the Near East from Africa. If the ancestors of small-brained and diminutive Dmanisi went all the way to the Caucasus approximately 1.8 mya, why not previous hominin groups, especially when the Sahara desert was a luscious forest, about 3.0–3.5 mya[19]? Currently, there is a tendency to question or even ignore any evidence predating the traditionally held date for the first migration of a hominin group.[20] In addition, any undated Oldowan artifacts (see next section for description) are routinely assigned to periods subsequent to the arrival of *Homo* to the Near East, 1.9 mya. In other words, these findings are invariably attributed to *H. erectus* without giving a second thought to the possibility that an earlier hominin such as *H. habilis* could have been responsible. Furthermore, previous dates associated with a given cultural tradition in a certain locale are used to tendentiously assign the antiquity of sites lacking reliable and specific direct dates. Thus, the possibility exists that the Oldowan artifacts from Erq el-Ahmar in present-day Israel (see section on Erq el-Ahmar below) were not contributed by *H. erectus* but by some earlier group such as *H. habilis* or a species of australopithecines. This is a real possibility considering that the Oldowan industry was initiated in Africa by *H. habilis* and *Australopithecus garhi* and was responsible for the lithic tools in Northeast Africa about 2.6 mya.

Also, although the traditional view is that *H. dmanisi* and *H. habilis* did not contribute genetically by introgression to the modern human population, scientists are beginning to uncover evidence that suggests that modern humans coexisted and interbred with other *Homo* species. In the case of *H. habilis,* it seems that it was able to interbreed with modern humans on several occasions. Investigators argue that *habilis* and *sapiens* interbreed extensively and regularly in Africa. The genetic data suggest several thousands of introgression events.[21,22]

Homo erectus

H. habilis was the first member of the genus *Homo*. It descended from an australopithecine group about 2.6 mya, and it became extinct about 1.4 mya. *H. erectus* is thought to have evolved from *H. habilis* about 2.3 mya in East Africa. Thus, *erectus* and *habilis* lived side by side in Africa for almost 1 million years. This cohabitation likely extended into Eurasia. *Erectus'* tenure extended to as recent as 70,000 or 60,000 ya if *H. floresiensis* was in fact a regional subspecies on the island of Flores. With a temporal range of approximately 2.0 my, as compared with just 200,000 for modern humans, the group was very successful having survived 10 times longer. During its existence, *erectus* underwent a number of anatomical, cultural, and technological transformations. For example, early *H. erectus* groups had bodies and brains not much larger than australopithecine species, whereas more recent forms possessed larger brains, about 1100 cm^3 or around 69% the size of the modern human brain and higher statures, reaching 180 cm or 6 feet. During its evolution, this extremely successful group improved on toolmaking technologies and discovered how to initiate and control fire and utilizing it for cooking. There is even evidence for compassion and empathy for the needy as is evident in one of the hominin remains from Dmanisi that belonged to an elderly male who had lost all but one of his teeth some years before his death. This individual could not have remained alive without the support of other clan members.

As a species, *erectus* rapidly dispersed within and out of Africa.[23] Remarkably, the orthodoxy believes that *erectus* reached Dmanisi in Southwest Asia about 1.85 mya, a date contemporaneous with the earliest African *erectus* specimens. This migration is known as the first Out of Africa I dispersal. Notably, by approximately 1.7 mya *erectus* arrived in northern China. The range of *erectus* was considerably more extensive than the territory occupied by Neanderthals. It included Africa, Europe, most of Asia, and parts of Near Oceania (Fig. 2.12 in Chapter 2). Since the original discovery of the species in Java by the Dutch anatomist Eugène Dubois in 1886, numerous specimens have been discovered throughout its range. For the purpose of the discussion in this chapter, *H. erectus* is the same species as *Homo ergaster* and the direct ancestor of *H. heidelbergensis, Homo neanderthalensis,* and modern humans. The most informative archeological sites in the Near East are described below. For a description of *H. erectus'* characteristics see Chapter 2.

Dmanisi

As far as premodern human occupation in the Near East, Dmanisi represents the flagship of sites. In a very important geographical area, at a nexus of continents that is relatively void of archeological evidence, Dmanisi, in the Caucasus, stands out as an area that provides a richness of artifacts as well as faunal and hominin remains unparallel in the Near East. Near the border with Armenia, the town of Dmanisi and adjacent areas have provided an abundance of prehistorical and historical relics since the excavations began in the early part of the 20th century. In 1984, the site acquired prominence when primitive stone tools and hominin remains were discovered. Intense interest in the settlement has dramatically increased since then.[24]

During the span of 14 years, between 1991 and 2005, the dig has provided evidence of an early *H. erectus* settlement dating back to 1.85 mya. The human remains were dated directly by paleomagnetic (record of the Earth's magnetic field in materials), potassium–argon (relative proportions of radioactive potassium-40 and its decay product, argon-40), and argon–argon (proportion of argon-40 to argon-39) dating methods.[25] The hominin occupation at Dmanisi lasted about 80,000 years. Significantly, this age estimate makes the settlement in the Caucasus contemporaneous to the genesis of *erectus* in Africa and makes us wonder about the migrational speed of *erectus,* or earlier hominins (e.g., *habilis*), through Arabia *en route* to the northern regions of the Near East.[26] Five skulls and postcranial skeletons have been unearthed. These remains belonged to one or several hominin groups at an evolutionary phase just subsequent to the transition from the genus *Australopithecus* to the genus *Homo* (Fig. 2.4 in Chapter 2). As previously discussed in Chapter 2, the hominins found in Dmanisi represent a collage of ancestral and derived characteristics seen in earlier and later hominin species. Some of the observed anatomy reflects the *Australopithecus* condition while others the *Homo* morphology. For example, these early *erectus* individuals exhibited ancestral brain cases and upper body structures but a vertebral column and lower extremities more similar to modern humans. As previously argued in Chapter 2, this medley of combinations of primitive and advanced morphological traits suggests an organism still retaining an upper body and hands for climbing from trees while below the waist the adaptations for upright posture were already apparent and progressing.

Furthermore, the skeletal heterogeneity observed among the Dmanisi remains (Fig. 5.5) is so profound that scholars are divided on their opinion on whether the individuals found in the same

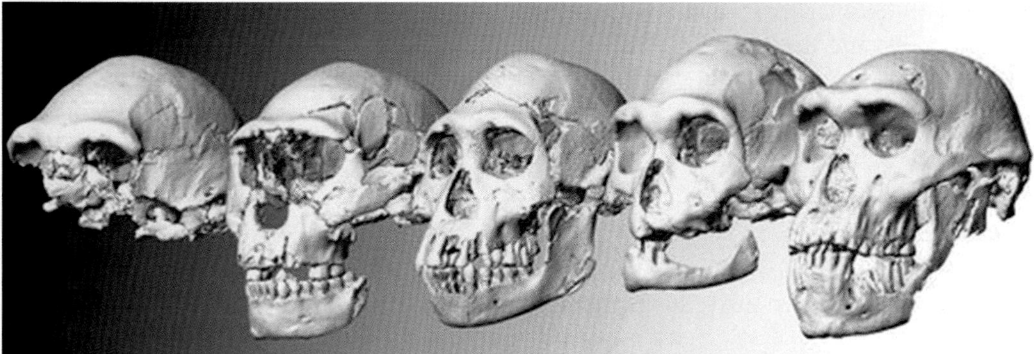

FIGURE 5.5 The five Dmanisi skulls. *From http://www.sci-news.com/othersciences/anthropology/science-dmanisi-human-skull-georgia-01474.html.*

stratum, within the span of about 20,000 years, were members of a single population exhibiting an unprecedented degree of anatomical diversity or, conversely, they represent hominins belonging to different cohabitating species, possibly interbreeding and exchanging DNA. Furthermore, as previously mentioned in this chapter, it is possible that some of the residents at Dmanisi were *habilis* or australopithecines. In the context of the remains found throughout its extensive geographic range, *H. erectus* (Fig. 2.12 in Chapter 2) exhibited a high degree of morphological diversity.

The last cranium found at Dmanisi, known as skull 5, initiated a debate that still fuels interest in the study of hominin evolution. Skull 5 exhibits a pointed face, thick brow ridges, and a brain case only 546 cc, about the size of an australopithecine male's brain and one-third the size of a modern human skull. It is also only 75% of the size of the largest skull found at Dmanisi.[14] It is significant that although skull 5 is diminutive, the size of the individual was within the range limit of modern variation. Altogether, the morphological diversity observed among the five skulls is so high that it encompasses the variability observed among a number of early *Homo* groups currently classified as distinct species, such as *H. habilis* and *Australopithecus (Homo) rudolfensis*. For example, the encephalization quotient (a measure of brain size in relation to body size) for the Dmanisi specimens ranges from 2.4 to 3.13, values at the lower end for African *H. erectus* and more similar comparable with *H. habilis* or australopithecines.[14] If indeed anatomical heterogeneity in time and space was an *erectus* hallmark, then it is possible that various other early *Homo* groups and even some australopithecines, throughout their ranges, that have been assigned as independent taxa may, in fact, be members of only one very polymorphic *H. erectus* species, possibly comprised of multiple geographical races.

Erq el-Ahmar

A number of locations signal the occupation of the Near East by *H. erectus* or an earlier species such as *H. habilis*. One of them is the Erq el-Ahmar site in which the human artifacts have been dated to about 2.0 mya. This settlement may have been older than Ubeidiya (see paragraph below), and both may represent examples of the earliest hominin presence in the Near East.[27] This dig encompasses archeological remains from the Early Pleistocene period belonging to the Oldowan lithic tool industry (Fig. 5.6). This toolmaking tradition is characterized by very simple cores derived from chunks of rock usually made up of quartz, quartzite, basalt, obsidian, flint, or chert. The resultant debitage (broken flakes produced by striking the core with another rock) was used to produce a large range of sharp-edged cutting and chopping tools, and the remaining core may not have been purposely shaped into a certain shape but probably came about as a by-product of generating the flakes. From this location, chopping tools, flakes, and pebble artifacts have been recovered.

Ubeidiya

Ubeidiya was a lakeside settlement in proximity to the Jordan River. The Ubeidiya site is located near Erq el-Ahmar, a few kilometers to the north, in the Jordan Rift Valley in present-day northeast Israel. Curiously, this region is regarded by some clergies as the location of the "the garden of Eden." The old age of the settlement, the abundance of artifacts, and the fossil remains found suggest that Ubeidiya signals one of the significant initial sorties of hominins out of Africa into Eurasia.

FIGURE 5.6 Lithic tool from the Oldowan industrial tradition. *From http://www.aggsbach.de/2014/08/chopping-tool-from-israel/.*

FIGURE 5.7 An Acheulean hand axe. *From https://en.wikipedia.org/wiki/Acheulean#/media/File:Biface_Cintegabelle_MHNT_PRE_2009.0.201.1_V2.jpg.*

Based on the location of the artifacts, the animal remains present, and pollen data, it seems that the settlement was located by the shore of a lake in a habitat that consisted of a hilly Mediterranean mixed forest and open grasslands (Fig. 5.8).[28] Over 80 occupation levels and a total of six major assemblages have been identified with a fauna that included hippopotami, crocodiles, deer, horses, elephants, rhinoceroses, pigs, gazelles, primates, baboons, and a number of carnivores. There are 45 animal genera represented in the assemblages including species originally from the Near East (n = 11) and East Africa (n = 6). The rest were imports from sub-Saharan Africa, North Africa, Europe, and Asia. This kind of fauna profile was found throughout the Near East all the way north to the Caucasus and would necessitate a luscious flora for survival.[29] Clearly, this type of fertile landscape and the organisms present indicate a much more benign environment compared with the arid conditions of today and would have been conducive to hominin settlements. The initial age estimates of Ubeidiya were quite inaccurate encompassing a wide range of dates (0.8–2.6 mya). The area is considerably disturbed with contamination from both older and younger strata as a result of tectonic stress (movement of the Earth's crust). It was not until fauna comparisons were made between animals in Africa and Europe that a more clear time period of about 1.4 mya was assigned.[30]

Curiously, a thin layer of sediment embedded with bones and artifacts characterizes the campsite at Ubeidiya. Scholars have labeled these thin pavement strata a "living floor." The name "living floor" hints at a surface with a utilitarian function, such as sitting or sleeping, made by hominins. Over the years since the discovery, this covering in certain areas of the settlement has been the subject of debate. On one hand, many researchers believe that the deposition of the overlay is purely accidental with no predetermined purpose while on the other hand, others find in the film a planned design. It is made up of basalt, limestone, and flint with clay overlaid. The tools on or within these sediments were of the same type as the ones throughout other parts of the settlement. Another telling detail is that these unusual surfaces are found

FIGURE 5.8 Ubeidiya habitat, fauna, and flora. *From https://www.google.com/search?q=ubeidiya+image&tbm=isch&tbo=u&source=univ&sa=X&ved=0ahUKEwj2zJCb25naAhVOJKwKHX34ANIQsAQIKA&biw=1440&bih=740#imgrc=Gf5MVtnR-bx0xM:.*

close to the shores of the extinct lake. This body of water retreated and expanded as a function of precipitation, and therefore, it is difficult to assess the location of the so-called living floors in relation to the shoreline at any given time. However, it is unlikely that hominins would set up camp and build the living floors close to the lake, near the water, in an area where they would be exposed and ambushed by predators such as carnivores and crocodiles.

The early fossil findings at Ubeidiya included human remains, specifically four cranial pieces and two teeth.[31] These were attributed to *H. erectus* and were dated to about 1.4 to 1.0 mya and thus are somewhat younger than the faunal material. More recently, a right lower incisor was discovered dating to the same period and may have belonged to *H. erectus* or *H. habilis*. This tooth exhibits striking similarities to incisors from 1.5 mya fossils from Lake Turkana in Ethiopia and to the 1.8 mya Dmanisi skull 2. Thus, they may belong to the same *habilis* or early *erectus* species that had a geographical range including Northeast Africa, the Near East and the Caucasus.

The artifacts at Ubeidiya included 10,000 lithic tools, including hand axes, chopping tools, denticulates (fine blades having a toothed margin), and scrapers, belonging to the early Acheulean style.[32] These tools were made from basalt, flint, and limestone. At Ubeidiya, specific types of stones were preferred by the inhabitants to make certain utensils. For example, bifaces were made from basal and chopping tools were mainly made from flint. This tool tradition is thought to derive from the more ancient Oldowan technology about 1.76 mya in Africa by *H. habilis*. Acheulean hand axes are typically oval and pear shaped (Fig. 5.7) and are associated with *H. erectus* remains across Africa, West Asia, South Asia, and Europe. Yet, the use of fire was not identified at this locale. At the time of the *H. erectus* occupation, the area was a luscious landscape that included a lake. The Ubeidiya tools are significant as well as controversial because they represent one of the oldest hominin occupation in the Near East and predate by as much as 500,000 years any record of early Acheulean artifacts in Africa.[32] If these dates hold true, they may suggest that *H. erectus* originated outside Africa, possibly in Southwest or East Asia.

Nahal Zihor

This locality is in the vicinity of a dry lake system in the Negev Desert, south of the Dead Sea in Israel. Pollen types date this site to 1.8 to 1.3 mya. Today the area is extremely arid. At the time of the hominin settlement, the region possessed a Mediterranean humid climate.[33]

The artifacts found at Nahal Zihor were part of three assemblages near the shoreline of the ancient lakes. They total about 40 pieces including 29 hand axes, 11 picks, and a number of chopping tools made in the Acheulean tradition and resemble the material found at Ubeidiya using limestone or flint. No hominin remains have been found at this site. Nahal Zihor may be contemporary to Ubeidiya.

Dursunlu

The importance of the Dursunlu site in West Turkey stems from the scarcity of hominin material from this part of the Near East and the discovery of early *Homo* species further northwest from the point of entrance from Africa. Like several of the other locations indicating hominin presence, it was a lakeside settlement with a habitat similar to the present day, with grassland savannas typical of the Mediterranean basin. Although no hominin remains have been found, the animal fossils found date the location to the wide range of the Early (Lower) Pleistocene (2.6 to 0.78 mya). The fauna consisted of hippopotami, pigs, deer, gazelles, various types of carnivores, and bison. A total of 175 artifacts including cobbles and flakes mainly made up of quartz, limestone, and flints have been recovered.

Kashafrud and the Darband Cave

Kashafrud and the Darband Cave are the two most informative archeological locations dating back to the *H. erectus* occupation of Persia. These sites are significant because Persia (present-day Iran) represents a natural bridge connecting the Near East to southern and central Asia and as such was a main route for hominin dispersal into Central Asia and further east. As a region, in between the Levant/Arabia and the Caucuses where Dmanisi is located, Persia is crucial to the understanding of hominin migration in Eurasia. Therefore, evidence from this part of the Near East is paramount. Although about 10 additional settlements have been identified in Iran along a number of waterways, including the Kashafrud (northeastern Iran), Karun (Kargar, Mashkid, Ladiz), Sefidrud (north), and the Mahabad rivers (northwest), most of the sites are represented only by surface gravel deposits. Most of the artifacts from all of these sites reflect Acheulean technology.

Kashafrud, located in what is today western Iran, is of particular interest because it is likely that hominins dispersed from the Levant into the region of Kashafrud and then beyond to the north, east, and west via northern Mesopotamia and along the southwestern border of the Zagros Mountains.[34] The Kashafrud sites have yielded about 80 artifacts mainly from seven assemblages corresponding to seven locations along the Kashafrud river basin. The material includes cores, choppers, flakes, chunks, hammer stones, and debris. The toolmaking tradition was pre-Acheulean, yet some pieces resemble Oldowan assemblages from East Africa and styles from Dmanisi, further north in West Asia.[35] The most abundant stone used was

FIGURE 5.9 A quartz core-chopper from Kashafrud, northeastern Iran. *From https://www.academia.edu/224537/Biglari_F._and_Shidrang_S._2006_The_Lower_Paleolithic_Occupation_of_Iran_Near_Eastern_Archaeology_69_3_4_160-168.*

FIGURE 5.10 Darband Cave. *From http://theiranproject.com/wp-content/uploads/2015/05/Darband-Cave.jpg.*

quartz (Fig. 5.9). The fact that these early hominins work with quartz, a very fragile stone and difficult to control, is a testament to their skill in the reduction process.

The Darband Cave (Fig. 5.10) is located in northwest Iran, close to the southwestern shores of the Caspian Sea. It is approximately 21 m long with an entrance of 7 m in diameter. Although no hominin remains have been discovered, a large number of animal fossils and tools have been recovered. The lithic artifacts belong to the Acheulean tradition and include flakes, choppers, cores, scrapers made of chert, volcanic rock, and silicified tuff (volcanic ash compacted into solid rock). All the artifacts have been retouched suggesting some degree of sophistication in toolmaking techniques.

Curiously, the animal fossils consist mainly of cave and black bears in addition to a minor number of hoofed mammals. The abundance of bear remains in the cave suggests that it served not only as refuge for hominins but also as a bear den, clearly, not both at the same time. Some of the bear bones have been dated to about 200,000 ya and thus are not contemporary to the Acheulean tools made by more ancient *H. erectus* or *H. habilis*. Although the presence of hominin remains and bear bones in the same location does not necessarily indicate predation and transportation to the cave, signs of burning on the bones may be indicative of use of fire as well as cooking and consuming the bears. Yet, natural mortality cannot be ruled out as the cause for the accumulation of bear bones in the cave. Natural death was a real possibility because these bears hibernate in caves during the winter and they often die during this period of time due to malnutrition during their active months and/or as a result of injuries sustained prior to hibernation. It is also possible that hominins preyed on disabled bears while hibernating in their caves. Bones from injured cave bears exhibiting various types of fractures are often found in caves.

In addition to the lower Pleistocene sites described above, attributed to *H. erectus* and/or *H. habilis*, early dispersals out of Africa are marked by a number of other lithic locations in the Near East. These locales are not discussed in detail because of their lack of absolute dating and hominin remains. These sites consist of Acheulean or the more ancestral Oldowan tool industries and many of the discoveries were by-products of commercial geological surveys in recent years. For example, over 200 mostly Acheulean locations pepper the territory of Saudi Arabia, especially in the central and western part of the country. Shuwayhittiyah and Najran in Saudi

Arabia are earlier settlements representing Oldowan technology. Acheulean stone tool artifacts have been discovered in Sitt Markho and Khattab in Syria, Abu Khas in Jordan, and Evron and Gesher Benot Ya'aqov in Israel, all credited to *H. erectus*. The older Oldowan lithic tradition is also seen in Bizat Ruhama in Israel and Humma in Syria, the former indirectly dated to 1.6 to 1.2 mya. Yet, both locations lack absolute ages, fauna, and hominin remains.

Homo heidelbergensis in the Near East

H. heidelbergensis originated in Africa about 800,000 ya, and its tenure extended to 200,000 ya. Currently, it is not known in what region of Africa it evolved (see discussion in Chapter 3). *Heidelbergensis* represents a transitional group exhibiting a collage of *erectus* and modern human characteristics, and at times, it is difficult to surmise whether a particular specimen belongs best with the former or the later. The group cannot be uniquely defined by any particular trait (Fig. 5.11).

H. heidelbergensis reached the Near East approximately 500,000 ya. This wave of migration is known as the third Out of Africa I dispersal. Subsequently, migrant groups dispersed into Europe and the rest of Asia (Fig. 5.11). In Europe, they evolved into Neanderthals, whereas in Asia, they gave rise to Denisovans. One or more of the populations that remained in Africa became the ancestors to modern humans about 130,000 ya.[36]

In general, the older specimens resemble *erectus* more, whereas the more recent fossils are more human-like. In addition, *heidelbergensis* displayed considerable regional diversity. It is thought that some populations, exhibiting geographical differences, developed into races and subsequently two separate species; in Africa it led to modern humans, whereas in Europe it gave rise to Neanderthals. Although some scholars are of the opinion that *heidelbergensis* is not a separate species from *erectus,* and others argue that the more recent ones are in fact early modern humans, most investigators currently think that it represents an independent species that descended from *H. erectus* and gave rise to *H. sapiens*. *Heidelbergensis'* range included Africa, Europe, and Asia

FIGURE 5.11 Geographical range of *Homo heidelbergensis* indicating important sites. *From https://www.historiaeweb.com/2014/04/11/homo-heidelbergensis/.*

FIGURE 5.12 Mugharet El-Zuttiyeh Cave. *From https://fr.wikipedia.org/wiki/Mugharet_el-Zuttiyeh.*

(Fig. 5.12). As might be expected of a species that originated in Africa, the older specimens are mostly found there. Within its territorial expanse, *heidelbergensis* were able to survive diverse habitats, some at extreme conditions. It is thought that some of the anatomical diversities observed among individual fossils are thought to reflect regional differences. In its modest 600,000-year tenure, *H. heidelbergensis* evolved into a number of geographical populations, likely as a response to unique environmental pressures emanating from different types of habitats.

For the most part, *H. heidelbergensis* exhibits a progression of characteristics that originated with earlier hominins such as australopithecines and *erectus*. For example, it displays an increment in brain size averaging 1250 cm,[3] slightly flatter projecting face, sloping forehead, wide nasal opening, long legs, and more massive facial features. Unlike AMHs, it lacked a protruding chin, but like humans, the teeth were small, and the lower jaw was parabolic in shape, curved at the front and spreading out posteriorly. The frontal and parietal lobes of the brain were larger than *erectus*, suggesting greater intellectual capacity.

Heidelbergensis individuals were tall and strong with some specimens of certain populations reaching a height of over 2.13 m (7 ft). Some geographical groups from South Africa, for example, dating to 500,000 to 300,000 ya, were characterized as giants.[37] In other regions, males of the species were on average about 1.75 m (5 ft 9 in) tall, whereas females were around 1.57 m (5 ft 2 in) tall. From remains of a total of 28 individuals excavated from Las Simas de los Huesos in Atapuerca, Spain, the richest source of *heidelbergensis* discovered thus far, the average height, based on size of both sexes, was calculated at 170cm (5ft 7 in).[38] These height values in various locations of their range clearly illustrate the anatomical diversity of this group.

Heidelbergensis was the first hominin species to construct shelters. Their primitive dwellings were made of stone, wood, and plant materials. Most of their lithic tools follow the Acheulean tradition and consisted of large bifacial stone tools created by removing flakes from both sides with another rock. With this technique, they created hand axes, cleavers, and carvers. Some of the latter-day groups of the species also created artifacts from antlers, bone, and wood. These items were assembled into scrapers, hammers, and throwing spears. With these spears, they routinely hunted in organized groups, targeting megafauna that included rhinoceroses, hippopotami, bears, horses, and different types of ungulates. There is evidence of controlled use of fire, and because their geographical range included areas with cold climate, it is likely that the individuals wore some kind of clothing. Yet, no garments have been recovered from their sites, possibly due to the nondurable nature of the material, such as hide.

DNA Analysis

H. heidelbergensis is the oldest hominin species successfully characterized by DNA sequencing. Prior to the original DNA sequencing report on *heidelbergensis*, the oldest specimens capable of providing genetic information were Neanderthal fossils not older than 45,000 years. The capacity to examine and compare nucleic acid from individuals and populations provides scientists with an extremely powerful tool to assess true phylogenetic relationships among organisms, including hominin groups, because it allows for direct examination of the genetic makeup of individuals, not the phenotype (the expression of the genes). In other words, characters such as anatomical characteristics (e.g., brain size or height) are criteria that are influenced not just by DNA but by the environment as well, and therefore, they do not represent the ideal type of phylogenetic marker to ascertain ancestral relationships because it is difficult to assess how much of the trait is due to genes or environment. In other words, markers on the DNA itself are expected to provide signals more true to ancestry. Clearly, someone's height could be heavily influenced by diet or disease, both of which are environmental factors. Thus, their phylogenetic value to assess ancestry and relationships is limited. In addition, DNA data provide for robust and objective statistical analyses of the data. Traits such as anatomical characteristics, for example, height, usually exhibit a spectrum of values or possibilities (a bell-shaped curve of values), from the very short to the very tall and in between. On the other hand, DNA sequencing results generate data that are statistically easier to analyze because traits usually are present in only two versions or discrete categories, for example, nucleotides adenine or guanine in a certain position of the DNA.

The results of the DNA study were published in 2013 and were based on mitochondrial sequences (mtDNA) from fossils unearthed from the Sima de los Huesos, Spain.[39] The study is significant in a number of ways. The remains are superbly preserved, and the site is unique due to the large number of specimens recovered. The dig is still an active site, and it is likely that additional individuals will be recovered from the location. In addition, the investigation technically pushed DNA analysis to a new limit. This single report increased the age of usable fossils from about 40,000 ya to 430,000 ya. For the first time, ancient DNA of an order of magnitude (10×) older was within the reach of DNA sequencing. Furthermore, the individuals who were sequenced using mtDNA (mitochondrial DNA) were more closely related to the Denisova of

Asia than to the European Neanderthals. Subsequent analyses on the rest of the genomic DNA (non-mtDNA) indicated that the *H. heidelbergensis* specimens from Sima de los Huesos were more related to Neanderthals.[40] It is not clear why the two types of DNA reflect different evolutionary histories. Yet, mtDNA, being strictly maternally inherited compared with biparentally derived genomic DNA (both parents contribute roughly equally to the genome of an individual), is subject to unique evolutionary forces such as selection and dropouts (certain mtDNA types may be deleted from the species at random during its evolution) and thus is not representative of the entire genetic material of a species. Genomic DNA, on the other hand, is more characteristic of the evolutionary history of lineages. Importantly, the data from La Sima de los Huesos also pointed out that the separation between Neanderthals and Denisovans must have occurred 430,000 ya (the age of the fossils). It is also noteworthy that *H. heidelbergensis* individuals in Spain at least 430,000 ya indicate that the migration from the Near East to Western Europe was fast because the settlement of the former occurred approximately 500,000 ya.

Mugharet El-Zuttiyeh

There are a limited number of sites attributed to *H. heidelbergensis* in the Near East. Probably the best known is Mugharet El-Zuttiyeh which translates to "Cave of the Robbers" (Fig. 5.12). This site is located in the Upper Galilee region in what is today Israel. Zuttiyeh has the distinction of being the first paleontological excavation in the region and the location in which the first hominin was found in Western Asia. In 1925, a skull was discovered by Francis Turville-Petre in a limestone cave and was nicknamed Galilee skull or Galilee Man (Fig. 5.14).[41] It was dated to 300,000–200,000 ya. Initially classified as *Paleoanthropus palestinensis*, currently it is considered the most definitive example of *H. heidelbergensis* in the Near East. The artifacts found in the cave belong to the Yabrudian style complex generally characterized by biface-rich axes (Fig. 5.13) thought to be derived from the Acheulean tradition and to be the precursor to the Mousterian industry. The Galilee skull exhibits a flat face, and the prognathism of the jaw (protruding jaws)

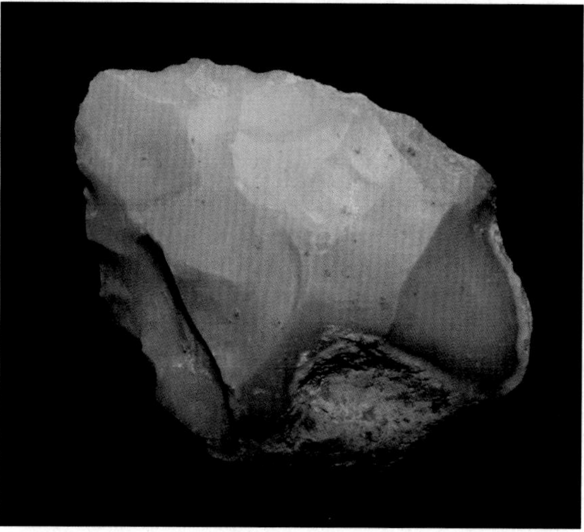

FIGURE 5.13 Yabrudian hand axe. *From http://www.aggsbach.de/2015/04/acheulo-yabroudian-handaxe-from-tabun-israel/.*

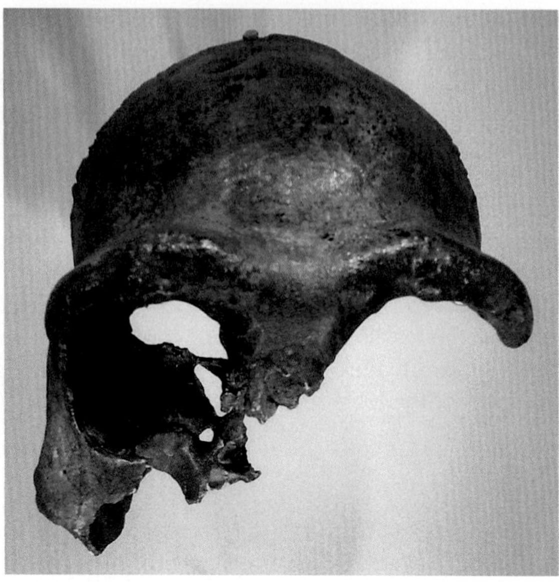

FIGURE 5.14 Galilee skull. *From https://en.wikipedia.org/wiki/Mugharet_el-Zuttiyeh#/media/File:IMJ_view_20130115_192522.jpg.*

is absent although it exhibits total prognathism of the face (an ancestral condition). Prognathism is considered an ancestral condition used to discriminate between Neanderthal and the more derived *H. heidelbergensis*. Together with the prominent brow ridge, Galilee skull is very much reminiscent of European and African Heidelbergs.

Tabūn, Gesher Benot Ya'aqov, and Other Potential Locations of *Homo heidelbergensis* Habitation

A number of additional *heidelbergensis* locations have been presumably identified in Israel, but many of the remains are fragmentary and it is not clear whether they belong to *erectus*, Neanderthals, early *sapiens*, or *heidelbergensis*. As *heidelbergensis* is a transitional group in the evolutionary progression from *erectus* to *sapiens*, exhibiting a mosaic of ancestral and derived traits, it is often difficult to decide if a particular fossil represents any of the abovementioned species. This situation is made more challenging considering the limited amount of fossils found in the Near East and the real possibility that the same cave or shelter was cohabitated by two or more species or consecutively occupied by various groups. Additional factors that contribute to the confusion in trying to identify specific species are the lack of absolute dates for fossils and potential intraspecies diversity and interbreeding.

To illustrate this predicament, let us consider the evidence at Tabūn and Gesher Benot Ya'aqov. The Tabūn site is a rock shelter in Mount Carmel facing the Mediterranean in northern Israel, and Gesher Benot Ya'aqov is located at the edge of an ancient lake in the northern region of the Jordan Valley. In Tabūn, archeologists have found part of a femur and a molar. The artifacts associated with the remains are Yabrudian in style (Fig. 5.13) and are comparable with tools discovered at Mugharet El-Zuttiyeh, thus the site has been dated at about 500,000 ya. The bones at Benot

Ya'aqov consisted of two femoral fragments, and the lithic tools belong to the Acheulean industry (Fig. 5.7). Thus, it is likely that this site is somewhat older than Tabūn. All three femurs exhibit shafts that are wider laterally, a condition characteristic of Heidelbergs. Yet, this condition is also seen among *erectus* and Neanderthals. Although the time of these remains fits the time line of *H. heidelbergensis*, the femurs cannot discriminate among the three potential species at these two sites.

THE MELTING POT

Currently, the general consensus is that *erectus* evolved into *heidelbergensis* and that *sapiens* evolved from the latter in Africa. The transition from the African *H. heidelbergensis* population into *H. sapiens* is thought to have occurred approximately 130,000 ya.[36] After the evolution of *H. heidelbergensis* into modern humans, *sapiens* move into the Near East in several waves within the period of 125,000 to 60,000 ya. This epoch is known as the Out of Africa II dispersals.[42] This *sapiens*-specific migration period occurred in two main phases. The first started about 125,000 ya, and it was made up of early *sapiens*. The second phase commenced about 70,000 ya and was made up of AMHs. Recent discoveries from a 300,000-year-old set of human fossils from Jebel Irhoud in Morocco and a 177,000–194,000 year-old *Homo sapiens* upper jaw from the Misliya Cave in Israel support the notion that *sapiens* probably began traveling out of Africa and settling the Near East at least 50,000 years earlier than previously thought. If these findings are substantiated, they also push back the genesis *sapiens* in Africa by at least 170,000 years (see Further Reading).

Subsequent to the transition from *H. erectus* into more recent groups within the genus *Homo*, the lines of demarcation between species become blurrier. Currently, for example, the evolutionary divides that discriminate *erectus* from early *sapiens* are not clear.[43] Similarly, the separation of AMHs from early *H. sapiens* is hazy. Experts in the field of physical anthropology oftentimes are in disagreement on several issues, including (1) whether a certain group of fossils deserves a unique taxonomic name, (2) the important diagnostic characteristics to consider, and (3) where to draw the lines among species. This lack of clearly distinguishable, well-defined *Homo* species after *erectus* depicts a nebulous view in recent hominin evolution. Furthermore, the number of remains exhibiting an intermediate combination of traits makes the fossil record look like a continuum with an unclear line of separation among populations. Also, the lack of direct dating for the remains in most sites prevents assessing temporal points of reference, exacerbating the confusion. This situation, although unsettling at times, is in fact expected. Considering the nature of evolutionary change, this continuous range of characteristics among species with intermediate forms, during evolutionary time, may reflect the typical and gradual transition among biological systems. In fact, it can be argued that the pre–*erectus* view of hominin evolution is artificially simple resulting from the limited fossil record dating back to more ancient times. In other words, the well-defined species distinctions observed in the pre–*erectus* epoch with a clear and staccato picture of hominin evolution may result from the limited information from transitional forms that have not been discovered yet. The greater number of representative specimens, from more recent times, provides greater details and resolution on evolutionary change, allowing us to discern the true complexity of the evolutionary processes involved.

Other factors may have contributed to this lack of definition among post–*erectus* hominins in the Near East. Geography may have played a role in generating the observed lack of discrete partitioning among hominin remains and their artifacts. As the Levant and Arabia represent a rather physically restricted passageway through which hominins most have traveled in their route to East Asia and Europe, it may have acted like a geographical funnel in which various populations from different locations and migration waves converged promoting various types of interactions. For instance, the narrowest portion of the Sinai Peninsula that connects Africa to the Near East (the Northern Corridor) is less than 150 km wide. The Sinai was a necessary region needed to be traversed by hominins as part of their trek across the Levantine passageway. It is noteworthy that this region connects directly to the region rich in archeological sites to the north, in current-day Israel, that have yielded most of our knowledge of hominin settlement in the Near East (see previous sections).

Another factor that may have contributed to the lack of distinction among post–*erectus* hominins was the technological advances that may have facilitated migration to and survival in extreme habitats. *H. erectus* was an innovator in many ways, and the new technological advances allowed this group to better control the environment. Among these advances, *erectus* practiced a hunter-gatherer existence, band societies based on extended family groups, communal hunting; cared for the disabled; controlled fire; cooked; started toolmaking traditions; and possibly used clothing. In addition, there are sites, such as Terra Amata in southern France, that indicate that *erectus* built oval shelters made up of stone foundations and branches. In addition, the anatomy of the neck indicates that they were able to articulate words, probably using some sort of protolanguage. The use of some type of verbal communication was certainly helpful during communal hunting, especially megafauna. Compared with australopithecines' short stature and slow locomotion, *erectus*' tall body frame, long legs, and slender physique allowed for running and traveling in difficult terrains. All of these cultural, technological, and biological characteristics freed them from many climatic and environmental restrictions, facilitating regular continuous dispersals out of Africa instead of rare sporadic incursions into the Near East. These features would have increased the number of incursions, migrants, and genetic diversity into the small region of the Levant.

The physical proximity of these various groups of migrants in the Near East as they moved northward to Asia and Europe must have provided ample opportunity for cultural and technical exchange as well as interbreeding. Technology transfer has in fact been documented involving Neanderthal and modern human populations in Central and Western Europe as well as in the Near East.[44] Specifically, technology such as bone utensils and grooved and perforated animal-tooth necklace amulets seems to have been transferred from early modern humans practicing a Chatelperronian/Aurignacian tradition to Neanderthals.[45] Conversely, some scholars see the Chatelperronian technology transfer in the opposite direction as well, from Neanderthal to recent anatomical modern human migrants.[46] Other clear evidence for technological transmission from Neanderthal to modern humans has also been found. Specifically, in a 50,000-year-old camp in Southwest France, lissoirs or leather smoothers were made from deer ribs by Neanderthal, long before AMHs acquired such know-how.[47] This procedure to tenderize hide is still in use today. Thus, currently, we may be using 50,000-year-old instruments and methodology learned from Neanderthals, a population considered by some to be technologically and intellectually limited. Although the lissoirs and the leather-softening technique could have been invented independently by both species, it is likely that technology transfer occurred. In fact, scholars have pointed out that this type of technology transmission is expected based on contemporary contacts between primitive and more advanced modern human populations.[48] The recent discoveries of paintings on the wall of La Pasiega cave in Spain, made over 64,000 years ago by Neanderthals, long before *sapiens* arrived in Europe, not only indicate that AMHs were not the first in this type of artistic expression but it suggest that Neanderthals may have transfer this art form to AMHs (see Chapter 6). This richness in biological and technological diversity in close geographical proximity was bound to be transferred due to sexual attraction and the advantages of adopting superior technology. It is likely that as groups came in contact with each other, cultural interactions led to the creation of a melting pot of various populations. These conditions may have contributed to the lack of definition and abundance of intermediate forms among fossils and artifacts in the Near East.

Although organisms, especially extinct ones, often cannot be partitioned into cubicles encompassing discrete species, the expectation of evolutionary change is that at some point, related populations become unique enough so that they become reproductively isolated. Reproductively segregating populations may become sister species when genetic flow between them stops completely. Cessation of genetic flow between populations may be the result of geographical barriers, such as a wide and deep river or mountain range. Alternatively, reproductive isolation can be achieved from genetic incompatibility that prevents the birth of fertile offspring. As soon as the reproductive barrier is established, the different populations of a species begin to mutate, differentiate, and evolve independently in response to different environmental pressures and selection or simply by random drift (mutations that accumulate randomly). In effect, this process starts changing the sister species from each other and eventually will generate enough differences between them that would allow the discrimination of one from the other. Thus, in the event that the populations do not experience genetic flow or reunite by interbreeding, prior to complete separation and genetic incompatibility, they will evolve into authentic species. Species that have experienced reproductive isolation for a considerable amount of time may exhibit unique anatomical characteristic evident in the fossil record providing for clear identification of groups as distinct. Of course, reproductive isolation and distinct characteristics do not necessarily mean a lack of trait overlap. In other words, completely unique but related species might retain aspects of their anatomy in common. And even more confusing is the possibility that totally evolutionarily unrelated species may exhibit similar anatomy due to convergent evolution, the process by which similar morphologies are generated as a response to similar selection pressure. Notorious examples are the flippers in species of the order cetacea (whales, seals, and dolphins) and the fins of fishes. These structures, although they resemble each other and perform similar functions, are not related embryologically.

From Early *Homo sapiens* to Anatomically Modern Humans in the Near East

Generally speaking, early *H. sapiens* are a heterogeneous group of populations that lived between 300,000 and 100,000 ya and coexisted with AMHs for at least 50,000 years in Northeast Africa. The largest number of specimens including the Omo and Herto groups was discovered in Northeast Africa, suggesting that the cradle of *H. sapiens* was there.

Similarly to the transitions among groups of pre–*sapiens* hominins described above, there is no clear line of demarcation between early humans and anatomical modern humans in the Near East. This dearth of diagnostic criteria is reflected in the confusion observed in the classification of fossils and species by different scholars. For instance, some investigators draw the line between early humans and AMHs with the Herto remains from the Afar region in the northwest of present-day Ethiopia. Of note, the Herto site is located on the African side of the Bab-el-Mandeb Strait ("Gate of Tears"), in the Horn of Africa, one of the two presumptive gateways to the Near East. These specimens share anatomical traits and technology with

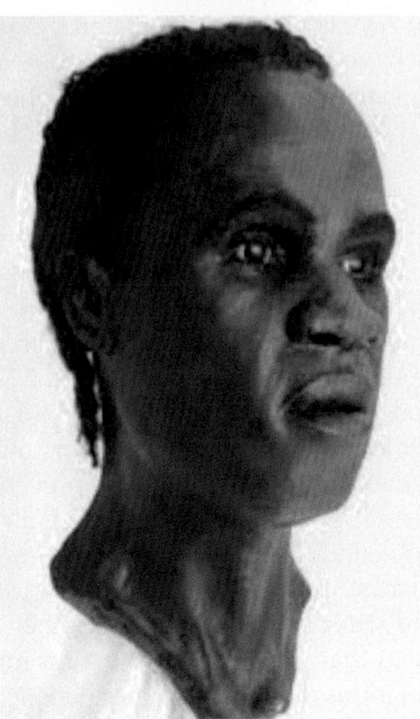

FIGURE 5.15 Reconstruction of Qafzeh 9 (female). *From http://ma.prehistoire.free.fr/qafzeh9.htm.*

a number of finds from the Near East such as Skhūl and Qafzeh (Fig. 5.15), and all may have been members of the same population occupying a geographical range encompassing Northeast Africa and the Near East. The Qafzeh population is considered by some to be at the threshold of becoming AMHs. It exhibits a collage of ancestral traits such as a robust body, thicker skull, prominent brow ridge, and less pronounced chins compared with contemporary humans. Yet, it exhibits AMH characteristics such as high round skull, flat face, and a brain capacity of 1450 cm^3, well within the contemporary human range. Contemporary humans have an average brain size of 1350 cm^3. Other investigators even include species such as *H. heidelbergensis* as part of the early human group. And still other scholars incorporate Neanderthals within early humans as well. This absence of consensus in terms of what the different names stand for makes comparisons among hominins difficult and prone to misunderstanding, especially among the laypeople. The situation is confusing to say the least.

EARLY *HOMO SAPIENS* IN THE NEAR EAST

The arrival of early *H. sapiens* to the Near East is thought to have occurred approximately 120,000–90,000 ya. Although arguments center on whether that migration wave contributed to the gene pool of contemporary humans, it is clear that early *H. sapiens* penetrated the Near East during that time period, dispersal known today as the first phase of Out of Africa II. The tool tradition, time range, and anatomy indicate that sites such as Skhūl and Qafzeh, in current-day Israel, belong to early modern humans. More recently, an assemblage of tools dating to about 125,000 ya was discovered at Jebel Faya in United Arab Emirates (UAE) as well as a 177,000–194,000 year-old *Homo sapiens* upper jaw from the Misliya Cave in Israel (see Further Reading) suggest an even earlier settlement of early humans in the Near East. In addition, a number of recent discoveries of early *H. sapiens* remains in China, including the Daoxian, Zhirendong, and Xujiayao sites dating to 120,000, 100,000, and 125,000 ya, respectively, indicate that the first phase of Out of Africa II started prior to the date previously estimated and may have contributed to populations in the Far East that possibly evolved into AMHs in situ. Although this notion is currently highly contested, genetic studies have detected early *H. sapiens* DNA in contemporary human populations, suggesting that these early migrations were not sterile evolutionary dead ends.

Skhūl

The Skhūl or es-Skhūl Cave is located in the slopes of Mount Carmel, 20 km south of the city of Haifa and about 3 km from the Mediterranean Sea in present-day northern Israel. The site is remarkable in that it contains three main layers of sediments designated as A, B, and C corresponding to Natufian, Mousterian, and Aurignacian traditions, respectively. The late Paleolithic Natufian tradition dates from approximately 14,500–11,500 ya in the Levant and represents the threshold to the Agricultural Revolution in the Near East. The Mousterian culture is associated with Neanderthals in the Middle Paleolithic from 160,000 to 40,000 ya, whereas the Aurignacian period is linked to the earliest migrations of AMHs from the Near East into Europe about 43,000 ya during the Upper Paleolithic. The Aurignacian eventually lead to the Gravettian tradition around 28,000 ya. Examples of all of these cultures are evident in the Near East. Thus, Skhūl provides a sequential record from early to AMH occupation in the region. The Skhūl site is also significant because it provides very early evidence of signs associated with humanity such as compassion, intentional interments, personal art, and artistic expression.

The skulls found at the site possess a mixture of ancestral and derived characteristics, and the specimens are considered a transition state toward AMHs. In general, the Skhūl skulls are robust indicating that they did not belong to fully AMHs but to an early population. Specifically, the brain cases are rounded with vertical foreheads as in contemporary humans, but they still exhibit brow ridges, jutting jaws, and a projecting facial profile reminiscent of earlier hominins. One of the brain cases has a capacity of 1518 cm^3. Although Neanderthals shared, in time and space, the same complex of caves, most authorities believe that the collage of ancestral and derived traits found at Skhūl are not the result of introgression but a retention of ancestral hominin anatomy of African origins. The Skhūl site was discovered between 1929 and 1935 and includes seven adults and three children. Electron paramagnetic resonance and thermo luminescence date the remains at about 80,000–120,000 years old.[49]

Some of the remains at Skhūl show signs of intentional interments. For instance, one of the skulls (number 5) that belonged to a 30- to 40-year-old male was found associated with the mandible of a wild boar positioned on the chest suggesting some sort of symbolic meaning in connection with the afterlife. Also, collections of perforated marine shells (*Nassarius gibbosuslus, Acanthocardia deshayesii, Laevicardium crassum*, and *Pecten jacobaeus*) collected in the Mediterranean were found in close proximity to the remains (Fig. 5.16). These collections of punctured shells may represent one of the earliest artistic expression and ornamentations. In addition, ochre, a source of yellow and red pigments, was found on the walls of the cave. Ochre was routinely employed by Paleolithic populations to paint and as food preservative, and it is also associated with burial sites.

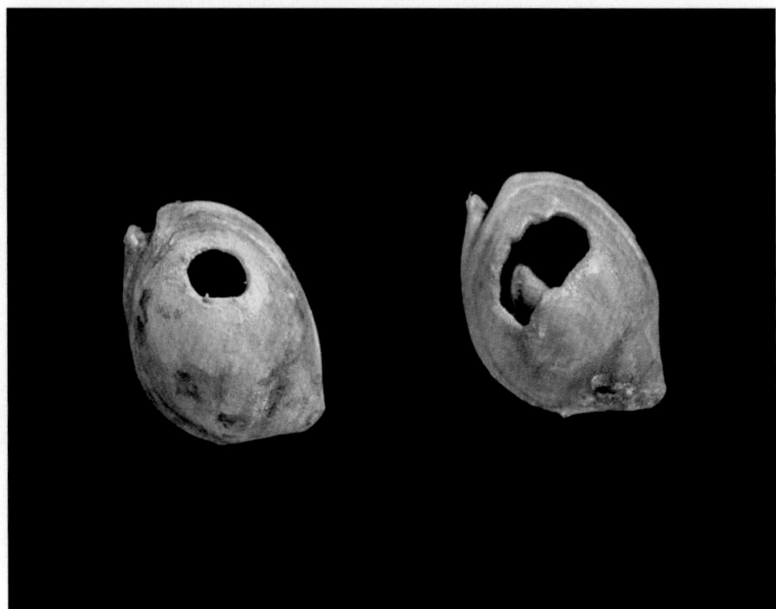

FIGURE 5.16 Skhul's beads. Assemblages of perforated *Nassarius* shells (a marine genus) significantly different from local fauna have also been recovered from the area, suggesting that these people may have collected and employed the shells symbolically as beads. *From https://s-media-cache-ak0.pinimg.com/736x/c9/ba/02/c9ba029c75d859c0 432d5fb9699b60e5.jpg.*

A rather sudden interruption in the early human fossil record in the Near East has prompted some to suggest that at around 80,000 ya these first-phase Out of Africa II migrants became extinct or returned back to Africa. It has been speculated that climatic changes to colder and drier weather and/or competition with Neanderthal groups triggered the disappearance of early humans from the Near East. The artifact assemblages at Skhūl, found in the context of the early human remains, belong to the Mousterian tradition, typically attributed to Neanderthals throughout their range in Eurasia. This association of species and a specific tool industry may be indicative of coexistence of Neanderthal and early *H. sapiens* as well as technology transfer from the former to the later.

Qafzeh

Located in the southern extreme of the Sea of Galilee, a few kilometers southeast from the town of Nazareth and 45–50 km from the Mediterranean Sea in current northern Israel, the Qafzeh Cave has provided important information on early *H. sapiens* in the Near East. Some of the oldest layers at Qafzeh date to 92,000–115,000 ya, and the lithic artifacts found are Mousterian in tradition.[50] Qafzeh is a multicomponent (different time periods are found) site with Middle (200,000–45,000 ya) and Upper (40,000–10,000 ya) Paleolithic as well as Holocene (last 11,700 years) occupations. Some of the sediments at Qafzeh are contemporary with some of the layers at Skhūl.

Fossils from five individuals were unearthed in 1934, and since then an additional 22 more remains have been recovered for a total of 27. Eight constitute partial skeletons while Qafzeh nine (Fig. 5.15) and ten are almost complete. Eight of the fifteen individuals were children and six of the fifteen were clearly interments. The Qafzeh cave contains one of the earliest unequivocal examples of intentional entombments in the world. Among the entombments, one is a double grave of a young adult female and an infant. Another burial belongs to a 12- to 13-year-old individual holding an anther of a large deer close to the chest. It is not clear the significance of the close association of various game animals with human remains, but this theme is repeated in various early modern human Near Eastern settlements. Their inclusion in entombments may have been for protection purposes during the transition to the afterlife or simply it may be indicative of the cause of death of the individuals. This same individual (Qafzeh 11) shows signs of a serious head injury about 8 years prior to his/her death. The trauma most have impaired the adolescent dramatically, likely affecting his/hers intellect, overall behavior, and capacity for self-reliance. Thus, for 8 years or so, this individual was cared and assisted in almost all aspects of his/her life, and then at the end of life, this individual received a ceremonial burial. In our current-day society, where the disabled are routinely marginalized into nursing homes, this level of devotion exhibited by early humans is noteworthy. A fourth tomb belongs to a 3-year-old child displaying several skeletal malformations consistent with hydrocephaly. This finding is highly compelling because it argues for substantial parental and community care and investment to an infant with a very serious medical condition. The survival to the age of three of a hydrocephalic child living in a cave suggests active and careful monitoring and assistance. It may even be expressions of love, an emotional commitment, and empathy associated with *sapiens*.

In general, anatomically the early *H. sapiens* from Qafzeh resemble the remains from Skhūl. They possessed derived modern human characteristics such as a high forehead and a prominent chin, but they still retained ancestral hominin traits such as projecting faces and jaws. Even though these skeletons were found in the context Mousterian artifacts, a Neanderthal industry, these remains were clearly early *sapiens* because they lack the occipital bone at the back of the skull.

A number of fire pits and tools including scrapers, disc cores, and points were also recovered indicating the control of fire, likely for comfort, protection, and cooking. The site shows evidence that some of the remains were buried. A number of animal bones from gazelle, horse, fallow deer (fallow, woodland, and red), European wild ox, rhinoceros, and microvertebrates were found as well. In addition, land snails and freshwater bivalves were used for subsistence. Similarly to Skhūl, clumps of ochre and numerous seashells collected from the Mediterranean shores were part of the assemblage. Some of the shells were painted with red, yellow, and black ochre and seem to have been hung as a pendant.[51] These findings indicate that these early *sapiens* were capable of artistic expression.

Tabūn

The Tabūn site is located adjacent to the Skhūl cavern and both are part of a group of four rock shelters (Tabūn, Jamal, El-Wad, and Skhūl) known as the Mount Carmel Caves. Positioned on the western slope of Mount Carmel, the cave entrances are clustered side by side along the south side of the el-Mughara valley a few kilometers away from the Mediterranean Sea in present-day northern Israel.

Although Tabūn is considered contemporary to Skhūl, not surprising because both caves are adjacent, signs of habitation in the former is about 20,000 years older than the later. Tabūn is remarkable in exhibiting one of the longest continuous hominin prehistoric cultural sequences in the world. The archeological assemblages include Acheulean remains dating from about half a million years ago to the Mousterian industry of 250,000–45,000 ya and into the Natufian culture of 15,000–11,500 ya. This extensive archeological record presents a story of intermittent and alternating occupations of Neanderthal, early *H. sapiens*, and AMHs. Depending on the sedimentary layer in question, at times Neanderthals and early *sapiens* seem to have coexisted at the site in the context of the same Middle Paleolithic Mousterian tradition, sharing the technology. It is difficult at times to access if specific Mousterian artifacts were made by Neanderthals or early *sapiens*.[52] In other periods, Neanderthals appear to have dispersed southward into the Levant, possibly from Anatolia and/or the Caucasus, during cold epochs while early *sapiens* migrated northward from Africa during the intervening warmer periods.

The Tabūn site contained less hominin remains compared with the adjacent Skhūl Cave. The entire bone assemblage consists of one skeleton, a lower jaw, a number of teeth, and some skull fragments. Most of the specific remains discovered in Tabūn belong to Neanderthals. These include the approximately 120,000-year-old burial of a female and a ~82,000-year-old teeth (Tabun BC7). Of particular interest among the hominin fossils found at Tabūn is a 120,000-year-old isolated mandible of a male individual that exhibits affinities with both the Neanderthals and the early *H. sapiens* of the region. As expected, opinions are divided as to the identity of this piece with some experts suggesting that the bone correspond to a hybrid.

In terms of artifacts, during the period of 1967–1972, excavations generated over 1900 bifaces (two-faced lithic tools or weapons), most of them belonging to the Late Acheulean (seen in the Near East about 350,000ya) and Yabrudian industries. The Levant-specific Yabrudian period started about 215,000ya. The Tabūn Mousterian deposits, in the upper layer and over the Yabrudian deposits, are more recent dating to about 200,000–45,000 ya. In these Mousterian layers, flints predominate as well as flakes of various shapes and sizes that were used to cut off meat. In addition, long points were found. These were presumably employed for opening holes in hides and possibly sawing.

Jebel Faya

Jebel Faya is a composite site containing about 5m of strata encompassing the Paleolithic, Neolithic, and Bronze Age as well as Iron Age layers. This dig is geographically located in what is today UAE, in Southeast Arabia, 62km from the Persian Gulf.[53] Jebel stands out from Skhūl, Qafzeh, and Tabūn for two reasons. The oldest assemblages in the rocky shelter date to approximately 150,000 ya, about 5000years older than the three Levant sites, and it may represent an earlier phase of the Out of Africa II. In addition, this location may signal a migration across Bab-el-Mandeb Strait (the Horn of Africa) using the Southern Route as opposed to the three Levant sites discussed above which probably resulted from dispersal via the Sinai Peninsula or the Northern Route. Both paths likely used coastal routes along the southern Arabian coast and the Eastern Mediterranean shore, respectively, to reach their destination. Unfortunately, this site is entirely made up of tools, no hominin or fauna remains have been found in Jebel yet. Geographically and chronologically, Jebel is closer to Herto and Omo in Northeastern Africa. In addition, there are evident parallelisms in tool styles seen in the two Northeastern African sites and Jebel. The population responsible for the lithic artifacts at Jebel may represent one of the earliest waves of early *sapiens* that venture out of Africa after they evolved from *H. heidelbergensis* about 130,000 ya. Alternatively, just as with the enclaves in the Levant, Jebel may correspond to a population in the northern fringes of a continuous range that included the Northeastern African populations. At the time, the Bab-el-Mandeb Strait connecting Arabia and Africa was just 4 km wide. Climatically, the Paleolithic settlement at Jebel corresponds to a wet time period in the coastal southern portion of the Arabian Peninsula and in Northeastern Africa.[54] Thus, ecologically, Arabia and the Levant would have been an extension of Northeast Africa at these times.

A number of Middle Paleolithic artifacts recovered from the 125,000-year-old sediments (most ancient layer) were made using the Levallois tradition (two-sided blade produced by striking flakes off from the edges). These artifacts strikingly resemble the lithic assemblages found in Northeast Africa. The orthodoxy is of the opinion that at some point during the Late Upper Paleolithic, the hominin populations that practiced this tradition evolved into AMHs. This transition likely occurred in Northeast Africa, not in Arabia. These artifacts consisted of tools such as handheld axes, foliates, scrapers, and denticulates (having small teeth generated by striking flakes off) cores as well as leaflike blades that might have been attached to spears. Unlike some of the artifacts recovered from the Levantine caves, these tools were clearly made by early *sapiens*, not Neanderthals.

Did Early Anatomically Modern Humans Survive?

The Jebel site is particularly important because of its location poised at the entrance into Persia and beyond into Far Asia. This strategic geographical location would have allowed these Arabian–Persian Gulf populations to move across a very shallow body of water known today as the Strait of Hormuz and then travel further east. This early *sapiens* populations would have been forced to move on eastward once the very arid conditions returned to the Arabian Peninsula. These putative dispersals eastward may be responsible for the ancestors of the early human East Asian settlements of Daoxian, Zhirendong, and Xujiayao dated at 120,000, 100,000, and 125,000 ya, respectively.[55] Settlements of early *sapiens* in the Far East would explain the ancestral traits such as massive molar teeth with very robust roots and complex grooves observed among the Chinese remains dating back to 125,000 ya. The geographical continuity of early *sapiens* groups extending from the Near East to the Far East argues against the notion of an extinction of the first phase of the Out of Africa II dispersal. Further, it is feasible to contemplate the survival of early *sapiens* DNA in contemporary human populations. The recent genetic results from the Metpaslu group in Estonia pointing to at least 2% of early human genetic material in contemporary humans[56] corroborate the archeological findings. Another important piece of evidence arguing for this notion is the trace amounts of *H. sapiens* DNA detected in a 100,000-year-old female Neanderthal from the Altaic region (Central Siberia) of north central Asia.[57] The implication of this finding is that early *sapiens* interbred with Neanderthals somewhere in Asia, possibly in the Near East or as far east as Central Siberia, prior to the arrival of AMHs to the Near East, approximately 60,000 ya. In other words, the findings from present-day China and the genetic studies suggest that the Near Eastern early humans were not an evolutionary dead end. Future findings from the Persian side of the Strait of Hormuz should shed light on this contentious issue.

The Near East: A Story of Cohabitation

Today the Near East is associated with the birth of three major religions, a volatile region of political turmoil, and conflicts since before biblical times. This land was also the scene of power shifts long before recorded history. Well-differentiated Neanderthals were the residents of Europe and Southwest Asia about 200,000 ya. In the region of Skhūl and Qafzeh in the Levant, evidence of Neanderthal occupation dates back to 130,000 ya. Of note, no Neanderthal artifacts were detected in Jebel on the Persian Gulf coast. Neanderthals were not inferior hominins. In fact, they were technologically and possibly intellectually equivalent to the early *sapiens*. In fact, at this time, it is difficult to ascertain whether Neanderthals or early humans made specific tools. Moreover, early *sapiens* were physically less endowed and not as well adapted to cold climatic conditions as Neanderthals. For a period of at least 40,000 years, these two groups of hominins lived together in the Near East. At times it seems they sheltered in the same cave. Then, after about 80,000 ya, all the hominin enclaves stopped exhibiting *sapiens* remains. This is clearly seen in the absence of early *sapiens* artifacts and fossils in sediment layers corresponding to about 80,000 to 65,000 ya. This corresponds to an epoch lasting about 15,000 years of uncontested Neanderthal occupation. Thus, Neanderthals prevailed in the Near East and early humans did not. If this outcome resulted from belligerent encounters between the two groups or competition for resources, the winners of the confrontation were the Neanderthals.

In what is known as the second phase of Out of Africa II, AMHs migrated into the Near East approximately 65,000 ya. As a result, Neanderthals were confronting human invaders for a second time. These migrants were not the same type of *sapiens* that reached the Near East about 115,000 (lower jaw from Misliya Cave in Israel) to 55,000 (Jebel) years earlier. These newcomer *sapiens* were not only physically and behaviorally different but also equipped with more advanced technology, superior set of tools, and a different outlook of the universe. It is likely that the early *sapiens* populations that remained in Africa gave rise to the AMHs that eventually migrated to the Near East. Neanderthals shared the land with AMHs for about 25,000 years until the formers became extinct in the Near East 40,000–35,000 ya. Again, as during their first encounter in the Near East, it is not clear how belligerent or passive were their interactions during this second period of cohabitation. If the disappearance of Neanderthals resulted from confrontational interactions between the two groups, AMHs won this second round. Thus, the relationship of Neanderthals and humans in the Near East could be characterized by periods of succession interspersed with epochs of communal living.

What Happened to the Early *sapiens* Populations of the Near East 80,000 ya?

AMHs reached the Near East approximately 60,000 ya during a brief warm epoch in the midst of the last ice age. Unlike Africa, archeological evidence indicates a lack of continuity between early *sapiens* and AMHs in the Near East. In Africa, both groups actually coexisted for thousands of years. In the case of the Near East there is no record of early human populations subsequent to about 80,000 ya.

The orthodoxy, over the years, has proposed that the arid cold weather conditions in the region forced the early *sapiens* populations to return to more benign habitats in Africa. Other suggestions for their disappearance include direct and/or indirect competition with Neanderthals or even the extinction by dilution of the pure human forms by interbreeding with the more numerous Neanderthal. This last mechanism could have the effect of drowning and erasing any early human genome component in a vast and overwhelming large Neanderthal population.

Even more controversial is the potential negative effect of the Toba supervolcano eruption that occurred about 75,000 ya in the island of Sumatra in Indonesia on early *sapiens* populations compared with Neanderthals in the Near East. It is known that this supereruption was followed by a global winter period that lasted at least 10 years and an approximately 1000-year-long cooling period. The volcanic winter resulting from the Toba eruption generated vast amounts of ash that blocked the sun light promoting a dramatic drop in temperature and sulfuric acid that contributed to the disappearance of a vast amount of the flora and fauna. How complete was this devastation is difficult to ascertain today, but judging from assessments of genetic diversity, our species suffered a dramatic drop in variability due to a reduction in the number of breeding couples to as few as 1000 or 3000 or 10,000 surviving individuals.[58] Although the Toba catastrophe theory is still controversial, the devastation would have resulted in what population geneticists call a bottleneck event. The impact of such an environmental and ecological catastrophe on a population on the fringes of its geographic range, possibly struggling to survive in competition with Neanderthal, would have been devastating. It would have had disastrous consequences because vegetation and, therefore, most fauna died in mass extinction worldwide. It is possible that Neanderthals were able to fare this debacle better, considering their physiological and anatomical advantages over *sapiens* in colder weather. In addition, it is likely that the number of early humans in the Near East was small adding to the strain on the population. None of these possibilities are mutually exclusive. Yet, as previously discussed, the worldwide existence of these early humans may not have stopped with their disappearance from the Near East. Recent evidence from East Asia and Oceania indicates that their DNA survived in what is today China and still present in populations of Papua New Guinea.[56]

Even though *sapiens* were absent from the Near East about 80,000 ya, there was a span of time in which early humans and Neanderthals lived together in the region. Recent discoveries in the Levant have demonstrated the presence of Neanderthal approximately 120,000 ya, long before 80,000 ya. Thus, these findings clearly provide for temporal and geographic overlap between the two hominin groups of at least 40,000 years, plenty of time for DNA to flow between the two.

Anatomically Modern Humans in the Near East

The settlement of AMH in the Near East dates back to about 55,000 ya. This occupation resulted from the second phase of Out of Africa II, which was facilitated by a short temperate period in the midst of the last ice age. These new migrants were anatomically different from the early *sapiens* that traveled to the Near East during the first phase of the Out of Africa II dispersal and subsequently disappeared 80,000 ya.

In general, these recent newcomers to the Near East were more slightly built or gracile (nonrobust physique) than the early modern humans. Their bones and musculature were less massive. This trend toward a less husky bone structure and musculature started about 50,000 ya,[59] and some scholars argue that this progression was the result of a greater dependency by AMHs on their technology and less on their body strength over a period of time. Overall, AMHs have progressed further toward the reduction of ancestral traits such as brow ridges, protruding faces, and occipital buns found on the back of the skull and the appearance of contemporary human characteristics such as a more rounded head, larger forebrain (with the brain above the eyes instead of behind them), high and vertical foreheads (not sloped like earlier hominins), smaller lower faces, and pointed chins. It is important to indicate that even some contemporary humans exhibit brow ridges, but they differ from those of early *sapiens* in that they have a supraorbital notch creating a split through the ridge above each eye.

Since this second phase of the Out of Africa II dispersal of AMHs, the occupation of the Near East has been pretty much continuous. During the late Middle Paleolithic (55,000–40,000 ya), Upper Paleolithic (40,000–10,000 ya), and the Neolithic (10,000 to present), AMHs have lived and evolved uninterruptedly in the Near East. Although AMHs have experienced some anatomical changes during these last 55,000 years, most of the changes experienced by the human population in the Near East have been cultural and technological in nature. The evolution of humans during this last portion of our tenure in the Near East is characterized by dramatic cultural changes involving modes of survival as well as population growth. Advances in tool manufacturing including the emergence of regional stone tool industries, such as the Perigordian, Aurignacian, Solutrean, and Magdalenian, are seen in the continuous occupation of the region. Improvements in food procurement and processing are also evident during this period. All of these changes allowed humans to dramatically modify their environment. The start of the Neolithic period is particularly pivotal for the human species because it marked the beginning of the Agricultural Revolution in the Near East. Humans stopped being hunters and gatherers and became farmers. This single event allowed for a mode of subsistence that was less impacted by environmental fluctuations such as climate changes as well as less dependent on the movement of game. Although the high concentration of people in the limited spaces provided by homesteads promoted the spread of infectious diseases, the Agricultural Revolution allowed people to devote more time to think, imagine, and create. These breaks from manual labor provided time to explore and become more conscious of themselves and their soundings. It provided time intervals for intellectual pursues. Some of the important AMH settlements are described below.

Manot

After the disappearance of early *sapiens* from the Near East approximately 80,000 ya, there is a gap in the presence of human fossils and artifacts in the area that lasted for 25,000 years. The oldest remains of AMHs in the Near East, in fact outside of Africa, have been found in the Manot Cave in Western Galilee in current-day Israel, a location about 50 km north from the Mount Carmel Caves (Tabūn, Jamal, El-Wad, and Skhūl). Excavations of this composite site have uncovered deposits and hearths from several different periods of hominin evolution starting with the Early Paleolithic. Stone tools such as points, burins (chisels used for engraving or carving wood or bone), scalpels, blades belonging to the Levallois tradition, as well as Aurignacian artifacts such as scrapers have been recovered. The fauna included deer, gazelle, horse, aurochs (wild cattle), hyena, and bear. In addition to numerous lithic tools and bone artifacts, this location is known for an AMH skull dated to 55,000[60] (Fig. 5.17). The skull clearly belonged to an AMH. The broadest part of the skull is high at the upper portion of the head, and the parietal bones (on the sides of the skull) are parallel to each other and position vertically such as in contemporary humans. The importance of the Manot skull is its early age that corresponds to the right time frame following

FIGURE 5.17 Manot skull. *From www.nature.com/nature/journal/v520/n7546/full/nature14134.html.*

the second phase of the Out of Africa II migration of modern humans that left Africa about 65,000 ya. It has been pointed out that although the Manot skull closely resembles more contemporary groups, such as the Cro-Magnons of 35,000 ya, it is not clear whether it belonged to the group of AMHs that were the direct ancestors of the individuals who first populated Europe about 45,000 ya. Alternatively, the Manot skull population may represent a vanguard wave of the second phase of the Out of Africa II that remained in Asia or became extinct, and thus do not represent the direct ancestors of contemporary Europeans.

Skhūl

This site is the same Skhūl Cave where fossils and artifacts from Neanderthals and early modern humans from the first phase of the Out of Africa II were discovered dating back to the more ancient dispersal and occupation of about 125,000–120,000 ya. More recent animal teeth were found in this cave corresponding to the 40,000–45,000 time period.[61] In addition, the anatomical characteristics of the fossil remains indicate that two distinct modern human populations (one less morphologically "modern" than the other) are represented at Skhūl, likely representing the two phases of Out of Africa II. In addition, it has been postulated that the *Nassarius* (mud snails) shell beads were created by individuals who arrived to the Near East 55,000 ya, not the occupants of the cave during the 125,000–90,000 time period.[62] This more ancient occupation is thought to be responsible for the perforated ochre-stained shells of the genus *Glycymeris*.

Kebara

Located in the western wall of the Mount Carmel range in Northern Israel, the Kebara site is best known for its numerous Neanderthal remains dating to a period between 60,000 and 48,000 ya. The cave is a multicomponent archeological site exhibiting three important constituents. The earliest goes back to 60,000 ya during the Middle Paleolithic. The second oldest component dates to 40,000 ya during the Upper Paleolithic, whereas the most recent period corresponds to the Natufian culture from the Epipaleolithic period (14,500–11,500 ya).

The earliest settlement at Kebara is characterized by artifacts belonging to the Aurignacian tool tradition associated to a Neanderthal habitation. As AMH were using basically the same lithic technology and styles, and hunted the same game, it is not possible to definitively rule out a modern human occupation of the cave during this same epoch. These layers contain numerous animal bones mostly of gazelle and deer, some showing scars indicating butchering and burned bones suggesting cooking. An abundance of stone tools and a number of junk piles in the strata are consistent with a permanent camp.

Kebara has the distinction of being the source of the most complete Neanderthal skeleton recovered worldwide thus far. Although the remains do not include the head, this fossil, nicknamed "Moshe," has been dated to 60,000 ya and seems to have been ceremoniously entombed. Typical of European Neanderthals, this individual was massive. Moshe also possessed an intact hyoid bone. The hyoid is located at the base of the tongue, and it is necessary for speech. This finding suggests that Neanderthals were able to speak. The capacity to speak not only would have facilitated communication among members of the same species during communal activities such as hunting dangerous megafauna but also may have permitted conversations with early *sapiens* and AMHs during the two periods of known cohabitation.

The second period of occupation at Kebara corresponds to the Initial Upper Paleolithic, an epoch between the Middle and Upper Paleolithic spanning roughly from 50,000 to 35,000 ya. The layers associated with these deposits correspond to an AMH occupation and contains a number of Mousterian artifacts manufactured using the Levallois technique (Fig. 5.18). In addition, hearths have been uncovered. Recent estimates indicate that AMH lived at Kebara from 49,000 to 46,700 ya.[63] This new set of dates obliterates any time gap between the Neanderthal occupations of the Kebara Cave during the Middle Paleolithic and the AMH habitation during the Upper Paleolithic to a few thousands of years.

FIGURE 5.18 Levallois hand axe. *From http://www.pasthorizonspr.com/index.php/archives/03/2015/a-carpet-of-stone-tools-in-the-sahara.*

The third epoch (14,500–11,500 ya) of occupation at Kebara is linked to AMHs practicing an Epipaleolithic Early Natufian culture just prior to the Agricultural Revolution in the Levant. Natufian populations were semisedentary, and it is thought that they were the precursors of the agricultural homesteads in the Near East. In addition to hunting game, Natufians are credited for the utilization of wild cereals for subsidence. From those wild varieties, artificial selection by humans led to the domestication and cultivation of grains, such as rye, and the genesis of agriculture. It has been suggested that Natufian-speaking populations such as the one from Kebara were the ancestors of the modern Semitic-speaking populations of the Levant.[64]

The site at Kebara is particularly important due to a large communal cemetery containing the remains of 17 people, 6 adults and 11 children. The bodies were entombed with a number of artifacts including crescent-shaped microlithic utensils, sickle blades, and mortars and pestles for grinding cereals. Some of the tools buried in the mass grave were likely important private possessions related to food processing and/or subsistence, likely intended to aid the deceased in his/her travel and survive in the afterlife. The type of precious tools included in the grave represents a radical shift in the importance of food processing items as opposed to hunting weapons. Just like in the case of the El-Wad site (see below), the bodies were interred sequentially. In other words, they were not buried at the same time as the result of a mass fatal accident or catastrophe. Of particular interest in relation to this grave is that it contained a middle-aged male with a blade inserted into the seventh or eighth thoracic vertebra (middle of the back), likely the result of a violent confrontation with another human, which possibly caused his death. The embedded artifact was a Helwan lunate (sickle-shaped blade) that is typically used as a cutting tool. Furthermore, two of the other mature individuals, or three out of the six adults corps, found in the massive grave exhibit signs of injuries resulting from physical aggression. It is interesting that this type of violence is rarely seen in preagricultural settlements. It is not clear if a correlation exists between aggressive behavior and communities larger than extended families (e.g., farming villages).

El-Wad

Located on the western slope of Mount Carmel, El-Wad is the largest of four neighboring caves (Tabun, Jamal, El-Wad, and Skhūl). Its entrance and terrace open to the coastal planes of the Mediterranean. The oldest layers at El-Wad date to the Late Paleolithic approximately 45,000 ya, whereas the most recent deposits represent a Natufian occupation of 13, 000–10, 500 ya.[64] The Late Paleolithic remains at El-Wad correspond to the migration of AMHs to the region during the second phase of the Out of Africa II dispersal.

El-Wad is a cave made up of six chambers; only one compartment, number III, contains Late Paleolithic material. In addition, an Early Natufian occupation is evident in chamber III while the most recent sediments (Late Natufian) are seen in the terrace. In the terrace, the remains of stone houses, pits, and living surfaces indicate several stages of the Natufian habitation. The population at El-Wad used the cave as well as the terrace outside as a permanent hamlet. This "hybrid" style of living accommodation was unique in that it combined the old practices of seeking protection in a natural shelter with human-made stone structures. These two lifestyle elements together in a single location represent a transition from the typical temporary living quarters of hunter-gatherers to more permanent hamlets used by farmers that is consistent with several family units sharing the site from which they venture to hunt, gather food and supplies.

The tools and artifacts recovered from the site include carefully retouched and exquisitely crafted microliths. These tools differ from previous traditions in the workmanship and refinement of the stonework and the elegance of the products. Some of the utensils were used for processing animal hides. Other artifacts such as points and drills were employed to perforate wood, rocks, and bones to make clothing and decorations. Blades for cutting meat and lunate-shaped lithics for arrowheads, harpoons, and fishhooks were abundant at the location. Sickle blades as well as pestles, mortars, and grinders were recovered. It is likely that the sickles were employed to harvest grain and the grinder and pounders to pulverize the seeds. These items signal the transition to an agricultural way of life.

The El-Wad site is significant in a number of ways. It exhibits the longest time sequence of Natufian deposits in the Near East encompassing at least 3000 years of occupation. The Early Natufian is a period of paramount importance in human evolution representing a transitional period from a nomadic lifestyle to one of agriculture and animal domestication. The genesis of stone architecture in the Near East is seen in the terrace signaling the transition from cave dwelling to human-made hamlet construction. Of particular importance in El-Wad are the burial grounds involving at least 100 individuals.[65] This mode of entombment may also mark the transition from the individual burials characteristic of the Upper Paleolithic to the cemetery-type interments involving large number of individuals seen in some Natufian communities. It is not clear whether this shift in burial practices derives from the permanent nature of the settlements, the larger number of individuals at the Natufian hamlets or both.

The entombments at El-Wad are found in three locations: inside the cave, in the cave mouth, and at the terrace. Curiously, the skeletons inside the cave and at the entrance to the cave were laid on their back in an extended position. The skeletons in the terrace, on the other hand, were buried in a flexed compressed position. The bodies were forced into very small spaces. No specific position was used to jam the bodies into the grave. Some of the corpses were in seating positions, in some cases the knees were pushed up to the side of the skull while in other the knees were close to the chin. In several instances, the bodies must have been bound prior to pushing them into minute holes. At times, more than one individual was pushed together into the same tiny space. Personal items were found in some graves in all the three locations including possessions made of stone, bone, or shells. All three groups of interment are pretty much contemporary dating to the Early Natufian period. Although it is possible that the entombments inside the cave and at the entrance of the cave belong to individuals of higher social status, considering the greater amount of space allocated for their burials to accommodate extended bodies, the lack of patterns in the many different positions of corpses in the small graves in the terrace argue for randomness that may have been simply dictated by convenience.

From the remains in the cemeteries, anthropologists have deduced that the Natufian population at El-Wad was anatomically characteristic of the Eastern Mediterranean populations of the Paleolithic–Neolithic transition period. Their crania are typically large with wide and low foreheads. The average size of the people who died at El-Wad ranged from 1.58 to 1.65 m.

How Sporadic Were the Settlement Events in the Near East?

It is possible that AMHs did not populate the Near East entirely in discrete and sporadic events with intermittent periods of complete isolation from Africa. Considering the similar climatic conditions, it is likely that at times some of the human populations that settled the Near East may have been within the geographical range of the species extending from Northeast Africa to the Arabian Peninsula. In addition, as the passage from Africa to Arabia and the Levant was not an insurmountable task, due to lower sea levels and a land bridge in the Sinai Peninsula, the movement of hominins back and forth may have been more of a routine.

Along these lines, according to genetic evidence using uniparental (found on the Y chromosome or the mitochondrial DNA) genetic markers, the Middle Paleolithic AMHs have used the Levantine or the Horn of Africa passageways in numerous occasions to get to the Near East.[66] These genetic markers suggest that settlement events were more frequent than just two dispersals during the first and second phases of Out of Africa II (i.e., 125,000 and 65,000 ya, respectively). Furthermore, the genetic markers indicate preferential use of the Levantine Corridor in the period between the Upper Paleolithic and Neolithic in contrast to an overwhelming preference in favor of the Horn of Africa for the intercontinental expansion during the Middle to Upper Paleolithic. In addition, a higher frequency of sub-Saharan mitochondrial DNA (maternal inheritance) compared with Y chromosome lineages (paternal inheritance) in the Near East contemporary human populations indicates that proportionally more migratory episodes of maternal markers occurred across the African–Asian corridors since the first African exodus of AMHs. It is also possible that more maternal lineages have survived bottleneck events during the settlement of the Near East compared with paternal Y chromosome markers. Although genetic markers can be very informative in delineating migratory events, DNA types usually represent an underestimation of the number of specific dispersal events because they are based only on DNA types transported by the migrants who survived the trip and the ensuing period of time to the present. In other words, different migrations of individuals with identical DNA types could go undetected as well as migrations carrying genetic markers that were dropped from the population and never made it to the present.

Where Hominis Went Next?

Subsequent to the settlement of the Near East, hominins likely traveled into Eurasia using three major routes, including a coastal route crossing the Strait of Hormuz into what is now Persia and continuing bordering the shores around the Indian subcontinent. This trajectory is usually referred as the Coastal Route and was used by AMHs on several occasions[66] and likely by other members of the genus *Homo* such as *habilis* and *erectus* and possibly by some australopithecines as well. These waves of AMHs eventually colonized the islands of Southeast Asia, Near Oceania, Australia, New Zeeland, and the islands of the Pacific Ocean as well as Madagascar, amazedly moving back west across the Indian Ocean. Chapter 7 delineates these trajectories in more detail. Another dispersal course went transversally from the area currently occupied by northern Pakistan to Central Asia and beyond to Northeast Asia. These migrants eventually traveled toward America across the Bering Strait. Chapter 10 describes this migration across Asia to Tierra del Fuego at the tip of South America. A third expansion led to the peopling of Europe (see Chapter 12). It is likely that the AMHs in the Near East moved northward and westward colonizing Eastern and Western Europe by way of the Caucasus and Anatolia, respectively. These three main migrational routes taken by AMHs occurred approximately at the interface between the Middle and Upper Paleolithic (~40,000 ya).

This was a critical and pivotal period in hominin evolution involving a number of major technological and cultural developments that facilitated the penetration and the adaptation to various habitats and climatic conditions.

Several of the following chapters of this book will explore the various paths taken by hominins out of the Near East that led to the peopling of the world.

References

1. Albright WF. *The archaeology of Palestine*. Great Britain: Penguin Books; 1954.
2. Groucutt HS, Petraglia MD. The prehistory of the Arabian peninsula: deserts, dispersals, and demography. *Evol Anthropol* 2012;**21**:113–25.
3. Parton A, Farrant AR, Leng MJ, et al. Alluvial fan records from southeast Arabia reveal multiple windows for human dispersal. *Geology* 2015;**43**:295–8.
4. Castañeda AS, et al. Wet phases in the Sahara/Sahel region and human migration patterns in North Africa. *Proc Natl Acad Sci USA* 2009;**106**:20159–63.
5. Bar-Yosef O, Belfer-Cohen A. From Africa to Eurasia–early dispersals. *Quat Int* 2001;**75**:19–28.
6. Alimen H. "Les "Isthmes" hispano-marocain et Sicilo-Tunisien aux temps Acheuléens. *Law Anthropol Int Yearbk Leg Anthropol* 1975;**79**:399–436.
7. Wood B. Fifty years after *Homo habilis*. *Nature* 2014;**3**:31–3.
8. Templeton AR. Haplotype trees and modern human origins. *Am J Phys Anthropol* 2005;**48**:33–59.
9. Tattersall I, Schwartz IJH. *Extinct humans*. New York: Westview Press; 2001.
10. Schrenk F, Kullmer O, Bromage T. The earliest putative *Homo* fossils. In: Henke W, Tattersall I, editors. *Handbook of paleoanthropology*. Berlin: Springer Verlag; 2007. Chapter 9.
11. Villmoare B, Kimbel H, Seyoum C, et al. Early *Homo* at 2.8 ma from Ledi-Geraru, Afar, Ethiopia. *Science* 2015;**347**:1352–5.
12. Spoor F, Leakey MG, Gathogo PN, et al. Implications of new early *Homo* fossils from Ileret, east of Lake Turkana, Kenya. *Nature* 2007;**448**:688–91.
13. Dembo M, Matzke NJ, Mooers A, et al. A complete skull from Dmanisi, Georgia, and the evolutionary biology of early *Homo*. *Proc Biol Sci* 2015;**7**:282.
14. Lordkipanidze D, Ponce de León MS, Margvelashvili A, et al. A complete skull from Dmanisi, Georgia, and the evolutionary biology of early *Homo*. *Science* 2013;**342**:326–31.
15. Schwartz JH, Tattersall I, Zhang C. Comment on "a complete skull from Dmanisi, Georgia, and the evolutionary biology of early *Homo*". *Science* 2014;**344**:360.
16. Baab KL. The place of *Homo floresiensis* in human evolution. *J Anthropol Sci* 2016;**94**:5–18.
17. Morwood MJ, Jungers WL. Conclusions: implications of the Liang Bua excavations for hominin evolution and biogeography. *J Hum Evol* 2009;**57**:640–8.
18. Brown P, Maeda T. Liang Bua *Homo floresiensis* mandibles and mandibular teeth: a contribution to the comparative morphology of a new hominin species. *J Hum Evol* 2009;**57**:571–96.
19. Dowsett H, et al. Joint investigations of the Middle Pliocene climate. I: PRISM palaeoenvironmental reconstructions. *Global Planet Change* 1994;**9**:169–95.
20. Dennell R, Roebroeks W. An Asian perspective on early human dispersal from Africa. *Nature* 2005;**438**:22–9.
21. Cox MP, Mendez FL, Karafet TM, Pilkington MM, Kingan SB, Destro-Bisol G, Strassmann BI, Hammer MF. Testing for archaic hominin admixture on the X chromosome: model likelihoods for the modern human RRM2P4 region from summaries of genealogical topology under the structured coalescent. *Genetics* 2008;**178**:427–37.
22. Hammer MF, Woernera AE, Mendezb FL, Watkinsc JC, Walld JD. Genetic evidence for archaic admixture in Africa. *Proc Natl Acad Sci USA* 2011;**108**:15123–8.
23. Dennell RW. Dispersal and colonisation, long and short chronologies: how continuous is the early pleistocene record for hominids outside East Africa? *J Hum Evol* 2003;**45**:421–40.
24. Garcia T, Féraud G, Falguères C, et al. Earliest human remains in Eurasia: new 40Ar/39Ar dating of the Dmanisi hominid-bearing levels, Georgia. *Quat Geochronol* 2010;**5**:443–51.
25. Gabunia L, et al. Earliest pleistocene hominid cranial remains from Dmanisi, Republic of Georgia: taxonomy, geological setting, and age. *Science* 2000;**228**:1019–25.

26. Ferring R, et al. Earliest human occupations at Dmanisi (Georgian Caucasus) dated to 1.85–1.78 Ma. *Proc Natl Acad Sci USA* 2011;**108**:10432–6.
27. Braun D, Ron H, Marco S. Magnetostratigraphy of the hominid tool-bearing Erk el Ahmar formation in the northern Dead Sea Rift. *J Earth Sci* 1991;**40**:191–7.
28. Bar-Yosef O. Prehistory of the levant. *Annu Rev Anthropol* 1980;**9**:101–33.
29. Webb SG. *The first boat people*. Cambridge, UK: Cambridge University Press; 2006.
30. Tchernov E. *H. erectus* in the near east. The age of the 'Ubeidiya formation, and early Pleistocene hominid site in the Jordan River Valley, Israel. *Isr J Earth-Sci* 1987;**36**:3–30.
31. Tobias PV. Fossil hominid remains from ubeidiya, Israel. *Nature* 1966;**211**:130–3.
32. Repenning CA, Fejfar O. Evidence for earlier date of 'Ubeidiya, Israel, hominid site. *Nature* 1982;**299**:344–7.
33. Hanan G, Zilberman E, Saragusti I. Early pleistocene lake deposits and Lower paleolithic finds in Nahal (wadi) Zihor, Southern Negev desert, Israel. *Quat. Res.* 2003;**59**:445–58.
34. Biglari F, Shidrang S. The Lower Paleolithic occupation of Iran. *Near E Archaeol* 2006;**69**:160–8.
35. de Lumley MA, Gabunia L, Vekua A, et al. Human remains from the upper Pliocene-early pleistocene Dmanisi site, Georgia (1991-200). Part I: the fossil skulls (D2280, D2282, and D2700). *Law Anthropol Int Yearbk Leg Anthropol* 2006;**110**:1–110.
36. Reid GB, Hetherington R. *The climate connection: climate change and modern human evolution*. Cambridge, UK: Cambridge University Press; 2010.
37. Burger L. Our story: human ancestor fossils. *Naked Sci* 2007. Science Interviews http://www.thenakedscientists.com/HTML/content/interviews/interview/833/.
38. Carretero J-M, Rodríguez L, García-González R, et al. Stature estimation from complete long bones in the Middle Pleistocene humans from the Sima de los Huesos, Sierra de Atapuerca (Spain). *J Hum Evol* 2012;**62**:242–55.
39. Matthias M, Qiaomei F, Ayinuer A-P, et al. A mitochondrial genome sequence of a hominin from Sima de los Huesos. *Nature* 2014;**505**:403–6.
40. Matthias M, Arsuaga J-L, de Filippo C, et al. Nuclear DNA sequences from the Middle Pleistocene Sima de los Huesos hominins. *Nature* 2016;**531**:504–7.
41. Cartmill M, Smith FH. *The human lineage*. London: John Wiley & Sons; 2009.
42. Meredith M. *Born in Africa: the quest for the origins of human life*. New York: PublicAffairs; 2011.
43. Dawkins R. *Archaic Homo sapiens. The Ancestor's Tale*. Boston: Mariner, Boston; 2005.
44. Bar-Yosef O. *The geography of Neandertals and modern humans in Europe and the Greater Mediterranean*. Cambridge, Massachusetts: Peabody Museum, Harvard Univ; 2000.
45. Mellars P. Neanderthal symbolism and ornament manufacture: the bursting of a bubble? *Proc Natl Acad Sci USA* 2010;**107**:20147–8.
46. Bar-Yosef O, Bordes J-G. Who were the makers of the Châtelperronian culture? *J Hum Evol* 2010;**2010**(59):586–93.
47. Soressia M, McPherronc SP, Lenoire M, et al. Neandertals made the first specialized bone tools in Europe. *Proc Natl Acad Sci USA* 2013;**110**:14186–90.
48. Mellars PA. The Neanderthal problem continued. *Curr Anthropol* 1999;**40**:341–50.
49. Lewin R, Foley RA. *Principles of human evolution*. Malden, MA: Blackwell Publishing; 2004.
50. Coqueugniot H, Dutour O, Arensburg B, et al. Earliest cranio-encephalic trauma from the Levantine Middle Palaeolithic: 3D reappraisal of the Qafzeh 11 skull, consequences of pediatric brain damage on individual life condition and social care. *PLoS One* 2014;**9**(7).
51. Bar-Yosef Mayer DE, Vandermeersch B, Bar-Yosef O. Shells and ochre in Middle Paleolithic Qafzeh Cave, palestine: indications for modern behavior. *J Hum Evol* 2009;**56**:307–14.
52. Stewart TD. Form of the pubic bone in Neanderthal man. *Science* 1960;**131**:1437–8.
53. Armitage S, Jasim SA, Marks AE, et al. The southern route out of Africa: evidence for an early expansion of modern humans into Arabia. *Science* 2011;**331**:453–6.
54. Groucutt HS, Petraglia MD. The prehistory of the Arabian peninsula: deserts, dispersals, and demography. *Evol Anthropol Issues News Rev* 2012;**21**:113–25.
55. Liu W, Martinón-Torres M, Cai YJ, et al. The earliest unequivocally modern humans in southern China. *Nature* 2015;**526**:696–9.
56. Luca Pagani, et al. Genomic analyses inform on migration events during the peopling of Eurasia. *Nature* 2016;**538**:238–42.
57. Kuhlwilm M, et al. Ancient gene flow from early modern humans into Eastern Neanderthals. *Nature* 2016;**530**:429–33.

58. Ambrose SH. Late Pleistocene human population bottlenecks, volcanic winter, and differentiation of modern humans. *J Hum Evol* 1998;**34**:623–51.
59. Hawks J, Wang ET, Cochran GM, et al. Recent acceleration of human adaptive evolution. *Proc Natl Acad Sci USA* 2007;**104**:20753.
60. Hershkovitz I, Marder O, Ayalon A, et al. Levantine cranium from Manot Cave (Israel) foreshadows the first European modern humans. *Nature* 2015;**520**:216–9.
61. Grün R. Direct dating of human fossils. *Am J Phys Anthropol* 2006;**131**(Suppl. 43):2–48.
62. Zilhao J. Origins of human innovation and creativity. In: Elias SA, editor. *Personal ornaments and symbolism among the Neanderthals*. Amsterdam, The Netherlands: Elsevier; 2012.
63. Rebollo NR, Weiner S, Brock F, et al. New radiocarbon dating of the transition from the Middle to the Upper Paleolithic in Kebara Cave, Israel. *J Archaeol Sci* 2011;**38**:2424–33.
64. Bar-Yosef O. The Natufian culture in the levant, threshold to the origins of agriculture. *Evol Anthropol* 1998;**6**:159–77.
65. Mastin BA. The extended burials at the Mugharet el-Wad. *J Roy Anthropol Inst G B Ireland* 1964;**94**:44–51.
66. Rowold DJ, Luis JR, Terreros MC, et al. Mitochondrial DNA gene flow indicates preferred usage of the Levant Corridor over the Horn of Africa passageway. *J Hum Genet* 2007;**52**:436–47.

Further Reading

Hublin J-J, Ben-Ncer A, Bailey SE, et al. New fossils from Jebel Irhoud, Morocco and the pan-African origin of *Homo sapiens*. *Nature* 2017;**546**:289–292.

Hershkovitz I, Weber GW, Quam R, et al. The earliest modern humans outside Africa. *Science* 2018;**359**:456–459.

CHAPTER

6

Neanderthals, Denisovans, and Hobbits

Being human means asking the questions of one's own being and living under the impact of the answers given to this question. And, conversely, being human means receiving answers to the questions of one's own being and asking questions under the impact of the answers. ***Paul Tillich, Systematic Theology, vol. 1.***

SUMMARY

Of all archaic human genera and species, Neanderthals are by far the most familiar to the public. Neanderthals are often portrayed as cavemen and brutish dullards in movies, media, and books, and the term Neanderthal is often used as a pejorative when describing someone's cognitive ability or poise. Regardless of public perceptions, Neanderthals are the best studied of all extinct hominins and are our closest extinct relatives. Although time estimates of Neanderthal extinction and fossil aging differ among scientists and archeologists, detailed studies of Neanderthal lifestyle and their genetics have begun to reveal that they lived a life similar to that of humans of the time. However, even after anatomical reconstruction and archeological findings, there is still disagreement as to whether the Neanderthals were a subspecies of *Homo sapiens sapiens* (*Homo sapiens neanderthalensis*) or belonged to their own separate species *Homo neanderthalensis*. The argument over the placement is even more complicated today because of the classical definition of species and recent genetic discoveries of admixture between Neanderthals and humans. Remarkably, Neanderthals made beneficial, as well as detrimental, contributions to the genomes of some modern humans, further calling into question whether Neanderthals were, in fact, a distinct species from *H. sapiens*.

A closely related species to the Neanderthals are the Denisovans. Fossil remains of this group were discovered in 2008, in the Denisova Cave, in the Altai Mountains of Siberia. DNA isolated from the remains was expected to be of modern human or Neanderthal origin because both groups were predicted to have been in the vicinity during the time period. Instead, the DNA differed from both humans and Neanderthals. The DNA sequence was found to be more closely related to Neanderthals, and although the public does not commonly recognize

Ancestral DNA, Human Origins, and Migrations
https://doi.org/10.1016/B978-0-12-804124-6.00006-9

this group of hominins, scientists have established them as a significant group that contributed to the evolution of modern humans. Genetic studies have indicated that they interbred with humans and contributed a significant amount of DNA to the genomes of modern people in Southeast Asia, Australia, and parts of Oceania.

An even lesser recognized group of hominins, *Homo floresiensis,* comes from the discovery of skeletal remains in Indonesia on the island of Flores. Today the fossilized remains are commonly referred to as the hobbit, because of this group's small stature, and in reference to a fictional humanoid group portrayed in a well-known trilogy by J. R. R. Tolkien. The remains continue to be controversial and an enigma with regard to their status as a species and their phylogenetic placement in human evolution.

It is not entirely clear when Neanderthals, Denisovans, and humans diverged from one another; however, many scientists agree that the common ancestor was *Homo heidelbergensis.* One current estimate of divergence based on DNA evidence predicts that humans split from *H. heidelbergensis* between 550,000 and 765,000 years ago, whereas Neanderthals and Denisovans diverged from each other between 381,000 and 473,000 years ago.[1] Other estimates suggest *H. heidelbergensis,* inhabited Africa, Europe, and Asia from around 600,000 to 200,000 years ago. The later model assumes that after *H. heidelbergensis* left Africa they evolved into *H. neanderthalensis* in Europe and Denisovans in Asia, whereas *H. heidelbergensis* populations that remained in Africa gave rise to *H. sapiens,* around 200,000 years ago.[2,3] The idea that *H. heidelbergensis* was the sole species that gave rise to modern humans has been heavily disputed recently, and there is mounting evidence that modern humans may have arisen from the interbreeding of a variety of hominin ancestors from East, West, and Southern Africa, before what we know of as anatomically modern humans left Africa. The different hominin groups could have been subspecies of *H. heidelbergensis* or a different *Homo* species. The topic is discussed further in Chapter 2.

ANTHROPOLOGICAL AND GENETIC TOOLS TO STUDY HUMAN EVOLUTION

Anthropologists use a variety of tools to help study evolution. There are several important subdisciplines under the heading of anthropology that are used when studying evolution. These are as follows: archeology, paleontology, and linguistics. Archeologists study ancient peoples and their culture by analyzing artifacts, graves, tools, inscriptions, architecture, and cultural and environmental landscapes, in an effort to learn more about human evolution, past human life, and the structure of prehistoric societies. Paleontologists study fossils to determine how organisms evolved and how they interacted with their environment. Paleontologists use tools from biology, geology, mathematics, physics, and engineering to determine the evolutionary history of life especially with regard to origin and identity of specimens. Historical comparative linguists study how languages have evolved over time. Morphology and semantics of language structure can be useful in following the migration of individuals and the evolution of language.

For many years, anthropologists used fossilized remains, ancient stone and bone tools, linguistics, artifacts (pottery, jewelry, etc.), and radiocarbon dating as the only means of investigating our evolutionary past. These were very useful for looking at body shape, functional anatomy, diet, energetics, and culture but told us nothing about the molecular genetics of the different hominin groups. A major breakthrough in molecular biology and molecular anthropology was the ability to isolate ancient DNA from fossilized remains and sequence the DNA of extinct animals and hominins. The first successful isolations of ancient DNA were done in the late 1980s by isolating and sequencing mitochondrial DNA (mtDNA). mtDNA is found in mitochondria (small subcellular structures) known as the powerhouses of the cell. The mtDNA is a small circular molecule that is typically passed down from females to all of their offspring and is more easily isolated than DNA from the nucleus. The process of isolating DNA has been vastly improved since the first attempts, and today we are able to isolate nuclear DNA from ancient remains and sequence the genome (all the DNA from a single individual) of ancient people.

The ability to sequence the genomes of ancient hominins has resulted in our ability to compare the genomes of modern-day humans with ancient human-like species to identify the differences in the DNA sequence that occurred over time. The DNA mutations (changes in the DNA sequence) can be helpful in identifying what led to our current-day status as modern humans and help us understand why other closely related species went extinct. In this chapter, we will discuss both the anthropological and molecular tools that have been used to study three different *Homo* species (Neanderthals, Denisovans, and Hobbits).

NEANDERTHAL ORIGIN AND FOSSIL DISCOVERY

Studying the fossil remains of hominins can provide information about the evolution of body shape, functional anatomy, diet, energetics, culture, and genetics. Often times, fossil records consist of only fragmented specimens. Today, advances in technology, especially X-ray computed axial tomography (CT scan) and digital technology, have allowed investigators to capture the shape of a specimen using only fragments to measure cranial morphology and compare different specimens for identification. A CT scan uses computer-processed combinations of X-ray images taken from different angles to reconstruct a 3D image. This technology has been used to scan rare specimens or parts of specimens embedded in rock, and to compare the fossils of Neanderthals and other extinct hominins with modern humans, to produce a virtual image and compare different morphologies. The process can also be used to share virtual data among scientists without ever touching the original specimen.

Paleontologists do not agree when Neanderthals can first be recognized in the fossil record, and estimates for their appearance range from 200,000–300,000 years ago. The skeletal remains of more than 500 Neanderthals have been uncovered, with approximately half of them children. Neanderthal remains have not been found in Africa with the closest discoveries to Africa in the Levant and Gibraltar, adding evidence to the theory that Neanderthals evolved outside of Africa (Fig. 6.1). The geographic distribution of the fossil remains suggests that Neanderthals inhabited western and central Europe, Ukraine and Western Russia, and Western, Central, and Northern Asia up to the Altai Mountains.[1,4]

The first Neanderthal remains were discovered in 1856 near the city of Dusseldorf, Germany in a gorge known as the Neander Valley. While excavating in a limestone formation for the steel industry, miners discovered a skullcap, ribs, and part of a pelvis. When examined by a local science teacher, the bones were immediately recognized as coming from

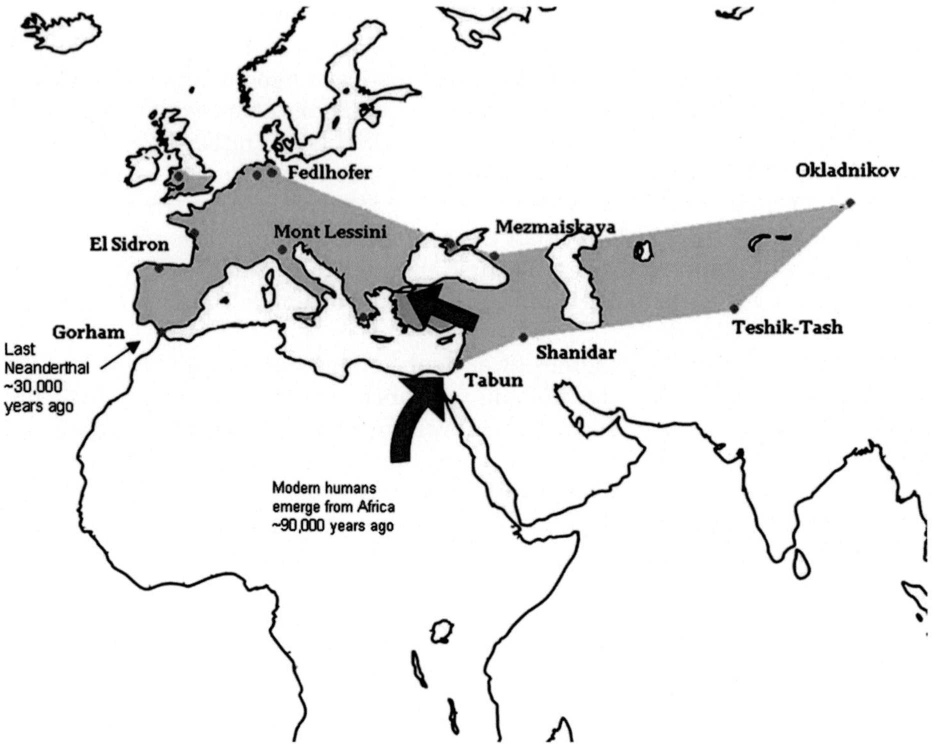

FIGURE 6.1 Neanderthal domain. Red dots represent regions where Neanderthal fossils have been found. *From http://www.ncbi.nlm.nih.gov/pubmed/23872234.*

a human-like source. Although the skull was clearly similar to a human in morphology, it had unusual eyebrow ridges and a retreating forehead that made the specimen look almost pathological, and for many years, most scientists regarded the specimen as a contemporary human pathological specimen. In 1863, William King, a professor at Queen's College in Ireland, suggested the remains be recognized as a different species, *H. neanderthalensis*, thereby acknowledging the fossils' link to humanity, yet distinguishing it from *sapiens*. Although he later retracted his statement, the specimen type kept the genus and species name.

Several years later (1848), in Belgium a Neanderthal jawbone was revealed that had been discovered in 1829 by Philippe-Charles Schmerling (before the type specimen described above), and in 1886 two Neanderthal skulls were found along with the remains of ancient mammoths, woolly rhinoceroses, and stone tools. These fossils, along with ensuing discoveries, began to convince 19th-century scientists that the Neanderthal bones were ancient and might represent robustly built humans or human forebears.

In the early 1900s, a large number of Neanderthal fossils were found in the Dordogne region of southwestern France. The region contained a number of almost complete Neanderthal skeletons that were reconstructed to form the first impression of a Neanderthal.[5]

The discovery of a female skeleton during the early 1930s in Israel at Mugharet et-Tabun, and a second find at the Mount Carmel site, Mugharet es-Skhul, produced remains of 10 individuals. The find was significant because it represented individuals that resembled Neanderthals but with some modified features (delicate jaw, smaller brow ridge, more pronounced chin, a more modern shaped cranium). These individuals clearly represented a transitional middle ground between humans and previously discovered Neanderthals. Later analyses of some of these individuals suggested that these specimens represent early *H. sapiens*.[6]

More recently, archeologists returning to the Neander Valley site in 1997 found the remains of three Neanderthal individuals, stone tools, and animal remains. Radioactive dating of the material indicated that the fossils were from Neanderthals that had lived 42,000 years ago.[7,8]

NEANDERTHAL SKELETAL MORPHOLOGY AND BODY TYPE

Fossil remains of Neanderthals reveal that they had a stocky build, with a wide short pelvis, heavily muscled limbs, prominent brow, large head, retreating forehead, and short shinbones. Their skulls showed a large nose, chinless lower jaw, and mid-facial prognathism (condition where the jaw protrudes resulting in the top and lower teeth misaligning).[8]

The Neanderthal cranial capacity of 1200–1700 cm^3 is notably larger than that of the modern human average of 1350–1400 cm^3 and larger than that of the proposed Neanderthal predecessors *H. heidelbergensis*[5]. Studies of 17 skulls from the Sima de los Huesos in northern Spain were dated to 430,000 years ago and thought to be from Neanderthal precursors. The skulls were much smaller than those of Neanderthals. The studies indicate that the Neanderthals developed larger skulls than their ancestors and that there was room for other neural development during the evolution of the Neanderthal brain.[9] Studies of juvenile and adult skull shape in humans and Neanderthals indicate that the brain of Neanderthals started with a similar developmental pattern in the womb but diverged after birth. However, it has been argued that humans who lived during the same period as Neanderthals had roughly the same cranial capacity as Neanderthals and that our contemporary cranial capacity evolved to accommodate our smaller stature.[10]

Based on skeletal reconstructions, the height of Neanderthals was approximately 165–168 cm for males and 150–156 cm for females. In comparison, fossils of Cro-Magnon (early European humans) ranged from 163 to 183 cm. Although average height has varied considerably over time, geographic region, nutritional source, and social status, the average height of modern humans at the time is thought to have been larger than that of the Neanderthals.[11]

Isotope studies of Greenland ice have shown that during the Neanderthal occupation of Europe, there were dramatic fluctuations in climate involving glacial and interglacial conditions. This unstable climate 65–25 kya is referred to as oxygen isotope stage 3 (OIS-3). OIS-3 was an interval between the two cold maxima—near-interglacial and peak-glacial conditions. It began with mild conditions and ended with ice covering northern Europe.[12] Neanderthals were the only hominins in Europe at the beginning of OIS-3 with modern humans entering Europe toward the end. The OIS-3 studies point out that the climate was increasingly unstable toward the end of the last glacial maximum, which would have brought on dramatic ecological changes. Neanderthals would have needed to adapt to these conditions to survive.[13] Large husky body structures are better at conserving heat, and because of the cold European ice age, the large husky build of Neanderthals could be a result of strong positive selection, allowing them to survive the extreme cold.[14] (This theory goes along with Allen's rule of body shapes and proportions varying by climatic temperature and Bergmann's observations that species of larger size are found in colder climates. The idea is that large, stockier masses minimize the ratio of surface area to weight and reduce heat loss). Data indicating that Neanderthals may have contributed keratin genes to the modern human gene pool suggest that keratin may have also contributed to their ability to adapt to the cold.[15] Keratin provides toughness to skin, hair, and nails and can be beneficial in colder environments by providing thicker insulation.

Neanderthal thorax structure indicates a large lung capacity. This may have helped with high energy demands and could have had biomechanical consequences for stronger muscle attachment.[16] Their small stature, broad shoulders, and large hands may have been ideal for hunting large mammals that lived in the region.[5,9] The development of the large brow and head is not easily explained by selection but instead may have been a result of random genetic drift.

Neanderthal birth and maturation rate have been proposed by studying skeletal remains. For example, the Neanderthal superior pubic ramus (pubic bone located at the front and base of the pelvis) was longer than that of modern humans. Researchers have suggested that the bone was longer than that of modern humans because of a larger birth canal in Neanderthals needed to accommodate the relatively large head size, or alternatively due to a difference in Neanderthal locomotion and posture.[17] No one knows for sure what the larger pubic ramus bone indicates, but reconstruction studies on infant skeletons have indicated that the gestation period and head size of Neanderthals was probably similar to those of modern humans.[18] The rate of head growth differed after birth leading to the larger Neanderthal head and other anatomical differences.[19] In addition, evidence from the development of human tooth enamel and comparisons with the enamel from Neanderthal fossils suggests that Neanderthal development was about 15% faster after fetal development than that of modern humans.[4,20] This could suggest that Neanderthals sexually matured at a faster rate leading to earlier reproduction. Faster maturation could have evolved as a way to overcome high mortality rates; however, it is still unclear if Neanderthals sexually matured at a faster rate than humans.

Three-dimensional reconstruction of skeletal remains has allowed scientists to point out small anatomical differences emphasizing the importance of reviewing the remains of old excavations; however, whereas skeletal remains indicate clear anatomical differences between Neanderthals and modern humans of the time, it is important to remember that both *H. sapiens* and *H. neanderthalensis* had diverse body types and that the ranges provided are those of individual remains. Just like today, we would not say all human males are between 1.7 and 1.9 m tall, weigh 80–85 kg, and have a cranial capacity of 1350–1375 cm; we cannot say all species of the time were identical to those whose remains we have uncovered, but the fossils do provide some clue to species anatomy.[20,21]

Over the years, full-sized reconstructions of skeletal remains have provided a complete picture of Neanderthal anatomy and bodily features.[4] After reconstruction of archeological findings there is still disagreement as to whether the Neanderthal's were a subspecies of *H. sapiens sapiens* (*H. sapiens neanderthalensis*) or belonged to their own separate species *H. neanderthalensis*. The argument over the placement is even more complicated today. This complication arises from the classical definition of species and some recent genetic discoveries, which will be discussed in the section on introgression.

The classical definition of species is a population of individuals capable of interbreeding and producing fertile offspring (i.e., different species cannot interbreed). Over the years, the term "species" has been loosely defined in many instances, and many scientists perceive species as a moving evolutionary target (i.e., a snapshot in time that is sometimes hard to pin down). An extension of this idea is the notion that species can be a population that has been geographically and temporally isolated but not necessarily reproductively isolated. The problem with the current definition of species with regard to ancient fossils of the genus *Homo* is that the species label has always been based on skeletal morphology. Based on skeletal morphology, Neanderthals are considered by most biologists as a separate species and not as a subspecies or race of *H. sapiens*. Yet, as we will see later in the chapter, based on the classical definition of species and recent genetic findings, the morphological labeling of Neanderthals and Denisovans as separate species is questionable because there was breeding among human, Neanderthal, and Denisovan populations.[4]

NEANDERTHAL LANGUAGE

Language is a social invention, meant to serve social or public purposes and convey ideas through sounds or symbols. Language can be thought of as complex and multifaceted when used as a mechanism for discourse or as a simple method of categorizing objects and basic ideas (similar to what we see in very young children or some mammals). Modern human language is thought to have slowly evolved into a mechanism for discourse through both biological and culture evolution, so any initial linguistic abilities of Neanderthals and early modern humans must have been a very simplistic method of communication. A study that looked at genetics, paleontology, and linguistics concluded that humans and Neanderthals may have shared some components of speech and proposes that not only did Neanderthals use language but that *H. heidelbergensis* may have also attained some form of speech.[22]

Initial investigations of Neanderthal remains led several authors to argue that Neanderthals did not have the physical characteristics suitable for speech. However, several anatomical features of Neanderthals suggest that they had the ability for speech. An anatomical feature that may be related to speech is the mental protuberance, which makes up the point of the chin and is responsible for the shape and size of the chin. The mental protuberance contains the mentalis muscle, which helps to move the lower lip. The mental protuberance has been seen in several Neanderthal fossils.[23] The large Neanderthal brain reflected by endocranial casts resemble that of modern man in an area important for speech, suggesting Neanderthals had the neural development necessary for language.[24] In addition to brain size, a 3D X-ray imaging of a Neanderthal hyoid bone (horseshoe-shaped bone in the neck) revealed that the bone was indistinguishable from modern humans. The hyoid allows for the proper interactions between the tongue and larynx to articulate sounds, which supports the notion that Neanderthals possessed the ability to develop speech.[25]

Although language acquisition is a multifactorial trait (i.e., no single gene is responsible for language, and it is dependent on environment) scientists have tried to identify single genes that contribute to linguistic ability. In 2001, the *FOXP2* gene was implicated in the development of language externalization and speech, and mutant forms of *FOXP2* were found to lead to defects in orofacial movements, language processing, and/or reading.[26] The gene codes for a protein that is common in mammals and is expressed more in females. The form of FoxP2 protein unique to humans has two small changes in the building blocks of the protein (amino acids) that appear to have been fixed in the human lineage since our split with chimpanzees (i.e., FoxP2 protein found in humans is unique among mammals and is not found in nonhuman primates).

Studies of ancient DNA have shown that Neanderthals had the same *FOXP2* gene that is found in modern humans. Krause et al. found these same two amino acid substitutions in two Neanderthal samples.[27] The findings of Krause et al. were challenged as possibly being the result of modern human contamination;[28] however, since the initial sequence was published, it has been confirmed that the *FOXP2* genes from Neanderthals and humans are the same, but the genes are not located on the same chromosome and not regulated in the same way.[29] Further studies by Vernot and Akey confirmed that the Neanderthal version of *FOXP2* is not found in modern human genomes, suggesting that the Neanderthal version of *FOXP2* was lost in Neanderthal–human hybrids by natural selection. They also suggest that *FOXP2* may not have functioned as an adequate component for modern language in Neanderthals.[30] Most scientists believe that although *FOXP2* is a possible contributor to speech, it is not the only source of linguistic ability.

The *CNTNAP2* gene has been associated with language development in children, and mutations in ASPM gene and the MHC1 gene are associated with tone deafness. All three genes are absent in the Neanderthal and Denisovan genomes, suggesting that if Neanderthals did have language it may have differed from human language.[31]

Although there is speculation on many fronts regarding the evolution of language and whether Neanderthals could speak, the truth is that all the conclusions are based on assumptions and in some cases weak archeological and genetic evidence. There must have been some form of communication that would allow Neanderthals to hunt, reproduce tools, and pass on some forms of knowledge; however, studies of nonhuman animals, archeological evidence, genetics, and modeling have not provided sufficient data to come to any concrete conclusions. This lack of solid evidence should not be surprising because currently we know very little about the genetics and evolution of modern human language let alone that of Neanderthals.[32]

NEANDERTHAL DIET

Diet is a significant component of any cultural analysis, and food-centered activities influence even modern human cultures. A group's diet can determine its activity, group size, predation risk, social structure, and its potential cognitive skills.[33] To infer the source of ancient diets, scientists use a variety of methods, including habitat reconstruction, floral and faunal collections at fossil sites, dental wear, dental tartar, and isotope analysis. Habitat reconstructions can provide limited information about the climate, the variety of flora and fauna at the time, and the overall environment at the time, because there may be no modern analogues for comparisons. Even when there is ample evidence of flora and fauna in the region during a specific period (through pollen and bone analysis) the potential food resources are not necessarily those used in the diet.[33]

Most analysis on diet is the result of findings at a fossil site. The accumulations of various animal bones, the distribution of bones, and cut marks on the bones have all been used to determine diet; however, there is often little consideration as to how the animal remains got to the site (washed in, brought in by predators, already at the site when hominins arrived, etc.) and little consideration that food may have been prepared at the scavenger or hunting site and bones left behind, and/or meat brought back to the living area. The same can apply for plant remains. Plant material, mostly pollen and seeds, found at fossil sites may be the result of windblown activity or animals that deposited the material. Plant material found in coprolites (fossilized feces) is a better indication of ingestion. However, even coprolites can contain pollen and phytoliths (plant microscopic silica structures) blown in and attached to coprolites, and it is difficult to determine the age of coprolites.[34]

Tooth wear can also be used to predict diet, but only specific types of dental wear can be used to infer diet because teeth can be worn when used as tools, and wear can result from dust during food ingestion.[35] Optical 3D topometry is a noninvasive technique for 3D scanning of the tooth surface to detect roughness or other characteristics. The characteristics of the scans can be compared with those of modern people whose diets are known. This technology has been used to determine that Neanderthal diets varied between warmer and colder climates with colder climates being high in protein.[33,34] Looking at dental micro-wear, El Zaatari et al.[36] found that most Neanderthals had micro-wear similar to populations that eat mostly meat, but that some populations found in forest habitat had a mixed diet (both plant and animal).

Dental tartar (a hardened deposit of biofilm and oral fluids i.e., mineralized plaque) can also be used to study mineral deposits left by food consumption. These studies have suggested that Neanderthals had a very broad diet that included meat and a variety of plants. Mineral deposits left on tooth tartar suggest that plant material was a source of food or possibly medicine.[37] These studies show various lipids and protein products of a variable diet. However, it has not been clearly shown that these plant deposits could not have come from chyme (the stomach contents of prey) that may have consumed the plant material and is considered a delicacy in several cultures.[33]

The stable isotope ratios $^{13}C/^{12}C$ and $^{15}N/^{14}N$ in collagen recovered from fossils is used for diet reconstruction. ^{13}C and ^{18}O values from tooth enamel can sometimes discriminate among terrestrial and marine foods. To make predictions, the stable isotope ratio from two different isotope pairs in a sample is measured relative to a standard producing an accurate estimate of food source. Bone chemistry from Neanderthal sites and the predicted cold habitat of the time suggests that large mammals were a primary source of food.[38] Fossil evidence from Gorham's Cave and Vanguard Cave reveal that Neanderthals consumed dolphins, seals, shellfish, birds, and rabbits. The isotopic analysis suggests that Neanderthals were at the top of the food chain consuming large mammals.[39,40] The problem with isotope analysis is that it only measures protein, therefore it is difficult to measure plant consumption. The isotopic ratios can be used to measure plant proteins, but it is not clear if the plant proteins were directly consumed by the Neanderthals, or if plants were consumed by animals, followed by subsequent consumption of the animals by Neanderthals.

Most anthropologists agree that the primary food source for Neanderthals was most likely large mammals. These would have supplied a high protein diet needed to survive in the cold climate. A healthy diet, however, would have required the consumption of carbohydrates in addition to animal protein (because high protein diets can cause ketosis and ammonia toxicity). Carbohydrates could be derived from seasonal fruits and other plant material or through the stomach contents of the plant-eating prey.[33,41]

NEANDERTHAL BEHAVIOR

Since the discovery of Neanderthal remains, the study and speculation of their behavior and cultural characteristics has been an ongoing debate. All of the evidence for Neanderthal behavior lies in their skeletal remains, the artifacts they left behind, and their genomes. Some of these sources provide direct evidence for behaviors, whereas others provide indirect evidence, or are open to speculation. Skeletal remains and artifacts can provide direct evidence for use of fire, shelters, diet, environment, use of tools, indications of violence, some diseases, and possible hunting strategies. The genome can provide evidence for disease, some external phenotypes, and some physiological characteristics. Less evident are the group and/or community interactions, water and land navigation, family structure, communication, and day-to-day activities. Cognitive, innovative, and emotional abilities of Neanderthals are almost purely speculative because of the inability of those traits to be fossilized and our lack of understanding of the genetics of these complex traits.

A number of behavioral differences have been noted when comparing modern humans with Neanderthals. With regard to lifestyle, fossil evidence suggests that Neanderthal males, females, and juveniles participated in a narrow range of activities that centered on obtaining large prey. The absence of role differentiation in the Neanderthals is consistent with the distinctive anatomical features of Neanderthal populations, the small groups found at archeological sites, and their perceived lifestyle.

The absence of role differentiation and division of labor could have decreased the competitiveness of Neanderthals. The ability to share decision-making especially in a democratic process can be useful in forming social bonds and strengthening group interactions and cooperation. The cooperative economic systems of humans may have given them a demographic advantage over Neanderthals, facilitating the rapid expansion of human culture throughout Eurasia. Kuhn and Stiner[42] argue that cooperative systems are more likely to have evolved first in the tropics or subtropics. They argue that although we take for granted today the division of labor by sex and age, it is not clear when it evolved; however, there are definite advantages to division of labor, and the advantages may have made modern humans more competitive than Neanderthals.

The small Neanderthal groups may have resulted in, or may have been the result of, interbreeding among close relatives, or matrilocal, or patrilocal behavior. Sequencing of the hypervariable regions of mtDNA and Y chromosome DNA analysis from 12 individuals from the El Sidron site in Asturias, Spain revealed low genetic diversity and suggested patrilocal mating behavior.[43] Genetic analysis of a female toe bone excavated from the Altai Mountains revealed that the parents of the woman were closely related, possibly half-siblings, an uncle or an aunt. It is not clear that inbreeding or patrilocal practices were common among Neanderthals or whether these were special cases that were unavoidable because of the small number of Neanderthals who lived in the area.[1]

Several Neanderthal remains show signs of violence. These include deeply grooved ribs showing signs of healing, misshapen bones, and broken hands and arms.[5] However, it is not clear if the damage was done by humans, other Neanderthals, or by accident. There is evidence that some Neanderthals may have practiced cannibalism. The evidence comes from 100,000-year-old remains from a cave at Moula-Guercy, near the Rhone River. The scattered distribution of Neanderthal bones, burned bones, marks on bones showing defleshing, hacked off arms and feet, and splintered bones for marrow extraction indicates consumption of the dead.[6,44,45] It is not clear if the individuals were killed or just consumed after death and/or if this was an isolated incident because of limited food resources.

Some authors have speculated that child mortality rates may have been high among Neanderthal children because of the number of juvenile remains uncovered. They have associated these high mortality rates with the aggressive life style of Neanderthals and the focus on collecting large prey.[8] Comparison of Neanderthal remains with modern groups with comparable life styles (hunter-gatherers, nomads, etc.) found no evidence of excessive or unusual trauma and no evidence that trauma played a more significant role in the lives of Neanderthals.[4,46] This is not to say that Neanderthals lived without trauma, but it indicates that the trauma they faced may not have been any more severe than that seen in ancient human skeletal remains. Research shows that throughout human history, cannibalism was widespread; mass killings, homicides, and assaults have been clearly documented.[47]

Despite potentially similar rates of violence among Neanderthals and modern humans, behavioral differences between humans and Neanderthals have been noted in the Levant. In spite of potential coexistence within the same region or even the same cave, modern humans and Neanderthals may have differed in habitat lifestyle. In one example, it appears that the two groups used different patterns of occupying caves. Animal remains (winter prey, mainly herbivores, such as gazelles) found in association with modern human fossils at the Skhul, Qafzeh, and Tabun caves, suggest single-season occupation, and that modern humans shifted their living quarters seasonally, including camping outdoors when the weather was favorable. Neanderthal fossils, from about the same time, are found in the context of local animals that were abundant during the entire year, suggesting Neanderthals practiced yearlong occupation.[48]

The hunting strategies of Neanderthal and modern humans were different. The more robust anatomy and body proportions of Neanderthals are thought to have allowed a more close-proximity style of hunting. It is thought that they thrusted and forced heavy spears with triangular stone points into the bodies of large prey.[49] This type of hunting would have necessitated constant replacement of spear points because of the force used during the penetration of the prey. Therefore, the expectations were that a number of point replacements would be found at Neanderthal caves. This was indeed the case where stone flakes shaped into various types of tools and spear points are found in caves occupied by Neanderthals. Many of the spearheads and flint tools from this type of archeological culture are referred to as Mousterian because of the type site in Moustier, France (Figs. 6.2 and 6.3). The evidence that Neanderthals used stone-tipped spears to hunt comes from impact scars on stone tips. Retouched Mousterian points from several sites in Western Europe have impact scars that suggest they were probably used at close range.[47] The weapons associated with modern human remains are more fragile wooden spears made for throwing at prey as projectiles from a distance. Neither the stone points found in Neanderthal caves nor the features of Neanderthal anatomy support the notion that they used long-distance projectiles.[50,51]

Microscopic analysis of tools from the Abri du Maras site in southern France found evidence for the use of twisted plant material used to make cords that could ultimately be used to form nets. The site dates to 90,000 years ago and has also been associated with the consumption of birds, aquatic mammals, and fish.[39]

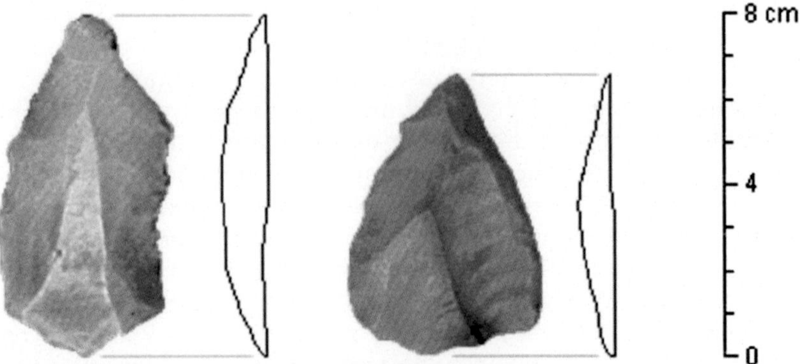

FIGURE 6.2 Mousterian tradition unifacial hide scraper (left) and spear point (right) made by late *Homo heidelbergensis* (both were made from Levallois flakes). *From http://anthro.palomar.edu/homo2/mod_homo_3.htm.*

FIGURE 6.3 Neanderthal stone tools. *From http://www.nhm.ac.uk/about-us/we-are-redeveloping-our-website.html.*

Analysis of ancient tools indicates that some Neanderthals traveled into central parts of Europe. In Western Europe, the material to make stone tools was local; however, in central Europe, the raw materials used to make the tools were not local, indicating that the material traveled and simultaneously providing a clue to the range of Neanderthal mobility.[52]

Researchers have speculated for many years about the cognitive ability and personality of Neanderthals. The difficulty in measuring cognition and personality lies in the inferences and indirect pieces of evidence that are based on fossilized material. For example, Wynn and Coolidge[53] attempted to address some of these questions by looking at the quality and quantity of Neanderthal working memory. They defined working memory as the ability to control decision-making, attention to tasks, storage of speech-based information, and storage of visual and spatial information. They defined two types of working memory: (1) long-term, which allows storage of knowledge and skills and (2) enhanced, which allows cultural creativity, complex communication, and experimentation. Based on the following premises: (1)

the fact that Neanderthals were replaced by humans even though at the time they both had similar technology and were exposed to similar resources, (2) the fact that the brain size and other anatomical characteristics were not significantly different enough to warrant entirely different behaviors, and (3) the fact that many modern behaviors and thinking evolved long ago, Wynn and Coolidge concluded that the lack of Neanderthal innovation and experimentation (as evidenced by archeological findings) ruled out enhanced working memory. Instead, they concluded that Neanderthals only had long-term working memory that may have led to more directive and crucial modes of restricted speech with limited vocabulary. In addition, they suggested that limited speech and experimentation would have helped to infer personality traits such as bravery, stoicism, and a lack of cost-benefit analysis. Other investigators have used eye socket measurements to estimate the amount of visual cortex in Neanderthal brains. They found that the orbital volume of Neanderthals was much larger than that of humans suggesting less neural tissue would be left over for other functions such as social networking.[10]

Recently, investigators have argued that Neanderthals were cognitively sophisticated and displayed many of the traits that are part of modern human intellect, including language.[54] Brain studies from fossilized skulls indicate that the frontal lobes, involved in problem solving, were similar in humans and Neanderthals. Studies of limbs, tools, and teeth indicate that Neanderthals were mostly right handed, a trait that distinguishes modern humans from chimpanzees and corresponds to brain asymmetries that are related to language.[9]

The clearest examples of cognition are innovation, symbolism, and the development of culture; for many years, there was very little evidence that Neanderthals had anything more than simple tools and hunting strategies. Recent evidence is beginning to point to a more sophisticated lifestyle than previously thought. Polished eagle talons with markings, found in Krapina in modern-day Croatia, show that they were a piece of jewelry. The talons were dated 80,000 years before modern humans entered Europe.[55] Similarly, anthropologists at sites in France and Italy have uncovered eagle talons dating from 40,000 to 90,000 years ago. Across Europe, decorative feathers, seashells bearing traces of pigment (red ochre), and seashells with holes (suggesting they were used as jewelry) are indications that Neanderthals had symbolic thinking.[9] In addition to decorative items, there are indications that Neanderthals also used art as a method of expression. Geometric-like engravings in a Gibraltar cave estimated to be more than 39,000 years old indicate that at least one Neanderthal tried to use art to convey an idea or message.[56,57] Even more recently cave art found in Spain point to sophisticated drawings that are discussed at the end of this chapter. This is important because many scientists believe that symbolic thinking is a precursor to language.

Neanderthal tool technology has also been found to be more advanced than previously thought. Early discoveries of Neanderthal sites revealed stone Mousterian flake tools. In 1961, Francois Bordes divided the tools into Mousterian points, denticulates (those with multiple notches), backed knives, bifaces, and convex and convergent scrapers.[58] Materials used for stone tools were mostly local, but there is evidence that some of these materials were transported into Western Europe within 5 km of the source) indicating some foresight, planning, and mobility.[52]

More recently, bone tools referred to as lissoirs (used today by leather workers to treat animal hides to make them tougher and waterproof) were discovered at two different sites in Dordogne, southwest of France. The tools were made of deer ribs and pressed against animal hides judging from the wear marks on the bone similar to modern-day wear. The site where the tools were found dates to around 51,000 years ago suggesting that they were an independent invention of the Neanderthals or indicating modern humans influenced Neanderthals earlier than previously believed.[59] Some have speculated that Neanderthals were able to plan their subsistence strategies and adapt their toolmaking to these strategies indicating that they had the ability to be flexible and solve problems. Although the evidence for Neanderthal use of symbols, ornaments, and sophisticated tools provides some indication of their potential ability for higher-order thought, some argue that Neanderthals might have acquired the technology from modern humans.[60] If this were the case, the use of these symbols, ornaments, and tools would merely indicate that the Neanderthals who interacted with modern humans were capable of mimicry rather than complex thought.

A key feature of modern humans and their capacity for caring and emotion is exemplified in their burial of the dead. Archeologists have argued that several Neanderthal sites show the signs of ritual burial; however, others have disputed those claims as speculation. Some of the evidence presented for burial is as follows: tools and animal corpses found with remains, red ochre pigment in Middle Eastern graves, and Neanderthal corpses found in the fetal position.[61] Skeptics claim that the evidence probably represents non-symbolic and nonreligious corpse disposal. One study found large amounts of pollen within the grave suggesting flower or medicinal plant deposits.[62] The pollen was identified as coming from Yarrow, Bachelors button, Ragwort, Grape hyacinth, etc., all of which have showy flowers and medicinal properties.[63] Although this represented strong evidence for ritual burial, rodent deposits of large numbers of flowers, and not human activity, later explained the pollen at the grave site.[64]

In 2013, new evidence for burial came from a 12-year reexcavation of the La Chapelle-aux-Saints site. Rendu et al.[65] found features that suggest the remains of an individual were modified for burial, a grave was dug, and the remains were buried quickly to avoid being eaten by carnivores. The site dates back 50,000 years ago, before any interaction with modern humans could have taken place, suggesting that the practice was original and not copied from modern human practices.

In addition to burial, there are a couple of examples of caring for the sick that may have been practiced by Neanderthals. The remains of an elderly man were found at Shanidar Cave in Iraqi Kurdistan. The remains, referred to as Shanidar 1, were one of four almost complete skeletons. Shanidar 1 was a 40- to 50-year-old whose remains displayed debilitating abnormalities, including a crushing fracture to the left side of his face, a withered right arm that caused the loss of his right hand, and deformities in his lower legs and foot that would have resulted in painful walking. All injuries showed signs of healing long before death, inferring that he was supported during his lifetime.[61,66] In addition, the man of La Chapelle-aux-Saints was older than average, and his skeleton revealed that he was severely bent over by arthritis and lost all but two teeth. The age and disabilities of the Shanidar 1 and La Chapelle-aux-Saints' individuals suggest that Neanderthals may have cared for the sick and aged.[61]

COEXISTENCE OF MODERN HUMANS AND NEANDERTHALS

For many years, the coexistence of humans and Neanderthals in Europe and the Levant was thought to have few or no consequences except possibly the demise of Neanderthals due to the occupation of Europe by modern humans. But the common anatomical, behavioral, and possible cognitive similarities that have been recognized recently have raised questions about the coexistence of these groups and their social interactions. One question is the time of coexistence, which has recently been disputed.

Initial estimates of the time period where Neanderthals and humans geographically overlapped in Europe ranged from 15,000 to 20,000 years. This time period was based on dating the youngest Neanderthals fossils found in Southern Iberia where Neanderthals were thought to have gone extinct (initial estimates of Neanderthal extinction based on the dating techniques were 28,000 years ago). In 2015, a new method of dating Neanderthal fossils was developed using pretreatment chemistry and a particle accelerator to measure radiocarbon. The new process was used on the youngest Neanderthal bones from Southern Iberian sites,[67] and the bones were determined to be 39,000–41,000 years old (about the same time frame as humans entered southwestern England and parts of Italy). These new measurements suggest that Neanderthals went extinct around 40,000 years ago (not 27,000) and place the Neanderthal–human overlap in southern Europe at approximately 5000 years, instead of the previously estimated 15,000–20,000 year overlap. Nevertheless, there was still a significant amount of time for spatial and temporal interactions and for introgression between the two groups.

Two windows of opportunity for admixture and/or cohabitation between humans and Neanderthals presented themselves when contemporary humans ventured out of Africa.[68] Modern human migration was most likely linked to improved climatic conditions outside Africa. Geological records from alluvial fan deposits indicate that increased rainfall and vegetation occurred in Arabia during Marine Isotope Stage 6 (approximately 160–150 kya), Stage 5 (approximately 130–75 kya), and early Stage 3 (55 kya). These climatic changes resulted in fresh water and increased vegetation and provided multiple opportunities for anatomically modern human dispersal out of Africa.[69] Anatomically, modern humans are thought to have undertaken at least two incursions out of Northeast Africa; the first one approximately 100,000–177,000 years ago reached the Levant (age range related to the dating of fossils found in the region).[70] Yet, after about 80,000 years before the present, human remains disappeared from the fossil record in the Levant. It is speculated that some modern humans could have retreated from the Levant back to Africa as a result of increasing cold weather or even competition with Neanderthals. However, modern human bones from the Zhiren Cave in China dating to about 100,000 years ago suggest that some of the early humans in the Levant may not have died in Arabia or retreated back to Africa but reached into the Far East.[71] The second movement of modern humans out of Africa resulted in a European settlement between 35,000 and 45,000 years ago.

Geological and archeological evidence suggests that Neanderthals inhabited the Near East during two separate occasions, being pushed south due to glaciers and cold climatic conditions in the north. The first incursion occurred from 120,000 to 100,000 years ago and a second one 61,000 to 48,000 years ago, with Neanderthals going extinct between 39,000 and 41,000 years ago.[72] The timing of modern humans leaving Africa and Neanderthal migration suggests that humans and Neanderthals could have cohabitated twice in parts of their ranges, in Europe and the Near East. The specific nature of the interactions between humans and Neanderthals is not clear, yet considering the options of conflict or collaboration, it is difficult to imagine living side by side for the estimated amount of time without some social interaction.

Indications of shared living quarters in caves in the Near East have been documented in present-day Israel. This cohabitation between Neanderthals and humans could have occurred during the emergence of the first modern humans out of Africa. The Mount Carmel range near the Sea of Galilee in northern Israel includes the caves of Tabun, Jamal, el-Wad, and Skhul. A series of discoveries of early modern human remains were made in the western slopes of the range. The remains were from the early middle Paleolithic age with dates ranging between 100,000 and 120,000 years ago. Of specific interest, these caves contain the some of the earliest modern human remains outside Africa, and the samples belong to the same strata as Neanderthal bones in the adjacent Tabun Cave.[73] Recently a human jaw bone dated to 177,000 years ago was found at Mount Carmel, Israel, representing the oldest AMH found outside Africa.[70] Dating techniques indicated that the Tabun Neanderthal samples were coexistent with the modern human remains found in the Mount Carmel range, with the Neanderthal samples dated to about 120,000 years ago. This proximity in time and space of modern humans and Neanderthal fossils suggests a possible cohabitation of the two groups. In addition, the remains showed modified skeletal features (delicate jaw, smaller brow ridge, more pronounced chin, a more modern-shaped cranium) that many claim clearly represent a middle ground between humans and Neanderthals (Figs. 6.4 and 6.5). Discoveries like this one led to what is known as "the admixture theory" which assumes that Neanderthals and modern humans mated and produced offspring. Supporting the admixture theory, the skeletal remains of a 4-year-old that represented a mosaic between a modern human and Neanderthal were discovered in Portugal. The cranium, mandible, dentition, and post crania all showed intermediate features suggesting admixture of Neanderthals and early modern humans. The authors point out that the find helps refute the replacement model of modern human origins.[74] (The replacement model suggests that there was no introgression between humans and Neanderthals and instead that humans eliminated Neanderthals as they moved into Europe and East Asia.)

In another example of possible hybridization, the remains of modern humans of Es-Skhul and Qafzeh caves exhibit a mix of modern human and archaic traits. The remains have been

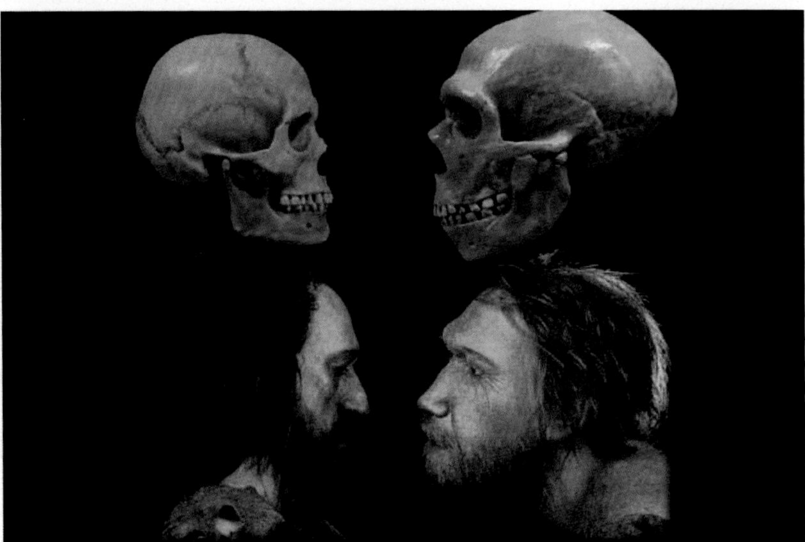

FIGURE 6.4 Comparison of Modern Humans and Neanderthal facial profiles. *From http://www.ilnavigatorecurioso.it/2014/04/24/i-neanderthal-e-i-sapiens-sono-geneticamente-simili-al-9984-allora-dove-sono-le-differenze/.*

FIGURE 6.5 Neanderthal recreations. *From http://www.npr.org/tags/133465492/neanderthal.*

dated to around 80,000 to 120,000 years ago using electron spin resonance. The brain cases of these individuals show both modern human traits and Neanderthal brow ridges and projecting jaws. More recently, Hershkovitz et al.[75] described a partial skullcap discovered at Manot Cave in Galilee dated by uranium–thorium techniques to be 50,000–60,000 years old. The shape and morphological features of the partial skull showed that it is modern and similar to African and European skulls from the Upper Paleolithic, but different from anatomically modern humans in the Levant, suggesting that it may be related to the first modern humans who later colonized Europe (i.e., the Manot people may be the forefathers of Paleolithic Europeans). The anatomical features suggest that they may have been inherited from earlier Levantine populations that resulted from an interbreeding event that took place between Neanderthals and humans.

INTROGRESSION

There are no known reproductive barriers between any living humans, regardless of geographic location and distance, and when different human populations encounter one another, genetic admixture episodes are a common occurrence. Although many species are reproductively isolated (unable to breed), there are examples of related primate species interbreeding. For example, bonobos, baboons, and chimpanzees interbreed with their closely related species and subspecies resulting in fertile offspring.[76] Many scientists have speculated that interbreeding between different primate species can occur as long as the species shared a last common ancestor within the last 2 million years.

A number of archaic, human-like species, such as *Homo erectus, Homo heildelbergensis, Homo habilis*, and *H. sapiens*, are known to have populated Africa, and *H. neanderthalensis, H. sapiens*, and Denisovans populated Europe and the Middle East (Fig. 6.6). It has long been suggested that the temporal and spatial overlap between these different species could have led to interbreeding. Of course, because of potential biological reproductive barriers (physiological or

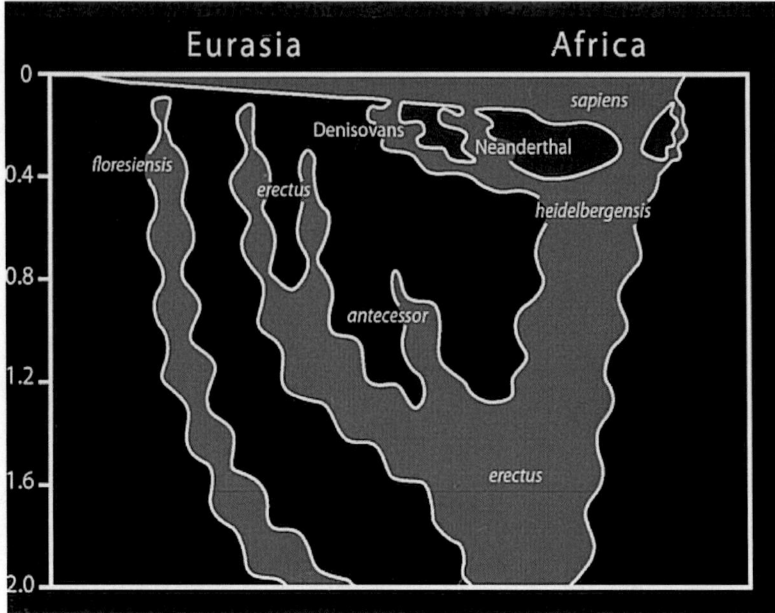

FIGURE 6.6 Tree representing the relationship of various *Homo* species and their geographic locations.

behavioral differences), fertile offspring may have been compromised or not have resulted. Several scientists also assumed that breeding would not have occurred because of perceived social and geographical differences (groups occupying different regions of Africa and Europe and/or interactions resulting in hostile outcomes).

Other scientists assumed that genetic variation between humans and Neanderthals was no different than the variation between subspecies of chimpanzees and subspecies of baboons that interbreed, and that if there was genetic evidence to support interbreeding and fertile offspring, Neanderthals and humans would be considered the same species.[4,77]

Interbreeding results in genetic admixture where alleles (different forms of the same gene) are obtained from each of the parental populations' respective genotypes. The frequency of alleles transferred is dependent on the population allele frequency, the amount of admixture, and the frequency of alleles in the respective parents. The genotypic results from random mating can be an increase or a decrease in specific alleles depending on mating preferences and whether the allelic combinations result in hybrid vigor, hybrid sterility, selective advantage, or disadvantage.

As previously noted, a major breakthrough in molecular anthropology was the ability to isolate ancient DNA and sequence the DNA of extinct hominins. The first successful isolations were done in the late 1980s isolating and sequencing mitochondrial DNA (mtDNA). mtDNA is found in mitochondria (small subcellular structures) known as the powerhouses of the cell because they provide the fuel for cell activity. The mtDNA is a small circular molecule that is typically passed down from females to all of their offspring and is more easily isolated than nuclear DNA. mtDNA is one of the most widely studied DNA sequences used in evolutionary research. mtDNA is particularly useful because of its maternal inheritance and consistent mutation rate (the rate at which the building blocks of DNA change over time). If the rate of DNA change over time is consistent, then the number of changes can be used to estimate the time it has taken for the changes to occur. For example, if one change in the DNA sequence occurs every 1000 years and this rate is constant, then five changes would indicate that the sample was 5000 years removed from the standard sample (standard being the sample with no changes). The use of DNA mutation rates to estimate the date of divergence since the last common ancestor and the relatedness between populations is an essential component of molecular evolution. Knowing the rate of mutation to estimate divergence is referred to as the molecular clock, and it has been used extensively to measure times of species divergence.[78]

Initial isolation of ancient mtDNA from Neanderthals was done from the remains of a Neanderthal-type specimen discovered in 1856 in the Feldhofer Cave, in the Neander Valley, Germany. The researchers drilled into the bone and isolated fragments of DNA from the bone powder. They sequenced the mtDNA and compared the Neanderthal mtDNA sequence with chimpanzee and human mtDNA. The comparison of Neanderthal with modern human mtDNA showed that the Neanderthal sequence fell outside the range of variation of modern humans. Modern human mtDNA sequences differed by an average of 8 substitutions, whereas the Neanderthal DNA differed by 27.2 substitutions, suggesting that no introgression occurred at the mtDNA level between humans and Neanderthals. The study also used the changes in mtDNA to determine that the age of the common ancestor of the Neanderthal and modern human was approximately 550,000 to 690,000 years ago.[79–81]

One year later, a second Neanderthal mtDNA sequence was derived from a geographically distant population from Mezmaiskaya Cave in Russia. The result of that mtDNA sequence was slightly different than that of the first described (with 23 substitutions instead of 27.2), but the research confirmed the distinction between modern humans and Neanderthals. The authors of the study propose that differences between human and Neanderthal in the two mtDNA sequences from different regions and different time scales suggest that Neanderthals went extinct before contributing mtDNA to modern humans.[82] Alternatively, Neanderthal female–human male copulations may have not survived and so no Neanderthal mtDNA would be seen in AMH populations, or that the interregional gene flow between modern humans overwhelmed the Neanderthal contribution to any admixture[81] (i.e., so much human mtDNA vanquished any Neanderthal contribution).

In 2010, Svante Paabo's group successfully isolated nuclear DNA from Neanderthal remains to compare the DNA with modern humans and look for similarities, differences, and signs of introgression with modern humans.[83] The group turned to archeological evidence that confirmed that Neanderthals had lived outside of Africa for 300,000 years where they were assumed to have evolved from *H. heildelbergensis* outside of Africa. This assumption is supported by the fact that no known Neanderthal remains have ever been found in Africa. This assumption would mean that if introgression had occurred between modern humans and Neanderthals, it would have occurred outside of Africa. Based on this hypothesis, the researchers predicted that Neanderthal DNA would not be detected in modern-day aboriginal humans from Africa, but only humans outside of Africa. To test their hypothesis, the scientists first compared 176 nuclear genomes in present-day humans from Africa with 846 nuclear genomes from non-African individuals. The comparison revealed 12 regions of variation between Africans and non-Africans that became candidates for Neanderthal DNA introgression. Comparisons of the Neanderthal DNA with the 12 DNA regions in non-African humans showed 10 of the 12 containing Neanderthal DNA (none of the African groups showed Neanderthal DNA). These studies were the first to suggest admixture and interactions between Neanderthals and non-African modern humans. The discovery of Neanderthal DNA in modern humans was a surprise. Because mtDNA from Neanderthals was not seen in modern humans, it was expected that Neanderthal nuclear DNA would also not be found in modern-day humans. Because of this finding, the experiments were redone, and the same results were found. Initially the public and other scientists were skeptical, and the results were assumed to be due to experimental error, human contamination of the samples, or the result of DNA that had been inherited from the human–Neanderthal common ancestor before the two diverged.[84] Although the inheritance of DNA from a human–Neanderthal common ancestor remains a remote possibility, subsequent experiments in other laboratories later confirmed the initial findings of Paabo's group that Neanderthal DNA did exist in the genome of some humans. Since then various estimates of admixture between the Neanderthals and humans have been reported ranging from 0% to 4%. In addition, the identification of unique genes, missing genes, and gene rearrangements (genes in different places) has confirmed that there was some level of introgression between Neanderthals and humans.

Overall, researchers have shown that about 1.5%–2.1% of DNA in humans outside of Africa is from Neanderthals; however, because the Neanderthal DNA sequences are not the same in all individuals, it is estimated that approximately 20–30% of the Neanderthal genome could survive in the human population outside of Africa.[1,85] Not all of the Neanderthal DNA is functional, but there is increasing evidence that humans have accumulated a number of Neanderthal sequences. The size of these fragments inside modern humans suggests that the introgression occurred as early as 37,000–86,000 years ago[87] or as late as 7000–13,000 years ago.[86] It has also been speculated that some of the DNA retained by modern humans from the Neanderthals may have helped humans survive.

Since the initial discovery of human–Neanderthal admixture, a number of different scientific groups have begun to identify the genes that are unique to Neanderthals and humans. Sankararaman et al.[87] showed that nine human genetic variants that affect lupus, biliary cirrhosis, Crohn's disease, optic disk size, type 2 diabetes, keratin production, and some disease risks likely came from Neanderthals. In addition, the group found that some areas of the modern, non-African human genome were rich in Neanderthal DNA, which may have been helpful for human survival. Other genomic areas they referred to as "deserts" with little or no Neanderthal DNA suggest that some Neanderthal genetic contributions that were harmful were removed by natural selection. Deserts were grouped in two parts of the modern human genome: (1) regions most active in the male sperm cells and (2) genes on the X chromosome that are specifically activated in testes. These large regions of Neanderthal X chromosome that are missing in modern humans suggest that genes in these regions may not have functioned properly in hybrids. The authors concluded that some Neanderthal–human hybrids may have shown hybrid infertility, where the offspring of a male from one subspecies and a female from the other subspecies have low or no fertility (i.e., Neanderthal–human hybrid males carrying the Neanderthal X chromosome were sterile, and females carrying one Neanderthal X chromosome may have been partially sterile). The overall conclusion is that modern humans have relatively few Neanderthal genes located on the X chromosome or expressed in the testes. This is consistent with the fact that a large proportion of male infertility is the result of faulty genes on the X chromosome.

The authors of this study further suggest that because of the hybrid infertility, humans and Neanderthals may have been close to reaching overwhelming reproductive barriers resulting in reproductive isolation.[85] An alternative explanation for the loss of the Neanderthal X chromosome in modern humans could be male hybrids may have not been able to reproduce because of sociological or cultural reasons (killing male hybrids, undesirable phenotypic outcomes, etc.); however, this would seem to have been rare.

Vernot and Akey[85] found similar results to Sankararaman and colleagues. Using a computational approach and whole genome sequencing of 379 Europeans and 286 East Asians, they found evidence that Neanderthal skin genes persisted in the modern human groups. Examples include the Neanderthal version of the skin gene POU2F3 found in approximately 66% of East Asians, whereas the Neanderthal version of the BNC2 gene, which affects skin color, was found in 70% of Europeans. In addition, they found that certain chromosome arms in modern humans are missing Neanderthal sequences, suggesting that some genomic regions of Neanderthals resulted in detrimental outcomes. The authors suggest that regions of Neanderthal DNA not found in modern humans may have helped humans adapt and evolve away from other *Homo* species. In other words, modern human DNAs involved in sperm movement, and cognition, are some of the genes that may have separated us from Neanderthals.[13]

One example of Neanderthal–human admixture that may have helped modern humans survive can be found in the genes of the immune system. The major histocompatibility complex (MHC) is a series of genes that define self and respond to antigens (foreign substances) in the bodies of vertebrates. (These are the genes that are activated when you receive another person's organs and your body tries to reject the transplant.) The MHC class of molecules present pieces of foreign proteins to T cells, which then trigger an immune response. The majority of MHCII molecules are composed of alpha (α) and beta (β) subunits from HLA (human leukocyte antigen) genes. Temme et al.[88] found that the first 90 amino acid residues of the beta 1 domain of HLA DPB1 from Neanderthal aligned with the human allele DPB1-0401. Because the 0401 allele is rare in sub-Saharan African populations, but frequent only in European populations, the data suggest it may have arisen from introgression of modern humans with Neanderthals. Yet, there is a probability that the similarities in HLA genes may be due to ancestral inheritance (before humans and Neanderthals split) and not admixture.[89]

An example of a Neanderthal gene that is a candidate for positive selection in modern humans is the innate immune gene STAT2. STAT2 is important in interferon-mediated responses and is potentially associated with autoimmune disorders. Mendez et al.[89] found that many Eurasians, but not sub-Saharan Africans, carried a STAT2 gene with a sequence that closely matches the Neanderthal STAT2. The gene is found in approximately 5% frequency in Eurasians and 54% in Melanesians. The group discounted genetic drift as the cause for the Melanesian frequency. Although they were not able to pinpoint the precise target of positive selection, they pointed out that STAT2 is also associated with the ERBB3 gene. ERBB3 plays a role in cell growth, differentiation, and suppression of cell death.

An example of a Neanderthal allele that was detrimental to modern humans is the SLC16A11 allele, which is a common risk factor in type 2 diabetes and occurs in high frequency in Mexico. SCL16A11 alters the way we process fat, causing an increase in cellular triacylglycerol levels. Analysis of the Neanderthal genome sequence indicated that the risk form of the gene introgressed into modern humans via admixture.[90] Recent studies have also found Neanderthal genes in modern humans related to LDL Cholesterol concentrations and schizophrenia, discussed further at the end of the chapter.

EXTINCTION

The enigma of Neanderthal extinction has been a topic for many conversations and speculations. They had survived for hundreds of thousands of years under varying climatic conditions and geological events, and then suddenly disappeared after seeking out isolated refuges in the southernmost regions of Europe.[91]

For many years, it was thought that Neanderthals became extinct in Southern Iberia, about 24,000–28,000 years ago; however, it is now accepted that Neanderthals went extinct around 39,000–41,000 years ago.[67,92] Because anatomically modern humans entered southwestern England and parts of Italy 35,000–45,000 years ago (during the second migration out of Africa), the new time estimates suggest that Neanderthals and humans overlapped in southern Europe for up to about 5000 years. Some have suggested this short overlap of 5000 years indicates that interbreeding between Neanderthals and humans most likely occurred in the Levant during the first migration of modern humans out of Africa, approximately 120–177 kya.[67] They suggest that the 5000 year time frame may not have been enough for significant interbreeding to occur. Yet, the 40,000-year-old remains of an early modern human found in Romania contain Neanderthal DNA, which indicates that at least some introgression occurred during the 5000-year overlap of humans and Neanderthals in Europe.[72]

There are three theories regarding the extinction of the Neanderthals as follows: (1) Extinction due to inability to adapt to new climate and geography in southern Europe, (2) The movement of anatomically modern humans into Neanderthal territory and subsequent competitive advantage of modern humans, and (3) Neanderthals became unfit due to interbreeding with humans.

The first theory suggests that Neanderthals could not adapt to the dramatic environmental and ecological changes associated with climatic fluctuations and eventually went extinct. Climate change around 55,000 years ago would have led to ecological changes in flora and fauna. The changes would have led to fluctuations in grasslands and the development of forests. This, in turn, may have led to significant changes in prey and other food sources. In addition, the cold-adapted physiology of the Neanderthals may have also made it difficult to adapt because of insufficient time to recover from the changes.[13,93] There is some evidence for new adaptations based on changes in tool types observed during this time period. Archeological finds have shown changes from crude stone flakes to tools using tree resin to bind stone points and crafting tools from bone and wood. These changes in tools may reflect the attempt to adapt to new prey and hunting strategies.

The first theory also assumes anatomically modern humans had already adapted to a variety of environments after leaving Africa and were better suited to tolerate changing conditions. Computer simulations of alternating climate change integrating archeological and chronological data show that the geographical range of Neanderthals contracted, whereas the human ranges expanded during the climatic phases of the Upper Paleolithic. This suggests that geographic expansion and competition from the larger populations of humans may have contributed to the Neanderthal demise.[94,95]

The second theory also suggests that climate changes pushed Neanderthals into southern Europe but proposes that Neanderthals could not compete for food or space with modern humans once humans occupied Neanderthal territory. The theory points to cultural, mental, and technological differences between humans and Neanderthals. It also suggests that humans may have eliminated Neanderthals because of their inferiority either by violence or by pushing them into regions that were too difficult to survive (similar to the effects of some invading humans on aboriginal groups).

One explanation for the slow demise of Neanderthals after humans left Africa is the occurrence of geological events that may have deterred modern humans from moving into Neanderthal territory. Although modern humans had already left Africa, there is some speculation the Campanian Ignimbrite volcanic eruption in Southern Italy, about 40,000 years ago, may have created a semi-desert buffer zone that prevented modern humans from entering Southern Iberia, allowing Neanderthals to have safe refuge in the area until modern humans arrived. This theory proposes that the Neanderthals were not as cognitively advanced as humans and may have had a more difficult time competing for similar resources. In this scenario, a failure to adapt and exploit small game would have resulted in a less diverse diet, and this inability to shift their prey focus and hunting strategies would have been instrumental in Neanderthal extinction.[93,95]

In addition to Neanderthals failure to adapt when humans arrived, several investigators have proposed that humans were more socially advanced and had created a culture where division of labor was the norm. Division of labor in the modern human population created a safer environment for children and the group and used the skills of hunters and gatherers as opposed to a more individualized effort used by Neanderthals for survival. It has been shown that in extant hunter-gatherer groups, complex social structure consisting of kin and metagroups can lead to advanced behavior and support from related and unrelated individuals within a group. This complex behavior in extant hunter-gatherers can be explained when men and women have equal influence in selecting group members, which may have had a transformative effect on social organization and social networking in early modern humans.[96]

Some have speculated that when humans did make advances into Neanderthal territory in Western Europe around 45,000 to 35,000 years ago violence and genocide led to the Neanderthal demise. The debate centers on which group of humans arrived in Western Europe at the time Neanderthals were present. The Aurignacian culture is an archeological culture located in Europe and Southeast Asia, which lasted between 47,000 and 41,000 years ago. Bone or antler points, personal ornaments, and perforated shells characterize the Aurignacian tool industry. The Protoaurignacian appeared about 42,000 years ago in southwest and south-central Europe and is thought to reflect the movement of modern humans from the Near East. Recent studies on dental remains of the Protoaurignacian revealed mtDNA of a modern human type. The teeth were dated at 41,000 years before present and represent the oldest remains of the Aurignacian culture in southern Europe. The DNA studies provide evidence that people of the Aurignacian culture were in the region at the time of Neanderthal extinction, and if the second theory is true, suggest that they could have triggered the demise of the remaining Neanderthals in the region either through violence, or the advanced human cognitive or social structure advantage.[97]

There is a lack of evidence in the archeological record that Neanderthals were outcompeted by modern humans in terms of subsistence and weaponry. Therefore, a third alternative explaining Neanderthal extinction is that the Neanderthals became unfit because of interbreeding with humans. The idea assumes that anatomically modern humans may have benefitted from some introgression, but that Neanderthals did not, or that interbreeding caused the demise of Neanderthals through assimilation, where those individuals who were not hybrids (totally Neanderthal) could no longer compete for a variety of reasons, including disease, discrimination, displacement, etc.[98]

It will never be clear exactly how Neanderthals went extinct and if it was a result of one of the theories or a combination of two or all three theories. Nevertheless, the radiocarbon data put the disappearance of Neanderthals in Europe, approximately 39,000–41,000 years ago in Southern Spain.

DENISOVANS

In 2010, the tree of modern human life gained a branch. An ancient finger bone (Fig. 6.7) discovered in 2008, in the Denisova Cave, in the Altai Mountains of Siberia (Fig. 6.8), led to the discovery of a new human species. The bone was dated between 30,000 and 48,000 years old. mtDNA isolated from the bone was expected to be of modern human or Neanderthal origin because both groups were expected to have been in the vicinity during that time period. Instead, the mtDNA differed from both humans and Neanderthals and was estimated to have diverged from their common ancestor almost a million years ago.[99] That same year, it was

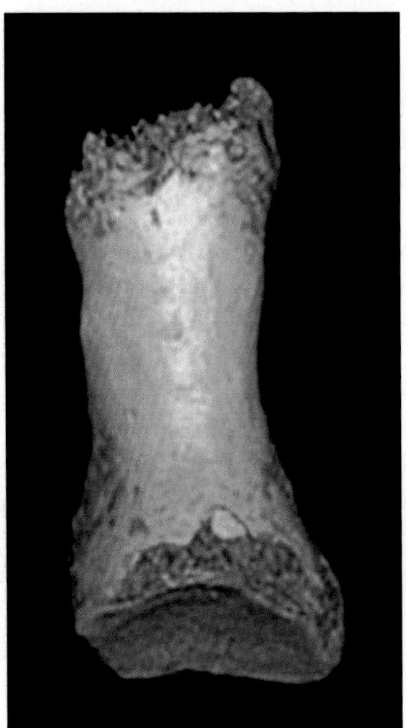

FIGURE 6.7 Denisova sample used to extract DNA. *From http://www.hominides.com/html/actualites/neandertal-denisova-tres-proches-0786.php.*

FIGURE 6.8 View of Denisova Cave. *From https://en.wikipedia.org/wiki/Denisovan.*

reported that the nuclear genome from the finger bone had been sequenced.[100] The genome sequence was compared with the Neanderthal and human genomes, and it was found to be more closely related to Neanderthals. The new species was designated Denisovan after the name of the cave where it was discovered.

In 2012, a new single-stranded DNA library preparation method allowed researchers to reconstruct a high-coverage Denisovan genome approximately 30 times greater than what could be done for modern humans at the time.[101] The high-coverage sequencing allowed Meyer and colleagues to determine that Denisovans had 23 chromosomes, the genome was from a girl with dark skin, brown hair and eyes, and that there are about 100,000 recent changes in the modern human genome that occurred since we split from the Denisovans. The comparison of the high-coverage sequence to 11 present-day human populations showed that approximately 6% of contemporary Papuan DNA comes from Denisovans. Australian aborigines and people from Southeast Asian islands also show Denisovan DNA, and about 0.2% of the DNA of mainland Asians and Native Americans is Denisovan. The DNA evidence points to a patchwork of people with and without Denisovan DNA, and estimates from the DNA conclude that interbreeding between anatomically modern humans and Denisovans occurred in Asia about 40,000 years ago.

Comparison of the X chromosome from Papuans and Denisovans revealed that there is less Denisovan DNA on the X chromosome of the human New Guinea population compared with the rest of the genome. One possible explanation is that mating was predominately between male Denisovans and female humans (because the male only contributes one X chromosome), or alternatively predominately human female migrants encountered Denisovans, thereby diluting the Denisovan component of the X chromosome (both explanations assume intermixing between the two female X chromosomes, one human and one Denisovan, may have eliminated some of the Denisovan X chromosome DNA while retaining the human components). There could also have been detrimental genes on the Denisovan X chromosome that would have been removed by natural selection.[101]

Later, 2M were found from the same Denisovan population, and their DNA was also sequenced. In 2014, Prufer et al.[1] sequenced the genome of a Neanderthal from the Caucasus and compared archaic genomes with 25 present-day human genomes. They demonstrated that several interbreeding events occurred among Neanderthals, Denisovans, and early modern humans, possibly including gene flow into Denisovans from an unknown archaic group. The unknown archaic group contributed between 2.7 and 5.8% of the Denisovan genome. The researchers estimated that the archaic DNA contributor split from humans, Neanderthals, and Denisovans between 900,000 and 4 million years ago, long before the groups diverged from each other. They suggest that the ancestor was possibly *H. erectus* because it is believed to be the predecessor of all modern humans. There are no known indications that the archaic group that introgressed with Denisovans interbred with humans or Neanderthals.[1] The data further suggested that at least 0.5% of the Denisovan genome came from Neanderthals; however, no Denisovan DNA was detected in Neanderthals. The group was not able to determine whether any gene flow from modern humans to Neanderthals or Denisovans happened, most likely because of the fact that the Neanderthal and Denisovan DNA that have been sequenced lived at a time before modern humans were around.

As discussed previously, introgression and adaptation appear to have significantly shaped some modern human immune systems. In one study, Abi-Rached et al.[102] used whole genome comparisons and identified introgression from Denisovans to modern humans. The analysis of human leukocyte antigen class I (HLA-I), a set of immune system components, revealed how modern humans acquired the HLA-B*73 allele in West Asia through admixture with Denisovans. They identified archaic HLA haplotypes from Denisovans and Neanderthals carrying functionally distinctive alleles that were introgressed into modern Eurasian and Oceanian populations. Several of the alleles, which encode unique or strong ligands for natural killer cell receptors, represent more than half the HLA alleles of modern Eurasians, and they speculate that the alleles were introduced later into Africans.[102]

A 50,000-year-old skull and rib and a 35,000-year-old finger bone found in the Strashnaya Cave in the Altai Mountains of Central Asia in August of 2015 may shed more light on the links between Denisovans, Neanderthals, and modern humans. The finger is thought to be from modern humans, whereas the skull and rib may be from a Neanderthal or Denisovan. So far, genetic analysis has not been done, but the addition of specimens over time, as well as improved genetic analysis methods, will continue to enhance our understanding of the interactions between these groups of early hominids.[103]

Because only a few fossilized fragments have been found, we still do not know what Denisovans looked like, but by comparing modern human, Neanderthal, and Denisovan genomes, researchers have identified more than 31,000 genetic changes that distinguish humans from Neanderthals and Denisovans. The changes may be associated with human survival, and a number of the genetic changes have been linked to brain development.[104] Although it remains uncertain when anatomically modern humans, Neanderthals, and Denisovans diverged from one another, current estimates predict modern humans split from the common ancestors of all Neanderthals and Denisovans between 550,000 and 765,000 years ago, and Neanderthals and Denisovans diverged from each other between 381,000 and 473,000 years ago, with Neanderthals evolving in Europe and Denisovans evolving in Asia.

HOBBIT

In 2003, a partial female skeleton just over 1 m tall and a complete skull was found in Liang Bua Cave on the Indonesian island of Flores. The skeletal remains were from an adult with a cranial capacity one-third the size of modern humans. The specimen is called LB1 and because of its size and features is often referred to as the hobbit. Partial skeletons of nine other individuals were also found. Because of their cranial *Homo*-like features the group was designated *H. floresiensis* and initially thought to be descendants of *H. erectus*. The remains were dated at 18,000 years ago and, the group has been estimated to have lived in the region as early as 12,000 years ago, long after Neanderthals, Denisovans, and *erectus* had disappeared.[105]

It is interesting to speculate that a possible form or descendant of *H. erectus*, *H. floresiensis*, was alive as recent as 18,000 years ago in the island of Flores in Indonesia. Yet, the taxonomic status of this fossil has been the subject of considerable controversy, partly because of its complex features. The specimen is small (only 1.06 m tall with a brain size of 380 cm) similar to *Australopithecus afarensis* because of height and brain size, but the feet are not ape like. Instead, the foot length is as long as the femur, which is atypical of humans and more similar to bonobos. The pelvis, leg, and foot indicate that the individual walked upright, but the short legs and long feet suggest that the individual must have had a high step while walking. One of the bones in the wrist, called a trapezoid, is shaped like a pyramid similar to that of apes, whereas the clavicle is short and curved, unlike humans, which have a straight clavicle. The pelvis resembles the pelvis of *H. erectus*, and *Australopithecus*.[106] Brumm et al.[107] determined that the tools found near the site were very primitive, suggesting the tools resemble those found in Olduvai Gorge dating 1–2 million years ago and most likely made by *H. habilis*.

Since modern humans are thought to have reached Flores about 45,000 years ago, *H. floresiensis* and *H. sapiens* were together on the same island for almost 30,000 years. Because of its recent existence, some investigators have speculated that the Ebu Gogo local traditions and myths of a forest humanoid creature derive from the existence of *H. floresiensis*.[108]

Argue et al.[109] used cladistics (the study of morphological forms to determine relatedness) to determine if *H. floresiensis* belonged to another *Homo* species, instead of descending from *erectus*. She determined there were two possibilities: (1) *H. floresiensis* evolved after *Homo rudolfensis* but before *H. habilis* and (2) *H. floresiensis* emerged after *habilis* but before *erectus*. She determined that there was no support for *H. floresiensis* and *erectus* having a close relationship, suggesting *H. floresiensis* was not a dwarf of *erectus*. Brain size was a determining factor because *erectus* had a relatively large brain, but the brain of *H. floresiensis* was larger than the brain of *habilis*. It is still not clear if brain size was the result of dwarfing or it represents the typical size of the species. Overall, the results support the idea that *H. floresiensis* is a different species from *H. sapiens*.

Most theories suggesting that *H. floresiensis* was a diseased representative of modern humans with cretinism, Laron syndrome, microcephaly, dwarfism, etc., have been disproven. Yet, some have argued that the diagnosis of various diseases has been hampered by limited access to the specimens and the inability to isolate DNA from the specimens. It is important to note that the designation of *H. floresiensis* as a new species has been heavily disputed, and recent data suggest that the size and characteristics of *H. floresiensis* were due to Down syndrome in a modern human.[110,111]

RECENT UPDATES

New evidence of the world's oldest cave art suggests that it was produced by Neanderthals. The investigators discovered more that a dozen paintings in a Spanish cave, dated to 65 kya, which predates AMHs in Europe. The investigators also discovered perforated seashell beads and pigments at Cueva de los Aviones, a cave in Southeastern Spain, that are at least 115,00 years old. The seashell beads and ochre pigments are the oldest ornaments ever found and predate by 20–40 thousand years anything found in Africa. The authors of the study suggest that these findings indicate that Neanderthals had a cultural competence that mimics the ancient artifacts of modern humans. Until now the artistic work of Neanderthals was much younger in age and thought to be mostly copied from the work of AMHs.[112]

Another recent finding are additional genes that have been contributed by Neanderthals to AMHs. DNA sequence analysis from a 50,000-65,000 year old Neanderthal woman found in Vindija, Croatia suggests that Neanderthals lived in small somewhat inbred small groups with low genetic diversity. In addition, the investigators found that the woman was more closely related to Neanderthals that mixed with the ancestors of present day humans, than previously sequenced Neanderthals from Siberia.They found variants of genes involved with LDL cholesterol concentration, schizophrenia, and other diseases and suggest these variants remain in the modern Eurasian gene pool.[113]

More recently investigators sequenced five Neanderthal genomes from bones and teeth found in Belgium, France, Croatia, and the Russian Caucasus. The remains were dated from 39,000–47,000 years old. The sequences were compared to a prior Neanderthal genome from the Caucasus region and indicated that Neanderthals had split from a common ancestor around 150,000 years ago and experienced a major population turnover approximately 38,000 years ago, just prior to their extinction. The researchers suggest that this could have been due to extreme cold periods, leading to extinction of local populations and then recolonization from southern Europe or western Asia. Overall the research also suggests that Neanderthal populations showed a geographical population structure (different characteristics in different geographic regions). The researchers point out that the genomes they looked at lacked modern human DNA although four of the genomes were from Neanderthals that lived at a time when modern humans were in Europe.[114]

References

1. Prüfer K, et al. The complete genome sequence of a Neanderthal from the Altai Mountains. *Nature* 2014;**505**:43–9.
2. Stringer C. *The origin of our species*. London: Penguin Books; 2011. p. 26–9.
3. McDougall I, et al. Stratigraphic placement and age of modern humans from Kibish, Ethiopia. *Nature* 2005;**433**: 733–6.
4. Neubauer F. A brief overview of the last 10 years of major late Pleistocene discoveries in the old world: *Homo floresiensis*, Neanderthal and Denisovan. *J Anthropol* 2014:7. Article ID 581689. https://doi.org/10.1155/2014/581689.
5. Campbell BG, et al. *Humankind emerging*. 9th ed. Boston, MA: Pearson; 2006. p. 358–62.
6. Campbell BG, et al. *Humankind emerging*. 9th ed. Boston, MA: Pearson; 2006. p. 367–80.
7. Schmitz R, et al. The Neanderthal type site revisited: interdisciplinary investigations of skeletal remains from the Neander Valley. *Proc Nat Acad Sci USA* 2002;**99**:13342–7.
8. Seddon C. *Humans: from the beginning*. San Bernardino, CA: Glanville Publications; 2015.
9. Wong K. Neanderthal minds: analysis of anatomy, DNA and cultural remains have yielded tantalizing insights into the inner lives of our mysterious extinct cousins. *Sci Am* 2015;**312**:36–43.

10. Pearce E, et al. New insights into the differences in brain organization between Neanderthals and anatomically modern humans. *Proc Roy Soc B* 2013;**280**:1758. https://doi.org/10.1098/rspb.2013.0168.
11. Campbell BG, et al. *Humankind emerging*. 9th ed. Boston, MA: Pearson; 2006. p. 370.
12. Dansgaard W, et al. Evidence for general instability of past climate from a 250-yr ice-core record. *Nature* 1993;**364**:218–20.
13. Wong K. Our inner neanderthal. *Sci Am* 2012;**22**:76–81.
14. Mayr E. Geographical character gradients and climatic adaptation. *Evolution* 1956;**10**:105–8.
15. Sankararaman S, et al. The genomic landscape of Neanderthal ancestry in present-day humans. *Nature* 2014;**507**:354–7.
16. Bastir M, et al. The relevance of the first ribs of the El Sidron site for the understanding of the Neanderthal thorax. *J Hum Evol* 2015;**80**:64–73.
17. Rak Y, Arensburg B. Kebara 2 Neanderthal pelvis: first look at a complete inlet. *Am J Phys Anthro* 1987;**73**: 227–31.
18. Ponce de Leon MS, et al. Neanderthal brain size at birth provides insights into the evolution of human life history. *Proc Nat Acad Sci USA* 2008;**105**:13764–8.
19. Rozzi FVR, Bermudez de Castro JM. Surprisingly rapid growth in Neanderthals. *Nature* 2004;**428**:936–9.
20. Gomez-Olivencia A, et al. La Ferrassie 8 Neanderthal child reloaded: new remains and re-assessment of the original collection. *J Hum Evol* 2015;**82**:107–26.
21. Will M, Stock JT. Spatial and temporal variation of body size among early *Homo*. *J Hum Evol* 2015;**82**:15–33.
22. Dediu D, Levinson SC. On the antiquity of language: the reinterpretation of Neandertal linguistic capacities and its consequences. *Front Psychol* 2013;**4**:397.
23. Schwartz JH, Tattersall I. The human chin revisited: what is it and who has it. *J Hum Evol* 2000;**38**:367–409.
24. Le May M. The language capability of Neanderthal man. *Am J Phys Anthro* 1975;**42**:9–14.
25. Ruggero D'Anastasio, et al. Micro-Biomechanics of the Kebara 2 hyoid and its implications for speech in Neanderthals. *PLoS One* 2013;**8**:e82261. https://doi.org/10.1371/journal.pone.0082261.
26. Fisher SE, Scharff C. FOXP2 as a molecular window into speech and language. *Trend Genet* 2009;**25**:166–77.
27. Krause J, et al. The derived FOXP2 variant of modern humans was shared with Neanderthals. *Curr Biol* 2007;**17**:1908–12.
28. Coop G, et al. The timing of selection at the human FOXP2 gene. *Mol Biol Evol* 2008;**25**:1257–9.
29. Maricic T, et al. A recent evolutionary change affects a regulatory element in the human FOXP2 gene. *Mol Biol Evol* 2013;**30**:844–52.
30. Vernot, Akey. Resurrecting surviving Neanderthal lineages from modern human genomes. *Science* 2014;**343**:1017–21.
31. Berwick RC, et al. Neanderthal language? Just-so stories take center stage. *Front Psychol* 2013;**4**:1–2.
32. Hauser MD, et al. The mystery of language evolution. *Front Psychol* 2014;**5**:401. https://doi.org/10.3389/fpsyg.2014.00401.
33. Buck LT, Stringer CB. Having the stomach for it: a contribution to Neanderthal diets? *Quat Sci Rev* 2014;**96**:161–7.
34. Kullmer O, et al. Hominid tooth pattern database (HOTPAD) derived from optical 3D topomentry. In: Marfat B, Delingette H, editors. *Three-dimensional imaging in paleoanthropology and prehistoric archeology. Liege Acts of the XIV UISPP Congress, BAR International series 1049*. 2002. p. 71–82.
35. Lucas PW, et al. Mechanisms and causes of wear in tooth enamel: implications for hominin diets. *J R Soc Interface* 2013. https://doi.org/10.1098/rsif.2012.0923.
36. El Zaatari S, et al. Ecogeographic variation in Neandertal dietary habits: evidence from occlusal molar microwear texture analysis. *J Hum Evol* 2011;**61**:411–24.
37. Hardy K, et al. Neanderthal medics? Evidence for food, cooking, and medicinal plants entrapped in dental calculus. *Naturwissenschaften* 2012;**99**:617–26.
38. Richards MP, Trinkaus E. Isotopic evidence for the diets of European Neanderthals and early modern humans. *Proc Nat Acad Sci USA* 2009;**106**:16034–9.
39. Hardy BL, Moncel M-H. Neanderthal use of fish, mammals, birds, starchy plants and wood 125-250,000 years ago. *PLoS One* 2011;**6**:e23768.
40. Stringer CB, et al. Neanderthal exploitation of marine mammals in Gibraltar. *Proc Natl Acad Sci USA* 2008;**105**:14319–24.
41. Florenza L, et al. To meat or not to meat? New perspectives on Neanderthal ecology. *Am J Phys Anthropol* 2014;**156**:43–71.
42. Kuhn SL, Stiner MC. The division of labor among neandertals and modern humans in Eurasia. *Curr Anthro* 2006;**47**:953–81.

43. Lalueza-Fox C, et al. Genetic evidence for patrilocal mating behavior among Neanderthal groups. *Proc Natl Acad Sci USA* 2011;**108**:250–3.
44. Defleur A, et al. Neanderthal cannibalism at Moula-Guercy, Ardeche, France. *Science* 1999;**286**:128–31.
45. Monnier G. Neanderthal behavior. *Nat Educ Knowl* 2012;**3**:11.
46. Estabrook VH. Is trauma a Krapina like all other Neanderthals trauma? A statistical comparison of trauma patterns in Neandertal skeletal remains. *Period Biol* 2007;**109**:393–400.
47. Walker PL. A Bioarcheological perspective on the history of violence. *Annu Rev Anthropol* 2001;**30**:573–96.
48. Lieberman DE. *Neandertal and early modern human mobility patterns*. Cham, Switzerland: Springer International Publishing; 2002.
49. Villa P, Soriano S. Hunting weapons of Neanderthals and early modern humans in South Africa: similarities and differences. *J Anthropol Res* 2010;**66**:5–38.
50. Rhodes JA, Churchill SE. Throwing in the middle and upper Paleolithic: inferences from an analysis of humeral retroversion. *J Human Evol* 2008;**56**:1–10.
51. Shea JJ, Sisk ML. Complex projectile technology and *Homo sapiens* dispersal into western Eurasia. *PaleoAnthro* 2010;**2010**:100–22.
52. Féblot-Augustins J. Mobility strategies in the late Middle Paleolithic of Central Europe and Western Europe: elements of stability and variability. *J Anthropol Archaeol* 1993;**12**:211–65.
53. Wynn T, Coolidge FL. The skilled Neanderthal mind. *J Hum Evol* 2004;**46**:467–87.
54. Johansson S. The thinking Neanderthals: what do we know about Neanderthal cognition. *WIRE Cogn Sci* 2014;**5**:613–20.
55. Radovcic D, et al. Evidence for Neanderthal jewelry: modified white-tailed eagle claws at Krapina. *PLoS One* 2015;**10**:e0119802.
56. Rodriguez-Vidal R, et al. A rock engraving made by Neanderthals in Gibraltar. *Proc Nat Acad Sci USA* 2014; **111**:13301–6.
57. Pike AWG, et al. U-series dating of Paleolithic art in 11 caves in Spain. *Science* 2012;**336**:1409–13.
58. Campbell BG, et al. *Humankind emerging*. 9th ed. Boston, MA: Pearson; 2006. p. 374–5.
59. Soressi M, et al. Neanderthals made the first specialized bone tools in Europe. *Proc Nat Acad Sci USA* 2013;**110**:14186–90.
60. Appenzeller T. *Neanderthal culture: old masters. Nature news (may)*. 2013.
61. Campbell BG, et al. *Humankind emerging*. 9th ed. Boston, MA: Pearson; 2006. p. 376–9.
62. Solecki RS. Shanidar IV, a Neanderthal flower burial in northern Iraq. *Science* 1975;**190**:880–1.
63. Lietava J. Medicinal plants in a Middle Paleolithic grave Shanidar IV? *J Ethnopharmacol* 1992;**35**:263–6.
64. Sommer JD. The Shanidar IV 'flower burial': a re-evaluation of Neanderthal burial ritual. *Camb Archaeol J* 1999;**9**:127–9.
65. Rendu W, et al. Evidence supporting an intentional Neanderthal burial at La Chapelle-aux-Saints. *Proc Natl Acad Sci USA* 2013;**111**:81–6.
66. Edwards O. The skeletons of Shanidar cave. *Smithsonian Magazine, March* 2010.
67. Higham TFG, et al. The timing and spatio-temporal patterning of Neanderthal disappearance. *Nature* 2014;**512**:306–9.
68. Reyes-Centeno H, et al. Genomic and cranial phenotype data support multiple modern human dispersals from Africa and a southern route into Asia. *Proc Natl Acad Sci USA* 2014;**111**:7248–53.
69. Parton A, et al. Alluvial fan records from southeast Arabia reveal multiple windows for human dispersal. *Geology* 2015;**43**:295–8.
70. Hershkovitz, et al. The earliest modern humans outside Africa. *Science* 2018;**359**:456–9.
71. Liu W, et al. Human remains from Zhirendong, South China, and modern human emergence in East Asia. *Proc Natl Acad Sci USA* 2010;**107**:19201–6.
72. Fu Q, et al. An early modern human from Romania with a recent Neanderthal ancestor. *Nature* 2015. . https://doi.org/10.1038/nature14558.
73. Coppa A, et al. Newly recognized Pleistocene human teeth from Tabun cave Israel. *J Hum Evol* 2005;**49**: 301–15.
74. Duarte C, et al. The early Upper Paleolithic human skeleton from the Abrigo do Lagar Velho (Portugal) and modern human emergence in Iberia. *Proc Natl Acad Sci USA* 1999;**96**:7604–9.
75. Hershkovitz I, et al. Levantine cranium from Manot Cave (Israel) foreshadows the first European modern humans. *Nature* 2015;**520**:216–9.

76. Bergmann TJ, Beehner JC. Social system of a hybrid Baboon group (*Papio anubis* × *P. hamadryas*). *Int J Primatol* 2004;**25**:1313–30.
77. Herrera KJ, et al. To what extent did Neanderthals and modern humans interact? *Biol Rev* 2009;**84**:245–57.
78. Rieux A, et al. Improved calibration of the human mitochondrial clock. *Mol Biol Evol* 2014;**31**:2780–92.
79. Krings M, et al. Neanderthal DNA sequences and the origin of modern humans. *Cell* 1997;**90**:19–30.
80. Krings, et al. DNA sequence of the mitochondrial hypervariable region II from the Neandertal type specimen. *Proc Natl Acad Sci USA* 1999;**96**:5581–5.
81. Relethford JH, Harding RM. Population genetics of modern human evolution. *eLS* 2001:0001470.
82. Ovchinnikov IV, et al. Molecular analysis of Neanderthal DNA from the northern Caucasus. *Nature* 2000;**404**:490–3.
83. Green RE, et al. A draft sequence of the Neandertal genome. *Science* 2010;**328**:710–22.
84. Lowery RK, et al. Neanderthal and Denisova genetic affinities with contemporary humans: introgression versus common ancestral polymorphisms. *Gene* 2013;**530**:83–94.
85. Vernot B, Akey JM. Resurrecting surviving Neandertal lineages from modern human genomes. *Science* 2014;**343**:1017–21.
86. Fu Q, et al. Genome sequence of a 45,000-year-old modern human from western Siberia. *Nature* 2014;**514**:445–9.
87. Sankararaman S, et al. The date of interbreeding between Neandertals and modern humans. *PLoS Genet* 2012;**8**:e1002947.
88. Temme S, et al. A novel family of human leukocyte antigen class II receptors may have its origin in archaic human species. *J Biol Chem* 2014;**289**:639–53.
89. Mendez FL, et al. A haplotype at ATAT2 introgressed from Neanderthals and serves a candidate for positive selection in Papua New Guinea. *Am J Hum Genet* 2012;**91**:265–74.
90. Williams AL, et al. Sequence variants in SLC16A11 are a common risk factor for type 2 diabetes in Mexico. *Nature* 2014;**506**:97–101.
91. Finlayson C, et al. Late survival of Neanderthals at the southernmost extreme of Europe. *Nature* 2006;**443**:850–3.
92. Higham, et al. The earliest evidence for anatomically modern humans in northwestern Europe. *Nature* 2011;**479**:521–4.
93. Fa JE, et al. Rabbits and hominin survival in Iberia. *J Hum Evol* 2013;**64**:233–41.
94. Banks W, et al. Neanderthal extinction by competitive exclusion. *PLoS One* 2008;**3**. https://doi.org/10.1371/journal.pone.0003972.
95. Seddon C. *Humans: from the beginning*. San Bernardino, CA: Glanville Publications; 2015. p. 112–5.
96. Dyble M, et al. Sex equality can explain the unique social structure of hunter-gatherer bands. *Science* 2015;**348**:796–8.
97. Benazzi S, et al. The makers of the Protoaurignacian and implications for Neanderthal extinction. *Science* 2015;**348**:793–6.
98. Villa P, Roebroeks W. Neandertal demise: an archaeological analysis of the modern human superiority complex. *PLoS One* 2014. https://doi.org/10.1371/journal.pone.0096424.
99. Krause J, et al. The complete mitochondrial DNA genome of an unknown hominin from southern Siberia. *Nature* 2010;**464**:894–7.
100. Reich D, et al. Genetic history of an archaic hominin group from Denisova cave in Siberia. *Nature* 2010;**464**:1053–60.
101. Reich D, et al. Denisova admixture and the first modern human dispersals into southeast Asia and Oceania. *Am J Hum Genet* 2011;**89**:1–13.
102. Abi-Rached L, et al. The shaping of modern human immune systems by multiregional admixture with archaic humans. *Science* 2011;**334**:89–94.
103. Stewart W. Could a skull found in Siberia give new clues about human evolution? 50,000-year-old remains may be a Neanderthal or Denisovan. *Dly Mail (Lond Engl)* August 14, 2015.
104. Meyer M. A high-coverage genome sequence from an archaic Denisovan individual. *Science* 2012;**338**:222–6.
105. Morwood MJ, et al. Archaeology and age of a new hominin from Flores in eastern Indonesia. *Nature* 2004;**43**:1087–91.
106. Morwood MJ, Jungers WL. Conclusions: implications of the Liang Bua excavations for hominin evolution and biogeography. *J Hum Evol* 2009;**57**:640–8.
107. Brumm A, et al. Early stone technology on Flores and its implications for *Homo floresiensis*. *Nature* 2009;**441**:624–8.
108. Forth G. Hominids, hairy hominids and the science of humanity. *Anthropol Today* 2005;**21**:13–7.

109. Argue D, et al. *Homo floresiensis*: a cladistic analysis. *J Hum Evol* 2009;**57**:623–39.
110. Eckhardt RB, et al. Rare events in earth history include the LB1 human skeleton from Flores, Indonesia, as a developmental singularity, not a unique taxon. *Proc Natl Acad Sci USA* 2014;**111**:11961–6.
111. Henneberg M, et al. Evolved developmental homeostasis disturbed in LB1 from Flores, Indonesia, denotes Down syndrome and not diagnostic traits of the invalid species *Homo floresiensis*. *Proc Natl Acad Sci USA* 2014;**111**:11967–72.
112. Hoffman DL, et al. Symbolic use of marine shells and mineral pigments by Iberian Neanderthals 115,000 years ago. *Science Adv* 2018;**4**. https://doi.org/10.1126/sciadv.aar5255.
113. Prufer K, et al. A high coverage Neanderthal genome from Vindija Cave in Croatia. *Science* 2017;**358**:655–8.
114. Hajdinjak M, et al. Reconstructing the genetic history of late Neanderthals. *Nature* 2018;**555**:652–6.

CHAPTER

7

Dispersals Into India

All of the interesting questions that I can see in science, and for the most part in scholarship, are based on the topic of origins. ***Lawrence Krauss***[1]

SUMMARY

As the geographical nexus between West and East, the subcontinent of India has been a crossroads of hominin dispersal since the genesis of the genus *Homo* in Africa. Second only to Africa, the subcontinent is also the region with the second highest genetic diversity in the world. Its elevated levels of genetic variability suggest that anatomically modern humans (AMHs) migrated to India via the Near East shortly after exiting Africa. The subcontinent has also served as an incubator and reservoir of human genetic variability as well as an epicenter of multidirectional clinal dispersions to Europe and East Asia. It is likely that the high degree of genetic heterogeneity in India was triggered by population expansion episodes experienced by AMHs when the environment became humid and more inviting after the Last Glacial Maximum (LGM).

In this chapter, a number of topics regarding the role of the Indian subcontinent in hominin evolution will be discussed. Starting with the earliest hominin migrants of the Early Pleistocene (2.6 million years ago [mya] to 800,000 years ago [ya]), we evaluate the putative migration routes as well as the impact of geographical barriers and available natural resources in the context of archeological, anthropological, and genetic data. Specifically, the support for the coastal route along the Persian Gulf and Indian Ocean south to the littoral of India is contrasted to the likelihood of an internal penetration and crossing of peninsular India. As a prelude to the description of specific sites, the Oldowan and Acheulean lithic traditions are contrasted. Most of the key important archeological sites, such as Riwat, Pabbi Hills, Dang Valley, Masol, in Northwest India, and Attirampakkam, in the southeast coast of India are described in some detail to provide an idea of the types of evidence currently available from the epoch in the subcontinent. In the context of these locations, we discuss the importance of the sub-Himalayan plains and the Siwalik corridor in hominin dispersal.

This chapter also reviews the archeological and genetic evidence that signals the appearance of AMHs in the subcontinent, including fossils, artifacts, and uniparental

Ancestral DNA, Human Origins, and Migrations
https://doi.org/10.1016/B978-0-12-804124-6.00007-0

(e.g., mtDNA- and Y chromosome–specific) and biparental (e.g., whole genomic DNA) genetic markers. Of particular interest is the general theme that modern human fossils from South Asia represent a collage of Archaic and *sapiens* characteristics. The geographical location of South Asia at a crossroads of hominin dispersals and its consecutive waves of migrants have been cited as potential reasons for the persistence of a number of unique ancestral characters in fossils that should otherwise be, because of their young age, fully developed AMHs. It is possible that hybridization involving archaic and AMHs may have introduced ancestral attributes into otherwise modern populations. This blend of ancestral and modern traits in *sapiens* argues against population replacement and for absorption or fusion of populations in South Asia. In addition, unlike Africa and the Near East, where the transition from archaic populations and AMH behavior was more clear and delineated, in South Asia the changes were gradual and occurred late, generating ambiguity and blurry lines of demarcation among populations and species. Furthermore, the technological transitions were also marked by regional differences in their time lines. This phenomenon may be the result of geographical isolation in a sparsely populated subcontinent, limiting communication from the site of the technological origin in Africa, as well as poor interactions among separated indigenous populations creating states of sociocultural stagnation. Incidents of violence were associated with the beginning of the Mesolithic in South Asia, which may have resulted from new sociocultural demands imposed on the populations in a short period of time. In the Mesolithic, for the first time, AMHs had to change their living routine from hunter-gatherers to structured communal work, with interpersonal interactions capable of generating occupational stress, interpersonal conflict, aggression, and, in some instances, violent death. The subsequent Indian Neolithic is thought to have been influenced by the Agricultural Revolution of the Near East and is marked by similar issues as in Europe. Genetic markers suggest that demic diffusion of individuals brought about a new way of life to South Asia. With the Agricultural Revolution, a number of positive and negative changes were introduced into India. Among the positive outcomes, there was an increase in food production as farming centers grew into larger homesteads and eventually into cities. Food surplus tended to prevent starvation during winter and harsh times. And very importantly, it provided time for humans to think, imagine, and create novel things, launching a new era of cultural development. On the negative side, however, the Agricultural Revolution promoted the genesis and transmission of diseases and pandemics. The shear large number of people in close quarters and the lack of sanitation and adequate waste disposal systems as well as the commensal cohabitation of rodents and livestock living in close proximity to humans greatly facilitated the spread of infection diseases.

FROM THE NEAR EAST TO INDIA

Geography and Routes

As discussed in Chapter 5 and supported by the findings at Dmanisi, it is possible that the first group of hominins that migrated out of Africa and settled in the Near East was not *Homo erectus* but its predecessor *Homo habilis* or even the more primitive australopithecines. Whatever their taxonomic affiliations may have been, these hominins arrived at Dmanisi as early as 1.8 mya. The path that took them to Dmanisi was very demanding due to the topography of the territory. Dmanisi is embedded in between the Lesser and Greater Caucasus, and a number of mountain ranges, with peaks as high as 4000m above sea level, separate the locale from the Arabian Peninsula. These massive ranges include the Taurus, Elburz, and Zagros Mountains, as well as the Lesser Caucasus. Thus, hominins must have traveled through difficult terrain to reach and populate Dmanisi. Whether the settlement was made by *Australopithecus habilis* and/or *erectus*, these first migrants must have been technologically equipped to endure the trip and must have possessed the flexibility to adapt to different habitats and environments along the way.

It is likely that similar or the same populations of hominins that populated the Caucasus range from the Arabian Peninsula also migrated eastward in the direction of India at about the same time. A pertinent question is what route they used to travel to the Indian subcontinent? What was the exact path that these hominins took? In addressing these issues, we must keep in mind the topology of the land, the path of least resistance or barriers, and the resources available to the migrants throughout the putative paths. In matters involving hominin migrations, geography is a crucial determining factor. Just as with the migration out of Africa in which hominins picked the easiest route across the narrow Bab-el-Mandeb Strait and the Horn of Africa, it is likely that the migrants crossed the Strait of Hormuz into present-day Iran on their way to India.

Presently, the Strait of Hormuz is about 200m deep and 54km wide. During the time period when *erectus* or *habilis* were colonizing the Near East, about 2mya, global temperatures and sea levels began to experience pronounced oscillations (Fig. 7.1). This volatility in sea level became more pronounced during the Terminal Pleistocene from about 600 kya to the LGM 24,500 ya (see Table 7.1 for the time line of geological ages, types of stone tools, and methods of subsistence). Around 2mya, at the interface of the Pliocene and the Pleistocene, the sea was about 50m below the current level. This change in sea level occurred within a span of just 100,000years or less. Periodically, during the periods of maximum glaciations the sea level reached lows of about 100m below the present level. It is likely that under these climatic conditions the Strait of Hormuz was less of an obstacle for a crossing. In fact, as depicted in Fig. 7.2, it is likely that during epochs of maximum glaciations the Strait of Hormuz was a continuous land bridge

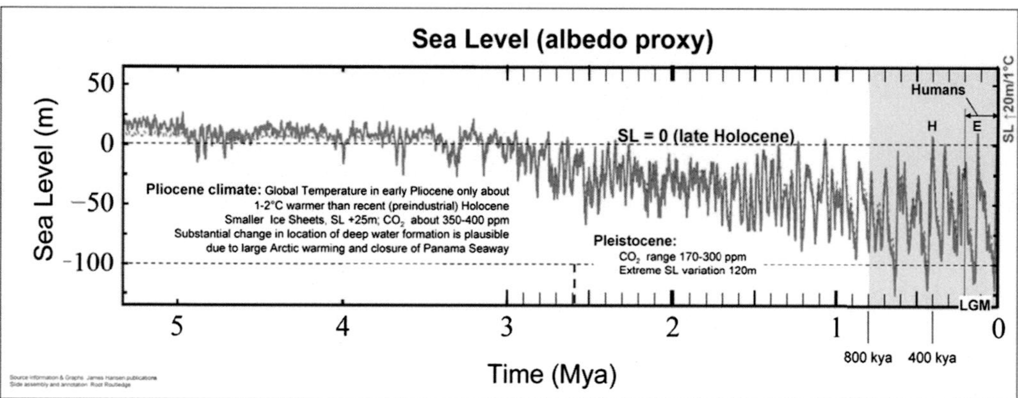

FIGURE 7.1 Global sea level fluctuations during the last 5 million years. *From https://www.e-education.psu.edu/earth107/node/901.*

TABLE 7.1 Cultural Period, Geological Age, Stone Tools, and Mode of Subsistence in the Indian Subcontinent

Terminology	Geological Age	Typical Indian Stone Tool Types	Main Subsistence Base
Lower Paleolithic	Lower Pleistocene	**Pebble** and **core tools** such as hand axes, **cleavers**, and **chopping tools**	Hunting and gathering
Middle Paleolithic	Middle Pleistocene	Flake tools, including those made by core techniques such as the **Levallois technique**	Hunting and gathering
Upper Paleolithic	Upper Pleistocene	Blade tools made on flakes, e.g., parallel-sided blades and burins	Hunting and gathering
Mesolithic	Holocene	**Microliths**	Hunting, gathering, fishing, with instances of animal domestication in a few places
Neolithic	Holocene	**Celts** (ground and polished hand axes)	Food production based on animal and plant domestication

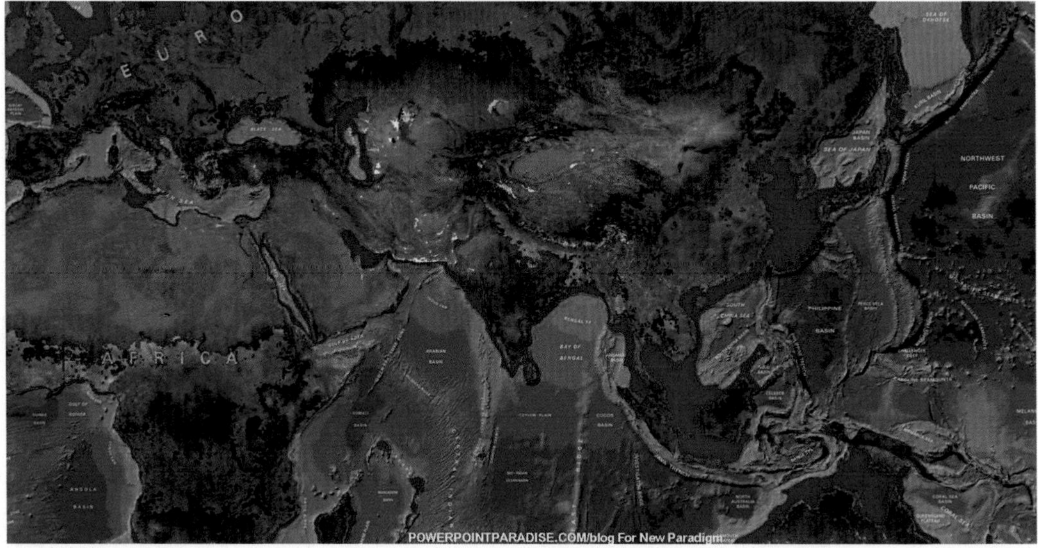

FIGURE 7.2 Coastal areas that emerge at low sea level of 100 m. *Reproduced from Ancientpatriarchs.wordpress.com.*

reaching from the Arabian Peninsula to West Asia. Once in the Arabian Peninsula, the other possible alternative route eastward was crossing the Zagros Mountains, a 1600-km-long and 240-km-wide range with peaks reaching 4000 m high. Although the Dmanisi colonization is a testimony to the capacity of *erectus* and/or *habilis* to travel through difficult terrains and survive in extreme habitats, it would have been less challenging to reach Persia via a coastal route.

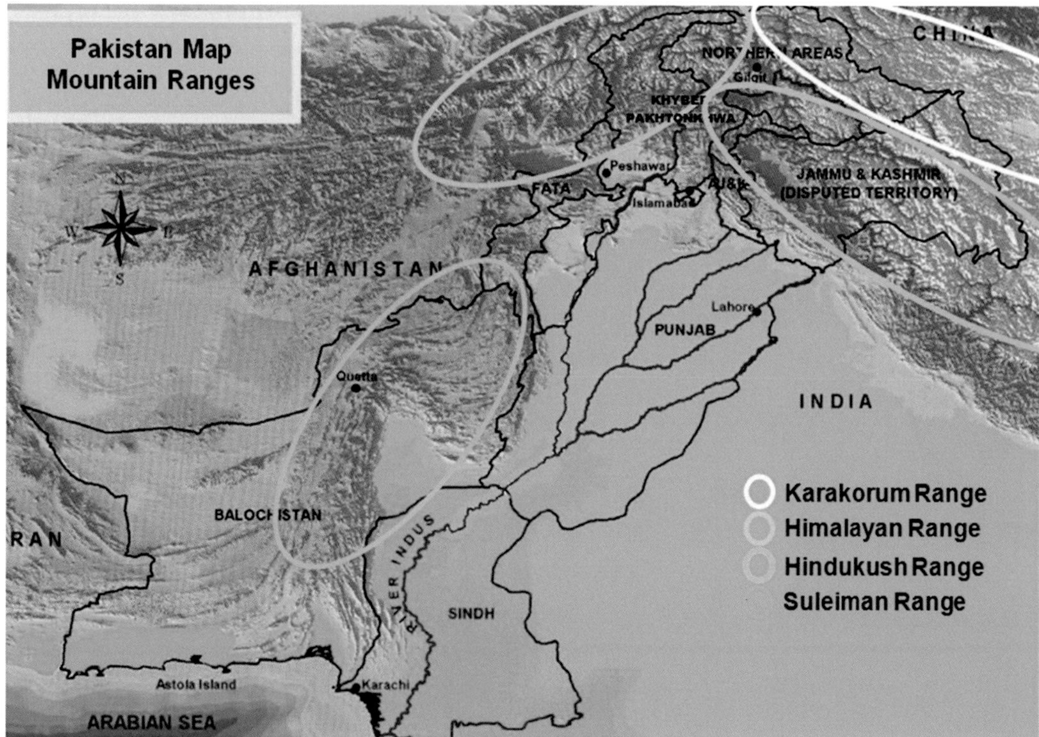

FIGURE 7.3 Mountain ranges encapsulating India. *From http://www.wildlife.pk/pakistan/images/Map-of-Mountain-Ranges-of-Pakistan.png.*

Once on the east side of the Strait of Hormuz, hominins had the option of moving into the interior. Selecting to move into the interior would have brought them again face to face with the mighty Zagros massifs or the Central Mehran Range (Fig. 7.3). An alternative route eastward to India would have been a coastal trajectory bordering the shoreline of present-day south Iran and Pakistan, a 1200 km distance and by far a shorter passage to the subcontinent. Such a course would have provided a coastal plane about 50 or more kilometers wide (Fig. 7.2), depending on the region, with minimal barriers (Fig. 7.3) and the resources available from the sea, including mollusks, shellfish, and fish. As the sea was shallow for several kilometers from shore, the width of this coastal plane would have varied considerably depending on the sea level at the time of the journey. Eventually, these hominins faced the Indus River Delta, just west of the current Pakistani-Indian border, an important obstacle for these travelers. It is not clear how a hominin, 2 mya, was able to cross the Indus waterways by the coast. Did *habilis* and/or *erectus* possess the technology to build rafts? Accidental crossings would not account for the sizeable groups of individuals needed to provide for a genetically healthy and sustainable population.

An inland approach into the subcontinent from the north or west would have been challenging, considering the number of ranges that literally encircle India. These impressive mountains are the Suleiman, Hindu Kush, Karakorum, and Himalayan Ranges (Fig. 7.3).

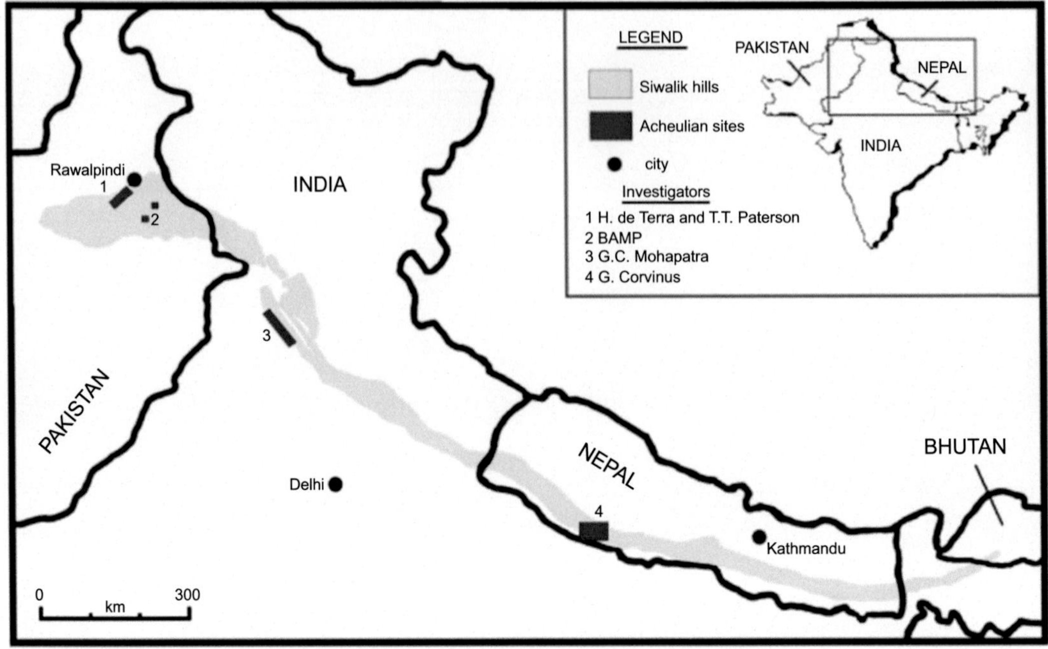

FIGURE 7.4 Siwalik corridor and Acheulean sites. *Reproduced from Archaeologydataservice.*

The Suleiman and Hindu Kush ranges block the passage into India from the west and northwest, respectively, whereas the Karakorum and Himalayan mountains obstruct entrance from the north. Of these four mountain chains, the Suleiman Range is the least impressive with several peaks over 3000 m high. The combined enclosure provided by these four cordilleras makes noncoastal dispersals very difficult, especially for early hominins with limited technological resources. The Suleiman Range stops approximately a couple hundred kilometers short of the current shoreline prior to the Indus River Delta (Fig. 7.3). Thus, the Suleiman Range would not have presented a barrier for a coastal migration into India.

Three Alternative Routes to Southeast Asia

Once hominins found themselves on the western banks of the Indus River (see Fig. 7.3), three main alternative routes were available to them for their migration eastward. One potential path would have avoided the Indian Peninsula altogether by dispersing eastward along the sub-Himalayan floodplains. It is likely that this putative passageway known as Siwalik corridor was employed by the earliest hominin travelers (pre–Acheuleans) when they traversed into South Asia (Fig. 7.4). The Acheulean period is characterized by oval- and pear-shaped hand axes, whereas the pre–Acheulean or Oldowan tradition is defined by choppers, scrapers, and pounders that were made from and resemble pebbles. The absence of pre–Acheulean implements in the subcontinent proper suggests that *erectus* and/or their ancestors used this northern passage as they traveled to Southeast Asia and Near Oceania about 1.8 mya. Yet, clearly this trajectory does not account for *all* the dispersions into East Asia because plenty Acheulean sites have been uncovered in the interior of the Indian subcontinent. A second passageway to reach Southeast Asia may have involved moving across the subcontinent; however, the lack of pre–Acheulean remains in the interior of the subcontinent indicates that such a route was not employed or was not successful during the earliest phase (phase one) of the Out of Africa I dispersal. It is possible that Oldowan or equally ancient sites simply have not been discovered yet. A third scenario involves a littoral route, which posits a circum-Indian trip around the subcontinent using the shoreline. Although some sites have been discovered in relative proximity to the coast, very limited hominin presence has been detected by the present-day coastline. Higher sea levels resulting from the current interglacial period may account for the failure to identify coastal settlements because they presently lie deep under water. Fig. 7.2 illustrates the extent of land currently under water that was exposed during periodic dry epochs of glacial maxima in West Asia and India. During glacial maximum periods, sea levels have dropped over 100 m. In Fig. 7.2, note the extensive coastal areas of Iran and Pakistan along the Arabian Sea and by the shores of the subcontinent that become exposed and available for habitation during minimum sea level periods. It is highly likely that a number of sites along the West Asian and Indian littoral route are now submerged and out of reach of conventional archeology.

Oldowan Versus Acheulean Traditions

The Oldowan is the archeological name given to the Earliest Stone Age. It is considered the earliest tool-manufacturing tradition used by hominins. It is thought to have originated approximately 2.6 mya by *H. habilis* or even earlier by an australopithecine group in East Africa. It lasted until 1.7 mya when it was gradually replaced by the more sophisticated Acheulean tradition practiced from about 1.6 mya to 200,000 ya. Although the Acheulean industry is mainly associated with *erectus*, it has been established that *Homo heidelbergensis*, in existence between 600,000 and 200,000 ya, also practiced the technology. Yet, in some locales (e.g., the Near East) both industries were practiced during the same time period. In general, the Oldowan tradition is characterized by the absence of bifaces, whereas the Acheulean is made up of cores flaked on both sides to remove most of the cortex and generate an ovate or pointed implement. Depending on the degree of workmanship and retouching, both Oldowan and Acheulean technologies are subdivided into early or late periods. These two traditions did not develop in situ in the subcontinent of India but arrived from the west with migrating populations. Acheulean assemblages are detected in Africa, South and Southwest Asia, and Western Europe but are not seen in Central, East, and Southeast Asia, where the artifacts are made up of unrelated simpler flakes and cores. The Movius Line, based on fieldwork conducted by Hallam Movius in Myanmar (Burma) in 1938, is routinely used as a line of demarcation to illustrate the geographical extent of the Acheulean industry in Asia. Yet, the lack of consistency in the longitudinal distribution of the Acheulean tradition has prompted some experts to question the distribution paradigm, considering it an oversimplification. Scholars often classify populations and cultures based on the presence/absence of these tool traditions.

The distinction between the Oldowan and Acheulean traditions is paramount in discussions on the peopling of the Indian subcontinent and the Out of Africa dispersals because there is a complete absence of pre–Acheulean implements in the interior of the Peninsula. Considering the relatively greater antiquity of Oldowan tools, this void may suggest that the very early Out of Africa migrants never traveled through the interior of India on their way to East Asia. On the other hand, because Acheulean sites are abundant within the subcontinent, it is possible that the limited technological sophistication of Oldowan hominins, such as *erectus*, *habilis*, and possibly australopithecines, was not enough to survive the harsh interior of India with limited lithic resources. It is possible that the environmental challenges provided by the interior of the subcontinent were just too much for the early hominins.

EARLY HOMININS DURING THE EARLY PLEISTOCENE, 2.6MYA TO 800,000 YA

The history of paleontology studies in the Siwalik foothills dates back to 1879 with the discovery of the first fossil ape in a locale near the Indus River. Since then, a number of other extinct fauna have been unearthed from the Siwalik. In addition, numerous early hominin sites have been identified along this range in Pakistan, India, and Nepal. A number of pre–Acheulean archeological sites have been discovered, especially in the foothills of the Siwalik range in northern Pakistan since the early 1980s. These remains provide the best evidence of early hominin occupation practicing an Oldowan-style industry in South Asia. The term "early hominins" in this context is used to denote groups that migrated eastward from the Near East as part of the early phases of the Out of Africa I dispersal. In other words, these were pre–Acheulean populations that eventually dispersed eastward.

It is noteworthy that *all* of the well-documented early homimin sites that practiced an Oldowan tradition in South Asia are found along the Siwalik trail within northern Pakistan and northwestern India. In other words, all reported well-studied locations are inland. No location has been found to the south in peninsular India. Thus, no pre–Acheulean coastal hominin settlements have been found that would confirm circum-Indian dispersal eastward. And yet, populations practicing Oldowan industries eventually colonized East Asia. The Oldowan penetration into East Asia is evident by a number of settlements, including Xlhoudu, Xiaochangliang, He County, and Donggutuo, all of which exist in what is today Northeast China. This dearth of information signaling an Oldowan presence in peninsular India underscores the need to reevaluate the paradigm of a coastal Indian dispersal eastward by the earliest Out of Africa I migrants.

The three well-studied and informative pre–Acheulean sites from the Siwalik region of North Pakistan (Riwat and Pabbi Hills) and Northwest India (Masol) are highlighted in the following case studies.

The Riwat Site

In the context of this discussion, it is noteworthy that one of the oldest hominin sites outside of Africa is located in northern Pakistan and not in the littoral. This location is Riwat in the province of Punjab. Riwat is located about 1000km from the Arabian Sea coast. The artifacts at this site have generated two dates: 1.9 mya and 2.5 mya.[2] If a coastal route was, in fact, used by the earliest hominins to reach India, then why is Riwat so far away from the shoreline? A plausible explanation is the proximity of the Riwat site to the Soan River basin. The Soan River is an important tributary of the Indus that empties into the Arabian Sea. It is possible that some of the hominin populations traveling east, by way of the Arabian Sea shore, when confronted with the Indus delta and the difficulty in crossing it, decided to move north along the Indus estuaries, water networks, and valleys. The Indus and Soan rivers must have provided a fertile landscape for these migrants as they traveled north. This trajectory likely provided abundant floral and faunal resources for subsistence during the trip and settlement of the region. Another possible reason for the lack of evidence of coastal habitation is the current higher sea levels that may be obscuring potential archeological sites.

The Lower Paleolithic Riwat settlement was discovered in 1983. The implements at Riwat included flakes, chopping tools, and cores made of quartzite according to the Oldowan tradition. An example of a typical tool found at Riwat consisted of a cobble flaked in three planes that usually involved the removal of about seven flakes.[3] Of the 23 artifacts collected at Riwat, only 3 are still considered authentic tools manufactured by a hominin, the other may be the result of natural forces. Although the orthodoxy considers Riwat a *H. erectus* dwelling, it is possible that an earlier hominin such as *H. habilis* could have been responsible for the implements. In addition to the 1.9 million-year-old remains, Riwat also provides evidence of a more recent occupation dating to 45,000 ya.[4]

Pabbi Hills

Pabbi Hills, a site close to Riwat in the Soan Valley in northern Pakistan, is a fast-eroding location made up of soft fluvial sediments deposited from about 2 mya to 100,000 ya.[5] It is likely that this location served as a tool-manufacturing station for a population of *erectus* or *habilis*. Although this site is less well known compared with Riwat, it is much richer in remains. The assemblage at Pabbi Hills is made up of 607 artifacts, most of which (96%) are fabricated from quartzite. Most of the tools are cores (41%) and flakes (58%). These implements are very simple and made in the Oldowan tradition. The more advanced tool industry, the Acheulean, has not been detected at Pabbi Hills, although numerous examples have been discovered throughout the subcontinent. Many of the pieces were found scattered in the area. And although a large number of bones were found at the area, they were not associated with the lithic remains, and no cutting marks were detected in the fossils. While paleomagnetic dating (magnetic field determinations of rocks or sediments lock-in at the time of formation) points to an age of around 1.9 to 2.1 mya of the sediments, the lack of absolute assessments have generated some skepticism among scholars.

Dang Valley

A less studied site, also in the foothills of the Siwalik range, is located in the Dang Valley along the Babai River in western Nepal. Dang is about 600 km southeast of the Masol site discussed below. The Riwat, Pabbi Hills, Dang, and Masol sites lie diagonally in a southeasterly direction on the Siwalik corridor. Dang dates to the early Paleolithic, around 1.8 mya. Yet, this age estimation is based entirely on the context of the geological strata. There is no direct dating based on fossils or associated fauna. The early Paleolithic component of the site has yielded a number of Acheulean tools, including 20 artifacts, all of which were made from quartzite.

It is noteworthy that the 1.8 mya tools at Dang reflect an Acheulean style developed by *H. habilis*, a direct ancestor of *erectus*. The Acheulean industry originated in Africa from the more primitive Oldowan tradition about 1.76 mya. It is not clear how a tradition that allegedly originated in Africa would be found contemporaneously in South Asia. The assemblage includes two hand axes, an unfinished hand axe, a cleaver, a pick, nine large- and medium-sized flakes, two large and one small core, one discoid, and two hammer stones.[6] The early Paleolithic hominins at Dang lived in what are today the banks of the ancient river, using the quartzite cobbles of the fluvial gravel as raw material for the tools. In addition, this location and a number of other nearby sites exhibit assemblages from the late Paleolithic, Neolithic, and Mesolithic periods.

Masol

The Masol site is part of the Siwalik deposits near the town of Chandigarh in Punjab, northwestern India. It is located about 450 km southeast of the Riwat site in northern Pakistan. Excavations at Masol started in the 19th century, and since then a number of stone tools and flakes associated with fossil remains of vertebrates have been discovered by different groups of investigators.

The original reports by Singh in 1988[7] and 2003[8] on the Khetpurali and Masol locations indicated the discovery of mandibular and postcranial fragments belonging to hominins in association with stone tools. The mandible still contained the lower right first molar and the third and fourth premolars. Further expeditions by Singh and collaborators recovered more hominin remains, including a second mandibular fraction with two premolars (P3, P4), a molar (M1), part of a canine, the proximal end of a left femur, the distal end of a left femur, and a right patella. Paleomagnetic dating results on the Upper Siwalik Formations indicated an age of 3.4 my. In addition, 50 vertebrate remains were collected. Quartzite pebbles and unifacial and bifacial choppers were also recovered. The conclusions derived from these reports have been questioned on the basis of the lack of evidence confirming that the fossils are from hominins and not from various large vertebrates, as well as the lithic material being much more recent.[9]

Additional excavations of the site started in 2007. An Indo-French research team under the auspices of the Siwalik Program is conducting the fieldwork, and it has yielded additional artifacts, most of them made of quartzite. Masol was redated by paleomagnetic methods and found to be about 2.6 mya.[10] The fossils and implements recovered were found under the Gauss–Matuyama paleomagnetic reversal (geologic event, approximately 781,000 ya, when the magnetic field of Earth last underwent reversal) layer, and it is thought that they were exposed recently by the dismantling of the old sedimentary layers.

The region of Masol, 2.6 mya, was a subtropical savannah with a diverse fauna made up of elephants, giraffes, hippopotami, buffalo, equids, and a number of carnivores, as well as giant terrestrial turtles. The Himalayas were in place but not the Siwalik Hills that today extent to the south and run parallel to it. The habitat at Masol was transformed periodically with floodplains and a multitude of rivers flowing down from the Himalayan range.

A total of 1469 lithic remains have been recovered from the site in a period of about 7 years (2008–15) from an area encompassing 0.5 km^2. The utensils recovered included quartzite pebbles and unifacial and bifacial choppers (Fig. 7.5). The artifacts do not reflect any given lithic technical tradition, such as the Acheulean. The fauna recovered includes 3 orders of reptiles (crocodilian, turtle, and lizard), 2 families of carnivores (Hyenidae and Felidae), and 10 families of herbivores. Masol clearly represents an abundant assemblage of fauna and stone artifacts in a fluvial landscape. Unfortunately, the material recovered was scattered on outcrops and their slopes as opposed to excavated out from undisturbed buried layers (in situ). Most paleoanthropologists only feel constable in calling material undisturbed if it is found at least 1–1.5 m below the surface. The location of the site at the foothills of the Himalayas is constantly subject to monsoon-related floods. It is highly likely that the unexpected torrential downpours periodically surprised herds of herbivores, reptiles, and carnivores, drowning them as the deluges flowed down. This repeated phenomenon may have led to the accumulation of a large number of animal cadavers at the Masol site. Among the fossils found, 12 exhibit marks resulting from carnivore activity and 3 bovine bones show cuts best explained

FIGURE 7.5 Cobble tools from layer 2 of the trial trench B1 in Masol. 1–3 represent three different cobble tools showing different angles. *From www.sciencedirect.com/science/article/pii/S1631068315002286.*

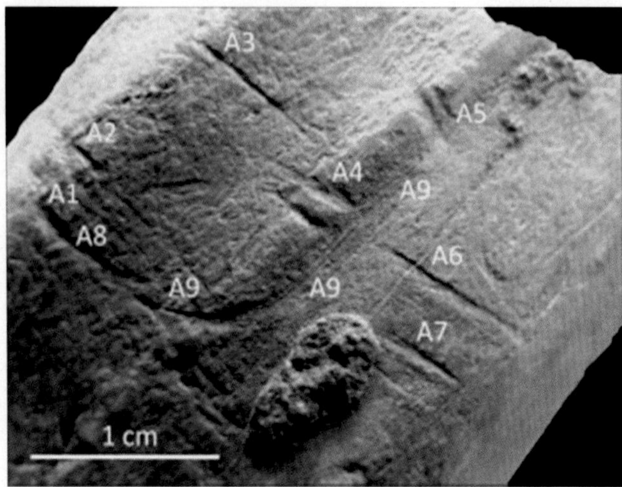

FIGURE 7.6 Cut marks made on a tibia. A1–A9 indicate different cut marks. *From https://doi.org/10.1016/j.crpv.2015.09.019.*

by hominin activity (Fig. 7.6). It seems that the cuts were designed to slide off the meat from the carcass and to break the bones to reach for the bone marrow.

The discovery of these indentations on bone remains immediately generated arguments among taphonomists. The extraordinary claims of the nature and age of the bone cuts, as expected, generated a number of criticisms. Some comments have alleged that land predators and scavengers could have made the marks. Others attribute the cuts to crocodile, rodents, and weathering. Crocodile remains were found among the bones, providing an explanation for the disarticulated fossils found. Even the laboratory experiments trying to duplicate the fossilized cuts have been criticized on the basis that pig bones were used instead of bovine bones. The unexpected very ancient age of the remains was also a concern considering that hominins capable of creating the cuts were not supposed to inhabit South Asia around 2.6 mya. Scholars studying these sites needed to contend with the long-held paradigm that posits that the genus *Homo* was the only one with the dexterity capable of constructing the necessary tools and dexterity to butcher animals.

Recent experimentation has shown that three bones from Masol clearly exhibit butchery marks.[11] The results of laboratory recreations of the cutting marks were published in 2016.[12] These studies show that the incisions exhibit traces of mineral deposits, suggesting intentional dismembering of the animal with stone tools. The residual mineralization left at the cuts is consistent with traces of quartzite, the same stone from which the flakes or cobbles from the site are comprised. Experimental cuts made at the laboratory confirm that the cuts left on the bones are consistent and typical of the grooves made by a sharp-edged object, such as a flake or cobble of quartzite. The dimensions and spatial orientation of the indentations suggest that the incisions required strong force and agile wrist action with precision and knowledge of the bovid anatomy. Considering the age estimates of the fossils with marks, the sophistication of the dismembering techniques, and the geographical location of the site, it is not clear whether the ancient inhabitants of Masol were australopithecines or members of the genus *Homo*.

Cut marks have been observed previously from the Lower Awash Valley in Northeast Africa, dating to 3.4 mya, the oldest known lithic activity visible on bone.[13] This location on the Afar depression in Ethiopia is near the spot where a 3.3 million-year-old juvenile *Australopithecus afarensis* fossil nicknamed "Lucy's Baby" was found suggesting that this species was dismembering animals 800,000 years prior to what previous evidence indicated. It is likely that *afarensis* did not venture out to the plains of Africa to hunt and kill an impala-like animal, but they were definitely capable of scavenging predatorial kills. The age of these indentations from Africa clearly shows that *erectus* was not the first to butcher game. In other words, dismembering game precedes the genus *Homo*. In connection with the Afar finding is the recent discovery of a tool workshop dated to 3.3 mya at Lomekwi 3, Kenya.[14] In addition, a number of other sites older than 2.0 my exhibiting cut marks have been reported at Kada Gona (Ethiopia) and Kada Gona (Bouri) and dated to the early Lower Pleistocene, 2.53 mya[15,16] and 2.5 mya,[17] respectively. These locations allegedly assigned to the genus *Homo* are contemporary with an assemblage of implements discovered in Longgupo, South Central China,[18] dated at 2.48 mya credited to *erectus* and tools from Renzidong, Southeast China, of a similar age (2.58 mya).[19–21] Masol is located at the same latitude as Longgupo and Renzidong, both in the Yangtze River basin at a distance of 3150 and 3800 km from Masol in Central South China.

Attirampakkam, a 1.5-mya Site Near the Southeast Coast of India

The early hominin settlement of Attirampakkam in the Kortallayar River basin in Tamil Nadu is rather unique in being close to the coast (about 45 km) in southeastern peninsular India. Other than the northern sites along the Siwalik corridor discussed earlier in the northern regions of the subcontinent, Attirampakkam is probably the most significant site dating to the Early Pleistocene.[22] The archeological complex is made up of several multicultural sites with in situ Lower, Middle, and Upper Paleolithic deposits. The inhabitants of the settlement practiced the Acheulean tradition, and the site has been dated 1.5 to 1.0 mya by paleomagnetic measurements. Attirampakkam is significant in that it is the oldest Acheulean site in the subcontinent proper, and as such, it may represent a second dispersal wave (phase two) of Out of Africa I. This second dispersion from the Near East likely brought with it the more advanced Acheulean technology that allowed the colonization of the harsher environs of peninsular India.

The Acheulean artifacts discovered are mainly hand axes made of quartzite. In addition, large bifaced cleavers, large flaked utensils, small tool components, cobbles, and devitage demonstrate refinement of the lithic instruments on the site. It is noteworthy that quartzite is not found in the region, and the scarcity of the cores suggests that early hominins transported the partially finished lithics to the site because of the available game in the area. Subsequently, on location, likely *erectus* finished tools by reducing them further. The discovery of three molars and footprints from the Lower and Middle Paleolithic belonging to a water buffalo, a horse, and a medium-sized bovid demonstrates that the habitat was an open and wet field.

Considering the age of the Attirampakkam location, it is expected that the residents were a band of *H. erectus*. The hominin occupation at Attirampakkam indicates a continuous stay during the Lower, Middle, Upper Paleolithic and Neolithic representing a number of species, including *H. erectus*, *Homo sapiens*, and possibly *H. heidelbergensis*. The Attirampakkam site also demonstrates that, at some point, hominin dispersal in South Asia included the continent proper and not just the northern Siwalik corridor.

RECONSIDERATION OF THE OUT OF AFRICA I PARADIGM BASED ON SOUTH ASIAN FINDINGS

Although South Asia as a territory suffers from limited fieldwork, poor recording of data, lack of systematic surveys, and undated sites, it is a region undoubtedly rich in archeological remains. In addition, as a strategic location in between the Near East and East Asia, it represents a pivotal middle stage in the hominin dispersals that led to the peopling of the world. Furthermore, its singular anthropological history has repeatedly challenged many of the premises and theories that have been accepted over the years. As such, the findings from the subcontinent of India allow for the reevaluation, refinement, or even rejection of theories addressing hominin evolution.

For example, the prevailing notion today is that hominins, other than Neanderthals, originated in Africa and then dispersed out of the continent into Eurasia. The orthodoxy also believes that *H. erectus* was the first species to venture out of Africa a little over 2 mya and expeditiously dispersed into Asia and Europe. Yet, a growing number of discoveries do not fit into this view. Masol provides an excellent case study.

As expected, the remarkable findings in Masol have revitalized the interest in early hominin habitation and dispersal in Southeast Asia, a crucial region in the Out of Africa I dispersal toward the Far East. The term early hominins in this context is used to indicate pre–*sapiens* populations, such as australopithecines, *H. habilis*, *H. erectus*, *H. heidelbergensis*.

Who Were the 2.6-mya Inhabitants of Masol?

If the age estimate of 2.6 mya for Masol holds, the findings indicate that an early hominin preceding *H. erectus* found its way to what is today Northwest India after exiting Africa. Alternatively, the emergence of the genus *Homo* in Africa dates back to earlier than 3.0 mya. Another possibility is that the Masol population evolved in situ in South Asia and then dispersed to other regions of the world. This last scenario has the potential to revolutionize the field of human evolution by bringing back, in full circle, the old notion that the cradle of humanity may reside in South Asia and not in Africa. Considering the unexpected old age of the settlement at Masol and possibly Riwat (if the 2.5 mya date is correct), future findings from South Asia may reveal a more complex picture of hominin evolution than what holds true today. For example, it is possible that *H. erectus* was only one of several hominin species that had the capacity to migrate out of Africa. We can also contemplate a scenario in which the various hominin branches may have evolved in different locales instead of one single origin in Africa and a one-directional dispersal from Africa to Asia and Europe.

Before the dating of Masol, the earliest settlements of early hominins outside Africa were 1.8 mya in Dmanisi in the Caucasus in the Near East and 1.6–1.8 mya in the island of Java in Indonesia. These age estimations are rather early and necessitate impressive traveling speeds from Africa by hominins. Considering that *erectus* is thought to have evolved in Africa only about 100,000 years before the 2.6-mya settlement of Masol, it clearly cannot represent the occupation of descendants of those *erectus* that evolved in Northeast Africa. So, who were the inhabitants of Masol? Considering that *H. habilis* lived in East Africa from 2.1 to 1.5 mya, an occupation dated to 2.6 mya in Masol suggests a pre–*habilis* hominin, a population in the time range of late australopithecines, such as *Australopithecus garhi* or *Australopithecus africanus*. The presence of individuals exhibiting australopithecine morphological characteristics in Dmanisi and in the island of Flores (*Homo floresiensis*) in Indonesia lends credence to the possibility that pre–*habilis* groups had the capacity to travel out of Africa. In addition, recent evidence suggests that australopithecines used tools. Of note is the fact that the Masol hominins were at least partly omnivorous just like *garhi* and *africanus*, whose remains have been associated with butchered animal bones approximately 3.4 mya in Northeast Africa.

The Out of Africa I theory posits that the first hominins to disperse out of Africa were *erectus* groups shortly after their origin in Northeast Africa approximately 1.9 mya. It also postulates that a circum-Indian coastal route was employed by these early pre–*sapiens* hominins on their way to East Asia. Thus far the information obtained from Masol does not substantiate the abovementioned premises.

Similarly, the presence of Acheulean implements dating to 1.8 mya in the Dang Valley, Nepal, midway along the length of the Siwalik corridor, is not compatible with the Acheulean technology developing in Africa about the same time. One of the three scenarios is possible: (1) the Acheulean tradition originated earlier in Africa and then early hominin migrants transported it to South Asia; (2) its genesis was in South Asia; or (3) both originated independently, an unlikely scenario.

Location of Sites Along the Siwalik Corridor

The Siwalik Hills are rather young, as recent as 5.2 million year old, which is a blink of an eye in geological time. The range runs diagonally east to west for about 2400 km from the Indus River to the Brahmaputra River in Bhutan. The mountains are made up of sandstone rock formations, a consequence of the creation of the Himalayas.

The locations of the four major archeological sites exhibiting evidence of the earliest hominin settlements in South Asia (Riwat, Pabbi Hills, Dang, and Masol) are in the interior north of the subcontinent, just south of the Siwalik Hills running diagonally from northern Pakistan to Southeast Asia. With the exception of Masol, where fauna has been found, these locales are made up entirely of lithic remains; no hominin fossils have been unearthed yet. However, because these early hominins were tool-dependent creatures, the stone artifacts constitute evidence of their existence and provide a glimpse of the hominins' life as they traveled eastward during the first phase of the Out of Africa I dispersal.

The Siwalik findings do not corroborate the coastal Out of Africa I hypothesis, which postulates a littoral migration around the subcontinent. Although a coastal migration route to East Asia would have been longer, there are reasons to envision that such a path would have been intuitively easier, especially for early hominin populations with limited technology. Littoral resources would have provided for marine organisms such as fish, shellfish, crustaceans, and algae, as well as the occasional estuaries rich in terrestrial animal and plant life. Furthermore, the terrain close to the shoreline tends to be less mountainous and easier to transit. Yet, no evidence has been found of early hominin habitation in coastal peninsular India. It could be argued that failure to detect such sites results not from their absence but from their present location under water as the sea level rose from the melting of glaciers during the current interglacial period.

Together, the locations of these archeological sites seem to outline what is known as the Siwalik corridor. So, in spite of the relative ease of a coastal circum-Indian route, the available archeological evidence suggests that the earliest hominins in South Asia bypassed the subcontinent proper and employed a peripheral northern path along the Siwalik corridor just south of the Himalayas to get to East Asia.

ABSENCE OF THE EARLIEST MIGRANTS IN THE INTERIOR OF THE SUBCONTINENT

A route along the foothills of the Himalayas would have provided certain advantages to the earliest migrating hominins. The Siwalik Hills corridor possesses one of the richest fossil sites for large animals anywhere in Asia. The megafauna recovered from the foothills were diverse and included a number of extinct species of sloth bear, giraffe, and giant tortoise. To the south of the hills, an alluvial zone gives way to level plains. Particularly pertinent to the northern traveling route of the earliest hominins is the regular torrential rains, especially during monsoon season, which flow from the Himalayas and the Siwalik into the foothills and planes to the south. These regular downpouring events created a luscious habitat of springs in hillsides and marshes rich in vegetation that allowed different types of animals to thrive.

In addition, there is ample evidence that these frequent drenching episodes literally drown animals, as massive water torrents wash the fauna to the rolling hills and plains south of the Siwalik Hills. These colossal precipitation occurrences start when strong storms from the north collided with the Himalayas triggering dramatic deluges and mud avalanches to the south. These massive mudslides killed and pushed down with them herbivores, carnivores, and residents of ponds and lakes of the valleys in between the Himalayas and Siwalik Hills. The high concentrations of fossils found today at the foothills of the Siwaliks represent killing zones and are testament to these dramatic events. These episodes paint a picture of periodic incidents of accumulation of large numbers of carcasses available to hominins for the taking. It is highly likely that these earliest hominin migrants help themselves to this bounty of meat. Thus, early *erectus* or *habilis* practicing ancient pre–Acheulean lithic technology only needed to move to high grounds to avoid the approaching rolling mud, then step to where the dead animals are located, select what type of meat they wanted to eat, bend and reach out for pieces of the flesh that they fancy. No hunting was required; just scavenging. It would have been convenient and easy picking.

The scenario described above contrasts sharply with the adverse conditions present within the subcontinent. Although a number of drainage basin systems permeate peninsular India, internal peninsular travelers would have confronted a much drier environment in which hunting, as opposed to scavenging, was a necessity. In addition, lithic resources are scarce in certain regions of India. Specifically, it has been proposed that internal routes, such as the Indo-Gangetic river system in northern peninsular India, were not used by the earliest pre–Acheulean migrants during the Early Pleistocene, 2.6 mya to 800,000 ya, because of the limited availability of workable stones in these plains. The scarcity of lithic resources would account for the absence of hominin sites in the interior of the subcontinent during the Early Pleistocene. Furthermore, it is thought that it was not until depositions of boulder conglomerates in the Middle Pleistocene about 800,000 ya that hominin locations are seen in the interior Ganges region.

There is also evidence that the Siwalik corridor was used not only by hominins but also by extinct apes in earlier times (Fig. 7.7). When the locations of fossil great apes and hominin sites are plotted on a map of South Asia, the course of the Siwalik corridor is clearly discernible. These types of data suggest that long before hominins, organisms such as apes, with limited ways and means of survival, employed the north sub-Siwalik corridor to move from west to east Asia and possibly vice versa. It is possible that these extinct apes survived partially as scavengers feeding on dead animals as well.

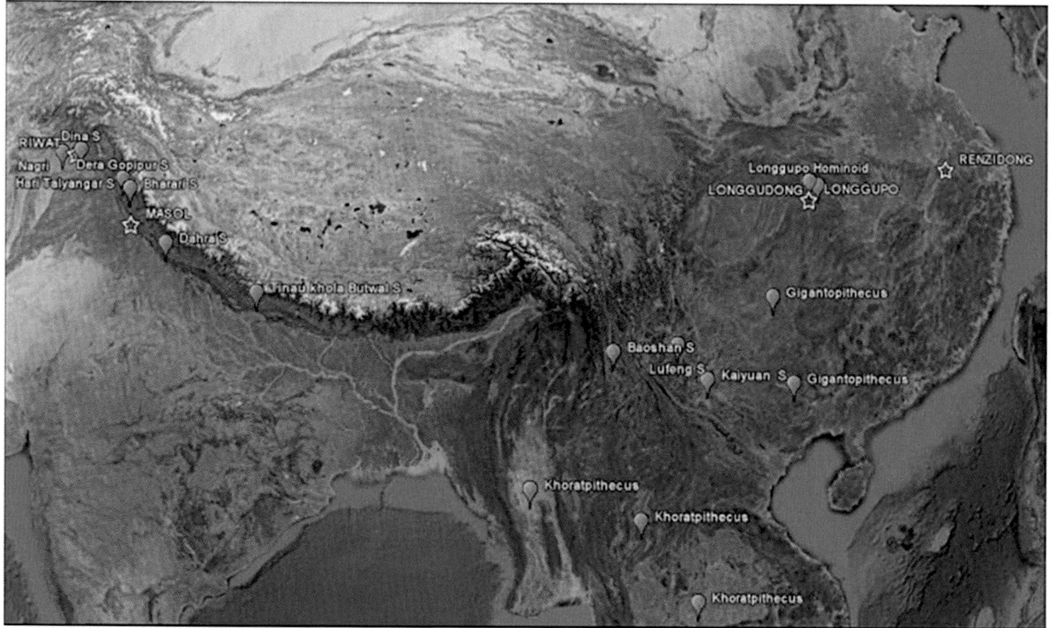

FIGURE 7.7 Locations in South Asia of Mio-Pleistocene fossil great apes and Plio-early Pleistocene paleo-hominin sites (stars): Riwat, Masol, Longgupo, Longgudong, and Renzidong. *Reproduced from Elsevier (https://www.sciencedirect.com/science/article/pii/S1631068315002286).*

SCARCITY OF THE PRE–ACHEULEAN INDUSTRY IN THE INDIAN SUBCONTINENT

Currently, limited evidence exists of the pre–Acheulean industry in peninsular India.[23] Other than the Riwat and Pabbi Hills sites in northern Pakistan, the Siwalik Hills in the northern part of Northwest India, and Narmada in Central India, no evidence of pre–Acheulean assemblages have been recovered from the subcontinent. In contrast, locations exhibiting evidence of Acheulean technology have been found deep into the subcontinent. This is somewhat perplexing, considering that the peninsula is the natural passageway for migrations of *erectus* and *heidelbergensis* on their way to Southeast Asia from Southwest Asia.

With the exception of Narmada, the absence of pre–Acheulean remains in the subcontinent proper is contradictory to the Out of Africa paradigm that posits peninsular India as a pivotal route in the dispersal eastward of early hominins. The meager evidence supporting a pre–Acheulean presence in peninsular India may simply stem from failure to identify sites and remains in a region of the world that is poorly surveyed. In connection with this scenario, it should be pointed out that few contemporaneous vertebrate remains that would be associated with Oldowan stone tools have been found.

Another scenario that may explain this dearth of pre–Acheulean remains in the subcontinent is the possibility that the pre–Acheulean populations that occupied northern South Asia did not disperse southward into the peninsula because of a number of potential reasons. In other words, the Oldowan early hominin populations (e.g., *habilis* and/or *erectus*) may not have continued migrating to lower latitudes in the subcontinent, and it was not until later, during the Middle Pleistocene approximately 780,000 ya, a time marked by the Brunhes–Matuyama reversal, that more advanced hominin groups practicing an Acheulean industry migrated south. If after additional survey of the subcontinent no additional pre–Acheulean industry is discovered, it is likely that the first widespread hominin occupation of the subcontinent proper is more recent, dating to about 800,000–700,000 ya.

Another explanation is that populations practicing Oldowan traditions did not permanently settle in India, and their presence in peninsular India was transitory, sporadic, and one-way. It is possible that some of these dispersals failed in their attempts to colonize the subcontinent because of the difficulties imposed by the harsh terrain. A brief occupation by these early hominins in peninsular India, just enough to allow passage to East Asia, may have resulted from minimal natural resources and raw materials (e.g., lithic substrates for artifacts) in the region, as well as their limited technology. So, these early migrants may have been forced to use coastal India as a fast passage to Southeast Asia. Unfortunately, coastal settlements may be currently under water because of higher sea levels, as we are in an interglacial period. This possibility of a brief stay in peninsular India is compatible with the early settlement of East Asia by *erectus*. In other words, as *erectus* originated in sub-Saharan Africa about 1.9 mya and they reached Sangiran in Java around 1.8 mya, their migration must have been rapid. In these speed-of-travel considerations, it must be kept in mind that these early hominins moved in different directions without any particular destination. With over 10,000 km ground distance separating Northeast Africa from the island of Java in Near Oceania, *erectus* dispersed at a speed of about 1 km every 10 years.

Alternatively, the populations of early hominins practicing a pre–Acheulean industry that colonized Riwat and Pabbi Hills in northern Pakistan and Masol in Punjab, northern India may have dispersed eastward along the Siwalik corridor or using a northern path through Central and East Asia into Southeast Asia. Of these two possibilities, the Siwalik corridor is the most likely because the terrain at the Siwalik foothills was a lowland region characterized by grasslands, scrub savannah, and abundant fauna. A northern passage, on the other hand, through Central and East Asia into Southeast Asia would have involved traveling through the harsh and inhospitable Tibetan Plateau. A passage through the foothills of the Siwalik range south of the Himalayans would have bypassed the subcontinent and explains the dearth of pre–Acheulean remains in the peninsula. This contention is congruent with the presence of a number of *erectus* sites in Northeast Asia that practiced an Oldowan industry, such as Xihoudu, Xiaochangliang, He County, and Donggutuo. The peopling of Northeast Asia by Oldowans is suggestive of a Siwalik Corridor passage to East Asia.

This scenario advocates for the putative coastal circum-Indian route being used at a later phase of the Out of Africa I and during the Out of Africa II dispersal involving hominins practicing the more sophisticated Acheulean technology. In other words, it is possible that the earliest hominin dispersals to Southeast Asia responsible for the settlement of Java were more direct and utilized the Siwalik corridor and planes of northern India. The single *heidelbergensis* presence in Narmada, Central India, stands out as an exception.

HOMININ ASSEMBLAGES AND FOSSILS FROM THE MIDDLE PLEISTOCENE, 800,000–126,000 YA

The findings at Isampur and Siwalik in Northwest India and Hathnora, Odai, and Bhimbetka in the interior of the subcontinent may signal a later Acheulean expansion of advanced *erectus* and/or *heidelbergensis* as part of the third phase of Out of Africa I. This third wave of Out of Africa I started about 0.8 mya and consisted of various cleaver-producing Acheulean groups. Based on the time period, it is likely that these waves of travelers were *H. erectus* and/or *H. heidelbergensis,* which likely interacted culturally and technologically with each other and possibly interbred in India. The widespread distribution of the Acheulean sites indicates some degree of adaptation to the interior of the subcontinent. Yet, the low density or spotted distribution of locations of these Acheulean sites within the peninsula may be indicative of more isolated groups composed of smaller numbers of individuals.

By about 250,000 ya, the subcontinent experienced a gradual technological transition from the large cutting lithics of the Late Acheulean to the prepared core and flake tools of the Middle Paleolithic (Table 7.1). Evidence of this technological shift is seen in sites such as Bhimbetka in the Raisen District of Madhya Pradesh, Central India. This transition seems to represent an in situ process resulting from adaption to local conditions within the subcontinent. This capacity to adapt to habitats in the interior of the subcontinent by the migrants of the third phase of Out of Africa I signals the potential for transcontinental dispersals across India by hominins on their way to East Asia and Oceania.

Isampur

At Isampur in the Northwest of South Asia, for example, about 15,000 Acheulean implements have been recovered. This site is of importance because of the stratified and continuous context of the material recovered. In addition, the remains of bovid, deer, and turtle have been unearthed. The date of this location is approximately 350,000 ya. At this site the occurrence of associated game and cutting tools in situ suggests that this location was used to butcher animals. Isampur represents a complex settlement made up of numerous camps and assemblages, each of which may have been occupied by an extended family group, with several groups together forming a population aggregate.

Hathnora

Hathnora in the Narmada Valley of Central India and Odai on the west coast (1 km from the shore) of South India are the only sites containing human remains older than 30,000 y in the subcontinent.[24]

In the Narmada Valley where Hathnora is located a number of other archeological sites, such as Netankheri, Sardarnagar, Shahganj, Ramnagar, Mahadeo-Piparia, Baneta, Devakachar, Khirighat, Gurla Beach, and Dhansi, have been discovered. It is postulated that this region of Central India has been a crossroads for game migration and attracted hominins to settle. Although some experts contest that the Hathnora fossil brain cap (Fig. 7.8) was an advanced

FIGURE 7.8 Narmada human's skull. *From http://www.frontline.in/static/html/fl2925/stories/20121228292512404.htm.*

erectus, it is likely that it was a *H. heidelbergensis* specimen, representing the easternmost enclave of this species. Hathnora also signals the easternmost reach of the Acheulean tradition. Further east, the hominins of the period practiced variations of a pebble core technology not related to the Acheulean industry found to the west.[25]

The Hathnora calvarium (upper portion of the skull without the lower jaw and face) was found in situ (not transported from a previous location) and belonged to a female individual approximately 25–30 years old. Based on the skull's features and paleomagnetic estimations, the Hathnora hominin was dated to about 500,000–600,000 ya. In addition to the calvarium, in the same area, two clavicles, a partial ninth rib, and a right and left collarbone were found. In the nearby site of Netankheri, 3 km from Hathnora, an arm and a thighbone were discovered, likely contemporaneous to Hathnora remains. A number of lithic artifacts are part of the assemblage recovered from this composite site, including cores, bifaces, flakes, thinning flakes, blades, scrapers, burins, denticulates, microliths, angular fragments, and microdebitage. Within the layers of this site, manufacturing changes are evident in the lithic material used for the tools as a function of time. For example, during the Lower and Middle Paleolithic (Table 7.1) the stone of preference was quartzite, whereas the more recent assemblages are made of chert, chalcedony, or quartz. The faunal evidence included mammalian teeth, fossils of cattle (*Bos*), elephant (*Elephas hysudricus*), and an ostrich (*Struthio camelus*) eggshell. Unfortunately, these remains were not found in association with the lithic implements, and therefore it is not clear if the tools were used to slaughter the animals.

Odai

The Middle Pleistocene Odai site is a stratified compound location containing Mesolithic and Upper Paleolithic artifacts as part of fluvial and wind-blown deposits. It is located 1 km inland from the coast (Bay of Bengal), a few kilometers from Bommayarpalayam, and about 260 km from the north coast of the island of Sri Lanka. White sand makes up the surface of the landscape; however, below this sand, at a depth of 6 m, there are seven alluvial and four wind-blown strata representing alternating sediments of wet and dry phases. Within the alluvial layers belonging to the Mesolithic and Upper Paleolithic, implements were found. This indicates that hominin occupations occurred only during humid periods.

Among the archeological material recovered from Odai, a skull and three cervical vertebrae were detected entombed in ferricrete (iron-rich soil cemented together into rock). The skull nicknamed "Laterite Baby" is about 312cc in size and still retains its maxilla and mandible with milk teeth. It is noteworthy that even as an infant, prognathism (protrusive jaws) as an ancestral trait is evident in the fossil. The fossilized skull belonged to an infant about 5months old, and it is likely that additional postcranial remains are embedded in the rock waiting to be recovered.[26] Although lithic artifacts have been found within the various layers at Odai, no tools were found in the ferricrete deposits where the skull was recovered.

This finding is unique in the sense that the ferricrete seems to have permeated the internal anatomy of the infant's brain soon after death, preserving its fine ultrastructure. The level of brain mass detail provided by the scanning electron microscopy (SEM) is impressive. The SEM images reveal an assortment of different tissues, including doughnut-shaped red blood cells, vessels, and membranous connective tissues (Fig. 7.9). The Odai infant has been dated to about 166,000 ya and is the second oldest hominin fossil discovered in peninsular India so far. Considering the anatomical features and the age estimates, the infant must have been an advanced *erectus* or a *heidelbergensis* from the late Middle Pleistocene or in the early Upper Pleistocene age.[24]

The location of Odai indicates that hominins from the third phase of the Out of Africa I dispersal had an enclave in the littoral of West India, highlighting the possibility of a coastal migration by *erectus* or *heidelbergensis* toward East Asia during the Upper Paleolithic.

Bhimbetka

The rock shelters at Bhimbetka are located in the geographical center of the subcontinent in the Raisen District of Madhya Pradesh, India, within the Ratapani Wildlife Sanctuary at the southern edge of the Vindhyachal hills. Bhimbetka is also about 40km north of the Hathnora site discussed above. Some 750 caves have been identified so far in the general area, and over 500 of these caves exhibit rock paintings. About 240 sites are located in the vicinity of the town of Bhimbetka. Presently, only 11 shelters have been excavated, and many of these only partially. A number of shelters possess well-stratified lithic assemblages, including Upper Paleolithic tools extending all the way to the Neolithic. The caves are compound settlements that chronicle hominin changes through time. Some of the shelters are indicative of occupations by advanced *erectus* dating to more than 100,000 ya, whereas others illustrate continuous habitation from medieval and historic times. Stone artifacts from the late Acheulean to the late Mesolithic have been identified along with some of the oldest stonewalls and living floors on record. Bhimbetka possesses important archeological value because it affords protractile and extended evaluation of the progressive changes in the sophistication of tools and thus provides an account of the technological and social evolution of hominins in the interior of peninsular India from crude, unworked lithics to highly specialized utensils that were made to address specific needs. In addition, the caves are famous for their numerous and exquisite rock paintings, covering a time period from 30,000 ya to historical times. The art at Bhimbetka will be discussed in the Artistic Expressions of Anatomically Modern Humans Section.

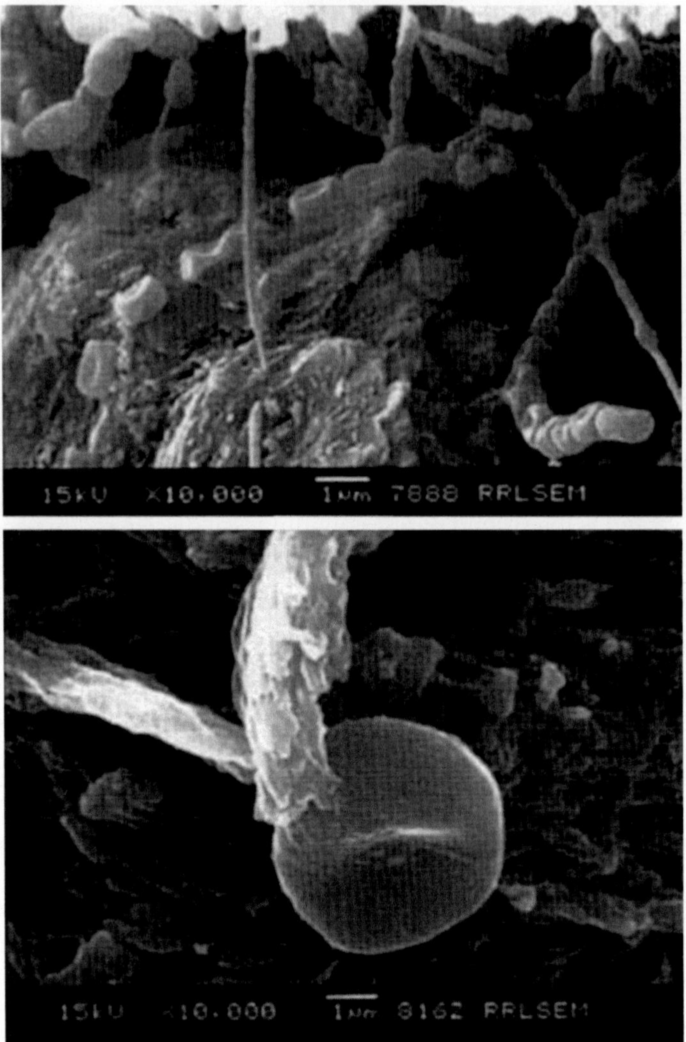

FIGURE 7.9 Images of the ultrastructure of the Odai infant's brain. Top, membranous connective tissue. Bottom, close-up of a red blood cell. *From http://www.ancient-asia-journal.com/articles/10.5334/aa.06102/.*

Attirampakkam

In a controversial recent report, Middle Paleolithic tools from Attirampakkam dated at 385,000 ya have been attributed to a population of technically advanced hominins, perhaps a group of *sapiens* (see Further Reading). These tools are significantly different from the older Acheulean technology known from the site. Traditionally, these advanced lithic tools have been associated with the incursions of *sapiens* into India from Africa about 125,000 ya but if this date and the characterization of tools are correct, the possibility of an earlier dispersal out of Africa and subsequent colonization of the subcontinent need to be considered. This earlier date of *sapiens* in India corroborate and validate the old *sapiens* sites recently discovered in East Asia (see section "A chronological conundrum" in Chapter 8).

HOMININS FROM THE MIDDLE LATE PLEISTOCENE, 126,000–70,000 YA

The archeological sites from the Middle Early Pleistocene are numerous and well distributed throughout the subcontinent. During this epoch, flakes, cores, and prepared cores dominated the lithic industry with different regional styles (Table 7.1). Yet, overall, there is a paucity of temporal information, and no hominin remains have been discovered. The Middle Late Pleistocene was a time of climatic oscillations, with alternating periods of dry and wet conditions and a progressive tendency for incremental aridity. In addition, tectonic activity was still affecting the subcontinent. Specifically, the Ganga plains were still geologically active during this time period. Furthermore, volcanic events, such as the Toba supervolcanic eruption of 75,000 ya, deposited layers of tephra (clastic volcanic material produced as dust during the eruption) throughout the river valleys of peninsular India[27] attesting to the impact of the devastation produced by a volcanic winter. In these fluctuating and demanding environmental conditions, two hominin species inhabited the subcontinent: *erectus* and *heidelbergensis*.

The 16R Dune in Rajasthan

In terms of artifacts, one location, 16R Dune in Rajasthan in the Thar Desert of northwestern India, stands above the others in the subcontinent. It is the only Middle Pleistocene site in the subcontinent, which exhibits a rather complete continuous stratified chronicle starting with a 19-m sequence older than 390,000 ya to the Upper Paleolithic sediment, 26,210 ya, including layers dating to 150,000 ya. Most of the other sites from this period only provide single implements or result from assemblies that are not part of stratigraphic profiles. The lithic tools from the Middle Late Pleistocene include scrapers, points, and cores (Fig. 7.10).

Patne

Patne in Maharashtra, East Central India, contains Middle Late Pleistocene artifact deposits at a depth of 10 m.[28] At this site, technological transitions of lithic tools are evident starting with the Middle Late Pleistocene (Fig. 7.11). Although, in general, the changes in the manufacturing of implements in the subcontinent resulted from differences in core reduction strategies that characterize the various traditions, at Patne, knapping techniques remain rather constant from Middle Late Pleistocene to the end of the Pleistocene, 11,700 ya.

ANATOMICALLY MODERN HUMANS IN INDIA

From archeological data, the earliest known arrival of AMHs in Australia dates back to at least 50,000 ya, and the estimated genetic coalescent time from contemporary human populations indicates that *H. sapiens* got to South Asia as early as 70,000 ya. The early presence of

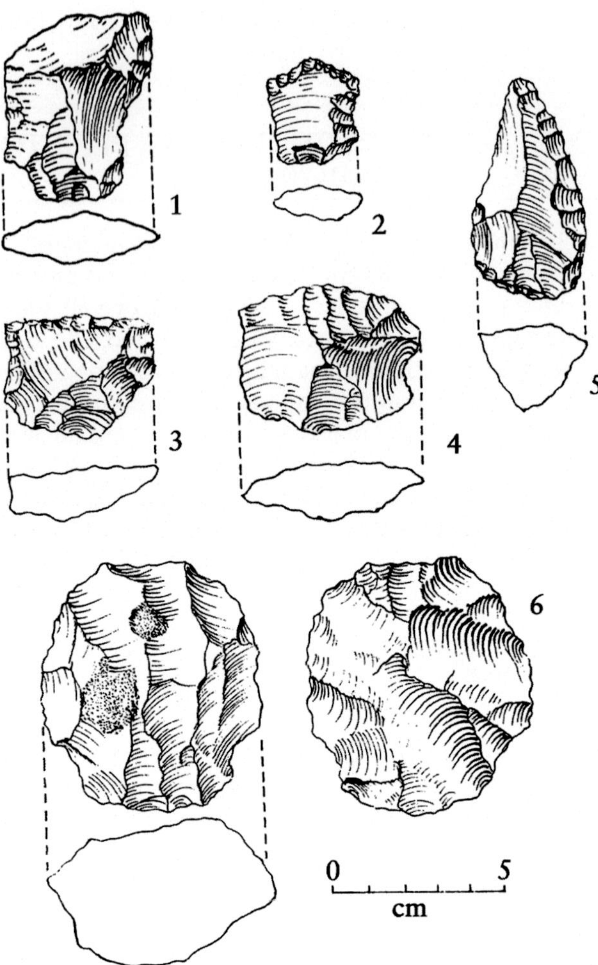

FIGURE 7.10 Middle Late Pleistocene tools from 6R Dune in Rajasthan. 1–6 represent drawings of various stone tools. *From http://www.journals.uchicago.edu/doi/full/10.1086/444365.*

sapiens based on archeological records sets a minimum time of penetration into the continent. So, if AMHs got to Lake Mungo (site of the oldest AMH fossil on the continent) in southern Australia 760 km due west from Sydney by at least 50,000 ya, *sapiens* must have entered India much earlier, maybe as early as 70,000 ya, as the genetic data suggest or even 385,000 ya as theorized from the tool findings in Attirampakkam (see above).

It is likely that as AMHs moved into South Asia from the Near East they encountered archaic hominin groups, such as *erectus* and *heidelbergensis*. As all three species practiced similar Acheulean technologies during the time period subsequent to their encounter, it is very likely that there was technological transfer and social interaction, possibly even leading to interbreeding and gene transfer. Unfortunately, no genetic studies have been conducted to date to ascertain possible introgression events among these species because investigators have not been able to sequence the genomes of *H. erectus* and *H. heidelbergensis*. Furthermore,

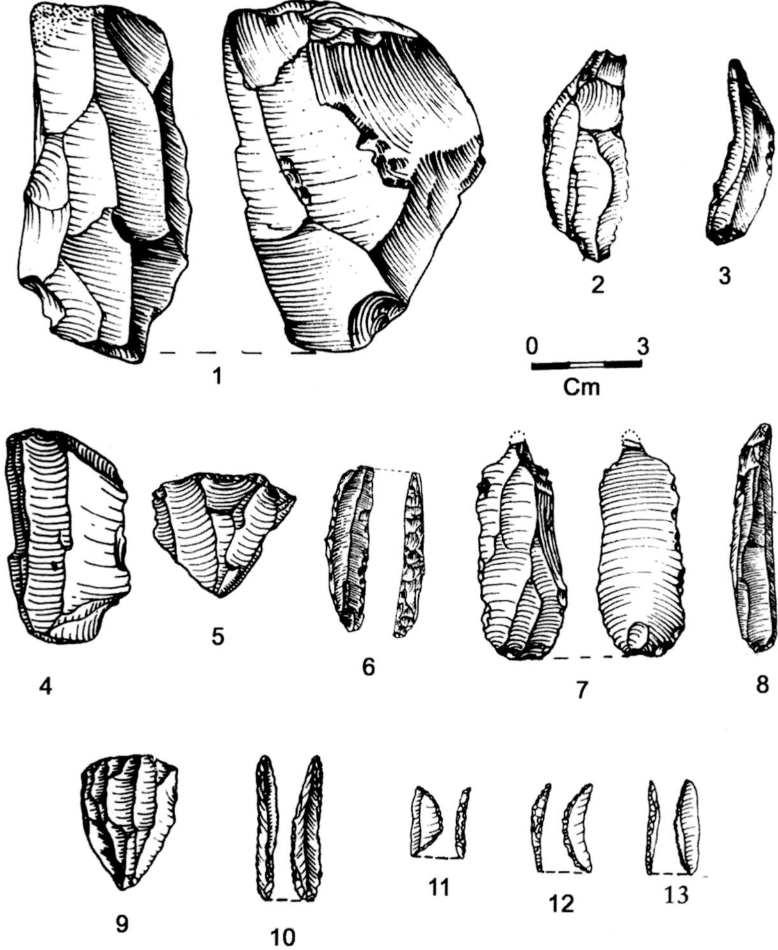

FIGURE 7.11 Technological stages of lithic tools at Patne. Top: advanced Middle Paleolithic (phase I); middle: early Upper Paleolithic (phase IIB); bottom: late Upper Paleolithic (phase IID) *(after Sali, 1989: Figs. 19, 21, 23)*. 1, 4, 5, 9, blade cores; 2, 3, 7, 10, retouched blades; 6, backed blade; 8, blade; 11, 12, 13, lunates. *From http://www.journals.uchicago.edu/doi/full/10.1086/444365.*

it is not known how much of the interactions among *erectus, heidelbergensis,* and AMHs was belligerent or amicable in nature. Technology and interbreeding among these three species could obscure the assessment of the time AMHs colonized the subcontinent. For example, transmission of toolmaking technology from *erectus* to *sapiens* could mislead scholars to assess an incorrect recent colonization of India by AMHs.

In terms of how AMHs got to South Asia, although a coastal route is currently favored among investigators as the most likely passage, no evidence exists today supporting a littoral-bound dispersal for AMHs. The absence of signals for near-shore migration may be actually artifactual resulting from the remains being under water because of high sea levels (Fig. 7.2). Most of the archeological signatures of AMH locations are from the interior of the subcontinent. Nevertheless, it is also possible that rapid transit along the coastline could also explain the minimal number of sites found close to the shoreline. In addition, the impressive speed that is required to account for the very early settlement of Southeast Asia, Australia, and Near Oceania during the second phase of the Out of Africa II dispersal by AMHs may suggest a transitory presence along the coast as opposed to permanent, long-lasting settlements.

The Arrival

The date of *sapiens'* arrival to the Indian subcontinent remains highly contentious. One of the reasons for this incertitude is the limited amount of available data that prevent meaningful correlations and time assessments. Overall, there are a meager number of well-dated sites and a small number of in situ undisturbed and stratified assemblages. In addition, because of the coexistence of three hominin species practicing similar technologies, it is difficult to determine which hominin made a given implement and to ascertain the authenticity of AMH-made implements. There is also disagreement on what constitutes a reliable marker for AMH behavior in the context of the specific habitats and conditions of the subcontinent. Furthermore, there is a tendency to interpret Indian material using African and/or European standards. In other words, oftentimes, scholars analyzed South Asian artifacts in light of technological and cultural events in other parts of the world. Drawing conclusions based on data from different evolutionary histories could be misleading and potentially generate incorrect conclusions. Also, the genetic data suffer from limited scope in terms of the ancient and contemporary human population sampled, lack of uniformity of DNA marker systems employed for meaningful and direct comparative studies, and the absence of commensal species studies (based on animal/plant systems) that could serve to corroborate or refute more traditional archeological data.

The Genetic Perspective

Genetics provides a unique view of human evolution. Genetic studies provide information about extant populations and extinct groups by examining ancient DNA. As a result of a number of revolutionary new techniques, ancient DNA studies are rapidly pushing back the ages of nucleic acid that scientists are capable of using to generate reliable data. In the field of human evolution, it is now possible to generate DNA sequences of Archaic hominins dating back to almost half a million years ago. Today, algorithms are capable of assessing that Denisovans and Neanderthals contributed to the DNA of present-day non-African AMHs, and a third *undiscovered* Archaic species provided some of its DNA to contemporary *sapiens* of Oceania. In addition, the large number of individuals whose DNA can be examined from minimal amounts of biological materials, such as a single hair strand, allow for robust statistical studies. This last characteristic is particularly useful in instances where fossil remains are so small that they are not identifiable morphologically. And although genetic clocks are gene or sequence specific, they can provide a powerful tool to ascertain age and coalescent time especially when used in conjunction with carbon isotope and archeological data.

The Indian subcontinent represents one of the first regions settled by AMHs. Furthermore, subsequently, the subcontinent served as the main epicenter of dispersal to east and west Eurasia. Although the exact arrival time of AMHs to the subcontinent as part of the second phase of the Out of Africa II spread is still uncertain, it is known that this dispersal originated in Northeast Africa and crossed the Bab-el-Mandeb Strait about 80,000 ya into the Near East. The range of departure time has been estimated to be between 50,000 and 100,000 ya.[29] Once in the Near East, the migration continued to South Asia. It is likely that this dispersal wave occurred serially in a stepwise fashion. And, as a result of founder effects (genetically nonrepresentative group of individuals establishing new settlements), elimination of genes through genetic drift (genes left out randomly from the gene pool), and accumulation of de novo mutations, the genetic distance between the source population in Africa and the descendant populations throughout the world increased as the geographic distances between them grew. Mitochondrial DNA evidence provides time estimates of 75,000 to 50,000 ya for the arrival of AMHs to the subcontinent.[30] A genetic study detected a male individual residing near the city of Madurai in South Central India possessing a Y chromosome derived from the second phase of the Out of Africa II exodus.[31] In terms of ancient remains, autosomal DNA sequencing (biparental genetic material) has been done on a 34,000-yo individual from Sri Lanka.[32] Yet, this sample is not likely a representative of the earliest AMH migrants.

Uniparental Markers

The term "uniparental genetic marker" denotes those DNA sequences that are inherited strictly through the maternal or paternal line of descent. Specifically, they are those pieces of DNA or genes that are located on the mitochondria (mt) or the Y chromosomes. As such, as mitochondria are passed only from the mother to her children and the Y chromosome only from the father to his sons, from generation to generation, these markers serve as signatures of maternal or paternal inheritance. Therefore, except for de novo mutations that occur constantly throughout time, paternally related hominin males possess similar Y chromosomes to the ones from their ancestors. This is because they are related by descent. Likewise, mtDNA from individuals who are maternally related have similar mtDNA. In addition, because de novo mutations accumulate with time, individual Y chromosomes or mtDNAs that belong to the same specific lineage differ from each other as a function of time or number of generations. For example, male individuals sharing the same paternal ancestry separated by 100 generations would exhibit Y chromosomes that differ from each other more than those of a father and son, which are only one generation apart. This gradual increase in DNA differences with time, known as a molecular or biological clock, is the basis for assessing the time since a common ancestor.

Uniparental markers provide a number of advantages in phylogenetic and anthropological studies. One of the most important advantages is that these markers allow for unequivocal assessment of lines of descent. For example, even though human mtDNA is a large molecule about 16,569 bp (base pairs) containing 37 genes and the Y chromosome is even larger with around 60 million base pairs (bp are the subunits of the DNA molecule) and over 200 genes, each behaves as a unit. Except for the extreme ends of the Y chromosomes, both mtDNA and Y chromosomes go undisturbed by recombination from generation to generation and thus can be used to trace ancestry and evolutionary connections unequivocally. So, it is possible to determine, for example, whether a certain type or lineage of Y chromosome, characterized by a given mutation that originated in Africa, is also found in the subcontinent of India, suggesting that the Indian population has received gene flow from Africa. The same sort of analysis is possible with mtDNA, but as its DNA is smaller, the ancestral connections tend to be less definitive. Unfortunately, because they behave as single genes, uniparental inheritance only reflects the passage of one marker during evolution and lacks the large numbers of independently segregating genes necessary for robust statistical analyses. In addition, uniparental markers can drop out from populations because individual carriers of a given mtDNA or Y chromosome type may fail to reproduce and thus the lineage ceases to exist in the population. The impact of these random dropout events is the creation of false negative scenarios in which investigators fail to detect phylogenetic relationships that indeed exist.

The Y Chromosome

As a paternal-specific marker, the Y chromosome has an important story to tell regarding the peopling of Southeast Asia by AMHs. The oldest Y chromosome types (haplogroups) are A00, A, and B and are found exclusively in Africa, whereas all other types (E, F, J, K, E, P, R, I, etc.), suprahaplogroups, and subhaplogroups are distributed worldwide in different proportions. All Y chromosome lineages outside Africa descend from three founder lineages: C, D, and F. Furthermore, haplogroups C and D are restricted to India and to regions to the east of it. This has suggested to some scholars that AMHs dispersed from Africa eastward moving through India on their way to East Asia and Oceania.

Specific mutations (nucleotide changes) on the Y chromosomes define each haplogroup and occur sequentially with the oldest mutations characterizing the earliest (oldest) haplogroups. For example, mutation M91 separates haplogroup A from the non-A types, and mutations M168 and M294 discriminate haplogroups A and B (African-specific) from all other (non-Africa-specific) Y chromosomes. Mutations M168 and M294 are shared by all non-African Y chromosomes and characterize the common ancestor(s) of all Eurasian and Oceanic populations. The defining M168 and M294 mutations predate phase two of the Out of Africa II dispersal of AMHs and are thought to have occurred in Northeast Africa. The P143 mutation that separates haplogroups A, B, D, and E from the rest also likely arose in Northeast Africa, and it was subsequently transported by AMHs into Eurasia and eventually introduced into South Asia and beyond. This progression of sequential nucleotide changes illustrating the earliest mutations originating in Africa is one of the strongest arguments in support of the Out of Africa theory, the proposition that AMHs originated in Africa.

The current contention is that after AMHs left Africa, they entered the Near East and promptly thereafter reached India. This view is supported by the fact that many of the extant haplogroups in India today likely date back to the original phase two of the Out of Africa II dispersion.[33] The contemporary Indian populations are known to have derived from ancestral groups from the Late Pleistocene, with minimal genetic input from outside the subcontinent since the Holocene. In other words, no major population replacement events have been detected since the initial peopling of the subcontinent during the second phase of the Out of Africa II. This is observed in all Indian populations regardless of language, caste, and religion. Recent exploration of the contemporary populations of India indicates that next to Africa, the subcontinent possesses the highest genetic diversity worldwide. Furthermore, three of the oldest major Y chromosome haplogroups, C, F, and K dating approximately to 60,000, 62,000, and 47,000 ya, respectively, exhibit the highest genetic diversity in India. This genetic variability is assessed by measuring the overall heterogeneity within Y chromosomes belonging to a given haplogroup. Genetic diversity usually correlates with the place of origin. The highest levels of genetic diversity may indicate place of origin because in those locations individuals have lived for longer periods of time accumulating mutations or diversity. In the cases of haplogroups C, F, and K, their high genetic heterogeneity signals the Indian origin of the mutations that define the haplogroups. In addition, a number of subhaplogroups of F and K, such as H, L, R2, and F*, are found predominantly in India.[34] The most abundant Y chromosome haplogroups in India are the descendants of F, R (mostly R1a1, R2, and R2a), L, H, and J (mostly J2).

Based on the abovementioned observations, it is likely that Y chromosomes carrying the M168 mutation arrived in South Asia soon after originating in Africa and differentiated in situ into different lineages in different regions of India. Because most of the Indian-specific Y haplogroups are very ancient and differentiated there, predating the Mesolithic and Neolithic, it is likely that many of the derived sublineages currently found worldwide originated in India and then spread out in different directions by demic diffusion. Thus, it is safe to say that the Indian subcontinent has been the birthplace of considerable genetic diversity and the epicenter for clinal dispersions to the east and west. In this way, the subcontinent has acted as an incubator of genetic variability and a conduit for genetic dissemination.

All these observations together indicate that India became populated by AMHs soon after *sapiens* left Africa. Moreover, Y chromosome information, based on haplogroup C (Fig. 7.12) and R (Fig. 7.13), supports a coastal circum-Indian migration dating to about 50,000 ya that eventually reached the Far East.[35] The distribution of the R haplogroup within South Asia (Fig. 7.13) illustrates its wide-range dissemination within the subcontinent and the regions of highest concentration, particularly in northwestern India.

It is noteworthy that some haplogroups derived from the major haplogroup F, such as J, R1b, and T, seem to have dispersed into Africa after originating in Southwest Asia coining the phrase "Back to Africa."

MtDNA

All the mtDNA lineages outside Africa are derived from three deep-rooted (old) founder haplogroups: M, N, and R. This is reminiscent of what is seen in relation to the Y chromosome in which all haplogroups in Eurasia descend from three ancient haplogroups, C, D, and F. In addition, both uniparental genomes (genetic makeup) in the populations of India exhibit little recent mtDNA and Y chromosome impact from non–Indian-Eurasian groups, and no evidence of extinction or replacement of the original settlers has been observed. Moreover, it is likely that both sets of relic haplogroups (maternally and paternally derived) accompanied each other in the departure and trek from Northeast Africa, arriving in India as part of the same second phase of the Out of Africa II dispersal about 65,000 ya. The very similar ages of haplogroups M, N, and R, 61,300, 64,100, and 65,500 ya, respectively, are congruent with a single early migration, possibly made up of several hundred migrants. Also, it is noteworthy that several subhaplogroups derived from the M, N, and R parent mtDNA types exhibit dates of origins very similar to the parent haplogroups themselves. This condition suggests that the mutations that define the subhaplogroups of M, N, and R occurred soon after the arrival of AMHs to the subcontinent. It is also likely that population expansion events took place soon after the colonization of South Asia by AMHs. These dispersals clearly extended beyond the borders of the Indian subcontinent and into the rest of Eurasia. These initial population expansion events in India occurred about 52,000 ya and resulted in a fivefold increase in the population.[36] Yet, signals of additional secondary expansions from the Near East to India involving lineages W, U7, and R2 (haplogroups descendants from N and R) are evident, dating to more recent time periods (about 30,000 to 20,000 ya). These younger population expansion episodes

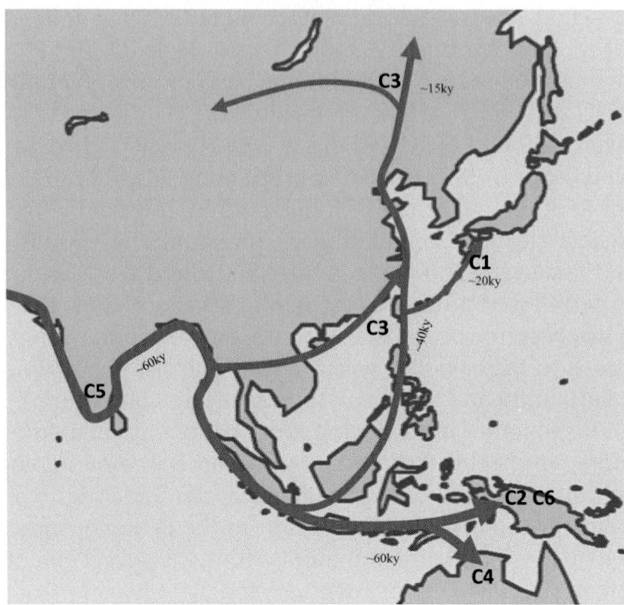

FIGURE 7.12 Estimated migration of haplogroup C along the coast of South and East Asia. *From https://en.wikipedia.org/wiki/Haplogroup_C-M130#/media/File:Migration_of_the_Y_chromosome_haplogroup_C_in_East_Asia.png.*

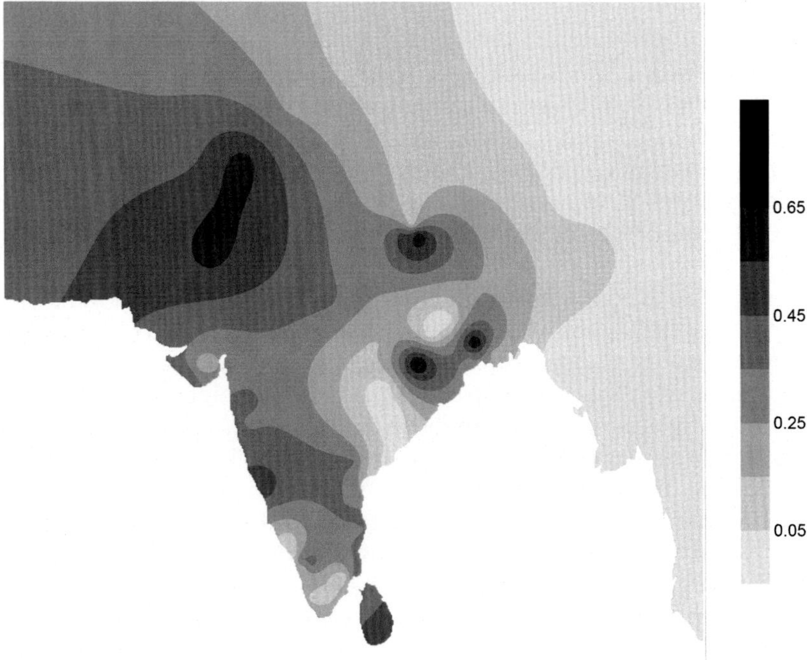

FIGURE 7.13 Distribution of Y chromosome haplogroup R within India.

coincide with humid epochs prior to the LGM 18,000 ya. Also this radiation and increment in population density coincides with the emergence of a novel, more refined, and sophisticated lithic tradition in India known as the geometric microlithic technology.

There are a number of other parallelisms between Y chromosome and mtDNA inheritance in AMHs with regard to the peopling of South Asia. For example, both sets of uniparental genetic systems indicate that the dispersals that led to the peopling of South Asia occurred soon after *sapiens* exited Africa in a speedy migration toward India and beyond to the east. The absence of nucleotide differences in the coding (gene-containing) mtDNA among South Asian, Southeast Asian, and Oceania groups is congruent with a scenario of a brisk dissemination eastward occurring during a time span of thousands of years rather than tens of thousands of years. If the dispersal had been slow, the DNA would have been able to accumulate mutations during the trip.

As with the Y chromosome haplogroups, the mtDNA lineages generally exhibit genetic uniformity among extant Indian populations across language, caste, and tribal groups. This suggests that the arrival of the primal mtDNA types took place before the creation and partitioning of caste and tribal groups. Also, the mtDNA M lineage characterizes populations of East Eurasia, including South Asia, whereas West Eurasian populations feature mtDNA haplogroups N and R and their derivatives.[37]

Moreover, similar to the founding Y chromosomes of India, studies dating back to the mid-1990s reported affinities between the ancient mtDNA of contemporary Indian groups and other Eurasian populations.[38] At that time some investigators proposed that population expansion from the subcontinent gave rise to the mtDNA genetic diversity observed throughout Eurasia. It is as though the ancestral mtDNA haplogroups differentiated into subhaplogroups after the arrival to the subcontinent and before spreading all over Eurasia. Thus, similar to the case with the Y chromosome, it is envisioned that India acted as an incubator of mtDNA variability and a source of gene flow. This parallelism between the two independent uniparental genetic systems may be the result of the same demographic events that occurred during the second phase of the Out of Africa II dispersal.

It is worth noting that a coastal route is also supported by both uniparental genetic markers. Specifically, the absence of mtDNA haplogroup M in contemporaneous Levantine populations suggests that AMHs carrying the mitochondrial M type departed Northeast Africa via the Southern route (the Horn of Africa) and continued through the littoral of Iran, Pakistan, and India to the east. The other suprahaplogroup, type N, predominantly of West Eurasia, could have traveled with migrants using the southern (Horn of Africa) or northern (Sinai Peninsula) route, which then moved into the Levant and westward.

Today the most common mtDNA types in the subcontinent are M, R, and U. Haplogroup U is a descendant of R. The ancient M haplogroup and its sublineages constitute about 60% of the overall Indian populace. M is found at 58% among the cast groups and 72% amid the tribes, with a demic increase toward the south and east of India. As a suprahaplogroup, M contributes considerably to the genetic diversity of the subcontinent. The other 40% of mtDNAs in India belong to suprahaplogroup R.

Genome-Wide Biparental Markers

As the name indicates, biparental markers derive genetic and ancestral information from both parents. In addition, biparental inheritance is subject to a number of evolutionary forces

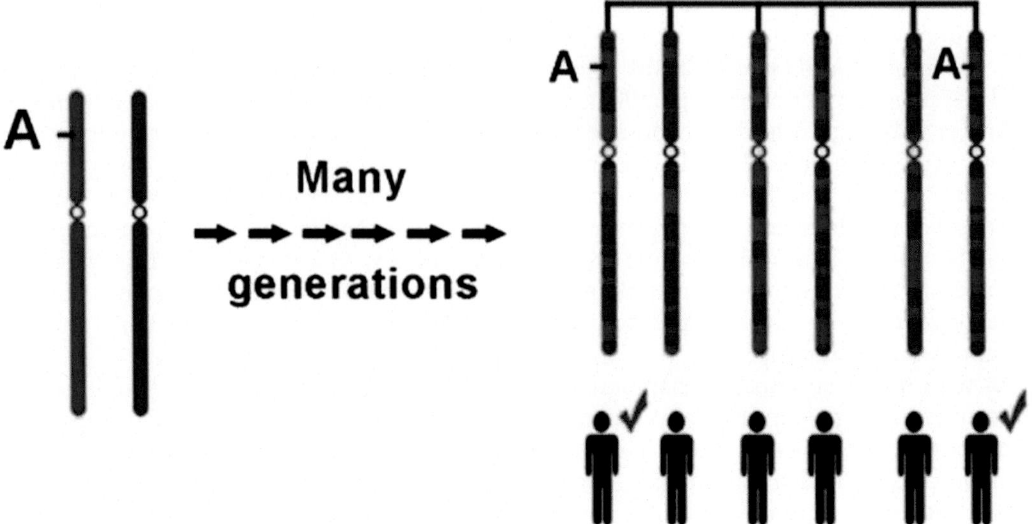

FIGURE 7.14 Consequence of DNA recombination. *From http://www.nanomed.unige.it/Projects/GHM.htm.*

(e.g., DNA recombination and independent assortment of chromosomes) not experienced by uniparental DNAs. In human females, both parents contribute equally to all of the genetic material in the nucleus of the cell whereas in males less DNA is provided by the father since only 1% of nuclear DNA is contained within the small Y chromosome only transmitted from fathers to sons and therefore uniparental. Thus, the vast amount of male and all female nuclear DNA obeys the rules of biparental inheritance, which includes equal contribution of both parents to the genetic makeup of the offspring. In addition, biparental DNA is present in double: one set of genes from the mother and a corresponding set from the father. Also, biparental DNA experiences recombination. The process of recombination involves the exchange of corresponding genetic material (e.g., equivalent genes) between the maternal and paternal sets of chromosomes. This mechanism, generation after generation, has the effect of mixing the maternal and paternal genetic components of chromosomes. And after several generations, the chromosomes end up being collages of the original chromosomes, the size of the maternally and paternally derived segments of DNA being indirectly proportional to the number of generations passed. Fig. 7.14 illustrates this process.

The consequences of recombination are profound in relation to the transmission of genes, ancestry, and evolution. For example, in biparental inheritance there is no unbroken line of descent as seen with the Y chromosome or the mtDNA. As the DNA of both parents is exchanged, there are no intact paternal or maternal lineages to follow, and, as a result, it is more difficult to assess the origins of specific genetic markers. Yet, because biparental DNA represents most of the genetic material in hominins, it provides a more representative picture of the genetic changes during evolution. Furthermore, with the duplicative nature of biparental inheritance, with DNA present in two copies, markers are less susceptible to dropouts. Markers are also less prone to random deletion from the gene pool. Moreover, the huge number of markers that genome-wide studies usually probe allows for a wider-ranging and more comprehensive examination of the genome and the capacity for more robust statistical analyses.

Currently, biparental inheritance is explored by examining specific genes of interest, scoring a large number of polymorphisms or mutations, such as single nucleotide polymorphisms (SNPs) and short tandem repeats (STRs), as well as sequencing large portions of (or the entire) genome of selective individuals. At present, sequencing entire genomes is expensive and time-consuming, and thus, it is usually reserved for key potentially informative individuals and only occasionally used in population studies that examine large numbers of people. With prices continually going down and technology advancing, complete genome sequencing on larger numbers of individuals should become more widely used in the near future.

So, short of complete genome sequencing of entire groups of individuals, many research teams favor the option to detect a massive number of mutations on the order of 500,000 SNPs or more. This approach has proven very informative in recent years, but it brings with it several shortcomings. One of these limitations is the shallow time span that the method allows for reliable timing information. For example, computer simulation studies using 600,000 markers (SNPs) have demonstrated that events (e.g., migrations, dispersals, etc.) dating back to more than 500 human generations ago (25 years per generation) or about 12,500 ya are difficult to detect, and the older the events get, the more difficult it is to observe and date the signals.[39] This is due to the increasing numbers of recombination events that accumulate as a function of time between the maternal and paternal copies of biparental DNA. In other words, recent events (e.g., timing of a genetic bottleneck) are reflected in longer stretches of undisturbed DNA resulting from few recombinations, whereas ancient incidents would be reflected in shorter length of unrecombined DNA. Because recombination splits and separates the original maternal and paternal DNA molecules as a function of time, generation after generation (Fig. 7.14), it decreases the sizes of uninterrupted linear sets of markers. In other words, the size of uninterrupted DNA (non-recombined DNA) is indirectly proportional to time. This allows scientists to measure the length of undisturbed maternal or paternal DNA segments and calculate the date of specific evolutionary events (e.g., the separation of the ape and hominin lineages). Given enough time, recombination would progressively mix and eventually totally scramble the original parental DNAs (Fig. 7.14) to a point when the original linear arrangement of markers disappears (becoming undetectable) making dating unreliable.

As the reliability of this type of SNP methodology only extends to events 12,500 ya or younger, dependable exploration of the timing of hominin evolution with it only goes back to the Mesolithic, a period just predating or coinciding with the Agricultural Revolution in the Near East. As a result, the events connected with the initial colonization of Southern Asia by AMHs is out of reach and would require a higher number of biparental markers to extend the early time limits of the technology. In other words, a higher density and concentration of markers would be needed to ascertain older incidents. Eventually, scholars would have to consider the point of diminishing returns for this SNP technology and adopt the strategy of complete (at least for part of the genome) sequencing of the DNA. In connection with this topic, it is noteworthy that Metspalu et al.[39] failed to detect the signals of the Indo-Aryan migration of about 3500 ya. As this putative date falls within the limits of this methodology, it suggests that the Indo-Aryan dispersals into Southern Asia must have occurred earlier.

During the first two decades of this century, assessments based on genome-wide studies indicate that India's generic diversity is second only to Africa.[40] Typically, these genome-wide studies probe over 500,000 recombining markers representing a comprehensive portion of the genome. Yet, unfortunately, the currently available genome-wide DNA data from the subcontinent is limited. This high level of genetic diversity in India is a strong argument for a rapid coastal route to South Asia soon after AMHs departed from Africa as part of the second phase of the Out of Africa II dispersal. The general maxim of Indian population genetics is that the current groups in the subcontinent are the result of different levels of admixture between two ancestral populations, the autochthonous Dravidian-speaking tribes likely dating back to the original migrants from Africa and the Indo-European-speaking nomads who invaded from the northwest some time during the Holocene, 11,700 ya to present. Genome-wide DNA analyses have provided support to this contention.[41] An independent study also detected two genetic components present in the contemporary Indian populations, the oldest more restricted to South India and accounting for more than 50% of the DNA of the southern region.[42] This study also concluded that the admixture event(s) between the two components were older than the Indo-Aryan invasion conventionally thought to date to about 3500 ya. Thus, in general, social history and the linguistic partitioning observed today parallels the genome-wide results. Thus, the biparental data are congruent with the hypothesis that the autochthonous people of the subcontinent became, at some point in time, Dravidian speakers; subsequently, Indo-European speakers invaded the subcontinent, allowing for admixture between the two groups. A review of the literature indicates that the current genetic composition and regional partitioning of the Indian populace is the result of complex protractile processes that cannot be simply explained by a recent Indo-Aryan penetration but requires multiple gene flow events into the South Asian gene pool from West and East Eurasia during an extended period of time.

Fossil Evidence

Although extrapolations from genetic studies indicate that AMHs arrived in India approximately 65,000 ya, shortly after their departure from Africa, the lengthy persistence of Late Acheulean (typical of Archaics) tools, the delay in the appearance of clearly diagnostic AMH lithic evidence, as well as the cooccupation in time and space of several hominin species on the subcontinent makes it difficult to empirically assess the time of the first settlement by *sapiens*. Furthermore, the fossil record of original AMHs in India is limited and difficult to reconcile among different sites and with genetic data. Also the locations exhibiting early AMHs remains are few, and among those, there is ambiguity in terms of age and species classification.

Darra-i-Kur

Although ancient AMH fossils are few, they are distributed in different locations within South Asia. For example, going from north to south, the Darra-i-Kur cave site in northeastern Afghanistan is at the juncture of West, Central, and South Asia and as such it would have been impacted by various dispersions throughout prehistory. This rock shelter is a stratified silt deposit derived from a nearby stream. The compound site is made up of multiple layers as recent as the Neolithic.

An almost complete temporal bone (compound bone that encases the inner ear located at the base of the skull) has been recovered from Darra-i-Kur. The temporal bone was found in the vicinity of an assemblage of late Middle Paleolithic artifacts dated to about 30,000 ya.[43] The temporal bone has been directly dated by AMS (accelerator mass spectrometry), which assessed the $^{14}C/^{12}C$ ratio and thus its age to about 4530 to 4410 ya. Age estimates based on genetic analyses are congruent with this carbon dating. This fossil was initially thought to belong to the same age as the tools described earlier from the late Middle Paleolithic, but recent direct dating shows that they are in fact Neolithic. It is likely that the temporal bone was introduced from a Goat-Cult interment into an older lithic layer as an intrusion as part of a Neolithic burial. A number of postholes in the surrounding also suggest tents from Neolithic times. Genetic time estimates corroborate the recent (Holocene) time of the specimen. It is noteworthy that one of the early anatomical reports of the temporal bone concluded that although it was from a modern human, it bore some resemblance to Neanderthals. In other words, some of the morphological parameters were within the range of Neanderthals and outside the bounds of AMHs. More recently, the consensus based on anatomical studies is that the mixed (archaic and modern) characters observed in the Darra-i-Kur specimen derived from in situ South Asian origins.[44] In other words, South Asia, being a contact zone for archaics and AMHs, may have been a site of interbreeding and retention of ancestral traits.

In addition, approximately 800 Mousterian (industry associated mainly with Neanderthals dating from 100,000 to 40,000 ya in Europe) lithic utensils and items have been recovered from the site, including axes, scrapers, pounders, blades, simple jewelry, pieces of pottery (some decorated), tin, and bronze. Also, several faunal remains were found, such as fish, rodents, wild horses, Asiatic wild asses, domesticated sheep, and goats. Several goat burials were also identified. It is likely that these interments had mystical value and were connected with the Goat Cult of the Neolithic.

The Darra-i-Kur fossil is unique in being the only ancient human specimen from Afghanistan whose DNA has been sequenced. The DNA analyses indicate that the temporal bone belongs to a modern human male with an H2a mtDNA type. It is interesting that this haplogroup is thought to originate in the Caucasus and dispersed eastward into South Asia as part of the Neolithic Agricultural spread.

Bhimbetka

From Central India, the Bhimbetka site provides hominin fossil evidence dating back to the beginning of the Upper Paleolithic. Although no direct dating of the remains has been generated, the assemblages found at Bhimbetka are clearly stratified. Fossils were found associated with early Late Pleistocene fauna and below these layers with sediments containing Upper Paleolithic tools, including a Late Acheulean axe. The hominin remains were recovered from a single interment of a male individual represented by a skull, mandible, teeth, and some postcranial bones. The anatomical traits and morphometrics indicate that the fossil was from an AMH. Yet, a number of archaic characteristics have been identified in its anatomy, such as thick cranial bones, a large mandible with molars that increase in size from M1 to M3, and no crowding. In addition, the skull possesses a sloping forehead and a low frontal height, both of which are ancestral traits. Owing to the sturdy nature of this individual's general anatomy as well as its ancestral characteristics, it was initially suggested that it possessed Neanderthal traits. Today, scholars recognize that the Bhimbetka remains represent a collage of ancestral and modern traits, and as discussed for Darra-i-Kur, it is not clear what mechanism(s) were responsible for the mosaic nature of their characters. We do know that different *Homo* groups, such as Neanderthal and Denisovans, interbred with AMHs, so it is possible these morphologically admixed specimens represent hybrid or transition individuals with features of both subspecies.

Batadomba-lena and Other Sites in Southern Sri Lanka

From the southern region of the subcontinent in Sri Lanka, a number of archeological sites containing ancient AMH remains have been discovered. These include Batadomba-lena, Fa Hien-lena, Beli-lena (Kitulgala), Fa Hien, and Bellanbandi Pellasa, all locations near the southern tip of the island of Sri Lanka, south of the capital Colombo. When considering the hominin fossils of the island, it is important to posit that Sri Lanka was connected to peninsular India by a land bridge several times during the Pleistocene, and only recently after the LGM was the land connection last severed. This intermittent land bridge undoubtedly allowed hominins and fauna easy access to Sri Lanka from southern India (and back) during extended periods of time throughout the Pleistocene, but especially since 600,000 ya when the sea level was at least 125m below its current mark on several occasions. This land strip became inundated recently, closing land access after the LGM when the ice caps partially melted about 7000 ya. Today the channel that separates the island from the subcontinent is known as the Palk Strait, and it is just about 70m deep at its maximum depth. Today, a man-made bridge connects Sri Lanka to peninsular India.

The Batadomba-lena location has been dated to about 38,000 ya, whereas hominin skeletal remains at the site generated an age of around 30,000 ya. From a nearby contemporaneous dig at the Fa Hien Cave, 28,500 ya hominin fossils were recovered as well. From Batadomba-lena, archeologists have recovered assemblages of geometric microlithics dating to approximately 28,500 ya. The discovery of microlithic artifacts in association with hominin fossils has suggested to many scholars that anatomically and specifically behaviorally modern humans reached and settled the subcontinent earlier than previously thought. The geometric microlithic tradition marks a new step in the progression of tool tradition industries after the blade types made from flakes during the Upper Pleistocene. The early settlement of behaviorally modern humans in India is underscored by the fact that together with African geometric microlithics, these Sri Lankan microtools are considered the oldest discovered thus far worldwide. This fact is often invoked to argue for the possibility that the microlithics of South Asia, and specifically Sri Lanka, developed in situ resulting from local environmental demands and necessities, rather than being imported by migrants from Africa or the Near East. At Batadomba-lena and Beli-lena (Kitulgala), the collections of microliths were located in stratified layers dating from the Late Pleistocene to the Early Holocene in continuous sequences. Charcoal, ocher, and bones were found among the sediments. In Batadomba-lena and Fahien-lena, some of the hominin bones were burnt, suggesting cremations that were probably ritualistic in nature. The location of hominin fossils in burial pits, the use of ocher to paint the remains, and the possibility of cremation strongly suggest that these individuals practiced ritual interments, a trademark of modern human behavior.

In the context of the hominin occupation at Batadomba-lena, a number of plant materials and animal fossils were discovered, including banana, breadfruit, fish bones, seashells, and shark bones. These remains were likely part of their diet. Other items, such as shell pendants, beads, and shark vertebra beads, were found at the site, indicating that these people traveled to the seashore some 40km away or were involved in trading with littoral populations. At the Beli-lena site, for example, residues of sea salt indicate that these hominins transported and utilized salt, possibly for cooking and/or meat preservation. In addition, it has been suggested that these hominins also hunted fauna, such as deer, wild cattle, and sambur, and collected wild cereals in the nearby central high plains and forests of the island.

Like the specimens from Bhimbetka and Darra-i-Kur, the anatomy and the morphometric values and nonmetric traits of the Batadomba-lena and Fahien-lena hominins demonstrate a unique combination of modern and archaic traits. The specimens from these two sites possess a number of ancestral characteristics, such as robust skulls with strong muscle markings, massive jaws, pronounced supraorbital ridges, depressed noses with large nasal openings, short necks, prominent nuchal crest (posterior lower part of the skull where the neck muscles attach), large teeth exhibiting considerable prognathism, and prominent occipital protuberance (bone projection located at the posterior inferior lower rear part of the skull). The mandibles retain a broad and massive ascending branch, or rami, characteristic of some Middle Pleistocene hominins from Africa and Europe. The permanent and deciduous dentitions exhibit stronger similarities to late Archaics, specifically to Neanderthals. The postcranial skeletons of these fossils are equally robust. These individuals were tall in comparison with the modern-day Sri Lankan populace. Males were about 174cm in height, and females around 166cm tall. Altogether, the observed anatomical characteristics indicate that the Late Pleistocene hominins of Sri Lanka were AMHs retaining a good number of archaic traits, likely the result of interbreeding.

It is interesting that morphometric studies comparing these fossils from the various strata corresponding to continuous time periods demonstrate a biological continuity from the Late Pleistocene to the present. These continuances of traits over long periods of geological time include several nonmetric dental characteristics heavily controlled by genes. This argues against invoking just environmental factors to explain morphological similarities. In addition, this continuum is seen extending to specific present-day aboriginal people of Sri Lanka, such as the Veddas.[45] In light of the relative geographical isolation of hominin populations in Sri Lanka during prehistoric times, it is not unexpected to find a biological continuum among groups. Furthermore, the anatomical and cultural continuity between early AMHs dating to 38,000 ya and contemporary *H. sapiens* in Sri Lanka suggests that at least some of the original Late Pleistocene migrants who dispersed into South Asia incorporated populations that migrated into India more recently by interbreeding and absorbing them. Thus, in the case of the Sri Lankan hominins, the data negate the model of population extinction by belligerent altercation, or any other means, followed by repopulation. The scenarios involving introgression, absorption, and fusion of different types of hominins to explain the observed biological continuity in South Asia prognosticate that when the genomes of contemporary Indian populations are investigated, evidence of hybridization among hominins will be uncovered.

Mosaic of Ancestral and Derived Traits in Anatomically Modern Humans in the Subcontinent

The data suggesting absorption of newcomer populations by resident archaic groups are related to the observation of ancestral and AMH characteristics in the South Asian fossil record. A general theme that underscores most of the modern human fossil sites described earlier from South Asia is the collage of archaic and *sapiens* characteristics seen in the fossil remains. Although retention of ancestral traits has been observed in a number of hominin species throughout recent human evolution (see Chapters 2 and 5), it is in South Asia that certain combinations of ancestral pre-AMH characteristics persist in rather recent populations, well after the arrival of AMHs. The geographical location of South Asia at the crossroads of hominin dispersals and its place as a region receiving consecutive waves of migrants have been cited as potential reasons for the persistence of a number of unique ancestral characters in fossils that should otherwise be, because of their young age, fully developed AMHs. The diversity of habitats and ecological niches has also been posited as important contributing factors that selected for the retention of ancestral traits in Indian *sapiens*. It is also possible that introgression among cohabitating species would have led to hybrids with various combinations of ancestral and modern traits. Hybridization involving Archaics and AMHs would have introduced ancestral characters into otherwise modern populations. The prolonged use of antiquated technology (Late Acheulean) throughout South Asia long after the time that AMHs arrived may be indicative of the lengthy occupation of Archaic groups, such as *erectus* and *heidelbergensis*, in the subcontinent, thereby providing the opportunity for gene flow among the different species.

Pre– or Post–Toba Arrival?

In relation to the AMHs' colonization of South Asia, the topic of the Toba volcanic super-eruption around 75,000 ya is often used as a landmark or point of reference to delineate *sapiens'* time of arrival. Chapter 5 describes some of the events resulting from the catastrophe on the island of Java in Near Oceania. Scholars are divided with regard to the impact of the Toba supereruption on the world's climate, habitats, and AMH populations. Some authorities believe that the volcanic winter and the resulting devastation ushered in the extinction of a large number of plant and animal species. Genetic studies indicate that our species suffered a dramatic drop in variability because of a reduction in the number of breeding couples to as few as 1000 surviving individuals.[46] This dramatic decrease in effective population size would have resulted in a marked decrease in genetic diversity. This drop in genetic heterogeneity is, in fact, observed in the genetic footprints of contemporary humans.

If the impact of the Toba supereruption in South Asia was as catastrophic as some investigators propose, it would have had dramatic effects on the behavior and evolution of the resident hominin populations. An obvious consequence would have been a pronounced bottleneck event. Yet, studies from the Jwalapuram site in southern India indicate that lithics attributed to AMHs have been recovered from sediments below and above the Toba ash level.[47] These data suggest that AMHs were already present in the southern portion of the peninsula before the Toba event. This discovery also indicates that the ecological consequences of the super-eruption were not damaging enough to cause the extinction of the resident hominins of the region. Yet, there are uncertainties about the actual age of the post–Toba tools recovered and the identity of the post–Toba species. In other words, subsequent dating of the site points to a more recent age (55,000 to 35,000 ya), raising the possibility that there was a time gap of habitation after the Toba eruption, allowing for resettlement of the area thousands of years after a putative regional extinction. Nevertheless, at this point in time, the presence of hominins before the Toba event indicates that *sapiens* reached India before 75,000 ya, much earlier than estimated from the DNA data.

Artistic Expressions of Anatomically Modern Humans

Reflections of humanity are also evident in the rock shelters and caves at Bhimbetka where artistic representations of everyday life, such as hunting and group dancing, decorate the rock walls (Fig. 7.15). The Bhimbetka paintings are one of the earliest artistic/spiritual

FIGURE 7.15 Rock painting at Bhimbetka. *From http://www.mpholidays.com/include/ckfinder/core/connector/php/images/WebPage/Bhimbetka_RockPainting.jpg.*

manifestations of human life in South Asia. This extensive archeological complex is located at the geographical center of the subcontinent in the state of Madhya Pradesh. The smooth rounded surfaces of the rocks that make up the caves indicate that, at one time, they had been under water. The name Bhimbetka means the seating place of Bhima, a hero-deity of the epic Mahabharata. Bhimbetka is not a single shelter but a conglomerate of about 750 rock caves. Some of the dwellings date back to 100,000 ya and were occupied by *erectus*. More than 400 rock shelters with images, most of the time overlapping or superimposed, have been identified. The depictions at Bhimbetka chronicle human existence during continuous periods of time, providing an enduring record of AMH's existence in South Asia.

Most of the paintings are in red and white, although some appear in green and yellow. The oldest paintings date back to the Upper Paleolithic about 30,000 ya and depict large images of animals, such as bison, tigers, and rhinoceroses, in green and dark red colors. Later illustrations from the Mesolithic depict anthropomorphic figures with body decorations, hunting tools, and scenes, as well as communal activities, such as dancing, drinking, interments, carrying animals, riding elephants, collecting honey, and performing household activities. Somewhat amusing is one representation of a bison chasing a hunter while two fellow hunters appear standing by viewing helplessly at the spectacle. Even drawings of childbirths have been identified. Most human figures are geometrical. The following period represented in the drawings is the Chalcolithic. In South Asia, this epoch dates to 7000–3300 ya, and the dwellers of the time painted scenes showing contact with the agricultural communities of the plains, exchanging goods with them. Some of the most recent paintings dating to medieval times illustrate trees, flowers, and even soldiers with weapons, including swords and kings with decorated horses.

Gradualism Instead of Sharp Replacement Characterizes the Archaic to AMH Transition

Unlike Africa and the Near East, where the transition from Archaic to AMH populations and behaviors was more clear and delineated, the changes in South Asia were gradual, late, and exhibited blurry lines of demarcation. In the subcontinent, the lithic traditions were so similar among Archaic species and AMHs that often it has not been possible to assess which group made a given implement. It was not until approximately 45,000 ya that gradual modifications and differentiation in toolmaking styles become discernible. For instance, the appearance of symbolism (e.g., beads as pieces of decoration and/or protection) in artifacts and the evolution to microlithic utensils is detected at a much later date in peninsular India than in Africa and the Near East. The unique dynamics of these cultural technological changes in different regions of the world suggests that the local *milieu* dictated, to some degree, the speed, style, and forms of hominin behavior.

Archeologists pay a lot of attention to the shape and size of lithics (especially shapes) as well as the context in which they were found and the techniques employed to produce them. Scholars rely heavily on these parameters to ascertain the chronology and evolution of hominins and their behavior. Thus, when the technocultural transitions are nebulous and variable in time and space, it generates ambiguity and lack of temporal demarcation. A good example of this type of uncertainty is seen in the subcontinent of India. For instance, the Middle Paleolithic technology of biface manufacturing was long-lasting in some regions of India, and in some areas it lasted until as recently as 34,000 ya. Yet, the same Middle Paleolithic technology of bifaces was less lengthy in other locales. As a consequence of this prolonged tenure and transition delay, the flake-dominated tradition that followed the Middle Paleolithic bifaces was short-lived.[48] In addition to being generally prolonged, the technological transitions were also marked by regional differences in their time lines. For example, in Central, Northwestern, and Western India, the diminutive biblade technology lasted until 120,000 ya, 95,000 ya, and 60,000 ya, respectively.[49]

The underlying reason(s) responsible for the above-described delay in technological progression and regional tenure differences within the Indian subcontinent is(are) not clear. In relation to the transition to microliths, for example, it is known that this tradition is associated with the creation of advanced composite tools and attention to detail. Microliths were used for the production of fine tools, such as spears made of a stone point and a long wooden shaft. This type of weapon is particularly made for hunting and fishing. The adoption of microlithics also coincides with the intensification of social and community-driven activities, as well as larger homesteads. It could be speculated that in the wide expanse of peninsular India, the land was sparsely populated and at the fringes of the technological innovations that originated in Africa. In other words, it is possible, for example, that the persistence of biblade technology was driven by limited communication flowing into the subcontinent from the site of origin in Africa, as well as poor interactions among indigenous populations creating isolation and sociocultural stagnation. Also, finite resources could have played a role by keeping population sizes down and geographically isolated, thus contributing to poor communication. In addition, it is also possible that because of the requirements imposed by the habitats, the hunting strategies used in the subcontinent were different from what was practiced elsewhere. In other words, because of the game available and the terrain, the natives did not require a change in technology or a switch to a new toolmaking tradition; there would be no need to adopt a new microlithic style when biblades did the job well.

Potential Interbreeding Among Archaics and Anatomically Modern Humans

As a result of the observed gradualism in the transition from the bifaces of the Late Acheulean technology practiced by Archaic species, such as *erectus* and *heidelbergensis*, to the microlithics of AMHs, it has been proposed that in South Asia there was biological continuity among these groups rather than extinction and population replacement by *sapiens*. In other words, *sapiens* may not have eliminated or displaced Archaic groups. It is possible that they fused into hybrid populations, yet no traces of Archaic DNA have been reported to substantiate introgression events. Failure to detect Archaic DNA may stem from infertile or reproductively compromised hybrids, in spite of cultural interactions. Furthermore, the limited DNA data available, and/or minimal amount of introgression in South Asia may account for detection failure. It is likely that *erectus* and *heidelbergensis* groups introduced into India the ancestral Late Acheulean tradition. And considering the late appearance of the microlithic tradition in South Asia, long after AMHs arrived, it is likely that all the three species practiced Late Acheulean technology at the same time and in the same places. In other words, the Archaic groups practicing the late-persisting biface technology could have interacted culturally and possibly biologically with AMHs.

The time line of these two lithic styles adds credence to this possibility of cultural and biological admixture. For instance, in several regions in South Asia, the Late Acheulean technology persisted and flourished, whereas the microlithic forms were extremely delayed. In regions such as Jwalapuram in the state of Andhra Pradesh, bordering India's southeastern coast, Late Acheulean bifaces were produced from 77,000 ya to 38,000 ya, and it was not until 35,000 ya that microlithic utensils appeared.[50] Similarly, in the north central state of Uttar Pradesh, the ancient biface technology was practiced as recent as 45,000 ya. These recent dates suggest that since AMHs arrived to South Asia about 65,000 ya, they adopted and continued using the technology developed by *erectus* and *heidelbergensis*. Thus, it is likely that both Archaic species and AMHs cohabitated and were involved in technology transfer.

Mesolithics and Violent Deaths in South Asia

By definition, the Mesolithic is the epoch in between the end of the Paleolithic and the start of the Neolithic. In northern India and Pakistan, the Mesolithic encompassed a period between 22,000 and 11,500 ya. In South India, on the other hand, the Mesolithic began somewhat later and lasted until 5000 ya. The following era, the Neolithic, lasted until 3400 ya in South India.

The start of the Mesolithic in South Asia experienced a dramatic increase in population density as evidenced by the increase in the number of sites in regions previously devoid of settlements. These higher population densities seem to coincide with wetter periods and a concomitant increase in the quantity and diversity of fauna. It is likely that the increased productivity of the land led to surplus food and the capacity to settle in previously uninhabited areas where water, flora, and fauna became plentiful.

In addition to the settlements, community cemeteries began to appear, with some of the more notable ones found in Sarai-Nahar-Rai, Mahadaha, and Damdama in the Ganga River Basin, dating back to a period between 10,000 and 4000 ya. The individuals buried in these graveyards were preceramic seminomadic foragers. Based on the unearthed skeletons, the human populations at these sites were made up of individuals with high stature, heavy build, and rather large crania.[51] Using morphometric analysis of 47 skeletons of males and females, researchers assessed that the average male height was about 180 cm and the average female stature was about 172cm, which are higher than the present-day average stature of the local inhabitants. The morphometric data derived from these skeletons reflect a distinct Mesolithic regional population that differed from contemporary humans in other regions of South Asia, as well as from current populations from the area. Although nutrition could account for anatomical differences, it is likely that these Mesolithic groups from the Ganga River Basin were genetically distinct and a certain degree of heterogeneity existed at the time among human groups.

Although the notion is still highly contested, the Mesolithic seems to be marked by an increase in violent deaths. This increase is evident in the frequency of physical trauma from interpersonal conflicts as seen in the skeletal remains from the Mesolithic in comparison with previous epochs. It has been suggested that this increase in violence resulted from territorial claims, as well as other economical, ideological, or social factors as population densities rose. It has been proposed that contacts between farming communities and foraging hunter-gatherers may have generated socioeconomic conflicts, or altercations between the "haves" (farmers) and the "have-nots" (hunter-gatherers). The increase in violence may have also been related to accumulation of precious surplus goods and the high demand during harsh winter times; prestige or social status; or centralization, stratification, and hierarchy of power within the communities. South Asia was not immune to this increase in interpersonal violence. At Sarai-Nahar-Rai in the Ganga Valley, for example, microliths were found embedded in the bones of some skeletons, suggesting violent death.[52]

The transition from the Paleolithic to the Mesolithic brought about a number of dramatic changes. Throughout evolution, hominins have been selected for a rather unstructured way of life. Hominins have been selected for an arboreal and then for a terrestrial existence, as scavengers, hunters, and gatherers. With the beginning of the Mesolithic, AMHs were suddenly confronting a new set of demands, having to cope with new challenges, both physical and mental. Humans were not only faced with having to plan and perform continuous scheduled labor during planting, irrigation, and harvest, but it is also likely that the gathering of people, living together in communities, as well as negotiating and trading with strangers may have resulted in new psychological issues with which to contend. Before the establishment of agricultural communities, humans enjoyed a relatively free, unscheduled existence; they scavenged or hunted whenever they needed food. This changed rather dramatically to routine communal work, with interpersonal interactions capable of generating stress, interpersonal conflict, aggression, and, in some instances, violent death. This impromptus way of life transformed humans in a very short period of time, providing no time for evolution to modify the hominin gene pool and make humans better adapted for the new survival mode. It could be argued that since the Mesolithic and the subsequent agricultural revolutions, humans have continued modifying their environment at a revolutionary pace that the species is not keeping up genetically.

Other events related to the lithic industry are associated with the Mesolithic, such as small blades designed to conform various geometric shapes. Artistic expressions in the form of human and animal images are abundant in caves and rock shelters. The Mesolithic was the age of the hunters and gatherers, which ended with the development of agriculture and domestication at the start of the Neolithic period. In certain parts of the world, including India, this transition was not always sharp. For a period of time, many of the pastoralist and agricultural societies lived side by side and traded with forager groups. Based on the violent incidents reflected in the skeletal remains, it is possible that some of the interactions between the two groups at times may not have been amicable.

The Neolithic

The orthodoxy is of the opinion that the Neolithic arrived to the subcontinent from the northwest approximately 9500 ya. It is likely that the Agricultural Revolution, with origins in the Fertile Crescent of the Near East, dispersed longitudinally eastward via the same route used by previous hominin migrants of the past. The first plant harvesters likely migrated eastward using the coastal route, eventually encountering the Indus River Basin and establishing agricultural settlements within the delta and tributaries. Subsequently, it is envisioned that these new migrants dispersed southward coastally and then into the interior of the Indian Peninsula.

It is also thought that these agriculturalists brought with them plants and animals from the West and incorporated them into the indigenous flora and fauna as domesticates. It appears that the domestication of plants and animals in South Asia was a complex protractile process in which some species and varieties arrived from the West, some from the East, and others were developed in situ. Further diversity was generated by interbreeding.[53] The contribution to South Asia domesticates from various agricultural centers to the West as well as from the East is not surprising considering that a number of agricultural revolutions emerged independently in various parts of the world; the subcontinent located geographically at a crossroads, in the middle, was prone to receive input from each. Specifically, it has been proposed that rice, water buffalo, and chickens were independently domesticated in East Asia and South Asia. It is also hypothesized that a number of strains and varieties of these species were introduced into South Asia at a later time. Other domesticates, such as the Zebu cattle (humped cattle), sheep, cotton, millet, and pulses (dried legume seeds), were generated and propagated autonomously in different regions of peninsular India, including the Ganga Plains, Indus Valley, Gujarat, and the cone of India. Furthermore, genetic evidence demonstrates that cattle were selected by humans in more than one region of India.[54]

Migration Versus Acculturation

It has been proposed that the Agricultural Revolution spread from the Near East to South Asia at an average rate of 0.65 km/yr.[55] Yet, as in Europe, the topic of the establishment of agriculture in India is plagued with a number of uncertainties. One of these controversial topics is the degree to which plant and animal domestication in South Asia derives from acculturation as opposed to actual movement of people. In other words, one might ask: Was the new more sedentary way of life of farmers introduced into India by sizeable number of migrating people, as part of a mass dispersion wave, or just by the dissemination of ideas including the novel subsistence system into the new land? Most of the available data from different fields support the contention that migration of individuals (not just communication) transmitted domestication and agriculture from the Near East to the Indian subcontinent.

For instance, based on genetic information, an acculturation model by itself would not explain the presence of DNA markers in India known to signal the movement of pastoralists and agriculturists from the Levant. Today the genetic signature of farmers and breeders from the Near East can be traced using Y chromosome–specific (Fig. 7.16) and mtDNA-specific lineages, as well

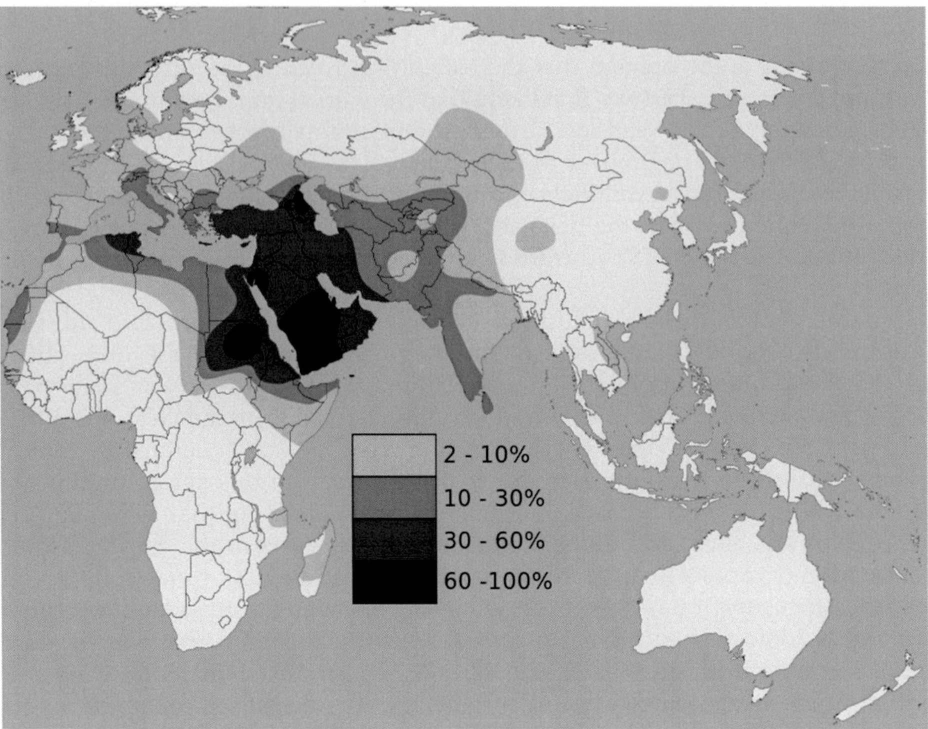

FIGURE 7.16 Distribution of Y chromosome haplogroup J. *From https://en.wikipedia.org/wiki/Haplogroup_J_(Y-DNA)#/media/File:Haplogroup_J_(Y-DNA).svg.*

as whole-genome genetic markers.[56] Y chromosome type J, for example, has a focus of high concentration within the Fertile Crescent and gradually diffuses along the Arabian Sea coast of Iran and Pakistan, as well as the littoral region of western India, eventually extending into Sri Lanka (Fig. 7.16). This is the expected genetic distribution pattern if haplogroup J males migrated into the subcontinent, disseminating their genes along a coastal route in peninsular India. Specifically, Y haplogroup J2a-M410 exhibits a pattern of gene flow from the Fertile Crescent during the Neolithic period about 10,000 ya into the Indian subcontinent.[57] More recent genetic studies suggest that the distribution of Y haplogroups J2a-M410 and J2b-M102 in South Asia indicates a complex scenario of multiple expansions from the Near East to South Asia.[58] Maternally derived mtDNA lineages also indicate that a number of the West Eurasian mtDNA haplogroups detected in the Indian populace are attributed to gene flow from the Near East about 9300 ya.[59] Whole-genome investigations also detected Eurasian gene flow from Iran and the Near East dating to the times of the Agricultural Revolution.[60] Additional recent studies based on specific genes, such as the one that controls lactose tolerance, suggest gene flow from Iran and the Middle East about 10,000 ya.[61] It seems that individuals in India carry the same lactose-tolerant gene mutation seen in the Near East and Europeans. Although there is always the possibility that the same gene variant (mutation) occurred in both places independently, it is more likely that a single lactose-tolerant gene originated in the Near East and then was transported to South Asia by migrating farmers. Altogether, these data are congruent with a demographic picture in which the lactose-tolerant

FIGURE 7.17 Female statuette from Mehrgarh. *From https://en.wikipedia.org/wiki/Mehrgarh.*

mutation dispersed in two directions from the site of origin in the Near East during the Agricultural Revolution. One branch moved into Europe, whereas the other moved into South Asia using a coastal trajectory following the Persian Gulf and the Indian Ocean where the mutation is found. It is highly likely that this lactose-tolerant mutation reached polymorphic levels throughout its distribution range as a result of positive selection generated by the consumption of milk and dairy products made by farmers from domesticates.

Perhaps, the most studied farming community in South Asia is Mehrgarh. The settlement of Mehrgarh is located in the fertile Kacchi Plain of Balochistan in central western Pakistan. The area is on the western rim of the Indus Basin by the eastern foothills of the Suleiman Range (Fig. 7.3). Mehrgarh is one of the earliest agricultural centers in South Asia dating back to 9500 ya. The site is of particular importance because it exhibits a continuous progression of stages from domestication and agriculture to developed civilizations. In addition, Mehrgarh is thought to be a forerunner of a number of Bronze Age urban centers, such as the Indus Valley Civilization, which first appeared in the northwestern regions of South Asia approximately 5000 ya and then spread throughout the subcontinent.[63] Mehrgarh had its beginnings as a small farming and herding community. Since its discovery in 1974, about 32,000 artifacts have been unearthed (Figs. 7.17 and 7.18). Habitation in the area extended to 4000 ya, the start of the Bronze Age.[64] Although it has been argued that Mehrgarh represents an in situ development of agriculture and domestication, evidence from various fields points to a connection with the Near East and the genesis of farming in the Fertile Crescent.[65] One line of evidence stems from studies on lactose tolerance in the subcontinent[61] mentioned in the previous paragraph.

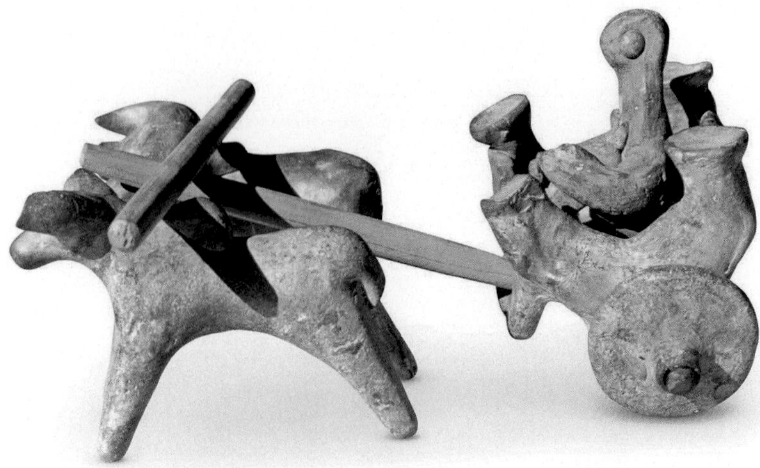

FIGURE 7.18 Depiction of oxens pulling man on cart from Mehrgarh. *From http://www.pakistansource.com/wp-content/uploads/2015/09/A-Corbis-LL002185_phf1cs.jpg.*

Parallelisms Between the European and South Asian Neolithic

As in Europe, the Neolithic had similar positive and negative consequences in South Asia. For instance, farming and domestication allowed the mass production of food. This new way of life proliferated throughout the subcontinent. The initial farming centers grew into larger homesteads and eventually into cities. For the first time, humans were able to save the surplus food from the summer harvest and keep animals during the winter. This storage of resources for the winter or harsh times brought about more free time to think and imagine and more time to create novel things. It is often argued that this spare time allowed humans to explore and innovate and extend their consciousness beyond their immediate surroundings, launching a new era of cultural development.

On the negative side, in South Asia, the development of cities ushered an era of malnutrition and diseases as compared with the more open and free living of hunter-gatherer groups of the Paleolithic. This is evident in the anatomical signs in skeletons found in the cemeteries of the Sarai-Nahar-Rai, Mahadaha, and Damdama settlements in the Ganga River Basin. It seems that in some settlements at some point in the growth of communities, the negative effects of overpopulation, including lack of proper sanitation and occupational stress, outweighed the benefits of farming and surplus food. From the skeletal remains at these Ganga Plain sites, it is clear that the incidents of dental cavities, diseases, trauma, and occupational stress augmented as the cities grew in size. Before urbanization the hunter-gatherer groups were generally healthy.[62]

Despite all the potential negatives mentioned above, it can be argued that overall the Agricultural Revolution led to positive outcomes for our species, although a number of other animal and plant groups have become extinct. Our life spans have increased substantially, and wars overall are down. Our continued rapid population expansion continuously extending our carrying capacity is also evidence that farming and domestication have benefited our species. Furthermore, humans have found ways to overcome many of the negative outcomes and potential threats, exceeding our carrying capacity on this planet several times over. It remains to be seen whether we can continue this trend, or if the Agricultural Revolution has afforded us only a short-term benefit to our long-term problem and detriment, as we use up our limiting resources and alter our planet beyond our own ability to adapt.

References

1. Lawrence K. *The beauty of science, the universe*. New York: The Chautauquan Daily; 2016. https://chqdaily.wordpress.com/2012/07/15/krauss-discusses-the-beauty-of-science-the-universe/.
2. Rendell H, Dennell RW. Thermoluminescence dating of an upper pleistocene site, Northern Pakistan. *Geoarchaeology* 1987;**2**:63–7.
3. Dennell RW, Rendell HM, Hailwood E. Late pliocene artefacts from Northern Pakistan. *Curr Anthropol* 1988;**29**:495–8.
4. Dennell RW, Rendell HM, Halim M, Moth E. A 45,000-years-old open-air Paleolithic site at Riwat, Northern Pakistan. *J Field Archaeol* 1992;**19**:17–33.
5. Fleagle JG, Shea JJ, Grine FE, Baden AL, Leakey RE. *Out of Africa I: the first hominin colonization of Eurásia*. New York: Springer International Publishing AG; 2010.
6. Corvinus G. A handaxe assemblage from Western Nepal. *Quartär Int Yearbook Ice Age Stone Age Res* 1989;**39/40**:155–73.
7. Singh MP, Sahni A, Kaul S, Sharma SK. Further evidence of hominid remains from the pinjor formation, India. *Proc Indian Natl Sci Acad* 1988;**54**:564–73.
8. Singh MP. First record of a middle pliocene hominid from the Siwalik Hills of South Asia. *Hum Evol* 2003;**18**:213–28.
9. Chauhan PR. Was there an Oldowan occupation in the indian subcotinent? A critical appraisal of the earliest paleoanthropological evidence. In: Schick K, Toth N, editors. *The cutting edge: New approaches to the archaeology of human origins*. Gosport, IN: Stone Age Institute; 2009.
10. Dambricourt AM. The first Indo-French prehistorical mission in Siwaliks and the discovery of anthropic activities at 2.6 million years. *Comptes Rendus Palevol* 2016;**15**:281–94.
11. Coppens Y. Human origins in the Indian sub-continent. *Comptes Rendus Palevol* 2016;**15**:279–80.
12. Malassé AD, Moigne A-M, Singh M. Intentional cut marks on bovid from the Quranwala zone, 2.6 Ma, Siwalik Frontal Range, Northwestern India. *Comptes Rendus Palevol* 2016;**15**:317–39.
13. McPherron SP, Zeresenay A, Marean CW, et al. Evidence for stone-tool-assisted consumption of animal tissues before 3.39 million years ago at Dikika, Ethiopia. *Nature* 2010;**466**:857–60.
14. Harmand S, Lewis JE, Feibel CS, Lepre CJ, Prat S, Lenoble A, Boës X, Quinn RL, Brenet M, Arroyo A, Taylor N, Clément S, Daver G, Brugal JP, Leakey L, Mortlock RA, Wright JD, Lokorodi S, Kirwa C, Kent DV, Roche H. 3.3-million-year-old stone tools from Lomekwi 3, West Turkana, Kenya. *Nature* 2015;**521**:310–5.
15. Semaw S. The world's oldest stone artifacts from Gona, Ethiopia: their implications for understanding stone tech- nology and patterns of human evolution between 2.6–1.5 million years ago. *J Archaeol Sci* 2000;**27**:1197–214.
16. Semaw S, Rogers MJ, Quade J, et al. 2.6-Million-year-old stone tools and associated bones from OGS-6 and OGS-7, Gona, Afar, Ethiopia. *J Hum Evol* 2003;**45**:169–77.
17. de Heinzelin J, Clark JD, White T, et al. Environment and behavior of 2.5-million-year-old Bouri hominids. *Science* 1999;**284**:625–9.
18. Han F, Bahain JJ, Deng C, et al. The earliest evidence of hominid settlement in China: combined electron spin resonance and uranium series (ESR/U-series) dating of mammalian fossil teeth from Longgupo cave. *Quat Int* 2015:1–9.
19. Hou YM, Zhao LX. An archeological view for the presence of early humans in China. *Quat Int* 2010;**223–224**:10–9.
20. Jin CZ, Zheng LT, Dong W, Liu JY, Xu QQ, Han LG, Zheng LL, Wei GB, Wang FZ. The early Pleistocene deposits and mammalian fauna from the Renzidong Fanchang, Anhui Province, China. *Acta Anthropol Sin* 2000;**19**:185–98.
21. Zhang SS, Han LG, Jin CZ. On the artifacts unearthed from the Renzidong Paleolithic site in 1998. *Acta Anthropol Sin* 2000;**19**:126–35.
22. Pappu S. Changing trends in the study of a Paleolithic site in India: a century of research at Attirampakkam. In: Petraglia MD, Allchin B, editors. *The evolution and history of human populations in South Asia. Vertebrate paleobiology and paleoanthropology series*. Dordrecht: Springer; 2007.

23. Anwar M, Dennell RW, Anwar M. *Early hominin landscapes in Northern Pakistan: investigations in the Pabbi Hills.* Oxford: Archaeopress; 2004.
24. Kumar RB, Rajendran B, Bhanu V. Fossilized hominid baby skull from the ferricrete at Odai, Bommayarpalayam, Villupuram District, Tamil Nadu, South India. *Curr Sci* 2003;**84**:754–6.
25. Ambrose SH. Paleolithic technology and human evolution. *Science* 2001;**291**:1748–53.
26. Rajendran P, Koshy P, Sadasivan S. *Homo sapiens* (archaic) baby fossil of the middle Pleistocene. *Ancient Asia* 2006;**1**:7–13.
27. Westgate J, Shane R, Pearce N, Perkins W, Korisettar R, Chesner C, Williams M, Acharyya S. All Toba tephra occurrences across peninsular India belong to the 75,000 yr BP eruption. *Quat Res* 1998;**50**:10712.
28. James HVA, Petraglia MD. Modern human origins and the evolution of behavior in the later Pleistocene record of South Asia. *Curr Anthropol* 2005;**46**:S3–27.
29. Liu H, Prugnolle F, Manica A, Balloux F. A geographically explicit genetic model of worldwide human-settlement history. *AJHG* 2006;**79**:230–7.
30. Appenzeller T. Human migrations: eastern odyssey. *Nature* 2012;**485**:7396.
31. Wells S. *The journey of man: a genetic odyssey*. Princeton: Princeton University Press; 2002.
32. Moorjani P, Thangaraj K, Patterson N, et al. Genetic evidence for recent population mixture in India. *Am J Hum Genet* 2013;**93**:422–38.
33. Chennakrishnaiah S, Perez D, Rivera L. Indigenous and foreign Y-chromosomes characterize the Lingayat and Vokkaliga populations of Southwest India. *Gene* 2013;**526**:96–106.
34. Kivisild T, Rootsi S, Metspalu M, et al. The genetic heritage of the earliest settlers persists both in Indian tribal and caste populations. *Am J Hum Genet* 2003;**72**:313–32.
35. Wang C-C, Li H. Inferring human history in East Asia from Y-chromosomes. *Invest Genet* 2013;**4**:11.
36. Mellars P, Gori KC, Carr M, et al. Genetic and archaeological perspectives on the initial modern human colonization of southern Ásia. *Proc Natl Acad Sci USA* 2013;**110**:10699–704.
37. Endicott P, Metspalu M, Kivisild T. Genetic evidence on modern human dispersals in South Asia: Y chromosome and mitochondrial DNA perspectives. In: Petraglia MD, Allchin B, editors. *The evolution and history of human populations in South Asia*. Berlin: Springer; 2007.
38. Mountain JL, Hebert JM, Bhattacharyya S, et al. Demographic history of India and mtDNA-sequence diversity. *Am J Hum Genet* 1995;**56**:979–92.
39. Metspalu M, Gallego Romero I, Yunusbayev B. Shared and unique components of human population structure and genome-wide signals of positive selection in South Asia. *Am J Hum Genet* 2011;**89**:731–44.
40. Majumde PP. The human genetic history of South Asia. *Curr Biol* 2010;**20**:R184–7.
41. Reich D, Thangaraj K, Patterson N, et al. Reconstructing Indian population history. *Nature* 2009;**461**:489–94.
42. Silva M, Oliveira M, D.Vieira, et al. A genetic chronology for the Indian Subcontinent points to heavily sex-biased dispersals. *BMC Evol Biol* 2017. https://doi.org/10.1186/s12862-017-0936-9.
43. Douka K, Slon V, Stringer C, et al. Direct radiocarbon dating and DNA analysis of the Darra-i-Kur (Afghanistan) human temporal bone. *J Hum Evol* 2017;**107**:86–93.
44. Bellwood P. Migration and the origins of *Homo sapiens*. In: Kaifu Y, Izuho M, Goebel T, Sato H, Ono A, editors. *Emergence and diversity of modern human behavior in Paleolithic Asia*. Texas: Texas A&M University; 2015.
45. Kennedy KAR. In: Abraham S, Gullapalli P, Raczek TP, Rizvi UZ, editors. *Connections and complexity: new approaches to the archaeology of South Asia*. Abingdon, United Kingdom: Routledge; 2013.
46. Ambrose SH. Late Pleistocene human population bottlenecks, volcanic winter, and differentiation of modern humans. *J Hum Evol* 1998;**34**:623–51.
47. Haslam M. A southern Indian Middle Palaeolithic occupation surface sealed by the 74 ka Toba eruption: further evidence from Jwalapuram locality 22. *Quat Int* 2012;**258**:148–64.
48. Blinkhorn J. A new synthesis of evidence for the Upper Pleistocene occupation of 16R Dune and its southern Asian context. *Quat Int* 2013;**300**:1e10.
49. Chauhan PR, Ozarkar S, Kulkarni S. Genes, stone tools and modern human dispersals in center of the Old World. In: Kaifu Y, Goebel T, editors. *Proceedings of the Symposium on the Emergence and Diversity of Modern Human Behavior in Palaeolithic Asia. November 29-December 1; Tokyo, Japan*. Texas: Texas A&M University Press; 2015.
50. Clarkson C, Jones S, Harris C. Continuity and change in the lithic industries of the Jurreru Valley, India, before and after the Toba eruption. *Quat Int* 2012;**258**:165–79.
51. Lukacs JR, Nelson GC. Stature in Holocene foragers of North India. *AJPA* 2014;**153**:408–16.

52. Misra VN. Microlithic industry in India. In: Misra VN, Bellwood PS, editors. *Recent Advances in Indo-Pacific Prehistory: Proceedings of the International Symposium Held at Poona, December 19-21, 1978*. New Delhi, India: Oxford; 1985.
53. Fuller DQ. Agricultural origins and frontiers in South Asia: a working synthesis. *J World PreHistory* 2006;**20**:1–86.
54. Chen S, Lin BZ, Baig M, et al. Zebu cattle are an exclusive legacy of the South Asia neolithic. *Mol Biol Evol* 2010;**27**:1–6.
55. Gangal K, Sarson GR, Shuku A. The near-eastern roots of the Neolithic in South Asia. *PLoS One* 2014. https://doi.org/10.1371/journal.pone.0095714.
56. Fernández E, Pérez-Pérez A, Gambá C. Ancient DNA analysis of 8000 B.C. near eastern farmers supports an early neolithic pioneer maritime colonization of Mainland Europe through Cyprus and the Aegean Islands. *PLoS Genet* 2014. https://doi.org/10.1371/journal.pgen.1004401.
57. Thangaraj K, Prathap Naidu NB, Crivellaro F, et al. . The influence of natural barriers in shaping the genetic structure of Maharashtra populations. *PLoS One* 2010;**5**:e15283.
58. Singh S. Dissecting the influence of Neolithic demic diffusion on Indian Y-chromosome pool through J2-M172 haplogroup. *Sci Rep* 2016;**6**:19157.
59. Quintana-Murci L, Chaix R, Wells SR, Behar DM, Sayar H, et al. Where West meets East: the complex mtDNA landscape of the Southwest and Central Asian corridor. *Am J Hum Genet* 2004;**74**:827–45.
60. Reich D, Thangaraj K, Patterson N, Price AL, Singh L. Reconstructing Indian population history. *Nature* 2009;**461**:489–94.
61. Gallego Romero I, Mallick CB, Liebert A. Herders of indian and European cattle share their predominant allele for lactase persistence. *Mol Biol Evol* 2012;**29**:249–60.
62. Lukacs JR, Pal JN. Skeleton variation among Mesolithic people of the Ganga Plains: new evidence of habitual activity and adaptation to climate. *Asian Perspect* 2003;**42**:329–51.
63. Sharif M, Thapar BK. Food-producing communities in Pakistan and Northern India. In: *Vadim Mikhaĭlovich Masson. History of civilizations of central Asia*. New Delhi, India: Motilal Banarsidass Publ.; 1999.
64. Jarrige J-F, Neolithic M. Paper presented in the International Seminar on the First Farmers in Global Perspective. Lucknow, India, 18–20 January 2006.
65. Coningham R, Young R. *The archaeology of South Asia: from the Indus to Asoka, c.6500 BCE–200 CE*. Cambridge, United Kingdom: Cambridge University Press; 2015.

Further Reading

Akhilesh K, Pappu S, Rajapara HM, et al. Early Middle Palaeolithic culture in India around 385–172 ka reframes Out of Africa models. Nature 2018;**554**:97–101.

CHAPTER

8

The Occupation of Southeast Asia, Indonesia, and Australia

We are so fortunate, as Australians, to have among us the oldest continuing cultures in human history. Cultures that link our nation with deepest antiquity. We have Aboriginal rock art in the Kimberley that is as ancient as the great Paleolithic cave paintings at Altamira and Lascaux in Europe. ***Kevin Rudd (Speech to the Australia Parliament)***

SUMMARY

The colonization of Southeast Asia and Australia represents the longest continuous dispersal of hominins from Africa. Remarkably, this migration occurred at least twice, involving two different hominin species. It first involved *Homo erectus* about 2.0 years ago or earlier and then *Homo sapiens* approximately 120 thousand years ago. It seems that the first wave of hominins managed to colonize the Indonesian archipelago but failed to cross a 90-km wide, deep-water stretch to reach New Guinea and Australia, at that time the last two land masses were connected by a land bridge. In the second attempt, the more technologically advanced modern humans were able to navigate all the way to Australia in sufficient numbers to establish a suitable genetically healthy population. It is not clear how *erectus* and early *sapiens*, with limited technology, were able to reach the Indonesian islands and Australia. It is postulated that these hominins somehow were able to build primitive fishing rafts, and accidental crossings occurred during storms, rip currents, or tsunamis. In any case, the presence of *erectus* throughout the Indonesian archipelago suggests that likely they were the first hominin open ocean navigators.

Today, it is clear that Southeast Asia played an important role in human evolution, yet over the years' research in the area has been neglected. And although continental and island Southeast Asia (ISEA) are represented by a number of *erectus* and *sapiens* specimens, it still remains a region full of controversy. The Southeast Asia fossil landscape is not only fragmentary, but also problematic to reconcile with the more extensive and comprehensive African data. Pre–*sapiens* sites as old as 1.7, 1.9, 2.0, 2.25 million years ago (mya) in Sangiran (Central Java), Mojokerto (East Java), Longgupo (Southwest China), and Renzidong (Central-East China),

Ancestral DNA, Human Origins, and Migrations
https://doi.org/10.1016/B978-0-12-804124-6.00008-2

respectively, are unexpectedly old and pose some important evolutionary implications. If *erectus* indeed evolved in sub-Saharan Africa, as the orthodoxy claims, these very early dates for the genus *Homo* in Asia requires that the genesis of *erectus* needs to be pushed back in time, much earlier than the traditional date of 1.9 mya.

Furthermore, these very old age estimates for hominins in Asia fuels the dilemma of whether *H. erectus* evolved in Africa or in Southeast Asia. Although the prevailing view on the presence of modern humans in Southeast Asia posits that *erectus* and *heidelbergensis* lineages became extinct in the region prior to the new dispersal and settlement by *sapiens*, the spatial and temporal overlap of hominin groups allows for the possibility of interbreeding and some degree of genetic continuity. In addition, it is pertinent whether *erectus* was the first hominin to arrive in Asia, or if more ancient groups, such as *Homo habilis*, settled the region first.

Another related issue that emerges from the anatomical features of Asian hominins is the diversity of combinations of intermediate archaic and modern traits seen in the fossil record. Starting with the oldest hominin remains that include *erectus*, *heidelbergensis*, archaic *sapiens*, and modern *sapiens*, a continuous progression of intermediate forms toward increasingly modern characteristics is observed in the region. In this chapter, various examples of fossils from the mainland and ISEA that exhibit this range of traits are described as case samples. The most parsimonious interpretation of the data is that pre–*erectus* hominins settled Southeast Asia, and *sapiens* derived from them subsequently in situ by a process of interbreeding and local evolutionary change. This does not rule out gene flow from Africa as well at some point.

A number of features set African and Southeastern *erectus* apart, as well as Chinese from Indonesian remains. Overall, robustness is the hallmark of *erectus* in Southeast Asia in comparison with its African counterparts. Indonesian-specific traits that differ from continental Southeast samples include a long and flat frontal bone (skull bone found in the forehead region), a protruding face with massive flat cheekbones, the presence of a zygomaxillary tuberosity (a small rounded projection at the base of the cheekbones), a rounded edge at the bottom of the eye sockets, and no clear delimitation between the nasal area and the lower face. These distinctions have prompted some to speculate that Indonesian and Chinese *erectus* had unique ancestral populations.

Similar to *erectus*' very old arrival date, the earliest time estimates of anatomical modern humans (AMHs) in Southeast Asia are older than expected. If *sapiens* originated and then exited Africa about 70 kya, as traditionally thought, it is not possible for the species to reach Southeast Asia 120 kya, 50 ky prior to its departure from Africa. As with *erectus*, one possibility is that AMHs migrated out of Africa much earlier. The recent findings of *sapiens* occupation from Attirampakkam in India, Jebel Irhoud in Morocco and Misliya Cave in Israel dated at 385,000 ya, 300,000 ya and 177,000–194,000, respectively, reinforce the contention of an earlier origin and exit out of Africa for AMHs (see Chapters 5 and 7 for additional information on these sites). Alternatively, *sapiens* may have evolved from *erectus* independently in different regions, in Southeast Asia earlier than in Africa, as Chinese investigators have been advocating for years.

The colonization of Australia by AMHs occurred approximately 60 kya, likely due to the limited navigational skills of *erectus*. Two routes have been postulated for the incursion. One would have taken humans from what is today the island of Sumatra, to Java, Flores, and Timor, eventually moving into what is today northwest Australia. The other route would have been more northerly, traveling from Sumatra to southern Borneo, crossing to the island of Sulawesi and then Halmahera, continuing to New Guinea and then northern Australia into the region that is today the Gulf of Carpentaria. At the time, the sea level in the region was about 150 m lower than today. The earliest AMH settlement sites in Australia, Madjedbebe, Nauwalabila, and Malakunanja, date to about 65, 50, 50 kya, respectively, and are located in the Northern Territory in between the two putative entrance ways.

Traditionally, it has been posited that at least two forms of aborigines existed in the continent, the gracile and the robust, reflecting independent migrations into the continent. The gracile and the robust types are described in this chapter using the Mungo and Kow Swamp sites, respectively. Mungo samples date to 40 kya, about four times older than the Kow Swamp material, and yet Kow Swamp exhibits a plethora of ancestral traits reminiscent of *erectus*, while the Mungo fossils look more like typical modern *sapiens*. The robust Kow Swamp samples suggest that a paradigm change is required regarding the peopling of Australia.

Genetic analyses using a number of marker systems, as well as contemporary and ancient populations for comparison, indicate that Australia possesses the deepest lineages of AMHs in the world. The DNA types that define Australian aborigines are found deep into the roots of human evolution, likely dating to the original settlement of the continent. In addition, the DNA evidence also indicates that, shortly after their entrance into Australia, the ancestors of the aboriginal Australians diverged from a single ancestral population and genetically differentiated as they migrated along the eastern and western coasts and then into the center of the continent. The deep-rooted DNA types in Australian aborigines suggest relative isolation subsequent to the initial settlement of the continent.

HOMININS IN SOUTHEAST ASIA

Despite the large number of *H. erectus* specimens found in Southeast Asia, the early hominin presence in the region is not only fragmentary but also problematic to date and difficult to reconcile with the more extensive and comprehensive African data. Herein the term Southeast Asia defines continental as well as ISEA (Near Oceania). Once early hominins arrived in East Asia from Africa via South Asia the archeological information becomes very fuzzy and perplexing. The limited number of archeological sites that are well dated and studied in East Asia is a major reason for this informational lacuna. Some scholars even contest that this overall disinterest in Asian anthropology coupled with limited accessibility over the years since the original discoveries of Java Man and Peking Man in the later portion of the 19th century and early part of the 20th century, respectively, have contributed to the greater number of discovered fossils and sites in Africa, shifting the interest and scientific opinion to the notion that the genus *Homo* is of African origins and that Asia played little or no role. This is an unfortunate situation considering that recent discoveries suggest that major technological developments, such as the Acheulean tool tradition may have originated in Southeast Asia.

The data contributing to this confusion started accumulating with the original findings of Eugène Dubois in 1891–92 that led to the characterization of Java Man in East Java, a *H. erectus* specimen currently dated at 0.7–1.0 mya.[1] Dubois, a Dutch surgeon, believed, unlike Darwin, that humanity had its start in Asia, and to prove this he embarked on a quest to find early human fossils in Indonesia. He succeeded soon after his arrival with the discovery of Java Man.

Subsequently, other remains were found at Sangiran (Central Java) and Mojokerto (East Java) in 1936–37, representing older *erectus* specimens with ages as old as 1.66 and 1.90 mya, respectively.[2] Later, in 1995, the surprisingly very old cranial remains from Longgupo in Southwestern China were described and dated at 2.0 mya. These very ancient fossils indicate that hominins, such as *H. habilis* and *Homo ergaster*, were possibly in the region prior to *H. erectus*.[3] More recently in 2003 Liang Bua and in 2014 Mata Menge, nearby locations on the island of Flores in Indonesia, have yielded diminutive early hominins fossils dated to 50–190 kya and 0.7 mya, respectively[4] (see Chapter 6 for information on *Homo floresiensis*). *H. floresiensis* is thought to have survived on the island as recently as 10 kya. These specimens from Flores also possess characteristics typical of *H. habilis* and *H. ergaster* (*Homo* species intermediate between its ancestor *habilis* and its descendant *erectus* that lived 1.9 and 1.4 mya). These findings, among others, have not provided a clear picture of hominin evolution in East Asia.

Altogether, these discoveries question whether *H. erectus* was indeed the first to arrive in Southeast Asia and whether *H. erectus* evolved in Southeast Asia or Africa, as considered by the orthodoxy. The traditional view in the field of human anthropology postulates that *H. erectus* originated in sub-Saharan Africa about 1.9 mya and then traveled and settled current-day China about 1 mya. Yet, a number of more recent discoveries of hominins during the last two decades in the Yuanmou Basin (currently southern China) and Gongwangling (present-day northern China) suggest that *erectus* arrived in East Asia earlier, about 1.7–1.6 mya and persisted until 400 kya.[5] It is also possible that a more ancient species, such as *H. habilis*, made the trek eastward as part of pre–*sapiens* migration waves known as Out of Africa I (Chapter 5).

The possibility that pre–*erectus* hominins existed in Asia about 2.25 mya (see discussion on the Renzidong site below) is consistent with the earlier diversification and speciation of early hominins within Africa and the very old dates of occupation of Dmanisi in the Caucasus about 1.8 mya[6] and Masol in Southwest Asia around 2.6 mya.[7] Chapter 7 provides a description of the Dmanisi and Masol locations with a characterization of the hominin groups, which are thought to have inhabited the sites.

The old dates of the Dmanisi, Masol, and Renzidong hominins were unexpected, and they pose some important evolutionary implications. The traditional view of hominin evolution indicates that early species of the genus *Australopithecus* evolved approximately 2.6–2.4 mya and *H. habilis* in turn evolved from one of the australopithecine species about 2.1–1.5 mya. In addition, some scholars are of the opinion that another species, *H. ergaster*, lived in an epoch in between *H. habilis* and *H. erectus*. *H. erectus* is thought to have derived from *habilis* around 1.9 mya in Africa and then dispersed into Asia. If indeed all of these speciation events occurred in sub-Saharan Africa, the very early dates of the genus *Homo* in Asia requires that the genesis of the genus and *erectus* to be pushed back in time, much earlier than the traditional date of 1.9 mya for *erectus*. An earlier emergence of *Homo* in Africa would accommodate an older migration and settlement of hominins, and specifically *erectus*, in Southeast Asia as part of the Out of Africa I spread.

In addition, the very old age of some of the Asian remains fuels the dilemma of whether *H. erectus* evolved in East Asia or Africa. Generally speaking, the African-Asian dilemma on the origins of *Homo* is basically drawn along Western and Eastern scientific schools of thought. Many Chinese anthropologists support an Asian genesis for *erectus* while their Western colleagues side with an African derivation for the genus. Another school of thought within many Chinese scientists is that the evolution of *H. erectus* in East Asia parallels an independent hominin evolution in Africa. This contention is in line with the candelabra theory of recent human evolution, which postulates independent hominin lineages in various parts of the world.

The arguments in China run deep and have inspired strong nationalistic fervor, prompting Chinese research agencies to fund projects specifically directed at these issues. These nationalistic feelings have motivated Chinese scholars to vigorously seek fossil evidence of ancient humans and reanalyze old remains. Even the government has been providing tens of millions of dollars per year into finding archeological sites and granting 1.1 million dollars to establish a state-of-the-art center designed to extract and sequence ancient DNA. Some Western archeologists believe these efforts are simply directed at demonstrating that hominins originated in East Asia.

Yet, independent of the continent of origin, what is becoming increasingly clear is that the genesis of the genus *Homo* needs to be pushed back considerably to accommodate the old age of the Asian materials. It is also evident that, in spite of the fragmentary, complex, and controversial nature of the available knowledge, both continental and island Southeast Asia has played an important role in early hominin evolution. Thus far, based on the available information, the most parsimonious interpretation of the data is that pre-*erectus* hominins settled Southeast Asia earlier than previously thought and *sapiens* derived from them subsequently in situ with some degree of interbreeding. Thus, it is likely that based on the available data, some details of recent human evolution would require a radical reassessment by the scientific community.

THE EARLIEST HOMININS IN EAST ASIA—2.0 MYA AND BEFORE

Archeological evidence indicates that early hominins were in East Asia by 2.25–2.0 mya or even earlier during the late Pliocene.[8] It is likely that when members of the genus *Homo* exited East Africa and entered the Near East, they continued dispersing into the Caucasus, West Asia, and South Asia (continental India), eventually reaching Southeast Asia (Fig. 8.1).

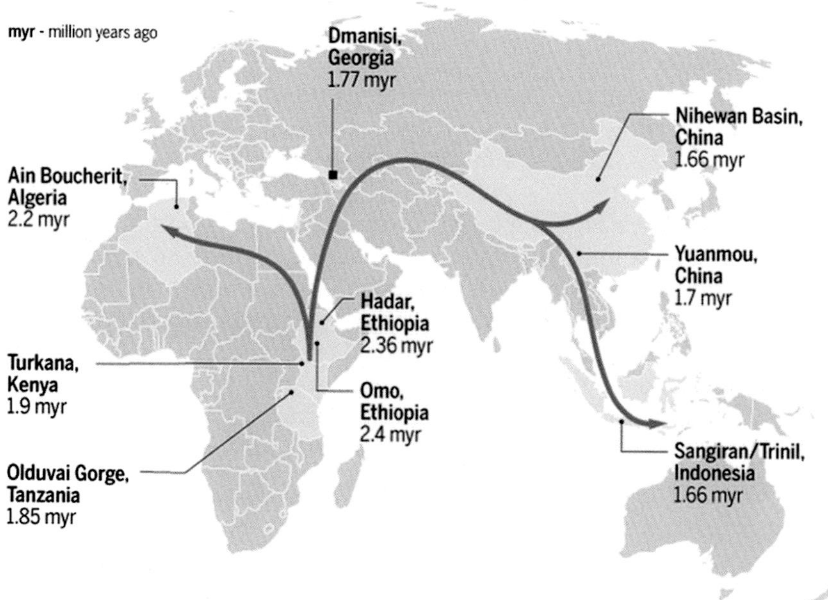

FIGURE 8.1 Trail of artifacts and fossils indicating possible routes for the spread of early *Homo* out of Africa to the Caucasus, West Asia, and South Asia (continental India), eventually reaching Southeast Asia. *From http://www.sciencemag.org/news/2016/11/meet-frail-small-brained-people-who-first-trekked-out-africa.*

It is also feasible that the primitive Dmanisi (Chapter 5) *H. erectus* and/or *H. habilis* from Georgia in Southwest Asia migrated further east and evolved into the East Asian early hominins unearthed in China, Java, and Flores. A description of Longgupo and Renzidong, the two best-known sites in Southeast Asia dated at 2.0 and 2.25 mya, respectively, is given below.

Longgupo

The oldest hominin fossils in East Asia include a mandible fragment with two very worn molars and a single maxillary incisor from the Longgupo (Dragon Hill) Cave site in the Three Gorges area of the Yangtze River in the southwestern Sichuan Province of China.[9] The remains were paleomagnetically dated to 2.0–1.8 mya. In general, the Longgupo jaw fragment resembles a pre–*erectus* hominin with similarities to *H. habilis* from Africa, and yet exhibiting affinities to Javanese *erectus*. To most non-Chinese experts the teeth exhibit characteristics seen in East African early hominins suggesting an Out of Africa I dispersal prior to 2.0 mya. Yet, to most Chinese investigators, these findings corroborate an independent origin for the genus *Homo* in Asia from Asian apes. Although a number of different interpretations of the Longgupo remains are possible, one scenario would require a dramatic change in the traditional hominin evolution paradigm, suggesting that *erectus* was not the traveler out of Africa, but australopithecines and/or *habilis* about 2.0 mya, which subsequently evolved in Southeast Asia into *erectus* and then dispersed into Africa.

Also found at the site was an abundant collection of fauna totaling 116 species that match the age of the hominin bones and artifacts. In addition to the fossil evidence, a spherical cobble hammer stone and a large flake were unearthed from the site. These lithic tools are of the ancient Oldowan-type typical of *H. habilis* and *H. ergaster* and not of the more recent Acheulean style associated with *H. erectus* (see Chapter 7 for a discussion of Oldowan and Acheulean lithic traditions). Remarkably, all of the archeological horizons from the Longgupo Cave were found in their undisturbed natural context within the original deposition layers. For most anthropologists the Longgupo findings provide evidence that *erectus* was not the first *Homo* species in Asia.

Yet, opinions differ with regard to the species represented at Longgupo and the age of the specimens. For example, while the upper lateral incisor is thought to be from *sapiens* and relatively recent, according to some scholars, the mandible fragment is considered by others to belong to an ape.[10] Since the fossils were found in association with Oldowan-type tools, the findings cannot be simply dismissed as of Ape origins. It is possible that Longgupo is a composite site containing a mixture of hominin and nonhominin primate remains. Although highly controversial due to contested species of origin and age, the general scientific opinion is that Longgupo suggests that *H. erectus* was not the first hominin to leave Africa and disperse into East Asia.

Renzidong

In addition to the Longgupo location, lithic implements and 5000 bone remains of about 75 species of animals mainly belonging to a type of elephant of the genus *Sinomastodon*, an extinct tapir, and the ancient monkey *Procynocephalus* were discovered at the bottom of a large rock fissure at Renzidong (Renzi Cave) in the Anhui Province of eastern China.[11] The material was dated at 2.25 mya. Apparently, the animals had fallen accidentally or were driven by humans or other predators to stampede into the deep crevice.

In Renzidong, some 100 lithic artifacts and fractured bones used as blades and scrapers were found. It is likely that the hominins went into the cavity to claim, dismember, and butcher the prey after they had fallen. Interestingly, the elephant-like bones were found together along one wall while the ancient tapir remains were located in a different area where they were apparently cut and partitioned for allocation to clan members.

The Renzi Cave is the oldest site among a growing number of hominin locations in East Asia, indicating the deep antiquity of the genus on the continent. The findings at Renzidong also suggest that *erectus* may not have been the first *Homo* group to inhabit Asia. With such an unexpected ancient age, the artisans of the tools found in the Renzi Cave were either early *erectus* or a more ancient *Homo* species, such as *H. habilis*. It is remarkable that the implements discovered are older than the ones made by *erectus* about 2 mya in Africa, but because the fossil records of pre–hominin species are so extensive in Africa, most authorities still support an African origin for the genus *Homo*. Yet, as previously indicated for Longgupo, some of the gaps in the Asian archeological horizons could be attributed to the limited interest or access to sites.

Just like with the Longgupo site, the Renzi discovery once again ignited the East–West divide on the birthplace of *Homo*. Both schools of thought posit that plate tectonics (movements of continental plates in geological time) are responsible for climatic changes that prompted habitat transformations from wet forest to arid savannahs, environmental changes that selected for bipedalism in previously arboreal primate populations (Chapter 2). In Africa, plate tectonics and the creation of the East African Range resulted in climatic changes that gave rise to aridity and the transformation of the land, east of the mountains, to the East African savannas (discussed in Chapter 2), whereas in Asia a similar phenomenon, in this case the convergence of the Indian and Eurasian continental plates, generated the Himalayans and, to the north, the dry Tibetan Plateau about 8 mya. Contrary to the East African hominin evolution scenario, many Chinese anthropologists contend that *erectus* derived in situ in Asia from Asian primates and then migrated west to Africa and/or evolved in parallel with an independent African hominin lineage. It has been proposed that hominins could have migrated out of Africa as australopithecines 3.0 mya or before, and in East Asia, they speciated into the genus *Homo* and species such as *erectus*, subsequently moving from Asia to Europe and Africa. Although this contention is not in favor by most scholars, in recent years an increasing number of specialists, including some Westerners, consider it a real possibility. A third scenario is that species, such as *erectus* or *habilis*, during their long tenure, migrated back and forth between Eurasia and Africa. Bidirectional movement of hominin species between Africa and Asia is not unprecedented. In the short existence of *H. sapiens*, for example, incidents of back migrations signaled by DNA markers are numerous.[12]

HOMO ERECTUS VERSUS HOMO ERGASTER

The scientific community is currently divided regarding whether the African and Asian specimens of *erectus* belong to the same species. One school of thought posits that only the Asian groups should be named *H. erectus* and the African forms should be referred to as *H. ergaster*. On the other side of the divide, there are anthropologists that attribute the observed anatomical differences to regional and time variation, but these scientists consider that both should be considered *H. erectus*. The rationale for this taxonomic split is the degree of morphological variation between the two geographical groups. Furthermore, some proponents argue for an ancestral relationship between the older versions of African *ergaster* and the Asian *erectus*. The theory also proposes that the Asian *erectus* represents only a side branch within the hominin tree not directly connected to *sapiens* and that *ergaster* (in Africa) is in fact the ancestor to AMHs. Because this issue remains highly speculative and contested, in this chapter, as in Chapter 2, the African and Asian forms are considered a single species. Based on the anatomical differences between the African and Asian groups, it is likely that this argument will continue until additional data provide clarity on the origins of the Asian and Indonesian *erectus*. In other words, a better understanding of whether *erectus* evolved in situ in Southeast Asia in parallel with the African forms or migrated from Africa is paramount to a resolution.

SOUTHEAST ASIAN *ERECTUS*

Ever since the discovery of Java Man by Eugène Dubois in 1891 (Trinil 2) a number of archeological sites clearly indicate a *H. erectus* occupation in Southeast Asia and Indonesia. Shortly after his findings, Dubois published the first report describing the new species giving it the scientific name of *Pithecanthropus erectus* (erect ape man). The genus was later renamed *Homo* based on Ernest Mayr's comments.[13] *Erectus* is the best-represented AMH relative in Southeast Asia with over 40 individual fossils excavated from Java and innumerable more from China. Just from Zhoukoudian, 40 km south of Beijing in China, 50 individuals have been excavated. Yet, in spite of this sizeable number of *erectus* remains, the subsequent evolutionary steps connecting them to archaic *H. sapiens* and then to AMHs are not clear. To illustrate the characteristics of *erectus* in Southeast Asia, the assemblages recovered from Trinil 2, Mojokerto, Sangiran 17, Talepu, and Mata Menge in Indonesia as well as Peking Man, Yuanmou Man are presented below in some detail.

Although it is not known if *erectus*, during its long tenure and geographical expanse, constituted a single species or several regional *Homo* groups, it is clear that as a taxon, it was anatomically highly variable in time and space, and very successful, with some populations in Indonesia surviving until as recent as 140 kya. Its longevity as a group is about 10 times greater than *sapiens* as of today. *Erectus* is also credited with the creation of a novel lithic tradition, the Acheulean, approximately 1.76 mya. These new stone tools were larger than the more ancient Oldowan-type characteristic of previous hominin species. *Erectus'* fresh arsenal of tools included novel large hand axes and cleavers designed to cut. During the later part of its tenure, it coexisted with other hominin species in Southeast Asia, such as *Homo heidelbergensis*, *H. floresiensis*, and, very possibly, *H. sapiens*.

Although it is not definitive that *erectus* lived in Europe, abundant specimens from Africa and Southeast Asia have been recovered since the original discoveries in Java. The most complete skeleton of *H. erectus* recovered to date is known as Turkana Boy from near Lake Turkana in Kenya (Fig. 8.2). Turkana Boy lived during the early Pleistocene, 1.5 and 1.6 mya. Probably the best-known examples of *erectus* in Southeast Asia are from Mojokerto and Sangiran, both in Java Indonesia, and Zhoukoudian or Choukoutien (Peking Man) in Beijing, China, and the island of Flores in Eastern Indonesia.

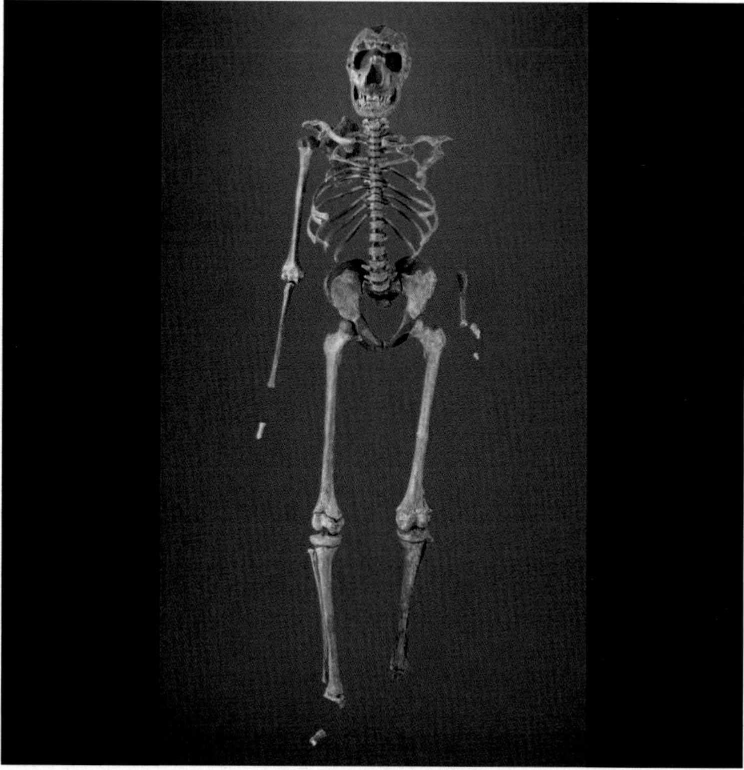

FIGURE 8.2 *Homo erectus* skeleton from Turkana, Kenya. *From http://humanorigins.si.edu/evidence/human-fossils/fossils/knm-wt-15000.*

How Different Were the Regional Populations of *Homo erectus*?

Chapter 2 provides an anatomical and cultural description of *H. erectus* over their geographical range and as a function of time. Over the years, as the number of *erectus* samples have increased, scholars have began to notice specific morphological characteristics that discriminate between African and Southeast specimens, as well as between Chinese and Indonesian remains. For example, samples from Mojokerto and Sangiran in Java and Zhoukoudian, just south of Beijing, are considered classic examples of the *erectus* group from East Asia, but a number of characters sets the Javanese and the Chinese specimens apart as well. In this section, the salient traits that differentiate between African and Southeast Asian types are described as well as some differences between continental Southeast Asian and Indonesian *erectus* groups.

Overall robustness is the hallmark of *erectus* in Southeast Asia in comparison with the African counterparts, which are more gracile. Even to the casual observer the Southeast Asian samples have very prominent orbital ridges above the eye sockets, and the skulls exhibit overall thickness of the walls. In addition, the Southeast Asian crania possess a prominent sagittal crest in the middle of the frontal bone at the sagittal suture. This bony

FIGURE 8.3 Early hominin skull from Sangiran 17 in Central Indonesia. *From http://whc.unesco.org/en/list/593.*

keel extends from front to back as a ridge and provides for the attachment of strong jaw muscles. The neurocranium or carvarium (part of the skull that protects the brain) and face are particularly massive in the Chinese and Indonesian samples, and their teeth are generally much larger.

The Indonesian specimens as a group are characterized by a number of traits that set them apart not only from African *erectus* but also from the mainland Asian *erectus* fossils. These Indonesian-specific traits include a long and flat frontal bone (skull bone found in the forehead region), a more protruding face with massive flat cheekbones, the presence of a zygomaxillary tuberosity (a small rounded projection at the base of the cheekbones), a rounded edge at the bottom of the eye sockets and no clear delineation between the nasal area and lower face (Fig. 8.3).

Although traditionally it was thought that a single dispersal wave gave rise to all *erectus* populations in Southeast Asia, the abovementioned anatomical differences between the Javanese and Chinese specimens have prompted some experts to propose that the two groups were not directly linked and that the mainland population arrived to Southeast Asia later from a different source population, possibly in Africa.

JAVANESE *HOMO ERECTUS*

Javanese remains, such as Trinil 2, Modjokerto, Sangiran 17, Sambungmacan, and Ngandong, group into a lineage that sets them morphologically apart from other Southeast Asian fossils, such as the Chinese and Flores *erectus*. According to proponents of the multiorigin theory of human origins, some of these Javanese-specific traits evolved in situ and are still evident in Javanese populations. Three representative samples of Javanese remains are described below.

Java Man

Eugène Dubois discovered the original Java Man; a skullcap, tooth, and thighbone in Trinil on the banks of the Solo River in East Java. The skullcap and the leg bone were discovered 1 year apart and approximately 15 m (50 feet) away from each other on the cave strata, and thus they may have been from different individuals (Fig. 8.4).

Dubois, whose full name was Marie Eugène François Thomas Dubois, was born on January 28, 1858, in Eisden and died on December 16, 1940, in de Bedlaer, Netherlands. Dubois was a rather eclectic scholar and as such played various roles during his professional life. His father, Jean Dubois, was an apothecary and wanted him to share in the same profession, but in 1877, Eugène enrolled as a medical student at the University of Amsterdam and became a lecturer of anatomy in different institutions in the city. During vacations, he invested much of his time in fossil hunting near the town of Rijckholt, where he discovered a number of prehistoric human skulls. After graduating in 1884, he did research for 6 years on the comparative anatomy of the larynx of vertebrates. This tenure as a comparative anatomist instilled in him an interest in the evolution of the organ, which led him to the study of human evolution. At the time, Darwin's book *On the Origin of Species*, published on November 1859, was in the spotlight, and evolutionary thought was circulating throughout Europe. Yet, unlike Darwin, who insisted on the African origins of humans, Dubois was a firm believer in the genesis of *sapiens* in Asia.

Without research funds, but with a keen interest in anatomy, human evolution on his mind and Darwinian ideas circulating in Europe at the time, he enlisted in the army as a military surgeon to serve in the East Indies, at the time a Dutch colony. Fascinated by the theory of evolution, he was convinced, somehow, that the origin of humans resided in that part of the world and not in Africa. His resolution was so strong that in 1887, he moved his wife and newborn daughter to the Dutch colony. This caused extreme consternation among family, friends, and colleagues. After arriving on current-day Indonesia, in his spare time, he started excavating for human remains in the island of Sumatra, north of Java. In Sumatra, he soon found a number of fossilized mammals, generating enough interest among his superiors to convince them in 1889

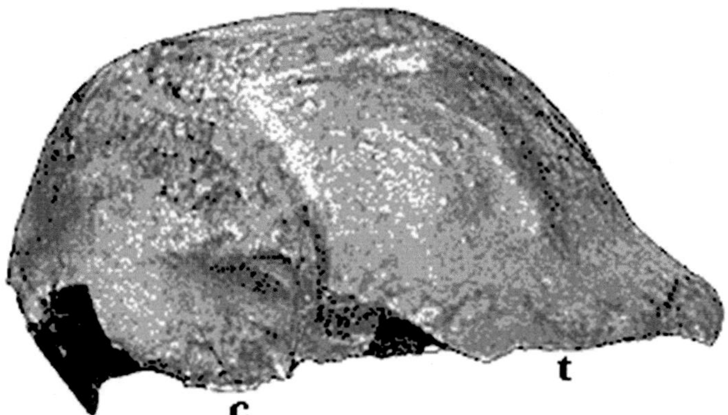

FIGURE 8.4 Skullcap of Java Man (*Homo erectus*): (c) postorbital constriction; (t) brow ridges. *From http://www.athenapub.com/13dubox1.htm.*

to be relieved of his military duties and to provide him with two engineers and fifty convicts to help him with his excavations. After failing to recover hominin remains in Sumatra, Dubois moved to the island of Java and initiated a series of digs in Trinil where he found Java Man. After returning to Amsterdam in 1894, he published his findings[14] and embarked in on a new career, teaching geology at the University of Amsterdam. Java Man became so controversial that the fossil was not reevaluated until 1923. Although his ideas on hominin evolution became increasingly accepted after the 1920s, he died a bitter man and was secularly laid to rest because of his thoughts on human evolution and antagonism with the Catholic Church.

The Trinil 2 remains have been dated at 0.7–1.0 mya and is the youngest of the three *erectus* Javanese sites described herein. The fossil likely belonged to a mature woman, and the skull is characterized as robust, with thick bones, protruding brow ridges, retreating forehead, no chin, and a massive jaw. The discovery immediately generated contradictory opinions regarding the evolutionary position of the find with respect to *H. sapiens*. Initially, some scholars thought it was the skull of an ape, some considered it an AMH, and many argued for an inconsequential side branch and dead end in the hominin tree. Only a few experts supported the contention of Dubois that the fossil belonged to an ancestor of AMHs. Interestingly, because Dubois was not actually present on site during the excavation of Trinil 2, he was not able to pinpoint the exact location of where the samples were found and thus had difficulty addressing a number of criticisms.

The Mojokerto Site

Mojokerto 1 is another example of Javanese *H. erectus*, and yet it is unique because the remains belonged to a child. Mojokerto 1 or Perning 1, named after nearby towns in East Java, Indonesia, represents a fossilized calvarium of a juvenile (2–4 years old) skull discovered in 1936. Soon after the finding of the Mojokerto Child, as it was nicknamed, it was the subject of controversy. The first dispute to emerge involved the scientific name. Initially, Ralph von Koenigswald described the fossil and named it *Pithecanthropus modjokertensis* after the previous discovery by Eugène Dubois of Java Man or *P. erectus*. Yet, Dubois insisted the name needed to be changed because it could not possibly belong to the same genus as Java Man. So the child skullcap's name was change to *Homo modjokertensis*. Finally, after Ernest Mayr's suggestion, the classification was changed to the present form of *H. erectus*.

Confusion also stemmed from the exact location of the fossil at the time of collection.[15] Notes on the dig were not clear enough to assess from where precisely the brainpan was collected. Furthermore, conflicting information was provided at different times regarding the depth at which the samples were located. Initially, Ralph von Koenigswald indicated that the fossil was found on the surface and later the account was changed to unearthing the fossil from a stratum 1 m deep. The exact location of remains is important since, for instance, it is generally agreed that surface material is less reliable for dating and association with other artifacts because of potential soil distortion and contamination with extraneous material.

Dating the sample also proved to be an issue. Initially, the skullcap was thought to be less than 1.0 mya. Then, in the 1960s, when the calvarium was redated by the potassium–argon method (based on the decay of potassium to argon), it generated an age of 1.9 mya. The material was further reassessed in 1994 using the then novel and more accurate ^{40}argon–^{16}argon method, providing an age of 1.8 mya.[15] This very early time estimates immediately caused controversy because it forced experts to reconsider the time of Out of Africa I and *erectus'* geographical origin in Asia versus Africa.

Although the digging and dating site was pinpointed by an Indonesian paleoanthropologist who participated in the original study group with von Koenigswald, it was later questioned as not being the precise site of excavation. One argument contested the validity of dates because the samples for dating were collected 60 years after the skull was unearthed and the two may have come from different layers. A more recent study conceded that the degree of confusion regarding the precise location of collection makes it difficult to ascertain the actual age of the Mojokerto specimen.[17] Currently, the age of the Mojokerto specimen is presented as a wide-ranging date of 1.9–1.5 mya.

The discrepancies regarding the Mojokerto child also include the age of the infant at the time of death. As the remains lacked teeth, the task of assessing the child's age is difficult. A number of quite different postnatal estimates raging from 0 to 8 years old have been reported, including 2–5, 1.5, 4–6, and 0.5–1.5 years.[18] The latest determination of 0.5–1.5 years was based on computed tomography scans of the calvarium with resulting images clearly indicating that the anterior fontanelle (soft membranous gap between the cranial bones that allows flexibility of the skull during birth) was still open, indicating that the child was less than 1.5 years old at the time of death. The anterior fontanelle normally closes between the ages of 12 and 18 months.

Brain Development in *erectus*

The discovery of the Mojokerto child was a unique and important event for evolutionary developmental biologists. Although only the brain cap of the individual was recovered, the fine structure provided by the interior cavity of the cranium of the *erectus* infant allowed the opportunity to ascertain the growth kinetics of *erectus* in relation to AMHs and Great Apes. One interesting area of inquest was whether *erectus* exhibited a *sapiens* brain developmental pattern or if the brain of a direct ancestor of AMHs lagged behind in the development of the central nervous system while other anatomical characteristics, such as *adult* brain size, general body plan, and locomotion, were quite advanced, similar to that of *sapiens*. Over the years since the discovery of Mojokerto, and with limited amount of supporting data, the scientific community has been divided into two groups: one is of the opinion that *erectus'* brain developmental kinetics is like the pattern seen in the Great Apes and the other argues that it resembled the development as seen AMHs.

In addressing these issues, it is important to consider that *erectus* was a particularly heterogeneous group of hominins that varied considerably in time and space. This fact has prompted a number of investigators to contemplate the notion that *erectus* was a conglomerate of sister species. An individual from Africa differed from its counterpart from Southeast Asia, and an early *erectus* dating to 2.0 mya was very distinct from one that lived 150,000 years ago. Thus, in any consideration of the growth kinetics of the species, one must first decide which adult specimen should be used for comparison. In other words, which adult sample is appropriate?

As expected, when the Mojokerto skullcap was compared with that of an adult *erectus*, the results were highly dependent on the adult employed in the collation.[19] For example, in a 2005 study, it was concluded that the *erectus* infant experienced brain development different from *sapiens*.[20] Specifically, the research group concluded that during its ontogeny, *erectus* experienced anterior–posterior flattening of the brain as well as anterior–posterior development of the frontal lobes, characteristics dissentingly different from AMHs.

In a more recent study in 2013, the internal anatomy of the calvarium was reexamined, this time employing a larger number of Javanese *erectus* samples older than 1.2 mya in an effort to provide a more appropriate set of adult parameter readings for comparison.[21] The results indicated that AMHs reach 62% and chimpanzees reach 80% of their adult brain volume at 0.5–1.5 years of age. In addition, AMHs attained 65% and chimpanzees 81% of the adult brain mass by the same age. Furthermore, when comparisons were performed with the adult *erectus* from Indonesia, the infant brain cap was 70% of its extrapolated adult size. Thus, this data suggest that, based on the Mojokerto individual, Javanese *erectus* exhibited brain growth kinetics intermediate between AMHs and chimpanzees. In addition, the results indicated that the growth pattern was within the range of both AMHs and chimpanzees. The study concluded that the Javanese *erectus* population of 1.2 mya and older exhibited a unique brain developmental pattern within the evolutionary continuum connecting the Chimpanzee's ancestor and AMHs. This conclusion was not particularly unexpected.

Sangiran 17

Another important Javanese *H. erectus is* Sangiran 17. Sangiran is the name of a city in Central Java, Indonesia, where nearby archeologists over the years have discovered one of the richest collections of *H. erectus* fossils in Asia. It seems that the region of Sangiran was first settled by *erectus* around 1.6 mya. Starting with the original findings by villagers on behalf of von Koenigswald during the period encompassing 1936–40 and continuing after World War II until the present, about 80 hominin fossils, including 10 partial skulls and 14 fragments of jaws were found. The ages of the hominin remains at the Sangiran complex range from 2.0 to 1.0 mya.

Specifically, the sample Sangiran 17 is a 1.2-million-year-old adult male skull, discovered in 1969 by S. Santono.[22] Over the years, some of the morphological characteristics of this cranium have attracted the attention of a number of scholars who are proponents of the multiregional theory of AMHs.[23] The reason for this interest stems from the fact that the Sangiran 17 fossil, together with the findings from Modjokerto, Sambungmacan, Ngandong, and contemporary Javanese exhibit anatomical characters that link them together as a lineage in time. Probably the best known of these persistent traits found in the present-day population are enlarged cheekbones, a long and flat frontal bone (forehead), and a zygomaxillary tuberosity (projection at the base of the cheekbones).

CHINESE *HOMO ERECTUS*

Peking Man

In continental Asia an increasing number of *H. erectus* fossils have been unearthed since the early 20th century. Probably one of the best-known groups of fossils is Peking Man discovered near the town of Zhoukoudian 30 miles southwest of Beijing, during a series of excavations between 1923 and 1927 (Fig. 8.5). The site was considered as the first indication that early hominins migrated all the way from Africa to what is today China. The remains date to 680–780 kya using $^{26}Al/^{10}Be$ isotope decay ratio.[24] Peking Man is not a single hominin fossil but a rather extensive collection of bones from at least 50 individuals collected from a quarry and caves. This population lived in the area for a period of 200,000 years, and it seems that the occupation of the area was not permanent, but sporadic. Both sexes are present among

FIGURE 8.5 Peking Man. *From http://www.nature.com/news/how-china-is-rewriting-the-book-on-human-origins-1.20231.*

the remains. These assemblages of remains included 6 complete or nearly complete skulls, 14 cranial pieces, 6 facial fragments, 15 jawbones, 157 teeth, 3 upper arms, and a number of other body parts. In addition, approximately 100,000 quartz and sandstone artifacts, including chopping tools and flakes, were found in the caves.

Although Peking Man postdates Java Man by approximately 250,000 years and the two are separated by a distance of about 5500 km, both sets of fossils share a number of important morphological features that group them together as *H. erectus*. These common traits include a flat and projecting profile, massive skull bones, small forehead, a keel along the top of the head for muscle attachment, prominent brow ridges, an occipital torus (bulge of the occipital bone at the back of the skull), a large palate (roof of the mouth), and a large chinless jaw. And yet, Peking Man differs from the Javanese samples by lacking a number of characters present in the Java Man specimens (see last paragraph of previous section). In addition, Peking Man possessed a larger cranial capacity averaging 1000 cm^3, a larger forehead, smaller teeth (similar to AMH in size and shape), and nonoverlapping canines as compared with Javanese specimens. These diagnostic anatomical differences have prompted some archeologists to propose that the Javanese and Chinese *erectus* populations were not directly connected and that the ancestors of groups, such as the Chinese Peking *erectus* arrived at East Asia later than the Indonesian population and both came from different sources.

Two interesting facts relate to these fossils. It is known that the first Peking Man samples were discovered by local people, who thought of them as "dragon bones" and sold them to medicine shops as traditional Chinese medicine. Since then, the site has been known as Dragon Bone Hill. In fact, fossil bones were recovered so often by ordinary people that archeologists working in the region routinely check the apothecary shops first for "dragon bones" to buy them and to get information on where to dig. Although excellent casts and detailed information exist, the actual Peking Man fossils disappeared during World War II. The fossils were supposedly hidden in 1937 as the Japanese troops invaded China before World War II. Yet, in 1941, when the American and Chinese governments mutually agreed to send them to the United

States for safe keeping, they vanished. According to records, the fossils were packaged and loaded on a military ship named President Harrison bound for the United States. As the ship was getting ready to depart on December 8, 1941, Japan declared war on the United States and attacked Pearl Harbor. The Japanese troops in China confiscated the cargo as the boat was getting ready to depart. Since then, the fate of the bones is unclear. The Chinese accused the United States, and the United States blamed the Japanese for the loss. It is likely that the Japanese opened the two boxes containing the bones, thinking it had valuables, and disappointed at the dusty content, discarded the precious fossils. In 2005, the Chinese petitioned the governments of the United States, Japan, and South Korea to investigate the whereabouts of the fossils, but no evidence has been found. In an effort to recover additional remains, excavations were reinitiated at Zhoukoudian with some additional fossils as well as core and flaked tools recovered.

Yuanmou Man

Another example of Chinese *H. erectus* is Yuanmou Man. The Yuanmou site is located in the vicinity of Danawu, a Village in Yuanmou County in the southwestern province of Yunnan, China. Two inside incisors from the upper jaw of a young male have been recovered. These incisors exhibit the classic shovel-shaped characteristic of *H. erectus*, which is also seen in East Asian and Native American contemporary populations. In addition to lithic artifacts, fragments of animal bones, some with indications of manipulation, and ash from what seem to have been campfires were identified.[25] The stone tools included quartz scrapers, a 90-mm spindle-shaped core, and a red sandstone flake core.

Two sets of dating originally indicated very different ages for the teeth. Estimates based on paleomagnetic readings of the surrounding rocks indicate an age of about 1.7 mya. This age makes the Yuanmou teeth one of the oldest hominin fossils in Southeast Asia. Other estimations using faunal layer sequences suggest a much younger age during the Middle Pleistocene in the vicinity of 500,000–600,000 ya.[26] The investigators involved in this second set of assessments argued that the strata had been disturbed by inversion with the older fossil animals in the upper layers and the younger ones at the bottom, contributing to the previous much older dates. Considering this large discrepancy, in 2008, a third group of researchers tested the soil from the exact location where the teeth were unearthed and again generated paleomagnetic date of 1.7 mya, confirming the original results.[27]

HOW *HOMO ERECTUS* GOT TO INDONESIA?

Although the exact classification of *H. floresiensis* is still highly contested (Chapter 6), it is clear that hominins managed to reach the island of Flores in the Lesser Sunda Islands in the eastern half of Indonesia. These hominins crossed treacherous deepwater channels to get to Eastern Indonesia. In addition, lithic artifacts from strata corresponding to the *erectus* timeline pepper the various deepwater islands of Indonesia. These include Sumatra, Java, Sulawesi, Roti, and Timor. The stone tools from Roti and Timor, at the extreme east of Wallacea, date to 781–126 kya. *H. erectus* was able to navigate from Alor to Timor, a distance of at least 60 km. Altogether, the archeological data suggest that *erectus* was able to island-hop throughout Indonesia, stopping short of reaching Australia. In fact, the more recent archeological data indicate that *erectus* was more widespread in Wallacea than initially thought.

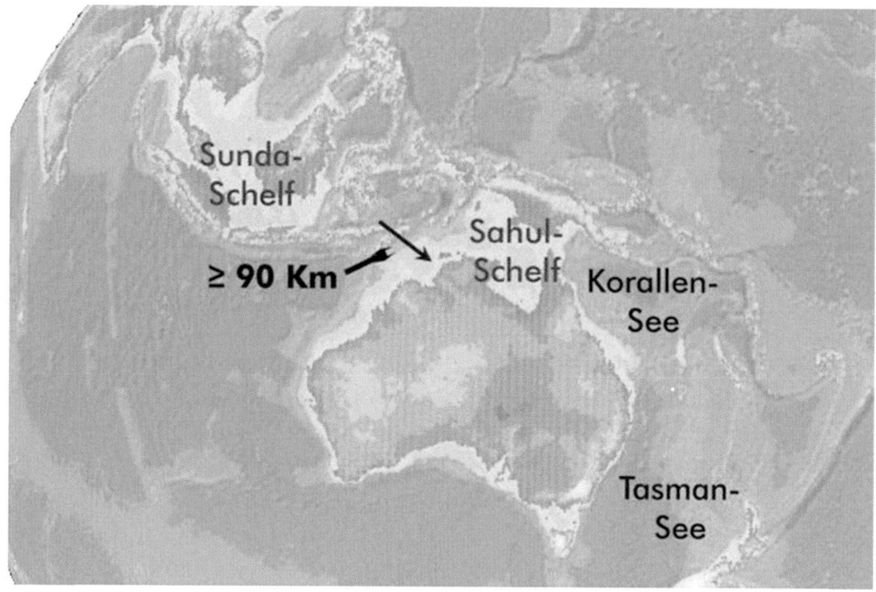

FIGURE 8.6 Lydekker Line (biogeographic line separating the island of Timor from the Australian continent) represents a 90 km-wide, deep and treacherous water gap that prevented *Homo erectus* from reaching Australia. *From https://en.wikipedia.org/wiki/Prehistory_of_Australia#/media/File:Sunda_shelf.jpg.*

Because no *erectus* stone tools or fossil remains have been identified past the Lydekker Line (biogeographic line separating the island Timor from the Australian continent), it is likely that the 90 km of deep treacherous water that sets Timor apart from Australia (Fig. 8.6), known as the Timor Sea, was too much of a distance for the early hominins to cross, preventing any further eastward dispersal of *erectus* to colonize the continent.

Mata Menge

Some experts adhere to the original contention that the diminutive hominins from the island of Flores represent a case of island dwarfism of a *H. erectus* type that reached Eastern Indonesia, whereas more recent anatomical analyses suggest that they were descended from the same australopithecine ancestor that evolved into *Homo habilis*, and thus they are not directly related to *erectus* and likely older.[28]

Recently, in an open-field site approximately 74 km east from where *floresiensis* was first discovered in the same island of Flores, additional hominin fossils were unearthed. This location, known as Mata Menge, generated hominin remains from individuals that arrived on the island at least 1.0 mya.[4,29] At this site, a mandible and six teeth were found from three individuals (one adult and two children) who predate *H. floresiensis* found in the original Liang Bua cave site by about 600 kya. The morphology of the Mata Menge fossils reflects a primitive *H. erectus*. The Mata Menge specimens are at least eight times older than the original *floresiensis* findings. In addition, Mata Menge yielded fossils of small elephant-like species (likely resulting from island dwarfism), Komodo dragons, crocodiles, and approximately 50 lithic artifacts.

Based on the size of the adult mandible, the Mata Menge population was also small, possibly as a result of insular dwarfism affecting *H. erectus* migrants from continental Southeast Asia.

What is perplexing about the Mata Menge population is that to reach Flores, in the Lesser Sunda island chain east of Java, *H. erectus* had to navigate across at least two channels. Even during glacial periods, when most of Indonesia was linked together into two large landmasses (Sunda and Sahul) and sea level in the region was as low as 150m below today's levels, these dangerous deep bodies of water needed to be traversed to reach Flores from the west. Traveling to the island of Java was not a problem because it was connected to the Asian continent during the glacial period. Yet, the 19-km deepwater gap of the Lombok Strait between the islands of Bali and Lombok clearly must have been a formidable barrier to eastbound *erectus* dispersals. The second crossing involved navigating between Sumbawa and Flores, a 9-km distance. These two water crossings may represent the most ancient seafaring events undertaken by hominins in known history. Although *erectus* was likely able to see the island of Lombok at a distance standing on Bali and looking east from the extreme end of the Sunda supercontinent (Fig. 8.7), the details of how *erectus* managed such a crossing are not clear. However, the fact that *erectus*

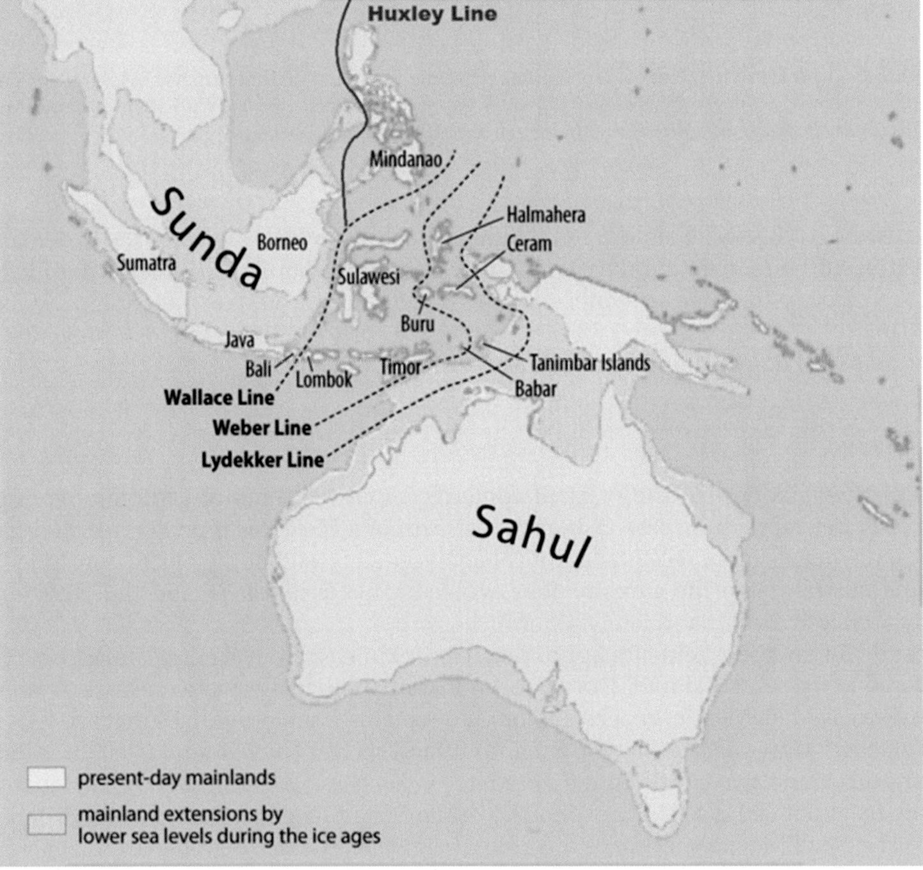

FIGURE 8.7 *Homo erectus* navigation across the channels of Wallacea, present-day Indonesia. *From https://en.wikipedia.org/wiki/Wallace_Line.*

migrated as far east as Flores implies that it colonized the intermediate islands of Lombok and Sumbawa as well. These crossings are testaments to the creativity, intellect, and spirit of exploration already present in the early hominins that remain traits of modern *sapiens*.

Talepu

Another site related to *H. erectus* in Indonesia is located on the island of Sulawesi, east of the island of Borneo. Borneo constituted the extreme east of the supercontinent of Sunda that was part of a continuous landmass with continental Southeast Asia during glaciations. During the ice age, Sulawesi was separated from Borneo by a 19-km body of water known as the Makassar Strait.

The original set of discoveries was made in the late 1940s and consisted of stone tools and megafauna dated to the Pleistocene (2,588,000–11,700 years ago). Subsequent work in the Cabenge region (Southern Sulawesi) of the island between 2007 and 2012 identified a number of additional sites with stratified undisturbed artifacts. One of these locations is Talepu where deep (at least 4 m) trench excavations were performed. Although no hominin fossils were found, some 270 lithic tools were found, including cores, choppers, and flakes.[30] The raw material from which the assemblage was made is silicified limestone in cobbles up to 130 mm in diameter. The cores were reduced by hitting the cobbles on one side or bifacially, then the flakes were further worked on one or both sides. There is no evidence that the artisans aimed at producing specific tool types; rather, they were interested in producing multiple flakes from which the best were selected for use. The megafauna remains consisted of Stegodon (extinct insular dwarf elephant), crocodile, anoa (miniature water buffalo), and a number of Celebochoerus (even-toed ungulates). Age estimates of the site using a number of methods, including multiple-elevated-temperature post-infrared infrared-stimulated luminescence (MET-pIRIR), indicate continuous habitation from before 200 kya until about 100 kya, likely representing an advanced population of *H. erectus*.

Were *Homo erectus* the First Hominin Mariners?

Thus, the question still remains how an early hominin such as *erectus* or *habilis*, with limited technological resources, made the crossings across Wallacea. Although Lombok Island is visible at a distance from Bali and would likely have enticed these hominins to reach it, most scholars express doubts on the navigational skills of *erectus*. Investigators question whether *erectus* purposely set out to move from one island to the next. It is likely that any deliberate attempt to reach the adjacent island would require building a raft with logs or bamboo (both plentiful in Indonesia at the time) and assembling them with plant fibers. If the crossings were made purposely, this would mean that *erectus* preceded Austronesians as Oceanic travelers by about 1 my.

Although the bodies of water within Wallacea were relatively narrow during ice ages, the channels were treacherous and the currents were strong, running north to south, perpendicular to the direction of the hominin crossings. This scenario would tend to push rafts or swimmers into the open sea north or south of the Indonesian archipelago. Could accidental crossings in primitive fishing rafts or rip currents explain this enigma? Possibly an unexpected tsunami was responsible? Could *erectus* survive a tsunami or escape a volcanic eruption by holding to vegetation only to be carried away into open ocean or, if lucky, to a nearby island? Considering that the islands of Indonesia have been highly seismic and volcanically active, it is not far fetched to envision that these types of sudden catastrophic events could have deposited hominins from one island to the next nearby landmass as unintentional swimmers or piggybacked on vegetation.

In addition, because the currents in the interisland channels of Wallacea run north to south, it would have been difficult to move longitudinally across the archipelago, but it would have been less problematic to colonize from Borneo to the island of Sulawesi to the North.

HOMO HEIDELBERGENSIS IN SOUTHEAST ASIA

The traditional view on *H. heidelbergensis* is that it originated in Africa approximately 600 kya and existed in Africa, Europe, and West Asia until 200 kya.[31] Most archeologists also think that *heidelbergensis* migrated from Africa to Europe about 400–300 kya where they evolved into Neanderthals. A separate branch is thought to have dispersed into Central Asia giving rise to Denisovans (40 kya or earlier), which interbred with East Asian *H. sapiens* populations. The orthodoxy also stipulates that populations of *heidelbergensis* that remained in Africa evolved into *H. sapiens* who subsequently dispersed into Europe and Asia between 125 and 60 kya as part of the first and second phase, respectively, of the Out of Africa II migrations (Chapter 5). Although some scholars are of the opinion that *heidelbergensis* represents a form of advanced *H. erectus*, most authorities consider it a separate transitional species with ancestral features and derived anatomical characteristics later seen in *sapiens*. Generally speaking, *heidelbergensis* exhibits *erectus* characters, such as a massive brow ridge and no chin, in combination with *sapiens* traits, such as smaller teeth and bigger braincase.

Until relatively recent, the presence of *heidelbergensis* in East Asia was arguable. Then, in the second part of the 20th century a number of *erectus–sapiens* transitional fossils began to show up in East Asia, and many Western investigators had no clue how to classify them. Presently, it is clear that during the period between 300 to 100 kya, Southeast Asia was home to several hominin populations exhibiting a collage of modern and ancestral traits. These intermediate hominins persisted as the degree of *sapiens* characteristics increased until about 100 kya when the number of modern traits dominated and anthropologists felt confident in considering them AMHs.

As these transitional fossils became more abundant, the traditional perspectives on the origins and migrations of *heidelbergensis* started to change and were challenged, especially among Chinese archeologists. Specifically, very ancient fossils dated to 300 kya or earlier exhibiting features more advanced than *erectus*, but with advanced traits seen in African and European *heidelbergensis*, started to appear in East Asia. Although these Asian hominins were similar to the African and European *heidelbergensis*, certain skull morphologies set them apart. Asian representatives had larger braincases, shorter flatter faces, larger angles between the nose and the forehead, flat nose bridges, rectangular eye sockets, and larger forward-projecting cheekbones.

Altogether, the various combinations of ancestral and derived characteristics of this transition taxon in Africa, Europe, and Asia do not allow assessing its geographical genesis. Yet, the very ancient dates of *heidelbergensis* in East Asia indicate that the Far East and not Africa could have been their place of origin. A Chinese-based genesis for a hominin directly ancestral to *sapiens* would definitely revolutionize the field of recent human evolution. Alternatively, it is possible that *heidelbergensis* left Africa and subsequently entered East Asia much earlier than traditionally thought. If this scenario is correct, it means that this species cohabitated with *erectus* during the later part of their existence in East Asia, potentially being the venue for interbreeding and gene flow.

H. heidelbergensis is an intermediate species with a mixture of *erectus* and *sapiens* characteristics, and as such, it is difficult to delineate unique *heidelbergensis* traits. If the group is examined as a whole, a spectrum of advanced traits is seen as a function of time. At times, it is difficult to ascertain if a given specimen is *heidelbergensis, erectus, or sapiens*. The ambiguity of this taxonomic group has persisted over the years since the initial discovery in 1907 of a *heidelbergensis* mandible near Heidelberg, Germany. Some archeologists still question its validity as a discrete species. Genetic continuity stemming from in situ ancestor–descendant relationships among East Asian *erectus, heidelbergensis*, and *sapiens* is the most parsimonious explanation for the observed morphological gradients exhibited by these three groups.

To illustrate the characteristics of East Asian *H. heidelbergensis* and its relationships to *erectus* and *sapiens*, two well-preserved key fossils (Dali and Jinniushan) are described below. Both the Jinniushan and Dali are of particular interest because both retain almost complete facial skeletons. Albeit the cranium is less robust than that of *erectus* and Neanderthals, the Dali specimen is nevertheless massive by modern human standards. It still retains protrusive brow ridges, receding forehead, median ridge, and a ridge along the posterior part of the skull. A pointed angle is formed by the occipital bone and nuchal plane similar to *erectus* (Fig. 8.8). Although highly circumstantial, the Dali and Jinniushan *heidelbergensis* remains have

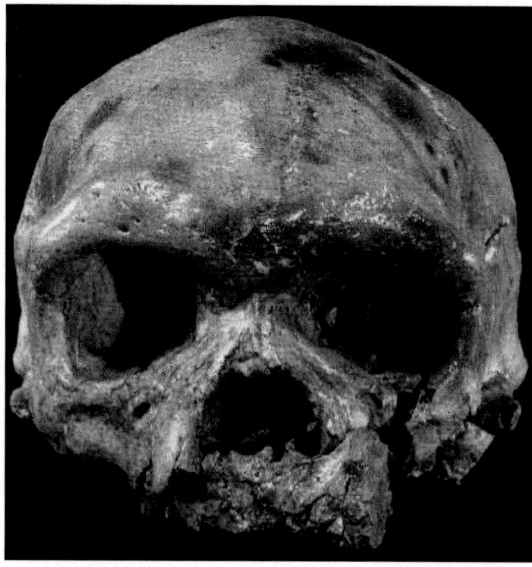

FIGURE 8.8 Dali skull. *From http://www.peterbrown-palaeoanthropology.net/dali.html.*

been link to the Denisovans of Central Asia of which only a finger bone fragment and few teeth have been recovered. Denisovans and a second uncharacterized species only seen by genetic analyses are known to have contributed DNA to the hominins of Australia, Papua New Guinea, and Polynesia (Chapter 6). The overall picture, illustrated by these examples and others beyond the scope of this chapter, is that during the transitional period between 300 and 100 kya a myriad of hominin types coexisted in East Asia. It is likely that this spatial and temporal cohabitation led to technological transfer and interbreeding. It is also likely that introgression among them generated the various permutations of anatomical combinations seen in the fossil record.

Dali

The Dali fossil was found embedded in a mud terrace near the city of Jiefang in the Dali district of the northwestern Chinese province of Shaanxi. In 1978, a cranium dated to about 200 kya was found exhibiting a number of ancestral characteristics reminiscent of *H. erectus* in combination with *sapiens*-specific traits.[32] In addition to the hominin skull, various small lithic artifacts, mainly scrapers, and megafauna remains, such as ox teeth, were also unearthed from the site. Although the cranium suffered some postmortem alterations and distortions of the palate and left maxilla, the skull was in fair condition.

In spite of the retention of ancestral features, the Dali facial anatomy is more like AMHs. In general, the facial bones are more modern-like, overall smaller, and flatter relative to *erectus*. As such, the face is short and although it still possesses relatively large cheekbones, they are less massive than in *erectus*. The nasal bones are flat and the nose of Dali would have been broad and low.

Also similar to *sapiens*, the maximum cranial width is located on the posterior–superior temporal region rather than the cranial base as in *erectus*. The mastoid process (conical pyramidal projection at the posterior base of the skull), which varies in size among contemporary humans, is small. At 1120 cm^3, its cranial volume was intermediate between *erectus* and *sapiens*.

The Dali discovery immediately elicited controversy due to its collage of ancestral and derived characteristics. Scholars were not sure how to classify it. Anatomically, it landed in a no man's land, with some experts characterizing it as an advanced *erectus* part of a continuous evolutionary path to *sapiens*, whereas others thought of it as a unique species similar to *H. heidelbergensis*. In retrospect, looking back into the dilemma, both schools of thought were right because these intermediate hominins were part of a continuum that eventually led to AMHs.

It is noteworthy that this almost complete skull has traits that set them apart from African and European *heidelbergensis*, such as the general craniofacial morphology and vault shape as well as bigger brain volume, a shorter face, and lower cheekbones. These characters in the East Asian specimen were also interpreted as representing more derived *sapiens*–like features.

Jinniushan

The Jinniushan remains were discovered in a limestone cave in the vicinity of the town of Sitian in the northeastern Chinese province of Liaoning bordering North Korea. The finding took place in 1984 and included a skull, the left ulna, the left innominate or hipbone (bones forming the sides of the pelvis each consisting of three joined bones: the ilium, ischium, and pubis), six vertebrae (one cervical, five thoracic), one complete left patella, two left ribs, and various bones of the hands and feet. From dental comparisons, it was estimated that the remains belonged to a young adult female about 20–30 years old. Noteworthy, the individual was approximately 78.6 kg (173 lb) in body weight representing the heaviest female hominin in the fossil record.[33] The large body size of the Jinniushan individual is not necessarily confounding because hominins reached maximum mass during the Middle Pleistocene and particularly the fossil was located at a high latitude and a cold climate location.

As part of the finding, a number of faunal remains were unearthed including species of *Macaca* (a genus of monkeys), Trogontherium (an extinct genus of giant beavers), *Megaloceros* (an extinct genus of deers), *Dicerorhinus* (a genus of rhinoceros), *Microtus* (a genus of voles), and several species of extinct birds.

Unfortunately, the skull was damaged during the extraction from the ground, and subsequent reconstruction led to damage and additional lost of bone.[34] A number of independent studies by Chinese archeologists have dated the fossils to the range of 310–200 kya.

The skull exhibits a remarkable resemblance to the Dali cranium (Fig. 8.9). They both have flat broad faces reminiscent of other East Asian *heidelbergensis* such as the specimens from Hulu Cave and Nanjing.[33] Yet, Jinniushan also possesses a unique medley of ancestral and derived traits. Just like the Dali skull, Jinniushan exhibits a number of *sapiens* characters. It has a rather delicate skull with overall *sapiens*-like features: a cranial volume of about 1400 cm^3 with the maximum cranial width present on the posterior–superior temporal region rather than the cranial base as in *erectus*. This greater brain capacity compared with that of Dali is due to overall thinner vault bones. It also has a median frontal ridge, which extends posteriorly. Like Dali, the mastoid process is small.

In contrast to Dali, the Jinniushan skull is overall less robust with thinner supraorbitals. It is not clear to what degree these gracile features could be explained by sexual dimorphism because the specimen was a female. The brow ridges in Jinniushan are less prominent and less massive in the midorbit area. The postorbital constrictions (gap in the cranium just behind the eye sockets or orbits) are more pronounced than in Dali. The occipital and nuchal planes do not meet at an angle as sharp as in Dali.

Most notably about Jinniushan is the anatomy of the foot representing a mosaic of *erectus* and AMH features. The foot provides evidence of increased walking stability present in the *sapiens*-like anatomy of the medial longitudinal arch. And yet, it retains a number of ancestral *erectus* foot characteristics such as lower arches and a less stable hallux metatarsophalangeal joint (joints between the metatarsal bones and the proximal bones of the toes) compared with AMHs. These unique combinations of foot bone orientations suggest that the Jinniushan individual walked differently but only slightly different from modern humans.[35,36]

HOMO SAPIENS IN SOUTHEAST ASIA

Although the prevailing view on the presence of AMHs in Southeast Asia posits that the *erectus* and *heidelbergensis* lineages became extinct in the region prior to the new dispersal and settlement by *sapiens*, the spatial and temporal overlap of hominin groups allows for the

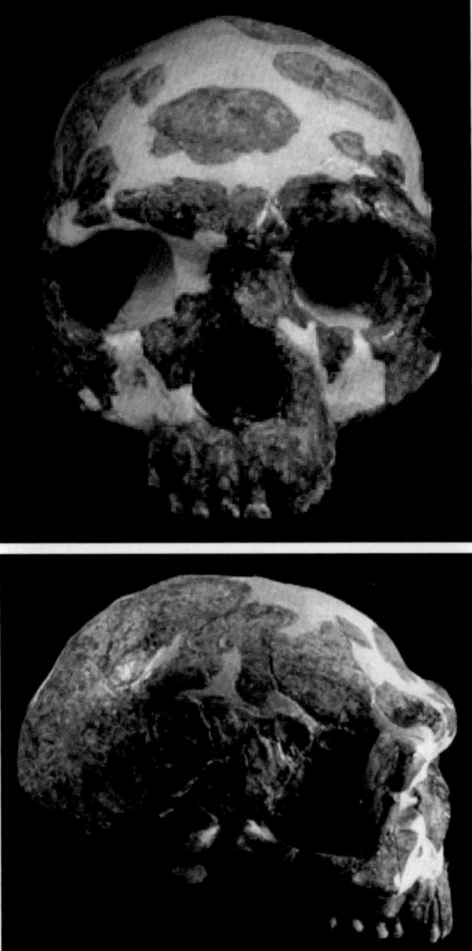

FIGURE 8.9 Jinniushan skull. *From http://www.peterbrown-palaeoanthropology.net/jinniush.html.*

possibility of interbreeding and some degree of genetic continuity. During the period that encompasses the tenure of these species in Southeast Asia, cohabitation of various hominin groups was the rule, not the exception.

During this time, anatomical features point to an epoch of progressive change from ancestral *erectus* features to modern *sapiens* characteristics. A casual observer would easily contemplate the possibility of genetic continuity beginning with *erectus*, and possibly earlier, leading to AMHs in Southeast Asia. Also, the Denisovan DNA component in Southeast Asian and Oceania provides genetic evidence of gene flow among hominin species. The main consideration regarding some degree of hominin interbreeding resides on biological reproductive barriers, population densities, and/or habitat ranges, the last two factors potentially keeping hominin species apart. In Southeast Asia, transitional hominins showing the entire gamut of ancestral and derived states are observed during the second half of the Middle Pleistocene (300–120 kya).

Notwithstanding these considerations, this section presents AMHs in Southeast Asia as the result of independent migrations from Africa as part of the Out of Africa II dispersal. The premise of the peopling of Asia by the Out of Africa *sapiens* is that small numbers of individuals migrated along the coast of the Indian subcontinent (see Chapter 7 for details), eventually arriving in Southeast Asia. In addition to this coastal route, it is possible that modern humans traversed the subcontinent via an inland migration to reach Southeast Asia.

Another route to East Asia would have taken modern humans through a northern passage diagonally from the Near East via Central Asia to Northeast Asia (see Chapter 7 for details). The well-documented genetic divide that exists between the southern and northern East Asian contemporary populations supports this double-pronged (i.e., a southern coastal and a northern inland) migration scenario. Specifically, mitochondrial DNA (mtDNA), Y chromosomal and genome-wide markers (see Chapter 5 for explanation of marker types) clearly illustrate this north–south dichotomy.[37] In terms of mtDNA, types A, C, D, G, Y, and Z almost completely characterize the mtDNA of Northeast Asians, whereas among the Southeast Asians, haplogroups B and F overwhelmingly predominate. In addition, mtDNA types C, Y, and Z are very uncommon among Southeast Asians. Based on the Y marker distribution, Southeast Asian populations are more diverse than Northeast Asian groups, and haplogroups O2-M95 and O1-M119 are most abundant in the south. This suggests gene flow from Southeast Asia northward.

This latitudinal partition is also evident in the archeological, anatomical, linguistic, and surname distributions.[38] Therefore, it is possible that the present-day East Asian populations are descended from two independent migrations, one populating the north, whereas the other populating the south.[39] Yet, it is not clear how far back the putative independent source populations go. In other words, did the split originate in the Near East, where one branch traveled coastally along the Indian subcontinent into Southeast Asia, whereas the other dispersed in a northeasterly direction toward Northeast Asia? Or do the two independent branches go all the way back to Africa?

Modern *Homo sapiens* Versus Archaic *Homo sapiens*

Generally speaking in African archeology, the term archaic *H. sapiens* is used to describe the earliest examples of modern humans or the transitional forms retaining a number of ancestral traits. In East Asia, on the other hand, the term archaic *sapiens* is employed to group transitional fossils that cannot be assigned to *erectus* or modern humans. Remains exhibiting a mosaic of features have a tendency of being assigned to the archaic human category. There is a critical difference between the two rationalizations because the African usage connotes archaics and moderns being part of the same modern human family, whereas the Asian definition just indicates a transitional state.

In practice, the line of demarcation between archaic *H. sapiens* and modern *H. sapiens* is, to some degree, an arbitrary one. Although the relative proportion of ancestral and derived characteristics normally are the ultimate criteria, a lot of what goes into the decision of whether a particular fossil falls into one category or the other lies on the professional opinion of the investigator, the importance given to specific diagnostic traits, the associated fauna, tool tradition, and age estimates. Some of these criteria engender subjective decisions from the scholars. Considering the transitional nature of these *sapiens* hominins, often the investigators need

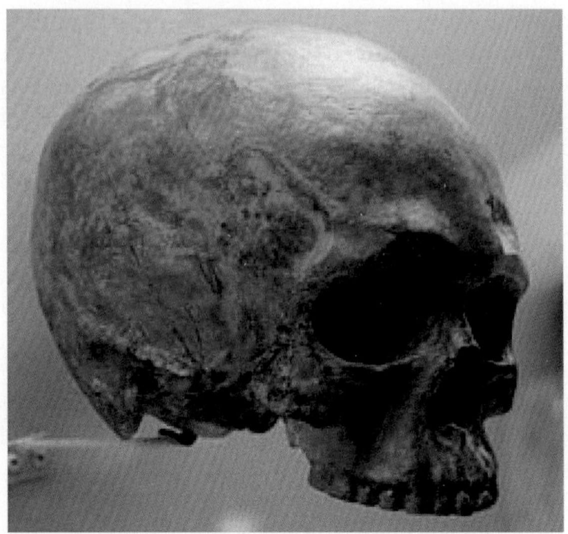

FIGURE 8.10 Liujiang man. *From https://en.wikipedia.org/wiki/Liujiang_man.*

to decide where to draw the line between archaic and modern forms within a continuum of anatomic variability. In general, the orthodoxy is of the opinion that *sapiens* properly dated between approximately 370 and 250 kya represent archaic *H. sapiens*, whereas fossils 150 kya and younger are usually considered modern humans.

In continental East Asia, a number of sites, including New Cave at Zhoukoudian dated at 248–269 kya and Chaoxian at 310–360 kya in eastern China, are now considered well-established archaic *sapiens*, whereas Huanglong Cave in Central China dated at 81–101 kya, Luna Cave at 70–127 kya, Daoxian at 80–120 kya, Liujiang at 111–139 kya (Fig. 8.10), and Zhiren Cave in South China with an age older than 100 kya are considered modern *sapiens*.[16]

Inconsistent Age Estimation of Hominin Remains in East Asia

A review of the East Asian human evolution literature illustrates that some East Asian archeological sites exhibit a lack of consistency in the age estimations when they are independently dated by different methodologies and/or investigators. This is not a phenomenon unique to the region, but results from technical and methodological limitations as well as the nature of particular digs and a number of other factors.

A notorious example of this condition is the Chinese site of Xujiayao in the Nihewan Basin (see below). Xujiayao was discovered in 1974, and for almost 5 decades since the finding, numerous conflicting age estimates have been reported, varying with dating technique, exact locations within the site and source (e.g., sediments or fossils).

The original dating in 1979 based on animal remains and sedimentary soil samples provided an age of at least 100 kya during the mid-Pleistocene. Subsequent uranium series dating of rhinoceros dentition provided an estimated age of about 104–125 kya. Then, in 1991, carbon 14 dating of bones and teeth generated dates older than 40 kya. Furthermore, environmental magnetic data in the 2000s recorded a time of about 500 kya, and more recently using $^{26}Al/^{10}Be$ isotope dating of buried quartz gave a mean age of 240 kya.[40] Not surprising, these highly incongruous results have caused consternation and generated alarm among archeologists. More recently electron spin resonance (ESR) studies have provided an average age of 260–370 kya for the hominin-containing sedimentary layer.[16] Overall these dates partition the assemblages into two groups. One set encompasses the dates around 100 ky while the three most recent estimations provide much ancient ages older than 240 kya. These two ranges of dates fall within the tenure of at least four possible hominins, *erectus*, and *heidelbergensis* as well as ancient and modern *sapiens*. In retrospect, considering all the dates reported for Xujiayao, the carbon 14 reading of greater than 40 kya should be invalidated because the isotope system employed is almost out of its useful range. Radiocarbon dating has a maximum reliable upper limit of about 50 ky.

Another example of age discrepancy is seen in the Yunxian remains where a geomagnetic age range of 830–870 kya and an ESR date on stratigraphically associated tooth enamel of 581 kya were independently obtained.[41]

As a result of recent technological advances and refinements during the last 10 years, age estimates have been revised and more accurate dates of hominin remains are now available. This new chronology based on U-series estimations of calcite rather than bones has provided more accurate ages for a number of important Late and Middle Pleistocene sites.[16] Fig. 8.11 illustrates the best-known archeological sites in continental East Asia indicating their corresponding updated age estimates.

Although it is difficult to attribute the time inconsistencies to any given cause, they are at times observed in regions where investigators confront complex cave depositions or layers that cannot be timed with potassium–argon techniques. As pointed out above for carbon 14, the procedure needs to be within its effective range. Inherited limitations of antiquated

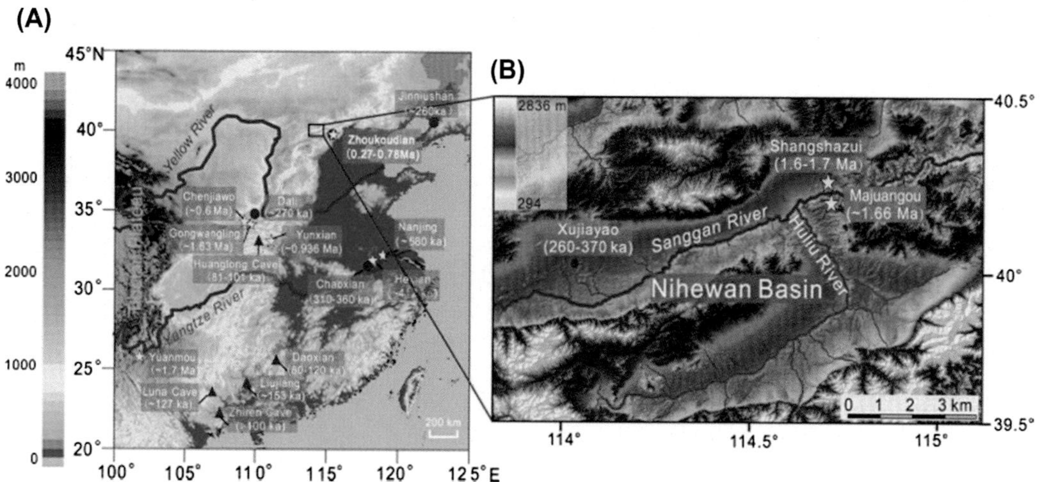

FIGURE 8.11 Location of archeological sites in continental Southeast Asia. (A) Location of main archeological sites in Southeast Asia. (B) Detail of the Nihewan Basin about 120 miles west of Beijing. *From http://www.sciencedirect.com/science/article/pii/S0047248417300349.*

techniques and poor record keeping as well as delineating the precise location of remains have been at times responsible for variability. Other sources of putative inconsistencies are exposed sediments subject to fluvial and/or wind erosion or distortion.

A Chronological Conundrum

Presently, it is generally accepted that the AMHs that eventually reached Southeast Asia exited Africa about 70 kya. This time estimate is based on archeological and genetic data, and, depending on the study in question, it tends to vary considerably from 100 to 60 kya. Furthermore, the traditional notion holds that *sapiens* evolved strictly in Africa starting about 200 kya.

The time of arrival of AMHs to Southeast Asia has been intensely contested since the earliest hominin finds of Eugène Dubois in Java in 1891. Over the years since the seminal discoveries, Southeast Asian archeology has been suffering from lack of exact documentation of site locations and wide discrepancies in age estimation, likely the result of diverse dating methodology. Notwithstanding these important issues, more recent findings have added fuel to the argument pertaining to the initial *sapiens'* arrival as AMHs. Dates of 120 kya and earlier have been reported throughout Southeast Asia. These very ancient AMHs in Southeast Asia pose chronological issues with the traditional views. How could *sapiens* be in Southeast Asia 120 kya, about 50 ky *prior* to departing from Africa? According to conventional thought the species was still within Africa 120 kya.

Because *H. sapiens* is thought to have originated in sub-Sahara Africa approximately 200 kya, the easiest way to reconcile this conflict is to adjust the time that *sapiens* departed from Africa after it evolved; that is, closer to its genesis. The most parsimonious solution to this apparent conundrum would be that modern humans dispersed out of Africa earlier to allow them to reach Southeast Asia and settle it before 120 kya. Yet, there are other more complicated scenarios. For example, it is possible that AMHs originated in Asia and not Africa. This notion is not new. Dubois was originally motivated to search for the origins of modern humans among the islands of Southeast Asia in the late 1800s. This theory has intensified more recently since the discovery of hundreds of very old modern human fossils in China. And although there is a divide between Western and Chinese scholars regarding the subject, an increasing number of investigators worldwide are beginning to give serious thought to the possibility of *erectus* leading to *sapiens* in Asia and then populating Africa and the rest of Eurasia.

An earlier migration of AMHs out of Africa is supported by the well-documented presence of modern humans in the Arabian Peninsula dating to around 100–125 kya or even earlier as suggested by the 177,000–194,000 year-old *Homo sapiens* upper jaw from the Misliya Cave in Israel (Chapter 5). These sites include Skhūl and Qafzeh, in current-day Israel, and an assemblage of tools dating to about 125,000 ya from Jebel Faya in United Arab Emirates (UAE). As part of the first phase of the Out of Africa II dispersal, this occupation of the Near East by *sapiens* is traditionally thought to represent a dead end because it did not contribute to the permanent peopling of Eurasia (Chapter 5). Yet a number of recent discoveries of early *H. sapiens* remains in China, including the Daoxian, Zhirendong and Xujiayao sites, dating to 120, 100, and 125 kya, respectively, indicate that the first phase of Out of Africa II may have commenced prior to the date previously estimated and may have contributed to the settlement of the Far East. Although this notion is currently highly contested, genetic studies have detected early *H. sapiens* DNA in contemporary human populations suggesting that these early migrations were not sterile evolutionary dead ends (Chapter 5).

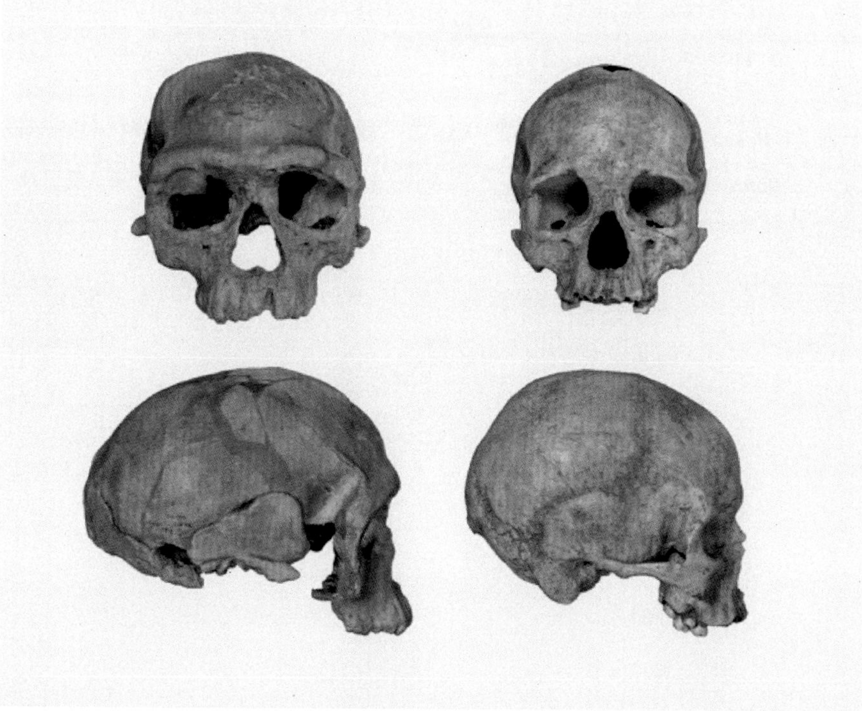

FIGURE 8.12 Skull of anatomical modern humans (AMHs) dated to 315 kya from Jebel Irhoud, Morocco. Early member of AMHs from Morocco (left) display a more elongated skull shape than modern humans (right). *From http://www.nature.com/news/oldest-homo-sapiens-fossil-claim-rewrites-our-species-history-1.22114.*

The likelihood of an early migration eastward to account for the very ancient modern humans in China has recently become more probable with the discovery of AMHs dated to 315 kya in Jebel Irhoud, Morocco (Fig. 8.12).[42] This finding of a 315-kya *sapiens* in Northwest Africa indicates that the premise that *sapiens* evolved only in East Africa is not correct. The date estimate is at least 100 ky older than previously thought, and the location was unexpected. Furthermore, a number of reports on mtDNA diversity outside of Africa argue for AMHs out of Africa much earlier than previously thought, about 270 kya.[43] These new dispersal estimations are based on the fact that the elevated DNA variation observed outside Africa (i.e., 70 kya) could not be accounted for because greater evolutionary time is needed to generate the mtDNA diversity observed in Eurasia.

Overall, the recent archeological and genetic data suggest that the origins and dispersal of modern humans out of Africa are more ancient than previously thought. This paradigm shift also accommodates the very old Chinese *sapiens*. Fig. 8.13 illustrates a number of well-dated modern human (orange) and archaic *sapiens* (blue) sites in China. Yet, a review of the issues pertaining to the origins of specific species and the precise timing of migrations clearly underscore the fact that the picture of hominin evolution in Southeast Asia is lacking crucial information to allow proper explanation. As case samples, three early sites from Southeast Asia are described below.

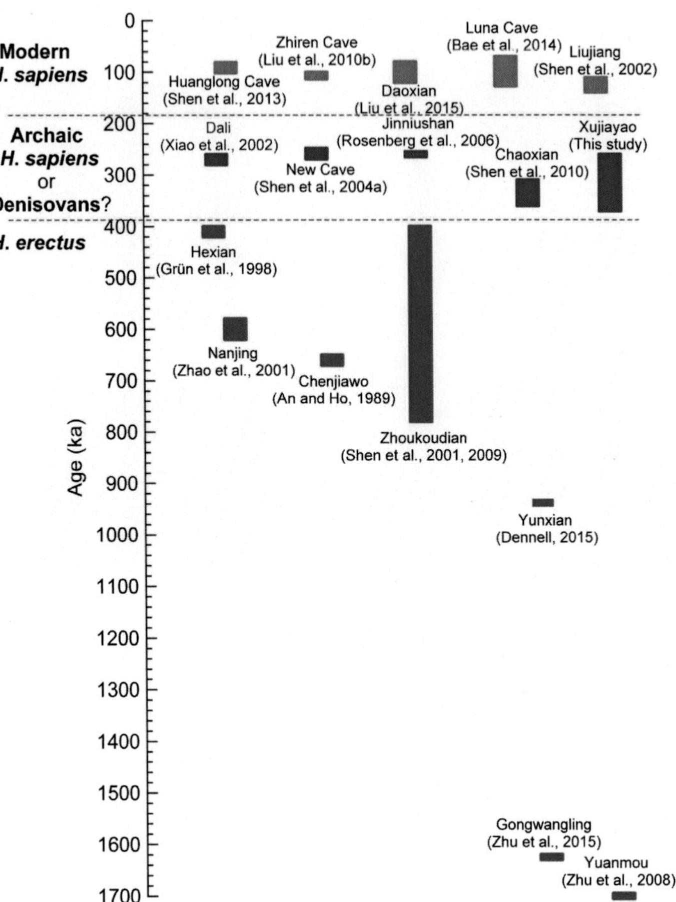

FIGURE 8.13 Chronological illustration of well-established dates of modern *Homo sapiens* (orange), archaic *H. sapiens* (blue), and *Homo erectus* (green) sites from China. *From http://www.sciencedirect.com/science/article/pii/S0047248417300349?via%3Dihub.*

Xujiayao, 125 kya

The Xujiayao site was discovered near the village of Houjiayao in the province of Hebei in Northern China. The open-field dig lies adjacent to the Liyigou River, 1 km from the village. A total of 27 fossilized bones have been recovered. The hominin remains found at the site represent an assemblage of cranial bones, including 12 parietals, some complete, one temporal, two large occipital fragments, one mandibular fragment, three isolated teeth and one maxilla, from a 6- to 7-year-old male infant. Typical of the time in East Asia, Xujiayao fossils have a medley of characteristics associated with European Neanderthals, Asian *erectus*, and modern *sapiens*.[44]

The Xujiayao fossils have their own unique complex combination of hominin transitional features making it difficult to precisely identify and classify into any given species and may even represent a unique taxonomic lineage. Specifically, the Xujiayao bones reflect individuals with thick cranial vaults relative to AMHs, an ancestral condition, while the maxilla is reminiscent of modern *sapiens*. Its dentition exhibits ancestral *erectus* features from East Asia as well as some Neanderthal characteristics, but overall the teeth do not fit any known species. According to some experts, the molars are massive with very large roots and complex pattern of grooves, reminiscent of those from Denisovans. For instance, the central incisor and canine exhibit marked shoveling. Also, as in *erectus*, the nuchal and occipital planes form an angle when viewed laterally, the maximum cranial width is located on the cranial base, and the temporal line (lateral skull line) is crested anteriorly. On the other hand, the inner ear is more typical of Neanderthals. Some scholars have postulated that Xujiayao is the result of interbreeding between cohabitating East Asian hominin species. Yet, based on the most recent and comprehensive age estimates, the Xujiayao fossils represent one of the oldest archaic *H. sapiens* in East Asia.

About 30,000 artifacts have been recovered from the site. Recently, an additional collection of over 10,000 stone tools have been discovered, but they have not been published yet. In this assemblage, large tools are not abundant. Some of them are made of stone, but a number of them were crafted from bones and antlers and they include scrapers, blocks or hammering surface, points, engravers, choppers/chopping tools, bolas, and spheroids.[45] The most abundant tool type found is scrapers, and most of the artifacts were reworked. The lithic remains found at Xujiayao belong to the primitive Oldowan-like tradition and contrast with the more advanced Acheulean stone tools from contemporary African sites. The retention of primitive Oldowan-like tradition is also characteristic of other East Asian locations of the time (mid-Pleistocene).[46] The recovered fauna consisted of approximately 5000 remains representing over 20 species. Judging from the large number of horse bones, the inhabitants of Xujiayao were excellent hunters of different types of *Equus*, including the Asiatic wild ass and the Przewalski's horse. Other animals that appear to have been hunted include an extinct species of woolly rhino, wolves, voles, wild cattle, antelopes, pigs, tigers, and a number of gazelles (e.g., Asian gazelles). The types of species found at Xujiayao suggest that at the time the habitat was a cool and open savannah with localized areas of forest.

Daoxian, 120 kya

The Fuyan Cave in Daoxian, southern China, has yielded an assemblage of 47 teeth from at least 13 individuals. The teeth are small with corresponding slender roots and flat unstructured crowns, a morphology typical of archaic or modern humans. In fact, the teeth are almost indistinguishable from contemporary *sapiens* specimens and fall within the range of variability found in AMHs.[47] On the basis of these striking similarities, these hominins are generally considered a population of AMHs living in Southeast Asia about 120 kya, similar to the archaic *H. sapiens* from Xujiayao. This temporal presence in East Asia raises the possibility of coexistence and introgression involving modern and archaic humans.

The Daoxian cave is a limestone system encompassing approximately 3000 m^2 of space located near the village of Tangbei in Hunan Province. No artifacts have been found suggesting that possibly the hominins and animals were dragged into the cave by predators for consumption as opposed to being a hominin dwelling.

The hominin dentition unearthed has been dated to around 120 kya using reliable uranium series dating from pure calcites. They possess moderate basal bulging and longitudinal grooves in the buccal surface of canines, premolars, and molars reminiscent of the teeth found at Xujiayao (see above). It is noteworthy that morphologically the Daoxian teeth resemble contemporary humans more than the teeth of Skhūl and Qafzeh in the Near East, which possess more ancestral characteristics. Skhūl and Qafzeh are thought to represent a failed incursion of AMHs during phase one of Out of Africa II, approximately 125 kya. Furthermore, the Daoxian dentition indicates that these East Asian humans possessed more derived traits than Late Pleistocene *sapiens* in central and northern China. Overall, the teeth are smaller than 120-ky-old African and European specimens and look more like the teeth of present-day humans.

The presence of *H. sapiens* dating to 120 kya in Daoxian does not support a scenario in which a 50–70 ky migration out of Africa resulted in the peopling of East Asia and strongly argues for AMHs arriving in Southeast Asia much earlier than conventionally thought or originating in East Asia. If indeed AMHs originated in Africa and then traveled to East Asia, then the available data suggest a separate Out-of-Africa migration 120 kya. This information by itself may indicate that East Asian AMHs may have evolved independently in Africa and Asia.

In addition to the hominin remains, fauna fossils, predominantly hyenas, extinct giant pandas, and dozens of other vertebrate species, including extinct members of the elephant family, giant tapirs, and pigs, have been recently recovered. The animal fossils found at Daoxian are typical of the Late Pleistocene, and their age coincides with the uranium series dating.

Zhirendong, 113 kya

In 2007, hominin remains were found in the Zhiren Cave at Zhirendong near the city of Chongzuo, Guangxi Zhuang Autonomous Region in Southeastern China. The fossils were discovered undisturbed in situ and included a partial mandible (anterior portion) and two isolated molars. The molars were fully erupted at the time of death and belong to two young adults; one molar and the mandible belonged to a single specimen. The fossils were solidly dated and assigned to the initial Late Pleistocene about 100 kya using U-series methodology.[48]

As in the Xujiayao and Daoxian remains and other Southeast Asian hominin fossils, Zhirendong exhibits its own unique merger of ancestral and derived anatomical traits. The mandible is small compared with other Late Pleistocene AMH mandibles and presents a number of features of modern humans such as a chin clearly projecting from the anterior surface of the midmandible junction. The anterior symphyseal (external faint ridge marking the line of junction of the two half of the mandible) is morphologically modern human. The two molars have crowns that possess diameters below the range of an East Asian modern human. They also exhibit five cusps and no midtrigonid crests (ridge that runs between main cusps), both are ancestral traits. Overall, the anatomy of the molars and mandible indicates an early modern human skull and dentition.

On the other hand, Zhirendong retains a number of archaic human traits. In general, the jaw is still robust. In fact, its massiveness exceeds the jaws of Middle Paleolithic modern humans. Also, the digastric fossa or oval depression on the internal surface of the body of the mandible for the attachment of the digastricus (a muscle that attaches to the internal side of the lower jaw for lowering the mandible) is barely apparent and the interdigastric (two halves) ridge, also on the internal side of the lower jaw, is absent. In addition, the lingual symphysis, or junction of the two halves of the lower jaw in the inside of the mouth, is an archaic trait reminiscent of anatomically Late Pleistocene archaic *sapiens*.

The faunal remains associated with the hominin fossils corroborate the age derived from the uranium series dating. These animal bone remains lacked Early Middle Pleistocene species such as *Gigantopithecus* (extinct giant ape), *Sinomastodon* (extinct elephant species from Late Miocene to the Early Pleistocene), and *Stegodon* while it contains a megafauna typical of the Late Pleistocene including elephant species such as *Elephas kiangnanensis* and *Elephas maximus*.

Thus, the Late Pleistocene hominins at Zhirendong argue for an early appearance of modern humans in Southeast Asia as well as coexistence and interbreeding with archaic human groups in the area.[49] As with the other two previously discussed Late Pleistocene hominins (Xujiayao and Daoxian), assuming an African origin for *sapiens*, it would require an earlier departure from Africa (prior to 120 kya) of AMHs. This poses no temporal conflict because according to the Out of Africa theory AMHs evolved about 200 kya in Africa, providing ample time for *sapiens* to reach the Far East. In addition, it would necessitate a process of absorption of regional late archaic humans into the East Asian breeding population and/or regional continuity with substantial gene flow to explain the complex and different (unique) combinations of ancestral and derived traits present in these three Late Pleistocene hominin sites described, among others, in East Asia. The cohabitation of archaic and moderns for over 50 ky in Southeast Asia would have accommodated the required time overlap. The most parsimonious mechanistic explanation for the diverse sets of anatomies seen in Late Pleistocene sites in East Asia is not a simple Out of Africa population replacement scenario. In fact, the traditional Out of Africa views are incompatible with the recent Southeast Asian findings from the last two decades. The various different combinations of ancestral and derived traits may indicate a situation in which several anatomically different populations were free to reproduce, generating a gamut of morphological amalgamations.

EN ROUTE TO ISLAND SOUTHEAST ASIA AND BEYOND

The traditional view pertaining to the colonization of Southeast Asia postulates that AMHs arrived from India approximately 70,000 ya.[51] The theory also stipulates that soon after, *sapiens* moved down from continental Southeast Asia through the Malay Peninsula, the Indonesian archipelago, and eventually reached Australia by 60 kya or earlier. The orthodoxy also states that this incursion was independent of the *erectus* migration out of Africa about 1.5 mya. Because *erectus* survived until 50 kya in the region, a temporal overlap of at least 20 ky occurred involving *erectus* and *sapiens*.

Yet, although recent findings throughout continental Southeast Asia point to a much older date of arrival from the Indian subcontinent approximately 120 kya, evidence of AMHs in ISEA prior to 60 ky is only supported by nonskeletal evidence.

Migrations Through Continental Southeast Asia Into Island Southeast Asia

Mainland Southeast Asia is a critical portion of the route connecting Africa to ISEA. Geographically, it fits into the great coastal arc of human dispersal from Africa to Australia. Thus, knowledge of migration routes through continental Southeast Asia is essential for a complete understanding of human evolution. The area represents a nexus between the Indian subcontinent and Near Oceania. This part of the arc has been a geographically intermediate zone for migrations between the last stages in the dispersal toward Australia.

Data from *sapiens* sites in continental Southeast Asia have been employed to pinpoint the putative migration paths from the Mainland to ISEA during the Late Pleistocene and Early Holocene. Three main passageways that are not mutually exclusive are considered. One model posits a central approach into what is today the Malay Peninsula by way of the Chao Phraya River Basin, a highly reticulated lattice of waterways of rivers and tributaries. The animal fossils discovered in this region and Indonesia support this putative approach. These water thoroughfares may have acted as causeways instead of barriers. Today, these waterways converge in a delta by the present-day coast and the Gulf of Thailand. Yet, at the time of Out of Africa II, during glaciations, this portion of Southeast Asia was fused with Indonesia forming the Sunda meta-continent. From present-day south Thailand, hominins could have traveled southeasterly into the rest of the Sunda meta-continent that included (in a west to east direction) Borneo, Java, Sumatra, Sulawesi, Bali, Lombok, Sumbawa, Flores, Sumba, Timor, the Alor archipelago, the Barat Daya Islands, and Tanimbar Islands. This central approach or migration route was, at the time, a fertile wide plane continuously well watered and replenished with sediment-rich monsoon rains from what is today northern Thailand. In addition, the mountain range to the west of the Chao Phraya Basin would have provided protection from inclement weather and vantage points to ambush megafauna during hunting. Also, the cordillera would have afforded different habitats with the corresponding diversity of food resources.[50] This type of favorable landscape would have allowed AMHs to transverse what is today Eastern Myanmar, Western Laos, and Thailand with little difficulty, supplying the migrants with plentiful natural resources necessary for survival. The rivers would have served as a source of ample freshwater, fish, and mollusks while the surrounding plains would have provided game.

A second coastal migration route would have taken humans through an easterly path along Vietnam. This path is supported by a number of recently discovered AMH archeological sites in Thailand, Vietnam, and Laos. For example, in Laos, at the Tam Pa Ling site, an AMH partial cranium and a number of teeth have been found. In Thailand (the Tham Wihan Naki site) and Vietnam (the Tham Kuyean site), dental specimens have been recovered exhibiting transitional characteristics of *erectus* and *sapiens*. Located in coastal Vietnam, Tham Kuyean corroborates a littoral route.

The third scenario envisions hominins traversing the region along the west coast of the Sunda meta-continent down the coast of South Asia and present-day Myanmar. This path represents a natural extension of the theorized coastal route that stretches from the Near East, along and around the Indian subcontinent into Southeast Asia. Unfortunately, archeological evidence for this potential passage may be currently under water and inaccessible due to elevated sea levels during the current interglacial epoch.

Homo sapiens in Sunda

The dates of *sapiens* sites in current-day Indonesia encompass a period of time that began about 60 ky later than the oldest continental sites in present-day China. The reason(s) for this temporal occupation gap are not apparent. The older settlement of Lida Ajer (63 kya), located in the Padang Highlands of western Sumatra, compared with 125 kya for Xujiayao in continental East Asia could be explained by unfavorable climatic conditions such as high sea level separating one or several Indonesian islands from the continent. It turns out that after a short period of time (few thousand years) in which waters were as high as today, sea levels started dropping about 120 kya and by 60 kya the sea was approximately only 50 m below the present (see Fig. 7.1, Chapter 7). Thus, it is not clear if the 60-ky gap of *sapiens* in Indonesia was the result of water barriers. Nevertheless, the early age of 63 kya for Lida Ajer in Sumatra underscores the need to reassess the timing of a fertile Out of Africa II that eventually led to the settlement of Indonesia by AMHs. Furthermore, the age estimated for Lida Ajer fits nicely with the time frame to allow the colonization of Australia by about 60 kya. Fig. 8.14 provides the location of the most important *sapiens* archeological sites within Sunda. The Lida Ajer, Niah, and Laili sites in Sumatra, Borneo, and Timor, respectively, are described below as case studies.

Lida Ajer

Lida Ajer located in the Padang Highlands of western Sumatra, was initially excavated by Eugène Dubois in the 1880s. At the time, Dubois discovered two human teeth and fossils from fauna. Since then, over the years, Dubois findings have been largely ignored due to controversies regarding the identity of the teeth and their age. Some critics argued that the dentition belongs to an ape species, such as orangutan, whereas others felt that they were from an early hominin and not *sapiens*. Yet, recently the teeth were dated between 73 and 63 kya.[51] This age

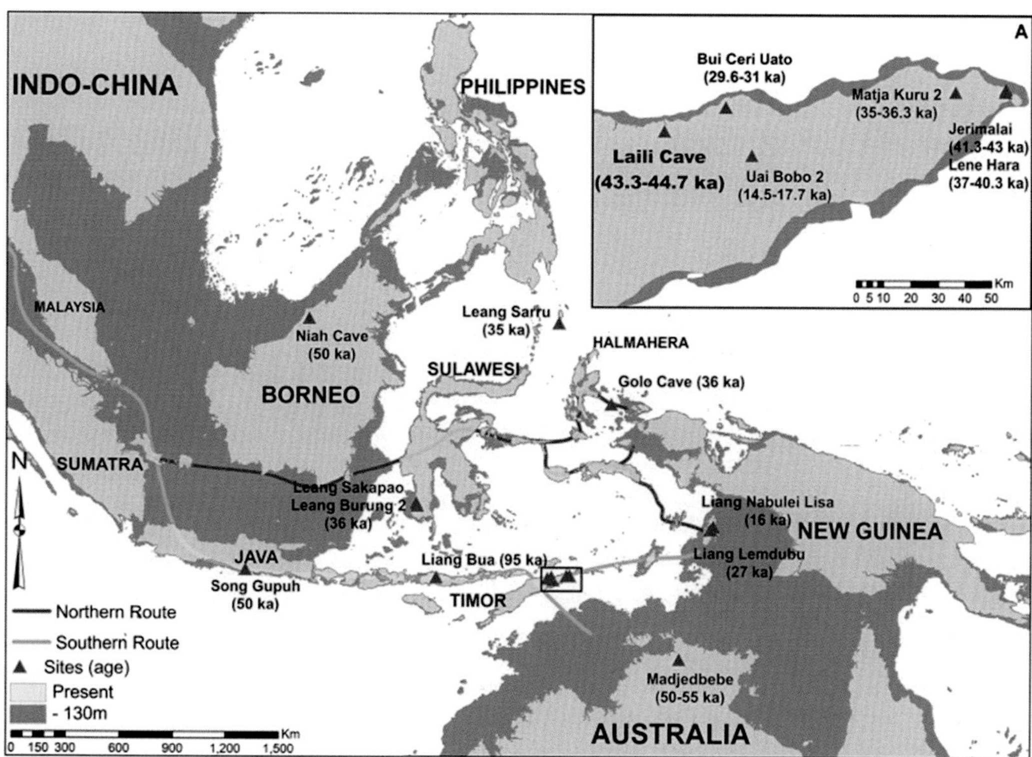

FIGURE 8.14 Location of anatomical modern human sites in Island Southeast Asia showing northern and southern migration routes into Australia. *From www.sciencedirect.com/science/article/pii/S0277379117302470.*

indicates that AMHs were in ISEA about 20 ky before previously thought. These early dates from Sumatra are also in line with the mtDNA data and overall paleoanthropological timeline from the region.[52] The age of Lida Ajer was generated using a number of reliable dating techniques, including luminescence, uranium series, and ESR on fossil-containing layers and mineral deposits in the cave. At the time, the habitat in the region was rainforest. Couple with the very early date of this site, finding *sapiens* in a rainforest was unexpected and suggested that these modern humans had already attained a level of technological sophistication that allowed them to survive in an environment with more limited resources, requiring hunting technologies and planning, as compared with the richness of river beds and sea shore lines.

The proposed age of Lida Ajer is also in line with a geographical and chronologically west-to-east dispersal of AMH into Australia considering that the Madjedbebe cave site in the Northern Territory is the proposed entrance into Australia and has been dated to 65 kya.[53] For a discussion of the putative regions of AMH incursions into Australia, see the *Routes into Australia* section below. In addition, the age provided by Lida Ajer lines up nicely with existing genetic evidence that suggests a date of about 50 ky or older for the arrival of AMHs to Australia[54] as well as favorable climatic conditions and low sea levels.

Although the 73- to 63-ky-old Lida Ajer site in Sumatra and the 65-ky-old Madjedbebe site in northern Australia push back the dates of modern humans in Sunda and Sahul, a time differential of at least 50 ky still exists from the earliest *sapiens* in continental Southeast Asia (Daoxian, 120 kya) and Indonesia (Lida Ajer, 73 kya). Considering that mainland Southeast Asia and Sumatra were part of the same continuous landmass (Sunda) and the speed hominins were traveling at the time, it is hard to imagine that such a trek would require such an extended amount of time. Therefore, it is likely that additional older sites exist and await discovery. Because several of the proposed routes from Southeast Asia to Australia are envisioned to have been coastal (see *Routes into Australia* section below), it is likely that some archeological sites are underwater due to current interglacial high sea levels.

Niah

The Niah limestone caves are located 17 km from the west coast of the island of Borneo in Malaysia close to the Batu Niah village. The site provides extensive stratified sequences of lithics and fauna dating back to about 46 kya and encompassing most of the occupation history since then. The caves were first excavated in the 1950s and 1960s, leading to the discovery in 1958 of an AMH cranium popularly known as Deep Skull.[55] This skull is currently the oldest AMH cranium from ISEA. The nickname derives from the fact that it was extracted from a deep, humid and hot pit known as "Hell Trench." The cranium was unearthed from an area in the cave where both a number of stone tools and charcoal were found. Recently, both the cranium and the charcoal were securely dated to about 45 and 39 kya by radiocarbon and uranium series, respectively.[56]

The cranium is the centerpiece of the Niah assemblage. It was put together from about 23 individual pieces, and unfortunately, it was painted over with shellac. Closure of vault sutures, considerable wear of molars and evidence of age-related degenerative changes at joints indicating possible osteoarthritis suggest that the individual was at least an adult or even in advanced age. Assessment of sexing criteria, such as weak muscle markings over the entire cranium and lack of cranial cresting, indicates that the remains likely belong to a female.[57]

The entire assemblage was also recently reexamined by Curnoe and collaborators[57] who concluded that even after accounting for sexual dimorphism and the expected less robust physique of a female specimen, the Niah individual was relatively delicate and gracile and in line with other Late Pleistocene *sapiens* from Southeast Asia. The Niah remains, specifically, exhibit a number of anatomical features typically associated with contemporary modern humans inhabiting ISEA, including a fully arched vault, maximum cranial width located high on the parietals, absence of a frontal sagittal crest, a vault that is relatively short and wide, mild postorbital constriction, a nonprominent glabella (flat area of bone between the eyebrows), poorly developed superciliary ridges and limited supramastoid crests, absence of occipital crests, minimal prognathism, and a flat midface. Furthermore, statistical analyses based on 18 anatomical characters indicated that the Niah skeleton partitions closest to Pleistocene East Asians.[57]

This new data contradict previous results that suggested morphological similarities between the Niah fossil and New Guinean, Australian and Tasmanian samples.[58] Instead, the Niah skeleton resembles more closely the current autochthonous population of Borneo with their diminutive anatomy and delicate build. Additional studies probing into the pre-Neolithic and Neolithic layers of the cave also suggest anatomical continuity through time between these two epochs.[59] Furthermore, these recent findings suggest that the original AMH population of Sunda goes back to about 45 kya in the Terminal Pleistocene and were not totally replaced by more recent migrations.

Laili

The Laili dig is part of a limestone cave system in the vicinity of the village of Laleia in northern East Timor or Timor-Leste at the eastern end of what is today the Indonesian archipelago. The site is 4.3 km from the northern coast of the island. It exhibits a continuous stratified sedimentary sequence that span approximately from 45 to 11 kya.[60] The Laili site is of particular importance because it currently represents the oldest human settlement in insular Southeast Asia and is geographically proximal to Australia and pertinent to the eventual colonization of the continent.

From Laili, over 28,000 flaked lithic artifacts have been recovered, the largest assemblage per unit volume in ISEA during the Late Pleistocene.[61] The collection includes 128 cores. Most of these tools were made from microcrystalline quartz. Only 42 were retouched flakes, and 17 flake fragments show evidence of retouching. Of particular note in Laili is the high incident of stone tools made by heat shattering the rocks into flakes and cores. The heat shattering technique produced a total of 6472 utensils, making up 23.8% of the entire collection. The significance of the preponderant use of the heat shattering method to produce tools in Laili is not apparent because it is not observed in other contemporaneous sites in ISEA. It may simply represent a local tradition developed and learned over generations in situ. These stone artifacts were found throughout the whole continuous stratified sequence from 45 to 11 kya.

A number of small fragments of human ribs and phalanges were identified in the different layers of the dig, yet these were too small on which to base any conclusion. Although rats of different sizes dominate the faunal remains, a highly diversified group of animal species was found at the site, including juvenile dog (*Canis familiaris*), fruit bats, insectivorous bats, birds, frogs, lizards, sea turtles, snakes, eels, mollusks, and crustaceans. It is likely that the diversified habitats in the form of coastal and river banks (the nearby Laleia River) contributed to a highly varied microfauna for AMHs to hunt. The absence of megafauna among these fossils signals lack of large terrestrial animals in Timor. During the Late Pleistocene, the region in the vicinity of the cave included marine, mangrove, riverine, grassland, and forested habitats. The fossil species recovered signal a combination of riverine, marine, and estuary conditions with a local environment made up of grassland, woodland, some forest as well as wetlands.

COLONIZATION OF AUSTRALIA

Routes into Australia

During the past several million years, global lower temperatures and glaciation periodically have resulted in reduced sea levels worldwide. These lower sea levels repeatedly exposed large areas of the surrounding continental shelf of Australia from New Guinea in the north through Tasmania in the south forming a single continuous landmass called Sahul. Yet, to reach this landmass, hominins needed to cross some 8 to 10 bodies of water or channels, depending on the potential route and the sea levels at the time of the dispersal.[62] It is generally thought that these barriers prevented early hominins, such as *erectus,* from reaching New Guinea and Australia. In fact, it is thought that the persistence of the final stretch of 90 km of deepwater separating the meta-continents of Sunda from Sahul was possible only as a result of the greater technological savvy of *sapiens*. This greater behavioral repertoire possibly allowed AMHs to build watercrafts for this extended voyage. Thus far the archeological evidence indicates that AMHs were the first hominins that colonized Sahul.

Two likely routes have been proposed for the colonization of Australia. In one, *sapiens* could have penetrated the meta-continent through a southern trajectory into the outspread Australian coastline.[63] This path would have taken humans from what is today the island of Sumatra, onward to Java, Flores, Timor, eventually moving into what is today northwest Australia in the area of present-day Wyndham/Derby in the present-day Kimberley region (Fig. 8.14). The other passage would have been more northerly, traveling from Sumatra to southern Borneo, crossing to the island Sulawesi and then Halmahera, continuing to New Guinea and then northern Australia in the region that is today the Gulf of Carpentaria/Cape York Peninsula.[60] Assuming a sea level of 150 m below the present level, both southern and northern routes would have included a most challenging crossing of 93, 69, 103, 98, or 87 km, depending on the specific islands used as stepping stones. It is noteworthy that the earliest AMH sites in Australia, Madjedbebe, Nauwalabila, and Malakunanja, date to about 65, 50, 50 kya, respectively, and are located in the Northern Territory in between the two putative entrance ways, substantiating the theorized locations of entry.

In these considerations of reaching Sahul, two important factors are (1) whether the crossing(s) were accidental or intentional; and (2) the number of migrants involved. It is likely that if the movement was unintentional, the number of travelers was small. It would not be expected for large number of people to drift in a log, floating vegetation, or a canoe, except for instance, if a tsunami thrusted a good portion of populations from villages to the shores of Australia. However, there are also the issues of survival as a population after arrival to the new land. Small populations always run the risk of not possessing a healthy gene pool that would lead to extinction due to the lack of genetic variability and evolutionary flexibility in the changing environment. Furthermore, there is the risk of population death resulting from being unable to sustain growth in a demanding habitat. The likelihood of these scenarios would be dependent on a number of factors, such as number of migrants, ratio of fertile males and females in the party, and age of members. It is interesting that in simulation studies, it has been assessed that although the survival of an isolated population depends on the size of the group, no population less than 60 individuals is immune from extinction and only rarely do they remain extant for more than 1000 years in isolation.[64] Based on these considerations, it is likely that the successful colonization of Australia by AMHs occurred by several successive trips and/or deliberate contacts between the pioneers and the source population. In both scenarios the probability of attaining a healthy gene pool and potential long-term reproductive success would be more probable.

Archeology of Australia

Archeological and anthropological data from Australia indicate that at least two forms of aborigines existed on the continent, the gracile and the robust.[65] Proponents of the candelabra theory, who advocate multiregional origins for modern human populations, suggest that the robust form is indicative of continuity with earlier *H. erectus* populations in Southeast Asia. Both gracile and robust remains have been recovered from fossils from the Pleistocene and early Holocene. The oldest human remains in Australia were found in Southeastern Australia in a dry lake known as Mungo, and they date back to at least 60 kya. The Mungo samples belong to the gracile type, suggesting that this type arrived prior to the robust type. These two distinct anatomical types have been linked to different continental penetrations. However, the anatomical parameters of the robust type are outside the indices of contemporary aboriginals. Considering the distance of the dry Mungo Lake, about 2700 km from the putative zone of initial incursion into northern Australia, it is likely that the first arrival by modern humans dates prior to the Mungo settlement by several thousand years.

The archeological evidence retrieved from the Australian continent is rather limited, and investigations involving aboriginal remains are difficult to perform. The discoveries, in general, suffer from poor descriptions and inaccessibility of discovered fossils and casts for further research. A policy of returning remains to the aboriginal groups (e.g., Mungo 1 and 3, Kow Swamp) tends to inhibit reexamination of fossils. Once fossils are returned to the aboriginal communities for reburial, exhumation for reexamination of remains by new technology, such as DNA sequencing, is very difficult. For example, the reburial of remains would prevent DNA sequencing with more advanced technology that minimizes the impact of degradation or contamination with contemporary DNA. Also, in recent times, some of the actual fossils have been locked up and are currently inaccessible to scientific inquiry. A good example of the situation involves the discovery in 1988 of a fossil of a child contemporary with Mungo 3. Tribal leaders blocked scientific studies of the specimen, and the fossil still remains in situ. While the wishes of aboriginal indigenous tribes should be respected, the loss of these specimens to reburial has prohibited a better understanding of the peopling of Australia

Mungo 1 and 3

The Mungo remains are made up of three main fossils: Mungo 1, 2, and 3. Mungo 1 and 3 are the best-known remains of the three and have been popularized as Mungo Lady and Mungo Man, respectively. The location of the finds is New South Wales in the Willandra Lakes region in Southeast Australia approximately 760 km west of Sydney. Today the location is a dry lake exhibiting three main sedimentary layers that started accumulating about 200 kya. The AMH remains were found in the middle layer that dates from 50 to 25 kya. Since then, as a result of glaciation and drier weather, the lake began to evaporate, increasingly becoming basic and salty and making the habitat unsuitable for most animals.

This midsediment layer was deposited during a wet epoch, and, at the time, the lake was home to fish and abundant megafauna, including giant wombats, marsupial tapirs, diprotodons or hippopotamus-sized marsupials most closely related to the wombat, various species of large kangaroos such as the meat-eating and the short-faced kangaroos, giant wallabies, and marsupial lions. Most of the Australian megafauna dramatically became extinct, and today most experts support the notion that the rapid extinction of the Australian megafauna resulted from hunting and burning of the landscape by *sapiens*.

Mungo 1 was discovered in 1969 embedded in a crescent-shaped sand elevation at the rim of the Mungo dry lake. It belonged to a female that died between 25 and 40 kya.[65] The fossil includes a cranium, teeth, and mandible. The integrity of the cranial and postcranial remains was compromised by two partial cremation events. The age of the individual at the time of death have been questioned based on the fact that the cranial sutures of the skull are all open, suggesting an early teen. Also, it seems that bones were broken prior to cremation. Her body was cremated twice and then covered generously with ocher (earth yellow/orange pigment made up of ferric oxide, clay, and sand). The use of ocher by ancient human populations during interments has been associated with ceremonial burial and suggests that *sapiens* practiced burial rituals at Mungo during the Late Pleistocene.

Mungo 1 is described as having a gracile build and a small vault. Yet, it exhibits a number of ancestral characters, such as prominent postorbital constriction (narrowing of the skull just behind the eye sockets) and decline of the frontal squame behind the orbital margin.

Mungo 3 was found in 1974 about 500 m away from Mungo 1 on the south end of the dry lake.[66] The body belonged to a middle-aged male who was intentionally laid on his back, with knees bent and his hands over the groin area with interlocked fingers (Fig. 8.15). Ocher was spread all over the corpse. As in the case of Mungo 1, these are signs that the corpse was ritually entombed, this time in a male-specific mode. Contrasting with the small size of Mungo 1, Mungo 3 was 196 cm (6 ft, 5 inches) in stature, relatively taller than contemporary aborigines.[67] Yet, in spite of its size, the skeleton exhibits an overall gracile anatomy. Although the reported age of Mungo 3 exhibits a wide range between 28 and 63 kya, most scholars consider the fossil to be approximately 40 ky old.

Anatomically, Mungo 3 is clearly an AMH.[68] Overall, it possesses modern features including a very thin cranial vault. Although its maximum cranial width is at the base of the cranium (a common feature among Australian aboriginal skulls) and the chin projection is unremarkable and the occipital bone (curved bone at the back base of the cranium) is thick, there is no evidence of a median frontal crest and the nuchal muscle insertion region is not prominent.

MtDNA analyses have shown that Mungo 3's genetic material was quite similar to present-day aboriginal sequences. Unfortunately, the results have been strongly challenged on the grounds of contamination with contemporary DNA.

Kow Swamp

This site is located in northern Victoria about 10 km southeast of Cohuna in the central Murray Valley in southeastern Australia. The dig is positioned at the edge of an irrigation

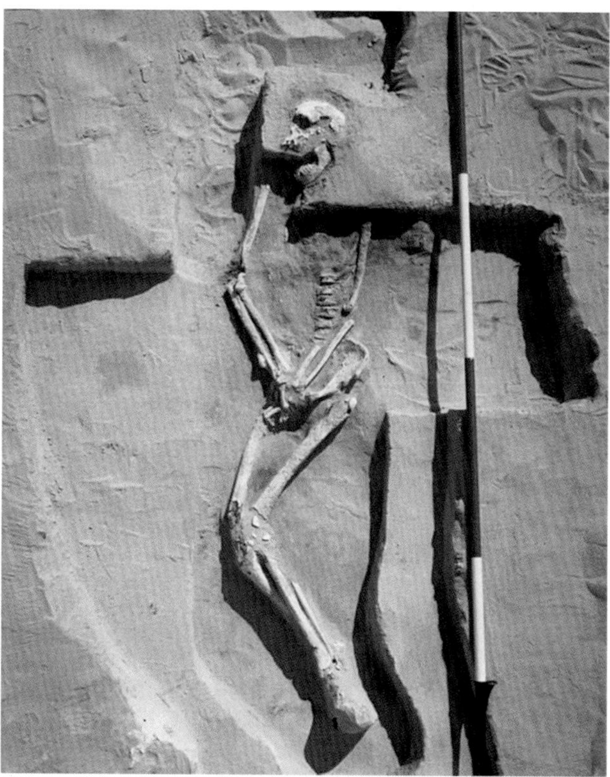

FIGURE 8.15 Mungo Man. *From https://www.theguardian.com/world/2014/feb/25/mungo-man-physical-reminder-need-for-indigenous-recognition.*

swamp. Alan Thorne carried out the excavations between 1968 and 1972. Since then, the remains have been returned to the aboriginal communities for reburial. Although the number of people recovered from the Kow Swamp site varies depending on the published sources, ranging from 22 to 40 individuals, it nevertheless represents the largest collection of characterized Terminal Pleistocene to Early Holocene remains in the world.

Significantly, among the fossils unearthed are infants, juveniles, adults, and midlife individuals of both sexes. The remains have been dated to a period of time between 13 and 9.5 ya. Curiously, the bodies were laid to rest in a number of positions. Some were extended on their backs. Others were buried on their sides while others were found in a flexed or bended position, facing up or on their sides. Still other individuals had their knees up and pressed onto their rib cage, while others were arranged covering their faces. One cremation was found. Some of the interments included artifacts clearly intended to accompany the deceased to the afterlife. Many of the goods accompanying the dead were of utilitarian nature. Among the items were animal teeth, quartz, shells, and ocher. Even a headband made of resin-glued kangaroo incisors was found wrapped around the head of a body. The significance of all of these positions, orientations, and differential distribution of offerings is not clear. Yet, it is possible that the different burial permutations are related to sex and/or status. If the various burial arrangements were related to status, it would suggest social stratification in this ancestral aboriginal population.

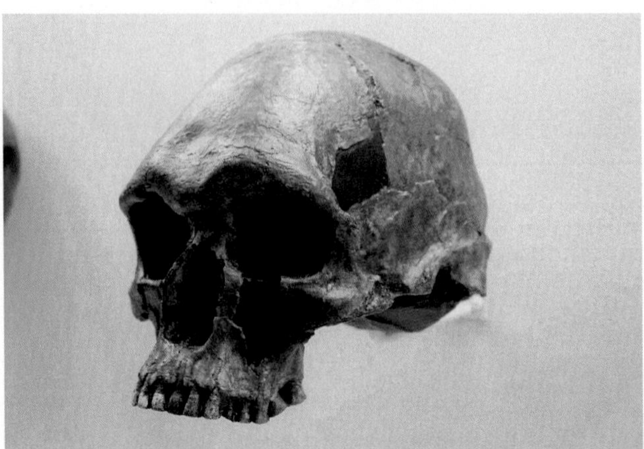

FIGURE 8.16 Kow Swamp skull. *From https://en.wikipedia.org/wiki/Kow_Swamp_Archaeological_Site#/media/File: Kow_Swamp1-Homo_sapiens.jpg.*

The Kow Swamp fossils are remarkable for several reasons. Unlike the gracile Lake Mungo specimens, the Kow Swamp bones are robust with thick cranial vaults. This assessment is rather unexpected considering that the Mungo samples date to 40 kya, about four times older than the Kow Swamp material. Specifically, the Kow Swamp population is characterized by having robust and large heads with thick bones, about 13 mm thick. Their faces were large, flat, and wide exhibiting some degree of facial prognathism (protruding face). In addition, they possess marked brow ridges, flat receding foreheads, and curvature behind the eye sockets. Their teeth and jaws are massive. All of these characteristics are ancestral traits and reminiscent of *erectus* (Fig. 8.16).

The population at Kow Swamp is not unique. Similar groups of robust individuals have been recovered from other Australian nearby locations, such as Cohuna and Talgai. Yet, the Kow Swamp fossils differ profoundly from the ones at Lake Mungo and Keilor. In general, as previously discussed, the Mungo specimens are gracile, and their skulls possess thinner bones and are void of marked brow ridges and massive jaws with large teeth and pronounced facial prognathism. Their faces are not flat and lack a tapered forehead. And yet, the 40 Kow Swamp individuals are usually classified by the orthodoxy as just an ancient population of Australian *H. sapiens*.

The Enigma of Younger *sapiens* Exhibiting More Archaic Features Than Older *sapiens*

Over the years, Alan Thorne has argued that the gracile and robust sets of fossils known as Mungo Lake and Kow Swamp, respectively, represent two independent and distinct source populations and migrations of *H. erectus* into Australia. According to Thorne, one population of *erectus*, the gracile group, dispersed from continental East Asia while the robust traveled from Java. This theory is not currently in favor by mainstream archeologists. Furthermore, the presence of a population of robust individuals dating to the Terminal Pleistocene in Kow Swamp is generally attributed to pre- or postmortem cranial deformations and distortions, such as intentional altering of the skull during life for the former and geological events for the latter. Genetic defects also have been invoked to explain the unusual fossils. Yet, in the absence of logical explanations, the reality of the presence of archaic features in a relatively young aboriginal Australian population, as compared with older samples exhibiting more modern human characters, still poses problems for the archeological community. As discussed above, repatriations and reburial of specimens have complicated recovering the fossils for further examination and testing.

There are several problems with the typical explanations of the unusual Kow Swamp individuals. The genetic defect hypothesis can be easily debunked because the robust prototype is not exhibited by a single individual but by a good number of individuals, about 40 specimens. It is highly unlikely that a sizeable group of people lived together suffering from the same condition. Even if neutral mutations are invoked as the cause of the anatomical phenotype, it is unlikely that such a non–wild-type allele or syndrome would generate such a diverse group of *ancestral* anatomical traits and become fixed in a population. Furthermore, there is no reason to expect that skull deformations, congenital, genetic, and/or environmental, should mimic a whole set of specific ancestral traits.

So, how did the Kow Swamp unexpected features came to be? Based on the well-documented diminutive *H. floresiensis* on the island of Flores, which lived as recently as 50 kya, is it possible that a 13-ky-old population exhibited a combination of archaic features in Australia? This is possible. *H. floresiensis* possesses several ancestral features and is considered by some specialists to be a miniaturized form of *erectus* resulting from living on a small island. Ironically, *floresiensis* was also thought by some scientists to be the result of Down syndrome. The genetic defect theory suffered a set back when a number of small individuals with the same characteristics were unearthed from the same site and from other locations in the same and different Indonesian islands. It seems that the case of Flores is the result of a well-documented evolutionary phenomenon observed in a number of species known as island dwarfism.

Also, the possibility of a syndrome generating such a unique combination of well-characterized ancestral characteristics is quite unlikely considering that such a condition has not been observed in other cases or reported in the medical literature or anywhere. Yet, the most challenging fact to explain is the occurrence of a number of known ancestral characters in a sizeable number of individuals that were part of a community. In other words, how traits such as a massive general anatomy, thick vault, large teeth and mandible, facial prognathism, marked brow ridges, flat receding foreheads, and curvature behind the eye sockets can be explained invoking geological forces, a phenomenon that tends to alter or destroy nondirectionally, aimlessly? This scenario is not likely. Geological disturbances do not create specific structures. Independent of whether one or two migrant populations of *sapiens* settled Australia, it is likely that the Kow Swamp population is the result of evolutionary processes, such as introgression and/or gene flow involving unique ancestral populations.

In addition, the archeological record on Southeast Asia is permeated by examples of cases of intermediate fossils with various combinations of ancestral and derived traits. This precedence makes interbreeding and gene flow among archaic and modern groups possible providing a mechanism to explain the mixture of characters found in Kow Swamp.

The robust Kow Swamp samples suggest that a paradigm change is required regarding the peopling of Australia. Yet, it seems that the conservative mainstream archeological community is ignoring the Kow Swamp fossils.

DNA Data From Australian Aborigines

A number of policies instituted by past European colonial governments of Australia have obscured the relationships and distribution of the native populations. This makes the assessment of their genetic history difficult. The enforced population relocation and child removal policies not only affected the cultural identity and connection of the aborigines to the land, but it also promoted mixing of the gene pools of the various tribal populations. In spite of these issues, recent genetic studies are beginning to provide useful information on the past of this remarkable and very ancient people.

Uniparental genetic marker systems (mtDNA and Y chromosome sequences), for example, have delineated some specific details on the time of arrival and spread of migrants within the continent (for a description of the various types of genetic markers, see Chapter 5, section on *DNA analysis*, and Chapter 10, section on *Three main types of DNA*). Precolonial ancient mtDNA analysis, which bypasses potential mixing of tribes by Europeans, also indicates that Australia possesses the deepest lineage of AMHs in the world.[69] The mtDNA and Y types that define the aborigines are ancient and basal in the phylogenetic tree, likely dating back to the original arrival of AMHs on the Australian part of Sahul or shortly after. In addition, the DNA evidence also indicates that, shortly after their entrance, the ancestors of the aboriginal Australians diverged from a single ancestral population and genetically differentiated as they migrated along the eastern and western coasts and then into the center of the continent.

The deep haplogroups in Australian aborigines suggest relative isolation subsequent to the initial settlement of the continent. Thus, the uniparental genetic data support, for the most part, the view that the technological developments seen later in Australia, including the backed-blade lithic industry, are in situ developments and, for the most part, not introduced by more recent migrations.

Furthermore, mtDNA- and Y-specific DNA studies indicate that 40–70 kya, from a single genetically heterogeneous wave or several independent penetrations Australia was settled. Recent studies using 111 mtDNA mitogenomes (entire mtDNA sequences) extracted from *historical* aboriginal Australian hair samples have delineated the entrance into Australia to a narrow window of time between 43 and 47 kya.[54] This age estimate was ascertained employing the time elapsed from the most recent common ancestor of aborigines to the sampled individuals using a molecular clock calibrated with mtDNA mutation rates from known populations (well-established rates). These genetic time estimates nicely correspond to the arrival date of 48.8 kya, traditionally thought to be based on archeological data within Sahul,[54] but are considerably younger (about 15 ky) than the 65 ky, new minimum age set by the northern Australian site of Madjedbebe for the arrival of humans in Australia.[53]

Based on assessments from mtDNA- and Y-specific DNA marker systems, contemporary Australian aborigines are genetically quite heterogeneous and, for the most part, are similar to residents of Southeast Asia. The observed genetic similarities between Australian and New Guinean aborigines also suggest a common source of migration(s) and/or admixture events subsequent to colonization.

Overall, the genetic evidence indicates that Australian aborigines and Eurasia share a common ancestor in sub-Saharan Africa, a major bottleneck event as *sapiens* exited Africa restricting genetic diversity and several admixture events with a number of archaic hominins such as Neanderthals and Denisovans. The split between Eurasian populations and the ancestors of the Australian aborigines came about soon after modern humans departed from Africa.

In addition, Australian aborigines exhibit gene flow from an unknown hominin only seen in the genetic imprint that it left behind. When and where exactly these introgression events occurred is not clear. But, the presence of this mysterious DNA and Denisovans' in the descendant of the original settlers of Australia indicates that these admixture events likely occurred when the original AMHs were in East Asia prior to penetrating Sahul because Australian aborigines received little foreign DNA after their arrival in Australia.

mtDNA

A number of specific mtDNA types signal the original settlement of Australia. For example, haplogroups N14, a descendant of the African L3, was brought into Australia during the original migration. It is thought that N14 originated in Asia around 71 kya.[70] Another haplogroup, M, possibly of South Asian origins from around 60 kya, also marks the entrance into Australia.[70] It is interesting that most of the basal (ancestral) lineages within mtDNA types are found in northern Australia. This is likely indicative of founding basal mtDNA types arriving in the north of Australia with the original populations of AMHs with some individuals remaining there while others moving south along the coast where the more derived lineages evolved. Furthermore, the basal separations between haplogroups O, S and N13, P and R, M16 and M42 likely reflect an early split in northern Australia around 50 kya.[54]

mtDNA markers have also provided information on the migration patterns within Australia, indicating that basal ancestral types geographically partitioned soon after arrival (Fig. 8.17). For example, mtDNA types P, S, and M42a predominate in eastern Australia while haplogroup O is totally absent. Similarly, within the P types there is a separation between Northeastern and South Australia. Parallel splitting patterns are seen among the other basal mtDNA types indicating strong geographical partitioning of populations soon after arrival. Limited dispersal, mixing, and gene flow occurred among tribes after separations and initial coastal migration southward. Altogether, the geographical partitioning of mtDNA types within Australia suggests that the migrants followed two simultaneous routes, an eastern coastal path southward carrying haplogroups P, S, M42a and a west coast dispersal southward with haplogroups O and R (Fig. 8.17). The presence of types O and S in the center of the continent suggests a meeting of the two waves in South Australia. The close dates of occupation sites throughout Australia, including the age of putative entrance locations in the north, indicate that the internal dispersal occurred rapidly probably a few thousand years after arrival.

Y Chromosome

The Y chromosome–specific mutations linked with the arrival of aboriginals to Australia include C4-M347 and S-P60. Just like the mtDNA types, C4-M347 and S-P60 are tied to the original circum-Indian dispersion. About 99% of the Australian tribal population possesses either the C4-M347 or S-P60 Y types. C4-M347 is exclusively Australian, and its parent major haplogroup C evolved in Asia approximately 53 kya.[71] C4-M347 is the most abundant Australian aboriginal Y type with continent-wide frequencies of around 60% in the tribal populations.[71] The other major native Y chromosome, S-P60, is found in 40% of the aborigines continent wide.[72] The abundance of these two predominant indigenous Y chromosome types varies within the Australian continent. For example, in South Australia the frequency of S-P60 is approximately 60%, whereas that of C4-M347 is around 40%.[73] It is not clear if

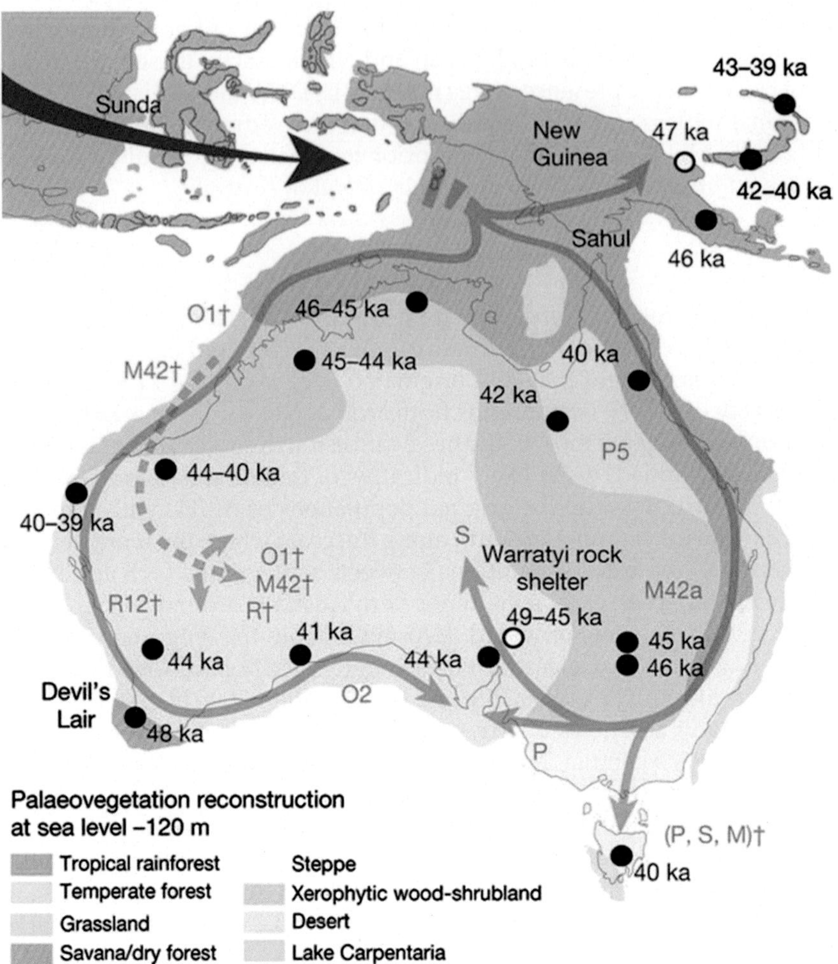

FIGURE 8.17 Potential migration routes within Australia based on mtDNA markers. Designations in brown indicate mtDNA types, whereas numbers in black followed by ka indicate age estimates. *From http://www.nature.com/articles/nature21416.*

these regional differences have resulted from genetic drift or independent migration waves and clinal expansions (e.g., migration of C4-M347 individuals as they moved southward into S-P60 territory). S-P60's parent haplogroup, S-P405, has its origins in Southeast Asia or New Guinea soon after its arrival to the region. It is not clear if S-P60 is of Australian origins or had its genesis in Southeast Asia or New Guinea like its parent haplogroup S-P405.

Genome-Wide Studies

Corroborating the Y and mtDNA results, recent genome-wide data have provided comparable ages for the out of Africa departure. It has been estimated that the aboriginal Australians and Papuans diverged from Eurasians about 51–72 kya.[74] Also, the genome-wide investigations support the coastal southern route from Africa to Southeast Asia with comparable arrival dates in Australia. These studies provide a date of around 50 kya for *sapiens'* initial incursion into the continent.[75] These investigations also indicate that subsequent to their arrival in Sahul, the populations of New Guinea and Australia started to diverge genetically approximately 25–40 kya, resulting in almost complete cessation of gene flow between the two populations.[74] As with the signals obtained from the Y chromosome and mtDNA uniparental markers, the genome-wide studies indicate a single founding Australian population, which differentiated and gave rise to all aborigines living today. According to the genome-wide data, this partition and continent-wide migration of groups took place about 10–32 kya.

Yet, although Australian aboriginal groups have been living in relative isolation since the initial colonization, some recent gene flow has been detected from India dating back to around 4230 years ago, which coincides with the introduction of certain technological advances, including new tool manufacturing traditions, food processing methods, and the introduction of the Australian dog or dingo (*Canis lupus dingo*).[76] These data demonstrate that Australia did not remain in complete isolation since the initial settlement about 60,000 ya.

SOUTHEAST ASIAN EVIDENCE FOR THE OUT OF AFRICA VERSUS CANDELABRA THEORIES

The original colonization of the Sahul meta-continent that included present-day Australia, Tasmania, New Guinea, Seram, and nearby islands is intimately tied to the origins of AMHs in Africa and their dispersion worldwide. Both proponents of the Out of Africa and candelabra models acknowledge that these first settlers represent the descendants of the oldest *H. sapiens* populations.[77] The ideas on how the settlement of Australia occurred are very much influenced by whether *sapiens* originated in various locations in situ from *H. erectus* or all humans have a single recent common origin in Africa.

The Out of Africa theory posits that AMHs exited East Africa and migrated to the Near East, likely across the Horn of Africa, about 120–70 kya. From there, these humans followed a coastal route along Southwest Asia (currently south Iran and Pakistan) and continued in a circum-Indian trajectory eventually reaching Southeast Asia and then Near Oceania (then Sahul). The data support a single wave but not necessarily belonging to a single genetic lineage of fully realized *H. sapiens*. The degree of anatomical variation observed in contemporary and fossil Australian aborigines is consistent with a single incursion followed by a subsequent geographically widespread dispersal within the continent. This diffusion is thought to have occurred fast, over the course of approximately 10,000 years.

The candelabra hypothesis, on the other hand, views the colonization of Australia as taking place by way of two separate migrations of hominins. One group was descended from Indonesian *H. erectus* and a second from Chinese *H. erectus*. According to this hypothesis, present-day aborigines derive from these two *erectus* types.[23] Whether Australian, Indonesian, or continental East Asian *sapiens* populations are considered, the main premise of the candelabra theory is *continuity* of anatomical traits and DNA in hominin populations. Continuity in the sense that *H. sapiens* is postulated to have evolved regionally and independently in situ from *H. erectus* with limited fresh gene flow from Africa. As discussed earlier in this chapter, a number of traits exhibit what appears to be persistence of ancestral traits beginning with *erectus*. Although the Out of Africa hypothesis is supported by an overwhelming amount of archeological and genetic evidence, certain continuities of anatomical and genetic characters in Southeast Asian hominins are difficult to explain without some degree of genetic continuum and lineage persistence.

A number of important scientific articles published since the 1990s have attempted to shed light on the validity of the Out of Africa and multiregional candelabra theories. The Out of Africa thesis (see Chapter 4) is currently the most widely accepted by the scientific community. It is supported by substantial archeological and genetic evidence. One of the strongest lines of evidence supporting the Out of Africa scenario involves mtDNA data. In 1987, in a landmark study of contemporary worldwide individuals, it was established that all the identified mtDNA human lineages derive from a common female ancestor that lived in sub-Saharan Africa approximately 140–290 kya.[78] This type of result strongly suggested that AMHs originated and migrated out of sub-Saharan Africa and populated the rest of the world replacing and not interbreeding with previous autochthonous populations. Similarly, for the most part, analyses of Y chromosome DNA, are in accordance with a recent out of Africa dispersal.

Continuity of Anatomical Traits

In terms of morphological traits, the data generally point to some degree of continuity of characters, although the studies differ in the number and the specific diagnostic traits exhibiting continuity. It has been pointed out that these morphological continuities are particularly observed in the hominins of East Asia as well as of Indonesia and Australia. In recent years, for example, Southeast Asia teeth sequences from Laos, Thailand, and Vietnam (e.g., Tham Wihan Naki in Thailand; Tham Kuyean in Vietnam) suggest transition stages between *erectus* and AMHs, possibly resulting from introgression.[50]

In Indonesia, for example, specific continuous sequences of characters are claimed to exist in the earliest fossils from Sangiran in Java through Ngandong from Solo River in Java and in prehistoric and recent Australian aborigines. In East Asia, the continuity of traits begins with the Lantian and Peking Man, persists in Dali and most recently are seen in Liujiang and present-day Chinese populations. Some of these populations were discussed in preceding sections of this chapter.

A number of studies have explored the continuity of anatomical features in extinct and extant Southeast Asia populations over the years. All of these studies share a comparative approach in which specific characteristics are followed in evolutionary time, from older specimens to younger ones. For example, in 1991, 17 nonmetric traits were examined, with the conclusion that 8 of the 17 traits exhibited continuity from Sangiran to modern Australians.[79] In a more recent study, it was concluded that only four characteristics exhibit continuity between Indonesia and Australian aborigines.[80] The observed facial prognathism similarities between Kow Swamp samples and contemporary Australians were noticed to be particularly striking. These anatomical features are described in previous sections of this chapter. In continental East Asia, a study involving 10 characters argued for continuity within the region.[81] Subsequent examination of the specimens claimed that only one combination of three features supported continuity. For example, a unique extreme form of shovel-shaped incisors has been reported to show continuity among ancient and modern East Asian populations.[82]

The traditional critics of the candelabra theory insist that no single anatomical human feature is characteristic to any given region. Some experts indicate that proponents of continuity exaggerate their claims because most of the features represent ancestral polymorphisms present in common ancestors.[83]

Continuity of Genetic Traits

In a comprehensive genetic assessment of 12,127 contemporary male individuals representing 163 East Asian populations, three Y chromosome mutations (YAP+, M89T, and M130T) were associated with a fourth mutation M168T, which originated in Africa approximately 35–89 kya. In this study, no input from East Asia was detected and the dispersal eastward points to a relative recent migration. These results are congruent with a recent out of Africa migration.[84] Yet, not all the genetic data available is consistent with a recent out of Africa migration of AMHs giving rise to all human populations worldwide. For example, in a study of 15 DNA sequences on the X chromosome, it was found that the DNA sites date back to various times and not a single time as expected by the single origin theory. Some of these DNA sites date back to about 2.0 mya when *erectus* originated in Africa.[85] These results also suggest that a population separation occurred at the time of *erectus* and not a recent split of *sapiens* as predicted by the Out of Africa theory. Furthermore, some of the DNA sequences exhibit more diversity in Asia, not in Africa as expected by the Out of Africa model. Greater variability is generally considered diagnostic of regions of origin because diversity is generated as a function of time; older locations possess more diversity.

A number of other sites distributed throughout the *sapiens* DNA also indicate inconsistencies with the notion that AMHs recently and uniquely originated in Africa in the absence of introgression. These studies point to old separation times of human populations dating to the origins of *erectus* in Africa.[86] Included among these deep-rooted diagnostic DNA locations are a ribonucleotide reductase (RRM2) gene, the microtubule-associated protein tau (MAPT) gene, an acetylneuraminic acid hydroxylase (CMP-N) gene, and a pyruvate dehydrogenase (PDHA1) gene. All of these studies are contradictory to the seminal mtDNA data that support a recent single Out of Africa dispersal. In addition, the notion of a recent single dispersion Out of Africa was challenged when 34 unique migration events involving Africa and Eurasia were detected by statistical analyses of genes. Nineteen of the thirty-four DNA sequences indicated that the gene flow occurred approximately 1.46 mya, three are correlated with the *erectus* migration out of Africa about 2.0 mya, only five were related to a recent expansion Out of Africa, and seven took place at intermediate time periods. In this study, the single migration from Africa with replacement was soundly negated with greater than 99% confidence.[87]

CONCLUSION

Based on the information presented here, it is clear that there is much work that remains to be done in the field of hominin evolution in Southeast Asia, a region traditionally neglected since the seminal discoveries of Eugène Dubois in the late 19th century. Despite the fact that there is still much we do not know and may never know, a number of common themes on hominin migration into and through Southeast Asia are evident. One of these is the continuity of certain traits seen among hominins during the past 2 million years. Some experts have interpreted this persistence of anatomical features as evidence in support of the multiregional origins of modern humans. Furthermore, in recent years, a plethora of fossil findings indicate that the presence of modern humans in Southeast Asia is more ancient than previously thought. These very old ages of human remains are not compatible with the traditional view that modern humans exited Africa recently, about 60 kya, and then populated the rest of the world. The early presence of modern humans in Southeast Asia also has been construed by some investigators to mean that modern humans evolved independently in Southeast Asia. Also, of particular interest are the gracile and robust hominin types from the Mungo and Kow Swamp sites, respectively, in Australia. The Mungo fossils date to 40 kya, about four times older than the Kow Swamp material and yet Kow Swamp exhibits a number of ancestral traits reminiscent of *erectus*, while the Mungo fossils look more like typical modern *sapiens*. Overall, the recent archeological and genetic data from Southeast Asia and Australia underscore a number of interesting incongruencies that are generally ignored. If additional new findings continue to support these new data, it may require a number of fundamental paradigm shifts in how we view the peopling of Southeast Asia and the entire world.

References

1. Shipman P. *The man who found the missing link*. New York: Simon & Schuster; 2001.
2. Dennell R. *The palaeolithic settlement of Asia*. Cambridge World Archaeology, Cambridge: Cambridge University Press, UK; 2009.
3. Huang W, Ciochon R, Yumin G. Early *Homo* and associated artifacts from Asia. *Nature* 1995;**378**:275–8.
4. van den Bergh GD, Kaifu Y, Kurniawan I, et al. *Homo floresiensis*-like fossils from the early middle pleistocene of Flores. *Nature* 2016;**534**:245–8.
5. Hong A, Chun-Ru L, Andrew PR. An updated age for the Xujiayao hominin from the Nihewan Basin, North China: implications for middle pleistocene human evolution in east Asia. *J Hum Evol* 2017;**106**:54–65.
6. Dennell RW. Dispersal and colonisation, long and short chronologies: how continuous is the Early Pleistocene record for hominids outside East Africa? *J Hum Evol* 2003;**45**:421–40.
7. Dambricourt AM. The first Indo-French prehistorical mission in Siwaliks and the discovery of anthropic activities at 2.6 million years. *C R Palevol* 2016;**15**:281–94.
8. Ciochon R. The earliest Asians yet. *Nat Hist* 1995;**104**:50–4.
9. Han F, Bahain J-J, Deng C. The earliest evidence of hominid settlement in China: combined electron spin resonance and uranium series (ESR/U-series) dating of mammalian fossil teeth from Longgupo cave. *Quat Int* 2017;**434**:75–83.
10. Schwartz JH, Tattersall I. Whose teeth? *Nature* 1996;**381**:201–2.
11. Mysterious cave stirring hopes for new clues to man's early evolution. *Bull Chinese Acad Sci* 2008;**22**:216–8.
12. Rowold DJ, Luis JR, Terreros MC, et al. Mitochondrial DNA gene flow indicates preferred usage of the Levant Corridor over the Horn of Africa passageway. *J Hum Genet* 2007;**52**:436–47.
13. Mayr E. Taxonomic categories in fossil hominids. *Cold Spring Harb Symp Quant Biol* 1950;**15**:109–18.
14. Dubois E. *Pithecanthropus erectus-Eine Menschenaehnliche Ubergangsform aus Java*. Landesdruckerei: Batavia; 1894.
15. Swisher III CC, Curtis GH, Lewin R. *Java man: how two geologists changed our understanding of human evolution*. Chicago: University of Chicago Press; 2000.
16. Ao H, Liuc C-R, Roberts AP. An updated age for the Xujiayao hominin from the Nihewan Basin, North China: implications for Middle Pleistocene human evolution in East Asia Author links open overlay panel. *J Hum Evol* May 2017;**106**:54–65.
17. Huffman FO, Zaim Y, Kappelman J, et al. Relocation of the 1936 Mojokerto skull discovery site near Perning, East Java. *J Hum Evol* 2006;**50**:431–51.
18. Coqueugniot H, Hublin JJ, Veillon F, et al. Early brain growth in *Homo erectus* and implications for cognitive ability. *Nature* 2004;**431**:299–302.
19. O'Connell CA, DeSilva JM. Mojokerto revisited: evidence for an intermediate pattern of brain growth in *Homo erectus*. *J Hum Evol* 2013;**65**:156–61.
20. Balzeau A, Grimaud-Hervé D, Jacob T. Internal cranial features of the Mojokerto child fossil. *J Hum Evol* 2005;**48**:535–53.
21. DeSilva JM, Lesnik JJ. Brain size at birth throughout human evolution: a new method for estimating neonatal brain size in hominins. *J Hum Evol* 2008;**55**:1064–74.

22. Sartono. Early man in Java: Pithecanthropus skull VII, a male specimen of *Pithecanthropus erectus*. *Proc Kon Ned Akad Wetenschappen Ser B* 1971;**74**:185–94.
23. Thorne AG, Wolpoff MH. Regional continuity in Australian Pleistocene hominid evolution. *Am J Phys Anthropol* 1981;**55**:337–49.
24. Shen G, Gao X, Gao B, et al. Age of Zhoukoudian *Homo erectus* determined with (26)Al/(10)Be burial dating. *Nature* 2009;**458**:198–200.
25. Li P, Chien F, Ma H-H, et al. Preliminary study on the age of Yuanmou man by palaeomagnetic technique. *Scientia Sinica* 1977;**20**:645–64.
26. Liu T, Ding M. A tentative chronological correlation of early fossil horizons in China with loess-deep sea records. *Acta Anthropologica Sinica* 1984;**3**:93–101.
27. Zhu RX, Potts R, Pan YX, et al. Early evidence of the genus *Homo* in East Asia. *J Hum Evol* 2008;**55**:1075–85.
28. Argue D, Groves CP, Lee M, et al. The affinities of *Homo floresiensis* based on phylogenetic analyses of cranial, dental, and postcranial characters. *J Hum Evol* 2017;**107**:107–33.
29. Brumm A, van den Bergh GD, Storey M, et al. Age and context of the oldest known hominin fossils from Flores. *Nature* 2016;**534**:249–53.
30. van den Bergh GD, Li B, Brumm A, et al. Earliest hominin occupation of Sulawesi, Indonesia. *Nature* 2016;**529**:208–11.
31. Mounier A, Marchal F, Condemi S. Is *Homo heidelbergensis* a distinct species? New insight on the Mauer mandible. *J Hum Evol* 2009;**56**:219–46.
32. Wu X, Athreya S. A description of the geological context, discrete traits, and linear morphometrics of the Middle Pleistocene hominin from Dali, Shaanxi Province,. *China Am J Phys Anthropol* 2013;**150**:141–57.
33. Rosenberg KR, Wu X, Chapter 3: a River runs through it: modern human origins in East Asia. In: Smith FH, editor. *The origins of modern humans biology reconsidered*. 2nd ed. London, UK: Wiley; 2013.
34. Wu R-K. The reconstruction of the fossil human skull from Jinniushan, Yinkou, Liaoning Province and its main features. *Acta Anthropologica Sinica* 1988;**7**:97–101.
35. Rosenberg KR, Zuné L, Ruff BR. Body size, body proportions, and encephalization in a Middle Pleistocene archaic human from northern China. *PNAS* 2006;**103**:3552–6.
36. Lu Z, Meldrum DJ, Huang Y, He J, Sarmiento EE. The Jinniushan hominin pedal skeleton from the late Middle Pleistocene of China. *Homo* 2011;**62**:389–401.
37. Wang C-C, Li H. Inferring human history in East Asia from Y chromosomes. *Investig Genet* 2013;**4**:1–10.
38. Jin L, Su B. Natives or immigrants: modern human origin in East Asia. *Nat Rev Genet* 2000;**200**(1):126–33.
39. Karafet T, Xu L, Du R, et al. Paternal population history of East Asia: sources, patterns, and microevolutionary processes. *Am J Hum Genet* 2001;**69**:615–28.
40. Tu H, Shen GJ, Li HX, et al. 26Al/10Be burial dating of Xujiayao-Houjiayao site in Nihewan Basin, northern China. *PLoS One* 2015;**10**:e0118315.
41. Chen T, Yang Q, Hu Y, et al. ESR dating on the stratigraphy of Ynxian *Homo erectus*, Hubei, China. *Acta Anthropol Sinica* 1996;**15**:114–8.
42. Callaway E. Oldest *Homo sapiens* fossil claim rewrites our species' history. *Nature* 2017. https://doi.org/10.1038/nature.2017.22114.
43. Posth C, Wißing C, Kitagawa K, et al. Deeply divergent archaic mitochondrial genome provides lower time boundary for African gene flow into Neanderthals. *Nat Commun* 2017;**8**:16046. https://doi.org/10.1038/ncomms16046.
44. Xing S, Martinon-Torres M, Bermúdez de Castro JM, et al. Hominin teeth from the early late pleistocene site of Xujiayao. Northern China. *Am J Phys Anthropol* 2015;**156**:224–40.
45. Wu X-J, Xing S, Trinkaus E. An enlarged parietal foramen in the late archaic xujiayao 11 neurocranium from Northern China, and rare anomalies among Pleistocene Homo. *PLoS One* 2013;**8**:e59587.
46. Bae CJ. The late Middle Pleistocene hominin fossil record of eastern Asia: synthesis and review. *Yearb Phys Anthropol* 2010;**53**:75–93.
47. Liu W, Martinón-Torres M, Cai Y-J, et al. The earliest unequivocally modern humans in southern China. *Nature* 2015;**526**:696–9.
48. Jin CZ, et al. The *Homo sapiens* Cave hominin site of Mulan Mountain, Jiangzhou District, Chongzhou, Guanxi with emphasis on its age. *Chin Sci Bull* 2009;**54**:3848–56.
49. Liua W, Jina C-Z, Zhang Y-Q. Human remains from Zhirendong, South China, and modern human emergence in East Asia. *PNAS* 2010;**107**:19201–6.
50. Marwick B. Biogeography of Middle Pleistocene hominins in mainland Southeast Asia: a review of current evidence. *Quat Int* 2009;**202**:51–8.

51. Westaway KE, Louys J, Awe RD. An early modern human presence in Sumatra 73,000-63,000 years ago. *Nature* 2017;**548**:322–5.
52. Fu Q, Mittnik A, Johnson PLF, et al. A revised timescale for human evolution based on ancient mitochondrial genomes. *Curr Biol* 2013;**23**:553–9.
53. Clarkson C, Jacobs Z, Pardoe C, et al. Human occupation of northern Australia by 65,000 years ago. *Nature* 2017;**547**:306–10.
54. Tobler R, Rohrlach A, Soubrier J. Aboriginal mitogenomes reveal 50,000 years of regionalism in Australia. *Nature* 2017;**544**:180–4.
55. Reynolds T, Barker G. Reconstructing Late Pleistocene climates, landscapes, and human activities in Northern Borneo from excavations in the Niah Caves. In: Kaifu Y, et al., editor. *Emergence and diversity of modern human behavior in paleolithic Asia*. Texas, USA: Texas A&M University Press; 2015.
56. Hunt C, Barker G. Missing links, cultural modernity and the dead: anatomically modern humans in the Great Cave of Niah (Sarawak, Borneo). In: Dennell R, Porr M, editors. *Southern Asia, Australia, and the search for human origins*. Cambridge, UK: Cambridge University Press; 2014.
57. Curnoe D, Datan I, Taçon PSC, et al. Deep skull from Niah Cave and the Pleistocene peopling of Southeast Asia. *Front Ecol Evol* 2016;**4**:1–17.
58. Matsumura H, Yoneda M, Dodo Y, et al. Terminal Pleistocene human skeleton from Hang Cho Cave, northern Vietnam: implications for the biological affinities of Hoabinhian people. *Anthropol Sci* 2008;**116**:201–17.
59. Krigbaum J, Manser J. The West Mouth burial series from Niah Cave: past and present. In: Majid Z, editor. *The Perak man and other prehistoric skeletons of Malaysia*. Pulau Penang: Universiti sains Malaysia; 2005.
60. Hawkins S, O'Connor S, Maloney TR. Oldest human occupation of Wallacea at Laili Cave, Timor-Leste, shows broad-spectrum foraging responses to late Pleistocene environments. *Quat Sci Rev* 2017;**171**:58–72.
61. Balme J, O'Connor S. Early modern humans in Island Southeast Asia and Sahul: adaptive and creative societies with simple lithic industries. In: Dennell R, Porr M, editors. *East of Africa: Southern Asia, Australia and human origins*. Cambridge, UK: Cambridge University Press; 2014.
62. Davidson I. The colonization of Australia and its adjacent islands and the evolution of modern cognition. *Curr Anthropol* 2010;**51**:S177–89.
63. Kealy S, Louys J, O'Connor S. Reconstructing palaeogeography and inter-island visibility in the Wallacean Archipelago during the likely period of Sahul colonisation, 65–45,000 years ago. *Archaeol Prospect* 2017;**24**:259–72.
64. Moore JH. 2001 Evaluating five models of colonization. *Am Anthropol* 2001;**103**:395–408.
65. Durband A, Rayner DRT, Westaway M. A new test of the sex of the Lake Mungo 3 skeleton. *Archaeol Oceania* 2009;**44**:77–83.
66. Thorne A, Curnoe D. Sex and significance of Lake Mungo 3: reply to Brown Australian Pleistocene variation and the sex of Lake Mungo 3. *J Hum Evol* 2000;**39**:587–600.
67. Bowler JM, Jones R, Allen H, et al. Pleistocene human remains from Australia: a living site and human cremation from Lake Mungo, Western New South Wales. *World Archaeol* 1970;**2**:39–60.
68. Barbetti M, Allen H. Prehistoric man at Lake Mungo, Australia, by 32,000 years BP. *Nature* 1972;**240**:46–8.
69. Adcock GJ, Dennis ES, Eastea S, et al. Mitochondrial DNA sequences in ancient Australians: implications for modern human origins. *Proc Nat Acad Sci USA* 2001;**98**:537–42.
70. Pierson MJ, Martinez-Arias R, Holland BR, et al. Deciphering past human population movements in Oceania: probably optimal trees of 127 mtDNA genomes. *Mol Biol Evol* 2006;**23**:1966–75.
71. Zhong H, Shi H, Qi X-B, et al. Global distribution of Y-chromosome haplogroup C reveals the prehistoric migration routes of African exodus and early settlement in East Asia. *J Hum Genet* 2010;**55**:428–35.
72. Scheinfeldt L, Friedlaender F, Friedlaender J, et al. Unexpected NRY chromosome variation in Northern Island Melanesia. *Mol Biol Evol* 2006;**23**:1628–41.
73. Hudjashov G, Kivisild T, Underhill PA, et al. Revealing the prehistoric settlement of Australia by Y chromosome and mtDNA analysis. *Proc Nat Acad Sci USA* 2007;**104**:8726–30.
74. Malaspinas A-S, Westaway MC, Muller C, et al. A genomic history of Aboriginal Australia. *Nature* 2016;**538**:207–14.
75. O'Connell JF, Allen J. The process, biotic impact, and global implications of the human colonization of Sahul about 47,000 years ago. *J Archaeol Sci* 2015;**56**:73–84.
76. Redd AJ, Roberts-Thomson J, Karafet T, et al. Gene Flow from the Indian subcontinent to Australia: evidence from the Y chromosome. *Curr Biol* 2002;**12**:673–7.
77. Wolpoff MH, Spuhler JN, Smith FH, et al. Modern human origins. *Science* 1988;**241**:772–4.
78. Cann RL, Stoneking M, Wilson AC, Allan C. Mitochondrial DNA and human evolution. *Nature* 1987;**325**:31–6.

79. Kramer A. Modern human origins in Australasia: replacement or evolution? *Am J Phys Anthropol* 1991;**86**:455–73.
80. Habgood PJ. A morphometric investigation into the origins of anatomically modern humans. In: *British Archaeological Reports, International Series 1176. Oxford, England.* 2003.
81. Wu X. The evolution of humankind in China. *Acta Anthropologica Sinica* 1990;**9**:312–21.
82. Frayer DW, Wolpoff MH, Thorne AG, et al. Theories of modern human origins: the paleontological test. *Am Anthropol* 1993;**95**:14–50.
83. Sanchez-Quinto F, Lalueza-Fox C. Almost 20 years of Neanderthal palaeogenetics: adaptation, admixture, diversity, demography and extinction. *Phil. Trans. R. Soc B* 2015;**370**. https://doi.org/10.1098/rstb.2013.0374.
84. Ke Y, Su B, Song X, et al. African origin of modern humans in East Asia: a tale of 12,000 Y chromosomes. *Science* 2001;**292**:1151–3.
85. Hammer MF, Garrigan D, Wood E, et al. Heterogeneous patterns of variation among multiple human x-linked Loci: the possible role of diversity-reducing selection in non-africans. *Genetics* 2004;**167**:1841–53.
86. Kim H, Satta Y. Population genetic analysis of the N-acylsphingosine amidohydrolase gene associated with mental activity in humans. *Genetics* 2008;**178**:1505–15.
87. Templeton AR. Haplotype trees and modern human origins. *Yearb Phys Anthropol* 2005;**48**:33–59.

CHAPTER

9

The Austronesian Expansion

The need of expansion is as genuine an instinct in man as the need in a plant for the light, or the need in man himself for going upright. The love of liberty is simply the instinct in man for expansion. ***Matthew Arnold***[1]

SUMMARY

The Austronesian Expansion represents the largest geographic dispersal ever undertaken by humans. It's reach expands from Madagascar to the west, Easter Island to the east, New Zealand to the south, and the Hawaiian islands to the north. This expansion is characterized by the Austronesian language, one of the sixth largest language families in the world, which now has over 386 million speakers of various sublanguages across the globe. Linguists have been able to follow this language dispersal along with cultural artifacts known as the Lapita Cultural Context (LCC). The patterns seen in the archeological record of the LCC align with Austronesian language acquisition. In addition, various genetic data, such as genomic-wide studies, Y chromosomal and mitochondrial DNA, have been used with the tools of population genetics to try to determine the migration patterns and interactions of the people of the Islands of Southeast Asia (ISEA). These tools have provided some conflicting, but mostly corroborative, data on this diaspora and the interactions of the people of ISEA. Most of the data point to a Taiwanese homeland and Taiwanese aboriginal people that left from there 5–6 thousand years ago (kya), moving south, carrying their culture, language, and genes into the remote regions of Oceania.

It has been postulated that this migration by Taiwanese tribal farmers initiated a dispersal throughout ISEA, first into the Philippines and Melanesia (mixing with Papuan speaking groups), then arriving in the Tonga and Samoan archipelagos, about 2.8 kya. Genetic and archeological evidence suggests that the initial Austronesian migrants "paused" in western Polynesia for approximately 500–1000 years, possibly allowing for the development of technology needed to sail further into the eastern Pacific. Once they attained the means to travel further, they reached the Cook Islands, the Hawaiian archipelago, French Polynesia, New Zealand, and finally Easter Island, about 1.2 kya. This initial dispersal into the Philippines, and ensuing island hopping into the Pacific Ocean, was likely driven by population growth and the need for more land. Thus, what most likely started as an agrarian acquisition of land

Ancestral DNA, Human Origins, and Migrations
https://doi.org/10.1016/B978-0-12-804124-6.00009-4

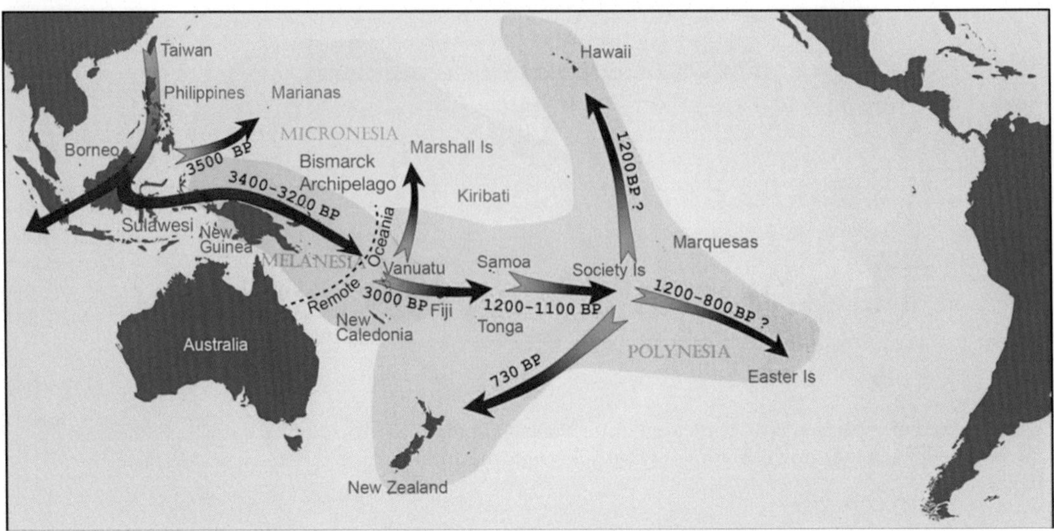

FIGURE 9.1 Region of the Austronesian Expansion. *Figure taken from E.A. Matisoo-Smith, Tracking Austronesian expansion into the Pacific via the paper mulberry plant, PNAS 112, 2015,13432–13433.*

in Mainland Southeast Asia resulted in the peopling of two-thirds of the circumference of the world, from the island of Madagascar in East Africa to Easter Island.

The Austronesian path through each of these island groups is mirrored by the residual aspects of their culture, language, and DNA. Each can be used as an indicator of the degree, nature, and location of interactions between these sea voyagers and the indigenous populations, if present.

INTRODUCTION

The Austronesian expansion Fig. 9.1 represents a good example of how linguistics, ethnobotany, cultural anthropology, molecular biology, and population genetics have been employed to assess human migration. Recent data have given us a deeper understanding of this human migration, but the traces of it were evident from an early point in history.

Captain James Cook and his traveling companions were one of the first to note portions of the migration patterns through Island South East Asia (ISEA). On his last voyage through the South Pacific, just before his death in 1779, Cook speculated on the origin of Polynesians, explaining they were descendants of Malaysia or from islands of the southwest Pacific. Similarly, a naturalist, Johaan Rheinhold Forster, on James Cook's second voyage (1772–75) recognized the linguistic and cultural unity of the islands of the central pacific and speculated that the similarities between populations reflected a short time since their dispersal. Forster correctly proposed a long-distance migration of people from ISEA to Polynesia as the ultimate source of Polynesian languages. This appeared to be the only explanation for the striking difference in phenotype that he observed between the peoples of the central Pacific and those of the intervening region. In 1832 the French explorer Jules-Sebastien-Cesar Dumont d'Urville, like Prichard, believed that the Pacific Islanders could not possibly be descendants of the dark-skinned individuals living in New Guinea. In addition, he divided the region into three groups: (1) Polynesia, the many islands of remote Oceania, (2) Melanesia, the diverse islands off the coast of Australia, and (3) Micronesia, the 2000-plus small islands of northwestern Oceania. The idea that the South Pacific was composed of two phenotypically distinct and unrelated people (the Melanesians and Polynesians) persisted until the 1900s.

By the early- to mid-20th century, anthropologists, biologists, and linguists began to examine the evolutionary relationships between the peoples of the Pacific.[2] Initially the linguists found two languages that showed no common features. These were the Papuan language of New Guinea and the Malayo-Polynesian or Austronesian languages of other Oceanian populations. The Austronesian languages are very diverse, are spoken by over 386 million people, contain about seven to eight dialects, and are thought to have arisen from the original Australoid settlers (indigenous people of Southeast Asia [SEA] and western Melanesia) more than 30 kya.[3] The Papuan languages are unique and do not appear to form one linguistic family as do the Austronesian languages, which have spread over 16,000 km across the Pacific.[4]

These languages represent the binary divide in the migration patterns of Southeast Asia. The Papuan branch denotes the first "Out of Africa" modern human dispersal. The settlement of the South Pacific Ocean by anatomically modern humans (AMHs) represents the final major migratory event of the global expansion that started in Africa. Human remains from archeological sites in the Philippines and Malaysia (Callao Cave and Niah Cave, respectively) suggest that SEA was first populated by AMH 50–80 kya. This migration followed the southern coastal route through SEA into Indonesia, reaching New Guinea and the Bismarck Archipelago around 33 kya, and eventually the Solomon Islands approximately 29 kya. The descendants of this initial exodus from Africa speak Papuan and are referred to as Melanesians. In addition, genome-wide studies concluded that all East Asians and SEAs originated from a single wave "Out of Africa" through a southern coastal route.

However, genetics and other concurring fields also show a second dispersal, the Austronesian Expansion, which started about 5–6 kya from SEA. It has been postulated that this migration by Taiwanese tribal farmers initiated a dispersal throughout ISEA, first into the Philippines, Melanesia (mixing with Papuan speaking groups),[5] then arriving in the Tonga and Samoan archipelagos, about 2.8 kya.[6] This second migration introduced the Austronesian language and the Lapita Cultural Complex (LCC) to ISEA. The LCC is represented by a variety of artistic works and tools used by people of this region. While the term Austronesian refers specifically to language and not to people or cultural traits, it has been used loosely to refer to all people and cultures that speak a dialect of the language. Because the current literature uses the term loosely, this chapter will follow that trend, using Austronesian to refer to both people and culture.

Because the Austronesian Expansion is so heavily characterized by language, linguistic data have been very effective in tracing the remnants of this migration. While there are many Austronesian speakers, each subgroup has their own individual language. To study relationships between languages, linguists create language trees and phylogenies to determine pathways of interaction between the subgroups. They use databases of words, such as the Austronesian Comparative Dictionary, to see the differences and changes to/in words, similar to how geneticists analyze DNA. It should be noted that DNA has a set mode of inheritance, whereas language can be more variable. Linguists use statistical modeling to see which language pair best fits with others and which is likely to be the root of the tree.

The spread of the Austronesian language coincided with the dispersal of the LCC. First and heavily noted in Near Oceania, specifically the Bismarck Archipelago, LCC is most often characterized by its elaborate, dentate-stamped, red-slipped pottery in various shapes; although, other artifacts also constitute the LCC assemblage.[7] These include adzes/axes, shell ornaments, tattooing, and bark cloth made from the paper mulberry tree. These artifacts were typically found at large coastal sites on small island beaches, which provided good launching points for their designer's ocean adventures. These tools suggested the Austronesian explorers were agriculturalists, but fish and poultry bones also found in the middens, or trash deposits, of the sites suggest a mixed foraging economy.[8] Archeological findings suggest that anything they would need to survive they brought with them in their outrigger canoes.

WHO WERE THE AUSTRONESIANS?

Determining the Homeland

Since the 1990s, investigations have been persistent in their efforts to determine the genesis of the Austronesian people. Each of the fields invested (linguistics, genetics, and anthropology) in this endeavor promoted a different initial homeland for these sea voyagers. However, as more data were collected, most of the fields have concluded Taiwan as the homeland for the Austronesian Expansion.

Before the conclusion by most investigators that Taiwan was the homeland for expansion, many scholars teetered between South China and Taiwan as the birthplace for the Austronesian Expansion. However, even with a South China origin, Taiwan was tied heavily into the narrative, usually being the first stop in the migration. Many of the excavated archeological sites provide links between southern China and Taiwan, suggesting the first settlers of Taiwan originated from the South China coast. There are several theories about Taiwanese origins; one is that Taiwanese aboriginals came from the Miao hill tribes (Hmong) or Kweichow aborigines from southern China, ancestors of the Yue ̈h people. This theory is partially based on the writings of Shen Ying (2300 ya) as interpreted by Chan Chi-yun, a Chinese historian. The documents assert that the Kweichow moved to the coast of China then crossed the Taiwan Strait, bringing their customs to Formosa (the ancient name for Taiwan). Several cultural practices of the Kweichow (including couvade) are still practiced by several Taiwanese aboriginal tribes.[9]

Other theories, based on phenotypic similarities, cultural similarities, and genetic studies, describe a different story. One states the Taiwanese aboriginals are descendants of people that occupied the Ryukyu Islands, a chain of islands off the southeastern coast of Japan that stretch from Kyushu (the most southwestern island of Japan) to Taiwan Fig. 9.2. The theory suggests that these people left Japan and moved down the island chain eventually finding their way to Taiwan 12 kya. An alternative theory suggests that Malayans followed warm ocean currents from the south eventually landing in Taiwan. These Malayans colonized the southern part of the island and expanded north, mixing with preexisting people that originated in Japan.[9]

Yet another study proposed that Taiwanese aborigines are genetically related to the Daic speakers that cover a substantial region of East Asia and SEA.[10] These coastal agriculturalists are considered to be the original inhabitants of China's southeast coast, and their origin can be traced to 20 kya. Today, the Daic is the second largest ethnic group in China, after the Han, and the population has strong presence in Thailand, Laos, Vietnam, Myanmar, and India. It is

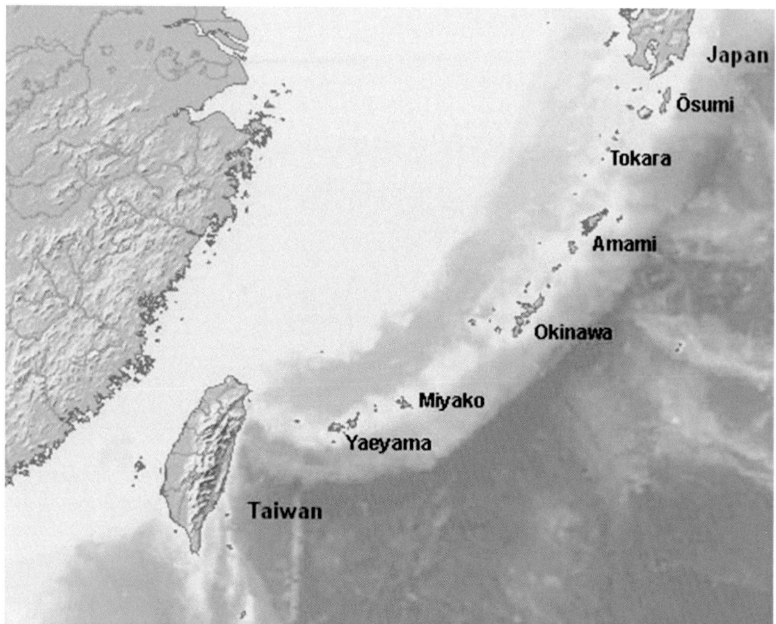

FIGURE 9.2 Location of the Ryukyu Islands in Japan. One theory suggests that Japan was the original source of contemporary Taiwanese aboriginals. The migrants moved south from Japan eventually reaching Taiwan.

likely they are the first inhabitants of Taiwan, bringing the Tapenkeng (TPK) culture and the foundation of the LCC. There are clear continuations of artifacts in the archaeological assemblages of the pre-TPK time period along coastal Southeastern China and the northern shores of Taiwan. Shell mound sites, stone bark cloth beaters, and coarse cord-marked pottery are seen in sites on both sides of the South China Sea between 6 and 4 kya. The middle Neolithic/Transition period, which started around 2 kya, displayed a continuation of the earlier TPK culture with fine cord-marked red-slipped pottery, polished stone knives, axes, sickles, and adzes found at village sites. These artifacts show evidence of fishing and hunting with some sustenance horticulture (meaning no surplus or animals).[11]

Archeology is not the only field that supports a Daic settlement of Taiwan. Geneticists have found a 9-bp deletion in mtDNA, which is unique to Polynesia (almost fixed in some regions) and occurs at lower frequency in Chinese populations and thought to have occurred 35–40 kya. Melton et al.[12] found 41.5% of the Taiwanese population contained the 9-bp deletion, including 47.4% and 46.6% of the Bunan and Paiwan aboriginal populations, respectively. This so called HSV1 haplotype also referred to as the "Polynesian motif" suggests two possible migration routes of the 9-bp deletion in east and SEA. Both proposed routes originated in China, with one leading to the Pacific Islands via Taiwan and the other to SEA and possibly the Nicobar Islands. Along both routes, a decrease in HVSI diversity of the mtDNA haplotypes is observed outside of Polynesia. The "Polynesian motif (16217T/C, 16247A/G, and 16261C/T)" and the 16140T/C, 16266C/A, or C/G polymorphisms appear specific to each migration route.[13]

Another mtDNA study from the skeletal remains of an 8-kyo male (Liangdao man) found on Liang Island of the Matsu archipelago, approximately 180 km from Taiwan Fig. 9.3, also align the Daic settlement narrative. The mtDNA analysis showed the Liangdao man to have haplotype E,

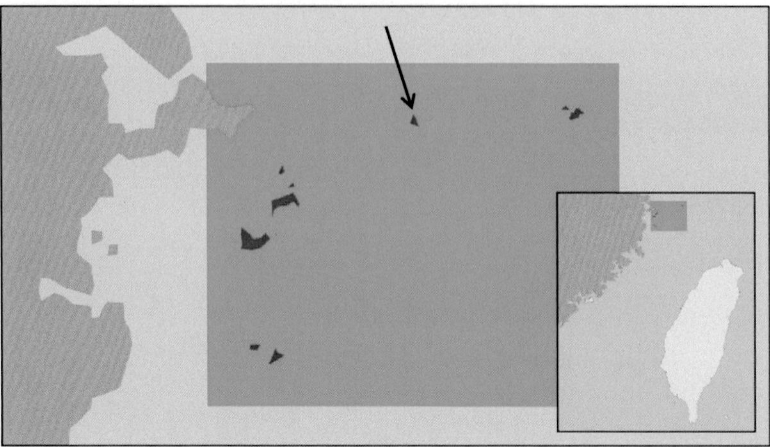

FIGURE 9.3 Figure shows the region containing the Matsu archipelago (the *arrow indicates Liang Island*). The insert shows the Matsu archipelago and its location relative to Taiwan (yellow). *From https://upload.wikimedia.org/wikipedia/commons/thumb/1/1b/Taiwan_ROC_political_division_map_Lienchiang_County.svg/800px-Taiwan_ROC_political_division_map_Lienchiang_County.svg.png.*

a sequence that is found in high frequency in Taiwan aborigines and also distributed in Maritime SEA and the Mariana Islands. Haplogroup E is a subclade of haplogroup M9 found along coastal China, Central China, and Tibet, but M9 has never been detected in Taiwan aborigines. This finding provided a link between southern China and Taiwan. Furthermore, the E lineages are not found in 84 populations in China, including the Austroasiatic, Daic, or Tibeto-Burman speakers, but they are prevalent among extant Austronesian speakers from the Philippines to Madagascar and east to the Bismarck Archipelago (the findings suggest that haplotype M9 converted to E and then individuals carrying E spread from Taiwan throughout Oceania).[14]

The mtDNA haplotype E data suggest the original descendants of Taiwan were from South China but also points out the differentiation between the first settlers of Taiwan and those that executed the Austronesian Expansion. Another genome-wide study indicated that Austronesians possess more affinity to tribal Taiwanese than to Melanesians or Mainland SEA populations. In addition, the data provided evidence for a mainland genetic component in western ISEA.[15] These data confirm previous data and suggest that Oceanic Austronesians' immediate ancestors are not mainland Chinese populations, like the Daic, but instead aborigines from Taiwan. The results suggest that insular SEA may have received gene flow from Taiwan, as well as from SEA mainlanders.

The Taiwanese aborigines were further proven to be the Austronesian descendants with the discovery that a subgroup of Y chromosome haplogroup O3, specifically O3a2, is widely distributed throughout the Islands of the SEA, Indonesians, and Polynesians.[16] The Dai groups, in particular, were missing the O3a2 haplogroup. More recently, a specific genetic relationship involving the O3a2c*-P164 subhaplogroup was detected between the Ami and Polynesian populations.[10] O3a2c*-P164 is found at very low levels and in only some Mainland East Asian populations. The Daic populations examined lacked O3a2c*-P164. These findings established a direct genetic link between a specific Taiwanese tribe and Polynesian groups previously undetected due to the minimal resolution of O3-derived Y chromosomes afforded by previous studies.

Linguistic data support the Daic-Taiwan and Taiwan-Austronesian connections. The origin of the Austronesian languages is considered by most linguists to have originated in Taiwan and then spread into Oceania. The language is thought to have begun with proto-Austronesian speakers spreading from early centers of rice cultivation in Central and South China, expanding to coastal China and across the Taiwan Straits. Some linguists have reported an ancient link between Austronesian languages and the Thai–Kadai language group (Kra–Dai, Daic, and Kadai) that may have arisen in Neolithic rice-cultivating communities in southern China, south of the Yangtze River, around 7–6 kya. Although controversial, there are records of southern Chinese mainlanders (the Yue ¨h) who may have spoken Austronesian languages, but Bellwood points out that evidence for this is weak.[17] Overall, there is little evidence to suggest an ultimate geographic origin for a pre-Austronesian language group, since today no Austronesian languages are spoken in South China. Likewise, Gray et al.[18] conducted some linguistic analyses of the Austronesian family, Taiwan or Formosan languages, and Old Chinese and Buyang, two ancient tongues spoken in China. They determined Formosan to be the base of the tree with deeper roots to Old Chinese and Buyang. These data overlap with the data collected before these linguistic studies, supporting the archeological record and genetic landscape of ISEA.

Taiwanese Aboriginals

Taiwan was inhabited by humans during the Paleolithic, long before the arrival of the current aboriginal populations. The earliest evidence for the first archaic *Homo* from Taiwan was a lower jawbone pulled from fisherman nets 25 km off the southwest coast of Taiwan in 2015. The site of the fossil was the Penghu Channel, which contains a series of ancient underwater islands. The Penghu jaw and teeth resemble a partial skull of *Homo erectus* from Longtan Cave, Hexian, mainland China. Because of seawater contamination and the lack of sedimentary deposits, it was difficult to date the fossil directly; however, it was found with an extinct species of hyena that suggests the specimen came from a human in the past 200–400 ky, suggesting that *H. erectus* or another type of human persisted late in this region. There is some speculation that the fossil may be a close relative of Neanderthals or a remnant of a Denisovan, although the fossil lacks some morphological features of these early humans[19] (see Chapter 6).

The early *Homo sapiens* inhabitants of Taiwan crossed the South China Sea, and/or the Taiwan Strait land bridge, as early as 27 kya, when sea levels were 140 m lower than today. Paleolithic sites in Taiwan have identified a stone pebble tool and flake industry and three cranial fragments and a molar tooth found at Chouqu and Gangzilin, in Tainan.[20]

The first settlement of Taiwan by AMH is thought to have occurred as a result of agriculture, although it occurred much later than the agricultural revolution associated with the Near East. The Asian agricultural revolution prompted a move from coastal Southeastern China to Taiwan, approximately 5–6 kya, as farmers expanded their domains within China and beyond to ISEA.[21] Evidence from ancient botanical remains indicate that rice was domesticated in southern China, as well as in the middle and the lower Yangtze River basins about 5 kya;[22] millets started in eastern Mongolia around 8 kya and soybeans in Northern China approximately 5 kya. The appearance of hand adzes used for land clearing in the early assemblages align with the botanical evidence of agriculture in coastal southeastern China.[11] It was these agricultural developments that are thought to have spurred first the settlement of Taiwan by the Daic and later the Austronesian Expansion.[23]

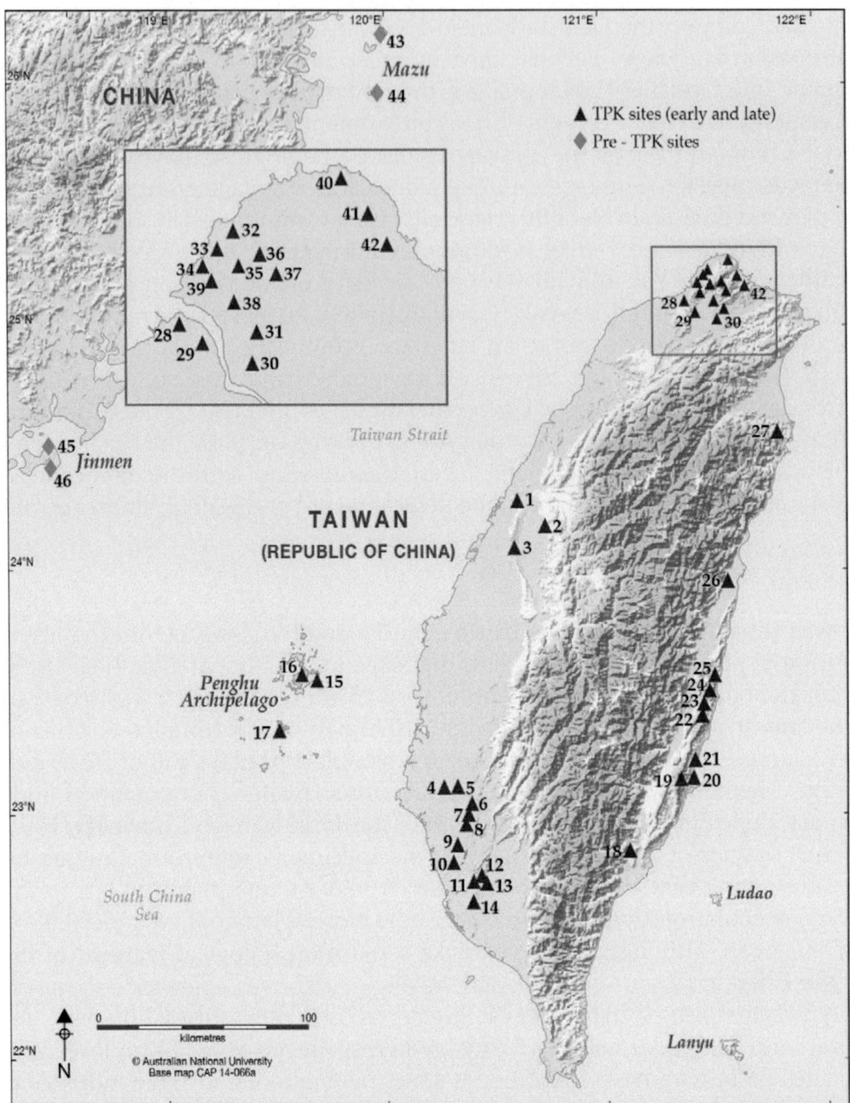

FIGURE 9.4 This shows the sites of the pre-Tapenkeng (TPK) culture, seen on the coast of southeastern China and the early and late TPK sites on Taiwan. Pre-TPK and TPK sites in the Jinmen and Mazu Islands, the Penghu archipelago, and Taiwan. (1) Niutmatou; (2) Huilaili; (3) Niupu; (4) Nanguanli; (5) Nanguanlidong; (6) Dachangqiao; (7) Qijia; (8) Bajia; (9) Xinyuan; (10) Gangkoulun; (11) Kongzhai; (12) Fudeyemiao; (13) Liuhe; (14) Fengbitou; (15) Guoye A; (16) Beiliao; (17) Liyushan; (18) Beinan; (19) Donghebei; (20) Xingang; (21) Zhitian II; (22) Changguang; (23) Chengzipu; (24) Zhenbing III; (25) Gangkou; (26) Yuemei II; (27) Xincheng: (28) Xiagudapu; (29) Dabenkeng (Tapenkeng); (30) Yuanshan; (31) Zhishanyan; (32) Xiaguirou II; (33) Yamuku; (34) Zhuangcuo; (35) Gezishan; (36) Linzijie; (37) Xiapidao; (38) Shuiduiwei; (39) Lanweipu; (40) Fuji; (41) Guizishan; (42) Wanlijiatou; (43) Daowei I; (44) Zhipinglong; (45) Jinguishan; (46) Fuguodun. *Taken from Hung H-c, Carson MT. Foragers, Fishers and Farmers: Origins of the Taiwanese Neolithic.* Antiquity *2014;88:1115–31; https://www.academia.edu/14545636/Foragers_Fishers_and_Farmers_Origins_of_the_Taiwan_Neolithic.*

FIGURE 9.5 Red slip pottery from Taiwan. The red color comes from aa combination of clay mixed with water and crushed red ocher that is used before firing. *Taken from https://www.taiwanteacrafts.com/wp-content/uploads/2016/05/Larger-Xi-Shi-Red-Clay-Teapot-Angled-View-1.jpg.*

These migrants brought their culture, TPK or Dabenkeng, with them. As previously stated, the TPK series shows continuations of the pre-TPK culture present on the coast of southeastern China, particularly the Pearl River Delta, and the early and late TPK sites in Taiwan Fig. 9.4. Taiwan's TPK sites see a change from small sites with short occupation in early TPK to more permanent village settlements. This change correlates with a reversal from a primarily coastal foraging economy with possible horticulture to an increased reliance on rice and foxtail millet farming with supplemental fishing and hunting. The Transition period continued this change, with a heavy reliance on agriculture shown in the increase in stone harvesting knives, sickles, and polished adzes at sites during this time period. The Xuntangpu site in northwestern Taiwan, in particular, epitomizes the Transition period and is possibly one of the earliest sites of farming.[11]

The pottery of the TPK series evolved from coarse cord-marked pottery to fine cord-marked pottery with a red slip Fig. 9.5. Red slip consists of water and crushed red ocher that is applied to the surface of the pottery before firing. Both the cord-marked and red-slip elements tie into the greater LCC narrative.[11,12] The LCC Taiwanese Corded Ware, excavated at Fengpitou (southwest Taiwan), TPK (north Taiwan), and other sites in Central Taiwan,[23,24] and predating 4.5 kya, is similar to Corded Ware of South China and north Indochina, dated to around 9 kya.[12,25] Lungshanoid pottery dated at around 4.5–2.5 kya is predominantly found on the west coast of Taiwan and is thought to have derived from Mainland China. These studies have suggested that multiple exposures to Mainland Chinese culture occurred in Taiwan, with the Corded Ware culture the most Austronesian, and found throughout the island of Taiwan.

The LCC is indicated by elaborately decorated red-slipped pottery, adzes/axes, shell ornaments, tattooing, and bark cloth. All the aspects of Lapita culture can be traced back to the

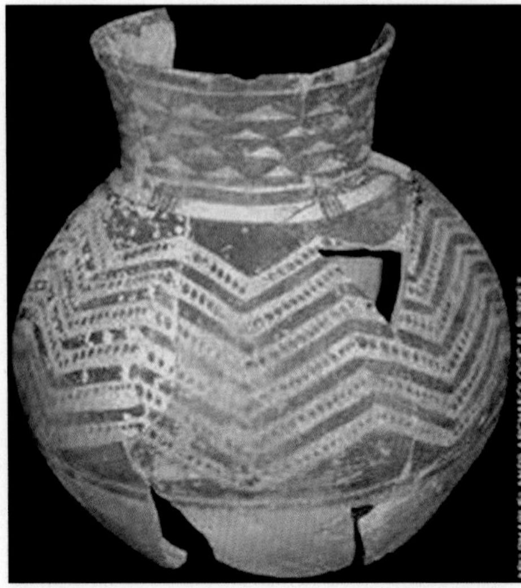

FIGURE 9.6 Figure shows a vase exhibiting Lapita designs. These designs are unique to the Lapita culture and their pottery. The discovery of the pottery in various regions of the South Pacific were useful for tracing the movement of individuals that bought, sold, or made the pottery. *From http://exploreoceania.weebly.com/melansian-lapita-art-and-poetry.html.*

FIGURE 9.7 The Lapita face. *From Carson MT, Hung H-c, Summerhayes G, Bellwood P. The pottery trail from Southeast Asia to remote Oceania.* J Isl Coastal Archaeol. *January 1, 2013;8(1):17–36. https://doi.org/10.1080/15564894.2012.726941.*

Transition period in the Tapenkeng culture of early Taiwan.[11,26] Specifically, the Lapita pottery is characterized by geometric dentate-stamped themes Figs. 9.6 and 9.7, which suggests a connection between the Taiwanese aborigines and the people of Oceania.[27] The finding of Taiwanese-mined nephrite within the Austronesian domain could also be interpreted as signs of ancestral kinship or even trade. Other similarities include a societal system based

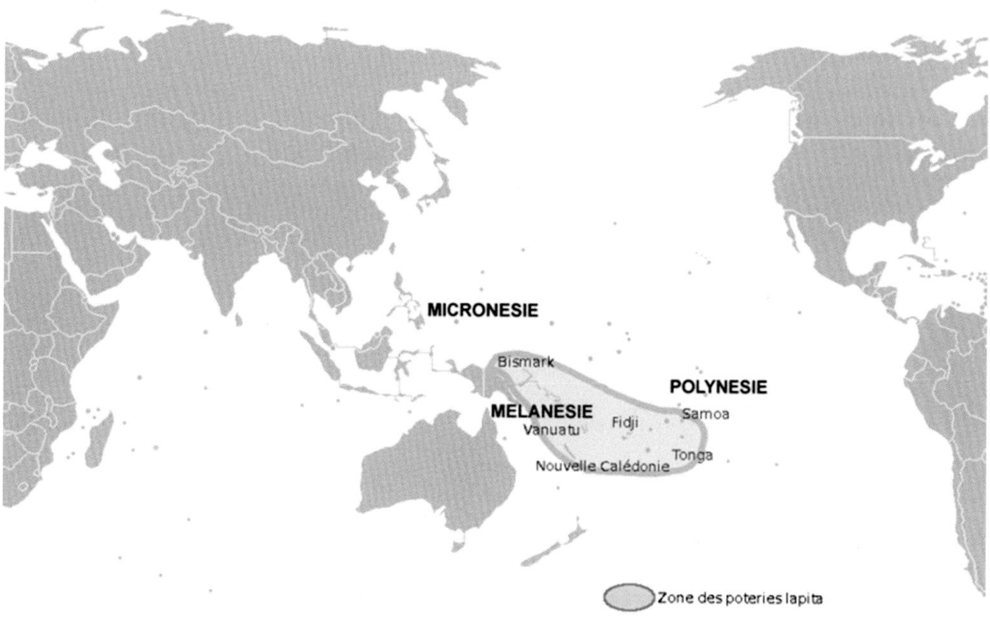

FIGURE 9.8 Figure shows the region where Lapita pottery has been discovered and thought to be related to the Austronesian Expansion. *From https://upload.wikimedia.org/wikipedia/commons/thumb/c/cb/Carte_lapita.png/800px-Carte_lapita.png.*

on agriculture and trade, a government by patrilineal chiefdoms, practicing pantheistic religions, the Austronesian-wide use of outrigger sailing canoes with shared designs, slit drums, and hula-type dancing. Of course, although these similarities suggest a Taiwanese origin, some could have been derived from Mainland SEA populations such as the Daic.[27]

However, until recently, the archeological record did not align with the "Out of Taiwan" narrative. The presumed origin of the LCC pottery was the Bismarck Archipelago because of its limited range only in Oceania. However, with more sites excavated in the Philippines and Mariana Islands (east of the Philippines), its range has expanded. A cross-cultural pottery series study involving the Mariana Islands, Philippines, and the early LCC in the Bismarck Archipelago determined Taiwan as the homeland for the red-slipped pottery. It was the unexpected link to Mariana that initially pulled the LCC homeland away from the Bismarck Archipelago. LCC pottery findings there mandate a non-Oceanian origin. This sparked the association of Taiwan's oldest red-slipped potsherds to the newer innovated LCC in Oceania. Undecorated red-slipped pottery was present in northern and southern Taiwan 4.5 kya. This pottery technique spread to the Philippines 4–3.8 kya and then simultaneously to Sulawesi in Indonesia and the Mariana Islands of Micronesia 3.5 kya Figs. 9.8 and 9.9. The Philippines continued to make red-slipped pottery but innovated with new designs and motifs. However, red slip dwindled in popularity in later Lapita populations in Oceania, but similar designs were continued. The Mariana pottery is an intermediary between an early Filipino pottery and the later Bismarck assemblages. Its presence in Mariana complicates the ease of travel from Taiwan to the Philippines to Indonesia down to Melanesia and Polynesia, but Carson and Kurashina[26] concluded it is possible different aspects of the Lapita culture took different ideas and incorporated them into the art.

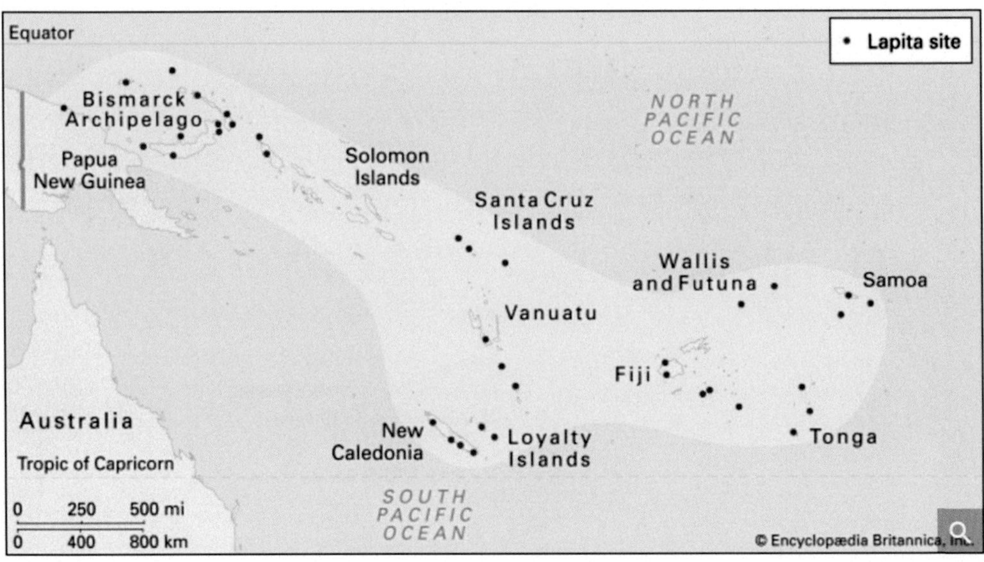

FIGURE 9.9 This shows all of the archeological sites with LCC assemblages in Melanesia. It highlights how the smaller islands were more effective, and the differential impact of Austronesians on the Bismarck Archipelago and Papua New Guinea. *From https://www.britannica.com/topic/Lapita-culture.*

Today, Taiwan is occupied by nine Austronesian indigenous groups whose populations have been in conflict with mainland Han Chinese invaders during the last several thousand years. Of the nine subpopulations (Saisiat, Atayal, Tsou, Bunun, Ami, Rukai, Puyuma, Paiwan, and Yami), eight are on the main island of Taiwan and one tribe, the Yami, inhabits Orchid Island off the southeast coast.[28] The tribes have been isolated from outsiders since the 18th century when the mainland Han Chinese began to occupy the western side of the island. The eight mainland tribes are distributed throughout the internal mountains and along the east coast, and all tribes have distinct languages, cultures, social organizations and show some phenotypic variation. This variation may be associated with the topography of Taiwan, where the large central mountain range stretching from the north to the south acts as an isolating factor. Studies by Chai[29] revealed that the morphological variation between the tribes was reflected in the geographical distances. For example, those tribes living in the east, namely the Ami on the east coast and the Bunun on the eastern mountain slopes, are the most different from the other groups based on discriminant function analysis (a statistical measure used to predict a group-dependent variable by one or more independent variables). Although the tribes are geographically isolated, several studies using nuclear and mtDNA have shown that gene flow did occur among some tribes that share borders. This was especially evident among the Atayal, Saisiat, and Tsou, as well as the Rukai and Paiwan.[12]

Taiwanese aborigines are unique in that 9 out of the 10 currently spoken Austronesian linguistic subgroups are found in Taiwan. The rest of the Austronesian-speaking world speaks only one subgroup, the Malayo-Polynesian or extra-Formosan branch. This high linguistic diversity is regarded as evidence that the Austronesian language family has its roots in the island. It is noteworthy to highlight that the natives of Orchid Island, 60 km off southeast Taiwan in the direction of Oceania, also speak the extra-Formosan Austronesian branch.[2]

Controversies surround the clustering of extant Taiwanese languages; For example, although these languages generally are divided into the Atayalic, Tsouic, and Paiwanic subgroups, there is uncertainty about whether Paiwanic and Tsouic languages were introduced later[25] and whether the language spoken by the Ami tribe should represent a separate branch from which extra-Formosan proto-Malayan languages arose.[8,12]

Similar to the language diversity, in genetic sampling of most islands of SEA populations, the Taiwanese aborigines have more genetic diversity between them than the other SEA language groups studied. The isolating geographical features of the island most likely caused this genetic divergence. This allows us to see the varying levels of impact each tribe had on the Austronesian Expansion. For example, the Ami and Saisiyat are thought to be the source of this population migration per Y-SNP and autosomal STR data. Ami, Bunun, and Saisiyat are genetically more linked to the Solomon Islands of Near Oceania and the Polynesians of the Society Islands of remote Oceania than the other tribes. On the other hand, Paiwan, Puyuma, and Saisiyat have the biggest contribution to the autosomal data of Polynesian and Madagascan Austronesians.[28]

Plant genetic diversity also gives us a window into the varying impact levels of each tribe. An ethnobotany study centered around the phylogeography of the paper mulberry plant correlates with the impact imbalance of the Taiwanese tribes. The paper mulberry plant was used by the many cultures in SEA, including the Taiwanese aboriginals and consequently the Austronesians as a nonwoven fabric. Owing to its reproductive limitations, the mulberry plant could not cross large bodies of water without human mediation. So, its presence in many ISEAs correlate directly to human migration. The paper mulberry plant was one of dozens the Austronesians carried with them to ensure their survival upon their arrival to new lands. Using DNA analysis Chang et al.[30] were able to trace the origins of the mulberry plants. Similar to mtDNA, chloroplast DNA can show the relationships between different species of mulberry plants. Specifically, on Taiwan there are separate species, and therefore specific chloroplast DNA sequences associated with the plants, from the northern, central to southeastern, and southern parts of the island. Using these differential haplotypes, Chang et al.[30] determined the southern mulberry plant was most prolific in Oceania with the presence of the cp–17 haplotype. This haplotype correlates with the southern tribes having the most impact on the Austronesian Expansion.

HOW THE DISPERSAL HAPPENED

The prolific sea voyagers of ISEA eventually moved from Madagascar to the west, Easter Island to the east, New Zealand to the south, and the Hawaiian island chain to the north Fig. 9.10, representing the largest geographic dispersal ever undertaken by humans. On the journey, they carried many things with them in their double-vaulted/hulled canoes Fig. 9.11: language, the LCC, livestock, their agricultural practices, plants, and, obviously, their genes. Each of these can be used to trace the path the migrants took across ISEA and Oceania. All of these also effected how the Austronesians interacted with the populations they encountered and their survival success. The "transported landscapes" they carried in their canoes increased their success rate as many islands in SEA are resource poor.[30] Hunter-gatherers could not survive with the limited fauna and flora, but the Austronesian brought all they needed to survive. This helped them dominate not only the resource-poor uninhabited islands but also those already inhabited with small hunter-gatherer populations. Their sustenance strategy gave the Austronesians a leg up on the competition and thus another angle through which to trace their

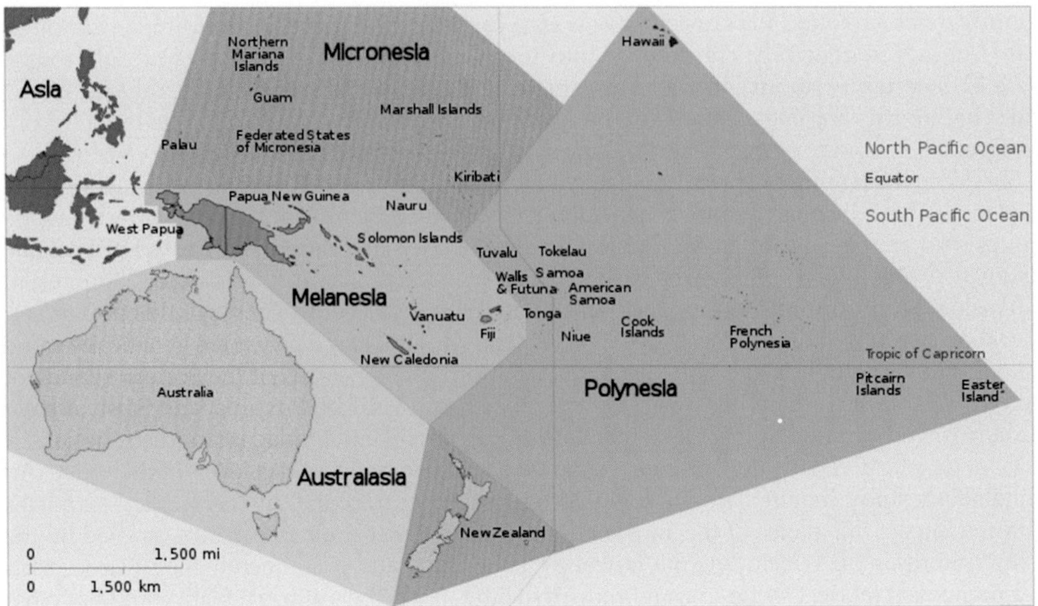

FIGURE 9.10 Figure shows the region in the South Pacific that contains Micronesia, Melanesia, and Polynesia. *From https://upload.wikimedia.org/wikipedia/commons/thumb/5/54/Oceania_UN_Geoscheme_Regions.svg/800px-Oceania_UN_Geoscheme_Regions.svg.png.*

FIGURE 9.11 An Austronesian outrigger canoe. *Taken from https://www.google.com/search?q=historical+austronesian+outrigger+canoes&rlz=1C1CHBF_enUS759US759&source=lnms&tbm=isch&sa=X&ved=0ahUKEwijoLCYj5XYAhWkdN8KHWWaCXgQ_AUICigB&biw=1422&bih=684#imgrc=wMGS72eD6NjcdM.*

movement: ethnobotany and animal speciation.[8] The differences between the indigenous species of plants and animals and those foreign species they carried help map the human dispersal.

Each of the carried components overlap geographically. Therefore, researchers have attempted to correlate the archeological record with the diversification of Austronesian dialects. As language moved, so did the artifacts associated with the people that used the language. For example, the use of geometric patterns and figures of plants and animals seen in the pottery moved into parts of Melanesia and Polynesia as the language intermixed with the indigenous tongue. However, it is a difficult task aligning data from each field that not only calculates time estimates differently but also has such different data sources. The molecular clock of nucleotide mutation will not always overlap with a carbon-tested time window of archeological specimens. So, the more sources used and the more data collected help to provide reasonable estimates of time and movement.

Data from each major field (linguistic anthropology, archeology, and population genetics) are not easily put on the same playing field. But each can give a different insight into human migration as they occur at different speeds. Archeology can detect immediate interaction with a population because it does not take long for a population to see a technique or design they like and copy it for themselves. This field can also give reliable dates, much more than the imprecision of the molecular clock. However, in the archeological record it is hard to differentiate between trade and acculturation. Pots may look similar because of a trade route or because one population assimilated another. Linguistic anthropology is the intermediary between the three fields. While full linguistic replacement requires many decades, populations can mix and share in a shorter period of time. Full genetic replacement, on the other hand, takes decades to complete, but again, residuals of shorter admixtures can still show interaction.

Interweaving this three-part narrative can give us a deeper understanding of the nature of the relationship and a time line. If it is a purely trade relationship, high amounts of material cultural overlap are to be expected, while shared terms will surround the techniques and items of trade, but genetic admixture might be limited. As the trade relationship lengthens and strengthens, more words and culture will intermix. The same can be true for colonization interactions. One might expect to see more genetic admixture and language acquisition, especially as interaction with the indigenous community increases.

The degree of interaction is a key component in understanding how the Austronesian Expansion occurred, as well as the rate of migration and extent and location of admixture pauses. Owing to the vastly different analyses techniques, each major field has its unique model. Overall, the greatest attention based on evidence from geneticists, linguists, and archeologists are given to five models, (1) the "express train," (2) "slow boat," (3) "pulse-pause," (4) Voyaging Corridor Triple I, and (5) the entangled bank.

The "express train" model, also referred to as "out of Taiwan" model, suggests that Austronesians originated in Taiwan and traveled rapidly through Micronesia and Melanesia with minimal genetic admixture with preexisting Melanesians before settling throughout Oceania.[31] It is mainly heralded by the genetics community as it explains the limited genetic residue left in these communities. This model was first supported linguistically, but in a more recent study, Gray and Jordan[2] concluded that the language patterns do not correlate to the "express train" model.

The "slow boat" model theorizes that the Austronesian migrants originated in Taiwan and traveled slowly through ISEA, influencing some of the preexisting genetic substrata and eventually reached Near and Far Oceania.[32] This model is almost the direct opposite of the

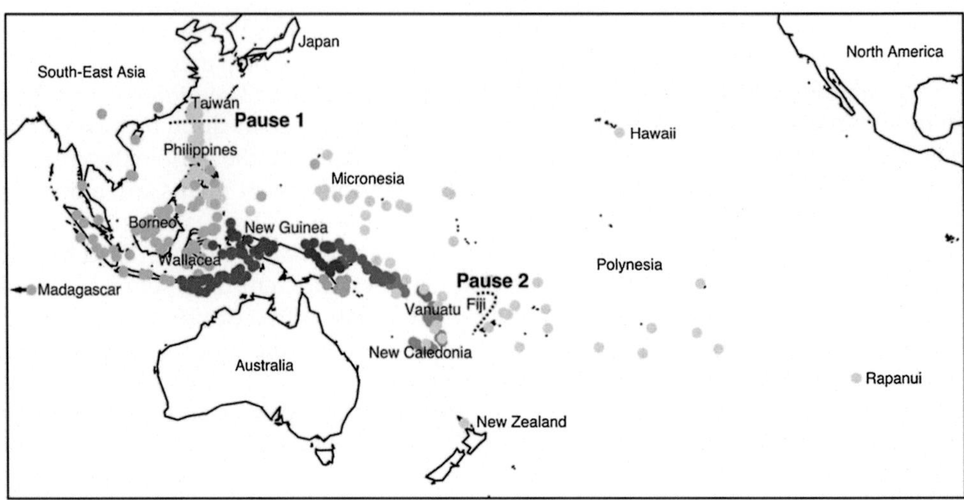

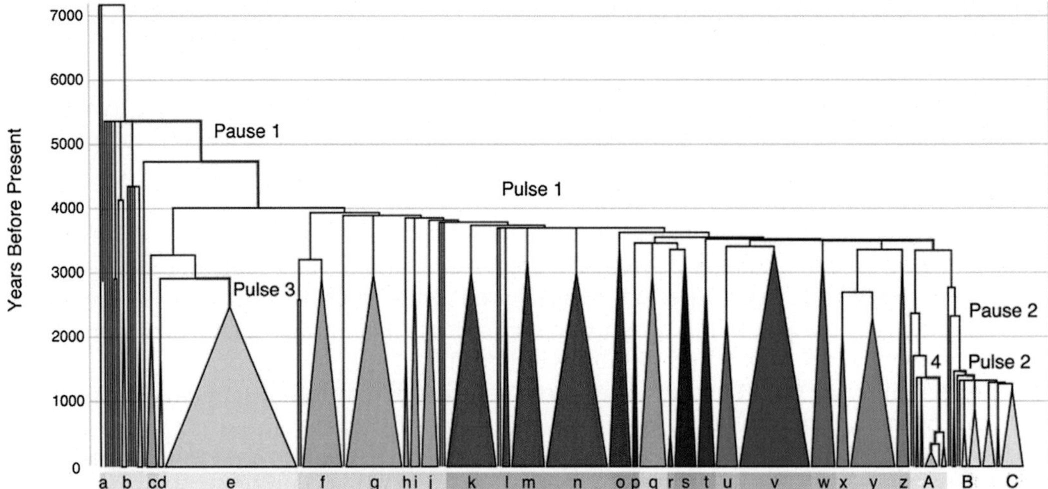

FIGURE 9.12 This relates to the pulse-pause theory and shows the different pulses per color and therefore different languages groups. The colors represent increased language diversification in areas where the pauses occurred and decreased language diversification during pulses. *Taken from Gray et al. Language Phylogenies Reveal Expansion Pulses and Pauses in Pacific Settlement.* Science *2009;***323***:479–83.*

"express train." The model aligns with the genetic admixture seen in Indonesia, the Bismarck Archipelago, and Polynesia. However, it still does not account for the limited residue left in New Guinea and other places.

The "pulse-pause" theory is an intermediary between both aforementioned models. It posits that the expansion came in waves. First was the settlement of Taiwan, then expansion into Polynesia and the Philippines, followed by a second pause in western Polynesia, and then a final push into eastern Polynesia.[33] This theory attempts to take into account the scale of interaction seen in the remnants of Austronesian language and genes Fig. 9.12. This is supported by increased language diversification in areas where the pauses occurred, such as Taiwan and western Polynesia, and decreased language diversification in areas of pulses, such as the Philippines. The "pulse-pause" theory considers the geographical and technological limitations that caused the pauses, such as the creation of the outrigger canoe and the ability to estimate latitude from the stars.[2] However, "pulse-pause" does not account for cultural considerations of each island.

The "Voyaging Corridor Triple I" theory attempts to explain the variation in degree, rate, and location of interaction through three concepts: intrusion, integration, and innovation. For example, Carson et al.[7] explained the pottery variation seen in the early collections through these concepts. First, intrusion was represented by the red slip, point impressions, and rectilinear shapes found widely in the pottery of the Philippines, Mariana Islands, and the early collection of the LCC in the Bismarck Archipelago. Integration included the expanded tool kit for dentate stamps and elaborative decorative motifs such as the labyrinth.[26] Both developed while the Austronesians were interacting with the community and then brought these developments as they continued their sea voyage. Once the Austronesians left, the communities of the Bismarck Archipelago continued to "innovate" with certain design motifs, such as curvilinear lines and the Lapita face. The Triple I theory attempts to situate the expansion in the contemporary cultural landscape. One of the biggest factors in the degree of interaction and assimilation is simply the presence of an indigenous population. Where no population is present the Austronesians can easily dominate, especially due to the factors already discussed, such as the ability of these agrarians to adapt to poor land quality.[8,26]

One intriguing example of walking the line between innovation and replacement is the changing of languages. Isolation can cause changes in basic lexicon and correspondences, the way the language sounds. Creolization or admixture with resident language can do the same. Donohue and Denham[34] characterized the East Indonesian dialect as no longer Austronesian because of its lack of overlap in bound morphemes, essentially meaning significant translations between words/phrases. Is this innovation of the Austronesian language or should it truly be its own language? These types of interpretations of related languages enforce the idea that language as a tool by itself can be misleading, thus the need for multiple sources of data to substantiate conclusions.

Another theory, the "entangled bank" model, is an important caveat to any of the proposed theories. It cautions that a dispersal from Mainland SEA into Oceania during the mid-Holocene (7–5 kya), along with continued and extensive gene flow with Melanesian natives throughout the trek is responsible for the current genetic characteristics of Austronesian groups.[35] This model highlights an important fact, that the current admixture may be blurring the lines for researchers to see the more ancient patterns. However, this is not to say that all trends found in the data are from recent migrations, since admixture data from Eastern Indonesia support an Austronesian Expansion.[36]

While these theories try to give the full picture of the Austronesian Expansion, several questions still arise: (1) What is the history of human occupation in Taiwan and where did the extant aboriginals of Taiwan originate? (2) Are the phenotypic and linguistic differences between the Taiwanese aboriginals the result of geographical isolation, or did these individuals arrive in Taiwan at different times from separate populations? (3) What was the impetus for their migration into Oceania? (4) How has this migration affected the culture and genetics of Oceania? (5) Are the phenotypic and linguistic differences between ISEA groups the result of the Taiwanese tribe that populated the island or have the isolation of islands caused this variation? (6) Are there more recent migrations that must be taken into account when outlining the Austronesian Expansion?

LEAVING TAIWAN: THE EXPANSION

According to the archeological and genetic data, people originating in Taiwan began to move southeast about 5 kya.[37] It is not clear what prompted Austronesians to venture out into the open seas. This sea voyage could have occurred first by chance; a fisherman drifted out to sea and was carried away with the prevailing currents. Yet, it is highly unlikely that most of the settlement events resulted from chance. As Austronesians were agriculturalists, it is also possible they were running out of land to cultivate and ventured out to find more resources. The topography of Taiwan contains a large central mountain range stretching from the north to the south, restricting agriculture to small farming villages on the coastal alluvial plains. Although it is unknown, changing conditions, such as poor soil, years of drought, inclement weather, internal conflicts between groups, feuds among families or tribes over land, may have made leaving a viable option.

Considering the distances, the dangers, and the duration of the excursions, it appears that Austronesians were highly driven, adventuresome, and saw the ocean as a highway for future opportunities attained in their double vaulted canoes. These outrigger canoes were first created so the Austronesians could take their first step in the colonization of SEA, and the Philippines. Gray et al.[18] propose the first settlement pause in Taiwan was due to a lack of sailing technology to cross the Bashi Channel. Linguistic evidence shows that outrigger terminology can only be traced back to Proto-Malayo-Polynesian and not Proto-Austronesian, conferring with the Austronesians first inventing this technology to populate the Philippines. These canoes were designed to provide stability for sailing long distances. The canoes could be fitted with huts for protection and were capable of carrying about 50 people. Journeys would include plants and animals that could all be used during and after reaching their destination. Fossil remains indicate that they transported dogs, pigs, chickens, rodents, and dozens of plants throughout the Pacific as they colonized various islands. The items they carried and their remnants are traces of their voyage, and these artifacts can be used to retrace their journeys and to determine the number of failed and successful colonization events. By comparing native and introduced species, it has been determined that a number of species of plants and animals were transported from islands in the Fiji, Samoa, and Tonga archipelagos in central Polynesia to the islands further east, north, and south in Oceania.[38]

THE PATH OF COLONIZATION THROUGH THE ISLANDS OF SEA

It has been postulated that the Austronesian migration by Taiwanese tribal farmers initiated a dispersal throughout ISEA, first into the Philippines, Melanesia (mixing with Papuan-speaking groups),[5] then arriving in the Tonga and Samoan archipelagos, about 2.8 kya.[6] Genetic and archeological evidence suggests that the initial Austronesian migrants "paused" in western Polynesia for approximately 500–1000 years, possibly allowing for the development of technology needed to sail further into the eastern Pacific. Once they attained the means to travel further, they reached the Cook Islands, the Hawaiian Archipelago, French Polynesia, New Zealand, and finally Easter Island, about 1.2 kya.[37,39]

The Austronesian Expansion was anything but uniform. It was an organic movement of people, and therefore not every island in SEA adopted the Austronesian language or culture nor did they do so in the same manner. Each assimilation took generations and some areas had a more substantial social pressure and/or incentive to adopt the language/culture. It was not a complete replacement on many islands. On top of the varied selection pressures, sea travel is not systematic. They did not stop at every island and consequently some were not introduced to the Austronesian language/culture. This only adds to isolating factors in the SEA islands as many have typological characteristic isolation rarely found across the globe.[34]

To see the varying level of assimilation and interaction, we walk through how each major group of islands was touched by the Austronesians.

Philippines

The Batan Island, the northernmost island of the Philippines, is likely the first island colonized by the Austronesians. The distance between Taiwan and the Philippine's Batan Island is just 190 km, which would have been navigable at the time. When the Austronesians reached the Philippines, they may have first encountered the Aeta, the indigenous people thought to be the earliest inhabitants of the Philippines. The Aeta are a large ancestral population found mainly in the Philippines where they live in the isolated mountain regions of Luzon. The population shows some unique physical characteristics, namely, their skin ranges from dark to very dark brown, they are small-statured, have curly to kinky hair, small noses, and dark brown eyes. Their religious beliefs range from monotheism to animism (all objects, creatures, and places are spiritually alive). Historically they are nomadic and build only temporary shelters, but modernized Aetas have been forced to move to villages due to continued deforestation and farming thereby decreasing their population size. Collectively they are known as Negritos, although the term describes several different populations that inhabit isolated regions of South Asia and Southeast Asia.[40,41]

While the Aeta are categorized socially as Negritos, they have adopted the Austronesian language. Not only do the Aeta speak an Austronesian dialect but all Filipino languages fall under the Austronesian family.[42,43] These languages are also the most conservative lexically and grammatically. Linguistic data show the Philippines are the dispersal center for the Malayo-Polynesian branch of the Austronesian family which spread into the Indo-Malaysian Archipelago.[8]

The populations of the Philippines assimilated culturally as well as linguistically with the presence of the LCC in its assemblages. The LCC appeared in the Philippines as early as 4 kya and specifically in Luzon, the northernmost of the three main island groups and where the Aeta lived/live, the pottery was noted 3.8 kya. The Lapita pottery consisted of mainly small jars with carinated shoulders with little decoration, suggesting utility rather than ornamentation. The Filipino assemblages propose a connection to the Mariana Islands. While the path to these far-flung Micronesian islands is unknown, the pottery is "virtually identical."[26] While the pottery shows clear cultural connections between the Philippines and the Austronesians, there are no contemporary populations of the paper mulberry plant on the islands. This may be due to a recent eradication of previous populations, but the mulberry plant was used heavily by the Austronesians as a cloth source.[30]

The geographical location and the acquisition of the Austronesian language correlates with new genetic evidence suggesting a close relationship between the people of the Philippines and Orchid Island, off the southeast coast of Taiwan (Bertrand-Garcia et al.

unpublished data). A study by Delfin et al.[14] came to the same conclusion. mtDNA haplotypes B4a1a, E1a1a, and M7c3c are associated with Austronesian Expansion and overlap with the time line of the migration.

Genes, language, and culture all align in the Philippines and show heavy assimilation of the population. Why was it so easy for the Austronesians to take over this island chain? According to Pawley and Ross[8] the pre-Austronesian population lacked the demographic and economic ability to overcome the impact of Austronesian languages in this region. As was previously discussed, the population structure of the island before Austronesian arrival impacts the amount of acculturation seen in the linguistic, archeological, and genetic residues. The Aeta were a small hunter-gatherer group. They had little power over the agricultural and oceanic force brought by the Austronesians.

Indonesia and Malaysia

Most migration models dictate the Austronesians moved from the Philippines to Indonesia and Malaysia, aligning with the dispersal centers for the Malayo-Polynesian language branch. However, acculturation was not as systematic or widespread in this archipelago as seen in the Philippines. This may be because of a different population structure of the residents. Peninsular Malaysia is comprised of several indigenous communities collectively known as "Original People" that make up approximately 0.6% of the Malaysian population. These indigenous communities are broken up into three groups: (1) Negrito (Semang), (2) Senoi, and (3) Proto-Malay based on linguistic, physical, and anthropological characteristics. These groups are also found in the islands of Malaysia and Indonesia.

Malaysian Negritos are Austro-Asiatic speakers inhabiting northern Peninsular Malaysia They practice the egalitarian and patrilineal descent system. Physically they have a relatively small body, dark skin, distinct facial morphology, and wiry hair. Malaysian Negritos are often grouped with other Negrito communities, and several studies have provided genetic evidence that Andaman islanders, Philippine Negritos, and Malaysian Negritos show some genetic affinity and similar phenotypes.[44,45] These similarities have led to some genetic evidence indicating that Negrito populations of SEA and Oceania are derived from the original ancient migration as opposed to the Austronesian Expansion.[46] Other genetic studies indicate that Negrito populations are closer to their non-Negrito communities.[42,47,48] There is likely no strict binary between the populations, but there are some trends across the islands. mtDNA and NRY data have shown that east Indonesian populations are of dual Papuan and Asian descent. Furthermore, genome-wide studies indicated that the Papuan ancestry gradually increased, while the Asian decreased, from west to east across Indonesia. The lowest proportion of Papuan ancestry was observed in south Sulawesi, whereas the population closest to New Guinea exhibited the highest proportion of Papuan ancestry.[36]

In other words, Papuan/Negrito genetic proportions increase as you move east, whereas Austronesian decreases. Therefore, Austronesian expansion occurred in an eastward direction, with proportionally greater genetic effects on the populations it initially encountered in Indonesia, and less genetic influence as Austronesians moved eastward. This correlates with the rejection of East Indonesian languages as Austronesian because of little contact with the expansion. Similar to other places such as New Guinea, the admixture was sex biased with a greater contribution from Austronesian women than men, showing the matrilocal culture of the Austronesians.[49]

One of the outliers in the genetic and linguistic landscape of Indonesia is Austro-Asiatic heritage, which occurs in those speakers on mainland SEA and in western Indonesia but nowhere else. Estimated dates of genetic admixture do not correspond to some of the linguistic and archeological studies, making it difficult to infer why there is a strong component of Austro-Asiatic ancestry in western Indonesia. While the admixture dates are inconclusive, there are three possibilities for the substantial Austro-Asiatic signal. First, during the Austronesian Expansion, these sea voyagers picked up the Austro-Asiatic component and brought it with them to Indonesia. However, if this was the case, we should see the two groups in similar proportions in each population, which does not occur. The second option is recent admixture which cannot be ruled out. Last, and most likely, is that the Austro-Asiatic lineage was already present in Indonesia, maybe through the migration of one of the indigenous populations (Negritos, Senoi, Proto-Malay) across Sundaland[49] (see Chapter 8).

Melanesia

The Austronesians moved from Indonesia to Melanesia, which includes islands from eastern Indonesia eastward to the Solomon Islands. Melanesians speak different forms of Papua, although some islands are linguistically Austronesian. Those that speak Austronesian languages exhibit a blend of Melanesian and Austronesian genetic elements. For example, in these admixed island populations, high frequencies of Asian mtDNA and Melanesian Y chromosome haplogroups were detected. Genome-wide results indicate that Melanesian groups are genetically very different from each other, with inland groups different from coastal groups. Inland populations are more homogeneous than coastal populations, which exhibit various degrees of Austronesian admixture. No Papua-speaking populations from the interior of islands exhibit Austronesian admixture. The low level of admixture seems to support the "express train" model;[50] however, mtDNA studies reflect a more complex story in which the genetic heritage of Austronesians appeared to be a combination of both Taiwanese aboriginal descent and Melanesians.[32] This suggests that Austronesians, before reaching Middle and Far Oceania, had a somewhat extended layover in Melanesia, with a stay long enough to allow admixture between the two groups. Thus, it looks as if there was not an "express train" but more likely a "slow boat."

Through the lens of the "Triple I" theory, we can see that both of the seemingly conflicting statements may be true. Genes, artifacts, and languages reveal that the Austronesians skirted their way around the edges of Melanesia, especially the big island of Papua New Guinea. Austronesian settlements occur near beaches as those are the best launching points for colonization voyages. It is therefore logical that Austronesians would keep to the exteriors of large islands. As we have seen in each of the other main ISEA groups, the presence of an indigenous population affected the dominance the Austronesians had over a population. Mainland New Guinea has very little residual Austronesian aspects, but its Bismarck Archipelago is one of the cultural centers for the LCC. Archeological data in this archipelago suggests a smaller population that is more easily culturally dominated.[8]

Polynesia

Genetically, Polynesia was one of the first island chains to be linked to the Taiwanese aboriginals and support the Taiwan homeland theory. The presence of the mtDNA haplogroup B4a1a1 (the so-called Polynesian motif) and its ancestral lineage B4a1a among

Taiwanese tribes as well as in Polynesian groups provided direct genetic evidence for a connection between these two groups of populations. Y chromosomal studies initially indicated genetic connections between Taiwanese tribes and Polynesians, but differences of opinion started when close genetic ties between Melanesia and Polynesia were reported.[32] The possible admixture of Melanesians and Austronesians during the Austronesian integration with certain islands could explain this occurrence of Melanesian DNA. However, this genetic signal could be coming from an indigenous Melanesian presence on the islands before they were populated by the Austronesians, as they are thought to be the original inhabitants of Oceania. But, autosomal DNA markers indicate Polynesians still segregate with Micronesians, Taiwanese tribes, and insular East Asians, not with Melanesians. In many cases, mtDNA and Y chromosome–specific haplogroups lack the resolution seen in autosomal markers that are less subject to being lost. For example, a family that has only female children will lose the Y chromosome markers in subsequent generations, whereas a family with only male children will lose the mtDNA marker in subsequent generations. Genome-wide studies, which assess recombining DNA as opposed to uniparental lineages, have detected minimal Melanesian DNA in Micronesian and Polynesian populations.[51] In other words, the genomes of contemporary Polynesian-speaking groups appear to be a mosaic of components derived from the coming together of long-diverged sources from ISEA and the region of northern Melanesia/ New Guinea.

Over time, the Austronesians reached all the islands of Polynesia, using them as centers for communication, commerce, and cultural exchange. The feat of reaching all of Polynesia is impressive, considering the vast area of Polynesia defined as the Pacific Islands located within a triangle containing New Zealand to the south, the Hawaiian archipelago to the north, and the Marquesas archipelago and Easter Island to the east (Fig. 9.10). The entire region is equivalent to the size of the entire North Atlantic, with less than 1% of the area occupied by land.[37] The center of ceremony on Ra'iatea (the second largest of the Society Islands in French Polynesia) can still be seen as an impressive complex that was used for departure in trans-island voyages.[50,52] Ra'iatea received the flow of goods and information from central Polynesia and in turn became a launching platform for the exploration and distribution to the islands in the north (Hawaii), south (New Zealand), and the Far East (Easter Island).[50,52] However, the barrier the sea presented is seen in the pulse and pauses of settlements. Gray et al.[18] proposed the second pause occurred in western Polynesia because the far-flung islands required more advanced technology. Austronesians first had to improve their sailing technology to reach their destination more efficiently.

Micronesia

This regional category of ISEA stretches from Palau and the Marianas in the west to the Marshall and Kiribati archipelagoes in the east, covering almost 5000 km. These islands are characterized by coral atolls, coral reefs completely or partially encircling a lagoon. On some islands, the lagoons are patterned with volcanic outcrops such as Chuuk. All Micronesian languages fall into the Austronesian family, but there is a split in which subgroup they fall into. The languages of the Marianas and Palau (Chamorro and Palauan, respectively) belong in the Western Malayo-Polynesian subgroup, found in the Philippines and Indonesia. Most linguists determine the Philippines as the source of Chamorro.[52] This pattern is reflected in the archeological record of the Marianas. As previously stated, pottery found in the Mariana Islands is identical to that found in the Philippines and represents the early design of the LCC later developed in the Bismarck Archipelago.[7] The two sites are clearly linked in timing and in pottery, but the interaction between the previously settled societies of Bismarck may explain the difference in pottery trends.[26] LCC assemblages are found on the larger islands of Marianas, Guam, Tinian, and Saipan.[27,50] Linguistic and archeological evidence supports an early voyage to the Mariana Islands, around 3.5 kya, the same time or possibly a little earlier than the habitation of the Bismarck Archipelago by the Austronesians.[26] Recent mtDNA studies have supported this early connection between Micronesia and the Philippines as most individuals had the haplogroup E found in the Philippines and Indonesia but do not have an abundance of the typical Oceanic mtDNA haplogroup B4.[53]

However, not all of the islands show an early expansion from the Philippines. Chamorro and Palauan may group with the Philippines and Indonesia, but the languages of the Carolines, Marshalls, and Kiribati fall under the Nuclear Micronesian category. The outlier, the language of Yap, located geographically between Palau and the Carolines, is an Oceanic language that has been heavily influenced by outside forces. Therefore, linguistically, the islands show a three-part dispersal: first, an initial long migration from the Philippines to Mariana and Palau; second, a quick island-hopping diaspora from the Solomons–Vanuatu region, creating the languages of the Carolines, Marshalls, and Kiribati; and finally, a possible direction colonization of Yap by the Bismarck Archipelago. Mariana and Palau show early LCC pottery, supporting the early colonization of those islands. The Mariana volcanoes are at a higher elevation than other Micronesian islands and so may have been one of the only islands significantly above the surface at the time of the Austronesian voyage.[26] This could be the reason why they were first colonized rather than the other eastern Micronesian islands. The Carolines also exhibit LCC assemblages with shell adzes and shell ornaments. These items are found in later LCC periods, which align with a later colonization time for eastern Micronesia, as this migration came after Austronesian contact with Melanesia and Polynesia. Yap has not been extensively excavated; therefore only historical linguistic data support the third stage of dispersal from the Bismarck Archipelago.[27]

Polynesian Outliers

As previously described, Polynesia, the last region to be inhabited by Austronesians, is typically considered to include the thousands of islands within the triangle bound by Hawaii, Easter Island, and New Zealand. However, there are about 23 islands that are culturally Polynesian, based on linguistic and archeological data, which lie geographically within Melanesia and Micronesia. These culturally Polynesian societies are referred to as Polynesian outliers and are scattered across five countries in Micronesia, Papua New Guinea, the Solomon Islands, Vanuatu, and New Caledonia Fig. 9.13. The people within these outliers not only demonstrate the cultural practices of Polynesians but also share the physical features of a large build, wavy hair, and brown complexion.[54,56] The outliers speak languages of the Samoic language family, a branch of the Polynesian languages, and many also speak the language of their local Melanesian or Micronesian population. Many researchers consider the Polynesian outliers to have been colonized by inhabitants of Tonga, Samoa, and Tuvalu.

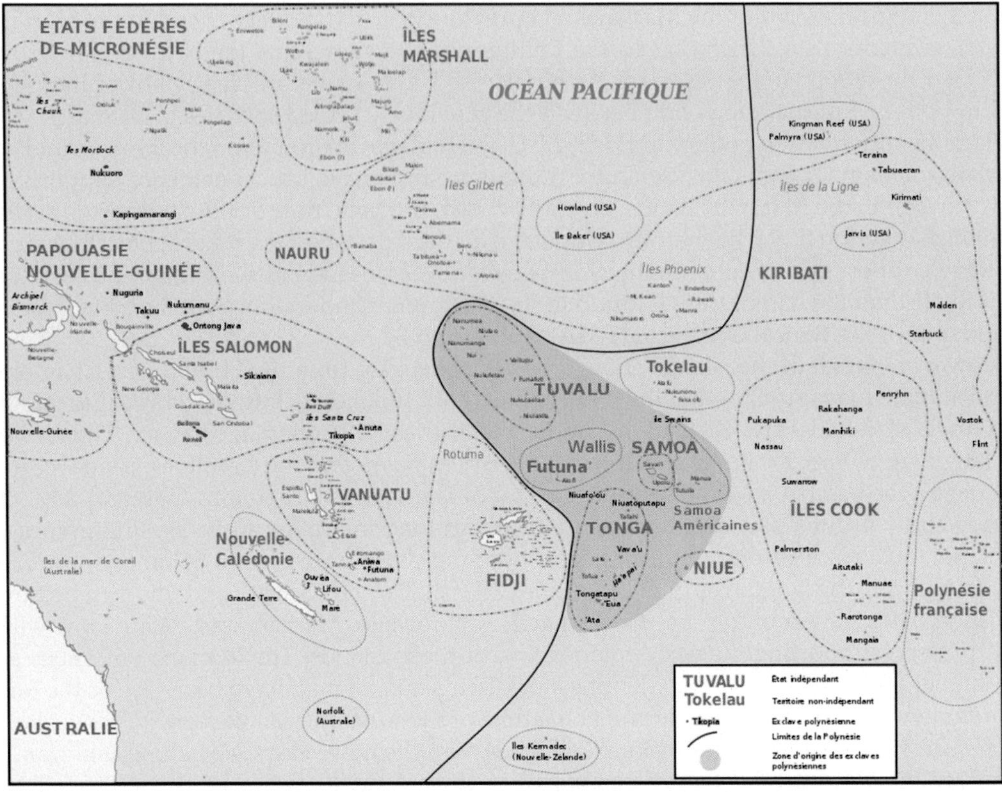

FIGURE 9.13 Figure showing the Polynesian outliers in red and the original Polynesian homeland in the pink zone. Outliers are shown in red within the five countries of New Guinea, Micronesia, the Solomon Islands, Vanuatu, and New Caledonia. *Taken from https://upload.wikimedia.org/wikipedia/commons/thumb/2/22/Western_Polynesia_and_Polynesian_Outliers_-_fr.svg/1024px-Western_Polynesia_and_Polynesian_Outliers_-_fr.svg.png.*

The existence of outliers and the fact that the genomes of contemporary Polynesian-speaking groups appear to be a mosaic of components derived from the coming together of long-diverged sources from ISEA and the region of northern Melanesia/New Guinea bring into question the origin of Polynesian society.[55] The origin of Polynesian society is a subject of debate, with three different hypotheses for the development of Polynesian cultural evolution: (1) advocates of isolation as the driving force, (2) those that invoke interconnectivity with external influences, and (3) an intermediate model with migrants from ISEA mixing with people from northern Melanesia before the settlement of western Polynesia.

Ancient DNA and population genetics indicate that genetic admixture occurred in western Polynesia significantly after settlement. In addition, recent results, based on haplotype analysis together with evidence from language, argue for greater consideration of outlier Polynesia in discussions of Polynesian origins.[56] Previously, outliers were not considered to be a source for the settlement of eastern Polynesia because of linguistic subgrouping and estimated dates of arrival into eastern Polynesia. However, genetic data from the Leeward Society Isles are consistent with a late settlement model for eastern Polynesia, which facilitates the consideration of the Polynesian outliers as a possible source region. These recent DNA and linguistic data are consistent with models of interconnectivity and support archeological dates for the late settlement of eastern Polynesia.[56] In addition, the data are consistent with a west-to-east movement of people, tracing the origins of eastern Polynesians to central northern outlier Polynesia, rather than Samoa. However, this does not necessarily exclude the possibility of an ultimate common origin shared with western Polynesians.

CONCLUSION

The islands of Southeast Asia were clearly populated in at least a two-step migration: the first, a continuation of the initial Out of Africa movement that built the foundations of most indigenous populations in the area and the second, the Austronesian Expansion. Many theories have been proposed to explain the pattern in the phenotypic variation, the level of interaction, and the remnants of Austronesian influence throughout the region. The "Triple I" theory explains the evidence we see most thoroughly, but while it can be readily applied to the Austronesians and fits the evidence well, in and of itself, it gives no specifics of the actual expansion. It is a framework in which to see the evidence. Other theories attempted to take the facts from various disciplines and turn them into a story, but missed certain key contributors to the variation patterns. For example, the "express train" and "slow boat" were both initially hailed by geneticists based purely on the level of admixture seen in the communities. However, the story line of the expansion falls in between these binaries, where the "pulse-pause" theory attempted to sit. This intermediary theory was first proposed by linguistics and tried to explain why some populations showed extended admixture (genetically, culturally, and linguistically) and others were barely touched by the migration. It took into account needed technologies to cross the immense expanse of the waters of Oceania but failed to correlate the presence of the indigenous populations structure as a factor in determining Austronesian influence. The "Triple I" theory looks at the Austronesian Expansion and human migrations in general in a beneficial light, but a theory that takes these concepts and data from all academic disciplines, and tailors them to the specifics of this human dispersal is needed for a complete and concrete outline of the events.

References

1. Arnold M. 1822–1888. www.brainyquote.com/quotes/matthew_arnold_148718?src=t_expansion.
2. Gray RD, Jordan FM. Language trees support the express-train sequence of Austronesian expansion. *Nature* 2000;**405**:1052–5.
3. Soares PA, et al. Resolving the ancestry of Austronesian-speaking populations. *Hum Genet* 2016;**135**:309–26.
4. Bellwood P. The Austronesian dispersal and the origin of languages. *Sci Am* 1991;**265**:70–5.
5. Wollstein A, et al. A demographic history of Oceania inferred from genome-wide data. *Curr Biol* 2010;**20**:1983–92.
6. Rieth TM, Hunt TL. A radiocarbon chronology for Samoan prehistory. *J Archaeol Sci* 2008;**35**:1901–27.
7. Carson MT, et al. The pottery trail: from Southeast Asia to remote Oceania. *J Island Coastal Archaeol* 2013;**8**:17–36.
8. Pawley A, Ross M. Austronesian historical linguistics and culture history. *Annu Rev Anthropol* 1993;**22**:425–59.
9. Goddard WG. *Formosa: a study in Chinese history*. Little Essex Street, London: Macmillan and Company Limited; 1966.
10. Mirabal, et al. Ascertaining the role of Taiwan as a source for the Austronesian expansion. *Am J Phys Anthropol* 2013;**150**:551–64.

11. Hung H-c, Carson MT. Foragers, fishers and farmers: origins of the taiwanese Neolithic. *Antiquity* 2014;**88**:1115–31.
12. Melton T. Polynesian genetic affinities with Southeast Asian populations as identified by mtDNA analysis. *Am J Hum Genet* 1995;**57**:403–14.
13. Yao YG, et al. Evolutionary history of the mtDNA 9-bp deletion in Chinese populations and its relevance to the peopling of east and southeast Asia. *Hum Genet* 2000;**107**:504–12.
14. Delfin F, et al. Complete MtDNA genomes of Filipino ethnolinguistic groups: a melting pot of recent and ancient lineages in the Asia-pacific region. *Eur J Hum Genet* 2014;**22**:228–37.
15. Lipson M. Reconstructing Austronesian population history in Island Southeast Asia. *Nat Commun* 2014;**5**:1–7.
16. Karafet TM, et al. Major east–west division underlies Y chromosome stratification across Indonesia. *Mol Biol Evol* 2010;**27**:1833–44.
17. Bellwood P. Early agriculturalist population diasporas? Farming, languages, and genes. *Annu Rev Anthropol* 2001;**30**:181–207.
18. Gray RD, et al. Language phylogenies reveal expansion pulses and pauses in pacific settlement. *Science* 2009;**323**:479–83.
19. Chang C-H, et al. The first archaic *Homo* from Taiwan. *Nat Comm* 2015;**6**:6037.
20. Chang K-c. The Neolithic Taiwan Strait. *Kaogu* 1989;**6**:541–50.
21. Jiao T. *The Neolithic of southeast China: cultural transformation and regional interaction on the coast*. Amherst, New York: Cambria Press; 2007.
22. Fuller DQ. Contrasting patterns in crop domestication and domestication rates: recent archaeobotanical insights from the old world. *Ann Bot* 2007;**100**:903–24.
23. Chang KC. *Fengpitou, Tapenkeng, and the prehistory of Taiwan. No. 73*. New Haven: Publications in Anthropology. Yale University; 1969.
24. Chang KC. Man in the Choshui and Tatu River Valleys in central Taiwan: preliminary report of an interdisciplinary project, 1972–1973 season. *Asian Perspect* 1974;**17**:36–55.
25. Bellwood P. The peopling of the Pacific. *Sci Am* 1980;**243**:174–85.
26. Carson MT, Kurashina H. Re-envisioning long-distance oceanic migration: early dates in the Mariana islands. *World Archaeol* 2012;**44**:409–35.
27. Kirch PV. Lapita and its aftermath: the Austronesian settlement of Oceania. In: Goodenough WH, editor. *Prehistoric settlement of the pacific*. Philadelphia, PA: American Philosophical Society; 1998.
28. Zeng Z, et al. Taiwanese aborigines: genetic heterogeneity and paternal contribution to Oceania. *Gene* 2014;**5**:240–7.
29. Chai CK. *Taiwan aborigines. A genetic study of tribal variations*. Cambridge, Mass: Harvard University Press; 1967. 238 pp.
30. Chang C-S, et al. A holistic picture of Austronesian migrations revealed by phylogeography of pacific paper mulberry. *Proc Natl Acad Sci USA* 2015;**112**:13537–42.
31. Diamond JM. Express train to Polynesia. *Nature* 1988;**336**:307–8.
32. Kayser M, et al. Genome-wide analysis indicates more Asian than Melanesian ancestry of Polynesians. *Am J Hum Genet* 2008;**82**:194–8.
33. Thomas T. The long pause and the last pulse: mapping East Polynesian colonization. In: Clark G, Leach F, 'Connor SO, editors. *Islands of Inquiry. Colonization, seafaring and the archaeology of maritime landscapes*. Terra Australis. Canberra: ANU Press; 2008. p. 97–112.
34. Donohue M, Denham T. Becoming Austronesian: mechanisms of language dispersal across southern Island Southeast Asia. In: Gil D, McWhorter J, editors. *Austronesian undressed. Pacific linguistics*. Canberra: ANU Press; 2015.
35. Oppenheimer SJ, Richards M. Polynesian origins: slow boat to Melanesia? *Nature* 2001;**410**:166–7.
36. Xu S, et al. Genetic dating indicates that the Asian–Papuan admixture through Eastern Indonesia corresponds to the Austronesian expansion. *Proc Natl Acad Sci USA* 2012;**109**:4574–9.
37. Moreno-Mayer V, et al. Genome-wide ancestry patterns in Rapanui suggest pre-European admixture with Native Americans. *Curr Biol* 2014;**24**:1–8.
38. Ponting C. *A green history of the world: the environment and the collapse of great civilizations*. New York: Penguin Press; 2007.
39. Wilmshurst JM, et al. High-precision radiocarbon dating shows recent and rapid initial human colonization of East Polynesia. *Proc Natl Acad Sci USA* 2011;**108**:1815–20.

40. Headland T, Reid L. Hunter-gatherers and their neighbors from prehistory to the present. *Curr Anthropol* 1989;**30**:43–51.
41. Reid L. Who are the Philippine Negritos? Evidence from Language. *Hum Biol* 2013;**85**:1–15.
42. Delfin F, et al. The Y-chromosome landscape of the Philippines: extensive heterogeneity and varying genetic affinities of Negrito and non-Negrito groups. *Eur J Hum Genet* 2011;**19**:224–30.
43. Aghakhanian F, et al. Unravelling the genetic history of Negritos and indigenous populations of southeast Asia. *Genome Biol Evol.* 2015;**7**:1206–15.
44. Jinam TA, et al. Evolutionary history of continental South East Asians: "early train" hypothesis based on genetic analysis of mitochondrial and autosomal DNA data. *Mol Biol Evol* 2012;**29**:3513–27.
45. Chaubey G, Endicott P. The Andaman Islanders in a regional genetic context: reexamining the evidence for an early peopling of the archipelago from South Asia. *Hum Biol* 2013;**85**:153–71.
46. Endicott P. Introduction: revisiting the "negrito" hypothesis: a transdisciplinary approach to human prehistory in Southeast Asia. *Hum Biol* 2013;**85**:7–20.
47. Wang HW, et al. Mitochondrial DNA evidence supports northeast Indian origin of the aboriginal Andamanese in the Late Paleolithic. *J Genet Genomics* 2011;**38**:117–22.
48. Scholes C, et al. Genetic diversity and evidence for population admixture in Batak Negritos from Palawan. *Am J Phys Anthropol* 2011;**146**:62–72.
49. Pugach I, Stoneking M. Genome-wide insights into the genetic history of human populations. *Invest Genet* 2015;**6**:6.
50. Herrera R, et al. *Genomes, evolution and culture: the past present and future of humankind.* NY: Wiley Blackwell; 2016.
51. Friedlaender J, et al. The genetic structure of Pacific Islanders. *PLoS Genet* 2008;**4**:e19.
52. Kirch PV. Peopling of the pacific: a holistic anthropological perspective. *Annu Rev Anthropol* 2010;**39**:131–48.
53. Hung H-C, et al. The first settlement of remote Oceania: the Philippines to the Marianas. *Antiquity* 2015;**85**:909–26.
54. Blake NM, et al. A population genetic study of the banks and Torres Islands (Vanuatu) and of the Santa Cruz Islands and Polynesian outliers (Solomon Islands). *Am J Phys Anthropol* 1983;**62**:343–61.
55. Soares P, et al. Ancient voyaging and Polynesian origins. *Am J Hum Genet* 2011;**88**:239–47.
56. Hudjashov G, et al. Investigating the origins of eastern Polynesians using genome-wide data from the Leeward Society Isles, *Sci Rep* 2018;**8**:1823. doi:10.1038/s41598-018-20026-8.

CHAPTER

10

From Africa to the Americas

Kinship with all creatures of the earth, sky, and water was a real and active principle. In the animal and bird world there existed a brotherly feeling that kept us safe among them... The animals had rights - the right of man's protection, the right to live, the right to multiply, the right to freedom, and the right to man's indebtedness. This concept of life and its relations filled us with the joy and mystery of living; it gave us reverence for all life; it made a place for all things in the scheme of existence with equal importance to all. ***Standing Bear***[1]

SUMMARY

In 1590, a Spanish Jesuit priest and naturalist by the name of José de Acosta posited the notion that Native Americans were descended from hunters who had come to America from Asia. It is not clear what observations or ideas prompted him to formulate such a theory, considering that over 400 years ago (ya) he had little scientific information on which to base such an assessment. Since the prevailing views at the time were that Columbus discovered China, it is possible that Father Acosta was influenced by the ideas of the time. Since then, scholars have tried to elucidate the exact provenience, route, and time of the migration(s) of Paleo-Natives to America, and a number of disciplines such as anthropology, archeology, linguistics, and genetics have contributed over the years trying to answer these questions. Today, more than four centuries after Acosta's initial declaration, investigators are still in search of explanations. Over the years we have learned that the origin of Proto-Americans can be traced back to the original Out of African migrants that settled in the Near East, probably in what is today northern Pakistan approximately 60,000 to 50,000 ya. From the Near East, this population dispersed in different directions. One branch traveled along a coastal route bordering Southwest Asia, the subcontinent of India, Southeast Asia, and finally into Oceania. Another dispersal wave migrated in a northeasterly direction, eventually reaching Central Asia about 42,000 ya. Investigators have proposed that the direct ancestors of extinct and extant Native Americans derive from the Altaic region of Central Asia. From the Altaic, Paleo-Natives traversed toward Northeast Asia, reaching extreme northeast Siberia approximately 35,000 ya. From there, humans moved into Beringia, a relatively flat land bridge that is periodically submerged every 100,000 years or so as glaciers melt during warm interglacial periods. It is thought that in Beringia humans experienced a layover of about 15,000 years that allowed the residents to genetically and anatomically differentiate and partition into distinct populations before moving on into the New World. This hiatus was likely prompted by severe weather conditions as glaciers encapsulated the captive Paleo-Natives on both sides of the landmass.

Ancestral DNA, Human Origins, and Migrations
https://doi.org/10.1016/B978-0-12-804124-6.00010-0

The migrational pause in Beringia provided the time for microevolutionary changes to occur, thereby generating genetic diversity that subsequently spread throughout America once a passageway, in the form of an ice-free corridor, opened up in North America at the end of the last ice age. Although the linguistic, archeological, cultural, and genetic evidence accumulated since the latter half of the 20th century supports an intercontinental migration from Central Asia to America, the available data accumulated over the years provide contrasting scenarios that differ in number of incursions, times, and speed of migration(s). Some of the proposed speeds are very fast. Some models suggest dispersals from Beringia to Tierra del Fuego that lasted only a few thousand years. These migrants are thought to have accomplished those feasts as pedestrians or coastal mariners. It is likely that some of the discrepancies today derive from the paucity of comprehensive studies using different genetic markers, which has prevented direct comparisons among populations. Another contributing factor for the lack of congruency is the use of data from contemporary populations to extrapolate ancient events. After many demographic changes over thousands of years, some of such assessments may not be realistic. It is expected that with better methods to extract ancient informative DNA, advances in DNA sequencing technologies, and improved analytical tools, a better picture of the peopling of America will emerge.

WHERE THE TREK BEGAN

Since our species originated in sub-Saharan Africa 200,000 to 150,000 ya, it could be argued that we all, including Native Americans, ultimately came from Africa. Yet, when we think of Native American's *immediate* ancestors, most of us envision Asians. Today, the orthodoxy believes that Native Americans are descendants from travelers that went across the Behring Strait from northeast Siberia to Alaska. Yet, the story of the peopling of America does not begin in extreme Northeast Asia. The general picture that emerges from different disciplines, including anthropology, archeology, morphometric measurements, and genetics, is that the ancestors of Native Americans are not initially from Northeast Asia or outmost Siberia but from further west in Central Asia (the Altaic region), about 4500 km south west from what is today the Behring Strait. This seems counterintuitive as the immediate land to the west of the Behring crossing is northeast Siberia.

In turn, travelers from the Near East, possibly from North Pakistan, are responsible for the peopling of Central Asia. And these people from the Near East were in fact the decedents of the original Out of Africa migrants who moved in a northeast direction in pursuit of migrating game.

THE ALTAIC

Specifically, the South Altaic region of Central Asia has been recently identified as the origin of the Paleo-Natives that migrated to America, eventually reaching the tip of South America approximately 15,000 ya or earlier.[2,3] The South Altaic is located at the boundary of Central Asia to the west and East Asia to the east. It occupies a relatively small geographic area at the nexus of Russia, China, Mongolia, and Kazakhstan (Fig. 10.1). A region of stunning

FIGURE 10.1 Geographical location of the Altaic region. *From https://www.pinterest.com/pin/113997434294392321/; https://s-media-cache-ak0.pinimg.com/736x/c4/b8/26/c4b8260cfbccec9d7dedc06432ce76d2.jpg.*

FIGURE 10.2 The Altaic in Central Asia. *From http://www.56thparallel.com/russian-regions/altai-republic/.*

beauty (Fig. 10.2), the Altaic has been the crossroads of a number of migrations and civilizations during humans' recent evolution. The Altaic range, with peaks as high as 4506 m, bisects the landscape into the North and South Altaic.[4] Thus, this cordillera may have served as a barrier to gene flow, allowing for diversification of populations observed to the north and south.

Climatic Conditions

Currently, our planet is experiencing a long-term period of low surface and atmospheric temperatures known as an ice age. In total, the earth has experienced five major ice ages (the Huronian from 2400 to 2100 million years ago [mya], Cryogenian from 720 to 635 mya, Andean-Saharan from 450 to 420 mya, Karoo from 360 to 260 mya, and Quaternary from 2.58 mya to present). The timing of these ice ages has been rather irregular and some so severe that made the earth look like a snowball (e.g., during the Cryogenian epoch). We are still into the current major ice age, known as the Quaternary Ice Age, which began at the start of the Pleistocene. Evidence of this recent cold stretch is seen in the Greenland, Arctic, and Antarctic ice sheets, as well as in the numerous alpine glaciers in the northern and southern hemispheres. Within the ice ages there are intermittent, less-pronounced fluctuations of warmer and colder periods, which are known as *glacials* and *interglacials*. During these temperature pulses, the temperature differential could be up to 10°C, and ice sheets several km in thickness have been known to expand during glacials, covering extensive regions of the planet. For example, during the Cryogenian period, an impressive glacial period generated ice sheets that reached all the way to the equator. More recently, 26,000 to 13,300 ya, during the current ice age, the earth experienced a less dramatic glacial period that generated ice slabs several kilometers thick covering what is today New York City. During this cold spell, the Eurasian ice plate bordered the Altaic and the region was spotted with numerous glaciers in the mountainous regions at lower latitudes. During this time, ice expanses also occupied the entirety of northeast Siberia, blocking passage to what is today the Bering Strait and the northern portion of North America. In the American continent, this glacial period is known as the Wisconsin epoch.

We are now in the Holocene interglacial, a relatively warm epoch within the Quaternary Ice Age. The Holocene interglacial started approximately 12,000 ya.[5] Before this warming, ice cores indicate that a glacial age that lasted approximately 60,000 years kept global temperatures about 8°C colder than today (Fig. 10.3). This lowering of temperature did not occur suddenly. The cooling process of this last glacial period began about 120 kya and it reached its lowest

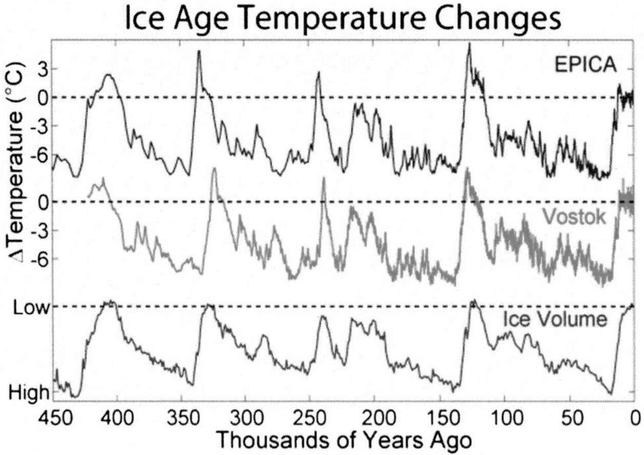

FIGURE 10.3 Glacial and interglacial temperature changes and ice volumes. *From https://en.wikipedia.org/wiki/Ice_age#/media/File:Ice_Age_Temperature.png.*

temperatures (last glacial maximum or LGM) around 20,000 ya. This glaciation, at its maximum, lowered the sea level in some regions of the earth by about 150 meters, and exposed a land bridge in what is today the Bering Strait. This ancient dry land is known today as Beringia, and it is thought to have facilitated terrestrial crossings by humans from Asia to America and back. Before this cold spell, there was a warm interglacial period from 115,000 to 130,000 ya. For the last 400,000 years, Earth has experienced a number of short interglacials lasting around 10–30 thousand years (ka) with interspersed glacial eras of 70,000 to 90,000 years in duration.

At the time of the original settlement of North Central Asia by AMHs as part of the second Out of Africa migration wave, about 42,000 ya[6] in the Upper Paleolithic, the world had experienced 80,000 years of a steady decline in temperature and was just short of the LGM, which would occur 20,000 years later. All the while, the temperature had already dropped 6°C from the previous warmest period, and the massive ice sheets were expanding rapidly covering the far north, turning it into a white wasteland and capturing much of the Earth's moisture. It is thought that these first travelers migrated from the region that is today the Near East and North Pakistan.[7] It is not clear why humans left the warmth of the south and moved north to the frigidity of North Central Asia as glaciers were advancing. Although the usual generic reasons for dispersals could be invoked, such as drought, tribal disputes, big game hunting, adventure, and/or curiosity about the unknown, no theories have received overwhelming support. Considering the hunting-gathering subsistence of these Paleolithic humans, it is likely that these people moved north following the big herbivores that preferred colder habitats. Thus, in such a scenario, the hunters travelled with the herds without a priori knowledge of where they were going, that is, without knowledge of a destination. They were just going where the food was. They had no planned route.

Early Settlement of North Central Asia

Notwithstanding the discovery of 45,000-year-old remains of an early human in the Ust'-Ishim district of Russia in western Siberia, the oldest encampment indicating occupation by modern humans in North Central Asia dates back to 32,000 to 33,000 ya at the Aldan River valley.[8] As the Ust'-Ishim sample consists of a single partial left femur found sticking out from the riverbank of the Irtysh River and no additional archeological artifacts are available, only DNA data exist from this mysterious population that may once have spanned Northern Asia. It is possible that the Ust'-Ishim femur represents the original out of the Near East migration to Northern Asia dating to about 45,000 ya.

Although the exact ages of the Aldan sites have been contested,[9] they clearly demonstrate a human presence in extreme weather at the height of the last glacial. Other more recent sites that are better documented and dated have been excavated above the Arctic Circle also in North Central Asia. One of these settlements was discovered in the Yana River delta dating to about 27,000 ya. The 27,000 ya time line corresponds closely to the pinnacle of the last LGM. Big game bones, fore shafts, and a variety of tools and flakes (Fig. 10.4) have been recovered from this site.[9] In general, these stone tools follow the Aurignacian tradition that extended from the Near East to Europe and Siberia. Thus, the style may signal a cultural continuity and link of Near East populations with Eurasians to the west and Siberians to the east. Notably, some of the tools resemble the Clovis tradition of Native American tribes. The wide diversity of animal remains left at these sites include horse, rhinoceros, caribou, musk, wooly mammoth, bear, boarder, and fox, among others. These animals were hunted with spears and

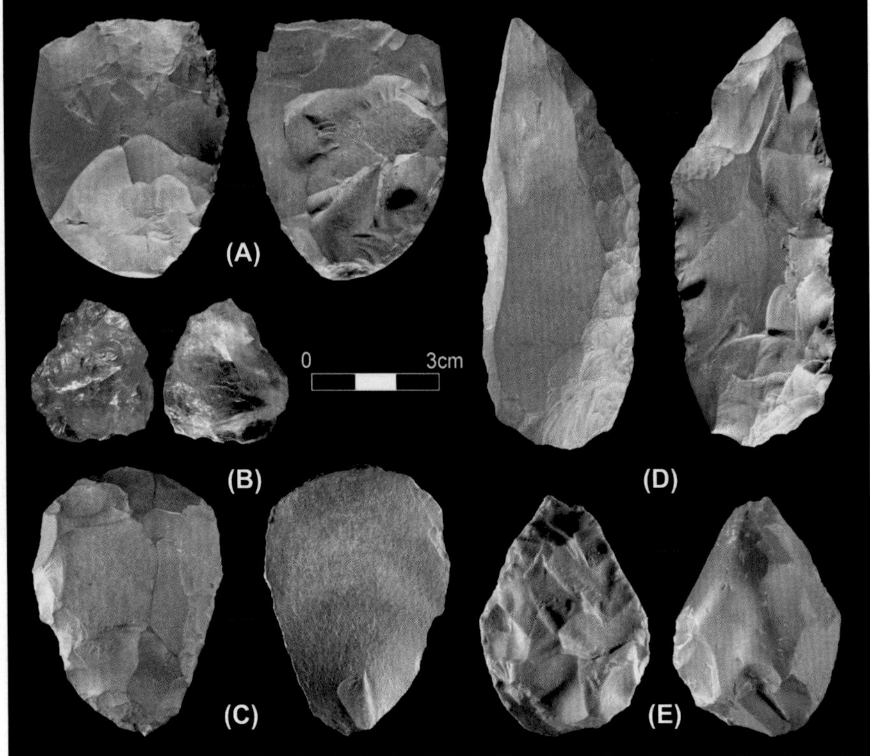

FIGURE 10.4 Central Asian stone tools. A–E represent various stone tools illustrating front and back views. *From http://www.ncbi.nlm.nih.gov/pubmed/?term=The+Yana+RHS+Site%3A+Humans+in+the+Arctic+Before+the+Last+Glacial+Maximum+V.+V.+Pitulko.*

lances without the assistance of implements like the spear throwers known as atlatls. These findings from the late Pleistocene demonstrate that humans somehow were able to adapt, hunt, and survive in the extreme Arctic and sub-Arctic conditions.

The People of the Altaic

Culture, Language, and Mythology

Altaic populations of today are very diverse physically, genetically, culturally, and linguistically. This high degree of heterogeneity is likely the result of gene flow and admixture involving ancient groups from East and West Eurasia that started converging in the region beginning in the Upper Paleolithic (around 42,000 ya) extending into the Mesolithic (about 11,500 ya).[4] Consequently, some Altaics possess oriental features, such as flat facial profiles, high cheekbones, epicanthic folds, thin eyes, and straight black hair. In other Altaics, these oriental characteristics are less prominent and reflect a more characteristic European physical appearance. In between, there are individuals that represent the entire spectrum of appearances. Also, undoubtedly, a considerable amount of gene flow has occurred in recent times, especially from European Russian settlers from the west.

Today, residents of the Altaic region speak a number of diverse, mutually unintelligible Turkic languages in addition to Russian. It has been postulated that the domain of the Altaic mountain range was the birthplace of the family of Turkic languages. Although their life style has change in recent times, traditionally, Turkic-speaking people have been nomadic pastoralists. Turkic is the largest branch of the Altaic family of languages. There are about 50 Altaic languages spoken by about 145 million speakers representing 2.5% of the world's population. The 30 or so Turkic tongues share a number of common words and roots for everyday things, such as *tau* for mountain, *kol* for lake, and *su* for water.[4] Many of the Turkic languages are only spoken and have no written system. Turkic can be further divided into four subbranches that correspond closely to geographic regions within Central Asia. These are the southwestern (e.g., Turkish), the northwestern (e.g., Kazakh and Rartar), the southeastern (e.g., Uzbek), and the northeastern (e.g., Yakut) groups. The northeastern Turkic group or Yakut-related Siberian languages are of particular interest to the topic of this chapter due to the geographical proximity of Siberia to Beringia and the American continent. It is noteworthy that the Yakut-related languages of Kamchatka and Chukotka in the extreme northeast Siberia are related to some Native American languages. These Siberian Paleo-Asians include the Chukchi, Koryaks, Itelments, Aleuts, and the Eskimos or Inuit. It is significant that the last two groups have representative populations currently living in Alaska, indicating relatedness among populations on both sides of the Bering Strait.

The traditional mythology of the South Altaic populations is based on Shamanism. In the practice of Shamanism, individuals, especially holy men or women, achieve an altered state of consciousness that allows interaction with the spiritual world, providing a mechanism for channeling power into the material world. The main deity in the Altaic region is Tengri or sky god with certain animals symbolizing specific lesser spirits with specific attributes. For example, the wolf exemplifies honor and represents the mother of the Turkic people. The horse is also revered by the South Altaics. The beginnings of shamanism involving worship of animal spirits and sky god reverence is thought to exist in North Central and North East Asia since the early Upper Paleolithic, about 40,000 ya. The shaman, as head of the spiritual life, occupied an elite social position and his/her privilege status could be seen as the beginning of sociopolitical stratification and elitism within these populations. Although a number of tribal groups worldwide practice shamanism, interesting similarities exist between Native American and Altaic mythos. In both groups of people, a state of drug-induced ecstasy allows shamans to depart the body and the material world and travel to communicate with spirits using key animal helpers for the purpose of obtaining knowledge and power and providing healing.

Genetic Characteristics

THE NEAR EAST

While the similarities in language, culture, and spiritual practices may suggest links between groups of individuals, genetic markers are also useful in establishing affinity. For example, genetic markers are increasingly providing information on the story of how modern humans traveled from East Africa to the Near East and beyond.

The immediate landmass encountered by modern humans after they left Africa was what we know today as the Arabian Peninsula. It is postulated that these people arrived in the Near East and then started a circum-Indian dispersal eastward along the coast of Iran, Pakistan, and the subcontinent of India. It is speculated that early in this migration, a group of travelers took a turn northward into Central Asia via Northern Pakistan.[10] Thus, the Near East and specifically Pakistan may have played a pivotal role in the peopling of Central Asia, Northeast Asia, Siberia, and the American continent.

The first modern humans that occupied the Altaic territory were migrants from Southwest Asia or the Near East, possibly northern Pakistan.[7] This part of the world is located within the putative southern coastal route travelled by modern *Homo sapiens* in their Out of Africa trek. It is thought that these modern humans reached this part of Southwest Asia approximately 60,000 to 70,000 ya or earlier (see Chapter 5), soon after the Red Sea and/or Levant crossing from Africa into the Arabian Peninsula.[11] Indeed, cave habitation has been detected in northwest Pakistan from the Late Paleolithic, but fossil remains are limited.[11]

Today, Pakistan is comprised of a very heterogeneous group of populations with very diverse origins that encompass at least 18 ethnic groups and speak about 60 very distinct languages.[12] As the settlements of many populations in Pakistan have occurred in recent times, in the past few hundred or 1000 of years (depending on the population in question) with highly diverse origins (e.g., Baluch from Syria, Balti from Tibet, Burusho from Alexander the Great's army, and Parsi from Iran, to name a few), the current genetic makeup of Pakistan does not reflect the gene pool when humans first ventured into North Central Asia about 45,000 ya. Even rather ancient groups like the Kalash, thought to derive from Greek migrants, date back to only 8000 to 12,000 ya in Pakistan.[13] Thus, it is highly likely that the ancient Pakistani DNA types have been largely replaced by subsequent migrations or gene flow from various locales. Unfortunately, only limited ancient DNA data are available from Pakistan and the Altaic regions. Most of the genetic composition in these regions is based on extrapolations from contemporary samples and absolute dates are rather circumstantial, especially when investigating remains dating to the early Late Paleolithic.

THREE MAIN TYPES OF DNA

Genetic data are derived from uniparental markers such as mitochondrial and Y-specific DNA passed down from mother or father, respectively, or biparental (inherited from both parents) sequences. The advantage of uniparental DNA as markers is that they provide direct lineage, nonrecombining information from generation to generation. In other words, in uniparental inheritance, maternal and paternal genes are not exchanged during reproduction, allowing straightforward assessment of ancestry along parental lines. In biparental inheritance, on the other hand, the genetic material from both parents is reshuffled every generation as a result of exchanges of maternal and paternal DNA within chromosomes. This reorganizes maternal and paternal DNA within individual chromosomes. Thus, the resulting recombined chromosomes are a mixture of both parents. This recombination of DNA makes it difficult to follow the maternal or paternal ancestry using biparental inheritance.

In the case of Y-specific genes, they allow assessment of the paternal ancestry of individuals and populations. Thus, for example, when investigators are interested in ascertaining the migration of male individuals along the Silk Road or from North Pakistan to the Altaic, they examine the distribution of Y chromosome–specific markers that occur within this geographic range. Then, they also would compare those distributions to those outside that range to see if the markers originate from those areas or whether they originate from other parts of the world. In addition, the Y chromosome is, for the most part, devoid of functional DNA. Almost the entire Y chromosome is made up of highly repetitive DNA sequences performing no obvious function. Thus, it is likely that these Y chromosomal sequences are essentially selectively neutral. Selective neutrality is paramount in evolutionary studies because the frequency of DNA markers is totally dependent on the occurrence of mutations and random frequency changes from generation to generation. Selection pressure, because of functional requirements and fitness, can obscure true phylogenetic relationships among individuals and populations. For instance, selection pressure could erroneously indicate close affinity between two populations when they are subject to similar environmental conditions and selection pressure. The converse may be the case as well if closely related populations are exposed to unique habitats, resulting in contrasting selection pressure. In such a scenario, phylogenetically related populations would *artificially* appear different.

Similarly, the maternal lineage is regularly examined by looking at genes or sequences on the mitochondrial DNA (mtDNA) located inside the mitochondrion, a cellular organelle. Unfortunately, the resolution afforded by mtDNA is limited because its size is only about 16,500 nucleotides (the building units of the DNA double helix) as opposed to the nearly 59 million nucleotides that make up the Y chromosome. As males and females of our species are subject to unique sociocultural conditions, and mtDNA and Y chromosome mutation rates as well as selection pressures are different, oftentimes the evolutionary panorama derived from both types of DNA markers differs. For example, in societies where males are involved in hunting, males may be exposed to certain types of dangers, which often result in premature death. These types of sex-specific biases may lead to differences in evolutionary histories based on mtDNA or Y chromosome sequences. Both male Y chromosome– and female mtDNA–specific lineages do provide DNA that is inherited intact from one's ancestors, only being altered by mutations that are accumulated within lineages over time.

Autosomal DNA, as previously discussed, experiences recombination and shuffling of DNA sequences, resulting in the mixing of maternal and paternal genetic material. Although autosomal DNA does not allow direct assessment of female and male lineages, it provides a more comprehensive picture of the genetic constitution of individuals and populations because the vast majority of the DNA in a cell is autosomal.

THE Y CHROMOSOME

Although Y chromosomes are usually examined using a number of genetic markers such as short tandem repeats (STRs), insertion/deletions (in/dels), and point mutations (one or a few nucleotide differences), they are ultimately characterized according to haplogroups or types. Scientists to characterize and compare extant and extinct human populations use these haplogroups as genetic markers. For example, the contemporary Y chromosome haplogroups of the general population of Pakistan are mainly R1a-Z93 (24.4%), J-P209 (15.3%), and L-M20 (13.1%). These three haplogroups represent over 50% of the Y types in Pakistan. In the Altaic populations, R1a-Z93 is the most abundant haplogroup, with frequencies ranging from 50.0% in the Southern Altaic to 11.8% in the Northern Altaic region. It is possible that these high R1a-Z93 frequencies represent the original migrations from the Near East via Pakistan to the Altaic. This is especially the case in relation to the Southern Altaic, considering the area's reported genetic affinities to Southwest Asia.[14] R1a-Z93 was transported by Bronze Age Indo-Europeans west of the Urals and into the plains and deserts of Central Asia and the metal-rich Altai mountain range. Thus, it is possible that R1a-Z93 signals an early Neolithic (approximately 12,000 ya) migration from the area of Pakistan into Central Asia and the Altaic. It is interesting that the R-M207, the ancestor haplogroup of R1a-Z93, is also thought to have originated in Southwest Asia, possibly before the LGM during the Late Paleolithic (26,800–34, 300 ya).[15] In a pertinent investigation, 24,000-year-old remains of a boy (the Mal'ta boy) belonging to haplogroup R-M207 were recovered from an area just east of the Altaic.[16] This person was a member of a group of people that hunted big game in North Central Asia extending into Siberia during the Late Paleolithic at a time of extreme glaciation. These studies may suggest that people carrying haplogroup R-M207 were part of the original population of Southwest Asia and Pakistan that migrated to Central Asia. Overall, these results could be interpreted as evidence for two migrations, one early during the Late Paleolithic and a second one more recent, dating to after the LGM from Southwest Asia via Pakistan to North Central Asia and the Altaic.

Surprisingly, when the complete genome (entire DNA) of the 24,000-year-old Mal'ta remains and an additional individual from the Afontova Gora mountains dating to 17,000 ya[17] were sequenced (genome-wide analysis), investigators found a surprisingly close affinity between Native Americans and modern-day western Eurasians that specifically link them to the Near East and no connection to contemporary East Asians. Furthermore, it was estimated that approximately 14%–38% of current Native American ancestry derives from the West Eurasian populations represented by these two ancient individuals. These data support the contention that at least some of the Paleo-Natives that migrated to America *initially* originated from West Eurasia and not from Northeast Asia or extreme northeast Siberia. This information and the genetic connections to the Near East and Pakistan by way of the Altaic could offer an explanation for why several crania of Native Americans exhibit morphological characteristics that do not resemble those of East Asians but rather those of West Eurasia.[18]

Another pertinent Y chromosome marker, J-P209, originated in the Near East about 42,900 ya and from there it expanded into Pakistan and then North Central Asia.[19] Although J-P209 is only present at 2.5% frequency in the Southern Altaic, this marker may represent genetic flow from the Near East as well. However, this marker does not appear in Native Americans, and thus, considering its low frequency in the Southern Altaic, it is possible that J-P209 dropped out via random chance (genetic drift) from the starting Asian Paleo-Native migrants before the Beringian crossing or it was lost from the founding populations after their arrival to America.

Pakistan and the Southern Altaic share a number of other Y chromosome markers as well. Of those, haplogroup Q-M242 and C-M130 are of particular interest as they are also found in America (Fig. 10.5).[20,21] Q-M242 is the original mutation that gave raise to lineage Q, all other types of Q haplogroups derive from it. Q-M242 possibly originated in Central Asia approximately 31,400 ya, and from there it expanded to the north and west, eventually reaching the Altaic, northeast Siberia, and then the American continent. At the South West extreme of its range in Northern Pakistan, the frequency of Q-M242 ranges from 2.1% to 5.2%[22] depending on the specific population in question. It is in Pakistan that the most ancient types of the Q haplogroup are seen, the less defined and older Q-M242 and Q-M346. The Q-M346 type derives from Q-M242 (the Q-M242 type is ancestor to Q-M346). Q-M346 originated in Central Asia. In the Northern Altaic, the frequencies of all haplogroup Q categories range from 68.0% to 40.7% and these individuals are almost totally of the Q-M346 type. Southern Altaic, on the other hand, possesses about 16.7% Q with all individuals belonging to the younger Q-L54

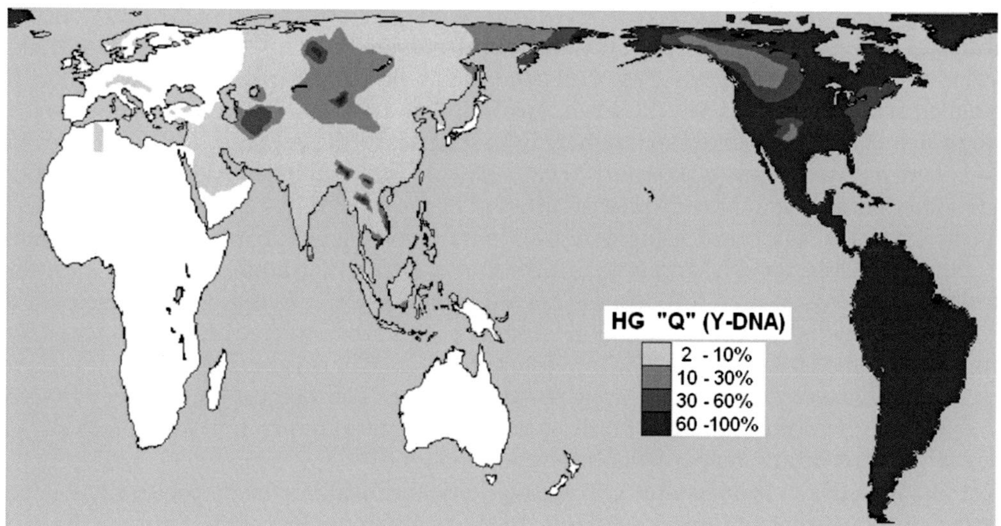

FIGURE 10.5 Worldwide distribution of haplogroup Q. *From https://en.wikipedia.org/wiki/Haplogroup_Q-M242.*

mutation.[14] Considering the easterly progression in place of origin and the sequentially younger age of haplogroups Q-M242, Q-M346, and Q-L54, in that order, it is possible that these mutations occurred as humans migrated northeastward from the Near East toward Central Asia and the Altaics.

C-M130, the other Y haplogroup of importance, in relation to the Native American diaspora, may be Indian in origin and dates back to about 53,000 ya. It is postulated that carriers of C-M130 initiated a northward dispersal to Central Asia around 40,000 ya. It is found at frequencies as high as 8.2% in Pakistan.[22] Its highest levels (75.5%) are observed in the Hazak of North Central Asia.[21] All Pakistani individuals belong to the basal or most ancient type of haplogroup C (C-M130) or to the more recent mutation C2-M217. It is thought that the C2-M217 mutation originated about 12,000 ya in Central Asia[23] and from that epicenter it spread in different directions. In the Southern Altaic, the frequency of C2-M217 is 20.0% while no haplogroup C has been detected in the Northern Altaic.[2] All of the C haplogroups in the Southern Altaic is of the C2-M217 type. As with the Q haplogroup, it is likely that the younger C2-M217 mutation occurred in Central Asia or possibly in Southern Altaic from the ancestral, older C-M130. It is thought that Na-Dené-speaking peoples transported the C2-M217 mutation into the northwest Pacific coast of North America. Na-Dené-speaking groups are thought to have arrived in America from Asia 6000–8000 ya. Alternatively, others contest that this language family, that includes Athabaskan, Eyak, and Tlingit languages of Native Americans, developed earlier in Beringia, between the two continents, during a migrational hiatus.

MITOCHONDRIAL DNA

Similarly to the classification of Y chromosome types discussed above, mtDNA can also be grouped into haplogroups and used as genetic markers for the purpose of following human migration and ancestry. All human females found outside Africa belong to only two mtDNA haplogroups (M and N) and their derivatives. Another way to express this fact is that all the mtDNA types outside Africa derive from M and N. In turn, both are descendants from haplogroup L3 that is African-specific and not found outside Africa. Currently, it is not clear whether the mtDNA M and N types originated in sub-Saharan Africa or in Arabia after the original crossing of the Red Sea (Strait of Sorrow). Some authorities are of the opinion that haplogroups M and N occurred somewhere in East Africa or the Persian Gulf.[24] Haplogroups M and N are approximately 60,000 and 70,000 years old, respectively.[25] Although it is not certain whether types M and N were part of different or the same initial migration out of Africa, the orthodoxy is in support of a single migration transporting both haplogroups into Arabia.

Although mtDNA types M and N define the Out of Africa migration, it has been proposed that type N was transported through the Levant route (north), whereas M dispersed via the Horn of Africa to the south.[25] The data in support of this scenario are the abundance of haplogroup N and the lack of the M type in western Eurasia, whereas M is observed at high frequencies in India and further east. Along these lines, data also suggest that dispersals across the Levantine corridor occurred not only to exit Africa but to return to it as well, as Back to Africa migrations during Upper Paleolithic and Neolithic.[26]

Just like with the Y chromosome–specific markers, the Near East and specifically Pakistan are rich in mtDNA diversity, and a lot of this diversity differentially partitions along the many ethnic groups of the area. In general, it seems that mtDNA types M and N, after leaving Africa, underwent a maturation phase in the Near East, subsequently branching out into a number of subhaplogroups, some of which went on to populate Central Asia.[27] In the northern regions of Pakistan, for example, the Pathans (the second largest ethnic group in Pakistan) are made up of 30.0% mtDNA type M (the most abundant subhaplogroup is M3 at 8.7%), 7.8% N and 61.3% R (the most abundant subhaplogroups are U7 at 11.3% and HV at 10.4%). The R mtDNA haplogroup is a descendant of type N that originated about 66,000 ya in South Asia not long after the initial Out of Africa crossing.[28] In other Pakistani populations, various haplogroups are detected, such as W6 (12.9%), M5 (11.7), and U2b2 (9.4%) in the Saraiki and L2a (15.0%) and R2 (6.0%) in the Makrani.

Approximately 40,000 ya, during the Upper Paleolithic, modern humans arrived in Central Asia from the Near East. The mtDNA types carried by these migrants included N and R. Specifically, the N haplogroup consisted of subhaplogroups N1, N2, and X, whereas type R was represented by subhaplogroups U1, R2e, R3, R4, R5, R7 and R8, TJ, and HV. At about the same time, there is evidence that Denisova (name of the cave where the remains were found) hominins were living in the Altai Mountains. Denisovans were members of an extinct *Homo* species (*Homo* sp. *Altai*) known only from teeth, and finger and toe bones (one each). Thus, the two *Homo* species may have cohabitated, providing for the possibility of interbreeding. This temporal and spatial overlap may explain the Denisova DNA detected in certain modern human populations (see Chapter 5 for modern humans and archaic *Homo* introgression).

Currently, Central Asia represents an amalgamation of Eastern and Western European mtDNA haplogroups.[27] The Altaic, in particular, is made up mostly of mtDNA of East Eurasian origin. Yet, Southern Altaic populations exhibit one of the highest proportions of West Eurasian mtDNA types (23.6%) in the region. Although both Northern and Southern Altaics possess the B, C, D, and U4 types, haplogroups C and D are the most abundant, with frequencies of 31.4% and 13.0%, respectively, in the Southern Altaic.[2] Type C mtDNA originated about 60,000 ya in Central Asia, and type D arose 48,000 ya in Asia from haplogroup M. Considering the distribution across Asia and the times of origin of types C and D, and their derivatives, it is likely that each matured into different subhaplogroups as modern humans migrated northward from the Near East to Central Asia.

In a recent study that examined populations from Central Asia and Sakha, to the northeast in Western Siberia, investigators found genetic continuity with western Eurasian populations.[29] This study was based on high-resolution mtDNA, Y-specific and biparental DNA markers. Although the major part of West Eurasian maternal and paternal lineages in Sakha could originate from recent gene flow from East Europeans, predominantly recent Russian settlers, mtDNA haplogroups H8, H20a, and HV1a1a, as well as Y chromosome haplogroup J, probably reflect ancient dispersals from West Eurasia (e.g., Near East) through Central Asia (e.g., Altaic) and south Siberia.

Haplogroups C and D are of special interest as they are also found in America.[30] Type B, although found at lower frequencies in Altaic populations compared with C and D, is also represented in America Natives. Haplogroup B is also linked to the Near East and the successful Out of Africa dispersal of modern humans since it originated approximately 50,000 ya in Asia and derives from haplogroup R.

BIPARENTAL DNA MARKERS

Starting about a decade ago, the field of human evolution experienced an increase in interest in phylogenetic relationships derived from biparental markers. Unlike genetic markers on the Y chromosomes and mtDNA, which derive only from father and mother, respectively (uniparental inheritance), biparental markers are passed on to offspring by both parents, and thus, are more representative and represent more comprehensively the genetic makeup of individuals and populations. These studies are sometimes referred to as genome-wide as the data are generated from most of the DNA content (genome) of individuals. In these investigations, DNA sequences are mined from throughout the genome and the data are used for individual and population comparisons. In other words, nucleotide differences in the DNA are scored as markers. In practice, these DNA differences or mutations are ascertained by determining SNP differences or scoring the number of repetitive sequences, such as STR. Considering that most of our DNA is not mitochondrial or Y chromosome–specific, genome-wide markers provide a more complete picture of the diversity and ancestry of populations.

Seminal studies using genome-wide markers based on SNPs and STRs have revealed genetic relationships consistent with the putative order of the modern human expansions; first Out of Africa, then to the Near East, Central Asia, North East Asia/Siberia and to America in that order.[31] This type of enquiry typically uses large numbers of markers in the hundreds of thousands of SNPs, affording extensive, representative, and uniform coverage of the entire genome. Typically, in a worldwide study, 650,000 SNPs were employed to explore the direction of modern human migrations.[31] In these studies, Near Eastern populations, including Pakistan and Central Asian populations, were found to exhibit genetic affinities to one another when examined with SNP and STR markers.[31,32] This is graphically represented in the orange component shared by the Pathan, Burusho, Sindhi, Hazara, Makrani, and Balochi populations from Pakistan with the Mongolian and Uygur groups of Central Asia in the structure analysis represented in Fig. 10.6. Structure analysis is a graphic representation (bar graph) that summarizes the genetic makeup derived from genome-wide markers of individuals in populations. This same genetic signature (i.e., the orange component) even extends into northeast Siberia (see the Yakut population) and America (Maya). This observation is consistent with the mtDNA and Y-specific data

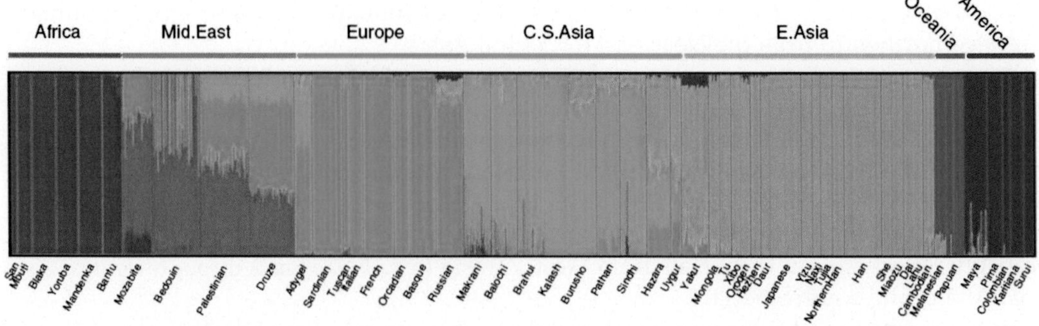

FIGURE 10.6 Genetic component analysis. *From http://www.zoology.ubc.ca/~bio336/Bio336/readings/Li2008.pdf.*

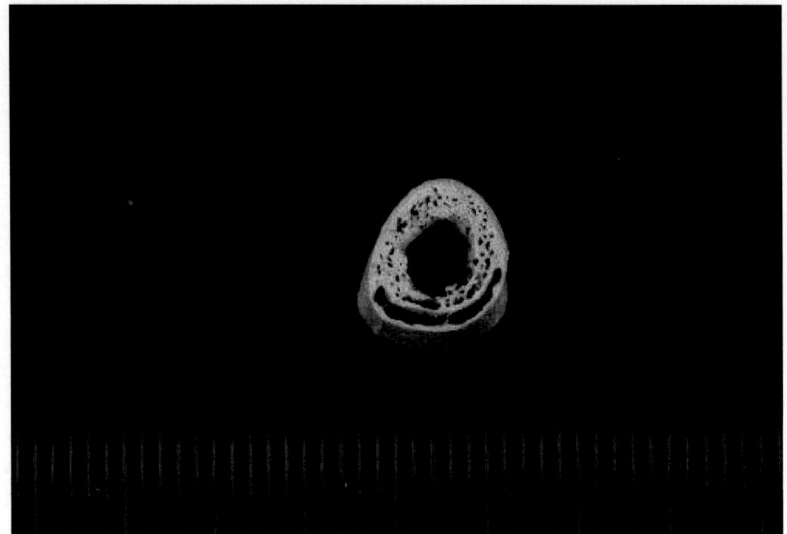

FIGURE 10.7 Cross section of Mal'ta boy's humerus used to extract DNA. *From http://phys.org/news/2013-11-ancient-siberian-genome-reveals-genetic.html.*

indicating connections between the Near East, Central Asia, and America. Also, the genomic data indicate a progressive reduction of genetic diversity following the putative migration route from Africa to America. In other words, high diversity is observed close to the origin of the migration (the Near East) and low diversity in the direction of America. This observation is consistent with the expected sequential bottleneck events involving small migrating groups of people as humans dispersed and branched out toward America.

Whole genome ancient DNA studies have also revealed western Eurasian and Native American connections.[16] The biparental genetic component of the 24,000-year-old Mal'ta boy, discussed above in connection with Y-specific DNA, has been completely sequenced. The DNA was extracted by grinding to a fine powder a small portion of bone from the boy's humerus (Fig. 10.7) and adding a solution to recover the nucleic acid within. The Mal'ta boy represents one of the oldest anatomically modern humans sequenced thus far. The Upper Paleolithic remains were recovered as part of a burial complex that included a Venus figurine, a hood, beads, and a number of other personal effects. To the surprise of the investigators, the sequenced whole genome was found to be quite different from the DNA of the contemporary human populations of the region, resembling more the gene pools of western Eurasians. These temporal differences in people indicate that extrapolating the genetic makeup of ancient populations based on data from contemporary people may be misleading. These results also indicate that some Paleo-Natives of North Central Asia migrated from the west. Furthermore, the Mal'ta boy's DNA exhibits close affinity to contemporary Native Americans, but remarkably, not to East Asians who are generally considered as being ancestrally closer to Native Americans.

THE PEOPLING OF NORTHEAST SIBERIA

Origins of Northeast Siberian Populations

It is well accepted by the scientific orthodoxy that Southeast and Northeast Asian populations differ morphologically and genetically. Since the late 1980s, investigators have noticed a partitioning in genetic markers[33] and anatomical features[34] when comparing Southeast and Northeast Asians. Northeast Siberian populations, for instance, are biphyletic in origin.[35] That is, they derive from two main sources. It seems that in addition to the Central Asian populations, including the Altaic, northeast Siberia received migrations from Southeast Asia, who were possibly the direct ancestors of the initial Out of Africa migrants. It is thought that these humans arrived to Southeast Asia via Middle Asia about 60,000 ya following the coastal route. Soon after this, humans migrated northward from Southeast Asia, reaching Japan about 30,000 ya.[36] Genetic age estimations based on distinct Y chromosome D-M174 lineages indicate that Out of Africa humans migrated from South Asia into Southeast Asia and Australia independently.[36,37]

Although it is possible that the Altaic ancestors of northeast Siberian populations also descended from East Asian groups, it is likely that they are of Southwest Asian ancestry via Central Asia, as postulated previously. These two sources for the genesis of the northeast Siberian populations are referred to as the northern (i.e., from the Altaic) and southern (from Southeast Asia) routes. Based on archeological and genetic data, as previously mentioned, humans travelling the northern route arrived to northeast Siberia during the Upper Paleolithic, about 30,000–40,000 ya. Genetically, this dispersal is evident in the Y-specific signature of the C-M217 and Q-L54 markers, as well as mtDNA haplogroups C and D in northeast Siberia. These genetic markers not only link Altaic and extreme northeast Siberian populations but also signal intercontinental connections across the Behring Strait. The ancestral populations that migrated from Southeast Asia to northeast Siberia (the southern route) are evident in the signature of the D-M174 and O3-M122 Y chromosome mutations as well as the mtDNA haplogroup B carried northward to northeastern Siberia by dispersals after modern humans arrived to Southeast Asia about 60,000 ya.

Habitat and Survival

It is estimated that humans got to northeast Siberia approximately 35,000 ya.[9] At that time, the earth's climate was about 15,000 years short of reaching the LGM (Fig. 10.3). The temperature 35,000 ya had already dropped about 6°C from the glacial minimum, 2° short of the LGM's utmost low. The ice sheet buildup was also at about 75% of its maximum. Although most of Northern Europe and a good portion of North America (to 45°N latitude or just south of New York City across the Hudson River) were covered with glaciers, some regions of northeast Siberia were mostly free of ice because of minimal snowfall, yet it was bitterly cold and inhospitable. Throughout the last glacial period, from around 120,000 to 15,000 ya, the land of northeast Siberia west of the Lena River and Lake Baikal (127.0°E longitude) was a grassland steppe almost devoid of trees. This plain, known today as the Central Siberian Plateau, dramatically changes northeast of the Lena River. Starting with the Verkhoyansk Range located between the Lena and Aldan rivers (129.0°E longitude) in an eastward direction, the terrain turns extremely mountainous. This cordillera is 1000 km across with peaks reaching 2500 m, encapsulating the easternmost portion of northeast Siberia. This rather elevated terrain continues into the eastern tip of Siberia. In certain regions, like the Kamchatka Peninsula, volcanic peaks reach over 4000 m high. This difficult terrain in combination with adverse climatic conditions certainly made any travelling very challenging for Paleo-migrants. For ancient Siberians heading east, the landscape again turned into plains and gently rolling hills in what is today submerged Beringia.[38] Even today, in a warm interglacial period, northeast Siberia is tundra, registering the lowest temperatures on the planet. Yet, in spite of these extreme geographical and climatic conditions, there is solid evidence that by 27,000 ya modern humans had coastal encampments above the Arctic Circle (71°N latitude) at the delta of the river Yana by the Arctic Ocean.[9]

The distance between the Altaic in Central Asia and northeast Siberia is approximately 2800 km of difficult real estate, a definite challenge for pedestrians. It is known that the people that dispersed northeastward to northeast Siberia were highly specialized hunters. Primarily, they were not gatherers. Among humans in the tundra of North Central Asia and Siberia, limited collection of plant products likely took place. Yet, a number of large mammals grazed the land in the Central Siberian Plateau and the mountainous valleys to the east. These animals included wolf, bison, horse, reindeer, caribou, musk ox, woolly rhinocero, lion, glyptodon, mastodon, and mammoth, among others. Individuals in coastal encampments would have had the opportunity to hunt seals and walruses as well as scavenge an occasional whale carcass. Also, it is likely that these groups of humans fished in the numerous streams and rivers that permeated the region. Under these conditions, the Paleo-Natives likely migrated without specific directional aims; rather they were just following food resources in the form of pasturing herds. Considering the arrival time of modern humans to northeast Siberia about 35,000 ya, the random dispersal from the Altaic must have taken at least 5000 ya at a rate of about 0.56 km per year. This is a reasonable speed, taking into account that the dispersal was widespread as an advancing wave, not unidirectional, and back and forth movements occurred at random. Thus, the only factors impacting the direction of travel were geographical obstacles, such as rivers, lakes, and mountains, the routes of the herds that they pursued, as well as any potential conflicts with neighboring clans that they likely avoided.

It is thought that these Paleo-Natives lived in small groups of one to a few dozen who carried on a communal existence that involved sharing in tasks as well as in the price and danger of the hunt. Considering the highly mobile mode of survival needed to succeed in the tundra habitat of the northeast Siberian landscape, it is likely that these modern humans built

FIGURE 10.8 Painting of Chukchi households by Louis Choris, 1816 Huts. Year 1816. *From https://en.wikipedia.org/wiki/Siberia#/media/File:Choris,_Tschuktschen.jpg.*

temporary tipi-like huts similar to the ones currently made by Siberian and North American natives (Fig. 10.8). This type of structure made out of small branches and/or large bones (Fig. 10.9) and wrapped in animal hide would have provided some protection and allowed for repeated dismantling, transportation, and reassembly as required during the trek. This type of shelter would have been essential in the rather structureless plains of the Central Siberian Plateau because natural dwellings, such as rock formations or caves, were not available to the nomadic hunters.

As wood was difficult to come by and limited to small amounts of driftwood, it is likely that these nomadic migrants utilized fat and oil from animals to keep warm, cook, and light up the nights. The discovery of burned mammoth bones in the remnants of archeological-site fireplaces indicate that massive bones (e.g., vertebra) of mastodons and mammoths were used as fuel for fire. This practice was logical and practical as the large bones burn longer and provided fat for fuel and food. We can envision Paleo-Natives sitting around the fire, performing some manual work, talking, eating, and throwing mammoth bones into the fire, mixing them with charcoal for comfort and cooking.

In terms of the clothing and footwear made by Paleo-Natives, few examples have been recovered as the material used, being skin and fur, deteriorates rapidly. Yet, judging by the garments of contemporary arctic people of Siberia and Alaska, such as the Eskimos, it is likely that Paleo-Natives wore layered animal hide and fur clothing. Specifically, reindeer and bear skin are particularly good in providing warmth. These animals were readily hunted at the time. Possibly, both clothing and footwear were made of a combination of fur for warmth and waterproof intestinal lining or sealskin for insulation to keep dry. Modern-day Yup'iks, for example, make boots known as mukluks using waterproof sealskin with fur inside and wear insulating socks made up of dry grass.

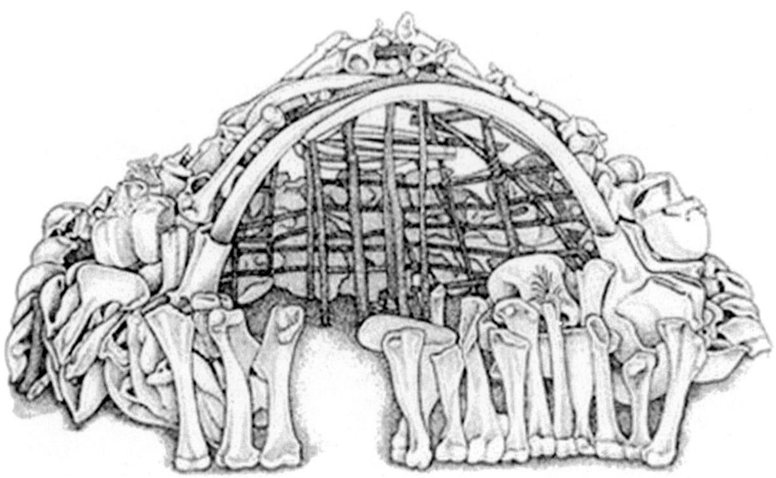

FIGURE 10.9 Drawing of a mammoth bone hut with cover made of wooly mammoth skin from the Paleolithic. *From http://grade3archaeology.weebly.com/ukraine.html.*

Hunting tools consisted of chipped stone tools employed for cutting meat or assembly into spears. Hunting strategies likely consisted of surprising blitzkrieg attacks in which animals were surprised and subdued by a number of clan members as spears were thrown as projectiles to penetrate the prey in their most vulnerable body parts, such as the neck, to cause maximum damage. These were not direct close-quarter attacks. The hunters positioned themselves at a relative safe distance from the dangerous large prey before launching their spears. Excavations at numerous sites suggest that the large animals were already in distress when corralled by humans. In many instances, it seems as if the megafauna were partially disabled as a result of being stuck in a mud pit of a swamp. Subsequent to the kill, animals were skinned, disemboweled, and partitioned into transportable pieces, which were then taken to camp to be distributed among the clan members.

Art

The artistic expressions of Siberian Paleo-Natives consisted of two main modalities. One category comprised articles of a personal nature. It included transportable items, such as beads, Venus figurines, hoods, articles of clothing, as well as bone decorations and amulets (Fig. 10.10A). The Mal'ta boy burial remains discovered in the Central Siberian Plateau and dating to at least 24,000 ya probably comprise the most well-preserved archeological ensemble thus far. Some of the pieces buried with the dead boy seem to be for decorative purposes and/or of religious significance, such as the string of beads, which probably was worn hanging by the neck (Fig. 10.10B). The Venus figurine is clearly of deistic significance. These statuettes are found throughout Eurasia, including northeast Siberia. The one found at this site belong to the Mal'ta–Buret' tradition. They usually portray a female figure, sometimes pregnant, with accentuated breasts, hips, backside, and genitalia.

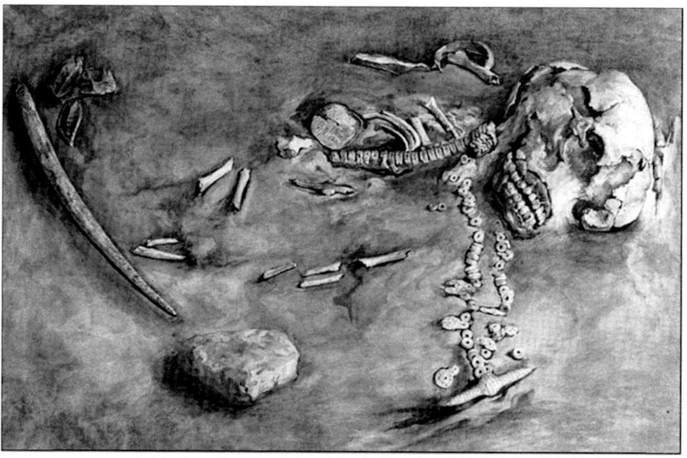

FIGURE 10.10A Mal'ta boy remains and details. *From http://siberiantimes.com/upload/information_system_38/1/2/0/item_1201/information_items_1201.jpg; http://donsmaps.com/images26/thehermitage185sm.jpg.*

FIGURE 10.10B Detail of necklace in Fig. 10.10A.

Compared with the European versions, such as the Venus of Willendorf (Austria) that dates to 27,000 ya, the Siberian counterparts tend to be less voluminous and usually are portrayed with their arms crossed (Fig. 10.11). Also, unlike the European counterparts, the Siberian Venuses exhibit facial features and are clothed, typically wearing a fur hood as well as decorations (notice the branch of a plant by the shoulder of the figurine in Fig. 10.11).

Some of the disproportionate body parts, such as the exaggerated breasts and private parts, as well as the state of pregnancy, symbolize fertility. These Venuses not only embodied human fecundity but also success in daily activities, such as hunting, which impacts indirectly on human fecundity. It is likely that these effigies embodied a deity that protected and provided for good luck in general, such as fortune in finding the herd, making the kill, not

FIGURE 10.11 They are the oldest prehistoric art ever found in Siberia, 22,000 ya. Carved out of mammoth ivory or reindeer antler. Fur hood. Central Siberia Plateau. Mal'ta deposits at Lake Baikal, Siberia. Comtemporary with Mal'ta-Buret'. *From http://donsmaps.com/images25/maltaimg_1530sm.jpg.*

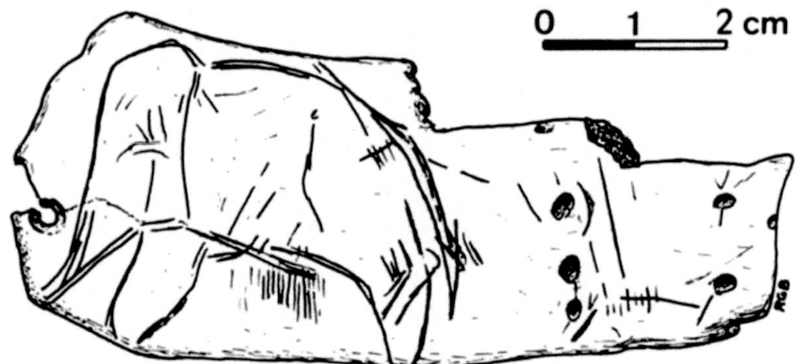

FIGURE 10.12 Mal'ta-Buret' culture engraving of a mammoth on a slab of mammoth ivory, from the Upper Paleolithic Mal'ta deposits at Lake Baikal, Siberia. Central Siberian Plateau. *From http://www.mdpi.com/arts/arts-02-00046/article_deploy/html/images/arts-02-00046-g007-1024.png.*

getting hurt, good weather, etc. The Venus figurine in Fig. 10.11, for example, with an orifice at its feet, was meant to hang, probably by the neck using some sort of fiber or leather string.

Other mobile art objects that seem to be connected to mysticism as it relates to successful hunts are slab or tablets depicting different types of game. The medium is usually ivory and the animal is etched to generate a silhouette (Fig. 10.12). The engraving in Fig. 10.12 is a

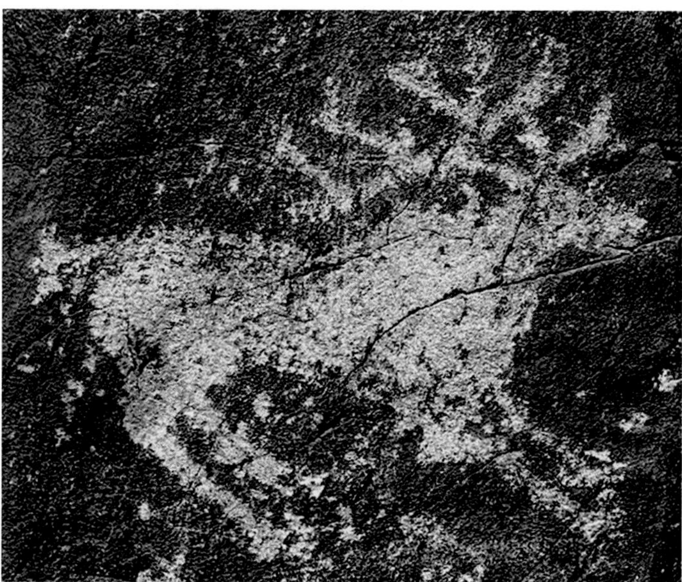

FIGURE 10.13 Petroglyph on volcanic rock. *From https://upload.wikimedia.org/wikipedia/commons/thumb/c/ca/Petroglyphs_on_Mount_Baga-Zarya_08.JPG/587px-Petroglyphs_on_Mount_Baga-Zarya_08.JPG.*

highly stylized portrayal of an adult mammoth. It was recovered from the Upper Paleolithic Mal'ta deposits. As the Venus in Fig. 10.11, this tablet possesses a hole at one end, possibly used to hang the amulet as a pendant. This particular lamina exhibits some curious marks at its narrowest end in between the perforations. It is noteworthy that these marks in the pendant display similarities and are reminiscent of some sort of counting system. Was the owner of the necklace counting the number of mammoths killed? These pieces of mobile art indicate that these people possessed a complex set of idolic symbols and were superstitious. It can be argued that these symbols represent the beginnings of religion or cult during the Late Paleolithic and Early Neolithic.

The inclusion of these mobile artistic expressions as part of the burial assembly, strongly suggests credence in a spiritual universe and a belief in an afterlife. The placement of these objects around the body was not random. The mystical objects were carefully placed in the vicinity of the body. These burial artifacts and their placement suggest that they were intended to provide the deceased with protection during the passage to the afterlife and/or eternal well-being.

The other mode of artistic expression in northeast Siberia takes the form of petroglyphs. The images are etched on exposed rock using lithic instruments and were not painted. Owing to the geological history of northeastern Siberia, the medium is usually volcanic rock (Fig. 10.13). The oldest of these petroglyphs date back to the Upper Paleolithic and are rather numerous, summing to several thousand in the region. Some of the images are thought to be 20,000 years old.[38] They are particularly abundant in the Olekma, Aldan, Lena, and Amur River valleys and in the region of Shishkino. The subjects and styles of these etches vary with time. The more ancient drawings, from the Upper Paleolithic, follow the Dyuktai tradition and typically illustrate silhouettes of megafauna. In addition, occasional anthropomorphic drawings are also observed. They are often found on cliffs along riverbeds or on massive volcanic boulders scattered in the field.

Contemporary Populations of Extreme Northeast Siberia

To attempt to understand the first Native Americans is crucial to examine the peoples which still live in the areas closest to the launching point of this intercontinental migration and which live in ways which may be similar to these first migrants. A number of distinct human groups currently inhabit extreme northeast Siberia, Chukotka. Chukotka today is mostly tundra (treeless arctic plains), with some taiga (plains with scattered trees) areas. The climate is bitter cold in the winter with temperatures dropping as low as −65°F (−54°C). Coastal regions are damp and foggy, whereas the climate is drier inland. The populations of the Chukotka peninsula include the Asian Eskimo, Chukchi, Koryak, and Itelmen. These populations sparsely populated the peninsula at the very utmost northeastern corner of Siberia. Chukotka, because of its geographic proximity to America, was the Asian territory that Paleo-Natives populated and saw last as they migrated before crossing or moving into Beringia on their way to the New World. Although it is highly likely that the present-day populace of Chukotka is not totally representative of the Paleolithic migrants, they may provide insight into the culture and genetics of Proto-Americans. The four populations that presently inhabit Chukotka do not represent a homogeneous group of people. In fact, culturally, some are dramatically different from each other.

Languages

All of the languages of Chukotka are either endangered, becoming extinct, or dead. This unfortunate condition is the result of migration of young natives to cities and sociocultural pressure by the dominant and dogmatic Soviet or Russian establishment, which has discouraged cultural differences, forced different groups of native to live together in single homesteads, and peppered the land with concentration camps. Yet, in spite of these external forces, the remaining natives persist in retaining their ethnicity and cultural heritage.

Linguistically, all four populations of the peninsula speak tongues that are mutually intelligible.[39] For example, there are three distinct Eskimo languages in Chukotka that are reciprocally incomprehensible. These are Chaplinski, Naukanski, and Old Sirenikski. Chaplinski and Naukanski belong to the Eskimo–Aleut family. The affiliations of Old Sirenikski are unclear. The largest, Chaplinski, is currently made up of at least four major idioms or dialects with several subdivisions each. It is interesting that some Naukanski-speaking Eskimos are bilingual, being also fluent in Inupiaq, the Eskimo language on the Alaskan side of the Bering Strait. This fact is indicative of the close cultural tides that still unite the Eskimo in Asia and America.

Eskimos are probably the most well-known group of people from Chukotka. This is possibly because of their relative large numbers and extensive range from East Asia to Alaska. In fact, a number of Asian Eskimo families have relatives living in the United States, separated by as little as 100 km or so of open ocean. They literally canoe back and forth between the two countries relatively freely. So, in terms of dispersion between Asia and America, Eskimos are still migrating back and forth, potentially contributing to gene flow.

Chukchi, the second most important language in the Chukotka Peninsula belongs to the Chukotko–Kamchatkan family and is very distinct from the Eskimo tongues. Although Chukchi can be subdivided into seven territorial dialects, a generalized form functions as *lingua franca* for all four populations in Chukotka. The language of the Koryak is also included within the Chukotko–Kamchatkan language family. The Itelmen language, on the other hand, is rather distantly related to Chukchi and Koryak, although it is still considered part of the same Chukotko–Kamchatkan language family. Today Itelmen is virtually extinct. It is interesting that although the Chukotko–Kamchatkan languages of Chukotka have no generally accepted relation to any other language family in Asia, it shares some similarities with a number of Native American dialects.[40] These linguistic congruencies suggest kinship between Chukotkan and Native American populations.

Subsistence

The inhabitants of Chukotka exhibit different forms of subsistence. Generally, all four populations partition into coastal and inland groups. The traditional homes of these populations are conical-shaped huts, called chum, similar to tipis of the American Plains Indians but less vertical. The groups in the interior survive mainly by pasturing reindeer, whereas littoral settlements are mainly involved in fishing and hunting marine mammals. Eskimos, especially the ones living in coastal areas, traditionally practiced a maritime existence, hunting whales and seals. The Chukchis, on the other hand, are mainly pastoralists surviving as nomadic reindeer herders. Chukchis, for example, regularly travel from the Kolyma River basin to the Bering Strait, several hundred kilometers apart, with their herds serving as liaisons or middlemen between the Eskimos and populations to the west and south, such as the Koryaks. All four inland Chukotka populations use the reindeer in a multitude of ways, such as a source of meat, milk, cheese, and blood (drink), as well as skin for clothing, footwear, bedding, covers, hut construction, and insulation. They also ride reindeer for transportation and sports (races) after they cut off their horns to prevent injuries.

The regular diet of the inland populations is mostly products of reindeer breeding. Other nourishments also include boiled venison, reindeer blood soup, reindeer brains, and bone marrow. One traditional dish, *rilkeil*, consists of semidigested moss from the stomach of a slaughtered reindeer cooked with blood, fat, and pieces of boiled reindeer intestine. The cuisine of coastal populations, on the other hand, relies heavily on boiled walrus, seal, and whale meat and fat, as well as seaweed.

Folklore and Religion

The folklore of the populations in the Chukotka Peninsula includes myths about the creation of the universe: earth, moon, sun, and the stars. They share stories about animals and enjoy ridiculing people they consider foolish. Some of the stories involve ancient tales of battles among the Eskimo, Chukchi, Koryak, and Itelmen. Some of the tales involve various animals and plants, as well as the moon, sun, rivers, trees, and other objects, which identify and possess specific spirits, good and bad.

Evil spirits are thought to be responsible for illnesses, as well as bad look, and shamans are responsible for curing the sick by summoning their extraordinary powers. For example, in a Chukchi folktale, several shamans were traveling together in a boat when the vessel develops a leak. The most powerful shaman in the group manages to stop the leak by plugging the hole with the assistance of seaweed spirits. When the seaweed spirits left, the leak reappeared, and he challenged the other shamans to fix the leak again. As their powers were weaker than his, they failed and drowned. The powerful shaman who was able to master the seaweed spirits was able to swim to shore and save himself. This epic myth illustrates how only the truly powerful shamans are perceived to be able to communicate and manipulate the spirits of nature.

As part of the rituals, shamans regularly fall into trances, oftentimes using hallucinogenic mushrooms to communicate with the spirits. People use shamans as mediums to interact with the underworld, speak through them, forecast the future, and cast spells. Since shamanism is practiced at home and not in congregations involving large numbers of people, the Russian authorities cannot easily persecute and prosecute this practice. Thus, it survives today unauthorized and underground.

Burials for all Chukotkans are generally simple. The deceased is cleaned and dressed in his/her best clothing and then is taken to the tundra where the body is burned or left there to decompose. Considering that wood is a prime commodity, funeral pyres tend to be small, which usually leaves most of the remains to rot or are eaten by animals in the field.

In addition to their strong religious/spiritual beliefs, the Chukotkan cultures are also intensely focused on hospitality. The virtue of generosity is valued highly among these peoples. Generally, the worst insult that could be bestowed to a Chukotkan is to say that he/she is not welcoming to strangers. They are expected to provide accommodation in the form of food and shelter to strangers and travelers.

Genetics

Overall, these present-day populations of Chukotka exhibit genetic similarities among themselves and to some neighboring populations. Yet, this genetic continuity is partially severed with groups just to the west in the adjoining district of Sakha.[29] It is not clear the reasons for this discontinuity as migrants must have crossed through Sakha to get to Chukotka. It is possible that genetic components from Southeast Asia, known to exist in Chukotka, contribute to the dissimilarities.

Early studies using classical blood groups, isozymes (proteins), and mtDNA markers show that the Chukchi and Asian Eskimo groups are genetically as similar to Alaskan Eskimos as to other populations of continental northeast Siberia.[41] In addition, coastal Chukchi are genetically as similar to Reindeer Chukchi from the interior of the peninsula as to Asian Eskimos. Furthermore, the Koryak were found to be closer to Reindeer Chukchi than to Asian Eskimos, whereas the Koryak and Itelmen are genetically equidistant to northeast Siberian populations and to Far East populations.[41]

The populations of the Altaic are characterized by five primary mtDNA haplogroups: A, B, C, D, and X. Although modern humans in the Chukotka Peninsula also possess haplogroups A, C, and D, these haplogroups are not identical to their counterparts in the Altaic. This is understandable considering that DNA likely mutated during the trek traversed by humans from the Altaic to the extreme northeast of Siberia and since they first populated Chukotka. The current day natives of the Chukotka Peninsula lack haplogroup B, possibly as a result of random genetic drift that lead to its drop-out from the gene pool. In addition, mtDNA haplogroups G and Y have been incorporated into the Chukotka autochthonous populations, except the Eskimos, from other regions.[42] Although the most prominent mtDNAs in Chukotka belong to major haplogroups A (Chukchi, 68.2% and Eskimo, 80.0%), D (Eskimo, 20.0%), and G (Itelmen, 68.1% and Koryak 41.9%), some recent investigations providing higher resolution have identified more specific mtDNA types, including C4b2, C5a2a, D2a, D4b1a2a, G1b, A2a, A2b, Y1a, and Z1a2.

In terms of the Chukotka Y chromosome diversity, three haplogroups have been identified in Chukotka: N1c, Q1, and C3. These haplogroups are defined by the mutations M46, L232, and M217, respectively. N1c possibly originated in what is today China about 25,000 ya and it parallels the spread of humans northward toward northeast Siberia and subsequently to America. Haplogroup C3 likely had its genesis in Central Asia approximately 15,000 ya and was carried to northeast Siberia by Late Paleolithic humans. Variations of this haplogroup are seen in present-day Native American populations. Haplogroup Q1 originated in Central Asia around 25,000 ya, and like C3, it was dispersed northeasterly in Asia by migrating Paleo-Natives. As previously mentioned, different forms of Q1 represent the most abundant haplogroups of American Natives. In other words, it is possible that all of the major Y chromosome haplogroups present in Native Americans today derive from Chukotka gene pools and are remnants of the populations that founded the New World via the Bering Strait.

Studies that focus on populations from Chukotka and Yakutia or Sakha (the district located just to the west of Chukotka) indicate a genetic discontinuity between the two regions.[29] This genetic divide is partially the result of and is clearly seen in the contrasting distribution of Y-specific haplogroup Q. The Q haplogroup, defined by the M346 mutation, is abundant in the Altaic region and Chukotka, and it is almost fixed in Native American populations but basically absent in most of the populations of Sakha. Also, in a recent report, haplogroup Q was found to be absent in the Buryat region by Lake Baikal.[3] The Buryat territory is located in between the Altaic and Sakha. Thus, it is noteworthy that Sakha and Buryat, a vast region that bisects the Altaic and the Chukotka Peninsula, is missing Q. This lack of Y haplogroup Q in Sakha and Buryat, areas that had to be traversed by humans on their way to northeast Siberia, Beringia, and America, poses an interesting dilemma: how did Paleo-migrants fail to leave behind the prominent Q imprint as they traveled? A number of scenarios can be imagined. It is possible that Q was present in Buryat and Sakha but was lost around Lake Baikal during or subsequent to the trans-Siberian migration northward to Chukotka and beyond. This random drift phenomenon is particularly pronounced in diffusely located populations because of their small effective (breeding) population size. Yet, although uniparental markers, such as Y haplogroups, tend to stochastically drop out from populations, disappearance over such a vast area is unlikely unless it occurred early in the trek. Furthermore, a similar discontinuity is observed when over 500,000 genome-wide autosomal (biparental) markers (SNPs) were examined.[29] With such large and comprehensive coverage of the entire genome offered by these markers, it is difficult to argue for random dropout to explain genetic discontinuity. It is also plausible that subsequent migrations into the area overwhelm in numbers haplogroup Q–containing populations, effectively erasing the signals from previous populations characterized by haplogroup Q. Another potential explanation is that the speed of the spread northward of these nomadic populations was so rapid that there was little time to allow the genetic signature of the migrants to linger in the traversed territory.

It is noteworthy that although a genetic discontinuity is seen between the contemporary populations of Buryat and Sakha, on one hand, as well as between Sakha and Chukotka, on the other, more ancient links were observed when deep-rooted genetic analyses were conducted.[29] In accordance with archaeological findings,[43] these results indicate direct cultural contacts between Chukotka and Sakha during the Late Paleolithic and Neolithic. Overall, the general picture that emerges from these findings suggests an initial time period of contact and gene flow, possibly during the initial dispersal from Central Asia to extreme northeast Siberia, followed by an era of cessation of contact, minimal gene flow, and isolation of Chukotka.

CULTURAL PARALLELISMS BETWEEN THE POPULATIONS OF NORTHEAST SIBERIA AND AMERICA

Examples of similarities between Northeast Asian and Native American populations are many. Striking congruencies exist between Northeast Asian and Native American populations not only in their anatomy and genetics but also in their artistic expressions, language, myths, and folklore. We will review some of the cultural parallelisms known and discuss the genetic and anatomical affinities in subsequent sections of this chapter when the American populations are covered. Although any given instance of cultural correspondence could be attributed to independent origins in Asia and America, the numerous examples taken together provide a compelling case for connections between the people from the two continents. In reviewing the subsequent examples, we must keep in mind that any association between the two groups of people on either side of the Bering Strait could represent migrations from Asia to America or vice versa as part of back migrations.

The Dyuktai culture is often mentioned as a clear case of similitude between the material world of American Natives and the people of northeast Siberia.[44] The scope of the similarities is extensive. The name Dyuktai derives from the cave of the same name in the Aldan River valley in northeast Siberia. Some of the Dyuktai assemblages from northeast Siberia date back to the Upper Paleolithic during the early Sartan Glacial period about 18,000 ya. These artifacts include Levallois and discoid cores, end scrapers, blades, choppers, mammoth ivory spearheads and bone needles.[45] The congruencies and chronologies in the Dyuktai traditions on both sides of Beringia are important considerations in timing the events leading to the peopling of America. In particular, some of the elements of this Dyuktai tradition, specifically the microcores and other tools, exhibit striking parallelisms with Native American artifacts. For example, the bifacial chipped knives and spearheads, some older than 18,000 ya, resemble the bifaces of the New World Paleo-Arctic and Clovis traditions. Also noteworthy is that these assemblages on the Asian and American side are from the right place and time in relation to the peopling of America by Paleo-Natives. Yet, some of this enthusiasm should be tempered since the Siberian Dyuktai bifaces lack fluting (groove or furrow carved into the blades running from the base to the middle of the tool), a distinctive trait of the American bifaces (see blades in Fig. 10.14). Considering the distances travelled from northeastern Siberia to America, it has been argued that cultural drift involving tool-making traditions could explain differences between some artifacts, such as the fluting of blades.

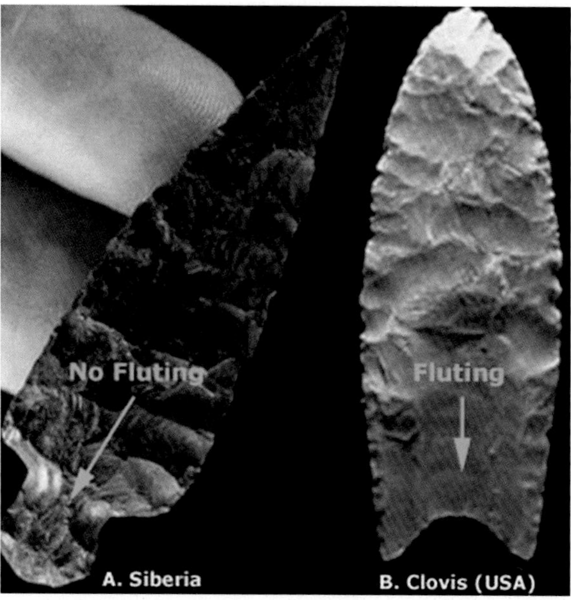

FIGURE 10.14 Similarities and differences between Dyuktai points from northeast Siberia (A) and Clovis (B) points from North America. *From http://www.viewzone.com/solutrean.blades.jpg; http://www.viewzone.com/solutrean.html.*

In addition to tool making, there are also similarities in artwork. The petroglyphs of northeastern Siberia often depict human images. Some of these human figures seem to be of religious significance. For example, the one illustrated in Fig. 10.15 seems to be wearing a hair dress with horns and may represent a shaman. Others, like the ones found in the Amur region, are more symbolic and display a mystic quality. For example, a favorite motif in Amur is a mysterious type of human mask (Fig. 10.16). These rigidly stylized rock drawings of the human face are reminiscent of the masks used by American Northwest populations in their art and remind us of the ritual masks worn by shamans in Northeast Asia and Northwest America. Some of the engravings depicting animals like the one in Fig. 10.17 from Baga-Zarya in the Central Siberian Plateau date to the Upper Paleolithic and are remarkably similar in style and composition to petroglyphs found in the southwest of the United States.

Other major sites with petroglyphs dating to the Late Paleolithic have been discovered in the eastern region of Chukotka. As previously mentioned, this land is the farthest northeast portion of Asia and the closest to America. Many of the petroglyphs in Chukotka consist of a number of megafauna including reindeer, elk, bison, horse and moose. Some are just heads, others whole animals. In addition, marine animals, boats with people inside and curious human footprints have been etched into volcanic rock. Some of the design types are realistic while others are rather schematic including triangular human bodies. This location is of particular interest due to the numerous anthropomorphic mushroom figurines found in this area (Fig. 10.18). Representative examples of this tradition are found in cliffs along the banks of the Pegtymel River in the rocks of the Kaikuul Bluff. The mushroom-looking hats are disproportionally large compared to the heads and bodies. The female figurines, at times dancing, seem to be involved in a ritual celebration. The species of mushroom

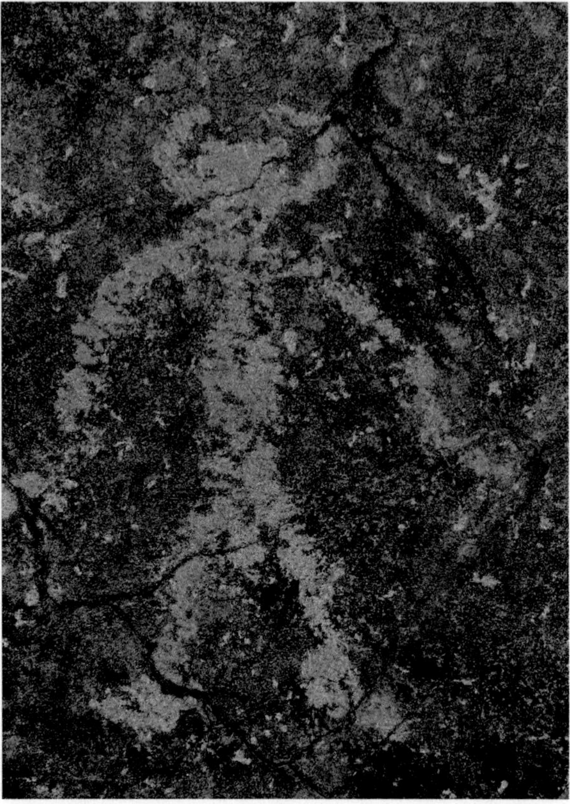

FIGURE 10.15 Mount Baga-Zarya. Shaman with horns? *From https://commons.wikimedia.org/wiki/File:Petroglyphs_on_Mount_Baga-Zarya_11.JPG.*

FIGURE 10.16 Mask petroglyph etched in volcanic rock. *From https://carolkiecker.files.wordpress.com/2015/07/dsc02556.jpg.*

FIGURE 10.17 Petroglyphs on Mount Baga-Zarya. *From https://commons.wikimedia.org/wiki/File:Petroglyphs_on_Mount_Baga-Zarya_04.JPG.*

FIGURE 10.18 Mushroom figurines and vessel. *Reproduced from https://www.pinterest.co.uk/pin/489133209510412298/.*

FIGURE 10.19 Mayan mushroom head figurines. Above are two of the nine miniature mushroom stones found buried together in a Maya tomb at Kaminaljuyu, along with nine miniature stone *metates* and *manos* (Soma stones?) used in the preparation of a ritual mushroom beverage. *From http://www.mushroomstone.com/somaintheamericas.htm.*

depicted is likely *Amanita muscaria*. *A. muscaria* is a psychoactive mushroom that is proposed to have originated in northeast Siberia and during Tertiary geological period (65–2 mya), before dispersing into other regions.[46] The prevalence of these drawings in ceremonial poses is suggestive of a cult, possibly a spiritual sect of schisms, based on the veneration and use of the mushroom for enlightenment. It is known that *A. muscaria* was used by a number of ancient populations of Siberia as an intoxicant and entheogen in connection with shamanic religious practices.[46] It is likely that this species of mushroom was used to help achieve an altered state by the shaman and possibly other tribal members. Interestingly, the use of mushrooms as head decoration is widespread among Native American populations, including the Aztec and Mayans (Fig. 10.19). In addition, the Mayans also used *A. muscaria* as a hallucinogenic after grinding the fungi to a fine powder with a mortar and pestle (Fig. 10.19). Considering the ubiquitous geographic distribution of this species of mushrooms and the numerous cultures around the world that employ mushroom-like symbols in their art (e.g., in India, China, etc.), it is not clear if the above noted parallelisms between northeast Siberian and Native American art and practices results from common cultural ancestry or independent origins.

Another artistic/religious expression exhibiting congruency between Far East Siberia and the Northwest of America are totem poles. On both sides of the Bering Strait, these statutes were created with a mystical intent using symbols to tell a story. The story may be creation, important events, genealogy or the history of a tribe or clan. A number of animals including the raven symbolizing the creator, the eagle as a representation of peace and friendship, the wolf as a provider of direction and leadership as well as protection and destruction and the beaver embodying determination and strong will are usually part of

FIGURE 10.20 Shigir totem pole. *From http://blogs.discovermagazine.com/d-brief/files/2015/09/big-shigir.jpg.*

the design in Native American totem poles. Totem poles dating as far back as the Mesolithic have been discovered in Siberia. The Shigir totem pole, for example, from Western Siberia has been dated to 11,000 ya (Fig. 10.20). Originally, this pole was about 5.3 m high and depicted a number of human heads. Other examples have been found in the Olkhon Island of Lake Baikal. In Far Eastern Siberia, in the Amur region, the Goldi totem pole is notorious for its remarkable resemblance to its American counterparts (Fig. 10.21). Images of various animals and anthropomorphic figures are seen along the length of the pole with a three-dimensional human head at the apex.

The myths of the Big Raven also provide a remarkable set of parallelisms between northeast Siberian and Native American populations. Ravens are remarkable birds. They are highly intelligent, capable of using tools and can be even trained to speak. And thus, it is not surprising that people of native populations portray them as seers. In general, Native Americans had great respect for these birds. Native American tribes such as the Tlingit and Tahltan people revere the raven as a bringer of light. Raven tales are common among Athabaskan-speaking tribes of the Pacific Northwest as well as in the Southwest and the plains of the United States. The Big Raven tradition is also particularly prominent among the Chukchi and Koryak populations of Chukotka. It is noteworthy that several populations from Chukotka in the far Siberian northeast share strikingly similar myths. Although raven stories may differ from population to population and often among clans, remarkably, certain specific elements are constant in Asia and America. The magical attributes remain consistent and sometimes even the titles of the tales are the same. The verbal stories are told generation after generation by the storytellers of each tribe. Yet, raven tales, as opposed to regular or children stories, are not considered entertainment, but cultural property of the clan or individual from which the story originates.

The raven is a hero and a creator. As a hero, the raven flies high toward the heavens, taking wishes and prayers from mortals to the spiritual realm and, in turn, brings messages back. As a creator, he is credited by the Haida people of northern British Columbia for discovering the

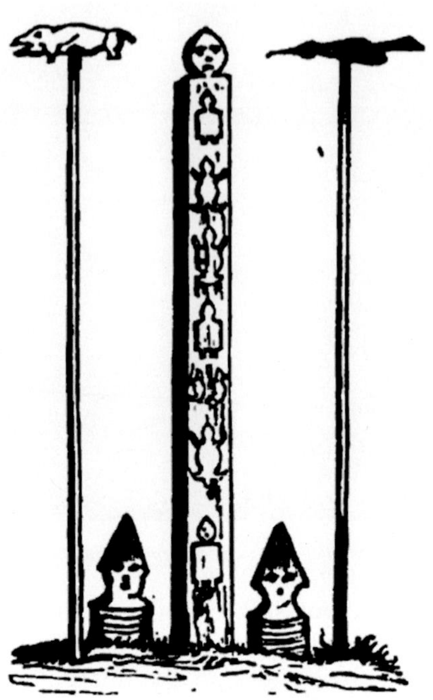

GOLDI IDOL POLES.

FIGURE 10.21 Goldi totem pole. *From https://en.wikipedia.org/wiki/Nanai_people#/media/File:Ravenstein-p377-Maack-Goldi-Idol-Poles.png.*

first human beings that were hiding inside a clamshell (Fig. 10.22). He then fed them with berries and salmon. The raven is also responsible for bringing light to where there was none. His creations are mostly driven by a desire to satisfy his own needs and not altruism. He is a keeper of secrets as well as a trickster. He is the protagonist of stories and invariably a source of both amusement and amazement. The raven is considered so intelligent among the Natives that it is able to transform into human form or adopt the shape of objects and animals. As a shape-shifter, the raven is capable of adapting to different situations as required. Representations of the raven are frequently found at the pinnacle of totem poles as crowns.

Music provides other illustrations of congruency in artistic expression among Altaic, Northeast Siberian and American populations. One example is overtone singing. Overtone or throat singing is an unusual type of singing in which vocalists, usually two, produce two distinct tones simultaneously. One performer sings low (25-20 Hz) at a fundamental pitch, while a second vocalist, at a higher tone, intonates flutelike harmonics. Overtone singers at times combine a normal glottal pitch with a low frequency, pulse-like vibration known as vocal fry.

Although throat singing is one of the oldest forms of music co-practiced by Altaic tribes, northeast Siberians and Native Americans, it is not known how ancient it actually is. Yet, overtone singing is uniquely linked to these three groups of people. It is noteworthy that these three populations followed the putative migration route from the Altaic into America. It is possible that this very unique style of singing may represent a musical signature of the expansion from Central Asia to America.

FIGURE 10.22 The creation of humans by the raven. *From https://mythicstories.files.wordpress.com/2015/03/haida-raven.jpg.*

Types of personal names given to children in extreme Northeast Asia and among Native Americans of the United States and Canada also exhibit strong similarities that suggest common origin. Personal names among Native Americans traditionally reflect natural events or animals in certain situations. The names of famous Native Americans, such as Crazy Horse, Red Cloud, Standing Bear and Sitting Bull, are good examples. Similarly, the names of people in extreme northeast Siberia take after natural occurrences often at the time of the individual's birth. For example, among the Chukchi people in the Chukotka Peninsula, names like Polar Bear and Sunrise are given to boys while Spring and Duck are used for girls. At times, names are assigned to indicate the traits and qualities that parents hope the child will possess. Examples are Robust Fellow for boys and Beautiful Woman for girls.

Most of the languages spoken in North America and Siberia are distantly related and thus offer no indication of links between the people of the two continents. Yet, in 1923 linguistic associations were proposed between Asian and Native American linguistic families. This notion was first advanced by Alfredo Trombetti based on the science of linguistic morphology.[47] Yet, it was not until 1998 that Merritt Ruhlen published a manuscript on the subject.[48] Specifically, Trombetti and Ruhlen noticed that both Na-Dené and Yeniseian languages fused together words and stems of words (e.g., prefix and suffix) to form complex words. Specifically, both language families exhibit a complex agglutinative prefixing verb structure. The similarities have compelled some linguists to advance the notion of a Dené-Yeniseian language family. This type of language structure is rare in Asia. At the turn of the 20th century, in the absence of genetic information and minimal archeological data, these linguistic connections between these two groups of people was not appreciated and it was received with skepticism. The speakers of these two language families live distant from each other. Yeniseian is spoken

FIGURE 10.23 Regions in Central Asia and America exhibiting similar language structure (agglutinative prefixing verb structure). *From https://en.wikipedia.org/wiki/Dené–Yeniseian_languages#/media/File:Dené-Yeniseian.svg.*

in southern Siberia, north of Lake Baikal while Na-Dené languages are found in Northwest and Southwest North America (Fig. 10.23). The approximate distance between the two extremes of the range is 15,000 km. It is noteworthy that the region currently occupied by Yeniseian speakers is an area just northwest of the location of origin of the presumptive Altaic migrants from Central Asia that colonized America.

More recent research on the observed linguistic similarities supports the notion that Yeniseian evolved from a Na-Dené dialect in the region of Beringia around 30,000 ya, 10,000 years prior to the LGM.[49,50] Statistical models based on Bayesian theory suggest that an early dispersal of people from the center of the geographical distribution of the Na-Dené language family, in the Beringia land bridge, took place with migrations dispersing south along the North American Coast and into the North American interior as well as westward, back into Siberia. In addition, this scenario argues for an early settlement of a Beringia refugium, long before the traditionally held contention of a crossing about 12,000 ya. Language development in Beringia also adds support to notions of a hiatus in this region based on genetic grounds. The Beringian Standstill Hypothesis posits that Paleo-Natives remained in the Beringia landmass for around 15,000 years. It is likely that climatic conditions trapped this population(s) within the land bridge, thereby preventing them from migrating to America proper or even moving back to Asia. It has been proposed that during this Beringia tenure, one of the Na-Dené languages differentiated and evolved into Yeniseian. Thus, although the linguistic and genetic data do not contradict the idea of hunter-gatherers entering the Americas through Beringia, they suggest that it was not a simple one-way trip to America; it involved back migrations as well.

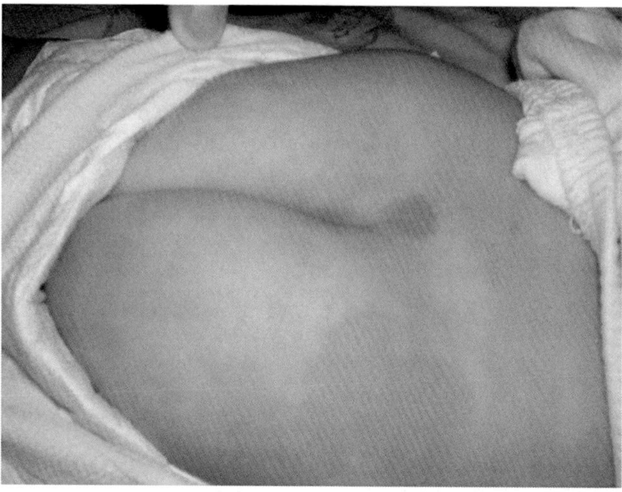

FIGURE 10.24 Mongolian spot characteristic of Asian populations. *From https://en.wikipedia.org/wiki/Mongolian_spot#/media/File:Mongolianspotphoto.jpg.*

The Na-Dené–Yeniseian linguistic connection is not the only one known, as the Eskimo languages provide another case of linguistic association between populations on both sides of the Bering Strait. Contemporary Eskimo tongues are divided into two major branches, Yupik and Inuit. Each one of these subfamilies is further characterized into more than a dozen dialects spoken in Siberia, Alaska, northern Canada, and coastal Greenland. Although for the most part Eskimo languages are mutually unintelligible, speakers of some specific dialects are able to assess the general meaning of conversations in other dialects. It is theorized that the Proto-Eskimo languages arrived relatively recently to America, about 5000 to 2000 ya, when the pre-Dorset culture, known for their Arctic Small Tool Tradition, crossed into Alaska. It has been proposed that the Eskimo culture and language further developed in Alaska and then returned to Siberia via back migration. In fact, Eskimos today continue to travel back and forth from Siberia to Alaska without impediments from the governments of Russia and the United States. Furthermore, although they now use motorboats for this task, it was accomplished using primitive boats within the last century and presumably for centuries or millennia prior.

THE BERINGIA HIATUS

The traditional view that modern humans in America are descendants of Asians is not new. In fact it is a centuries-old notion. The idea was first proposed in 1590 by José de Acosta (1539–1600), a Jesuit priest and naturalist from a small town near Valladolid, Spain, who posited that Native Americans were descended from hunters who had migrated to America from Asia.[45] Since then, this view has been generally accepted as the number of similarities between the native people on both sides of the Bering Strait accumulated. The similarities which were initially noted included anatomical variations such as the Mongolian Blue Spot (Fig. 10.24) found in high frequencies among Asian (~95%) and Native Americans (~85%)

(http://www.tokyo-med.ac.jp/genet/msp/about.htm) and the epicanthic folds. At first, it was not obvious how and when Asians managed to travel from Asia to America. Although dispersals from Europe have been proposed in recent years to account for the origin of Native Americans, a northeast crossing from Asia is still the theory which is most generally accepted by the orthodoxy. Yet, issues involving the time, duration, and route of the dispersion(s) from Asia are still highly disputed. These topics will be discussed in the following sections of this chapter.

Currently, it is generally believed that the Asian migrants crossed what today is the Bering Strait as pedestrians, although the possibility of a coastal trajectory for at least part of the trip cannot be formally ruled out. It is thought that the passage occurred after the LGM, about 12,000 ya. At the time there was a continuous ice-free landmass extending from northeast Siberia to what is today Alaska. This land is known today as Beringia, a term coined by the Swedish botanist Eric Hultén in 1937. This land bridge was not unique to the last glacial period. In effect, during the Pleistocene, starting 2.58 mya to the present, this landmass has risen from the shallow ocean on several occasions as a result of reduced sea levels because of water sequestration during glacial periods. During the LGM, sea levels in Beringia were 120 m below those of today. Every time after a glacial period, the dry land submerges as the temperature increases and the ice sheets melt. For example, it is known that since 740,000 ya there have been a total of eight glacial cycles with the corresponding exposures of the sea floor.[51] The traditional view regarding this intercontinental dispersal of Paleo-Natives is that it was only one way, from Asia to America, relatively fast, and relatively recent (approximately 12,000 ya) after the LGM. This was a logical assumption, as during glaciation the penetration into Alaska would have been difficult because of extreme cold weather and obstruction by ice. Thus, most scholars assumed that Beringia just acted as a transitory bridge, an available path. In other words, it was thought that the Asian migrants had no desire to stay and colonize such an inhospitable place, and therefore, they went across the landmass into America as fast as possible. Yet, starting about a decade ago, a different way of thinking about Beringia began to emerge in the scientific community[52] in which Paelo-Natives used the landmass not so much as a transitory passageway but as a refugium for several thousand years. This idea became known as the Beringian Standstill Hypothesis or the Beringian Incubation Model. It is likely that this migrational interlude was motivated by adverse conditions on either side of Beringia, a climate that blocked dispersal to the east (to America) and west (to Asia). Currently, many investigators envision that modern humans had an extensive tenure of 10,000 to 20,000 years encapsulated in a water-free, ice-free Beringia.[53]

For the last 12,000 years, we have been experiencing an interglacial period (the Holocene), and the Beringia land bridge is again submerged under a shallow body of water 30–50 m deep. This body of water, known as the Chukchi Sea, is a rectangular gulf of the Arctic Ocean, flanked on the west by the Chukotka Peninsula of northeast Siberia and to the east by the northwest coast of Alaska. Geologically, the landscape of dry Beringia during the last glaciations was a broad (about 600 miles wide), steppe-like tundra valley with isolated peaks (Fig. 10.25).[54] Core samples retrieved from the bottom floor under the sea indicate that western Beringia was more extensively forested than the eastern side. The region was dry with ice limited to sparse mountainous regions, mainly in the Brooks Range and the Alaska Range. The remaining landscape was an extensive, ice-free, relatively featureless flat land. The region was drained by a number of rivers that meandered through valleys emptying into the Arctic Ocean. It is thought that during cooling periods, as the sea level decreased, these valleys

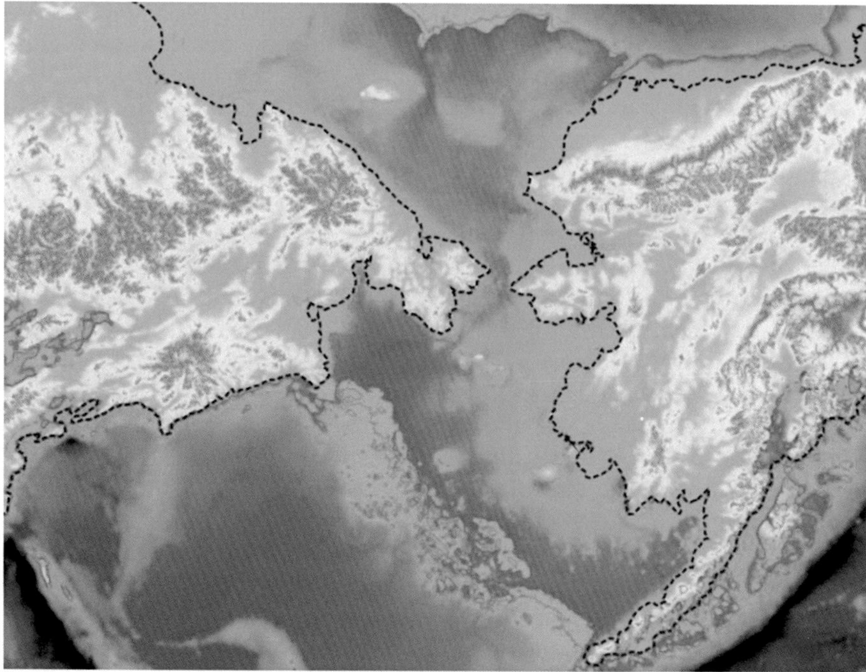

FIGURE 10.25 Beringia Refugium during last glacial maximum. Dark green represent land currently under 30–50 m of water. *From http://www.smithsonianmag.com/science-nature/ancient-migration-patterns-north-america-are-hidden-languages-spoken-today-180950053/?no-ist; http://www.smithsonianmag.com/science-nature/ancient-migration-patterns-north-america-are-hidden-languages-spoken-today-180950053/#RAfWU2Z7xX8UXZyq.99.*

turned into ice-covered estuaries. We can envision these wetlands being fresh water in the interior and brackish water closer to the coast. Taken together, it is likely that the dry and partially submerged Beringia was a highly productive ecosystem during the last glaciation and melting period after the LGM, about 20,000 ya.[55]

Currently, it is still not entirely clear whether the lowland bridge was covered with a shrub tundra that would not have provided much food for large grazing mammals, such as mammoths, or a steppe tundra with plenty of nourishing grass to sustain a megafauna for long periods of time. This, of course, would make a huge difference in the lives of humans trapped in this environment. Radiocarbon dating of organic materials retrieved via sea floor cores has provided evidence for the organic productivity of Beringia. For example, peat deposits from 50 m below current sea level (bsl) in the Chukchi Sea have been dated to 11,000 ya. Also, wood fragments older than 18,000 years have been recovered as well as sediment cores containing plant fossils and pollen from a wooded ecosystem.[54] In addition, sensing equipment has identified areas of potential human occupation, such as paleobeaches, terraces, and fluvial confluences.[54] In such a habitat, marine mammals, small game, and healthy megafauna likely lived and prospered. In turn, Paleo-Natives could have lived and thrived there for an extended period of time. In other words, it seems that humans had no need to rush through the Beringia landmass into America, especially when massive ice slabs covering much of Alaska and northeast Siberia blocked their way.

The time line of these glacial and interglacial periods as seen in the geological record indicates that approximately 125,000 ya our planet began to experience a gradual decline in temperature and a simultaneous increase in ice mass. The cooling process lasted about 100,000 years and culminated in the LGM around 20,000 ya. Our planet's temperature experienced a 10°C drop during this gradual cooling period. The interglacial thawing processes tended to be much faster, lasting about 20,000 years. According to the geological record, we should be initiating a cooling trend soon. Yet, man-driven global warming resulting from the greenhouse effect may counteract this natural progression in the current cycle, and it is difficult to predict how the two opposing forces will balance each other out. These freezing–warming cycles occur periodically every 85,000–100,000 years (Fig. 10.3).

Within the current glacial–interglacial cycle, about 9000 ya sea levels in the Bering Strait were down to around 44 m bsl but around 6000 ya they were up to only 12 m bsl.[56] This suggests that some areas of the seafloor in Beringia were exposed until about 8000 ya, allowing passage of game and humans. Based on the climatic, geological, and archeological information available, it is possible to reconstruct a putative scenario of the Paleo-Native layover and progression through Beringia. The arrival of modern humans to northeast Siberia from Central Asia (e.g., Altaic) about 35,000 ya infer a subsequent date for the dispersal into Beringia. This sets a limit on the age of the penetration into Beringia and then North America. It is also known that during the extreme weather of the LGM (28,000–15,000 ya), glaciers were present in western and eastern Beringia[57] and likely prevented the entrance of people into Beringia and the movement of migrants into America. The ice masses in the Verkhoyansk Range in Siberia and in the Mackenzie River valley in Alaska essentially isolated Beringia and its inhabitants. In addition, after about 12,500 ya, the flooding of the landmass because of the proactive melting of the ice caps limited pedestrian incursions into Beringia. These climatic conditions delineate a putative scenario in which Paleo-Natives could have reached Beringia 35,000–25,000 ya, a period before the LGM. Subsequently, in a span of 10,000 years (25,000–15,000 ya), glaciers to the west (northeast Siberia) and east (Alaska) would have trapped species, including humans, within the Beringian landmass. Following the LGM, during a melting epoch lasting 3000 years (15,000–12,000 ya), a window of opportunity for migration would have allowed the captive populations to move into America. After the complete inundation of the strait, further pedestrian crossing was not possible. Thus, this hypothetical scheme permits a prolonged layover of at least 10,000 years in Beringia. Alternatively, dispersals across Beringia could have occurred rapidly during two windows of opportunity with limited or no sequestration of humans and animals: an interval from about 35,000 to 25,000 ya and at a later period that spanned from 15,000 to 12,000 ya. In addition, the prolonged confinement and the fast-moving scenarios are not mutually exclusive. Hypothetically, some populations could have travelled rapidly across, using the land bridge solely as a conduit. These humans would have contributed to a wave of early arrival migrants, whereas groups that stayed trapped within Beringia moved on to America after the LGM, when the ice barriers partially melted creating a pedestrian corridor, and became part of more recent dispersals. A combination of these two types of travel may provide the basis for multiple and genetically unique migrations to America. Proponents of multiple migration waves to explain the currently observed anatomical and genetic diversities of humans in the New World see these possibilities as likely scenarios.

It has been argued that this incubation period in the Beringia refugium provided the opportunity for genetic diversification of the peoples that would later become the Native Americans and that subpopulation structuring via this genetic isolation may have contributed to this diversity.[58] If the incubation theory is correct, and substantial genetic diversification in fact occurred within Beringia during a protracted tenure therein, this variation would have contributed to the genetic heterogeneity observed today in the New World. An example of this Beringia-born genetic diversity is the Y chromosome Q-M3 mutation. It is likely that the Q-M3 mutation arose in Q-L54 Y chromosomes that travelled with Paleo-Natives from the Altaic all the way to Beringia. Haplogroup Q-L54 is thought to have originated in Central Asia, possibly within the Altaic region.[3] In Beringia, during the hiatus, within some of the founder Q-L54 Y chromosome carriers, the new Q-M3 mutation occurred. The Q-M3 mutation then dispersed with the Proto-Americans into the New World, becoming the preeminent Q lineage of Native Americans. Today, Q-M3 is geographically widespread, found from the northeast tip of Siberia (observed in the New Chaplino and Chukchi) to throughout the Americas. Considering the greater frequencies and genetic diversity within haplogroup Q-L54 found in America compared with northeast Siberia, it has been proposed that this haplogroup may have originated in eastern Beringia during the layover period, and its presence in northeast Siberian populations is the result of back migrations from Beringia or Alaska to Siberia.[59]

In terms of maternal inheritance, mtDNA analyses aimed to assess the genetic makeup within the ancestral northeast Siberian and Native American founder populations, indicate considerable differences between these two groups.[52] In other words, the American mtDNA founder lineages were found to be unexpectedly different from their ancestral Asian counterparts. Also, the founder American mitochondrial types exhibit unanticipated substantial differentiation from each other. The newly resolved phylogenetic structure suggests that the ancestors of Native Americans paused when they arrived in Beringia, providing enough time for the Asian sister clades to mutate and evolve from each other and from northeast Siberian ancestor mtDNA types. Specifically, the data suggest that mtDNA types A2, B2, C1b, C1c, C1d, C4c,D1, D4h3, and X2a underwent a maturation and diversification process during the Beringia standstill. Furthermore, in the same way the Y chromosome–specific Q-M3 mutation panorama supports recent bidirectional gene flow between Siberia and the Northwest America, the mtDNA displays not only signals of genetic continuity between Northeast Asian and American Paleo-Natives but also the back migration of Native American haplogroups C1a and A2a into Eastern Siberia.

COMING TO AMERICA

Significant questions about these first American migrations have yet to be answered. The duration of the Beringia layover, the possibility that at least some human groups having been just transient migrants through the landmass, as well as a number of other details regarding the peopling of the Americas remain elusive. Among them is the time of the incursion(s) into America, the number of such migrations, the direction of the dispersals within the New World, and the speed of travel. Today, after years of intense interest and research, our knowledge of the colonization of America, unfortunately, suffers from the limited scope of the studies often involving poorly anthropologically characterized populations and a lacuna of equivalent data incapable of affording direct comparison among extinct and extant populations. These situations, at times, have prompted some investigators to extrapolate and extent themselves based on limited data sets. In addition, because of technical limitations and dating of questionable artifacts (not the actual human remains), time estimate studies have been plagued with questionable data. Overall, this situation has often contributed to contradictory results in the literature such as, for example, the relationship of the Clovis people to other Native American populations.

The First Americans

The traditional view pertaining to the first arrival of humans to the New World involves one or several demographic expansion(s) induced by the onset of global warming after the LGM, about 12,000 ya. It is thought that this global thawing opened the Mackenzie corridor along the eastern flank of the Rocky Mountains, allowing Proto-Americans to disperse southward.[60] This notion remains generally accepted today. However, over the years, the number of penetrations and the times of the incursions have changed every few years as new data and analyses become available. Often, these fluctuations relate to spotty sampling, different marker systems and dating techniques, availability of higher resolution data sets, as well as allegations of disturbances of archeological strata, and contamination of archeological sites with contemporary material.

Time of Arrival

As previously outlined, the southbound migration from Beringia into America was feasible during two windows of time (35,000 to 25,000 ya and 15,000 to 12,000 ya). Humans may have used Beringia as early as 35,000 ya as an ecological refugium from extreme weather and remained isolated in the land bridge until climatic conditions improved after the LGM. Alternatively, Paleo-Natives may have used Beringia simply as a platform to transit across to America, or both. After the Beringian pause, it has been theorized that a small number of founders (about 100–2000 reproductively effective individuals) settled in America in a rapid expansion.[61,62] This southward dispersal from Beringia and Alaska could have occurred along a coastal Pacific route before 24,000 ya and/or after 15,000 ya.[63] Although a maritime dispersal could potentially increase the speed of the trek, no evidence of submerged encampments or sites along the American coast have been discovered. It is likely that if any evidence of a coastal trip still exists, it may be under water beyond reach, considering the rise in sea level subsequent to the LGM.

From purely logistical grounds, the timing and the speed of the human dissemination from Beringia to the extreme southern region of South America represents a conundrum, which is still under intense debate. In general, the proposed travelling times are too fast to account for the earliest sites discovered. Perhaps the most notorious case of time discrepancy involves the Monte Verde site in southern Chile. The Monte Verde dig is located in extreme South America about 58 km from the coast in the western Andean foothills. The distance separating Monte Verde from Alaska is about 16,000 km. For a pedestrian traveler, the trek would have been difficult, considering the numerous mountainous terrains located along the way. A path through the interior of South America would have necessitated crossing the massive Andes mountain range, which would have been a Herculean task. It has been calculated that at a rate of 200km per century, a very rapid pace for random dispersal (Proto-Americans were not going anywhere in particular and had no assigned arrival deadline), Native Americans would have travelled the 16,000km distance in approximately 7000years.[64] To reconcile these facts with the presence of archaeological remains in Monte Verde, Chile, dated at 14,800 ya,[65] the time line requires that the exodus from Beringia to Monte Verde at the cone of South America must have started at least 21,600 ya. This date coincides with the time of the LGM and considerably before the creation of a hypothetical ice-free passageway, known as the Mackenzie corridor, that extended from Beringia to the interior of North America for about 2000km. This corridor opened between 13,000 and 12,000 ya. Furthermore, even if the corridor was available, the question still remains concerning whether it was capable of providing at least minimal subsistence in the form of food and shelter during the travelling time of several thousand years at a rate of about 2km per year. In other words, 21,600 ya is a date about 8000years earlier than the traditional putative commencement of the migration (around 12,000 ya).

The time estimate of Monte Verde has come under considerable scrutiny over the years since its discovery in late 1975. A lot of the skepticism at the time stemmed from the reigning theory that pointed to a much later arrival of Proto-Americans. Allegations of contamination from differently aged strata generating older settlement dates have plagued the scientific literature over the years. The Monte Verde finding consisted of two fireplaces, wood tools, timber, fruit residue, marine animal remains, posts from about 12 huts, bones from game, hide and clothing fragments. It was estimated that about a couple dozen people lived at the site.[66] Even a piece of preserved meat was found within the confines of the camp further identified as belonging to a mastodon. Experts attribute the unique state of preservation of the settlement to flooding from the nearby river shortly after the establishment of the camp. Human coprolites (excrement) within small pits were identified as well. Clearly, the archeological orthodoxy was not ready for the quality and quantity of these findings. The debate intensified when the Chilean archeologist Mario Pino found fire pits at a lower sediment stratum estimated as being 33,000years old by radiocarbon dating. This finding today is still highly contested.[66]

Given that the archaeological evidence indicates that humans arrived at Monte Verde approximately 1800years before the time of the traditional putative trans-Beringian crossing, the simplest explanation for this perplexity is a much older dispersal from Beringia, contributing to the settlement of Monte Verde. In an attempt to reconcile the time data outlined above, scholars have suggested that the spread that established Monte Verde occurred very rapidly once the Proto-Native Americans left Beringia. Some investigators also suggest that this people colonized America from eastern Beringia to Tierra del Fuego, along a Pacific coastal route. A coastal dispersal along the western shore of America would have provided speed, sustenance from marine life, and a less challenging route that avoided huge mountain ranges, such as the Andes. Some investigators propose that such a trip bordering the coastline would cut down the traveling time to as little as 2000years. Yet, even with a very speedy coastal voyage, requiring some sort of unknown purpose for a prompt arrival, a 2000-year travel time after the Beringia departure 13,000 ya, would place the migrants in Monte Verde at approximately 11,000 ya, postdating the site's age by a few thousand years. Furthermore, thus far, no archaeological evidence has been found indicating Native American presence before 13,000 ya along a coastal migration route.

Other than Monte Verde, other Native American sites have been dated to as early as 33,000 years old. These include certain sediment layers in Topper, South Carolina, and Pedra Furada in Brazil. Yet, the radiocarbon dating of these sites remains highly contested. Moreover, there are a number of additional locations in southern Chile providing less contentious dates. They are comparable in age to Monte Verde. For example, Pilauco Bajo and the Pali Aike crater lava tube register dates ranging from 10,000 to 14,000 before the present. It is thought that Monte Verde was a rather permanent location of habitation, whereas Pilauco Bajo was a temporary hunting site where game was disemboweled, cleaned, and partitioned.

The Topper dig represents one of the oldest sites claiming human occupation in North America. It is located along the banks of the Savannah River in the state of South Carolina. The artifacts unearthed at the site are thought by some archeologists to date to about 20,000 ya.[67] Among the materials retrieved, archeologists found an axe exhibiting bifacial flaking of the edge. Other lithic materials are being disputed as the result of natural forces that were not man-made and the 50,000-year-old carbonized remains, the outcome of a natural fire. Other significant early archeological locations in North America comprise Cactus Hill in Virginia and Buttermilk Creek in Texas. The Cactus Hill dig has been dated from 18,000 to 20,000 ya and items have been recovered that include Clovis-like points (projectile points exhibiting a central furrow first discovered in Clovis, New Mexico), blades used for butchering and hide processing, phytoliths (microscopic silica structures derived from plant tissues capable of resisting fire and decay) that were estimated to be from carbonized hickory wood and animal remains, such as deer bones, turtle shells, and fossilized shark teeth. The dating has been strongly contested by mainstream archeologists on the grounds of strata disturbance. As the site is located in sand dunes, it is argued that wind action could have compromised the integrity of the depositions.

The Buttermilk Creek location in Salado, Central Texas, differs in that it is probably one of the most reliable sources of artifacts predating the traditional time of the colonization of the area by the Clovis people, 13,200–12,800 ya. It is dated at approximately 15,500 ya by optically stimulated luminescence and exhibits undisturbed strata and an extensive lithic tool assemblage.[68] The area represents a continuous, favorable climate, source of food and water, as well as an excellent source of raw material for tools. Specifically, within the general area, there are deposits of high-quality microcrystalline quartz rock known as chert. Based on the uninterrupted progression of artifacts, layer after layer, it is thought that the archeological record of the site is unbroken from the most ancient Buttermilk Creek Complex to the prehistoric Native Americans (before the arrival of Europeans). This continuity is significant because it clearly establishes that humans were in America to stay much earlier than the traditionally accepted date.

The assemblage of tools found at Salado encompasses 15,528 pieces, such as debitage (production debris and rejects derived from toolmaking), flakes, blades, bifaces, discoidal cores, edge-modified stones, radial break stones, and polished hematite. Debitage is produced when a rock is struck with another stone (functioning as a hammer) to reduce the chert into a bifacial shape. This process is also known as flint knapping. The tools found also exhibit evidence of use. A number of items show polish and markings typical of previous cutting, grooving, or incising. Also, the apparent connections and influences among the various cultural traditions discovered in the location make it particularly interesting. The continuity of the strata and the intermediary characteristics of some artifacts point to a scenario in which earlier traditions contributed to the development of subsequent cultures in situ. For example, the notion of core reduction as a technique to produce the assemblage seen in the Buttermilk Creek Complex suggests that the more recent fluted spear points seen in more recent layers developed from a more ancient style on location. In other words, the site may represent an uninterrupted continuation of lithic industries.

The Buttermilk Creek Complex is singled out by many archeologists as a definitive example of the arrival of Proto-Americans before the Clovis culture, traditionally thought of be the first colonizers of the New World. The oldest human material at the Buttermilk Creek Complex predates the Clovis tradition by about 3000 years. We will discuss Clovis in the next section.

Clovis First

The name Clovis derives from the town of Clovis in eastern New Mexico where this tool culture tradition was first discovered in the 1920s. Since then, numerous Clovis sites have been identified throughout North America, especially in the east portion of the United States, but there are also Clovis sites in South America.

Since the 1930s, the prevailing ideas pertaining to initial human habitation in America was that the first culture in the New World was the Clovis tradition and that most, if not all, Native American groups derived from them. For several decades, this paradigm was almost dogmatic among archeologists. Yet, toward the end of the last century, investigators started questioning and challenging the validity of the Clovis First hypothesis. This skepticism was primarily fueled by the increasing number of archeological sites generating dates presiding Clovis. The first in importance and chronological order was Monte Verde in Chile (see Time of Arrival section). The impact of the Monte Verde discovery on the Clovis First theory was profound, resetting the date of the initial Proto-American incursion back potentially by several thousand years from the time previously accepted. Given the multiplicity of Pleistocene archaeological sites in America, which have emerged over the subsequent decades, today most scholars do not support the Clovis First theory.

The Clovis people are characterized by a very specific style of spear-tip construction in which the blade is fluted to produce a furrow or grove in the middle of the blade (Fig. 10.14). The blades are made of stone, bone, or ivory. It is likely that these Native Americans carved this cavity to allow the point to attach more firmly to the wooden shaft of the spear. This tradition quickly spread throughout the continent. The current consensus today is that the Clovis directly gave rise to a number of other cultures such as the Folsom, Gainey, Suwannee-Simpson, Plainview-Goshen, Cumberland and Redstone traditions, among others. These subsequent industries were primarily characterized by the topology of the blades that varied, for example, by the length of the furrow in the middle of the point. It is thought that these differences are not just stylistic but also stem from the natural tendency of technologies to evolve with time in response to environmental pressures, such as changes in the type of game hunted. It has been postulated that the extinction of the New World megafauna may have contributed to the changes in point morphology as specific species became extinct and Paleo-Natives were forced to shift to different available game. It is noteworthy to notice that many of the North American large mammals became abruptly extinct shortly after the appearance of the Clovis people in the New World about 12,000 ya. It is likely that the melting of LGM glaciers changed megafauna living conditions as large areas of real state opened up to grassing and hunting. This coincidence has suggested to some researchers a cause and effect relationship between the rapid population expansion of the Clovis culture and the massive megafaunal hunting spree that led to megafauna extinction.

How Many Waves?

As scholars began to theorize on the notion that the ancestors of Native Americans were of Asian descent, they also began to question not only when they arrived and how they travelled but also how many times they independently reached America. Today, after many decades of intense controversy in the scientific community, the question remains largely unanswered. Investigators are still divided on issues such as the number of successful dispersals from Central Asia to extreme northeast Siberia, from northeastern Siberia to the Beringian landmass, and from Beringia to the New World. In addition to the archeological data presented above that signal a number of tool traditions, such as the Clovis and pre-Clovis, possibly reflecting unique dispersals, scientists have incorporated a number of other marker systems from different disciplines in their arsenal of parameters to ascertain the number of migrations. Many of the clues are linguistic, anatomical, and molecular.

When estimating the number of dispersal waves of Paleo-Natives, a number of concerns need to be addressed. One of these concerns is whether or not stochastic (random) changes mediated by genetic drift (i.e., bottleneck episodes, founder effects, and isolation) experienced by a presumptive single migrating group of people can sufficiently account for the geographical patterns and partitioning of linguistic, anatomical, and genetic variability observed in the extant and extinct Native Americans, or whether it is necessary to invoke separate migrations from different source populations to explain the diversity. In other words, can a dozen 1000 years or so of evolution and genetic partitioning generate the observed variability currently seen among extinct and existing Native American populations? If the answer is no, then the observed differentiation might be better explained in terms of multiple independent dispersals.

Climatic Considerations

As previously stated, the time limits of these migrations into America were set by a number of climatic conditions along the way. If the migrational process is divided into stages, three main phases can be delineated: (1) the human arrival to Northeast Asia from Central Asia (e.g., Altaic), (2) the tenure in the Beringia landmass, and (3) the dispersal into the New World. Thus, estimating the earliest time that humans got to Northeast Asia from the Near East and South Asia at about 35,000 ya sets the limit for a migration into Beringia. Similarly, the climatic conditions that forced the length of the layover in Beringia and the glacial barriers on the American continent dictated the time of the spread into the New World. If humans went across Beringia without stopping, in theory they would have reached Alaska a few thousand years after they arrived at the northeast tip of Siberia, which would likely be about 30,000 ya. During this time period, the planet was gradually cooling but was still short of the LGM by about 10,000 years. Thus, it is likely that glaciers in west Beringia (Asia) were not yet fully formed, and the ice sheets in North America were not blocking access to the interior of the continent yet. Under those conditions, Paleo-Natives could have migrated rapidly through to America from 30,000–20,000 ya. Thereafter, during the LGM (20,000 ya), dispersal would have been very difficult. With the beginning of thawing, at the end of the ice age (about 12,000 ya), pedestrian crossing would have become feasible again. Hypothetically, these windows of time would have allowed different populations of Paleo-Natives, possibly even from different sources and locations in Asia, to travel across without stopping in Beringia. Of course, a temporal hiatus in a dry Beringia was also possible in between the two periods of accessibility.

Linguistics

The linguistic data support either several dispersion waves or one highly multilingual group of migrants. Based on the elevated multiplicity of mutually unintelligible tongues currently spoken in America, the most parsimonious model is one invoking multiple migrations. Available data indicate about 150–180 characterized language families in existence in the New World today. It is estimated that at the time of the European arrival, thousands of dialects were spoken in America.[69] Although the relationship among Native American languages is not always clear, the linguistic diversity suggests different groups of Asian Paleo-Natives moving into the New World. Two main migrational scenarios have been inferred from the linguistic data: (1) a few linguistically distinct migrations or (2) a multilingual migration (a single migration with multiple distinct or related languages).[70] Currently, many linguists advocate for multiple migrations of Paleo-Native populations, speaking various dialects, dispersing along the Pacific coast of America and into the interior of the continent with a concomitant linguistic proliferation along the way.[70]

Greenberg's tripartite linguistic theory on Native Americans[71] has received considerable attention over the years in connection with the peopling of the New World, but most linguists today consider it rather simplistic, while other scholars argue that it is simply incorrect. In his model, Greenberg postulates three higher-level families or macrofamilies: the Eskimo–Aleut, Na-Dené, and Amerind. Greenberg thought that many of the Native American language groups could be amalgamated into these three superfamilies. Albeit its simplicity and shortcomings, the three partite model allows for the identification of three distinct Native American language groups that correlate with three distinct assemblages of populations, the Eskimos, Athabascans (Na-Dené), and the rest of the Native American groups (Amerinds).

Anatomy

Morphological/anatomical traits also provide useful information to address the issues involved in the peopling of America. Also, not surprising, there is considerable controversy in this field of inquiry as well. For example, marked cranial differences have been observed among Paleo-Natives from Washington State, Nevada, and Brazil dating back to 5600 to 10,000 ya compared with contemporary Native Americans.[72] The discrepancies observed in the anatomies of these populations are so striking that a number of investigators are convinced that random genetic drift emanating from a single migration cannot explain the observed differences. These physical anthropologists argue for at least two major migrations from different source populations in Asia.[73] Yet, this view is not universal among scholars.

For example, as indicated below, in the DNA sections of this chapter, this two-pronged source model featuring a late Pleistocene to early Holocene colonization of the New World followed by a temporally and genetically distinct proto-Native American settlement is not supported by some of the genetic data. In addition, other morphometric work suggests that random genetic drift and selective pressure acting within a single ethnic group over thousands of years could have brought about the degree of craniofacial and anatomical differences observed among fossil and contemporary Native Americans. These studies indicate that the variation observed falls into a morphological continuum, and thus, most likely reflects a single source population exhibiting a high level of genetic and craniofacial diversity. These conclusions can be reconciled with the proposed Beringia layover theory that postulates a process of genetic differentiation and population partitioning during a prolonged hiatus of up to 15,000 years within the landmass.

DNA

Owing to the quantitative nature of genetic and molecular traits, DNA-based differences or polymorphisms are particularly useful in addressing issues involving the peopling of the New World. Therefore, to address this question, researchers have used estimated inception times (i.e., times in which specific markers appear in populations) of specific genetic markers to assess the time line of migratorial events, as well as the likelihood of a single source population providing the genetic variation present in extinct and extant Native Americans.[74] Yet, over the years, these studies have provided a multitude of different answers, which we will explore.

Classical Markers

Before the biotechnology revolution that began in the late 1970s, seminal studies using classical genetic traits, such as blood groups, protein, and histocompatibility markers, have confirmed the long suspected Asian ancestry of present-day Native Americans.[75] In addition, these initial investigations provided insight into the underlying genetic differences among the people in the different regions of the American continent. This body of work allowed scholars to assess changes in gene frequencies as a function of location and distances. Yet, these marker systems are not sensitive enough to ascertain details and microevolutionary changes involving origin, timing, and minimal number of migratory waves. In other words, these studies were limited by the small number of markers and populations as well as by the fact that many of the genes examined are known to be under selection pressure, and thus, do not necessarily reflect true demographic events and phylogenetic relationships resulting from mechanisms such as dispersals, bottleneck episodes, and ancestry. Marker systems under neutral selection or no selection pressure provide more reliable phylogenic information on population relationships, including Native American ancestry.

mtDNA Types

More recently, in the early 1990s, researchers began to mine and compared the mtDNA of extant Native American populations and found that, for the most part, the data pointed to three migration waves,[76] corroborating Greenberg's tripartite theory based on the Amerind, Na-Dené, and Aleut/Eskimo language groups. Yet, the excitement of a one-to-one correspondence between language groups and mtDNA classes was short-lived. Subsequent studies in the late 1990s, also based on mtDNA markers, failed to corroborate the previous results. These new data indicated that the degree of genetic diversity and differences observed among the Native American populations could be explained with just a single major dispersal. In other words, there was no need to postulate multiple migrations to explain the amount of variability seen.[77]

Autosomal Markers

During the same time that mtDNA studies were starting to add to the data on migrations into America, autosomal STR studies began to support a single primary dispersal from Asia, with random drift of DNA types generating the observed genetic partitioning along geographic lines.[78] An example of the impact of this type of stochastic genetic change is seen in the geographic transition of diversity from North to South America. Along this north–south axis, the DNA markers reveal a marked and dramatic reduction in genetic diversity (Fig. 10.26). As an illustration of this change in genetic diversity from North and Central America to South America, note within Fig. 10.26 the relative color homogeneity exhibited by the Wai-Wai population (number 20 in the bar graph) of South America in comparison with the color multiplicity in the Mayan Kaqchikel population (population number 19 in the graph) of Mesoamerica. Overall, starting in Central America and continuing into South America, the genetic diversity drops compared with North America. It seems that as migrants traveled south, the already relatively limited diversity of DNA types in North America was further reduced in Central America and beyond. This is attributed by many scholars to major bottleneck episodes and founder effect events in North America and/or Mesoamerica before the peopling of South America.

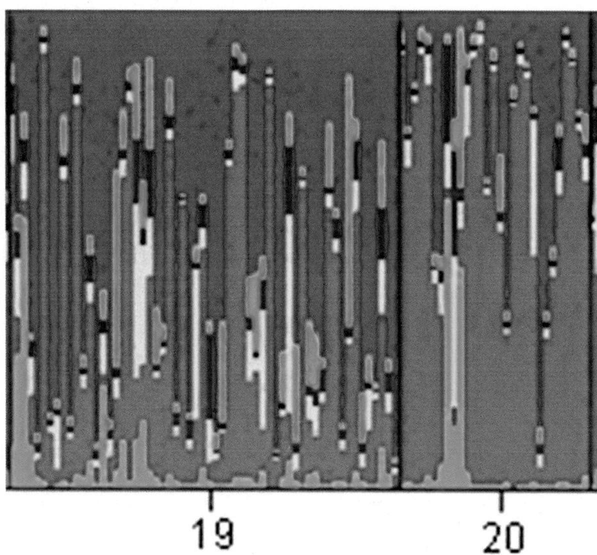

FIGURE 10.26 Transition between Mesoamerica and South America as seen from autosomal short tandem repeat analysis. Population 19, Mayan Kaqchikel. Population 20, Wai-Wai, South America.

As the technology in the field of molecular genetic analysis has improved, and investigators were able to probe larger number of genes and compare diversity among populations, the evidence started pointing to multiple incursions into America. For example, in an extensive study involving 364,470 autosomal SNPs in 52 Native American and 17 Siberian populations, the data align best with Greenberg's tripartite theory than with a single migration, yet the results also support the notion that the *majority* of extant Native Americans descend almost entirely from the first migratory wave of migrants.[60] In other words, these results suggest that the original migration contributed most of the Native American DNA in the New World.

More recently, in a study with greater genetic coverage involving approximately 600,000 SNPs (about double the number of autosomal markers examined compared with the previous study) and 21 Native American populations, a single *major* origin for most of Central and South America was indicated, which is consistent with the previously mentioned study[61] and the bottleneck episodes in Mesoamerica described earlier.[79] In other words, the genetic heritage of Central and South American Natives cannot be entirely explained by a single dispersal and suggests *at least* two genetically distinct migrations from Beringia and/northeast Siberia. In addition, various ancient genetic signals in certain Amazonian populations derived from a different Asian ancestral group, who are more closely related to indigenous Australians, New Guineans, and Andaman Islanders than to any present-day Eurasians or Native Americans, were detected. This genetic signature from an ancient ancestral population is not exhibited by present-day Northern and Central American natives or by a Clovis individual that lived about 12,600 ya. Thus, this extensive genome-wide study indicating Oceanian/Melanesian genetic input adds to a growing set of data suggesting multiple incursions into America.

The significance of a link between individuals that arrived to America as part of a distinct migration and present-day natives of Australia, New Guinea, and the Andaman Islands (Indian Ocean) is not clear. It is possible that the genetic commonalities observed in the study result from ancient gene flow between an ancestral Southeast Asian population directly derived from the Out of Africa migration to Asia and beyond (i.e., into Indonesia and Oceania) and populations that disseminated toward extreme Northeast Asia, Beringia, and the New World. This is a reasonable contention, considering that the ancient signals are detected among populations from Asia, Australia, New Guinea, and the Andaman Islands that are located along the putative coastal route followed by early modern humans during their trek out of Africa and into Asia. Also, this assumption is supported by the relative distant genetic affinities exhibited among the hypothetical ancestral population and Pacific and Indian Ocean islanders. The fact that this ancient signature in the Amazonian populations represents minor (not dominant) signals may be indicative of recent gene flow in the past few thousand years.

Y Chromosome–Specific Markers

The initial data based on Y-SNP and Y-STR studies failed to agree on the number of incursions into America.[80] Yet, in more recent investigations the general consensus has been in support of multiple dispersals.[81] In a detailed study of Y chromosomal Q types, the data suggested that humans entered America about 13,800 ya, a relatively recent date.[59] The study also indicated a latitudinal reduction of genetic diversity and Y chromosome types in contemporary populations of South America. Specifically, the authors noted a diminished presence of the Y chromosome Q-M346 type in South America, likely resulting from bottleneck and

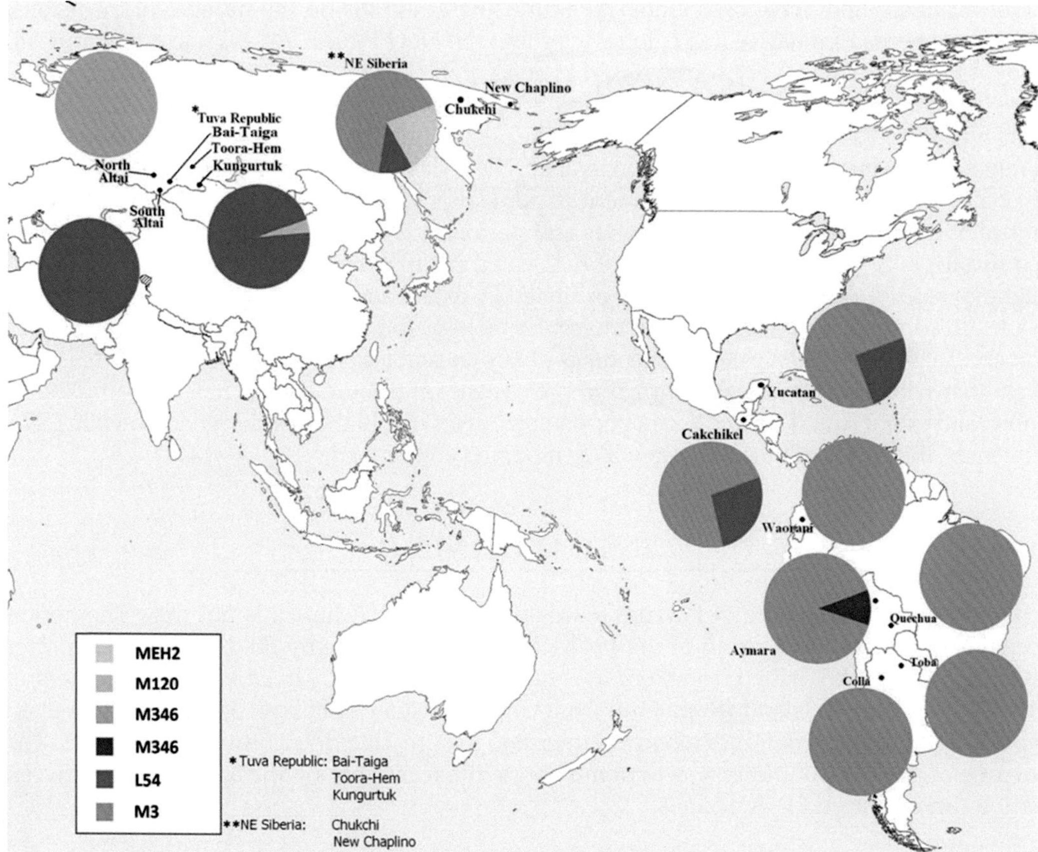

FIGURE 10.27 Transcontinental distribution of Q haplogroup types. *From Regueiro AJPA, 2013.*

founder effects during the Southward dispersal. Other reports have confirmed a second, more recent Circum-Arctic gene flow event as well.[82] Overall, these recent experiments confirm that the Q-M242 mutation that defines the entire Y chromosome Q lineage took place in Central Asia and it was transported, by groups such as the Altaics, in a northeastern direction to extreme northeast Siberia and beyond to America. The results also substantiate the notion that the Q-M13 Y chromosome type had its genesis in Beringia and was transported shortly thereafter to America as well as back into Asia.

In more recent investigations, a dramatic reduction of the ancient and ancestral Q L54 and MEH2 lineages of Central Asia have been described in America.[3] The near disappearance of these two ancient haplogroups in the New World is concomitant with an increase in the derived (from the L54 and MEH2 types) and younger Q-M3 mutation in Central and South American populations (Fig. 10.27). This increment in Q-M3 culminates with its fixation (100% presence) in South America. This study also detects a dramatic postdispersal population growth in Mesoamericans. Specifically, Proto-Mayan groups experienced a demographic expansion possibly fueled by ubiquitous and diverse agricultural- and trade-driven subsistence adopted during the preclassic age of their empire.

Various disciplines have contributed to addressing questions on the number of incursions, dates, and speeds of putative migration waves into the New World. Yet, it is clear that, overall, many issues still remain very controversial in spite of the efforts. Opinions have shifted, sometimes back and forth, since scholars started inquiring about the subject. Essentially, during the 400 years plus since Father Acosta postulated that the New World inhabitants were of Asian descent, highly valuable information has been generated. These data have been useful in delineating the diversity of many Asian and Native American populations. This information is necessary to formulate appropriate questions and theories and perform comprehensive comparative analyses. Specifically, over the years, scholars have delineated putative source populations, windows of migration opportunity, and the diversity exhibited by extant and extinct Native American populations. In addition, recent advances in molecular biology and bioinformatics (sequence annotation and analysis) have revealed the genomes of key ancient and contemporary individuals. It is likely that with additional technological improvements and more comprehensive sampling of extinct and extant American and Asian populations, involving high coverage genome-wide DNA sequences, the details of this diaspora will be modified and refined.

INTO OCEANIA

The idea that the people of Polynesia are migrants from America is not new. The notion became popularized as a result of the 1947 Kon-Tiki raft voyage by the Norwegian explorer Thor Heyerdahl who drifted from Peru to Tahiti and more recently in late 2015 by a successful trip reaching Eastern Island (Rapa Nui) from Chile, a distance of about 3, 690 km. The return trip had to be abandoned due to bad weather and progressive deterioration of the rafts. The purpose of these expeditions was to demonstrate the feasibility of ancient contacts between Native Americans and Polynesians.

Polynesian Chickens in Chile; South American Sweet Potatoes in Polynesia

Since the second half of the last century, anthropologists started noticing distributions of specific fauna and flora that suggested contacts between the people of South America and Polynesia. One of the best-known cases is the presence of the sweet potato (*Ipomoea batatas*) throughout Central and Eastern Polynesia.[83] This plant is not of Asian origins, so it could not have been transported during the Austronesian Expansion. Its genesis is in America. Specifically, the sweet potato was domesticated in the Andean region about 10,000 ya, during the in situ agricultural revolution of the region. Radiocarbon dating demonstrated that the plant arrived in the Cook archipelago and the Mangaia island in Central Polynesia about 1000 ya. After years of skepticism, currently the possibility that the plant got to Polynesia transported by American Natives or Austronesians is seriously considered by the orthodoxy. It is thought that this plant species was transported to Polynesia from South America around 1300 ya. These dates precede the European arrival in Oceania. Although it is possible that Native Americans reached these islands, taking the plant with them, it is likely that Polynesians, with their superior maritime technology, traveled to South America and back with the sweet potato as part of the cargo. This second scenario is also more plausible, considering the much larger geographical target that the American continent represented to Polynesian travelers compared with Native Americans finding tiny islands in a vast Pacific Ocean.

Chickens provide another case of imported domesticates. Near the coast of Chile, in a site named El Arenal, chicken bones were found. Yet, these were not ordinary chicken bones. Before this discovery, it was thought that Europeans first introduced chickens during the colonization period. Yet, radiocarbon dating and genetic evidence indicate that the bones got to Chile much earlier than the arrival of Europeans.[84] Although controversy still exists regarding the reliability of the radiocarbon dating and the origins of the mtDNA type found in El Arenal, proponents of the Polynesian–South American link affirm that chicken mtDNA haplogroup E was exclusively introduced from Southeast Asia into Oceania about 4000 ya, probably as part of the Austronesian Expansion and from there, some time later, into South America. At that time there were no European chickens in South America.

Native American DNA in Easter Island

DNA research has recently uncovered gene flow from South America to Polynesia.[85] This DNA flow was detected from the southern tip of South America toward Easter Island. Rapa Nui (the Austronesian name for Easter Island) was colonized by Austronesians around 800 ya. The small volcanic island of Rapa Nui (163.6 km^2) represents the easternmost Polynesian enclave in the Pacific Ocean, and it is thought to be the last land colonized by Polynesians as part of the Austronesian Expansion that started in continental Southeast Asia about 6000 ya. Europeans (the Dutch explorer Jacob Roggeveen) landed there for the first time in 1772. It is thought that the number of Polynesian inhabitants on the island reached a maximum of 15,000 from an original settlement of 50–100 people. Over the years, the native population has experienced dramatic declines resulting from a number of catastrophic events, such as self-inflicted ecocide, the Peruvian Slave trade of the later part of the 19th century, and a series of ensuing epidemics introduced by Europeans. These events reduced the number of natives to a low number of about 100 individuals in 1877.

Initial studies based on immuno-histocompatability genes (DNA responsible for acceptance/rejection of tissues during transplants) suggested connections between Easter Islanders and Native Americans.[86] Recently, with the implementation of more precise and sophisticated technology, these early connections have been confirmed. Comparing the relative proportions of nuclear DNA from contemporary Europeans, South Americans, and Polynesians present in individuals from Rapa Nui, averages of 6% and 16% of South American and European components, respectively, were detected in an Austronesian genetic background.[85] These frequencies were based on phylogenetic and statistical results obtained from structure analyses (Fig. 10.6) as well as the length of ancestry tract length distributions. The latter technique takes advantage of the indirect relationship that exists between the length of continuous nondisturbed (nonrecombined) DNA tracks and the time foreign chromosomes were introduced into Easter Island by interbreeding. These experiments, based on DNA track sizes, suggest that the time of the Native American incursion and admixture with the Rapa Nui natives was approximately 700 ya, about 100 years after the original Polynesian colonization of the Island. These dates are noteworthy as they delineate not only the time interval between the ensuing arrival of Austronesians to Rapa Nui and the rediscovery of Rapa Nui by the Europeans but also a time period before the cessation of long-distance travel by Polynesians about 575 ya. In other words, at the time of the putative contact of Native Americans with the Easter Islanders, Austronesians were still in their age of discovery. They were still in their expansion mode, hopping from island to island in the Pacific. By the time Europeans got to Easter Island, in 1772, the Austronesian maritime and canoe-building technology had deteriorated to a point that long oceanic crossings were not possible anymore. This corrosion of the Polynesian naval tradition among the Rapa Nui population may have been associated with their overall societal collapse and ecocide.

Considering the adventurous spirit of Austronesians, the proximity of Rapa Nui to the American continent and the relative large size of the New World extending from pole to pole, it is likely that sporadic contacts between Polynesians and Native Americans occurred. Also, taking into consideration that Polynesians were able to find and colonize little specks of land in the vastness of the Pacific Ocean, finding a whole continent should not have been more difficult. It is also plausible that any Polynesian genetic flow into Native American populations may go undetected as it may have been diluted out after its introduction into the vast expanse of Native American DNA. It is expected that greater coverage, probing more populations, individuals, and genetic markers in combination with more powerful analytical tools, may provide the resolution required to detect faint genetic signals of early Polynesian DNA in America.

References

1. Hohne K. *The mythology of sleep: the waking power of dreams*. Carnelian Bay, CA: Way of Tao Books; 2009.
2. Dulik MC, Zhadanov SI, Osipova LP, Askapuli A, Gau L, Gokcumen O, Rubinstein S, Schurr TG. Mitochondrial DNA and Y chromosome variation provides evidence for a recent common ancestry between Native Americans and Indigenous Altaians. *Am J Hum Genet* 2012;**90**:1–18.
3. Regueiro M, Alvarez J, Rowold D, Herrera RJ. On the origins, rapid expansion and genetic diversity of Native Americans from hunting-gatherers to agriculturalists. *Am J Phys Anthropol* 2013;**150**:333–48.
4. Mikhail S, Blinnikov A. *Geography of Russia and its neighbors*. New York: The Guilford Press; 2011.
5. Walker M, Johnsen S, Rasmussen SO, Popp T, Steffensen JP, Gibbard P, Hoek W, Lowe J, Andrews J, Björck S, Cwynar LC, Hughen K, Kershaw P, Kromer B, Litt T, Lowe DJ, Nakagawa T, Newnham R, Schwander J. Formal definition and dating of the GSSP (Global Stratotype Section and Point) for the base of the Holocene using the Greenland NGRIP ice core, and selected auxiliary records. *J Quat Sci* 2009;**24**:3–17.
6. Peregrine PN, Ember M. Encyclopedia of prehistory. *Artic and subartic*, vol. 2. New York: Kluwer Academic Press; 2001.
7. Baume CC. *The history of central Asia: the age of the steppe warriors*. London: Tauris Press; 2012.
8. Mochanov AY. Early stages of human occupation in NE Asia (Nauka, Novosibirsk, Russia, 1977 in Carlson, Roy L introduction. In: *Early human occupation in British Columbia*. Vancouver, B.C.: UBC Press; 1996.
9. Pitulko VV, Nikolsky PA, Girya EY, Basilyan AE, Tumskoy VE, Koulakov SA, Astakhov SN, Pavlova EY, Anisimov MA. The Yana RHS site: humans in the arctic before the last glacial maximum. *Science* January 2, 2004;**303**(5654):52–6.
10. Rakha A, Shin KJ, Yoon JA, Kim NY, Siddique MH, Yang IS, et al. Forensic and genetic characterization of mtDNA from Pathans of Pakistan. *Int J Legal Med* 2011;**125**(6):841–8.
11. Hussain J. *A history of the peoples of Pakistan towards independence*. Karachi, Pakistan: Oxford University Press; 1997.
12. Grimes BF. *Ethnologue: languages of the world*. Dallas: Summer Institute of Linguistics; 1992.
13. Ayub Q, Mezzavilla M, Pagani L, Haber M, Mohyuddin A, Khaliq S, Mehdi SQ, Tyler-Smith C. The Kalash genetic isolate: ancient divergence, drift, and selection. *Am J Hum Genet* May 7 , 2015;**96**(5):775–83.
14. Dulik MC, Zhadanov SI, Osipova LP, Askapuli A, Gau L, Gokcumen O, Rubinstein S, Schurr TG. Mitochondrial DNA and Y chromosome variation provides evidence for a recent common ancestry between Native Americans and indigenous Altaians. *Am J Hum Genet* February 10, 2012;**90**(2):229–46.
15. Karafet TM, Mendez FL, Meilerman MB, Underhill PA, Zegura SL, Hammer MF. New binary polymorphisms reshape and increase resolution of the human Y-chromosomal haplogroup tree. *Genome Res* May 2008;**18**(5):830–8.
16. Raghavan M, Skoglund P, Graf KE, Metspalu M, Albrechtsen A, Moltke I, Rasmussen S, Stafford Jr TW, Orlando L, Metspalu E, Karmin M, Tambets K, Rootsi S, Mägi R, Campos PF, Balanovska E, Balanovsky O, Khusnutdinova E, Litvinov S, Osipova LP, Fedorova SA, Voevoda MI, DeGiorgio M, Sicheritz-Ponten T, Brunak S, et al. Upper palaeolithic siberian genome reveals dual ancestry of native Americans. *Nature* January 2 , 2014;**505**:87–91.

17. Astakhov SN. *Paleolit Eniseia: Paleoliticheskie Stoianki Afontovoi Gore v G. Krasnoiarske.* Evropaiskii Dom; 1999.
18. Hubbe M, Harvati K, Neves W. Paleoamerican morphology in the context of European and East Asian Pleistocene variation: implications for human dispersion into the new world. *Am J Phys Anthropol* 2011;**144**:442–53.
19. Semino O, Magri C, Benuzzi G, Lin AA, Al-Zahery N, Battaglia V, Maccioni L, Triantaphyllidis C, Shen P, Oefner PJ, Zhivotovsky LA, King R, Torroni A, Cavalli-Sforza LL, Underhill PA, Santachiara-Benerecetti AS. Origin, diffusion, and differentiation of Y-Chromosome haplogroups E and J: inferences on the neolithization of Europe and later migratory events in the mediterranean area. *Am J Hum Genet* May 2004;**74**(5):1023–34.
20. Zegura SL, Karafet TM, Zhivotovsky LA, Hammer MF. High-resolution SNPs and microsatellite haplotypes point to a single, recent entry of native american Y chromosomes into the Americas. *Mol Biol Evol* 2003;**21**(1):164–75.
21. Zhong H, Shi H, Qi XB, et al. Global distribution of Y-chromosome haplogroup C-M130 reveals the prehistoric migration routes of African exodus and early settlement in East Asia. *J Hum Genet* July 2010;**55**(7):428–35.
22. Firasat S, Khaliq S, Mohyuddin A, Papaioannou M, Tyler-Smith C, Underhill PA, Ayub Q. Y-chromosomal evidence for a limited Greek contribution to the Pathan population of Pakistan. *Eur J Hum Genet* 2007;**15**(1):121–6.
23. Karafet TM, Osipova LP, Gubina MA, Posukh OL, Zegura SL, Hammer MF. High levels of Y-chromosome differentiation among native Siberian populations and the genetic signature of a boreal hunter-gatherer way of life. *Hum Biol* December 2002;**74**(6):761–89.
24. Torroni A, Achilli A, MacAulay V, Richards M, Bandelt H. Harvesting the fruit of the human mtDNA tree. *Trends Genet* 2006;**22**(6):339–45.
25. MacAulay V, Hill C, Achilli A, Rengo C, Clarke D, Meehan W, Blackburn J, Semino O, et al. Single, rapid coastal settlement of Asia revealed by analysis of complete mitochondrial genomes. *Science* 2005;**308**(5724):1034–6.
26. Rowold DJ, Luis TR, Terreros MC, Herrera RJ. Mitochondrial DNA geneflow indicates preferred usage of the Levant Corridor over the Horn of Africa passageway. *J Human Genet* April 2, 2007;**52**(5):436–47.
27. Comas D, Plaza S, Wells SR, Yuldaseva N, Lao O, Calafell F, Bertranpetit J. Admixture, migrations, and dispersals in Central Asia: evidence from maternal DNA lineages. *Eur J Hum Genet* February 11 , 2004;**12**:495–504.
28. Soares P, Ermini L, Thomson N, Mormina M, Rito T, Röhl A, Salas A, Oppenheimer S, MacAulay V. Correcting for Purifying selection: an improved human mitochondrial molecular clock. *Am J Hum Genet* 2009;**84**(6):740–59.
29. Fedorova A, Reidla M, Metspalu E, Metspalu M, Rootsi S, Tambets K, Trofimova N, Zhadanov SI, Kashani BH, Olivieri A, Voevoda MI, Osipova LP, Platonov FA, Tomsky MI, Khusnutdinova EK, Torroni A, Villems R. Autosomal and uniparental portraits of the native populations of Sakha (Yakutia): implications for the peopling of Northeast Eurasia. *Sardana BMC Evol Biol* 2013;**13**:127.
30. Volodko NV, Starikovskaya EB, Mazunin IO, et al. Mitochondrial genome diversity in Arctic Siberians, with particular reference to the evolutionary history of Beringia and pleistocenic peopling of the Americas. *Am J Hum Genet* 2008;**82**:1084–100.
31. Li JZ, Absher DM, Tang H, Southwick AM, Casto AM, Ramachandran S, Cann HM, Barsh GS, Feldman M, Cavalli-Sforza LL, Myers RM. Worldwide human relationships inferred from genome-wide patterns of variation. *Science* 2008;**319**(5866):1100.
32. Rosenberg NA, Pritchard JK, Weber JL, Cann HM, Kidd KK, Zhivotovsky LA, Feldman MW. Genetic structure of human populations. *Science* 2002;**298**:2381–5.
33. Zhao TM, Lee TD. Gm and Km allotypes in 74 Chinese populations: a hypothesis of the origin of the Chinese nation. *Hum Genet* 1989;**83**:101–10.
34. Zhang ZB. An analysis of the physical characteristics of modern Chinese. *Acta Anthropol Sin* 1988;**7**:314–23.
35. Chu JY, Huang W, Kuang SQ, Wang JM, Xu JJ, Chu ZT, Yang ZQ, Lin KQ, Li P, Wu M, et al. Genetic relationship of populations in China. *Proc Natl Acad Sci USA* 1998Vol. 95:11763–8.
36. Shi H, Zhong H, Peng Y, Dong YL, Qi XB, Zhang F, Liu LF, Tan SJ, Ma RZ, Xiao CJ, Wells RS, Jin L, Su B. Y chromosome evidence of earliest modern human settlement in East Asia and multiple origins of Tibetan and Japanese populations. *BMC Biol* 2008Vol. 6:45.
37. Hudjashov G, Kivisild T, Underhill PA, Endicott P, Sanchez JJ, Lin AA, Shen P, Oefner P, Renfrew C, Villems R, et al. Revealing the prehistoric settlement of Australia by Y chromosome and mtDNA analysis. *Proc Natl Acad Sci USA* 2007Vol. 104:8726–30.
38. Hopkins DM, Matthews Jr JV, Schweger CE, Young SB. *Paleoecology of Beringia*. New York: Academic Press; 1982.
39. Vakhtin N. Bicultural education in the north: ways of preserving and enhancing indigenous peoples' languages and traditional knowledge. In: Kasten E, editor. *Endangered languages in northeast Siberia: Siberian Yupik and other languages of Chukotka*. Munster: Waxmann Verlag; 1998.
40. Fortescue M. The relationship of Nivkh to Chukotko-Kamchatkan revisited. *Lingua* 2011;**121**:1359–76.

41. www.spri.cam.ac.uk/.../e.rockhill/mphil.pd.
42. Schurr T, Sukernik R, Starikovskaya E, Wallace D. Mitochiondrial DNA variation in Koryaks and Itel'man: population replacement in the Okhotsk Sea-Bering Sea region during the neolithic. *Am J Phys Anthropol* 1999;**108**:1–39.
43. Dikov NN. *Early cultures of northeastern Asia. Shared Beringian heritage program*. Anchorage: National Park Service; 2004.
44. Yi S, Clark G. The "Dyuktai culture" and new world origins. *Curr Anthropol* 1985;**26**(1):1–20.
45. Fiedel SJ. *Prehistory of the Americas*. London UK: Cambridge University Press; 1999.
46. Geml J, Laursen GA, O'Neill K, Nusbaum HC, Taylor DL. Beringian origins and cryptic speciation events in the fly agaric (*Amanita muscaria*). *Mol Ecol* January 2006;**15**(1):225–39.
47. Vajda EJ. A siberian link with Na-Dene languages in the Dene–Yeniseian connection. In: Kari J, Potter B, editors. *Anthropological papers of the University of Alaska, new series*, vol. 5. 2010. p. 33–99. Fairbanks.
48. Ruhlen R. The origin of the Na-Dene. *Proc Natl Acad Sci USA* 1998;**95**:13994–6.
49. Vajda E. Yeniseian, Na-Dene, and historical linguistics. *APUA new series*, vol. 5. 2010. p. 100–18.
50. Sicoli MA, Holton G. Linguistic phylogenies support back-migration from Beringia to Asia. *PLoS One* 2014. https://doi.org/10.1371/journal.pone.0091722.
51. Augustin L, et al. Eight glacial cycles from an antarctic ice core. *Nature* 2004;**429**(6992):623–8.
52. Tamm E, Kivisild T, Reidla M, Metspalu M, Smith DG, Mulligan CJ, Bravi CM, Rickards O, Martinez-Labarga C, Khusnutdinova EK, Fedorova SA, Golubenko MV, Stepanov VA, Gubina MA, Zhadanov SI, Ossipova LP, Damba L, Voevoda MI, Dipierri JE, Villems R, Malhi RS. Beringian stand still and spread of Native American founders. *PLoS One* 2007;**2**:e829.
53. Mulligan CJ, Kitchen A, Miyamoto MM. MM. Updated three stage model for the peopling of the Americas. *PLoS One* 2008;**3**:e3199.
54. Archaeological Assessment of Geotechnical Cores and Materials, 2011 Statoil Ancillary Activities, Chukchi Sea, Alaska June 2012 Prepared for Statoil USA E&P, Inc. 3800 Centerpoint Drive, Suite 920 Anchorage, Alaska 99503 Prepared by Jason Rogers, PhD Alaska Maritima 7009 Madelynne Way Anchorage, Alaska 99504.
55. Hopkins DM. Hard times in Beringia, a short note. In: Masters PM, Flemming NC, editors. *Quaternary coastlines and marine archaeology, towards the prehistory of land bridges and continental shelves*. London, UK: Academic Press; 1983.
56. Darigo N, Owen KM, Bowers PM. *Review of geological/geophysical data and core analysis to determine archaeological potential of buried Landforms, Beaufort Sea Shelf, Alaska*. Report prepared for U.S. Department of the Interior, Minerals Management Service (MMS). Fairbanks: Northern Land Use Research, Inc.; 2007.
57. Grosswald MG, Hughes TJ. Paleoglaciology's grand unsolved problem. *J Glaciol* 1995;**41**:313–32.
58. Kitchen A, Miyamoto M, Mulligan CJ. A three-stage colonization model for the peopling of the Americas. *PLoS One* 2008;**3**:e1596.
59. Malyarchuk B, Derenko M, Denisova G, Maksimov A, Wozniak M, Grzybowski T, Dambueva I, Zakharov I. Ancient links between Siberians and Native Americans revealed by subtyping the Y chromosome haplogroup Q1a. *J Hum Genet* 2011Vol. 56:583–8.
60. Reich D, Patterson N, Campbell D, Tandon A, Mazieres S, Ray N, Parra MV, Rojas W, Duque C, Mesa N, Garcıa LF, Triana O, Blair S, Maestre A, Dib JC, Bravi CM, Bailliet G, Corach D, Hünemeier T, Bortolini MC, Salzano FM, Petzl-Erler ML, Acuna-Alonzo V, Aguilar-Salinas C, Canizales-Quinteros S, Tusie´-Luna T, Riba L, Rodrıguez-Cruz M, López Alarcón M, Coral-Vazquez R, Canto-Cetina T, Silva-Zolezzi I, Fernandez-Lopez JC, Contreras AV, Jimenez-Sanchez G, Gómez-Vázquez MJ, Molina J, Carracedo A, Salas A, Gallo C, Poletti G, Witonsky DB, Alkorta-Aranburu G, Sukernik RI, Osipova L, Fedorova SA, Vasquez R, Villena M, Moreau C, Barrantes R, Pauls D, Excoffier L, Bedoya G, Rothhammer F, Dugoujon JM, Larrouy G, Klitz W, Labuda D, Kidd J, Kidd K, Di Rienzo A, Freimer NB, Price AL, Ruiz-Linares A. Reconstructing native american population history. *Nature* 2012;**488**:370–4.
61. Ray N, Wegmann D, Fagundes NJ, Wang S, Ruiz-Linares A, Excoffier L. A statistical evaluation of models for the initial settlement of the American continent emphasizes the importance of gene flow with Asia. *Mol Biol Evol* 2010;**27**:337–45.
62. Kumar S, Bellis C, Zlojutro M, Melton PE, Blangero J, Curran JE. Large scale mitochondrial sequencing in Mexican Americans suggests a reappraisal of Native American origins. *BMC Evol Biol* 2011;**11**:293.
63. Bodner M, Perego UA, Huber G, Fendt L, Ro¨ck AW, Zimmermann B, Olivieri A, Gomez-Carballa A, Lancioni H, Angerhofer N, Bobillo MC, Corach D, Woodward SR, Salas A, Achilli A, Torroni A, Bandelt HJ, Parson W. Rapid coastal spread of First Americans: novel insights from South America's Southern Cone mitochondrial genomes. *Genome Res* 2012;**22**:811–20.
64. Gibbons A. Mother tongues trace steps of earliest Americans. *Science* 1998;**279**:1306–7.

65. Dillehay TD, Ramırez C, Pino M, Collins MB, Rossen J, Pino- Navarro JD. Monte Verde: seaweed, food, medicine, and the peopling of South America. *Science* 2008;**320**:784–6.
66. Waguespack NM. Why We're still arguing about the Pleistocene occupation of the Americas. *Evol Anthropol* 2007;**16**:63–74.
67. Smallwood AM. Clovis biface technology at the topper site, South Carolina: evidence for variation and technological flexibility. *J Archaeol Sci* 2010;**37**:2413–25.
68. Waters MR, et al. The Buttermilk Creek complex and the origins of Clovis at the Debra L. Friedkin site, Texas. *Science* March 25 , 2011;**331**(6024):1599–603.
69. Campbell L. *American Indian languages: the historical linguistics of native America*. Oxford, UK: Oxford University Press; 1997.
70. R. Blench. Accounting for the Diversity of Amerindian Languages: Modelling the Settlement of the New World. Paper presented at the Archaeology Research Seminar, RSPAS, Canberra, Australia 2008.
71. Greenberg JH. *Language in the Americas*. Stanford, CA: Stanford University Press; 1987.
72. Powell JF. *The first Americans: race, evolution and the origin of native Americans*. Cambridge; New York: Cambridge University Press; 2005.
73. Raff JA, Bolnick DA, Tackney J, O'Rourke DH. Ancient DNA perspectives on American colonization and population history. *Am J Phys Anthropol* 2011;**146**:503–14.
74. Balter M. The peopling of the Aleutians. *Science* 2012;**335**:158–61.
75. Salzano FM, Gershowitz H, Mohrenweiser H, Neel JV, Smouse PE, Mestriner MA, Weimer TA, Franco MH, Simoes AL, Constans J, Oliveira AE, de Melo E, Freitas MJ. Gene flow across tribal barriers and its effect among the Amazonian Icana River Indians. *Am J Phys Anthropol* 1986;**69**:3–14.
76. Torroni A, Schurr TG, Yang CC, Szathmary EJE, Williams RC, Schanfield MS, Troup GA, Knowler WC, Lawrence DN, Weiss KM, Wallace DC. Native American mitochondrial DNA analysis indicates that the Amerind and the Nadene populations were founded by two independent migrations. *Genetics* 1992;**130**:153–62.
77. Bonatto S, Salzano M. A single and early migration for the peopling of the Americas supported by mitochondrial DNA sequence data. *Proc Natl Acad Sci USA* 1997;**94**:1866–71.
78. Wang S, Lewis CM, Jakobsson M, Ramachandran S, Ray N, Bedoya G, Rojas W, Parra MV, Molina JA, Gallo C, Mazzotti G, Poletti G, Hill K, Hurtado AM, Labuda D, Klitz W, Barrantes R, Bortolini MC, Salzano FM, Petzl-Erler ML, Tsuneto LT, Llop E, Rothhammer F, Excoffier L, Feldman MW, Rosenberg NA, Ruiz-Linares A. Genetic variation and population structure in Native Americans. *PLoS Genet* 2007;**3**:e185.
79. Skoglund P, Mallick S, Bortolini MC, Chennagiri N, Hünemeier T, Petzl-Erler ML, Salzano FM, Patterson N, Reich D. Genetic evidence for two founding populations of the Americas. *Nature* 2015;**525**:104–8.
80. Karafet TM, Zegura SL, Posukh O, Osipova L, Bergen A, Long J, Goldman D, Klitz W, Harihara S, de Knijff P, Wiebe V, Griffiths RC, Templeton AR, Hammer MF. Ancestral Asian source(s) of New World Y chromosome founder haplotypes. *Am J Hum Genet* 1999;**64**:817–31.
81. Schurr TG, Sherry ST. Mitochondrial DNA and Y-chromosome diversity and the peopling of the Americas: evolutionary and demographic evidence. *Am J Hum Biol* 2004;**16**:420–39.
82. Bisso-Machado R, Bortolini MC, Salzano FM. Uniparental genetic markers in South Amerindians. *Genet Mol Biol* 2012;**35**:365–87.
83. Van Tilburg JA. *Easter Island: archaeology, ecology and culture*. Washington, DC: Smithsonian Institution Press; 1994.
84. Storey AA, Matisoo-Smith EA. No evidence against Polynesian dispersal of chickens to pre-Columbian South America. *Proc Natl Acad Sci USA* 2014;**111**:E3583.
85. Moreno-Mayar V, et al. Genome-wide ancestry patterns in Rapanui suggest pre-European admixture with native Americans. *Curr Biol* 2014;**24**:1–8.
86. Thorsby E, et al. Further evidence of an Amerindian contribution to the Polynesian gene pool on Easter Island. *Tissue Antigens* 2009;**73**:582–5.

CHAPTER

11

The Bantu Expansion

The chief reasons why so many people are loath to admit the genetic variability of socially and culturally significant traits are two. First, human equality is stubbornly confused with identity, and diversity with inequality, as though to be entitled to an equality of opportunity, people would have to be identical twins. Human diversity is not incompatible with equality. Secondly, it is futile to look for one-to-one correspondence between cultural forms and genetic traits. Cultural forms are not determined by genes, but their emergence and maintenance are made possible by the genetically conditioned human diversity. ***Theodosius Dobzhansky, "The Problem of Human Evolution"***

SUMMARY

In spite of the homogenizing effect of the Bantu dissemination, sub-Saharan Africa is the most genetically diverse region in the world. It is this variability that actually points to this region as the cradle of humanity. It is estimated that sub-Saharan Africa possesses over 90% of the total human genetic diversity.

Bantu is a linguistic term; the word itself means *people*. The Proto-Bantu language had its genesis in West Africa on the border region of what is today Northern Cameroon and Western Nigeria about 5000 years ago (ya). The Bantu Expansion is the name given to a monumental migration that started 4000–3000 ya. By definition, Bantu is a branch of a family made up of approximately 500 languages. The Bantu branch within the Benue–Congo family is the most widely spoken group of languages within the Niger–Kordofanian supralanguage family (Fig. 11.1). About 60 million people speak Bantu languages today. The term Narrow Bantu is often used to discriminate between more inclusive Bantoid languages from the more strictly defined Bantu. Today, the Bantu language with the largest total number of speakers is Swahili, and it is thought of as a sort of *lingua franca*.

Signals of this major human dispersal were initially detected when linguists noticed language similarities among groups of people throughout sub-Saharan Africa. Subsequently, these vernacular parallelisms were supported by findings from a number of fields including art, archeology, architecture, anthropology, culture, genetics, mythology, and technology. Genetics clearly indicates that the observed shared characteristics among Bantu populations are not just the result of acculturation or transfer of knowledge but actual movement of people

Ancestral DNA, Human Origins, and Migrations
https://doi.org/10.1016/B978-0-12-804124-6.00011-2

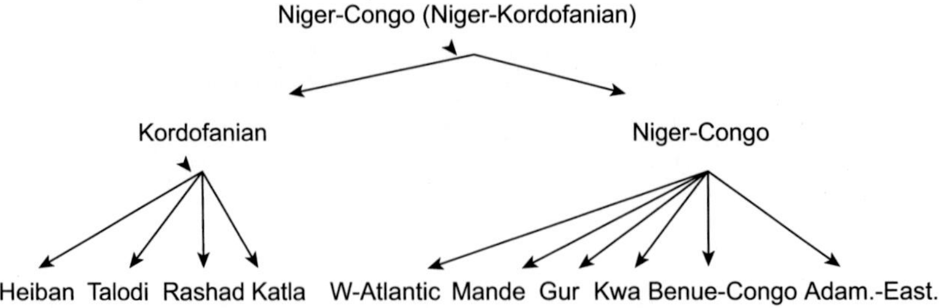

FIGURE 11.1 Niger–Kordofanian language family. *From http://tima.zonglist.zong.mine.nu/images/Family_tree.jpg.*

admixing with local groups and passing their genes to them. Although the Bantu dispersal is traditionally an intra-African migration, its impact is in fact felt outside the continent. It can be argued that the transatlantic slave trade that brought native Africans into America and the East African slave trade that supplied forced laborers to Muslim courts in West Asia are extensions of this Diaspora, considering that a good number of the captives were Bantu. According to the orthodoxy, the expansion may have been driven by overpopulation, the need for more land, famine, tribal conflicts, disease, and/or climatic changes such as drought. Yet, it is not clear to what degree the movement was self-conscious as opposed to opportunistic farmers just looking for better agricultural land and in the process displacing endogenous populations. A number of innovations including agriculture, new crops (bananas, jam, bulrush millet, sugarcane, etc.), and ironworking technology likely facilitated the spread. Although iron technology may not have been invented or introduced by Bantus and the travelers did not initially practice it, it is postulated that it allowed them to exploit forest niches in Central Africa. It is thought that iron technology and agriculture complemented each other because the metal allowed the clearing of the forest and that in turn provided hardwood for the purification of the ore. The outcomes of the expansion include permanent and larger settlements because of their capacity to store food for lean times, utilization of diverse ecological systems and crops at different times of the year. In addition, the Bantu Expansion resulted in the establishment of centralized government systems and kingdoms (e.g., The Great Zimbabwe) allowing for efficient trade within the Bantu domain and with Arab and Persian traders, as well as the assimilation and/or elimination of many native populations. The Bantus have impacted a number of autochthonous groups throughout Africa including the Cushitics, Pygmies, San, and Nilo-Saharans, among others. The scientific community actively debates the putative migration routes and the dispersal time line. A particular point of contention is whether the spread occurred through two independent paths, a western and an eastern corridor, from their homeland or just one southward along West Africa, traversing to East Africa recently (about 1500 ya). Another intriguing question is the nature of the interactions between the encrushing Bantus and the natives, especially during the initial period of contact. It is likely that the local groups were marginalized to less desirable lands by the technologically superior Bantus. Linguistic studies indicate that the Bantu language has impacted native languages differentially on an individual basis throughout the continent whereas genetics indicates that Bantu and non-Bantu populations have assimilated various proportions of each others' DNA depending on the populations in question.

THE BANTU CAUSATUM

Independent of whether the Bantu dispersion resulted from transmission of ideas and technology, actual movement of people or a combination of the two, the fact is that its aftermath has been profound in and outside Africa. Actively and passively, the Bantu phenomenon changed sub-Saharan Africa and beyond, forever. In general, it had a powerful homogenization effect reducing the amount of diversity. Most of the autochthonous populations south of the Sahara were impacted. To various extents, native groups lost their languages and culture. This wealth of heritage vanished almost completely in some instances and partially in others. The strength of the Bantu punch was strong and varied depending on the local sociocultural conditions at the time of the initial encounter and soon after. The end result of these processes was the assimilation of the Bantu way of life by the indigenous people and the elimination of native singularities.

Bantus disseminated the practice of agriculture throughout their domain. Novel crops were introduced and improved by human artificial selection. Furthermore, domestication of animals was spread into areas where it was not previously practiced. These changes brought with them profound shifts in sociocultural modes. As in other worldwide places where agriculture and domestication originated and flourished, in sub-Saharan Africa, it allowed for the storage of surplus food. This event had profound consequences. It freed humans from having to search for subsistence on a daily basis. The previously practiced foraging way of life demanded considerable amount of time and effort. As agrarians, people were now able to feed on previously accumulated supplies such as grains and dry meat as well as a continuous source of milk and dairy products.

In parallel to this agrarian revolution was the acquisition and propagation of iron technology. Although iron is more difficult and time consuming to purify than copper, as a material for tool manufacturing, it affords greater strength and durability. Iron was used by the Bantus to clear the forest, plow the fields, harvest the crops, cook, eat, build homes, create artistic expressions, and defend themselves. In general, iron greatly extended the ecological niches available to the Bantus and expedited their territorial advances. Therefore, developments in metallurgy complemented the advance of agriculture, and both were key in moving forward the Bantu civilization at unprecedented speed. In other words, iron tools facilitated the clearing of the forest, which in turn provided for hardwood essential for reaching the high temperatures require for the purification of the ore.

These changes, in turn, provided for a cascade of developments. It obliterated the need for foraging and humans had more leisure time to rest, think, and more importantly, be creative. Although time to meditate and conceive ideas is usually ignored as an important by-product of an agrarian society, it likely fuels the genesis of discovery. The additional food also allowed for population growth and the establishment of permanent settlements that then evolved into larger towns and cities. The physical landscape of these extended settlements contrasted distinctly with the foraging style or village way of living. Households went from mud–dung huts to impressive walled city-states, such as the Great Zimbabwe.

The development of large homesteads brought positive and negative outcomes. On one hand, it allowed for a centralized government that facilitated law and order. It gave its citizens a sense of community. It promoted technical specialization and individuals with expertise in specific trades and functions with the potential to excel. It stimulated local and distant trade and provided protection from invaders and enemies. On the other hand, it could have generated a tyrannical and corruptive ruling class that victimized lower classes. In addition, accumulation of people in confined areas facilitates the start and spread of infectious diseases that could kill in thousands or even millions.

In this chapter, we will explore these topics, among others, and provide insight into the nature of the Bantu from their origins, dispersals, kingdoms, and slavery into the New World and West Asia.

THE LINGUISTIC CUE

Bantu and Non-Bantu Languages of Sub-Saharan Africa

Bantu is a prominent group of languages in Africa. It represents one (Benue–Congo) of several branches within the Niger–Kordofanian family of languages. Furthermore, the Niger–Kordofanian classification is split into the Kordofanian and Niger–Congo divisions. Bantus encompass more geographical area than the rest of the Niger–Kordofanian family put together (see orange color in Fig. 11.2).

This degree of linguistic uniformity compelled linguists to consider mechanisms for such widespread distribution of a single language family. At the center of this apparent conundrum was the issue of whether the extensive Bantu domain was the result of acculturation (movement of ideas and knowledge *without* migration of people) and/or physical dispersal and settlement of people.

Niger–Kordofanians

In addition to the Niger–Kordofanian family, other major groups of language families are spoken in North Africa and sub-Saharan Africa (Fig. 11.2). Sub-Saharan Africa alone possesses the highest levels of linguistic differentiation in the world, in spite of the amalgamation and homogenization forces derived from the Bantu Expansion. Conservatively, it has been estimated that approximately 1000 languages and dialects are spoken in sub-Saharan Africa.[1]

Afro-Asiatic

Included among the major families of languages of Africa are the Afro-Asiatic languages from the Horn of Africa and North Africa. Within this family, there are a number of extensive branches such as the Semitic and the Chadic. The Semitic tongues are amid the oldest currently used languages in the world with records dating back to about 4500 ya.[2] Although some, such as Sumerian, Ancient Egyptian, Ancient Hebrew, are extinct, others such as Ethiopian Semitic and Arabic are pretty much alive. The Semitic branch is the only one spoken inside and outside Africa, specifically in Southwest Asia. The Chadic offshoot is used in Central and Western Africa. One of the Chadic languages, Hausa, exhibits a wide distribution. It is commonly used among the residents of Niger, Ghana, Togo, Benin, Cameroon, and Chad. Before meeting the Bantus, Semitic and Chadic tribes were traditional hunter-gatherers/pastoralists.

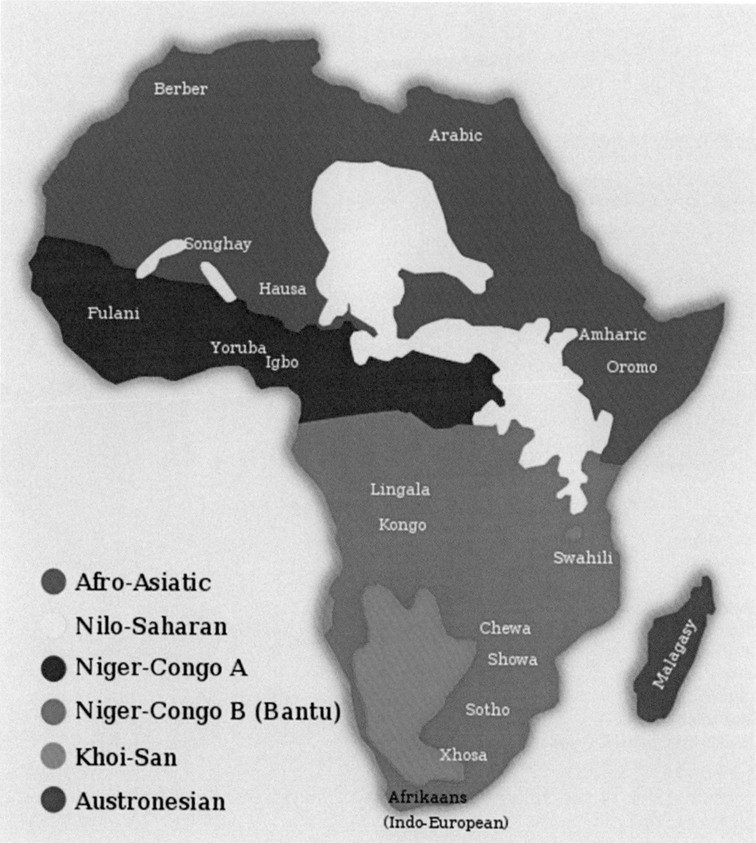

FIGURE 11.2 Languages of Africa. *From http://www.newworldencyclopedia.org/entry/File: African_language_families.svg.*

Khoisan

The Khoisan is another major family encountered by the dispersing Bantus. After the Bantu branch, the Khoisan is the second most widely spoken group of languages. Before interacting with the Bantus, the speakers of this family practiced a hunter-gathering/pastoral existence. Khoisan is another ancient family with distribution throughout Southern Africa and Southeast Africa. In Southern Africa, this family is represented by the Khoikhoi and San (Bushmen) people whose speech is characterized by clicks. In Southeast Africa, the Sandawe and Hadza are members of this language family. As the Khoikhoi and San, they also communicate using clicks. Together with the Bushmen, the Sandawe contrasts dramatically with the typical Bantus in physical appearance. The Sandawe are small in stature and thin framed with light yellow-brown skin, narrow lips, oriental-like epicanthic eye fold, and highly wrinkle skin at old age.

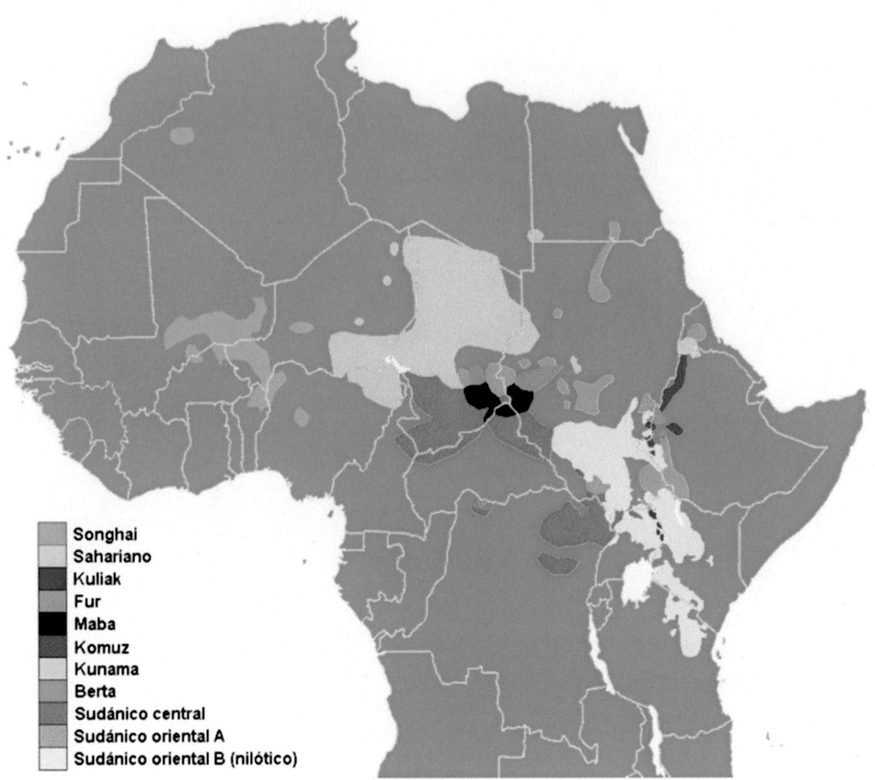

FIGURE 11.3 Subdivisions of the Nilo-Saharan family of languages. *From https://commons.wikimedia.org/wiki/File:Lenguas_nilo-saharianas.PNG.*

Nilo-Saharan

As the name implies, the languages within the Nilo-Saharan group are distributed mainly in the upper basins of the Nile and Chari rivers in North-central and Northeast Africa (Fig. 11.3). Fig. 11.3 illustrates the geographical location of the main subdivisions within this family. As seen in Fig. 11.3, it exhibits a rather spotted distribution within the continent. It is not a click language. It is spoken by about 60 million people in 17 contemporary nations in the northern half of Africa from Algeria to Benin to the west, from Libya to the Democratic Republic of Congo in Central Africa, and from Egypt to Tanzania to the west (Fig. 11.3). It is considered an ancient group of languages dating back to at least 7000 ya.[3] Genetic studies have shown that although Nilo-Saharan languages represent a mosaic classification and amalgamation of unique tongues that do not belong in the other groups, it has a genetic basis. This may result from ancient genetic sources going back to the late Upper Paleolithic, about 15,000 and 10,000 ya.

Pygmies

Contrary to common belief, Pygmies are not a homogeneous group of people. They do not share a common ethnicity or language, their geographical distribution is patchy within

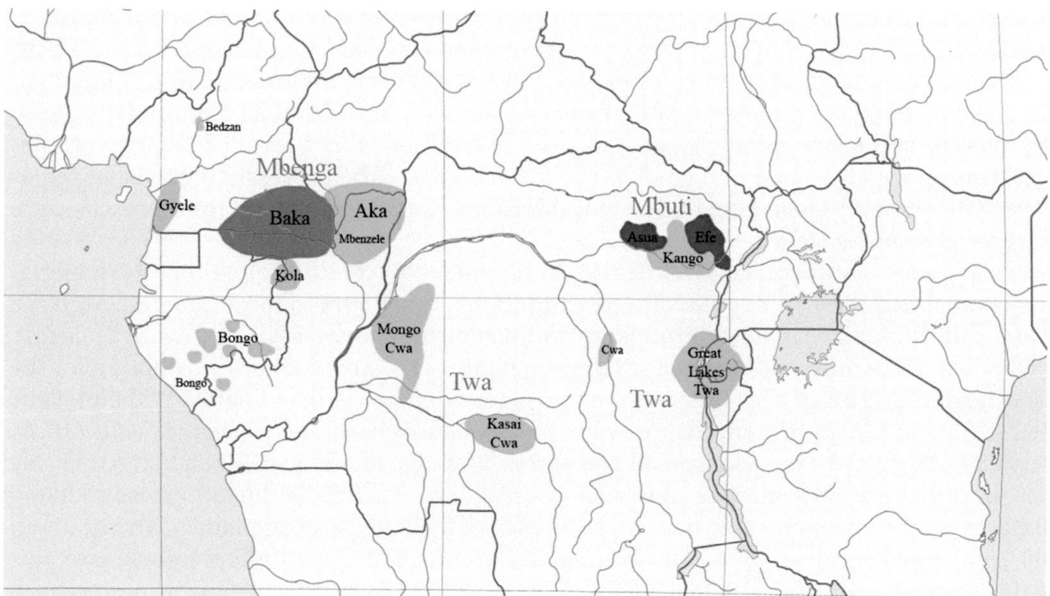

FIGURE 11.4 Distribution of Pygmy languages. *From https://en.wikipedia.org/wiki/Classification_of_Pygmy_languages.*

the rain forest, and they are isolated from each other (Fig. 11.4). The various colors of populations in Fig. 11.4 represent different Pygmy language groups. Although an ancestral Pygmy language has been proposed, current populations speak different Nilo-Saharan or Niger–Kordofanian languages with various proportions of Bantu mix in. The degree of Bantu linguistic admixture varies with specific populations and locations. It is likely that this diversity in the language that they speak is the result of interactions with neighboring tribes of various linguistic ancestries. The only possible signature of an ancestral language is the Mbenga forest vocabulary shared by the Aka and Baka Pygmies of Western Africa. This vocabulary relates to forest living and is primarily made up of words for plants and animals.

Yet, Pygmies share certain characteristics in common. They are forest people from Equatorial Africa. Their short stature is genetically determined. They practice a simple non-hierarchical societal existence based on hunting and gathering and exhibit subservient relationships with neighboring agricultural groups, such as the Bantus. They all express strong cultural and spiritual bonds with the forest. Their relationship with non-Pygmy populations is generally commercial in nature, and their languages have incorporated extensively the elements of Bantu. Occasionally, Pygmy females marry into Bantu families. It is interesting that the genes that determine their small size are not always the same in different Pygmy populations.[4]

The Surprising Widespread Distribution of Bantu Languages

Linguistics and the study of the history of languages are based on the premise that changes in languages stem from idiolects (a single person's speech) and dialects (the community's lingual). Thus, an idiolect, with time, may contribute to the creation of a dialect and dialects, in turn, may become part of a language when a number of communities adopt mutually understandable dialects. In other words, languages are made of a number of related dialects. In this way, dialects develop into full-fledged languages. This progression from individual linguistic diversity accumulating and dispersing among members of a community generating dialects, and subsequently, languages may be envisioned as a form of linguistic evolution. This process takes place in tandem in time and space. The process can be visualized as a tree with branches emanating in different directions from the trunk and from branches into increasingly smaller stems.

It was somewhat surprising to linguists in the mid-19th century to see that the majority of subequatorial African populations communicated using very similar tongues. A glance at the distribution of language groups in continental sub-Saharan Africa (Fig. 11.2) clearly shows that of the four major language families, Bantu is by far the largest accounting for the majority of the territory. This was first noted and reported by German linguist Wilhelm Bleek (1827–75) who coined the term *Bantu* in his 1862 book *Comparative Grammar of South African Languages*[5] to describe the widespread use of this language family in sub-Saharan Africa. He theorized that the large number of similar languages across the continent possess so many characteristics in common that they must be part of a single language family. Today, about 440–680 languages, over 400 ethnic groups and approximately 60 million people, are part of this linguistic branch.[6] The Malagasy language in the island of Madagascar represents a unique situation because its populace speaks an Austronesian tongue, a language family of Southeast Asian origin, likely Taiwan. How a Malayo-Polynesian language is spoken 200 km off the east coast of Africa is in itself a very interesting topic that will be discussed in the chapter on the Austronesia Expansion (Chapter 9).

Bantu languages are characterized by the use of tones and have various grammatical genders; the root of verbs stay unchanged for the different tenses and moods, whereas they lack case inflections. Linguists have been able to reconstruct the ancestral Proto-Bantu language dating back to around 5000 ya. By studying the similarities and differences among Bantu languages, as well as the evolution of contemporary lexicons and grammar, experts have been able to trace the expansion throughout the continent and beyond toward the Americas.

THE LINGUISTIC LANDSCAPE

A consensus among linguists is that all Bantu languages are closely related, and speakers must have dissipated quite fast to explain such acute similarities over such an extensive territory. According to linguistic studies, three main phases (Fig. 11.5) are recognized in the Bantu Expansion.[7] Within those three phases, successive dispersals occurred. In each step of the way during the dispersal, linguistic differentiation into new languages occurred with repeated fission events of dialects and languages taking place. It is highly unlikely that the spread of Bantu languages happened in a single large and continuous expansion. In fact it is speculated that several major diffusions and innumerable minor spreads occurred with intermittent periods of hiatus along the way. In addition, the language data do not signal dispersal along specific paths but suggest dispersal waves in certain directions. Of course, these current theories are likely the result of naiveté as future research and information may uncover specific routes.

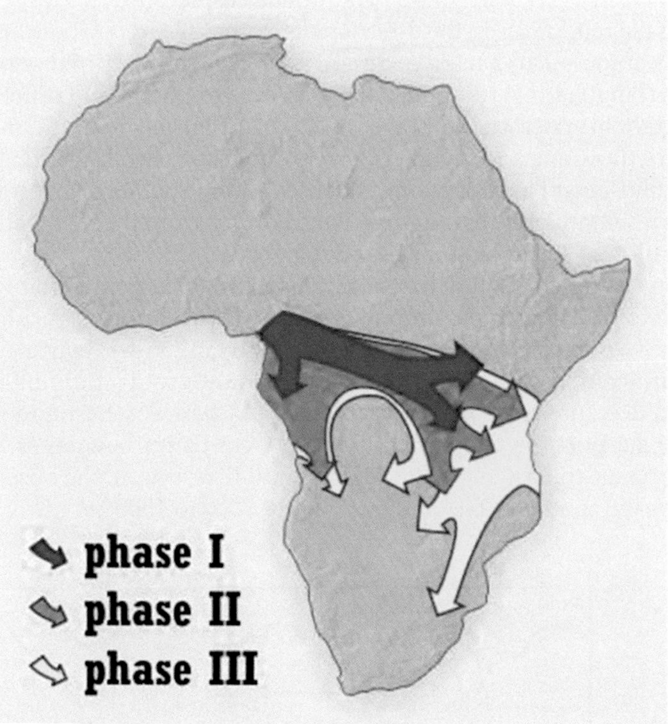

FIGURE 11.5 Three phases of the Bantu Expansion. *From http://www.uganda-visit-and-travel-guide.com/bantu.html.*

The 5000 years of migrational process has been somewhat arbitrarily partition into three phases (Fig. 11.5). It has been theorized that in the first phase, the original Proto-Bantu language differentiated from the Mbam–Nkam cluster of Bantoid languages in the western region of Cameroon.[8] Linguistic data indicate that the greater diversity within the Bantu family is located in this region, thus suggesting its origin in the area. It then expanded as a single language, mostly eastward toward the Great Lakes area of East Africa. It is likely that many of these movements of people transpired along the course of river basins. With time, the Proto-Bantus at the extreme of this language continuum began to differentiate and started speaking mutually indecipherable tongues. At the same time, it is thought that in the northwest domain of the range, West Bantu dialects began to deviate from each other evolving into unique languages. This ensued in the wetlands between the Sanga and Ubangi rivers. Then, there is evidence of dispersals along the northeast, and simultaneously, the northwest Bantu branches southward. It is conceivable that initially the locals became bilingual and gradually adopted the tongue of the invaders.

The next major dispersion phase involved movement along the West African route southward toward Southwest Zaire, Angola, and Namibia. A number of linguistic fission events happened in that general area of West Africa. It is speculated that subsequent to these differentiation events, two separate major migrations took place eastward into the heart of Zaire using river venues; one diagonally in a northeast direction and the second directly east.

It is likely that concurrently with these longitudinal displacements toward the east, Bantus were moving from the rainforests in the direction of Southeast Africa. There are indications that in the vicinity of the middle Zambezi river differentiation events occurred generating a number of dialects and then languages. A number of these languages underwent differentiation, fission, and dispersal events that generated a number of language clusters in northern Mozambique, in the Kilimanjaro area (the Caga and Thagicu clusters), in East-Central Africa (Shaba and Zambia language groups), the Great Lakes area of East Africa and in southern East Africa.

A graphical representation of the putative Bantu routes from their homeland based on the contemporary location of Bantu languages and dialects is provided in Fig. 11.6. Trees like this based on Bayesian (using characteristics as random variables possessing known probabilities of distribution) or maximum parsimony (simplest explanation) algorithms are generally congruent with the hypothesis that the northwest Bantu languages dispersed southward as a paraphyletic group (composed of some of the descendant groups but not all from the common ancestor) and does not support an *early simple* split between East and West Bantus. The dendrograms also support the contention that the West Bantu languages are equivalent in ranking and importance to East and Central branches. In addition, the observed distribution in the tree parallels the spread of farming from about 5000 to 1500 ya.

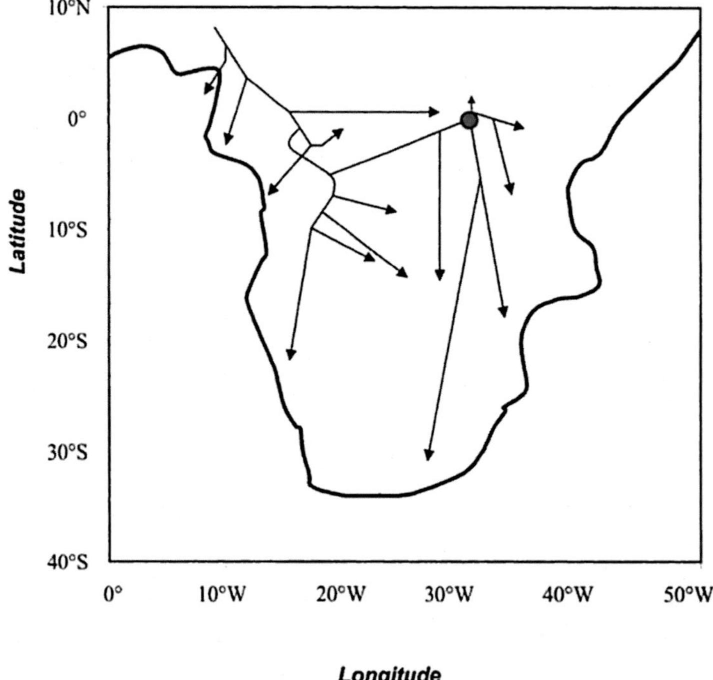

FIGURE 11.6 Theorized dispersal routes of Bantu from their homeland. Dark node indicates the location of the oldest Urewe pottery site in Northeast Africa and the putative origin of modern East Bantu languages. *From https://books.google.com/books?id=5YP_p5eS898C&pg=PA65&lpg=PA65&dq=bantu+eastern+stream&source=bl&ots=roda58XSpY&sig=QfJ36XkjOulnwySmYwscrmW_waE&hl=en&sa=X&ei=npGhVcrkDobX-QHa8pHQBg&ved=0CCQQ6AEwAQ#v=onepage&q=bantu%20eastern%20stream&f=false.*

The expansions described above represent only basic extrapolations from contemporary linguistic data. A number of microdifferentiation episodes are not accounted for, most of them lost in time. These movements, smaller in scale, many putative in nature, likely would reflect a profusion of migrational events accompanied by a myriad of linguistic interactions, such as separation and fusion of dialects and languages. No doubt these minor, regional dispersals were subject to the specific local conditions such as the technological level and social/cultural outlook of the local natives as well as the ecology of the region. Examples of complexity are recolonization by different Bantus speakers leading to the replacement of one Bantu language by another as well as dialects and languages created by the fusion of two or more parent tongues. For instance, a number of Malawi Bantu languages are the result of the fusion of Zambia and Yao-Mpoto clusters from neighboring Tanzania and Mozambique.[7] It is almost certain that all kinds of complex interactions are responsible for the extrapolated and current Bantu linguistic diversity and distribution. This is particularly reflected in the range of combinations of autochthonous and Bantu words in languages spoken in sub-Saharan Africa, today. Also, although genetics clearly indicates that physical movement of people with their DNA accompanied the Bantu language dispersals,[6] it is not unconceivable that minor spreads were the result of acculturation without interbreeding and altering gene pools.

ON EXPANSION MODE

The question of what motivates people to move to a new location is often times difficult to answer, especially when the population in question has remained sedentary for a considerable amount of time before the exodus. The typical generic reasons that are often provided include overpopulation, ecological collapse, famine, drought, wars, invasions, family feuds, internal conflicts, or just plain curiosity. These justifications seem, at times, to be general and unfounded. In the case of the Bantu dispersal, it has been postulated that overpopulation in their homeland resulting from dramatic climatic changes in the Saharan region was the main motivation. This represents probable cause considering that about the time of the initial move, 4000–3000 ya, the Saharan desert was experiencing a process of expansion into previously fertile regions to the south. Thus, it is likely that this dramatic increase in aridity pushed people southward. This displacement of humans generated an overpopulation in the cradle of the Proto-Bantu.[8] It is also consider that the settlement of new lands was made less difficult by a number of technological advances that provided the Bantu with a number of advantages over the locals. Technological know-how could have provided a sense of security and superiority over the locals. These included the practice of agriculture, the introduction of new crops, including bananas, bulrush millet, sugarcane, as well as the use of quality pottery.

All of these factors likely played a role in triggering the initial impetus for the move. Yet, what drove the Bantus to keep moving for about 3000 years, a dispersal so extensive that it managed to encompass most of sub-Saharan Africa? It would be hard to justify 3000 years of movement on the basis famine and the need for land to survive. Repeated episodes of overpopulation and acquisition of new land for agriculture would not explain such rapid migrations. It has been calculated that humans lacking a destination migrate at a rate of about 200 km per century.[9] Considering expected periods of hiatus, at this speed, the Bantu dispersal would have taken much longer. Then, what was motivating the Bantu to move? Did migrations become a way of life? Maybe, but Bantus were not nomads. The essence of Bantus is that they permanently settled the occupied territories. Were their dispersals driven by the desire to conquer and/or discovery or just acquisition of power, just like the Romans and other empires have done? It is likely that various motivations were at play at different times and places.

Iron technology, often cited as a Bantu creation, was not a Bantu invention.[10] Actually, Bantus procured this skill after the start of their dispersal in the Great Lakes area of East Africa about 2500 ya. Iron metallurgy was first developed in Turkey, and it is likely that it was reinvented independently in different parts of the world, including sub-Saharan Africa. Centers of iron technology developed around Lake Chad among the Chadics, in Nigeria by Yorubas, in Lake Victoria by Urewes, and in the Egyptian-Kushite Kingdom of Meroe along the Upper Nile. The assimilation of ironworking by Bantus could have been from one or more of these populations.[10] Although iron technology was not discovered by Bantus, their migrations contributed to its dissemination and facilitated their spread throughout sub-Saharan Africa.

AN AMICABLE TAKEOVER AND PEACEFUL COEXISTENCE?

Nature of the Interactions Between Bantus and Local Populations

Some of the early ideas concerning the Bantu dispersal portrayed the expansion as armies marauding, conquering, and subjugating the locals by brute force. One of the main proponents of this view was the English explorer and linguist Sir Henry Hamilton Johnston (1858–1927). With this perception, the Bantu advances were envisioned as migrating hordes of males, because females and children would slow them down, killing local males and procreating with their women. Even during the age of kingdoms, such as the Great Zimbabwe that flourished from 1100 AD to 1500 AD, and later starting in the early 19th century with the Zulu Empire, when military enterprises were undertaken, the Bantu dispersals were not violent in nature. The myth of an aggressive and brutal conquest by Bantus was gradually replaced in the mid-20th century when scientific data from archeology and historical linguistics presented Bantus as astute agrarians searching for new land.[7] Although some of the original ideas on the expansion have been recently challenged, such as the importance of iron technology in encouraging their moves, the fundamental premises and reasons for the dispersal still revolve around the thesis of opportunistic agriculturists.

There is no indication that the Bantu dispersal was particularly brutal or even belligerent. It is as if the Bantu "conquered" by peaceful assimilation of locals, maybe impressing them with their "superior" technology and culture. The dynamics of such a process could be envisioned by repeated series of separation, migration, and colonization events. For example, original Bantu hamlets split and one or several groups decided to penetrate and settle an uninhabited or sparsely populated nearby area in the rain forest to the south or the east. The genesis of separations may have been prompted by feuds or tribal disputes or localized ecological collapse trigger by soil exhaustion or simply overpopulation in a particular camp. The natives of the recently occupied lands and the Bantu arrivals should have known of each other because of their original proximity. Their interactions may have been peaceful coexistence and mutually beneficial trade, maybe intermarriage with the indigenous females. In this process, satellite villages would crop up in different directions at the periphery of the parent homestead. These sites were generally long-term settlements and often grew larger in size with time. It is possible that this sedentary living style had a positive effect on birth rate, promoting future migratory events. In the mean time, the fact that these new villages spoke the same language and were familiar with nearby settlements facilitated commerce and alliances with other Bantu groups. These communities were larger than the temporary encampments of the autochthonous tribes of the area providing the new comers with a sense of power and control. In addition, Bantu villages may have offered the locals an interesting place to visit, interact, trade, and even live. These interrelationships would have culturally enriched all the parties engaged, improved trade, and increased gene flow. With time, these interactions brought about linguistic, cultural, and genetic admixture of various combinations and proportions, depending on the populations involved. This process of separation, departure, and colonization was repeated resulting in the massive dispersal.

Bantu Hegemony in East Versus West Africa

The time line of Bantu migrational advance bringing agriculture and metallurgy along the Eastern and Western Streams down south was similar. In East Africa, the Bantu social values, culture, and philosophy were imposed from the very beginning of the encounters. Yet, in Western Africa the ideological and political forms, after the initial encounter with the autochthonous people, followed along Khoisan lines where Khoisan dictated assimilation terms with Bantus being incorporated into Khoisan social matrices by way of marriage and alliances.

This is evident in the archeological record as the Bantus dramatically appropriated the indigenous Khoisan forager and pastoralist systems in East Africa as they migrated south.[8] In the east, Bantus colonized the regions in numbers and quickly established a hierarchy of villages surrounding their main towns. In this imposed system, the groups practicing a pastoral and foraging subsistence had an economically subordinate role to the invading agriculturists. Yet, the incorporation of click sounds into a number of Bantu languages such as Zulu, Xhosa, and Sotho is indicative that the interactions between Khoisans and Bantus were intimate and complex, likely involving commerce.

In Western Africa, on the other hand, Bantu hegemony was not established until the mid-1800s. It seems that the indigenous Khoisan transferred cattle herding to the agrarian ironworking Bantus. Thus, in Western Africa the Bantus' ways were not initially and easily imposed.

MIGRATION ROUTES AND TIME LINE

Mitochondrial DNA (mtDNA) evidence indicates that the common ancestors of Pygmies and the future Bantu agriculturists separated about 70,000 ya.[11] Also from mtDNA data, we know that Pygmy women began to marry into Bantu families about 40,000 ya, a pattern that we still see today. Genetic flow from Pygmy males to Bantu females in Bantu communities was not detected. For the period of time between 70,000 ya and 40,000 ya, the two groups seem to have lived in relative isolation from each other. Linguistic reconstructions suggest that the genesis of the Proto-Bantu language occurred approximately 5000 ya in a region just below the Saharan desert in what is today the border of Northern Cameroon and Western Nigeria.

Genetic data indicate that the subsequent spread of the Bantu language, culture, and technology was not just an acculturation process but also involved massive movement of people and DNA flow. In other words, it was not a process in which a few individuals transmitted ideas and know-how, it included many humans traveling and taking their DNA to new locations. It is known that Bantus got to Gabon around 3500 ya and initiated a western dispersal by 3000 ya reaching Angola 2500 ya and Southwest Africa about 2000 ya. Signals of their presence are seen in the Central African rain forest by 2500 ya. Also, they arrived in East Africa approximately 2500 ya, and they have continued moving into Southeast Africa until as recent as 300 ya.[12] It is important to emphasize that the exact routes taken by the Bantus are not known and it is highly likely that they migrated rather randomly into fertile land without knowing their exact destination.

Currently, a highly debated issue is whether the movement toward East Africa took place around 4000 ya, or more recently, about 2000 ya. Another controversy relates to whether the crossing to East Africa occurred north or south of the Central African rain forest or alternatively across the rain forest establishing internal settlements in the jungle as they traveled. It is postulated that river systems were used in this process. This notion of aquatic transportation makes sense considering the speed necessary to occupy most of sub-Saharan Africa in about 4000 years. Independent of the time and location of the crossing to East Africa, today the orthodoxy recognizes that at some point in time there was an Eastern and a Western Stream wave of migration. Fig. 11.7 illustrates the location of Eastern and Western settlement sites within the Bantu domain and the boundaries separating the west and east branches of the migration.

THE MATERIAL EVIDENCE

Dwellings

Today, Bantus are primarily farmers cultivating a variety of cereals and other traditional crops as well as pastoralists raising a number of domestic animals such as cattle, sheep, and goats. They are known for their iron and copper works. They live in homesteads made up of simple single-room mud and cow dung huts (Fig. 11.8). Traditionally, these dwellings were of two types. One was the so-called Nguni or beehive style constructed as a circular room supported by long poles and covered with grass. The second style is the cylindrical-type single room made up of poles and covered with mud and dung. The floor of both types is made up of compact soil and the roof of tied-up sticks. Yet, during the time periods of the Bantu chiefdoms, from the 13th to the 17th centuries, and Kingdoms, from the 18th to the 19th centuries, more complex structures were built. An example of the former includes The Great Zimbabwe (Fig. 11.9), massive walled cities that stand in sharp contracts with the humble mud huts.

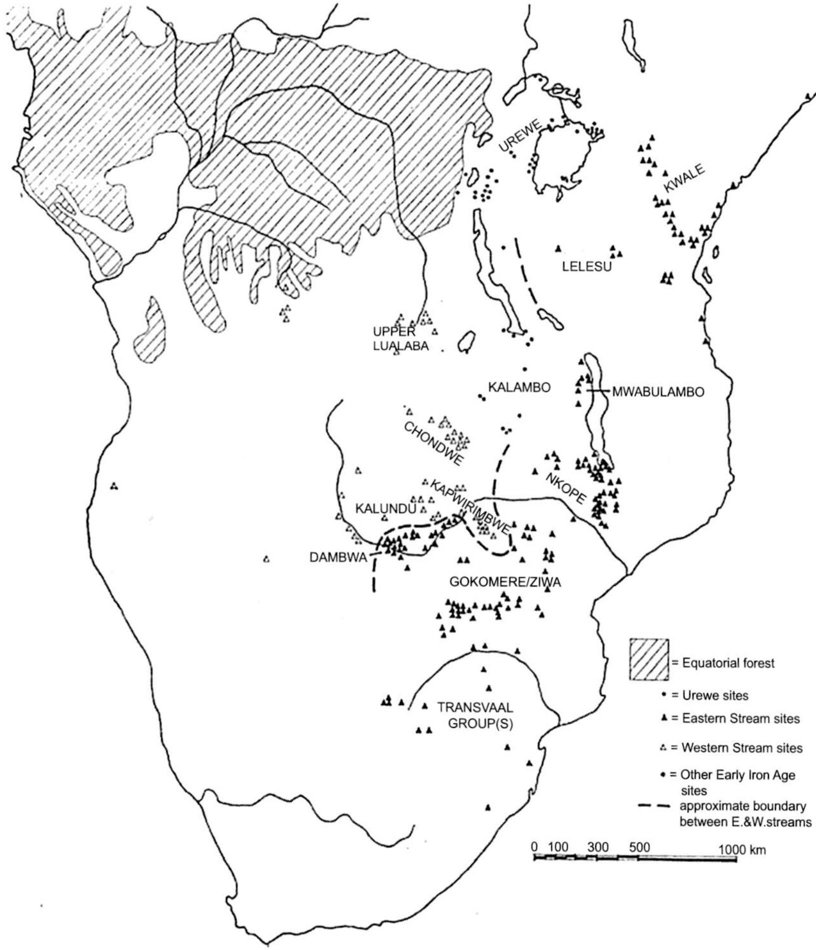

FIGURE 11.7 Eastern and Western Streams and their boundaries. *From http://links.jstor.org/sici?sici=0043-8243%28197606%298%3A1%3C65%3AAABL%3E2.0.CO%3B2-W.*

FIGURE 11.8 Bantu homestead and typical huts. *From https://www.emaze.com/@AOCTZWOW/The-Bantu.*

FIGURE 11.9 The Great Zimbabwe. *From https://thezimbabway10.files.wordpress.com/2011/01/img_4507.jpg.*

Pottery and Ceramic

In terms of art, it is logical to expect that correlations between general styles and population identity exist. Indeed, in the case of Bantus, the design repertoire in ceramic and pottery reflects group uniqueness and parallels dispersals during the Early and Late Iron periods (the last 2000 years). Pottery is a utilitarian form of art. The pottery and ceramic of Bantu origin is characteristic of an agrarian food-producing culture. Bantus value their pottery. They use it on a daily basis for crop collecting, processing, and cooking their food. As such, they decorated their pottery and ceramics with traditional symbolic patterns such as groves and stamped patterns to make them look beautiful. The patterns used reflected local traditions. Therefore, related populations made similar pottery, ceramic, and works of art. Thus, artistic traditions and motifs are used by archeologists to date settlements and connect populations.

One important example of pottery and ceramic tradition, among many, is the Luangwa pottery tradition (Fig. 11.10) from Zambia, Malawi, Mozambique, Zimbabwe and parts of Angola, the Democratic Republic of the Congo, and Tanzania. This style originated in Zambia in the mid-6th century and subsequently dispersed throughout the region. The hallmark of this tradition is its use to comb-stamped designs applied to necks and shoulders of pots and vessels. Common among the motifs are horizontal and diagonal bands made with delicate cuts, interlocking triangles, and jagged patterns. It most be emphasized that the specific designs are not universal among all Bantu groups.

Another important example of an earlier pottery tradition and culture is the Urewe. It originated in the region of Lake Victoria in Northeast Africa and dates to the Early Iron Age about the 5th century of the Common Era. The orthodoxy posits that the Urewe likely represents

FIGURE 11.10 Bantu Luangwa pottery tradition. *From http://www.jstor.org/stable/180367?seq=1#page_scan_tab_contents.*

the original migrants from northwest Africa. It is also thought that the Urewe gave rise to the eastern subfamily of Bantu languages. Subsequent to their arrival to the Lake Victorian area, Bantus migrated southward along the Eastern Track. In the process, the Urewe pottery tradition was spread to Southeast Africa. Some experts believe that the Urewe tradition is ancestral to subsequent styles such as the Luangwa pottery described earlier. This style incorporated distinctive elements such as dimples and concentric lines (Fig. 11.11). The term dimples refers to the indentations at the base of pots and bowls.

The Lydenburg Heads from South Africa are examples of ceramics with typical Bantu stylistic elements (Fig. 11.12). A set of seven heads dating to the sixth century of the Common Era was found buried near the town of Lydenburg in northeast South Africa close to Pretoria. Their location and arrangement suggest that the entombment was not accidental. The designs and patterns in these heads are found in different art media including beadwork, wood, and paintings. Typically, they are made of clay or wood and include modeled strips of clay, linear patterns, and animal figures. The heads exhibit a raised hairline, slightly protruding mouth, nose, and ears, and they seat in columnar necks. Some of these artistic elements persist in present-day Bantu art. For example, compare the stylistic similarities of the Lydenburg Head (Fig. 11.12) with the contemporary Bantu mask (Fig. 11.13) made out of wood and fabric from Southern Africa.

FIGURE 11.11 Bantu Urewe pottery tradition. *From http://www.tandfonline.com/doi/abs/10.1080/00672707109511552?journalCode=raza20#preview.*

Ironworking

Ironworking was developed for the first time in Anatolia, present-day eastern Turkey approximately 3500 ya. The making of iron is not as simple as the smelting of copper or tin, which are just melted and collected from the rock. In the case of iron, the impure metalliferous rock (see Fig. 11.14) that is mined needs first to be pulverized to increase surface area. Then, it requires to be heated in furnaces (see Fig. 11.15) at high temperatures to initiate a chemical deoxygenation (oxygen removal) reaction involving the ore and the added carbon in charcoal. The amount of carbon in the mixture is critical, too much or too little charcoal ruins the process. This reaction takes advantage of the carbon atoms' affinity to bind to oxygen, and in doing so, removing it from the ore. The products of this reaction are CO, CO_2, and a semi-refined form of iron. In addition, lime is usually added in the form of seashells to remove a number of rocky impurities in the ore. The heat provided by the boiler acts as a catalyst that transforms the limestone and the oxygen from the ore to calcium oxide. The solid contaminants are removed as slag from the bottom of the furnaces at the end of the smelting process. The process is initiated when the mixture of ore, charcoal, and lime is placed in the furnace, fire is initiated, and air is blown continuously with hand-powered bellows to increase the temperature. The smelting process continued for hours into the night.

FIGURE 11.12 Lydenburg Heads with typical Bantu motifs. *From http://web.uct.ac.za/org/cama/CAMA/countries/southafr/projects/artafric/RA3_10a.htm.*

FIGURE 11.13 Contemporary Bantu mask from Southern Africa. *Personal photo.*

FIGURE 11.14 Iron ore rock. *Reproduced from Wikipedia (https://commons.wikimedia.org/wiki/File:HematitaEZ.jpg).*

FIGURE 11.15 Bantu iron furnace. *Reproduced from Africa Unchained.*

Subsequently, the crude iron is retrieved from the boiler while hot and hammered to remove impurities further and mold it into shape and tools. Because of the rather complicate nature of this process, it was assumed for many years that ironmaking was an acquired knowledge or imported from Southwestern Asia (Anatolia). Yet, the time line of the appearance of the technology and its dissipation within Africa suggest that iron work may have been discovered in Africa independently, probably several times in different places. The separate genesis of knowledge has precedence in human history. Some of the best-documented cases are agriculture and domestication, which appeared singly in Anatolia (Fertile Crescent) 10,000 ya, China around 8000 ya, and in Central America approximately 5000 ya.

Iron technology was first seen in Africa in Egypt about 670 of the Common Era,[13] and traditionally it was thought that from there it spread all over Africa. This notion has changed because of the earlier dating of iron technology in sub-Saharan Africa. Although the dating of some iron sites has come into question, it is possible that the production of iron originated in several regions of Africa independently. Most researchers believe that iron smelting was practiced in the region between Lake Chad and the Great Lakes of East Africa starting about 3000 ya before its detection in Egypt. From those centers, the technology was likely transferred to the Bantus who adopted it and dispersed it eastward and westward approximately starting 2000 ya. Bantu iron artifacts can be classified as belonging to an Eastern or a Western Stream and correlate with the distribution and partitioning of Bantu pottery, ceramic, and art.

Although iron was more difficult to produce and mold into tools, it provided a number of important advantages. Compared with copper and bronze, iron is stronger, and sharper instruments can be created with it. In addition, iron ore is widely distributed and available in sub-Saharan Africa. To the Bantu, iron was paramount because it allowed them to penetrate and travel through the dense forest as well as clear the jungle for agriculture. Hardwood, in turn, provided them with the fuel to heat their furnaces to high temperatures, a necessary step in the purification of the ore. Thus, iron production and agriculture were activities that feed on each other.

It is thought that Bantus were not the inventors of ironworking. In fact there is no word for iron or ironworking in the ancestral Proto-Bantu language.[14] It seems that they merely assimilated the technology from other tribes and transmitted it as they dispersed in different directions and established their domain. For this reason, iron implements are very useful as markers to help trace the Bantu Expansion. Along these lines, it is not surprising that the location of iron-manufacturing sites in sub-Saharan Africa parallel very closely the Bantu dispersal routes (Fig. 11.16).

Because iron is highly corrosive, ancient tools are not usually preserved well as part of archeological finds. Some examples of ancient Bantu iron tools and weapons have been discovered (Fig. 11.17). These artifacts are mostly utilitarian in nature. Yet, in addition, iron was also used to make jewelry, artwork, coins, and various instruments. At times, iron was shaped into rods, presumably for ease of transportation and trade. This provided for flexibility of molding into any form. Evidence of iron-processing sites usually consists of waste slag from ovens and charcoal. These residues are routinely used for carbon dating.

FROM SCAVENGERS, HUNTER-GATHERERS, AND PASTORALISTS TO AGRICULTURISTS

The tools and artifacts made by humans are a reflection of their mode of subsistence. Starting in the Lower Paleolithic, sub-Saharan Africans developed a number of tool traditions, such as the Oldowan (2,5000,000 ya) and subsequently the Acheulian (1,500,000 ya). These styles were exported out of the continent relatively recent when humans migrated out of Africa during the Middle Paleolithic (about 100,000 ya) and then arrived in Europe some 45,000 ya. These were simple stone tools belonging to the Mousterian-like Levallois tradition. The

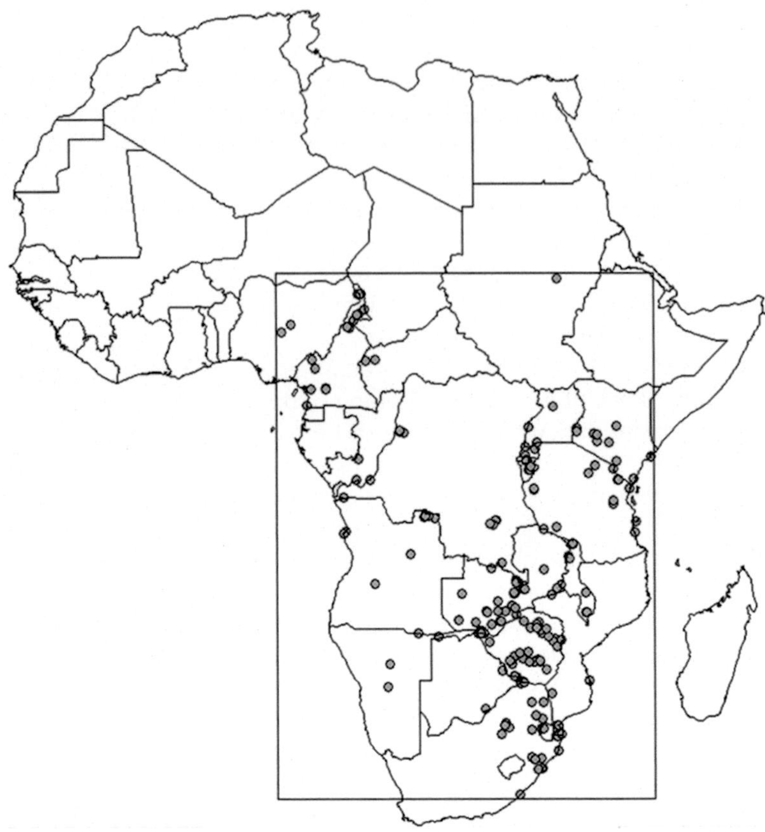

FIGURE 11.16 Distribution of iron technology sites in sub-Saharan Africa. *From http://www.ucl.ac.uk/~tcrnjst/RussellSteele2009.pdf.*

FIGURE 11.17 Typical Bantu iron tools and weapons. *From http://worldreligions13.weebly.com/african-history.html.*

creators of these implements were opportunistic hunter-gatherers. These stone instruments were used to hunt and cut meat. In West Central Africa about 75,000 ya, lances with dart-like bone points were used for hunting. And in South Africa by around 70,000 ya, heat-treated stones were used to make fine flakes for spears and cutting devices. These technologies were unknown to other hominin species and archaic humans. During the Upper Paleolithic, approximately 38,000 ya, beads of ostrich eggshells were being made, and animals were painted in rock refuges in South Africa about 28,000 ya.

More recently, during the Late Stone Age, about 15,000 ya, sub-Saharan populations were practicing hunting and fishing subsistence catering to the various niches of the subcontinent. These communities were the ancestors of the Cushitic, Pygmies, Nilo-Saharan, and Khoisan populations of today. Approximately 2000 ya, cattle and sheep were imported from the north, probably from Cushitic groups that practiced a mixed economy combining pastoralism with foraging for wild animals and plants. It is not clear whether Bantu speakers were responsible for the introduction of these life stocks, but it is known that as they migrated south, they took with them at least two species of cattle, the humpbacked (*Bos indicus*) and the flat backed (*Bos taurus*) as well as chickens and goats.[15] Together with the importation of cattle and sheep into the Bantu domain, a new pottery tradition known as Bambata emerged (Fig. 11.18). This early pottery tradition, characterized by thin walls and parallel lines, predates the Urewe and Luangwa styles discussed earlier. The Bambata pottery dispersed southward with the Bantu migrations.

These early Bantus also brought with them agricultural technology, which included crops such as banana, yam, coconut, sorghum, millet, cowpeas, and melons in addition to iron and copper metallurgy. Although these domesticated species were transported by the advancing Bantus along the Eastern and Western African Streams, many had origins outside Africa.

FIGURE 11.18 Ancient Bambata pottery. *From Huffman TN. The stylistic origin of Bambata and the spread of mixed farming in southern Africa.* Southern African Humanities *(2005);***17***:57–79. Pietermaritzburg, South Africa (www.sahumanities.org).*

For example, the humpbacked cattle is from India, whereas chickens, bananas, and coconuts were imports from Southeast Asia that were introduced into East Africa before the Bantu Expansion, possibly by various complex trading venues. Thus, Bantus were the transmitters of these forms of agrarian subsistence and metallurgy, but they were not their creators. It is likely that the spark that initiated husbandry and agrarian subsistence in sub-Saharan Africa arrived from the north, the regions that are part of the Okavango and Zambezi river basins. Additionally, crops of Asian origins could have been introduced into East Africa via Austronesian Madagascar into Mozambique, and more recently involving trade with Arabs and the Near East.

Unfortunately, the archeological evidence indicating the crops and food prepared by the early Bantu farmers is limited because of the warm and humid environment of sub-Saharan Africa. In spite of these conditions, the archeological evidence shows that Bantus moved south with an assemblage of tools, technologies, livestock, and plants not seen before among native populations. Evidence of this Iron Age package includes physical evidence of hamlets with repository pits for consumables, scorched huts, grain receptacles, ceramics, and pottery as well as animal remains in the form of feces, teeth, bones, and remnants of domesticated plants such as burn seeds, grindstones, soil scaffolds, and phytoliths. Phytoliths are microscopic silica structures that form in living plants and become fossilized. Different plant types possess characteristically unique phytoliths that allow archeologists to identify the species from where it came from. Residual by-products of metallurgy are also evident in the villages' furnaces, forge vesicles, and manual blowpipes.[16]

THE DNA EVIDENCE

Subsequent to the discovery of the widespread distribution of Bantus languages in sub-Saharan Africa, one of the major points of contention among scientists was to what extent it was the result of diffusion of ideas. As a marker, DNA is unique in being able to access if considerable movement of people accompanied the transfer of language and technology. The reason why DNA is capable of ascertaining if migrations actually occurred is because the genetic material is stable and is transmitted from generation to generation. In addition, the genetic material is not subject to sociocultural and environmental forces that can obscure or even erase linguistic and archeological signals. Any displacement involving considerable number of people moving will likely leave an imprint in the DNA. Of course, this would not happen if the population in question becomes extinct soon after the move or the migrants fail to settle the region. In addition to exchange of knowledge occurred in both directions, from Bantu groups to indigenous people and vice versa, the genetic evidence clearly indicates that DNA also moved in parallel with the dispersal of ideas.

Molecular anthropology is the scientific field that addresses questions of phylogenetic relationships among populations, ancient and contemporary. In trying to access if the Bantu Expansion was driven by people with their DNA or it resulted by acculturation or both, molecular anthropologists investigate if specific DNA markers are present in locations where people speak Bantu, presently or in the past. The technology involved in genetic testing has become so advanced that miniscule amounts of DNA can be examined from contemporary and ancient populations.

The major drawbacks of genetic technology as an anthropological tool are the tendency of DNA to degrade with time and the basic assumption that genetic diversity do not change during the passage of time. In other words, most genetic studies only examine DNA variation of existing populations, not ancient groups. Clearly the former provides a limitation to its use since the older the samples the greater the level of DNA degradation as well as the difficulty in obtaining results, and the later is truly a serious fundamental problem because the genetic constitution of populations changes with time. The first issue limits the use of ancient DNA. Although recent advances in molecular biology and bioinformatics have extended assessments of ancient specimens to about half a million years ago, and future technological refinements should extend the time threshhold even further into the past, genotyping still subject to contamination by contemporary DNA and age decomposition. The second limitation is more fundamental in nature. It is likely that present-day samples are not representative of populations dating back to the time of interest, and therefore, extrapolations from contemporary DNA could be at best misleading if not just incorrect. Although this issue is being alleviated by new discoveries extending the age of ancient samples examined, it represent a point of contention among scientists from different fields. Nevertheless, genetic data are increasingly becoming more central in anthropological studies.

This is a very eclectic discipline that is nourished by data from a number of seemingly unrelated sources. And it is particularly powerful when it corroborates other fields such as linguistics, archeology, anthropology, history, and art. In the event of congruency of results among data from different expertise, conclusions are more robust and carry more weight. Discrepancies among the results from various bodies of knowledge may signal technical problems, unknown variables, and/or incorrect assumptions. It turns out that, for the most part, the genetic data are consistent with the archeological and linguistic evidence.

mtDNA

The genome (the entire genetic makeup of an individual) is made up of various types of DNA. One way to characterize the genetic material is by the kind of inheritance the DNA represents. For example, the genetic material within certain cellular organelles (intracellular structures) called mitochondria (mt) is strictly maternally inherited. In other words, the mtDNA of all human beings is derived from our mothers and not passed down from our fathers, independent of whether we are male or female. The implication of such a mode of inheritance is that with mtDNA molecular anthropologists can follow maternal lineage singly. Another way of expressing this concept is that children have the same mtDNA as their mothers, grandmothers, great grandmothers, etc. Of course this DNA, like all other types, incurs de novo (new) mutations in each individual during a generation. These mutations may accumulate generation after generation becoming part of and characterizing specific mtDNA lineages. Thus, even though the individuals of a maternal lineage have very similar mtDNA, they are not identical. And, furthermore, the differences between individuals within an mtDNA lineage are greater as the number of generations increases. In other words, mutational differences increase directly proportional to the number of generations separating people. These small number of intralineage variations are useful because they allow for the discrimination among related individuals.

Bantus are characterized by mtDNA types (also known as haplogroups) L0a, L2a, L3b, and L3e, whereas haplogroup L1c is thought to represent ancient non-Bantu populations such as the Khoisan-speaking groups.[17] In general, the presence non-Bantu mtDNA haplogroups indicate considerable levels of hunter-gatherers DNA within the Bantu domain. This scenario is compatible with incomplete replacement of indigenous by Bantu mtDNA or genetic flow from autonomous populations into the Bantu gene pool or both. Initially, these results were interpreted by some experts to indicate that the Bantu spread was mainly a phenomenon of acculturation with limited genetic flow.

Examination of the proportions of Bantu and indigenous mtDNA across the Bantu realm demonstrated that the levels of both components differs depending on the region (e.g., East vs. West Africa), the specific populations involved, and the DNA type used in the analysis. This was not totally unexpected because each of the indigenous populations was socioculturally unique with different values and philosophy providing for different types of interactions. For example, at one end of the spectrum, a complete replacement of indigenous mtDNA by Bantu haplogroups is observed in Khoisan speakers in Southwest Africa.[18] In other Bantu-speaking groups, only traces of autochthonous mtDNA lineages have been detected.[11] In contrast, much higher frequencies of non-Bantu mtDNA lineages are reported in the Bantu groups from Gabon and Cameroon,[17] and Pygmy haplogroups, such as L1c1a, are present in Bantu agriculturalists, but Bantu mtDNA in Pygmy groups is minimal.[19] Thus, although the amount of Bantu maternal heritage differs from area to area, the data clearly indicate that acculturation by itself was not the mechanism for the presence of the Bantu way of life in the occupied territories. It is also evident that females were part of the dispersion, not just men, because the maternally derived mtDNA is detected throughout the Bantu domain today. In other words, the expansion was not a militaristic male-driven invasion but a family affair.

Y-Specific DNA

The male-derived counterpart of mtDNA is Y-specific chromosomal DNA. A chromosome is a structure inside the nucleus of cells that contains genetic material. Humans have a total of 46 chromosomes arranged in 23 pairs. The Y chromosome is one of those 46 chromosomes, and only males possess it. Females do not have a Y chromosome. In our species, as well as in other mammals, a gene on the Y chromosome determines the male sex of the individual. If you have the sex-determining region of Y (the SRY gene) on the Y chromosome, you develop into a male and if you do not, you become, by default, a female. Thus, the Y chromosome is inherited through the male lineage. That is, from father to son. Thus, the Y chromosome DNA of a human male child is almost identical (except for a small number of de novo mutations) to his father, grandfather, great grandfather's (etc.) DNA. The word *almost* is used because just like with mtDNA, Y chromosome–specific genetic material occasionally mutates to generate small differences among individuals of the same paternal lineage. So, just like mtDNA, which is used to trace the maternal line of descent, the Y chromosome is used by molecular anthropologists to follow male or paternal ancestry.

An examination of the Y chromosome signatures within the Bantu territory indicates that, in general, the paternal impact on the indigenous populations has been greater than the maternal one. A number of scenarios may explain this observation. For example, it is possible that the number of male migrants exceeded the female component. It is conceivable that some females stayed behind in the established settlements taking care of the infants, the disabled, and/or the elder component of the population. Similarly, it is also probable that sex-specific biases were in place that prevented or minimized the flow of genetic information from female

Bantus into the autochthonous tribes' gene pools. Sociocultural barriers and peculiarities among Bantus and non-Bantus groups could have acted in favor of genetic flow from male Bantus to indigenous females. For example, it is known that sociocultural taboos have been in place limiting marriages between female Bantu and male Pygmy.[20]

Another line of evidence supporting the physical migration of Bantus, as opposed to just acculturation, is the greater genetic diversity of the mtDNA component relative to Y chromosome DNA within the Bantu domain.[21] These data are congruent with a scenario in which large numbers of male Bantus, relative to females, migrated to the new territories reducing the indigenous non-Bantu DNA, especially the Y-specific DNA. Because these male Bantu travelers originated from a single source in their homeland in West Africa, they are expected to be more genetically homogeneous than the autochthonous populations. In a situation like this, the Bantu paternally derived Y DNA by shear number overwhelmed the native counterparts. In other words, the Bantu Y chromosomes may have had a homogenization effect on the paternal lineage composition of the land compared with the maternal mtDNA.

Specifically, several Bantu-specific Y chromosome markers have been identified. These mutations include E1b1a-V38, E2-M75, E1b1a1a1-M180, and B2a-M150. Conversely, non-Bantu haplogroups A and B2b-M112 represent the signature of hunter-gatherers groups. B2b-M112 is thought to signal Pygmy ancestry. The residual presence of these markers represents remnants of the paternal genetic composition before the Bantu dispersal. In general, the very minimal frequencies of these non-Bantu Y chromosome–specific markers in comparison with the mtDNA component suggest that replacement by paternal Bantu lineages was more complete than the maternal counterpart.

Yet, indigenous Y-specific signals indicate that low levels of ancient paternal lineages persist in a number of extant Bantu-speaking African tribes.[22] For instance, traces (1%) of the local hunter-gatherer Pygmy Y-haplogroups (B2b-M112 and A) are observed in the Bantu populations of Gabon and Cameroon.[17] Yet, in the Khoisan-speaking populations of Southwest Africa, there is a total displacement of the ancestral Y chromosomal lineages by Bantu-specific haplogroups,[18] a scenario very similar to present-day mtDNA distributions in these populations, as discussed above. Although the amount of residual indigenous Y chromosomes is small all over the Bantu domain, the observed regional differences in frequencies is indicative of diverse complex admixture scenarios. It is clear that the infusion of Bantu Y chromosome DNA as they advanced into new territory was massive and indicates that genetic flow was involved and not just acculturation.

Autosomal Bantu Signals

Autosomal DNA includes all genetic material that is not located on the X or Y chromosome and not mtDNA. Most of the DNA in our cells is of the autosomal type. Unlike Y and mtDNA, which are present in single copy in our genome, autosomal DNA exists in duplicate. In other words, humans have their autosomal DNA in a total of 22 pairs of chromosomes, half of the chromosomes (22 chromosomes) derived from mother while the other half (22 chromosomes) from father. Therefore, autosomal DNA is biparental in origin. When we are conceived, we receive one set of autosomes from mother (by way of the ovum) and the other set of autosomes from the father (by way of the sperm). An important characteristic of autosomal DNA is that it is capable of recombining. Recombination involves the exchange of DNA between the two chromosomes of each pair, the maternal and paternal members. The end result of recombination is that the original maternal and paternal DNA is reshuffled. This mode of inheritance is more complex than the typical mtDNA- and Y-specific uniparental transmission because individual genes can move back and forth between the two members of each pair of chromosomes generation after generation.

Although mtDNA and Y chromosome heritage offer powerful tools to ascertain the uninterrupted maternal and paternal lineages, respectively, it is limited in its analytical power because each (Y chromosome and mtDNA) behaves as a single genetic signal. In other words, a typical human male only has one Y chromosome and that single Y chromosome only contributes one set of genes (haploid). The same can be said about mtDNA. The advantage of examining autosomal DNA is that each set of autosomes has thousands of genes contributed by both parents (diploid) and each gene is a maker or signal that can be follow independently, generation after generation. Therefore, in autosomal DNA studies, investigators assess hundred of thousands of sequences or signals allowing robust analyses and conclusions. Thus, when studying autosomal DNA, the amount of information available for study is so much greater than with uniparental markers.

Currently, the most frequently examined autosomal markers are short tandem repeats (STRs) and single nucleotide polymorphisms (SNPs). STRs are repeats of 3–6 nucleotides (DNA subunits) in length that are repeated in tandem (for example, the sequence ATTCATTCATTC). SNPs are single nucleotide changes in the DNA such as substitutions, deletions, and additions. In addition to SNPs and STRs, there are other types of autosomal genetic markers available to the molecular anthropologist to study human populations such as repetitive DNA segments (hundreds to thousands of nucleotides long), insertions, and deletions of DNA. These two types are frequently used to supplement STRs and SNPs in genome-wide autosomal studies.

An examination of the autosomal STR profiles in sub-Saharan Africa indicates substantial DNA in common among the different regions (Table 11.1). As Table 11.1 illustrates, it is likely that the considerable amounts of DNA shared between sub-Saharan West Africa and the East African Great Lakes (25.4%) as well as Southern Africa (38.9) is mainly because of the spread of Bantus eastward and southward. In addition, a number of genome-wide autosomal investigations demonstrate strong correlations between linguistics and autosomal genetic markers.[23] In one of these studies, a prominent portion of autosomal DNA was shown to link populations from all over the Bantu dominion, including Western, Eastern, and Southern Africa (Fig. 11.19). Specifically, using structure analysis, a statistical tool designed to identify DNA in common among populations, Bantu West African groups such as the Mandinka share most of their autosomal DNA with Bantu speakers from Southern Africa, such as the Venda and Xhosa.[24] The autosomal data are also congruent with the archeological and linguistic

TABLE 11.1 Autosomal Short Tandem Repeat Genetic Links to the Tropical West African Region

Southern African	38.9%
Sahelian	26.2%
African Great Lakes	25.4%
Horn of Africa	8.8%
Other	0.7%

Modified from http://dnatribes.com/library.html.

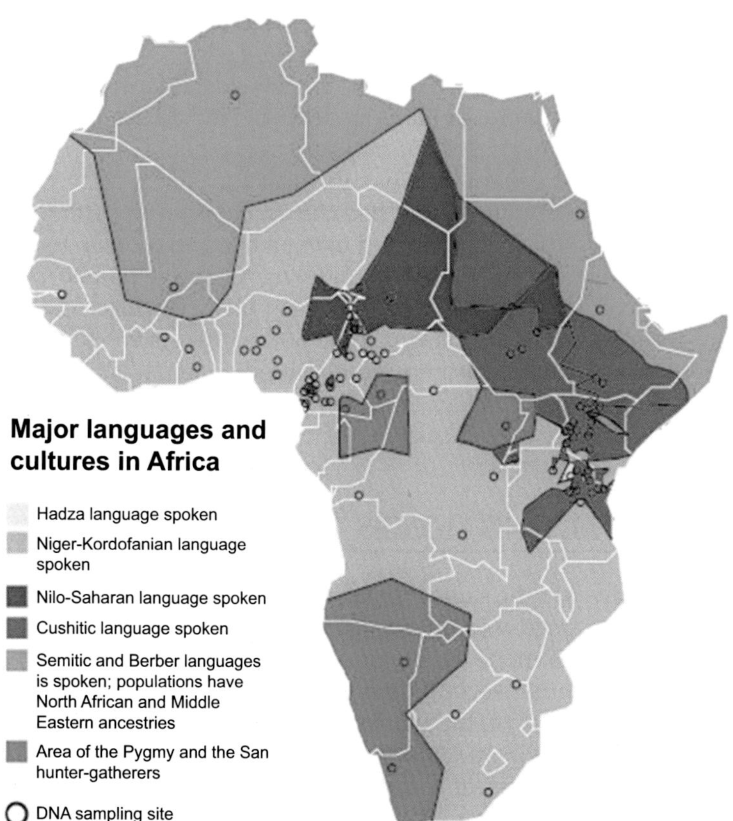

FIGURE 11.19 Genetic landscape of populations within Africa based on genome-wide autosomal markers. Different colors indicate regions of genetic continuity. Orange likely represents the genetic continuity resulting from the Bantu dispersal (tishkoff-et-al-on-genetic-structure-of.ht). *Reproduced from The American Association for the Advancement of Science. Science 2009;***324***(5930):1035–44.*

results regarding the East African Bantu dispersal signals into Southern Africa from an area west of Lake Victoria as well as the incorporation of Khoekhoe ancestry into several of the Southeast Bantu populations about 1500 to 1000 ya. In addition, the various levels of genetic flow from Bantus into autochthonous groups detected in uniparental mtDNA and Y chromosomes markers are also observed with the genome-wide autosomal markers.

Basically, just like the mtDNA and Y chromosome data, all of the genome-wide autosomal evidence supports a distinct Bantu migration from sub-Saharan West Africa.

BANTUS' ATTITUDE TOWARD THE LAND

It is likely that the early Bantus lived in small groups in villages made up of several huts. These simple single-room homes were located around the cattle kraal. This layout was designed for protecting the livestock from wild animals. Particular animals belonged to specific individuals and were a source of pride and traded as commodity. The livestock provided meat, milk, butter, and hide, and even the feces were used for building and fire. Animal skins were used for making clothes, thongs, bags, and shields. Intermarriage among families within the community was common. A village leader provided advice and made decisions in consultation with other elders.

It is thought that the Bantus generally only cultivated for themselves, that is, their immediate or extended families. Every household had the innate right to cultivate the land. Each family possessed a small field that men planted using a hoe and women tended. During the Age of Colonization, Europeans introduced ox plowing. Ox plowing allowed the farmer to plant in community land, increasing the harvest. This in turn, provided food not only for the family but also for the community or village. Some of the projects were surely communal in nature such as cooking on festive occasions and hunting by men. It is likely that community projects such as roads connecting villages and peripheral fences protecting the homestead required the efforts of the entire populace of the hamlet.

FROM COW DUNG HOMES TO GRANDIOSE CITIES

Today the rural Bantu lives in the same type of modest homes they lived during the expansion, circular single units framed with leafless branches and plastered with dung and clay (Fig. 1.8). The floor was and is made up of compressed soil and manure. With time, different types of settlements reflecting social status began to appear. For example, in Botswana, South Central Africa, about 1000 ya, the villages were social class-specific and differed in a number of characteristics. These included size, location, duration of occupation, size of livestock, imported items, and the level of social class structuring. At one end of the spectrum were the homesteads high in hilltops with large kraals and considerable number of foreign items suggesting extended trade and social stratification. The elevation of these "high-class" villages allowed for better visibility of hostile groups and provided better defensive positions in case of attacks. Other more modest villages were also located in elevated mesas but had less livestock, and the numbers of years of occupation were less. More humble settlements were much smaller, at prairies, kraals were minimal in size and numbers and large numbers of hut kills are found. The numbers of imported artifacts were meager. For the residents of these sites foraging was more important for survival.

The type of cattle sacrificed for meat concussion provides a glimpse at the existence of social classes and a hierarchy among these Bantu communities in these different types of villages. For example, in the larger more prominent sites and chiefdoms, archeologists observe the remains of the sacrificed select young specimens, whereas in the smaller hamlets older animals were killed for food.[25] The slaughtering of older livestock is more in-line with keeping a sustainable herd than with better quality meat because many of the animals sacrificed have passed their reproductive age. It is likely that the higher strata of the Bantu society not only were able to afford imported exotic items but also to acquire more tender desirable meat form the less affluent farmers from nearby more modest communities. Of course, for the underprivileged members of the Bantu society, this practice had negative effects on the maintenance of viable herd size and production of milk products.

It is not altogether clear how the modest, culturally stratified, agrarian, ironworking societies described above evolved into the builders of chiefdoms, kingdoms, and empires, a dramatic transition that occurred only in a few centuries. Albeit the Great Zimbabwe is the best known and largest of the Bantu city-states, about 200 other sites are recognized within the Bantu domain of Southern Africa. Other significant kingdoms include Bumbusi, Manyikeni, Kongo Lunda, Luda, and Zulu. Although the Lemba Bantus have been implicated with the construction of the Great Zimbabwe, it is the opinion of the archeological orthodoxy that the edifices were made by the Gokomere people who settled the current-day Masvingo Province. The Gokomere were the early Bantus who populated the area about 1600 ya. These Bantus were agriculturists in the nearby valley as well as iron and copper metallurgists. At these early stages, no stone edifices were built. Yet, about 1000 ya, Bantus started building the Great Zimbabwe complex. It took approximately 300 years to build. This city-state thrived from 1120 AD to about 1500 AD.[26] At its pinnacle, it occupied 7.3 km^2 and about 18,000 people lived there.[27]

Although the ruins of the Great Zimbabwe were discovered in 1531 AD by Europeans (the Portuguese Vicente Pegado), it was not until 1871 AD that the German explorer Karl Mauch performed minimal archeological investigations at the location.[28] Mauch, in fact, was more of an adventurer not an archeologist and initially he was just interested in finding the lost mines of King Solomon. On the rediscovery of the ruins, Mauch immediately indicate that the complex represented the ruins of the biblical city of Ophir and the source of the gold that Queen of Sheba gave to King Solomon. These declarations fueled a wave of ethnocentrism in the lay and scientific communities. Mauch could not bring himself to believe that primitive African natives could have been the builders of such impressive structures. The Great Zimbabwe has been a source of scientific controversy ever since as well as highly politicized as the white apartheid Rhodesian government pressured archeologist to negate its creation by the indigenous Africans. The public and especially the white Europeans, in general, embraced these ideas. Bigotry and xenophobia made it difficult to envision how any of the native African groups that were currently living in dung huts could build such magnificent stone-walled cities (Fig. 11.9). Even scientists were not able to put aside their Western European biases and contemplate the possibility that Black Africans could have made edifications such as the Great Zimbabwe.

Complexes such as the Great Zimbabwe served as palaces for the ruling class. One of the building known as the Great Enclosure has walls that stand 11 m high (Fig. 11.9). Structurally, the Great Zimbabwe is made up of three subunits: the Hill Complex inhabited from the 9th to the 13th century, the Great Enclosure active from the 13th to the 15th century, and the Valley Complex used from the 14th to the 16th century. These dates imply that when the Portuguese first encounter the complex, at least the Valley section still may have been active. It is not clear whether these subdivisions represent successive ruling dynasties or complexes of edifices for different purposes throughout the entire occupation of the area. Clearly, the Great Zimbabwe was a major trading center. Also mysterious were the causes for the decline and abandonment of the site. A number of theories have been put forward. In addition to the usual generic reasons of ecological collapse, drought, climatic changes, famine, and political instability, more specific factors such as depletion of gold and the exhaustion of salt mines have been invoked.[29]

Although trading and cultural connections with other civilizations, such as the Chinese and Arabs, through the African East Coast, may have triggered and jump-started the establishments and evolution of kingdoms such as the Great Zimbabwe, no such connections have been firmly supported. At the Great Zimbabwe excavations, a number of foreign items have been unearthed including fragments of Chinese ceramics, coinage from the Arabian Peninsula, and various other imported luxury articles. In addition, structural elements such as a prominent conical tower, thought to have been used for worship, resemble Arabesque architecture.[30] The impact of foreign influence in the construction of the Great Zimbabwe has received renewed interest with the claims that the Lemba people, architectural candidates of the complex, have Judaic roots. A list of parallelisms between Jewish and Lemba practices include observing Shabbat, self-identification as the chosen people, restriction of diet as indicated in the Torah as well as abstention from eating pig, ritual slaughters, circumcision, and restrain from marring outside the group. The Lemba's oral tradition tells that their ancestors were white and came from an area of large cities to the north looking for gold.[31]

The Jewish connection has been corroborated by genetic studies.[32] For example, the Lemba population has been shown to be genetically linked to the Jewish Cohenins, one of the tribes of Israel. The Cohenins are a priest class within the Hebrews. More than 50% of the Lemba male-specific Y chromosomes are of West Asian decent characteristic of Arabs and Jews.[33] In addition, a diagnostic fragment of DNA known as the Cohen Modal Haplotype is shared among Arabs, Jews, and the Lemba. In fact, this segment of the genetic material is detected in the Lemba at frequencies higher than in the Jewish population. On the other hand, other studies examining the maternally derived mtDNA failed to uncover Near Eastern ancestry among the Lemba suggesting that the putative migration of Cohenins to the region of the Great Zimbabwe was male specific.[34] Subsequent genetic research has demonstrated elevated levels of Y chromosomes of the T type of Near Eastern ancestry.[35] Because most of these genetic markers are Near Eastern and not Jewish-specific, it is highly possible that the observed genetic signals derive from Arab trading by way of the East African coast, that extended all the way south to Mozambique, before and after Islam.

BANTU MYTHOLOGY

No Creator or Creation Myth

Bantus believe in an eternal, nonchanging universe without a creator. Animal and plant symbols predominate for different aspects of their life. For example, certain plants and trees are sacred and cannot be cut unless permission is granted from specific deities. Elaborate ceremonies are performed before cutting certain plants. A specific case is the ceiba or silk-cotton tree (*Ceiba pentandra*). The Bantu captives brought this belief to the Americas where the taboo still persists today. Other examples of Bantu mythology include chameleons, which represent eternal life, whereas the lizard embodies death. In Bantu folklore, this duality of characteristics is known as "the double message." Other creatures such as the hare exemplify skill and cunning, whereas the hyena stands for sneakiness and deception. As expected, Bantu vocabulary is permeated with words that reflect their traditions and myths.

The Myth and Rituals of Iron

The production of iron was so important to the Bantu that a whole ritual myth developed from it.[36] The purification of the ore was performed by men and took place some distance from the settlements. These individuals were highly specialized in this trade. The presence of women during the ironmaking process or manufacturing of tools was considered taboo. Violations would compromise the output and quality of the iron. Having women in the vicinity of the furnaces would distract the men and promote sexual temptations.

These iron artisans not only were highly crafted in the technical aspects of the purification of the ore but also performed rituals to ensure good yield and purity. This spiritual aspect of the iron culture is permeated with taboos. These iron specialists were regarded as unique and mythical individuals by the rest of the community with positive and negative attributes. On one hand, they were capable of creating iron and wonderful tools from soil. And on the other hand, they desecrated the earth by robbing it from its precious possessions. In a way they performed magic by transforming soil into prized tools.

Their ceremonies were designed to cast bad spirits away and included chants, prayers and provided medicine and performed sacrifices. The offerings were burned during the purification of the ore. Evidence for these ceremonies has been excavated dating back from the Early Iron Age in Tanzania and Rwanda, East Africa.[37]

Some of the myths associated with iron production were linked with fertility through the incorporation of sexual symbols and elaborate rituals. In Bantu mythology, the purification of the ore was analogous to conception and birth. As part of this symbolism, the boilers were shaped in the form of a sensually curved woman with the chamber decorated with feminine features. Also the tips of the manual blowers were shaped in the form of pennies that were inserted into openings around the base of the furnaces. The process involved loading the cavity of the boiler with charcoal and lime and setting the charcoal on fire while rhythmically blowing air with the bellows to increase the internal temperature. Then, the crude ore was introduced into the chamber at the opening, on top of the furnaces, covered and sealed with branches and vegetation. The temperature was kept high by continuous synchronous blowing of air. This continued for hours into the night. Finally, when the vault was open, the semipurified iron was extracted and shaped into any number of utilitarian tools for farming or clearing the forest as well as decorative artifacts. This was done by reheating the metal and hitting it repeatedly into shape with a rock while still red-hot. The general idea was to create the appropriate setting for creation by representing the sexual act, conception, gestation, and birth, the purified iron being the newborn child.

BANTUS IN THE AMERICAS, A RECENT EXTENSION OF THE EXPANSION

The impact of Bantu people in the American continent is undeniable. Although their numbers vary depending on country, geographical area, and degree of admixture with other ethnic groups, the Bantu signature is clearly noticeable in current rural and cosmopolitan America. Their physical characteristics including broad facial features, wide nose, big lips, and short muscular physique[38] are evident in the American populace. Bantu tribes found themselves in the New World as a result of the transatlantic slave trade. During a period of more than 300 years (from 1525 to 1866) about 11 million Africans landed as slaves in the Caribbean, South, Central, and North America.[39] Captives were transported to America in the most horrendous circumstances, pack as sardines in slave ships (Fig. 11.21). In the United States, this extension of the Bantu dispersal continued beyond the period of the transatlantic slave trade. Bantu people went on moving further west into Native American territory and north into what is today New England and Canada via the Underground Railroad.

Sources indicate that Africans settling in the New World were from Angola, the Bight of Biafra[40] and the East Coast of Africa.[41] Among these, Bantu tribes were among the most represented in the slave trade. They included major groups such as the Bamoun, Fang, Kongo, and Xhosa among others, yet members of other West African non-Bantu tribes were enslaved and transported as well. These included the Bong, Fulani, Igbo, and Yoruba. These non-Bantu populations also belong to the Niger–Congo language branch, but they are cataloged with Group A, whereas Bantus are Group B (see Fig. 11.2). Even East African groups from present-day Mozambique and Madagascar were transported to America.[42]

The cultural impact of Bantus through the American continent is profound. Take for instance language. In the Spanish language, we find words such as Rumba (music), Mambo (song), and Konga (drum). In English, commonly used words such as ballyhoo (know about it), jive (the noise of talking), jambalaya (dish of tender cooked corn), jazz (cause to dance), and bodacious (to pulverize) are a few of many examples.[43] In Brazilian Portuguese, the Cafundó is an example of a whole dialect of Bantu ancestry.[44]

Music and arts in America also have been highly influenced by the Bantu. Afro-Cuban music and jazz have deep roots in Bantu rhythms and instruments. Musical styles such as Kimbisi, Mayombe, son, rumba, and mambo clearly demonstrate Bantu heritage. In the forms of Afro-Cuban music, these rhythms not only are played around the world, but in fact have returned back to Africa where the vocalists imitate the Spanish lyrics and the Cuban accent of Cuban music. That is, African bands sing using Spanish-sounding words, but in fact the words carry no meaning. It is just gibberish.

The Bantus brought their metallurgical and woodworking skills to the Americas. Their work is today evident, for example, in the fences and iron balconies of New Orleans, Charleston, South Carolina and Savannah, Georgia in the United States (Fig. 11.20). Their culinary styles are seen in cooking techniques such as deep fat frying. Specific dishes including millet and corn bread, boiled yams, okra or gumbo, and grits or eba are Bantu imports. Black peas, for example, were introduced into Jamaica in 1675. A staple dish made up primarily of smashed plantain called fufu (the name still used today in Cuba), mangu (Dominican Republic), and mofongo (Puerto Rico) is served in different parts of the New World.

Bantu DNA also came and stayed in America. Although West African non-Bantus were enslaved and brought to America, Bantu gene flow is evident in the frequencies of their genetic markers. Biparental autosomal as well as uniparental mitochondrial and Y-specific DNA signals of Bantus ancestry are detected among American populations all over the continent. Bantu heritage is seen in the presence of the Y chromosome–specific mutation or haplogroup E1b1a-M150.[45] This Bantu-specific DNA variant is found at frequencies as high as 62%, the most prominent component among the African-American population in the United States. Other Y-specific lineages such as B2b-M112, which is Pygmy in origin, is found at minimal amounts in the New World.[12] Similarly, Bantu mtDNA types such as L3b, L3e1a,

FIGURE 11.20 Iron fence work from Savannah, Georgia. *Personal photos.*

L1a, and L2a are abundantly detected in the Americas, whereas Khoisan-specific haplotypes (e.g., L1d) are not as frequent.[46] Yet, some autosomal genome-wide results from biparental DNA markers suggest that the African components in African-Americans from the United States derive mostly from West African non-Bantu Niger–Kordofanian-speaking populations such as the Yoruba.[47] As previously discussed, autosomal genome-wide studies assess most of the genetic material of cells and as such are more representative of the genome of the population.

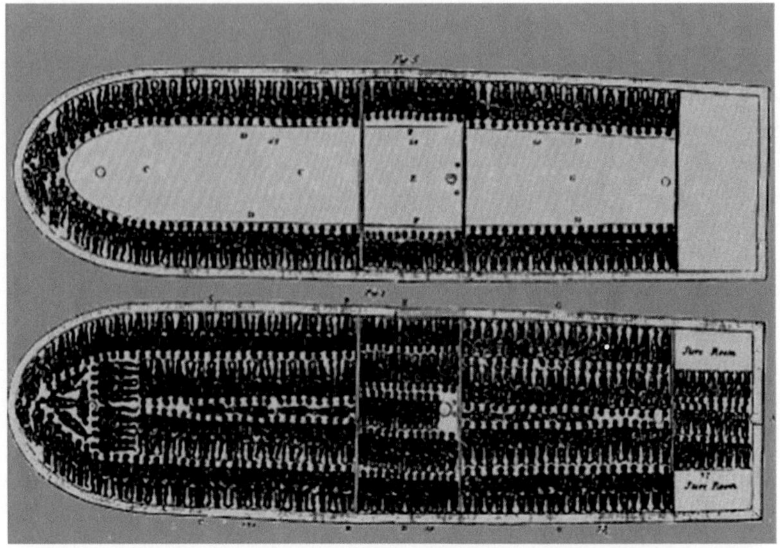

FIGURE 11.21 Illustration of slave ship with human cargo. *From https://en.wikipedia.org/wiki/Slave_ship.*

It is noteworthy that the dissemination of slaves in the New World was not uniform in terms of the location and tribe of origin of the captives. For example, most of the slaves transported into the United States came from tribes from Senegal down to Nigeria. This is in sharp contrast to other countries in the Americas, such as Brazil, where most of the slaves were from Angola and West Africans from the Bantu domain. Thus, this lack of homogeneity in source populations in Africa is responsible for the complex cultural and genetic scenarios when it comes to the topic of Bantus in the new continent. In other words, people of African ancestry in different parts of America reflect various regional and ethnic origins back in Africa.

References

1. Bowden R. *Africa south of the Sahara*. Mankato, Minnesota USA: Coughlan Publishing; 2007.
2. Brown K, Ogilvie S. *Concise encyclopedia of languages of the world concise encyclopedias of language and linguistics series*. Amsterdam, Netherlands: Elsevier; 2008.
3. Clark JD. *From hunters to farmers: the causes and consequences of food production in Africa*. California, Oakland, USA: University of California Press; 1984.
4. Perrya GH, et al. Adaptive, convergent origins of the pygmy phenotype in African rainforest hunter-gatherer. *Proc Natl Acad Sci USA* 2014;**111**:E3596–603.
5. Bleek W. *A comparative grammar of South African languages*. London, UK: Trübner & Co.; 1862.
6. Clark JD, Brandt SA. *From hunters to farmers: the causes and consequences of food production in Africa*. California, Oakland, USA: University of California Press; 1984.
7. Vansina J. New linguistic evidence and the Bantu Expansion. *J Afr Hist* 1995;**36**:173–95.
8. Phillipson DW. *African archaeology*. Cambridge, UK: Cambridge University Press; 2005.
9. Gibbons A. Mother tongues trace steps of earliest Americans. *Science* 1998;**279**:1306–7.
10. Stahl A. Intensification in the west african late stone age: a view from Central Ghana. In: Shaw T, Sinclair PJJ, B.Andah, Okpoko A, editors. *The archaeology of Africa: food, metals and towns*. London, UK: University College London; 1993.

11. Quintana-Murci L, et al. Maternal traces of deep common ancestry and asymmetric gene flow between Pygmy hunter-gatherers and Bantu-speaking farmers. *Proc Natl Acad Sci Unit States Am* 2008;**105**:1596–601.
12. Rowold DJ, et al. At the southeast fringe of the Bantu expansion: genetic diversity and phylogenetic relationships to other sub-Saharan tribes. *Meta Gene* 2014;**2**:670–85.
13. Shillingtog K. *History of Africa*. New York, NY, USA: Palgrave Macmillan; 2012.
14. de Maret P, Nsuka F. History of Bantu metallurgy: some linguistic aspects. *Hist Afr* 1977;**4**:43–65.
15. Ehret C, Posnansky M. *The first spread of food production to Southern Africa (Chapter 8) in the archaeological and linguistic reconstruction of African history*. Los Angeles, California, USA: University of California Press; 1982.
16. Huffman TN. The stylistic origin of Bambata and the spread of mixed farming in Southern Africa. *Southern African Humanities* 2005. Pietermaritzburg, South Africa.
17. Berniell-Lee G, et al. Genetic and demographic implications of the Bantu Expansion: insights from human paternal lineages. *MolBiol Evol* 2009;**26**:1581–9.
18. Beleza S, et al. The genetic legacy of western Bantu migrations. *Hum Genet* 2005;**117**:366–75.
19. Batini C, et al. Phylogeography of the human mitochondrial L1c haplogroup: genetic signatures of the prehistory of Central Africa. *Mol Phylogenet Evol* 2007;**43**:635–44.
20. Destro-Bisol G, et al. Variation of female and male lineages in sub-Saharan populations: the importance of sociocultural factors. *Mol Biol Evol* 2004;**21**:1673–82.
21. Wood ET, et al. Contrasting patterns of Y chromosome and mtDNA variation in Africa: evidence for sex-biased demographic processes. *Eur J Hum Genet* 2005;**13**:867–76.
22. Tishkoff S, et al. History of click-speaking populations of Africa inferred from mtDNA and Y chromosome genetic variation. *Mol Biol Evol* 2007;**24**:2180–95.
23. Scheinfeldta LB, et al. Working toward a synthesis of archaeological, linguistic, and genetic data for inferring African population history. *Proc Natl Acad Sci USA* 2010;**107**:8931–8.
24. Tishkoff SA, et al. The genetic structure and history of Africans and African Americans. *Science* 2009;**324**: 1035–44.
25. Shillington K. *Encyclopedia of african history*. London, UK: Routledge Publisher; 2004.
26. Beach D. Cognitive archaeology and imaginary history at Great Zimbabwe. *Curr Anthropol* 1998;**39**:47–72.
27. Kuklick H. Contested monuments: the politics of archaeology in Southern Africa. In: Stocking GW, editor. *Colonial situations: essays on the contextualization of ethnographic knowledge*. Madison, Wisconsin, USA: University of Wisconsin Press; 1991.
28. Shadreck C, Pikirayi I. Inside and outside the dry stone walls: revisiting the material culture of Great Zimbabwe. *Antiquity* 2008;**82**:976–93.
29. Holmgren K, Helena Öberg H. Climate change in Southern and Eastern Africa during the past millennium and its implications for societal development. *Environ Dev Sustain* 2006;**8**:1573–2975.
30. Gayre R. *The origin of the Zimbabwean civilization*. Paris, France: Galaxie Press; 1972.
31. le Roux M. *The Lemba – a lost tribe of Israel in Southern Africa?*. Pretoria, South Africa: University of South Africa Press; 2003.
32. Kleiman Y. *DNA and tradition – Hc: the genetic link to the ancient Hebrews*. New York, NY, USA: Devora Publishing; 2004.
33. Spurdle AB, Jenkins T. The origins of the Lemba "Black Jews" of Southern Africa: evidence from p12F2 and other Y-chromosome markers. *Am J Hum Genet* 1996;**5**:1126–33.
34. Thomas MG, et al. Y chromosomes traveling south: the cohen modal haplotype and the origins of the Lemba–the "Black Jews" of Southern Africa. *Am J Hum Genet* 2000;**66**:674–86.
35. Mendez FL, et al. Increased resolution of Y chromosome haplogroup T defines relationships among populations of the Near East, Europe, and Africa. *Hum Biol* 2011;**83**:39–53.
36. Schmidt PR. Iron technology. In: *East Africa. Symbolism, science and archaeology*. Oxford, UK: James Currey Publishers; 1997.
37. Childs ST, Herbert EW. Metallurgy and its consequences. In: Stahl AB, editor. *African archaeology: a critical introduction*. Oxford, UK: Blackwell Publishing; 2005.
38. Menkhaus K. Bantu ethnic identities in Somalia. *Ann Ethiopie* 2003;**19**:323–39.
39. Eltis D, Richardson D. *Extending the frontiers: essays on the new transatlantic slave trade database*. New Haven, Connecticut, USA: Yale University Press; 2008.
40. Parra EJ, et al. Estimating African American admixture proportions by use of population-specific alleles. *Am J Hum Genet* 1998;**63**:1839–51.

41. Childs M. *The Yoruba diaspora in the Atlantic world*. Bloomington, IN, USA: Indiana University Press; 2004.
42. Simms TM. The genetic legacy of the transatlantic slave trade in the island of new providence. *Forensic Sci Int: Genetics* 2008;**2**:310–7.
43. Vass WK. *The Bantu speaking heritage of the United States*. Los Angeles, California, USA: UCLA Press, Center for Afro-American Studies; 1979.
44. Fry P, Vogt C. *Cafundó, a África no Brasil: Linguagem e Sociedade*. São Paulo, Brazil: Companhia das Letras; 1996.
45. Hammer MF, et al. Population structure of Y chromosome SNP haplogroups in the United States and forensic implications for constructing Y chromosome STR databases. *Forensic Sci Int* 2006;**164**:45–55.
46. Pereira L, et al. Prehistoric and historic traces in the mtDNA of Mozambique: insights into the Bantu expansions and the slave trade. *Ann Hum Genet* 2001;**65**:439–58.
47. Bryc K, et al. Genetics genome-wide patterns of population structure and admixture in West africans and african Americans. *Proc Natl Acad Sci USA* 2010;**107**:786–91.

CHAPTER

12

Modern Humans in Europe

Europe is not based on a common language, culture and values.... Europe is the result of plans. It is in fact, a classical utopian project, a monument to the vanity of intellectuals, a program whose inevitable destiny is failure; only the scale of the final damage done is in doubt. ***Margaret Thatcher***[1]

SUMMARY

The origin of modern-day Europeans continues to be debated because of the complex prehistory of Europe. Evidence from archeological sites, linguistic studies, and DNA tells a story of humans in Europe that underwent relocation and resettlement, introgression, invasion, migration, and important technological developments, leading to vast population growth. Although many aspects of early European lifestyle remain a mystery, it is well known that their existence has had a profound impact on modern-day Europeans and the development of governments and social hierarchy. Hominin evolution and the eventual colonization of Europe took place over millions of years during the Paleolithic, Mesolithic, and Neolithic period, (these titles are commonly referred to as the Old, Middle, and New Stone Age, respectively), lasting from the early Pleistocene through the Holocene. During the Lower Paleolithic and into the Upper Paleolithic, 2 million years ago (mya) to 20,000 years ago (kya), Europe was populated by *Homo erectus*, *Homo heildelbergensis*, and *Homo neanderthalensis*. Then from the Upper Paleolithic through the Middle Paleolithic, 45 kya to 6 kya, Europe was settled by anatomically modern humans (AMHs), *Homo sapiens*. The earliest of these *H. sapiens* were the Cro-Magnon who were characterized by strong musculature, a broad face, and tall stature. The Cro-Magnon interbred with Neanderthals in Europe and the Middle East. However, during the Last Glacial Maximum (LGM), 24 kya to 17 kya, when Northern Europe was covered in ice, most *H. sapiens* left Europe and then resettled about 15 kya when the glaciers had receded. Fossil remains and extensive archeological discoveries reveal a large variety of innovations belonging to the early Europeans, including hunting tools and weapons, sophisticated art, ornamentation, and an awareness of time. This evidence, along with recent genetic discoveries, offers clues about their behaviors, cultural characteristics, rituals, and beliefs. Archeological evidence separates Europe into several ages including the Neolithic, Bronze, and Iron ages. Agriculture marked the last of the Stone Age periods and began the Neolithic

Ancestral DNA, Human Origins, and Migrations
https://doi.org/10.1016/B978-0-12-804124-6.00012-4

period, starting in Southeastern Europe about 9 kya and spreading to Northern Europe by about 5 kya. The Neolithic was followed by the European Bronze Age, which began around 5 kya in Greece. The Bronze Age is characterized by early forms of writing and the use of bronze (mixture of copper and tin) in the development of art, tools, and weapons. Following the European Bronze Age, the Iron Age began in Southern Europe around 3 kya, spreading to Northern Europe 500 years later. During the Iron Age, literacy was evident in Southern Europe along the Mediterranean and various religious beliefs, artistic styles, and diverse agricultural practices began to emerge; however, Northern Europe remained more primitive in writing, architecture, and related hallmarks of civilization until around 1.5 kya. The Iron Age was the last of the three archeological ages (Stone, Bronze, and Iron), and all three ages saw migration, invasion, and trade, leading to genetic admixture.

Following the Iron Age, an expansion of literacy emerges and historical documents, archeological evidence, and technological advances help separate European history into the Middle Ages, the Renaissance, and the Industrial Revolution. These last three periods showed dramatic improvements in agriculture, increased trade, improved infant care, increased nutrition, and reduced mortality, leading to increases in population density. Larger populations had certain advantages but also led to environmental pollution and an increase in insects, microorganisms, and various disease vectors. Today the study of European history and prehistory continues to be a focus of archeological excavations and modern-day genetic analysis trying to unravel the mysteries behind human evolution in the region. Contemporary studies not only are of anthropological and historical value but also have revealed various genes that may have been involved in selection, adaptation, and disease.

TIME PERIODS, HUMAN EVOLUTION, AND THE MOVE OUT OF AFRICA

Archeologists have discovered a variety of hominin fossil remains and artifacts that have helped to unravel African and European prehistory. They use this collection of artifacts and fossils to divide human history into three archeological periods, the Stone Age, Bronze Age, and Iron Age, according to the materials used to make many of the artifacts from each age.[2] The use of stone, bronze, and iron varied geographically and temporally, meaning the dates of these ages can differ considerably depending on location.

The Stone Age (named for stone tools with an edge or point) took place from about 3 mya to about 43 kya, ending with the advent of metal working and making it one of the longest periods in the history of humankind. Because of its length and variable activities that occurred during this time, it has been divided into the Paleolithic (Old Stone Age), Mesolithic (Middle Stone Age), and Neolithic (New Stone Age). Owing to significant changes in stone tool complexity that took place during the Paleolithic, it has been further divided into three different periods, the Lower Paleolithic, the Middle Paleolithic, and the Upper Paleolithic, which reflect developmental and technological changes.[2]

The length of the Stone Age in any particular region is dependent on cultural artifacts and on the transition from stone to metal. Because this transition time varies within and across continents, many investigators prefer to use geological timescales instead of archeological scales.[2] From a geological perspective, the Paleolithic period refers to characteristics of hominin activities of the late Pliocene to the end of the Pleistocene epochs. On the geological timescale, the Pliocene extends from 5.3 mya to 2.6 mya, and the Pleistocene followed the Pliocene and ended around 10 kya. In this chapter, we will use the archeological timescale and focus on Europe and the Middle East where the Paleolithic period extended from about 3.3 mya to around 10 kya.[3]

The Paleolithic was characterized by the development of the first stone tools and the evolution of humans from an ape-like ancestor. The evolution of modern-day humans from ape-like species occurred over a million years of the Paleolithic period, with the genus *Homo* originating about 2.5 mya.

The Lower Paleolithic extended from about 3.3 mya to roughly 300 kya and is associated with the small-brained hominins of the genus *Homo*, including *Homo antecessor* (living around 750 kya) which is believed to have later evolved into *H. heildelbergensis* (living around 650 kya) in various parts of Europe.[4,5] The Lower Paleolithic hominins were the first users of stone tools, which included choppers, hand axes, burins, and awls.[6] Fire, another significant tool in human evolution, was first used during this time, suggesting that early hominins may have begun to cook their food[7] and could move around and survive colder landscapes.

The Middle Paleolithic (which spans the period 300,000 to 30,000 years ago) brought about advanced tool-manufacturing techniques and the creation of more elaborate devices, such as the stone-tipped spear, which allowed hunters to access a greater variety of food sources.[8] The style of flint implements used during this period is associated primarily with Neanderthals and is referred to as Mousterian (or the Mousterian industry or tradition). The name Mousterian comes from the site of Le Moustier, in the Dordogne region of France, where Neanderthal remains and artifacts were discovered. Since the original discovery, Mousterian tools have been found throughout Europe and parts of Northwest Africa.

The early segment of the Middle Paleolithic also marks the evolution of modern humans (possibly from *H. heildelbergensis, Homo rhodesiensis, or H. antecessor, or a mixture of species*) and the migration of humans out of Africa (Fig. 12.1). There are several theories regarding the origin of modern humans, but fossil evidence strongly supports the theory of evolution occurring in tropical Africa about 200–300 kya. Although there is no direct genetic evidence from 300 kya fossils, some archeological evidence suggests that there may have been interbreeding with another species in Africa, leading to the *H. sapiens* that eventually left Africa. Evidence includes a 195-ky-old (kyo) modern human from Omo Kibish, Ethiopia, and three 160-kyo *H. sapiens* skulls from Herto, Ethiopia[9] (known as *H. sapiens idaltu*). The fossils from Omo Kibish, Ethiopia, appear as morphologically intermediate between archaic African and later anatomically human and most likely represent the ancestors of AMHs that left Africa. Although there is some disagreement about the timing and the routes taken out of Africa, most researchers agree the first migration occurred around 100–170 kya toward the Near East and was followed by a retreat back to Africa or extinction. The second migration occurred around 65–75 kya across the strait of Bab-el-Mandeb (gate of tears) that connects Ethiopia and Yemen. During that time, the strait was much shallower or possibly dry because of glaciation, allowing for emigration into Southeast Asia. Evidence for this migration comes from archeological artifacts and indicates a movement of humans along the coast of the Indian Ocean, with modern humans reaching Australia around 50 kya.[10,11]

In sum, according to a recent theory, humans may have crossed over into the Arabian Peninsula as early as 125,000 years ago. From the Middle East, migration continued into India

FIGURE 12.1 A depiction of *Homo heidelbergensis* potentially the ancestor of both Neanderthals and anatomically modern humans. *From http://humanorigins.si.edu/sites/default/files/styles/full_width/public/images/square/heidelbergensis_JG_Recon_head_CC_3qtr_lt_sq.jpg?itok=0E_UqEh9.*

around 70 kya and into Southeast Asia. Settlers then could have crossed over into Australia and New Guinea because they were one continent due to lower sea levels 55 kya. Migration into Europe took somewhat longer to occur.

UPPER PALEOLITHIC: THE AGE OF EUROPE

The first definitive archeological evidence of human settlement on the European continent was discovered in 1964 in the Grotta del Cavallo in Apulia, Italy. Investigators discovered infant teeth that were later determined to be 43–45 kyo and because of their unusual hybrid-like structure were some of the first evidence to suggest that Neanderthals and humans coexisted over thousands of years and the possibility of interbreeding between the two species. Genetic evidence also suggests that settlement of Europe may have occurred as early as 45 kya.

The appearance of AMHs in Europe and the nature of the transition from the Middle to Upper Paleolithic are matters of intense debate. Most researchers accept that before the arrival of AMHs, Neanderthals had adopted several "transitional" techno-complexes. Two of these are, the Uluzzian, a prehistoric stone tool technology used by Neanderthals and adopted by early modern humans from Southern Europe, and the Châtelperronian, a blend of tool types from the Middle and Upper Paleolithic from Central and Southwestern Europe. These industries are key to current interpretations regarding the timing of arrival of AMHs in Europe, their potential interaction with Neanderthals, the cognitive abilities of Neanderthals, and the reasons behind Neanderthal extinction.[2,3] However, the actual fossil evidence associated with the Uluzzian and Châtelperronian is limited, and recent work has questioned attributing the Châtelperronian industry to Neanderthals on the basis of mixed fossils analysis.[12,13] Uluzzian artifacts have been unearthed from caves in Italy and Greece and are considered the most likely transitional tool types following the Mousterian that then spread across Europe before the Aurignacian. Bennazi et al.[12] reanalyzed the deciduous molars from the Grotta del Cavallo in Apulia and associated them with the Uluzzian. Using two independent morphometric methods based on microtomographic data, they showed that the specimens can be attributed to AMHs. The teeth provided crucial evidence that the makers of the Uluzzian technology were not Neanderthals. In addition, new chronometric data for the Uluzzian layers of the Grotta del Cavallo obtained from associated shell beads and included in a Bayesian age model show that the teeth must date to 43–45 kya. The remains from the Grotta del Cavallo are the oldest known European AMHs, confirming a rapid dispersal of modern humans across the continent before the disappearance of Neanderthals.

Based on radiocarbon dating of fossil remains, modern humans entered Southeastern Europe around 46 kya and then migrated into Central and Western Europe between 42 and 45 kya, whereas other modern human groups migrated south into Spain around 41 kya and into the Iberian Peninsula from France about 28 kya. Cro-Magnons are the oldest known modern humans in Europe, and the term "Cro-Magnon" is used interchangeably with "European early modern humans," *H. sapiens*, and AMHs.[9,14] The scientific community most commonly uses the term European early modern humans (Fig. 12.2).

From a geographical perspective, the late migration of modern humans into Europe seems counterintuitive considering its close proximity to Africa; however, there were only two routes that scientists have concluded could have been taken into Europe from Asia because of climatic conditions and this may explain the late migration. The route early humans took

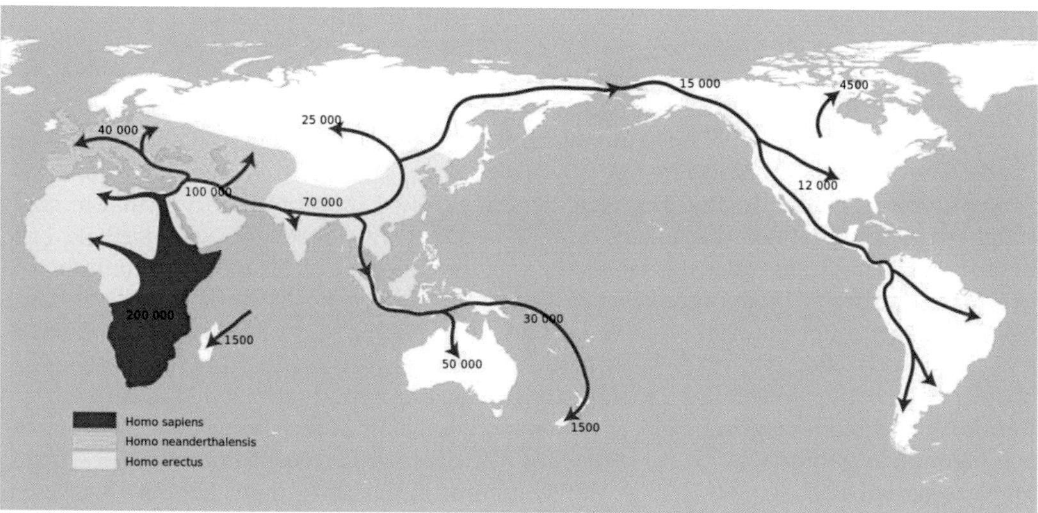

FIGURE 12.2 The route for human migration within Africa, and the route of human and Neanderthal migration out of Africa. *From https://upload.wikimedia.org/wikipedia/commons/thumb/2/27/Spreading_homo_sapiens_la.svg/2000px-Spreading_homo_sapiens_la.svg.png.*

to Europe is still unclear, but several hypotheses have been proposed that are discussed later in this chapter.

The Middle Paleolithic is also associated with the evolution of *H. neanderthalensis* (Neanderthals) from *H. heildelbergensis* about 300 kya and their subsequent extinction about 40 kya.[2,15] Neanderthals occupied Europe for approximately 200 kya, and the most recent radiocarbon dates indicate that the Neanderthals went extinct about 5000 years after the first modern humans arrived in Europe 45 kya.[15] Iberia was one of the last regions occupied by modern humans and the last stronghold of Neanderthals. The specifics of Neanderthal and modern human interactions are still unclear; however, genetic and archeological evidence has revealed they had some social interaction within the 5000 years they both inhabited Europe. One example is the skull of a 4-year-old Neanderthal from a cave in Gibraltar that showed characteristics of hybridization. Another example is a recent study that exposed evidence of shared living spaces in caves of present-day Israel, depicting cohabitation and shared technology of Neanderthals and modern humans during the first modern human migration from Africa into the Near East (See Chapter 5). The modern human remains found at this site displayed skeletal features that seemed to represent a mix of Neanderthal and early modern human anatomy. This evidence and numerous genetic studies support the admixture theory in which Neanderthals mated and produced viable and fertile offspring with modern humans (see Chapter 6). As we will see, the admixture of Neanderthals and early humans played a significant role in the evolution of modern-day Europeans and Asians, and Neanderthals may have contributed both useful and detrimental genes to the human gene pool.[15,16]

How and why Neanderthals became extinct is still a hotly debated question; however, continuing research on Neanderthal and modern human coexistence has provided potential clues for their extinction. There are three common theories regarding the extinction of the Neanderthals upon the arrival of the first modern humans in Europe.[14,15] First, it is commonly believed that the Cro-Magnons had a competitive advantage over the Neanderthals. Their advancements in tool manufacturing and weapons led to a larger variety of food sources, and their social structure placed the European early modern humans a step ahead of the Neanderthals who were competing for the same prey and other resources. Another widely accepted theory suggests the extinction of the Neanderthals because of their inability to adapt to the ecological changes that came with the fluctuating climates in Southern Europe. The harsh, cold environments may have led to significant changes and a decrease in food sources, potentially affecting Neanderthal survival. Another theory is that humans brought diseases with them (pathogens or parasites) that were deadly to Neanderthals. Last, scientists have suggested that the relatively small population of Neanderthals interbreeding with the relatively large population of humans led to genetically unfit hybrids. Because no Neanderthal mitochondrial DNA is found in humans, it suggests that female humans were able to interbreed with male Neanderthals, leading to fertile offspring, whereas female Neanderthals mating with human males may have led to hybrids that failed to thrive, were infertile, or unviable offspring. Selection was more efficient in removing deleterious genes from the human population. The evidence for this comes from recent findings that suggest that Neanderthal DNA sequences were slowly removed from the human genome by selection pressure, leading to the pattern of Neanderthal genes that we see today in modern humans.

The Upper Paleolithic period began around 40–50 kya and ended about 10 kya in Europe and the Near East. The Upper Paleolithic overlapped with some of the previous mentioned periods (the Middle Paleolithic), with considerable dating differences in various regions.[9]

The Upper Paleolithic is characterized by major changes in innovations and technology. The arrival of AMHs in Europe is one of the most compelling events that led to the extensive use of stone blades, decorative art, awareness of time, celebration of seasons, and the widespread manufacture of new tools that, in addition to stone, also contained bone, wood, antler, and ivory (Fig. 12.3). Primitive cave art, Venus figurines, rituals, and organized settlements were also noted during this period. Increasing intellect, creativity, abstract thinking, and greater awareness of the universe are assumed factors that led to such technological advances.[2,5]

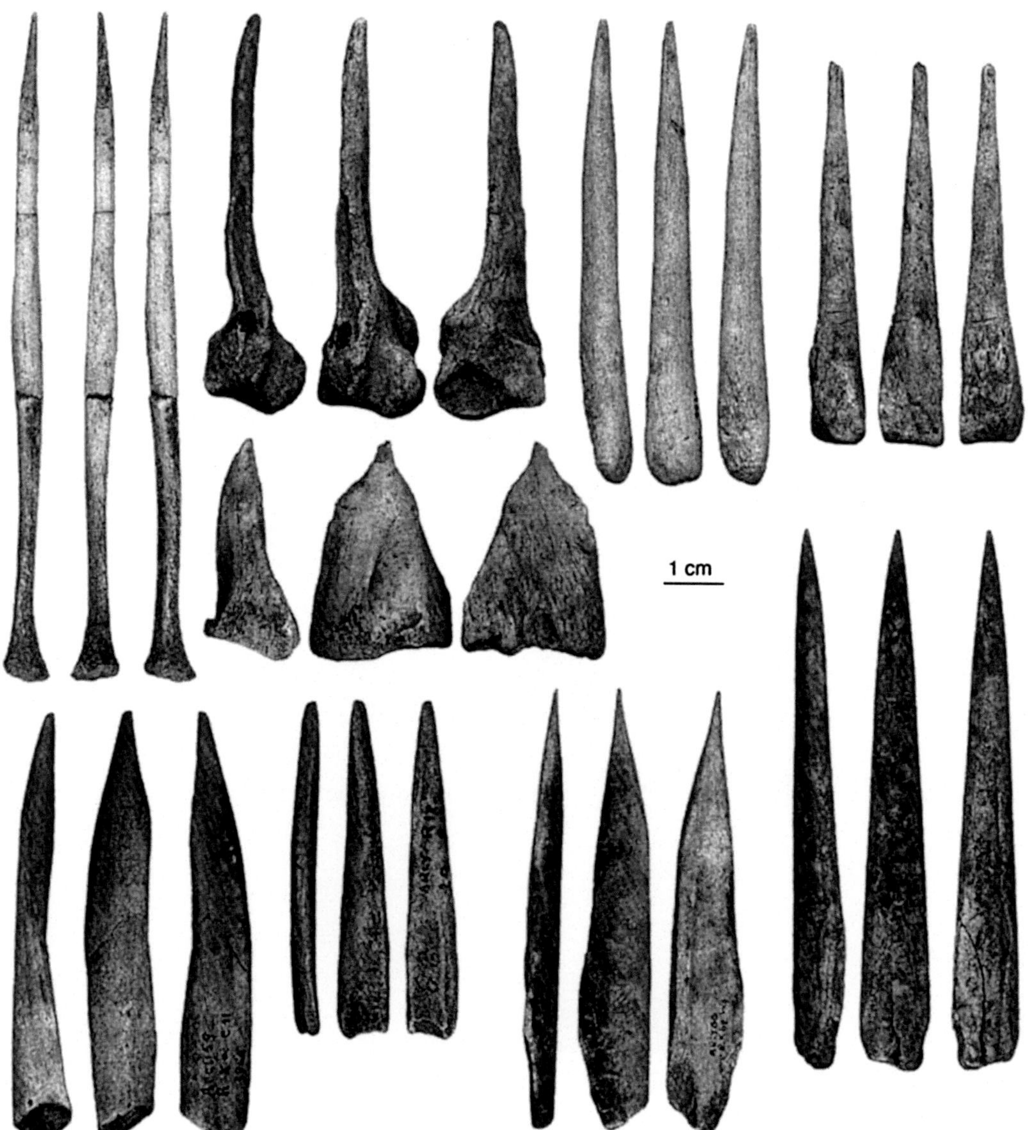

FIGURE 12.3 Bone tools from the Châtelperronian period, which refers to a tool industry of the Middle to Upper Paleolithic. *From https://www.thoughtco.com/guide-to-the-chatelperronian-173067.*

Numerous archeological materials from the Upper Paleolithic were found during the 20th century in Europe, especially southwestern France.[2] Excavations in caves and rock shelters in this area revealed layers of materials from the last part of the Pleistocene. Studies conducted on the contents of these shelters exposed distinctive tools and other artifacts that led to the division of the Upper Paleolithic into subperiods that arose from the earlier Mousterian industry. These new archeological subperiods are known as the Châtelperronian, Aurignacian, Gravettian, Solutrean, and Magdalenian, whose names come from famous French sites where distinctive cultural artifacts were first documented.[2] The Châtelperronian period was the time that Cro-Magnon existed in Europe. As we will see later, recent evidence shows these early modern humans disappeared from Europe and were replaced by several other groups that are represented in the other subperiods of the Upper Paleolithic.

CRO-MAGNON ORIGIN AND FOSSIL DISCOVERY

Railway workers first discovered Upper Paleolithic human remains at a rock shelter in southwestern France known as *Abri de Cro-Magnon* near Les Eyzies-de-Tayac-Sireuil, Dordogne in 1868[9]. Geologist Louis Lartet excavated the cave and uncovered five fragmentary human skeletons including an infant, a middle-aged man, two younger men, and a young woman.[9,17] They had round heads with mostly straight foreheads and only slight brow ridges and were anatomically similar to modern humans (Figs. 12.4 and 12.5). The first specimen discovered at the *Abri de Cro-Magnon* site was Cro-Magnon 1, a fossilized human skull

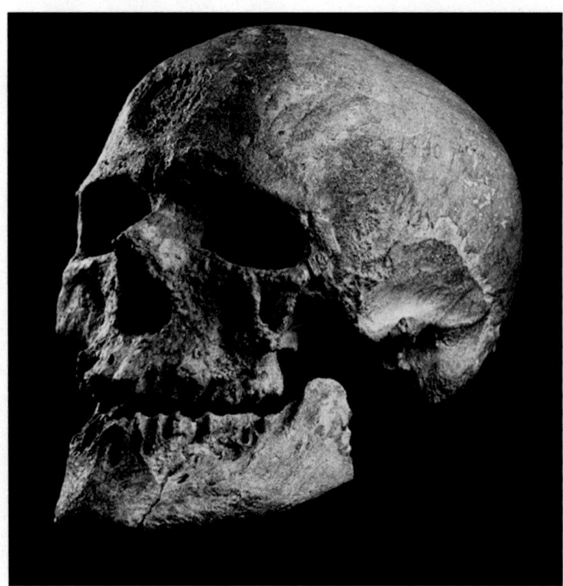

FIGURE 12.4 A Cro-Magnon skull showing the straight forehead, slight brow ridges, and round head typical of modern-day humans. These features are very different from those seen in *H. heidelbergensis* depicted in Fig. 12.1. *From http://exploregram.com/wp-content/uploads/2015/10/2-of-2-A-side-profile-of-a-Cro-Magnon-skull.-Evolution-HumanEvolution-@RobertClarkPhoto-@thephotosoc.jpg.*

FIGURE 12.5 An artist's depiction of Cro-Magnon based on skeletal remains. *From https://s-media-cache-ak0.pinimg.com/564x/33/81/41/3381413443fab50cc97e7e6c6c0e2dda.jpg.*

from a middle-aged male carbon dated to about 30,000 years old.[17,18] The location of the skull among a variety of shells and decorative pendants led researchers to believe that the individuals were buried intentionally and ceremoniously.[16] This would suggest that Cro-Magnons understood the importance of containing dead bodies possibly to prevent disease or possibly as spiritual rituals. Additional specimens from the site were skulls from an adult female, Cro-Magnon 2, and a male, Cro-Magnon 3. Several of the individuals contained severe injuries, such as infections, fused vertebrae in the neck, and a fractured skull.[14] Such life-threatening injuries imply that their lives were physically demanding. In addition, as some of these individuals were middle-aged, this could indicate that Cro-Magnons relied heavily on a sense of community support to aid each other in the case of illness or life-threatening injuries.[19,20]

In 1927, anthropologist Sir Arthur Keith reported finding an upper jawbone from a site in the United Kingdom—Kent's Cavern in South Devon.[21] The teeth were initially thought to be from a Neanderthal, but in 2011, Higham et al.[22] compared the three teeth of the maxilla to well-documented remains of Neanderthal and modern human teeth from other European sites and found that the remains belonged to an AMH. The maxilla was initially thought to have been no more than 36.4 kya until Higham et al. used a dating technique that eliminated chemical contamination from the bones to help determine the maxilla age between 43 and 42 kya, making it one of the oldest specimens of an AMH in Northwest Europe.[9] The other being Cro-Magnon teeth from the Grotta del Cavallo were discovered in Italy in 1964 and are dated to about 41.5–44.2 kya.[12] Other discoveries of Cro-Magnons in Italy and Britain found evidence of individuals that lived in Europe around 45,000 years ago.[14,22]

Cro-Magnons are the oldest known modern humans in Europe, and although the term "Cro-Magnon" is often used to describe the earliest European humans, it does not carry taxonomic or cultural status. Increasingly, genetic evidence suggests that there have been a variety of genetically different populations (other than the original Cro-Magnon) that inhabited Europe during the Stone Age.[23,24] Despite the extensive time and research that has taken place over the past century and a half, there are still unanswered questions about the first modern inhabitants of Europe. The cognitive skills of European early modern humans are thought to have developed at some point in Africa between 50 and 100 kya; however, very little is known about this time in their history.[11] After only a short time of being in the Near East, about 5000 to 10,000 years, AMHs moved into Europe.[18] Scholars still heavily debate the motives and means of this migration. One motive may have been a reduction in the plants and animals used for food and/or an increase in the population in the Near East. There are two proposed routes taken into Europe based on archeological and genetic evidence. One proposed route is that humans leaving Africa went around the Mediterranean, near present-day Israel, Lebanon, and Jordan, and continued west through Turkey into Europe. Evidence for this route, known as the "Levantine corridor," comes from stone tools and ornaments from sites in the Levant that are like those found in the earliest human sites of Europe. Another hypothesis is a route through Russia and then west into Europe. Evidence for this route comes from the genome of a 45-kyo human from Western Siberia, whose genome is similar to human remains found in the earliest sites of Europe.[24]

EUROPEAN EARLY HUMAN SKELETAL MORPHOLOGY AND BODY TYPE

Cro-Magnon fossil remains reveal that the species had powerful and heavy musculature, short and wide faces, slightly long and straight foreheads, prominent chins, and rectangular eye sockets.[21] They were straight-limbed and stood as tall as 163–183 cm, compared with Neanderthals who were approximately 150–168 cm tall (see Chapter 6). The average height of modern humans during this time was larger than that of the Neanderthals and other early human species. Furthermore, in comparison with Neanderthals, Cro-Magnons had significantly less prominent brow ridges, weaker jaws, smaller teeth, and moderate to no prognathism (protrusion) of the face or jaw. The changes leading to the unique skull of the Cro-Magnon may have arisen from a slight reduction in the length of the sphenoid bone, a central bone at the base of the cranium.[25–27]

The Cro-Magnon cranial capacity was approximately 1600 cm^3, similar in size to the Neanderthals and somewhat larger than the 1350 to 1400 cm^3 average for modern humans.[28] Studies have been conducted on the cranial capacity for the brain of these individuals, providing crucial insights into the morphological differences in brain organization and function over time. One study by Eiluned Pearce et al.[29] revealed that Neanderthals had significantly larger orbits and visual cortices than Cro-Magnons. This suggests Neanderthals invested more neural tissue to their visual and somatic systems, therefore leaving less neural tissue for areas associated with problem-solving and critical thinking such as the prefrontal cortex and parietal lobe. On the other hand, Cro-Magnon brains were organized differently, allotting more neural tissue to regions associated with social adaptations and problem-solving. These data support the notion that AMHs had increasingly more cognitive abilities and social networks than Neanderthals, potentially contributing to the extinction of Neanderthals after the arrival of Cro-Magnons in Europe. The small but continual decrease in human cranial capacity over time suggests that our current cranial capacity may have evolved to accommodate our smaller physique.[29]

CRO-MAGNON DIET

Diet greatly depended on the hunting techniques that the early modern humans brought with them to Europe during their migration from Africa and Southwest Asia. The open spaces and dispersal of both game and plant foods over long distances forced modern humans to acquire efficient hunting and plant identification skills in Africa. They became clever stalkers and used the landscape to help trap their prey. This combination of skills helped shape the Cro-Magnons into successful hunters. The treeless landscape of Europe, however, challenged the early Europeans to use a variety of materials other than wood to create hunting tools and weapons. They turned to stone, antler, bone, and ivory, which led to the advancement of manufacturing techniques for artifacts.[19]

The creation of the net, bola, spear, spear-thrower, and bow and arrow were critical as they were primarily big game hunters. They used these devices for hunting mammoths, cave bears, reindeer, wooly rhinoceros, horses, wolves, and steppe bison.[7] The spear-thrower was an essential tool that propelled spears quicker and at much longer distances than even the most skilled hunter could physically perform (Fig. 12.6). Harpoons were also created during this time, which introduced fish into the menu.[17]

The particular regions in which Cro-Magnons lived played a major role in their diet. Campsites were typically focused in locations along the migration trails of large herds

FIGURE 12.6 The spear-thrower helped to increase the force and distance that a spear could be thrown. This allowed hunters to take down larger game. *From https://mir-s3-cdn-cf.behance.net/project_modules/disp/79062014991619.562ab1034d026.jpg.*

of herbivores, signifying their ability to track the migratory patterns of their prey. The specific species targeted along these trails differed among regions. For instance, steppe bison and woolly mammoth were preferred in Eastern and Central Europe, reindeer and horses were targeted in the tundra regions of Western Europe, red deer and chamois were preferred in the more forested south, and reindeer were the prey of choice at sites in southwest France.[30]

The European climate had a profound impact on the Cro-Magnon diet as well. For example, the LGM (the last period when ice sheets were at their greatest extension) brought about harsh weather conditions including wind, ice, and snow that made hunting and survival much more difficult.[9,19] Fat was an essential part of the Cro-Magnon sustenance because it provided calories and energy when food was scarce. Late summer and early fall were therefore crucial times when Cro-Magnons worked hard to hunt big game animals that had accumulated large fat deposits.[18] Cro-Magnons additionally used the animal bones, fur, and hides to make huts for shelter. Evidence of large dugout pits implies that they used the permafrost to create natural ice coolers for their meat, allowing them to live off these reserves once animals migrated. The ability to store food and the creation of tiny bone and antler needles used to make clothing provided Cro-Magnons with the energy and ability to hunt in extremely cold conditions and harvest a wider variety of animals, therefore giving them a competitive advantage over the Neanderthals during intense winter months.[17,30]

EUROPEAN EARLY MODERN HUMAN BEHAVIOR AND CULTURE

The arrival of modern humans in Europe was marked by the influx of innovation, intellectual behavior, rituals, and sophisticated art. It was during this creative explosion that early modern humans intellectually set themselves apart from Neanderthals. In addition to the elaborate tools, natural freezers, and layered clothing that aided Cro-Magnon during harsh winters, cooperation and critical thinking skills were just as, if not more, important to their survival. For example, Cro-Magnons' keen sense of cooperation provided for some members of the community to look out for the needy of their group while others were hunting, gathering plants, traveling, or involved in warfare.[17] Lifestyle evidence also suggests that Cro-Magnons made decisions as a community and equally divided labor. Another characteristic of Cro-Magnon behavior was sharing resources. One example of this is reflected by the discovery of the skeleton of a single animal distributed among three campsites in France, all within hundreds of feet of each other and likely occupied by separate families of the same community. These are significant aspects that strengthened the group and made Cro-Magnons more competitive than the Neanderthals.[31]

Sophisticated art is one of the unique characteristics of European early modern human culture. It is assumed to be an essential part of their day-to-day lives and to have played a critical role in the success of their hunting, gathering, and overall survival because of its ability to communicate contemporary lifestyles and past culture.[31] Cave paintings, ornaments, decorative utensils, Venus figurines, and other portable art are some of their most common forms, greatly contrasting with the utilitarian culture of the Neanderthals.[32]

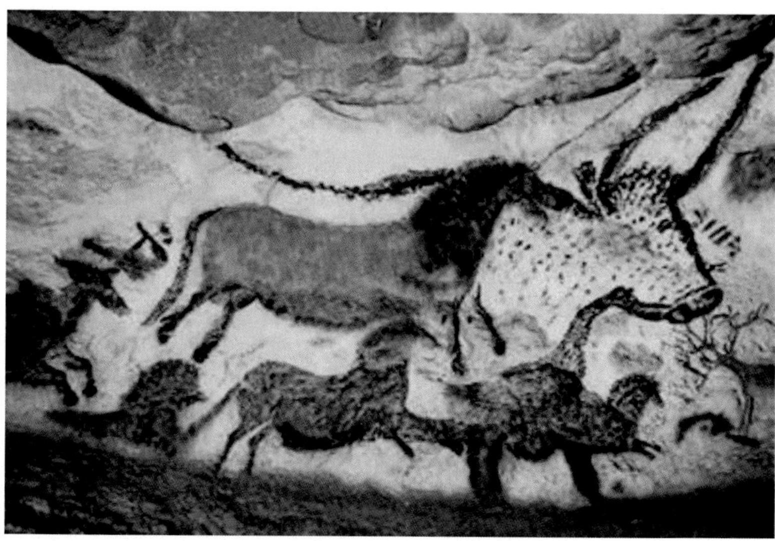

FIGURE 12.7 Cro-Magnon cave art depicting regional animals that were most likely food sources. *From http://www.davidpratt.info/americas/amer8-7.jpg.*

Cro-Magnons generally used antler points, as well as manganese and iron oxides, to paint images on the walls of caves, revealing details about their lives that otherwise would have remained unknown (Fig. 12.7). For example, by observing cave paintings of animals in different climates and scenarios, we now know that Cro-Magnons carefully monitored their prey over the seasons. Cave art has also exposed anatomical details of now-extinct animals including the physical attributes of woolly rhinoceroses and Irish elk. Some of the most spectacular cave art are located at Lascaux, Chauvet, Cosquer, Pech-Merle, and Niaux in France, as well as El Castillo and Altamira in Spain. Three hundred and twenty sites have been discovered in these regions since the mid-19th century, all of which help depict the arrival and behaviors of the modern humans in Europe.[9,31]

Ornaments are another popular form of art created by the Cro-Magnons. They were typically made of carved bone, ivory, and shell and used as pendants, necklaces, and headband decorations.[2] The elaborate details and designs on these objects suggest that Cro-Magnons put an extensive amount of effort into their artistic creations. A recent study found a collection of ornaments at a site near Dolní Věstonice, currently within the Czech Republic, that were originally from the Mediterranean, several hundred kilometers to the south. This reflects the Cro-Magnons' establishment of exchange networks in Europe, allowing some of the highest quality shells, ornaments, and ivory beads to be traded and circulated over long distances.[9,32]

Venus figurines are artifacts carved from materials including mammoth ivory, serpentine, limestone, or clay that were colored with ocher and characterized by exaggerated sexual features such as large pendulum breasts and enormous hips. The figurines range in size from 4 to 25cm tall and are often perceived as fertility figures or mother goddesses. These figurines are extensively present in Upper Paleolithic sites across most of Europe and Asia (Fig. 12.8), indicating their significance as icons of certain shared beliefs and behaviors.[2]

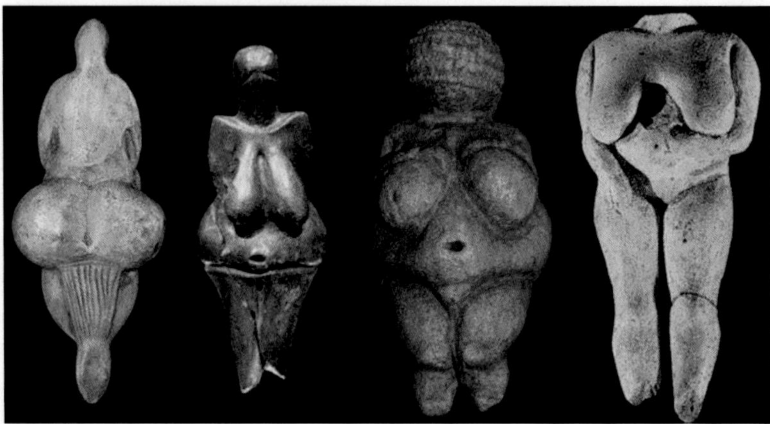

FIGURE 12.8 Venus figurines from Paleolithic Europe. Most figurines were female with exaggerated breasts; however, some were male or of undetermined sex. These were small portable items thought to be icons. *From http://www.ancient-origins.net/sites/default/files/field/image/venus-figurines-europe-paleolithic.jpg.*

Another key feature of Cro-Magnon behavior and culture was their burial of the dead. All that we know about the Cro-Magnons' rituals and beliefs comes from their art.[17] Individuals buried with shells, ornaments, and decorative pendants have been recovered from several sites including Cro-Magnon 1 from *Abri de Cro-Magnon* in southwestern France. Interment of the dead with decorative items symbolizes a belief in the afterlife. They may have believed that the artifacts would be useful to the deceased in the future, making them an essential part of the obsequies and therefore a significant aspect of early European culture.[31] In comparison, Neanderthal burials have been controversial, with most investigators agreeing they existed but were a rare occurrence. Among the Neanderthal burials that have been confirmed, some grave sites contained stone tools and animal bones but overall were less elaborate than those of Cro-Magnon.

All the surviving relics and material evidence of the European early modern humans' lives reinforces the conclusion that they were equipped with almost all the cognitive skills we associate today with humanity and were therefore our intellectual parallels.[33,34] Their effective technology, acute self-awareness, cooperation with one another, and exceptional knowledge of the world around them were the most instrumental weapons in life. Although they were the newcomers in Europe, they were capable of critical thinking, planning, and adapting to situations, which allowed them to flourish and increase their numbers significantly after the Neanderthal extinction and during some of the most difficult climates and environments in recent history.[17,34]

EUROPEAN EARLY MODERN HUMANS AND THEIR ARCHEOLOGICAL CULTURES

Archeologists have uncovered numerous layers of Upper Paleolithic artifacts, tools, and fossil remains across Europe. These distinctive antler, bone, and stone utensils; animal and human skeletal remains; and their indications of cultural and behavioral changes have been used to classify the findings and the populations associated with them into subperiods, also known as European archeological cultures. The five cultures whose names come from sites where their unique artifacts were first documented are the Châtelperronian, Aurignacian, Gravettian, Solutrean, and Magdalenian.[9,17]

The Châtelperronian culture is the earliest culture of the Upper Paleolithic, extending from 45 to 40 kya.[14] This culture was primarily located in Central and southwestern France as well as Northern Spain. The name is derived from the site la Grotte des Fées in Châtelperron, Allier, France, where a backed knife blade known as the Châtelperron knife was discovered. These artifacts were sandwiched between layers of Neanderthal artifacts, indicating a period of overlap and cohabitation between Neanderthals and the Cro-Magnons. The Châtelperronian culture is primarily associated with people wearing perforated animal teeth, as well as the creation of denticulate stone tools and some knifes formed from bone or ivory (Fig. 12.3).[14,17,31] Recent genetic studies indicate that these individuals that originated in Africa and moved into Europe died out and that their genetic material is no longer present in the modern-day European gene pool.[16]

The Aurignacian culture followed the Châtelperronian and was widely distributed over Europe and into the Near East, spanning from about 39 to 29 kya. The origins of this culture are thought to have arisen in what is now Bulgaria (proto-Aurignacian) and Hungary (first full Aurignacian). Despite low temperatures and glaciers covering Northern Europe from 25 to 19 kya, most of these Ice Age inhabitants never left the region. The exotic name is derived from a small town in the eastern foothills of the Pyrenees known as Aurignac, where a road worker exposed a cave with 17 human skeletons as well as flints and antler tools.[33] Flint utensils, grooved bone and antler points, and razor-sharp spears are some of the most distinctive artifacts of this culture. Some of the earliest known cave art, pendants, ivory beads, and figurines were also created during this time (Fig. 12.9). The Aurignacian used lunar calendars to document the phases of the moon and track the migration of game animals, and there is evidence they buried their dead.[11] These innovations, in addition to the evidence of Aurignacians' cognitive abilities and complex social interactions, have led some archeologists to consider the Aurignacians the first truly modern humans in Europe. In addition to these innovations, Aurignacians are thought to have had articulated speech. Evidence for this comes from recent genetic analysis and archeological evidence of their cooperative intelligence, which is thought to be a precursor of language.[34]

Although art and relics are very useful for distinguishing different populations and cultures and providing information on intellect and cultural sophistication, genome data can reveal subtle differences that may not be obvious in archeological studies. In 1860, the remains of a 35-kyo male was discovered by Edouard Dupont in Goyet, Belgium. Recently, the upper arm bone from the individual was used to extract DNA. The DNA sequence revealed that this early European ancestor contained the earliest evidence of a genetic contribution to modern-day Europeans and although the sequenced DNA did contain Neanderthal DNA, it was much less than that found in the remains of individuals from the Châtelperronian culture, indicating Neanderthal DNA had been diluted out of the individual through human-to-human breeding and the extinction of Neanderthals by this time.[16]

Genetic and archeological evidence suggests that the Aurignacians that dominated most of Europe between 34 and 26 kya were eventually replaced by the Gravettian culture. The Gravettian culture extended from about 29 to 18 kya, and archeological and genetic evidence

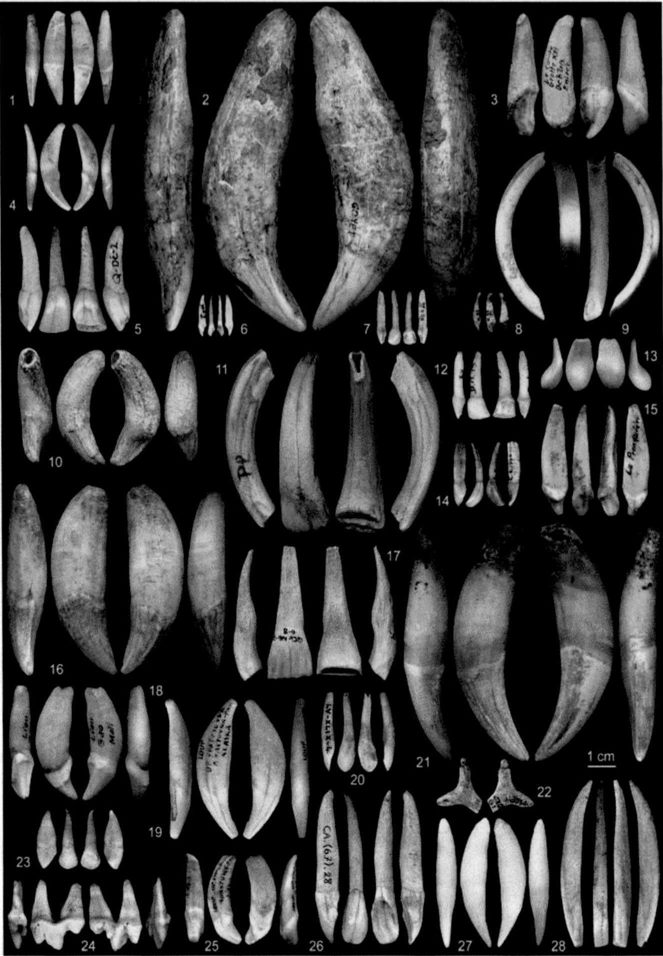

FIGURE 12.9 Animal teeth from the Aurignacian culture used as ornaments. *From Vanhaeren M, d'Errico F. Aurignacian ethno-linguistic geography of Europe revealed by personal ornaments.* J Archeol Sci *2006;33:1105–28; https://s-media-cache-ak0.pinimg.com/564x/2f/58/8f/2f588fcf89355f934e774dc5ef3e4806.jpg.*

suggest that the Gravettians interacted with the Aurignacians. This culture is named after the site, La Gravette in the Dordogne region of France, and is thought to have first developed from Aurignacian roots somewhere in Central Europe or Western Eurasia. Over 40 Gravettian burials are known among sites from France to Russia.[9,17] The Gravettians are associated with several innovations that aided in survival during the LGM. Some of these include small pointed blades and spears used for big game hunting and the creation of nets for small game, needles, layered clothing, natural freezers, and shelters made of mammoth bones, hides, and sod. They also typically formed campsites near gorges, blind valleys, and other natural traps where prey could easily be ambushed. Venus figurines and elaborate ornamentation are primarily associated with the Gravettian period (Fig. 12.10). A discovery by Hohle Fels

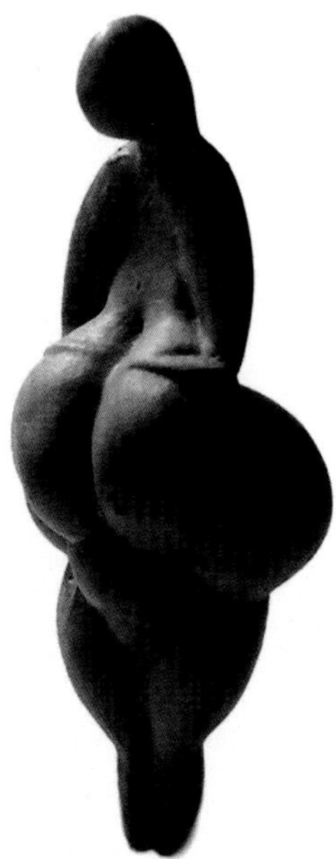

FIGURE 12.10 A Venus figurine (Venus of Lespugue) from the Gravettian culture. *From https://en.wikipedia.org/wiki/Gravettian#/media/File:Venus_de_Lespugue_(replica).jpg.*

confirmed an Aurignacian origin of Venus figurines and further suggested some degree of cultural continuity between the Aurignacian and Gravettian. Many of these innovations provide clues for the emergence of a decision-making hierarchy, individual specialization, and greater social complexity during this period.[9]

Genetic evidence suggests that the Gravettians and the Aurignacians were descended from the same ancient founder population; however, the genetic signature of the Aurignacians disappeared from Europe after the Gravettians arrived.[16]

The Solutrean culture spanned from about 18 to 17 kya and was in northern Spain and southwest France. Its name is derived from a site in the town of Solutré, near Mâcon, France. The people of this culture produced finely worked ornamental beads, cave art, and bifacial pointed tools, as well as advanced flint utensils with a pressure flaking technique that produced thin, laurel leaf points and spearheads. These tools may have been effective in killing prey such as reindeer and horses.[17] There is a considerable similarity in the European Solutrean and North American Clovis lithic technologies (Fig. 12.11). This has led some investigators to hypothesize that people from the Solutrean tradition migrated by boat to North America along the packed ice in the Atlantic. Most scientists do not support this hypothesis, and moreover, genetic evidence from Russian skeletal remains show a linkage between Europe, Asia, and North America, supporting the widely held contention that Asians originally crossed the Bering Strait (Chapter 9).[35] However, support for an alternate route of migration comes from recent studies, showing that the route across the Bering Strait did not become biologically viable for human migration until well after humans occupied the Americas. The evidence comes from studies of pollen, macrofossils, and metagenomic DNA (microbial DNA samples from the

FIGURE 12.11 Solutrean tools. These tools closely resemble the Clovis points made by Native North Americans. The similarity led to the theory that early Europeans migrated to North America along the ice-packed coast of the Atlantic. *From https://upload.wikimedia.org/wikipedia/commons/e/e4/Solutrean_tools_22000_17000_Crot_du_Charnier_Solutre_Pouilly_Saone_et_Loire_France.jpg.*

environment) that show the ice-free corridor became open for human migration approximately 12.6–11.5 kya, while it is known that humans inhabited North America by 14.7 kya.[36,37]

The Magdalenian culture was the last cultural period of Ice Age Europe. Evidence for this culture comes from Central Europe and northern Spain about 11–17 kya after the LGM ended and human populations began to expand into warmer climates. It was named after a Magdalenian rock shelter in Dordogne, France, called La Madeleine, which was more than 180 ft long and assumed to have been occupied for several months a year. The people of this culture are best known for their elaborately worked bone, antler, and ivory projectile points and burins. These weapons were ideal for hunting in open terrain with the aid of spear-throwers.[18] The harpoon also appeared during the end of the Magdalenian, indicating that fish was an important part of their diet during this period. Fossil evidence of the bow and arrow suggests that Magdalenians experimented with new weaponry that targeted smaller game such as birds.[9,17] In addition to the manufacture of new tools, Magdalenians are also known for their extensive artistic abilities. They engraved intricate designs on projectile points, antler, bone, ivory beads, figurines, and even carnivore teeth that they presumably wore as necklaces (Fig. 12.12). This sudden appearance of sophisticated art is assumed to be a reflection of individual and group identity. As the climates grew warmer, populations grew and hunting territories became overcrowded, therefore enhancing the need for individualization.

FIGURE 12.12 Tools and weapons from the Magdalenian culture. *From https://upload.wikimedia.org/wikipedia/commons/thumb/4/4f/Magdalenian_tools_17000_9000_BCE_Abri_de_la_Madeleine_Tursac_Dordogne_France.jpg/220px-Magdalenian_tools_17000_9000_BCE_Abri_de_la_Madeleine_Tursac_Dordogne_France.jpg.*

Remarkably, genetic analysis of the Red Lady of El Mirón, in a cave from Northern Spain, revealed she had DNA that resembled DNA from the Aurignacian group. This finding insinuates that some of the Aurignacians had retreated south during the North's cold climate (25–19 kya) and established a small population in Southern Europe that eventually adopted the Magdalenian culture.[16] Genes of the Magdalenian culture are still prevalent among many modern Europeans today and reflect the significant role they had in European history and evolution.[9]

The unique artifacts exposed from each culture reveal significant details about the evolution of Early Europeans over more than 40 ky. New innovations, technological advances, and behaviors were brought about by each successive culture, allowing populations to survive some of the coldest ice age settings in Europe. The steady increase in cognitive abilities and social complexity implies that Early Europeans were our intellectual equals, which many believe is a potential reason for the extinction of Neanderthals.

GENETICS OF EUROPEAN EARLY MODERN HUMANS

Unraveling the genetic history of Europe is a difficult task; however, sequencing of nuclear DNA, mitochondrial DNA, and Y chromosomal DNA from prehistoric human remains and fossil evidence have provided vital information about their origin, body shape, behavior,

FIGURE 12.13 The Pestera cu Oase (cave of bones) in Romania. The cave is the site of bones from an early modern human that contained 6%–9% Neanderthal DNA. *From http://phys.org/news/2015-07-scientists-early-modern-human-neanderthal.html.*

culture, rituals, diet, and evolution. A variety of AMH remains have been uncovered in Europe, and in 2002, Erik Trinkaus [38] retrieved specimens of three individuals from the Cave of Bones (*Peştera cu Oase*) in Romania (Fig. 12.13). The findings included a lower jawbone, the partial skull of a 15-year-old adolescent, and a left temporal bone dated to about 37–42 kyo exhibiting a mixture of Neanderthal and AMH features.[2,9,17] Trinkaus suspected that the specimen was a hybrid because of the unusual lower jaw and huge wisdom teeth. Recently, geneticists have confirmed Trinkaus's suspicions and determined that the genome of an Oase man contained approximately 6%–9% Neanderthal DNA and that the individual had a Neanderthal ancestor only four to six generations before.[23]

In 2016, Fu et al.[16] reported genome-wide data from 51 Eurasians that dated from 45 to 7 kya. The data confirmed that modern humans and Neanderthals had mated, but the amount of Neanderthal DNA decreased from 6% in the oldest specimens down to 2% in the younger specimens, suggesting that random drift or natural selection may have eliminated some detrimental Neanderthal DNA (Fig. 12.14). In addition, a comparison of the Oase man's genome to that of modern-day Europeans led the researchers to conclude that the Oase individual did not contribute a substantial portion of his genetics to later humans in Europe and that the population that the Oase man belonged to went extinct. In other words, the data indicate the earliest humans in Europe (those older than 37 kya) did not contribute genetically to later populations. The authors suggest the results may indicate that individuals older than 37 kya migrated out of Northern Europe during the Ice Age or went extinct. However, they infer that

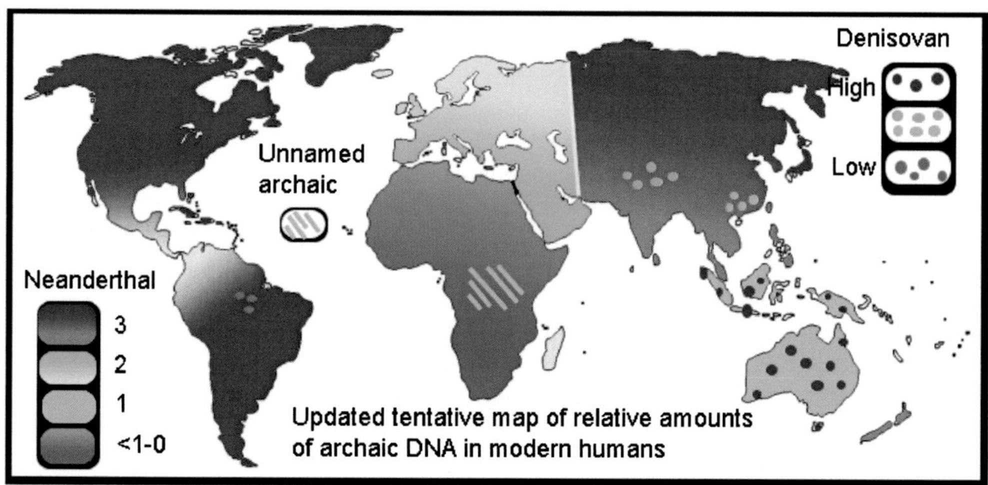

FIGURE 12.14 2016 map of Neanderthal and Denisovan DNA distribution in modern humans. *From https://evolutionistx.files.wordpress.com/2016/03/picture-12.png.*

a set of different European groups existed between 37 and 14 kya, which were descended from a single population that contributed significantly to the ancestry of modern-day Europeans. They propose the founding population of modern Europeans is represented from remains of a man found in Belgium (based on DNA analysis) that was part of the Aurignacian culture. They propose individuals from this mans population persisted through the Ice Age, with the peak of the ice age (including glaciers covering Scandinavia down to Northern France) from 25 to 19 kya. As the ice sheet began to retreat from the north about 19 kya, Northern Europe was repopulated by a group from southwest Europe, the data indicating they were most likely from modern-day Spain. A second migration into Europe about 14 kya occurred from the southeast (modern-day Turkey and Greece) replacing the first group of humans. Consequently, after 14 kya, Europeans became more genetically related to populations from the Caucasus and Turkey, which coincided with the end of the Ice Age. The expansion of people from the Southeast was possibly due to the agricultural revolution.

Interestingly, these genetic studies also found that early Europeans had dark skin and brown eyes until the beginning of the agricultural revolution (between 10 and 12 kya) when farmers from the Near East swept across Europe. As the farmers intermingled with the native populations, both light skin and blue eyes originated. Evidence for this admixture and the appearance of blue eyes in Europe comes from DNA analysis of a male hunter from Villabruna, Italy.[16,39] Blue eyes appeared in some Europeans around 14 kya, while light skin appeared around 7 kya.[39,40]

Investigators sequenced the genome of a boy who lived 24 kya in Southern Siberia. The sequence revealed genetic similarities to present-day Europeans and Western Siberians and indicated the boy was closely related to present-day Native Americans (Chapter 9). The data support the existence of genetically linked Upper Paleolithic populations extending from Europe into Central Asia, and the genetic similarities imply that present-day Europeans descended from these populations (Chapter 9).[41,42]

Mitochondrial DNA (mtDNA) is known to have a relatively rapid mutation rate, therefore providing a more accurate estimate of historical changes during smaller periods of time. Because mtDNA is passed from mother to children, it is useful for studying maternal lineage and reflects the historical movement of females in a region that came to be dominated by patriarchal cultures.

mtDNA is passed on as a single DNA sequence (unlike the two copies in nuclear DNA) so it constitutes a haplogroup. Most modern Europeans have been found to belong to eight main mitochondrial haplogroups and three major clades. Mitochondrial haplogroups are defined by differences in the DNA and are typically used to represent the major branches on the mitochondrial phylogenetic tree. Clades are groups of individuals that are believed to have descended from a common ancestor. The three major clades are (H,V), (U3, U4, U5, K), and (J, T), with H being the most abundant across Europe and Southwest Asia, with frequencies in Europe ranging from 40% to 60%. Haplogroup H is estimated to be about 20–25 kyo. The oldest mitochondrial haplogroup in Europe is U5, which is thought to have evolved in Europe about 25–50 kya and is consistent with an Aurignacian origin.[43,44] Ancient DNA studies of samples 32–22 kya (Gravettian culture) have confirmed *haplogroup U5* as the dominant female European lineage during the Mesolithic, although a few U2e and U4 samples were also found in northeast Europe.[16,40] Contemporary analysis show U5 is found in 8% of the Spanish population, with the highest frequencies of U5 found among the Basques (12%) and the Cantabrians (11%). Only one sample of ancient DNA from the Aurignacian period has been assigned to a mitochondrial haplogroup today. The full mitochondrial sequence was collected from 37.5-kyo remains from a human at Kostenki and identified as haplogroup U2, which is dated to around 50 kyo (U2 is found in very low frequencies across Europe, Central Asia and the Middle East).[41,42]

Another study assessed mtDNA from 821 contemporary individuals in Europe and discovered 6 clear lineage groups, indicating that Europeans were more genetically diverse than originally thought. They concluded the majority of the lineages dated between 30 and 50 kya, suggesting that 85% of mtDNA lineages in Europe were already present before the development of farming around 9 kya and therefore could be connected to one or several of the Cro-Magnon traditions.[45]

In 2016, Posth et al.[40] analyzed mtDNA from the remains of 55 different fossils from across Europe that ranged in age from 35 to 7 kya. Based on changes in the mtDNA sequences, geneticists have identified genetic populations that have the same haplogroup across a large geographical region (superhaplogroups) that shared common ancestors. They examined two superhaplogroups, M and N, and everyone from European descent has the N haplotype, whereas the M type is common among people from Asia and Australia. The study by Posth and collaborators determined that the M haplotype predominated in ancient Europeans until 14.5 kya then vanished and no longer exists in Europe. The group determined that the ancient Europeans shared a common ancestor with modern-day carriers of M, from around 50 kya before M disappeared from Europe. They speculate that the timing (14.5 kya) coincides with the end of the Ice Age and that the small populations that survived the Ice Age were not able to survive the climatic changes when temperatures increased, resulting in new populations carrying the N haplogroup replacing those with M. Although it is not clear where the new N group came from, it has been suggested they may have come from Southern Europe (where conditions were mild during the Ice Age) or possibly from the Near East. It is known from other studies that Europeans acquired DNA from people in the Near

East during the agricultural revolution that occurred at the start of the Neolithic. The data showing the loss of the M haplotype and replacement by N may coincide with the 2016 studies by Fu et al.[16] who determined that nuclear DNA which showed migration into Europe about 14 kya occurred from the southeast (modern-day Turkey and Greece), replacing the previous group of humans.

The picture painted by mtDNA and genomic studies have helped to clarify the movement of people in and out of Europe. Genetic evidence suggests that some individuals of the Aurignacian population (that was replaced by the Gravettian) did not disappear entirely but instead moved to northern Spain where they survived the LGM. The descendants of these Aurignacian individuals then became members of the Magdalenian culture that spread over Europe after the LGM.[16]

Y chromosomal DNA can tell us something about the dispersal of men in Europe. It is thought that Cro-Magnon's Y chromosomal haplogroups belong to C, F, and I. The analysis of Y chromosomal DNA from Neolithic Anatolia, Iran, Israel, Jordan, and various Neolithic cultures in Europe (Greece, Hungary/Croatia, Germany, Italy, France, and Spain) indicated that all sites, except those from the Levant, had a majority of haplogroup G2a individuals. G2a appears to be the dominant Neolithic male lineage of the Near East, with the only apparent absence of G2a in the southern Caucasus region. F, I1, and I2 were also found with I2 lineages found among early farmers in Siberia and southern France. Interestingly, the close haplotype relationship between Basques and Sardinians (I2a1a) indicate they share a common Neolithic ancestor.[46,47]

The highest genetic diversity within haplogroup G is found between the Levant and the Caucasus, which may indicate it is the region of haplogroup G origin. This would suggest that early Neolithic farmers expanded from Northern Mesopotamia westward to Anatolia and Europe, east to South Asia, south to the Arabian Peninsula and into North and East Africa.[46,47]

Despite the importance of genetic studies in uncovering details of European prehistory, there is still a lot that we may not fully understand, and further research may be beneficial in helping to uncover some of the remaining mysteries and gain a more holistic sense of how modern-day Europeans evolved.

THE EUROPEAN COPPER AND BRONZE AGES

When examining the different metallurgic ages, it is important to remember that they occurred at different times in different regions and are associated with the specific use of materials to make artifacts. Therefore, it is not possible to assign a specific age to a specific time in a region as large as Europe or the Near East. Instead, the time frames used for different ages are ranges and because one region may have advanced to using other materials for implements there could be overlap between periods, or entire periods could be skipped. In Europe and the Near East, the Neolithic period was followed by the Chalcolithic period (Copper Age), which specialized in the use of copper. Many archeologists believe that copper smelting may have originated in the Near East around 7 to 10 kya where copper ornaments have been found.[48] The oldest evidence for copper making in Europe comes from the archeological site in Belovode, Serbia, where copper residue and copper slag were found.[49] The copper deposits there were dated to 7 kya, and optical and chemical evidence suggests that the process of copper smelting may have differed from other regions, suggesting that the use of copper may have originated in different parts of Africa, Asia, and Europe independently instead of being spread from a single source. The Chalcolithic period in Europe is distinguished by the use of copper for tools and weapons, the use of ivory and ceramics for art, and increasingly complex stratified societies. However, in many cases it is difficult to find information on the European Copper Age because the period was relatively short (before it was discovered that the addition of tin could result in bronze). So, the European Chalcolithic is considered by most European archeologists to be part of the Bronze Age because information from the Copper and Bronze ages overlap.

The Bronze Age is loosely defined as a period where a civilization smelted its own copper and tin to make bronze or traded in bronze goods with other groups. Just like every other archeological age, the beginning of the Bronze Age varies from region to region, and the dates and places of origin are controversial. In many parts of the world, the Bronze Age followed the Neolithic (last period of the Stone Age), However, in some regions including parts of Africa, the Neolithic was followed by the Iron Age, and the Bronze Age was completely absent because of the lack of materials to make bronze and/or the civilizations remaining in the Stone Age for a longer period.

The European Bronze Age began around 5 kya in Greece and is referred to as the Aegean civilization because of its location around the Aegean Sea.[9] The term Aegean civilization covers three distinct regions including Crete, the Cyclades, and the Greek mainland. The region was the birthplace of two ancient civilizations, Crete, which is associated with the Minoan civilization from the Early Bronze Age, and the Mycenaean civilization. Crete situated between Greece, Turkey, and Libya is sometimes referred to as the birthplace of Western Civilization and has been extensively studied. The first inhabitants of the island were Neolithic farmers who are thought to have arrived from Anatolia 9 kya. These individuals formed the Minoan culture and the Minoans flourished in Crete as a literate society with extravagant palaces, the establishment of government administration, and trade with Mediterranean civilizations. The Mycenaean civilization, the conquerors of Homer's heroic poems, followed the Minoans and introduced innovations in engineering and architecture and a system of warrior aristocracy. The Mycenaeans came from Southern Russia around 4 kya and settled in Greece. They built a large navy, attacked nearby lands, and became the major civilization in the Aegean Sea. Recent evidence for the Mycenaean's wealth and culture comes from a grave site outside the acropolis of Pylos. In 2015, Jack Davis and Sharon Stocker, from the University of Cincinnati, led an expedition that uncovered a 3.5 kyo grave of a Mycenaean Griffin Warrior, which contained 2000 artifacts including objects from the Minoan culture on Crete.[50] It is not clear if these items had been forcibly taken by the Mycenaeans or if it represents an understanding and appreciation for the Minoan culture. Some have speculated that the findings represent a mixture of cultures and possible interbreeding.

The Mycenaean civilization marks the last phase of the Bronze Age in the Aegean region. The Aegean Sea and its islands played a significant role in European history as navigation through the sea was common and insular ports provided safe harbors where marble and iron were mined and traded with other regional civilizations. The Aegean Bronze Age established trade that imported tin and charcoal to Cyprus, where copper was mined to produce bronze artifacts that were then exported. Skeletal osteology of fossils in the region suggests that Aegean Bronze Age of Greece consisted of a very morphologically homogeneous population.[47,48] However, genetic evidence, from extant populations, using Y chromosome analysis has revealed a high degree of heterogeneity within the island of Crete when examining haplogroups A, DE, G2, I, J, and P. The data showed that although 96% of males in the Cretan regions of Chania, Rethymno, and Heraklion could be assigned to the tested haplogroups, only 82% of the markers in Lasithi were common. In another study looking at an isolated region in the Lasithi Plateau, Y-STR analysis demonstrated the close affinity that R1a1 chromosomes from the Lasithi Plateau shared with those from the Balkans but not with those from lowland eastern Crete. In contrast, Cretan R1b microsatellite haplotypes displayed more resemblance to those from northeast Italy than to those from Turkey and the Balkans.[51]

Greece arguably has the longest and richest archeological history in Europe, with the earliest evidence of habitation coming from a skull that was found in the Cave of the Red Stones (Petralona cave). Initial attempts to date the skull estimated it at about 800 kya; however, more recent estimates using uranium-series dating estimate the age to be between 160 and 300 kya. The species of the skull has also been disputed, with the most recent classification as either *H. erectus*, *H. heildelbergensis*, or *H. neanderthalensis*.[52] Some investigators have also proposed that it represents a hybrid between two of these early hominids or an early form of *H. sapiens*. Further evidence of habitation in Crete as early as 130 kya comes from stone tools found in the Franchthi cave. The tools have been dated to at least 130 kya and represent the Acheulean technology that originated in Africa about 1.7 mya, among prehuman populations. The findings suggest that *H. erectus* and/or *H. heildelbergensis* may have had high cognitive abilities and crossed the Mediterranean, indicating they may have been some of the first seafaring hominids. The earliest evidence of burials and agriculture in Greece also come from the Franchthi cave. Burials dated to about 95 kya contained stone artifacts, and 28,000 seeds from 27 different plant species were found in Mesolithic ground levels dating to 11.5 kya.[53]

Another island of interest in the Mediterranean is Sardinia. The early Bronze Age Sardinians are sometimes referred to as the Nuragic civilization. They built megalithic structures and maintained a flourishing trade arrangement with other Mediterranean people. Remains of amber, African animals, Mycenaean ceramics, and weapons from the Eastern Mediterranean indicate their vast trading abilities or their ability to attack and take objects from regions in the Eastern Mediterranean. Further evidence for the Sardinian adventurous nature comes from Otzi, the Tyrolean Iceman, a 5.3-kyo mummy found in 1991 in the Alps on the border of Italy and Australia. In 2014, investigators found that the Iceman's Y chromosome haplotype and whole genome sequence was linked to contemporary Sardinia. Archeological evidence suggests that he was a local farmer, yet his Y chromosome lineage and autosomal single-nucleotide polymorphism variation has vanished from Central Europe. In addition, examination of DNA from modern-day Europeans found that European farmers had a large share of their ancestry associated with modern Sardinians, suggesting that Sardinian ancestry may have once traversed Neolithic Europe. The authors propose that Sardinia may represent the genetic structure of the people associated with the spread of agriculture.[54]

Further influence of Sardinians can be seen in Southern Europe where marine mollusk shells of the cockle (Cardium edulis) found in Sardinia and regions of Northern and Southern Italy were used to imprint the clay pottery known as Cardial Ware (impressed ware). These Neolithic decorative items define the Cardial culture, which can be seen in Italy around

8 kya and eventually spread to the Atlantic coasts of Portugal and south into Morocco. Late Neolithic culture included the Eneolithic (copper) culture, with the Beaker culture marking the transition between the Eneolithic and the Bronze Age.

In addition to decorative pottery, Neolithic Italy and Sardinia also had statue menhirs, a type of carved standing stone that contains carvings of human figures and weapons. These types of sculptures lasted into the Bronze Age and may have been influenced by the Yamnaya culture (discussed later).

Early forms of writing, language, increased trade, and the use of bronze in the development of art, tools, and weapons characterize the European Bronze Age (Fig. 12.15). For many years archeologists debated whether these cultural changes resulted from immigration or were driven by the circulation of ideas. In other words, did the advances result either from acculturation or actual movement of people or both? Archeological and genetic evidence from two separate studies that excavated the remains of Bronze Age people in Europe and

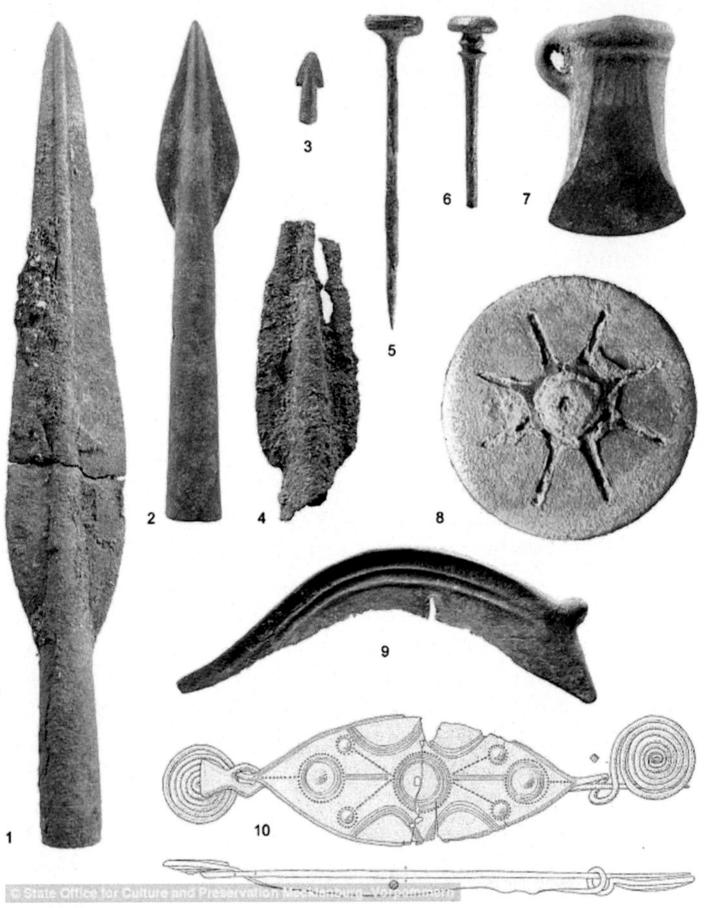

FIGURE 12.15 European Bronze. Figure shows an axe, spear points, ornaments, and tools. *http://i.dailymail.co.uk/i/pix/2016/04/14/17/332EC0A200000578-3509388-image-a-27_1460651456661.jpg.*

Asia suggests that immigration played the most significant role in cultural changes. In both studies, bones from across Europe (Italy, Romania, Poland, Denmark, Sweden, Germany, Russia, and Siberia) were analyzed. The bones ranged in age from 8 to 3 kya. In one study by Allentoft and collaborators,[55] DNA was isolated from 101 Bronze Age remains, whereas in the second study by Haak et al.[56] DNA was isolated from 69 Bronze Age people. The results from both groups concluded that there was a wave of migration about 5000 years ago by herders called the Yamnaya who entered Northern Europe. The authors suggest that this influx of immigrants to the north could explain some of the genetic differences between the people of Northern and Southern Europe. The Yamnaya came from the eastern Pontic–Caspian steppe, a region stretching from the northern Black Sea to the Caspian Sea, in present-day Ukraine and Russia. The studies indicate that the genetic makeup of the Yamnaya was an admixture of Eastern European hunter-gatherers and a population of Caucasus hunter-gatherers from the last ice age.

Archeological evidence indicates the Yamnaya people (also referred to as the Yamna culture, Pit Grave culture, or the Ocher Grave culture) brought major cultural changes to Eastern Europe, including organized family units, property ownership, villages, family burial mounds (Fig. 12.16), cultivation of livestock, and a chiefdom-type social stratification. Recent archeological and genetic studies also suggest the Yamnaya most likely spread Indo-European languages (the contemporary source of vernaculars spoken in Europe

FIGURE 12.16 Yamnaya pit grave showing the unique burial style. *From https://upload.wikimedia.org/wikipedia/commons/8/88/Yamna_culture_tomb.jpg.*

today) to the region.[55,56] Today over 46% of the world's population speak Indo-European languages including English, Spanish, Hindi, Urdu, Portuguese, Russian, Persian, and Panjabi. The root of these languages started in the Neolithic period and is referred to as Proto-Indo-European. Some linguists place the origins of the Proto-Indo-European languages in either Anatolia or Transcaucasia, in the Northern Caucasus mountains, and propose the Armenians and their migration into Greece were responsible for the spread of the language groups into Europe.[57]

Genetic analysis revealed that the Yamnaya were tall, dark-eyed, dark-haired, had skin color that was slightly darker than most modern Europeans, and were lactose intolerant.[56] The intolerance for lactose is not surprising as the absence of the lactase persistence allele has not been found in early Neolithic or Mesolithic skeletons and is thought to have arisen only in pastoral communities after the introduction of domestic cattle breeds.

Prior research of ancient human genomes has shown that Bronze Age Europeans were an admixture of three major ancestral populations that resulted from 3 ancient migrations, the original hunter-gatherers who entered 40 kya, the Near Eastern farmers from the fertile crescent who started penetrating Europe around 9 kya and introduced farming to Europe, and a northern Eurasian population (Yamnaya) that arrived from the east during the Bronze Age[58] The data showing that the Yamnaya were admixed (DNA from Eastern European hunter-gatherers [most likely from Siberia] and a population of Caucasus hunter-gatherers) disclosed the Caucasus as a fourth population that added to the genetic mix of modern-day Europeans. Genome analysis from two Caucasus hunter-gatherers from Georgia that are 13.3 and 9.7 kyo indicate they were probably the source of the DNA in the Yamnaya.[58] The Yamnaya people had an important impact on the genetics of Northern and Central Europe. For example, Norwegians show around 50% of their genetic ancestry from the Yamnaya, and R1b is also the most common Y-DNA haplogroup found in the Yamnaya and modern Western Europeans. Researchers also suggested that the Caucasus hunter-gatherers influenced populations further east, particularly in South Asia.[59]

It is still not clear what influenced the Yamnaya immigration west, but it may have been the result of a decrease in farming in Europe at the time, possibly due to illness or crop failure, which provided opportunities for the Yamnaya to inhabit the region. The discovery of plague DNA in a Yamnaya burial site and a population decline in Europe during the same period suggests their migration may have been influenced by the spread of disease or they may have brought the disease to the region.[60]

DNA evidence suggests that after the Yamnaya culture arrived in northeastern Europe, they may have also influenced social and genetic changes in the temperate regions of Europe by replacing the Neolithic farmers. The Yamnaya are thought to have brought the Corded Ware culture (named for corded pottery) (Fig. 12.17), the Bell Beaker culture, named for the pottery drinking vessels, and the Unetice Culture, named after the village of Únětice in the current Czech Republic of Central Europe (Figs. 12.17 and 12.18). In every case, ancient DNA analysis from skeletal remains associated with these cultures show various levels of Yamnaya DNA. The Corded Ware culture shows the highest levels of Yamnaya DNA, followed by the Bell Beakers, and the lowest levels are seen in the Unetice culture.

Recent studies of the Corded Ware culture using isotope analysis of bone and tooth enamel from two cemetery populations in Southern Germany revealed there was no dominant mode of subsistence and instead a great deal of variability in diet. The variability in diet consisted of

FIGURE 12.17 Pottery from the Corded Ware culture. The pottery clearly shows the corded imprints on the outer surface of the vases. *From http://storia-controstoria.org/wp-content/uploads/2015/08/eulau-dna-europa-cordata.jpg.*

FIGURE 12.18 Unetice pottery. *From https://en.wikipedia.org/wiki/Unetice_culture#/media/File:Uneticevessle.jpg.*

both animal and plant material and specific diets correlated with sex and mobility, especially at the Bergrheinfeld site, where nonlocal women had a different diet from locals and from nonlocal males. The study indicated that the women of this region were highly mobile and the authors suggest this may indicate a pattern of female exogamy involving different groups at differing economic levels[60]

The Yamnaya appear to have had a significant impact on both the genetics and cultures of Europe. One exception to the Yamnaya influence is the Copper Age culture of Italy, known as the Remedello culture. The Remedello culture does not show a genetic component from the Yamnaya or the Caucasus and was either not influenced by the Yamnaya or the Remedello predate the Yamnaya influence in Southern Europe.[58]

Part of the cultural legacy of Europe and the technological advances during the Bronze Age is the great monument Stonehenge found in Southern England. Although many questions remain as to what the monument represents (temple, calendar, etc.), it is known that the monument was constructed in multiple phases between 5.1 and 4.5 kya and was used as a burial site. The first group to start the construction of Stonehenge were the Windmill Hill people. These individuals built the large circular furrow and mounds which indicate a respect for symmetry. They practiced collective burials with most of their mass grave mounds facing east-west. The next group of people involved in the construction of Stonehenge were the Bell Beaker people, who are thought to have migrated from Spain and colonized Northwest Europe. The name Bell Beaker people is derived from their unique style of cups. In addition to their contributions to Stonehenge, the arrival of the Bell Beaker people gave rise to the Wessex culture exemplified by precious burial items, such as elaborately decorated daggers, gold and amber ornaments, and gold cups. The Bell Beaker people are thought to have brought bronze to Britain, and like many other Bronze Age cultural changes, it is associated with growing social complexity, cultivation of livestock, and economic growth.[61] The Bronze Age in Britain and Ireland occurred around 4.5 kya and lasted till around 2.7 kya. It is divided into the Early, Middle, and Late Bronze Age.

One reason for focusing on Britain's Bronze Age is because it is represented by one of the most well-preserved settlements in Europe. The Bronze Age settlement is a 1100 square meter site found on Must Farm in England's Cambridgeshire Fens. The site contains the remnants of a 3-kyo village settlement that was uniquely preserved in silt. The settlement was partially burned and then covered in mud from a tributary of the nearby Nene River, leaving much of the material preserved in pristine condition. The site reveals that people living there were much more sophisticated than originally thought. Almost 80 fragments of clothing were found that were made of finely weaved linen, along with beads made of glass, amber, and jet, that are thought to have originated in Anatolia. Bronze implements recovered included 15 axe heads, 5 sickle heads, a rapier, a sword, and 5 spearheads. Ash and oak wooden structures were also recovered that were preserved because of water logging. Eight dugout oak canoes ranging 3–9 m long, 4000 pieces of burnt timber, and the remains of 5 round houses were recovered at the site. Each house contained about a dozen handcrafted pots 2–35 cm wide, some still containing a porridge-like food. Wooden plates and boxes, a wooden wheel, carefully carved axe handles and posts indicate the inhabitants were skilled carpenters. Animal and plant remains suggest that the inhabitants were farmers that raised lamb and calves and that they also hunted wild boar, red deer, and freshwater fish.[61,62]

Archeologists have determined that the wooden beams for the houses were cut during the winter and the settlement was burned during the summer. The fire appears to have been started deliberately. Speculation on why the settlement was destroyed varies from attack by a nearby settlement, to the onset of disease within the settlement and attempts to rid the area of the disease. Whatever the reason, it is clear the settlers did not return and left a treasure of artifacts.[62]

THE END OF THE BRONZE AGE

The end of the European Bronze Age was both sudden and violent. The Helladic period is the term used to describe the culture of mainland Greece during the Bronze Age. Between 5 and 3 kya, the Helladic civilization became known as the Mycenaean civilization. The Late

Helladic (3.5–3.0 kya) was a time when Mycenaean Greece flourished; however, in the end none of the mainland palaces of Late Helladic Greece survived and the period became known as the beginning of the Greek Dark Ages.

During the Dark Ages, over a 40–50 year period, almost all the great cities were intentionally destroyed by fire, and many were never reinhabited. Most of the destructions were focused on palaces, administrative buildings, large homes known as corridor houses, and fortified urban villages. Fortified settlements covered areas ranging from 1 to 80 ha and contained as many as 15,000 people. These large towns are considered signs of a strong economy, a stratified and complex society with administrative and community organization, and an elite or upper class.[63]

The islands of the Aegean Sea were also sites of destruction. Excavations on the island of Paros have shown that during the Late Helladic a large complex was ransacked and burned and several skeletons of victims were recovered. A few major Mycenaean towns on islands in the central and western Aegean, including Phylakopi on Milos, Ayia Irini on Kea, and Grotta on Naxos, were spared until the end of the Late Helladic but were ultimately targeted as well. Other islands exhibit various degrees of destruction or abandonment, while some island settlements were unaffected. Destruction occurred all along the Eastern Mediterranean, and cities were burned throughout the Aegean, Anatolia, Cyprus, and the Levant. Cities that were rebuilt were mostly along the seacoast or high in the mountains.[64]

Theories for the catastrophe include: (1) climate change resulting in a little ice age, which caused crop failures and famine, leading the lower class to revolt in order to survive; (2) drought that led to social and economic problems for the lower classes; or (3) cultural differences led by the immigration of new ethnic groups and the disproportionate large number of politicians and elite individuals. The lopsided proportion of elites, the complex political structure, and a desire for a more equitable society were the most likely grounds for the destruction, especially because palaces, administrative buildings, and corridor houses were the main targets of destruction. Robert Drews[65] suggests that destruction began when there was a change in warfare strategy that gave a military advantage to the lower classes over established kingdoms. When the lower class recognized that the elite chariot-based forces of the kingdoms could be overwhelmed by infantrymen equipped with javelins, long swords, and simple defensive armor, they rose up to plunder the richest palaces and cities. These so-called barbarians of Northern Greece, Italy, Sicily, and elsewhere throughout Europe and the Near East used guerrilla tactics to outmatch and defeat the chariot armies.[65]

Overall the Late Bronze Age debacle resulted in lower literacy, enslavement, interrupted trade routes, a system collapse leading to new, less bureaucratic societies, and the slow but continuous spread of iron technology due in part to the relatively low cost and abundance of iron in the region, and the disruption of tin imports.[66] The gradual end of the Greek Dark Ages led to revitalization in Europe, the rise of Classical Greece, and the Iron Age.

THE IRON AGE

The Iron Age is the last of the three archeological ages (Stone, Bronze, and Iron). As with the other archeological ages, the Iron Age did not begin simultaneously around the world, and as mentioned previously, some regions of the world went right from the Neolithic to

FIGURE 12.19 Iron Age tools. *From http://1216.virtualclassroom.org/technology/iron_age.html*

the Iron Age. The technology for smelting iron was most likely first introduced through the Caucasus and because different techniques were used to mine and make steel (an alloy of iron and carbon), archeologists can differentiate between steel made in different regions of the world. The earliest evidence for making iron implements comes from Anatolia around 4.2 kya (in the middle of the European Bronze Age). The first signs of iron working in Europe come from Greece, during the Greek Dark Ages 3.2 kya, and lasted until 2 kya. The iron age spread from Greece into Central and northern Italy. Iron technology spread into Southern and Eastern Europe around 3.1 kya, Central and Western Europe around 2.8 kya, and Northern Europe 2.5 kya. Iron weapons, ornaments, pottery, utensils, tools, and specific flowing design patterns that were not previously seen in the Bronze Age characterize the European Iron Age (Fig. 12.19). Iron was abundant in many regions of Europe, and steel weapons and tools were stronger and more durable than bronze. The use of steel followed bronze because the purification of iron to make steel required special furnaces that could reach a temperature of 1500°C and the mining and production of steel required removal of more impurities.[66]

During the Iron Age, the Vikings had an important biological and cultural impact on many parts of Europe through raids, colonization, and trade. The Vikings expanded from Norway, east, south, and west. They moved throughout Europe to the North Atlantic islands, into England and south to Spain, Sweden, and Central Russia. Iceland was the last place to be settled and is still inhabited by Norse-speaking descendants, many of whom are thought to have originated in the British Isles. mtDNA studies in Iceland suggest that British women either voluntarily accompanied or were forced to accompany their invaders as they moved into Iceland.[67]

In 2014, investigators analyzed mtDNA from 80 individual skeletal remains in Norway. Forty-five of the sequences were verified as ancient Norwegians representing the Late Iron Age. The ancient DNA was genetically similar to ancient Icelanders and to present-day Shetland and Orkney Islanders, Norwegians, Swedish, Scottish, English, German, and French.

DNA from the Viking Age population had higher frequencies of K, U, V, and I haplogroups than their modern counterparts, but a lower proportion of T and H haplogroups.[68] Their analyses indicate that Norse women were important in the overseas expansion and settlement of the Vikings, and women from the Orkneys and Western Isles contributed to the colonization of Iceland. Other mitochondrial studies of contemporary populations in Iceland and the Faroes suggest that Norse settlers brought Gaelic women into the region, whereas islands near Scandinavia were settled by Norse men and women. In many cases, the Norse women and children moved with the armies, and Viking migrants contributed to the gene pools of their new homelands, spreading their genes as they moved from one region to another.[69]

Although it is clear the Vikings invaded Britain, genetic studies suggest that they did not mix much with the local populations. Research completed in 2015 looked at the genomes of 2039 people with strong ties to the UK before movements that occurred in the last century. The researchers compared the UK genetic data with genetic profiles from Europe and found the greatest genetic contributions to the UK came from Germany, Belgium, and France, with the Welsh populations showing the strongest link to the first settlers of Britain after the last ice age. Unexpectedly, there were very little genetic contributions from Danish Vikings, suggesting that Viking invasions encouraged pillaging but not rape. Some have speculated that the lack of rape may have been due to the Norse women that often followed their male counterparts into battle and that they may have discouraged the practice.[69]

The most important developments of the European Iron Age were the societal changes. For example, the dead were buried in an extended position, with some graves containing elaborate jewelry, as opposed to cremation practiced in the Bronze Age, the occurrence of different religious practices and beliefs, the development of more permanent homesteads, and even a change in attire. Differences in artistic expression occurred during the Greek Orientalizing period around 2.8 kya. The Greek art of this period was influenced by the art of the Eastern Mediterranean and the Near East because of increased cultural interchanges driven by the Silk Road, among others. The Orientalizing period brought a change from the geometric style of the Bronze Age to a more flowing style with animal and floral motifs. Many of the sculptures, pottery, and engraved gems contained religious or Greek mythological themes.

The Iron Age also marks the beginning of the Greek alphabet, which descended from the Phoenician alphabet. The Greek use of alphabetical characters associated with vowel sounds followed by the development of written language and the production of preserved manuscripts helped to spread written language to many cultures in Europe. This was also the age of the great Greek philosophers (e.g., Socrates, Plato, among others) and Alexander the Great who spread Greek culture to other states.

The utilization of iron for weapons and the subsequent distribution of weapons to the masses set off large-scale wars, including the rise and fall of the Roman Empire that changed the face of Europe. The various conflicts of the period, as with all wars, brought changes in culture and admixture of populations in various regions of Europe. Although there is still very little evidence and an ongoing debate persists on the subject, invasions of Rome by barbaric hordes of Goths, Lombards, and Slavs could have had an influence on the genetics of Southern Europe. Studies by Botique et al.[70] did show evidence for genetic contributions to Southern Europe from North Africa, likely partially due to the Moorish Berber conquest of

Iberia around 2.8 kya and Sicily in 827, as well as previous waves of migrations (Arabs and Berbers first invaded Iberia in 711). They showed that the highest levels of North African ancestry occurred in southwestern Europe and that the North African influence decreased in Northern European populations, with sharp differences between Iberia and France where Basques were less influenced by North Africa.

THE MIDDLE AGES

The Middle Ages began with the fall of the Western Roman Empire and the division in Western Christianity and concluded with the Renaissance, which heralded significant scientific advancement and the beginning of European overseas expansion. The Middle Ages lasted from 500 to 1500 ya and are denoted by the Holy Wars (instigated by the conflict between the Pope and the Emperor), the formation of nation states including France, Spain, Portugal, and England, each developing their own dialects, and the rise of feudal lords and serfdom.

The formation of nation states and larger cities along with improvements in agriculture, increased trade, improved infant care, increased nutrition, and reduced mortality led to increases in population density. The associated high population density lifestyle had certain advantages but also led to environmental pollution and an increase in parasitic insects, animal vectors (e.g., rats and mice), and human-associated microorganisms. This in turn ushered in an increase in the spread of disease.[71] One of the most significant and devastating events during the Middle Ages was the Black Plaque (Black Death) that occurred from 1347 to 1351 and resulted in the deaths of an estimated 100–200 million people (30%–60% of Europe's population). Analysis of the DNA from victims indicates that the malady was caused by a bacterial pathogen, *Yersinia pestis*. This infection is commonly transmitted by fleas carried by rodents and is thought to have originated in Central Asia and then been carried along the Silk Road into Crimea before spreading across to Italy and France by boat. Although there is evidence that the plaque was present during the Bronze Age it did not have the devastating effects seen in the middle ages when the population density had increased.[71,72] There were several penetrations of the epidemic during the Middle Ages as it spread to Sicily in October 1347 and rapidly disseminated within the island appearing in Genoa, Venice, and Marseille during January 1348.

Although Britain had been occupied by Celtic peoples, notably, the Britons and Picts, within the period designated as the Middle Ages and toward the end of the Iron Age, Britain was shaped by a series of immigrations including settlements by the Romans, Scandinavians (Vikings), the Normans, and the people from the North Sea coast known as the Anglo-Saxons. Starting with the exit of Roman support, the Anglo-Saxon period of dominance lasted from roughly 450 C.E. to 1066 C.E., when the French-speaking Normans took control. Normans held England as an elite dominion governing a predominantly Anglo-Saxon underclass until 410 ya, but the Anglo-Saxon invasion most significantly influenced both the linguistics and the genetic structure of Britain. A study by Schiffels et al.[73] examined the influence of immigration into Britain from the late Iron Age to the middle Anglo-Saxon period by examining whole genome sequences from 10 individuals exhumed in England. They compared the ancient DNA samples with DNA from modern-day British and Europeans and showed that on average the eastern English population derived 38% of their ancestry from Anglo-Saxons.

They also determined that the Anglo-Saxon samples were closely related to modern-day Dutch and Danish groups, whereas the Iron Age samples shared ancestors with Northern European populations, including Britain. The Anglo-Saxons are still the dominant population in England and parts of southern Scotland.

During the Middle Ages, the Near East grew more powerful and saw the rise of Islam, and with the emergence of their city-states, a concurrent proliferation of the arts and sciences, particularly medicine, math, astronomy, poetry, architecture, and philosophy. Many scholars today believe that these Islam-driven advances ushered in the Renaissance when they were introduced via Iberia and then transmitted throughout Western Europe. However, during the Middle Ages in Europe, often referred to as Europe's Dark Ages, scientific and artistic development were stifled by the rather young and very controlling Catholic Church. The European Middle Ages were followed by the Renaissance and Early Modern Europe, which began with the rediscovery of the New World and ended with the French Revolution. Toward the end of the 15th Century, the European Renaissance flourished in Italy and quickly dispersed throughout Europe. Borrowing much from Islamic scholars and their preservation of Classical Greek and Roman texts, the Renaissance brought about advancements in technology, secular politics, the economics of capitalism, and an increasing respect for the sciences, the scientific method, and objective thought exemplified by the new ideas of Voltaire. These changes eventually led to a decrease in the power of the Pope and an end to feudalism and serfdom. The end of early modern Europe was the French Revolution, which began because of political conflict between the monarchy and the nobility over tax reform. The revolution replaced the old monarchy government with a more democratic dictatorship. This revolution eventually started a decline in existing European monarchies and the development of republics and democracies (a so-called new modern society).[74]

THE INDUSTRIAL REVOLUTION

The Industrial Revolution followed the early modern period in Europe in the late 18th century. The Industrial Revolution marked the transition to new ways and the mass production of goods by machines. It started in England and Scotland and brought about profound changes in agriculture, manufacturing, mining, and transportation. The effects were significant and brought about long-lasting socioeconomic and cultural changes. Some of the technological advances included the use of coal for the development of large-scale iron manufacturing, the development of the steam engine, gas lighting, glass making, the development of machinist's instruments and metal implements, and the mechanization of textiles. Improvements in agriculture led to increased food production and improvement in nutrition as well as longer life expectancy. The development of roads and canals increased trade. The technology swept throughout Europe and North America and overall allowed for a higher standard of living. As with other technological advances, there were downsides to these advances. Increased environmental pollution, increases in population, labor abuses including child labor and poor working conditions, and loss of jobs due to mechanization and massive urbanization. Small cities grew into large urban centers and the number of people in rural communities declined.[75,76]

The accelerated population growth and the weak purifying selection from medical and other technological advances have led to a host of genetic diseases that are prevalent in

Europe and American Europeans. These include the LAMC1 gene associated with premature ovarian failure, LRP1, which is linked with both Alzheimer's disease and obesity, the CPE gene linked to hardening of the arteries, the C282Y gene linked to hemochromatosis, a common iron overload disorder, and mutations in the CFTR gene associated with cystic fibrosis. In addition, there are a growing number of rare functional genetic variants. Many of these rare variants are associated with complex genetic diseases (caused by more than one gene and influenced by the environment). These rare variants have complicated the search for disease alleles. This has led to a need for genome-wide-association studies that require large sample sizes to identify these rare allele variants.[75]

MODERN-DAY EUROPE

Genomic and archeological analysis has revealed at least 3 major ancient migrations that were influenced by climatic changes, disease, limited resources, and technological advances leading to modern day Europeans. Although there is a lot of information from Europes past Europe continues to be a focus of archeological excavations and has become a center of modern-day genetic analysis to further unravel the mysteries behind human evolution in the region. Contemporary studies are enhanced by the ability to sequence entire ancient and contemporary genomes and the ability to analyze genome-wide data and begin to look for subtle genetic differences within and among populations. This information is of not only anthropological, linguistic, and historical value but could also reveal various alleles that may have been involved in selection, adaptation, and disease. Modern genetic analysis has uncovered some of the Darwinian selection that was taking place during European evolution. Genetic analysis has disclosed genes selected for immunity (HIV resistance, pathogen recognition genes), morphology (hair, eye, skin color), and ability to metabolize lactose.[75]

Although at first it may seem, from historical and genetic data, that the genetic relationships in modern-day Europe are a hodgepodge admixture, studies by Veeramah and Novembre[77] and others[78] have demonstrated that the genetic relationships of Europeans are highly correlated with their geography, so that a summary of genetic data looks similar to a regional map of Europe. This is true even when considering that Europeans have shared hundreds of common ancestors over 5 ky and that there is relatively very little variation among modern European subgroups. The only major outliers in these genetic studies are the Sardinians and the Basques, which represent genetic isolates and are distinct from the general European pattern that may reflect older barriers to gene flow and genetic diversity dating back to the upper Paleolithic.[71]

Studies of European evolution have relevance today for forensic analysis, questions of ancestry, and genome-wide association studies, which compare genomes of diseased and healthy groups with the objective of finding genes associated with complex diseases.

THE BALTIC SEA REGION

Scandinavia was one of the last geographic areas in Europe to become habitable after the LGM. While the events that shaped Central and Southern Europe, have become increasingly clear the genetic evidence from Northern Europe surrounding the Baltic Sea and the genetics of the postglacial migrants of the North are just beginning to be revealed. Gunther et al., [79] sequenced the genomes of seven hunter-gatherers excavated across Scandinavia and dated from 9.5–6.0 kya. Their genetic data indicated an east–west genetic gradient that opposed the pattern seen in other parts of Mesolithic Europe. Their results suggested there were two different migrations into Scandinavia following the LGM. The first migration from the south, and a later migration from the northeast. The two groups met and admixed in Scandinavia, creating a population, adapted to the high latitude environments. These Mesolithic Scandinavians showed adaptations which included low amounts of pigmentation, a gene region associated with physical performance that is persistent in modern-day northern Europeans, and the TMEM131 gene that has been speculated to be involved in long-term cold adaptation.

In another recent study, Mittkin et al.,[80] isolated DNA from 38 North Europeans ranging from 9.5–2.2 kya. Their genetic evidence supports the idea that hunter-gatherers settled Scandinavia by two routes. They concluded that the Mesolithic Scandinavian hunter-gatherers are a genetic admixture of Western and Eastern hunter-gatherers along with substantial ancestry from ancient Northern Eurasians. In contrast, the two DNA samples from the Eastern Baltic Mesolithic Kunda population, that predate the Scandinavian hunter-gatherers, carried relatively low amounts of North Eurasian ancestry, indicating that North Eurasian influence was never widespread to the southeast of the Baltic Sea. In addition, they conclude that the Mesolithic Western hunter-gatherers extended to the east of the Baltic Sea, without geneflow from Central European farmers during the Early and Middle Neolithic. Further, their genetic analysis determined that following the early Neolithic period (about 4 kya) agriculture reached southern Scandinavia, followed by agriculture in Denmark and in western central Sweden. During the Late Neolithic shifts in economy toward marine resources began the spread of a new ancestry associated with the Corded Ware Complex in Northern Europe. The Corded Ware Complex has been shown to be associated with the pastoralists of the Yamnaya Culture which as previously discussed introduced a new genetic component into Europe that is seen in today's populations in a decreasing northeast to southwest gradient.

CONCLUSION

The prehistory of Europe and the evolution of humankind in the region is extremely complex, not only because of Europe's size but also because of the effects of climate change, migration, extinction of species, introgression, relocation and resettlement, and invasions. To help clarify the changes that have taken place over time, archeologists and geologists have divided the history of Europe and human evolution into various periods, with the earliest being the Paleolithic, Mesolithic, and Neolithic epochs commonly referred to as the Stone Age. During the Paleolithic, Europe was populated by *H. erectus, H. heildelbergensis,* and *H. neanderthalensis* and then later settled by *H. sapiens*. The earliest of these *H. sapiens* were the Cro-Magnon who were characterized by strong musculature, a broad face, and tall stature. These individuals interbred, to some degree, with Neanderthals in Europe and the Near East. However, during the LGM, Northern Europe was covered in ice, and most *H. sapiens* left Northern Europe and migrated south into a number of refugia, mainly in coastal Iberia, Catalonia, Aragon, and Southwest France then returned about 15 kya when the glaciers had receded. Southern

Europe has been shown to be more genetically diverse than Northern Europe, some of which may be explained by the LGM and retreat of *H. sapiens* from the north to the southern regions of Europe. The migration associated with the agricultural revolution also affected North and South Europe both genetically, socially, and politically through North African influence and the migration of farmers from the Near East.

The Cro-Magnons greatly contributed to the early period of European identity. Fossil remains and extensive archeological discoveries have revealed a large variety of innovations belonging to the early Europeans, including hunting tools and weapons, sophisticated art, ornamentation, and an awareness of time. Clues about their genetics, behaviors, cultural characteristics, rituals, and beliefs have also been uncovered. The Neolithic period marked the last of the Stone Age periods and began in southeastern Europe about 9 kya and reached Northern Europe by about 5 kya. The Neolithic was followed by the Bronze Age, which began around 5 kya in Greece. The Bronze Age is characterized by early forms of writing and the use of bronze in the development of art, tools, and weapons. Following the European Bronze Age, the Iron Age began in Southern Europe and spread to Northern Europe 500 years later. During the Iron Age, literacy was evident in Southern Europe along the Mediterranean, and various religious beliefs, artistic styles and differing agricultural practices began to emerge, however, Northern Europe, remained more primitive until around 1.5 kya. The Iron Age was the last of the three archeological epochs and was followed by the Middle Ages, the Renaissance, and the Industrial Revolution. These last three periods experienced dramatic advances in agriculture, increased trade, improved infant care, increased nutrition, and reduced mortality, providing for increases in population density. This lifestyle had certain advantages but also led to environmental pollution, and an increase in insects, microorganisms, and disease. Over the last 5 ky population increases led to increased numbers of random mutations some of which led to genetic disease. Disease alleles for cystic fibrosis, early-onset obesity, premature ovarian failure, phenyl ketone urea, and hardening of the arteries are all genetic diseases that occur in higher frequency in Europeans than in Africans and African Americans. Most are rare mutations but of growing concern.[75] Today the studies of European history and prehistory are still providing insightful information. Europe continues to be a focus of archeological excavations and has become a center of modern-day genetic analysis trying to unravel the mysteries behind human evolution in the region. Contemporary studies are not only of anthropological and historical value but can also reveal various alleles that may have been involved in selection, adaptation and disease.

From a genetic perspective, the recent changes in European social policies (BREXIT and Nationalism) could bring significant changes to European populations by isolating mechanisms, and a decrease in admixture seen throughout Europe over the ages. It will be interesting to see how social and political policies change the genetic makeup of Europe over the next millennium.

References

1. Thatcher M. *Statecraft: strategies for a changing world.* NY: Harper Collins New York; 2003.
2. Price DT. *Europe before Rome.* New York, NY: Oxford University Press; 2013.
3. Henke W, et al. *Handbook of paleoanthropology.* Berlin: Springer-Verlag; 2007.
4. Harmand S, et al. 3.3-million-year-old stone tools from Lomekwi 3, west Turkana, Kenya. *Nature* 2015;**521**:310–5.

5. Champion T, et al. *Prehistoric Europe*. London: Academic Press Inc.; 1984.
6. Semaw S. The world's oldest stone artefacts from Gona, Ethiopia: their implications for understanding stone technology and patterns of human evolution between 2.6-1.5 million years ago. *J Archaeol Sci* 2000;**27**(12):1197–214.
7. McClellan. *Science and technology in world history: an introduction*. Baltimore, MD: JHU Press; 2016. p. 6–12.
8. Marlowe FW. Hunter-gatherers and human evolution. *Evol Anthropol* 2005;**14**(2):15294.
9. Seddon C. *Humans: from the beginning*. San Bernardino, CA: Glanville Publications; 2015.
10. Rasmussen M, et al. An Aboriginal Australian genome reveals separate human dispersals into Asia. *Science* 2011;**334**:94–8.
11. Curnoe D, et al. Human remains from the Pleistocene-Holocene transition of Southwest China suggest a complex evolutionary history for East Asians. *PLoS One* 2012;**7**:3.
12. Benazzi S, et al. Early dispersal of modern humans in Europe and implications for Neanderthal behavior. *Nature* 2011;**479**:525–8.
13. Zilhao J, et al. Analysis of Site Formation and Assemblage Integrity Does Not Support Attribution of the Uluzzian to Modern Humans at Grotta del Cavallo. *PLoS One* 2015:0131181.
14. Higham T, et al. The timing and spatiotemporal patterning of Neanderthal disappearance. *Nature* 2014;**512**:306–9.
15. Longo L, et al. Did Neanderthals and anatomically modern humans coexist in northern Italy during the late MIS 3? *Quat Int* 2012;**259**:102–12.
16. Fu Q, et al. The genetic history of ice age Europe. *Nature* 2016;**534**:200–5.
17. Fagan B. *Cro-magnon: how the ice age gave birth to the first modern humans*. New York, NY: Bloomsbury Press; 2010.
18. Armesto FF. *Ideas that changed the world*. New York, NY: Dorling Kindersley; 2003. p. 400.
19. Fagan BM. *The oxford companion to archaeology*. Oxford, UK: Oxford University Press; 1996. p. 864.
20. Evolution: humans: origins of humankind. Pbs.org.
21. Mellars P. Palaeoanthropology: the earliest modern humans in Europe. *Nature* 2011;**479**:483–5.
22. Higham T, et al. The earliest evidence for anatomically modern humans in northwestern Europe. *Nature* 2011;**479**:521–4.
23. Fu Q, et al. An early modern human from Romania with a recent Neanderthal ancestor. *Nature* 2015;**524**:216–9.
24. Callaway E. Europe's first humans: what scientists do and don't know. *Nature* 2015. https://doi.org/10.1038/nature.2015.17815.
25. Campbell BG, et al. *Humankind emerging*. 9th ed. Boston, MA: Pearson; 2006. p. 370.
26. Velemínskáa J, et al. Variability of the Upper Palaeolithic skulls from Předmostí near Přerov: craniometric comparison with recent human standards. *HOMO* 2008;**59**(1):1–26.
27. Linares O. African rice: history and future potential. *Proc Natl Acad Sci USA* 2002;**99**(25):16360–5.
28. Campbell BG, et al. *Humankind emerging*. 9th ed. Boston, MA: Pearson; 2006. p. 367–80.
29. Pearce E, et al. New insights into the differences in brain organization between Neanderthals and anatomically modern humans. *Proc R Soc B* 2013;**280**:1758. https://doi.org/10.1098/rspb.2013.0168.
30. Tattersall I. *Becoming human: evolution and human uniqueness*. New York, NY: Harcourt Brace; 1998.
31. Kuhn SL, Stiner MC. The division of labor among Neanderthals and modern humans in Eurasia. *Curr Anthropol* 2006;**47**:953–81.
32. Formicola V, et al. The Upper Paleolithic triple burial in Dolní Věstonice: pathology and funerary behavior. *Am J Phys Anthrol* 2001;**115**:372–4.
33. Mellars P. Archeology and the dispersal of modern humans in europe: deconstructing the Aurignacian. *Evol Anthropol* 2006:167–82.
34. Evans V. *The crucible of language: how language and mind create meaning*. Cambridge: Cambridge University Press; 2015.
35. Yong E. Americas' natives have European roots. *Nature* 2013. https://doi.org/10.1038/nature.2013.14213.
36. Bradley B, Stanford D. the North Atlantic ice-edge corridor: a possible paleolithic route to the new world. *World Archeol* 2004;**36**:459–78.
37. Pedersen MW, et al. Postglacial viability and colonization in North America's ice-free corridor. *Nature* 2016. https://doi.org/10.1038/nature19085.
38. Trinkaus E. An early modern human from the Peştera cu Oase, Romania. *Proc Natl Acad Sci USA* 2003;**100**:11231–6.
39. Vercellotti G, et al. The Late Upper Paleolithic skeleton Villabruna 1 (Italy): a source of data on biology and behavior of a 14,000 year-old hunter. *J Anthropol Sci* 2008;**86**:143–63.
40. Posth C, et al. Pleistocene mitochondrial genomes suggest a single major dispersal of non-Africans and late glacial population turnover in Europe. *Curr Biol* 2016;**26**:1–7.

41. Krause J, et al. A complete mtDNA genome of an early modern human from Kostenki, Russia. *Curr Biol* 2010;**20**:231–6.
42. Raghavan M, et al. Upper Palaeolithic Siberian genome reveals dual ancestry of native Americans. *Nature* 2014;**505**:87–94.
43. Seguin-Orlando A, et al. Genomic structure in Europeans dating back at least 36,000 years. *Science* 2014;**28**:1113–8.
44. Torroni A, et al. Harvesting the fruit of the mtDNA tree. *Trends Genet* 2006;**22**:339–45.
45. Richards M, et al. Tracing European founder lineages in the Near Eastern mtDNA pool. *Am J Hum Genet* 2000;**67**:1251–76.
46. Mathieson I. Eight thousand years of natural selection in Europe. *Cold Spring Harbor* 2015. https://doi.org/10.1101/016477.
47. Lizaridis I. Genomic insights into the origin of farming in the ancient Near East. *Nature* 2016;**536**:419–24.
48. Fazeli R. Stone tool production, distribution and use during the late Neolithic and Chalcolithic on the Tehran plain. *Iran* 2002;**40**:1–14.
49. Radivojevi M, et al. Origins of extractive metallurgy: new evidence from Europe. *J Archaeol Sci* 2010;**37**:2775–87.
50. Mai-Duc C. Here's what archeologists found in a warrior grave that's been untouched for 3500 years. *Los Angeles Times* October 28, 2015.
51. Martinez L, et al. Paleolithic Y-haplogroup heritage predominates in a Cretan highland plateau. *Eur J Hum Genet* 2007;**15**:485–93.
52. Baryshnikov G, Tsoukala E. New analysis of the Pleistocene carnivores from Petralona cave (Macedonia, Greece) based on the collection of the Thessaloniki, Aristotle University. *Geobios* 2010;**43**:389–402.
53. Stiner MC, Munro ND. On the evolution of diet and landscape during the upper paleolithic through Mesolithic at Franchthi cave (Peloponnese, Greece). *J Hum Evol* 2011;**60**:618–36.
54. Sikora M, et al. Population genomic analysis of ancient and modern genomes yields new insights into the genetic ancestry of the Tyrolean Iceman and the genetic structure of Europe. *PLoS Genet* 2014;**10**(5):e1004353.
55. Allentoft ME. Population genomics of bronze age Eurasia. *Nature* 2015;**522**:167–72.
56. Haak W, et al. Massive migration from the steppe was a source for Indo-European languages in Europe. *Nature* 2015;**522**:207–11.
57. Herrera KJ, et al. Neolithic patrilineal signals indicate that the Armenian plateau was repopulated by agriculturalists. *Eur J Hum Genet* 2011;**20**:313–20.
58. Lazaridis L. Ancient human genomes suggest three ancestral populations for present-day Europeans. *Nature* 2014;**513**:409–13.
59. Jones ER, et al. Upper Palaeolithic genomes reveal deep roots of modern Eurasians. *Nat Commun* 2015;**6**:8912. https://doi.org/10.1038/ncomms9912.
60. Sjogren K-G, et al. Diet and mobility in the Corded Ware of Central Europe. *PLoS One* 2016;**11**:e0155083. https://doi.org/10.1371/journal.pone.0155083.
61. Adkins R, et al. *The handbook of British archeology*. London: Constable; 2008. p. 64.
62. Glass N. *History made: in an astonishing Bronze Age discovery a 3,000-year-old community has been unearthed*. 2016. CNN.com.
63. Forsen J. *The twilight of the Helladics: a study of the disturbances in East-Central and Southern Greece towards the end of the early Bronze Age*. Partille, Sweden: Paul Astroms Forlag; 1992.
64. MacSweeney N. Social complexity and population: a study in the early bronze age Aegean. *Papers Inst Archaeol* 2004;**15**:52–65.
65. Drews R. *The end of the Bronze Age: changes in warfare and the catastrophe Ca. 1200BC*. Princeton University Press; 1995. p. 19.
66. Stoia A. In: Stig Sørensen ML, Thomas R, editors. *The Bronze Age: Iron Age transition in Europe*. Oxford University Press; 1989.
67. Capelli C, et al. A Y chromosome census of the British Isles. *Curr Biol* 2003;**13**:979–84.
68. Krzewinska M, et al. Mitochondrial DNA variation in the Viking age population of Norway. *Philos Trans R Soc B* 2014;**370**. https://doi.org/10.1098/rstb.2013.0384.
69. Helgason A, et al. mtDNA and the islands of the North Atlantic: estimating the proportions of Norse and Gaelic ancestry. *Am J Hum Genet* 2001;**68**:723–37.
70. Botique LR, et al. Gene flow from North Africa contributes to differential human genetic diversity in southern Europe. *Proc Natl Acad Sci USA* 2013;**110**:11791–6.

71. Valtuena AA, et al. The Stone Age plaque and its persistence in Eurasia. *Curr Biol* 2017;**27**:3683–1.
72. Bos KI, et al. A draft genome of *Yersinia pestis* from victims of the Black Death. *Nature* 2011;**478**:506–10.
73. Schiffels S, et al. Iron age and Anglo-Saxon genomes from East England reveal British migration history. *Nat Commun* 2015. https://doi.org/10.1038/ncomms10408.
74. Livesay J. *Making democracy in the French revolution*. Cambridge, MA: Harvard University Press; 2001. p. 19.
75. Tennessen JA, et al. Evolution and functional impact of rare coding variation from deep sequencing of human exomes. *Science* 2012;**337**:64–9.
76. Feinstein C. Pessimism perpetuated: real wages and the standard of living in Britain during and after the industrial revolution. *J Econ Hist* 1998;**58**:625–58.
77. Veeramah KR, Novembre J. Demographic events and evolutionary forces shaping European genetic diversity. *Cold Spring Harbor Perspect Biol* 2014. https://doi.org/10.1101/cshperspect.a008516.
78. Stephan L, et al. The fine-scale genetic structure of the British population. *Nature* 2015;**519**:309–14.
79. Gunther T, et al. Population genomics of Mesolithic Scandinavia: Investigating early postglacial migration routes and high-latitude adaptation. *PLoS Bio* 2018;**16**:e2003703.
80. Mittnik A, et al. The genetic prehistory of the Baltic Sea region. *Nature Comm* 2018;**9**:442. doi:10.1038/s41467-018-02825-9.

CHAPTER

13

The Agricultural Revolutions

Agriculture... is our wisest pursuit, because it will in the end contribute most to real wealth, good morals & happiness. ***Thomas Jefferson, Paris Aug. 14, 1787.***

SUMMARY

The agricultural revolution is the name given to a number of cultural transformations that initially allowed humans to change from a hunting and gathering subsistence to one of agriculture and animal domestications. Today, more than 80% of human worldwide diet is produced from less than a dozen crop species many of which were domesticated many years ago. Scientists study ancient remains, bone artifacts, and DNA to explore the past and present impact of plant and animal domestication and to make sense of the motivations behind early cultivation techniques. Archeological evidence illustrates that starting in the Holocene epoch approximately 12 thousand years ago (kya), the domestication of plants and animals developed in separate global locations most likely triggered by climate change and local population increases. This transition from hunting and gathering to agriculture occurred very slowly as humans selected crops for cultivation, animals for domestication, then continued to select plants and animals for desirable traits. The development of agriculture marks a major turning point in human history and evolution. In several independent domestication centers, cultivation of plants and animals flourished according to the particular environmental conditions of the region, whereas human migration and trade propelled the global spread of agriculture. This change in subsistence provided surplus plant food that accumulated during the summer and fall for storage and winter consumption, as well as domesticated animals that could be used for meat and dairy products throughout the year. Because these new survival strategies no longer required relocation and migration in search of food, humans were able to establish homesteads, towns, and communities, which, in turn, caused rapid increases in population densities and lead to the emergence of civilizations. This dependence on plant and animal domestication entailed a number of other environmental adaptations including deforestation, irrigation, and the allocation of land for specific crop cultivation. It also triggered various other innovations including new tool technologies, commerce, architecture, an intensified division of labor, defined socioeconomic roles, property ownership, and tiered political systems. This shift in subsistence mode provided a relatively safer existence and in

Ancestral DNA, Human Origins, and Migrations
https://doi.org/10.1016/B978-0-12-804124-6.00013-6

general more leisure time for analytical and creative pursuits resulting in complex language development, and the accelerated evolution of art, religion, and science. However, increases in population density also correlated with the increased prevalence of diseases, interpersonal conflicts, and extreme social stratification. The rise of agriculture and the influence of genetics and culture (gene–culture coevolution) continue to affect modern humans through alterations in nutrition, predisposition to obesity, and exposure to new diseases. This chapter will cover the various regions that adopted early agricultural practices and look at the long-term positive and negative effects of agriculture on society.

HUNTING AND GATHERING

When anatomically modern humans left Africa around 45–50 kya, gathering food or foraging (the collecting of wild plants, fishing, and hunting) dominated their culture and lasted until the end of the last glacial period (LGP), approximately 12 kya.[1,2] Members of a hunter-gatherer society benefited from active lifestyles and a variety of foods containing a wide range of proteins, fats, and carbohydrates, derived from various meat and plant products. These individuals also shared resources and tool technologies associated with foraging, hunting, and food processing, practices unique to the hominins that underlie the cultural evolution of the genus *Homo*.[3]

However, these same individuals suffered from the inability to store food, the dangers of hunting, and seasonal changes that result in the need to adapt to new food sources. The hunter-gatherer communities were motivated to developed agriculture between 4.5 and 12 kya in several independent regions[3] to increase the amount and reliability of certain wild species, while consequently, reducing the risk and uncertainty associated with foraging.[4] For the majority of human societies the transition from foraging to agriculture was not rapid but gradual, especially among groups where hunting and gathering had previously been successful. Most hunter-gatherers initially adopted mixed economies that added certain crops or livestock to their hunting-gathering lifestyle.[4]

The transition to agriculture may have been influenced by several key factors[4]: (1) Hunting and gathering became less economically rewarding as resources (wild plant species, animal prey) became less abundant. (2) Climate changes during the LGP led to migration into new environments with unfamiliar resources. (3) Foragers developed the technologies of collecting seeds, food processing, and storage. (4) Accumulation of material resources (tools, weapons, etc.), which reduced mobility, favored child rearing and encouraged settlement instead of a nomadic lifestyle. These advancements were vital for the development of food production and enabled an easier transition to farming.

Although, agriculture is the dominant worldwide mode of subsistence, a few communities have retained a foraging-based economy. These groups mitigate the risk of uncertainty by storing food in preparation for hard times and creating kinships with nearby foraging groups.[4] An example of a modern-day foraging society is the San people of the Kalahari who occupy regions of South Africa, Botswana, and Namibia.[5] Within these difficult terrains, the San people have survived by hunting wild game and gathering roots and tubers. However, recently this group's economic existence has been threatened by reduction of territory due, in part, to increased farming encroachment.[5]

THEORIES ON THE ORIGINS OF AGRICULTURE

Hunting and gathering was the primary subsistence strategy for more than 95% of our existence as a species.[1] What motivated humans to abandon this apparently successful strategy in favor of cultivated food production? Several factors have been proposed both natural (climate change, scarcity of wild species and the attempt to protect them, population pressure from increased density) and/or cultural (lesser mobility of farmers favoring child rearing, deliberate displays of power through food accumulation, accumulation of other resources, property ownership, and eventually aggression to neighboring groups). Regardless of the contributing factors the process was gradual, and initially the two ways of subsistence were maintained (and still are, in some isolated human groups) as a dual strategy.

Domestication is defined as a selection process for adaptation to human agroecological niches and, at some point in the process, human preferences. The criteria for identifying domestication differ significantly for plants and animals. Plants can quickly show distinct morphological changes as a result of selection, whereas animals are much slower in presenting such developments mostly because of increased reproductive time. The archeological evidence for the domestication of plants comes from changes in plant morphology, seed shape, seed size, pollen structure, and tools used to process plants. In most cases, the evidence for plant domestication is clear, and the domesticated species clearly differ from the wild relatives. Evidence for animal domestication is much more difficult to establish because many of the changes during domestication were behavioral and not morphological. However, after long periods of selective breeding, animals can show skeletal changes related to domestication.[6]

Scholars have hypothesized and debated the motives behind the development of agrarian subsistence. Two major theories invoke climate change and increased population density as the most likely causes. The effects of climate change on cultural development have been the subjects of debate for many years; however, the most accepted explanation of the agricultural transition is that a more stabilized climate facilitated the origins of farming.[7–9] In addition, climate change and human hunting may have led to the extinction of megafauna that were used as a food source. Recent studies have shown that in South America, megafaunal extinctions coincided with human presence and climate warming, around 12 kya.[10]

The theories of early agricultural practices and the origins of agriculture have changed dramatically over the years as the number of archeologists and archeobotanists have increased. The increased number of investigators has led to an increase in the number of excavations and subsequently the number of proposed centers for the origins of agriculture. In 1971, 3 centers of origin were identified and by 2010, the number had risen to at least 24 centers of origin. As these numbers increased, debate over chronology and the identity of the wild species that were domesticated increased. One area of agreement is that although domestication of plants and animals caused a fundamental change in the way humans lived, it was a very slow process. In contrast to early investigations and experiments that suggested that domestication of some species could be done in 20–100 years, recent archeological evidence for rice domestication in China is estimated to have taken at least 3000 years to complete.[11] As more archeological, botanical, and genetic evidence accumulates, it appears that other crops would have also undergone a gradual change and may have taken more than 100 generations to domesticate. The process was slow because wild species do not always have the desirable traits needed for domestication. Traits such as plant size, seed size, seed dispersal, yield per plant, ease of harvesting, nutrition, taste, etc., all needed to be selected over time. Many of these desirable traits came about because of random mutations, and mutants were then used in subsequent plantings. In addition, cultivation practices (land clearance, soil preparation, land maintenance, irrigation) would all have taken experimentation along with the development of baskets, the mortar and pestle, and various agricultural tools. As selection, cultivation practices, and tools evolved, agriculture gained momentum and gradually replaced the nomadic hunting and gathering existence with sedentary food production and a more sedentary lifestyle.[4,11]

The same slow process of domestication would have also occurred with animals, and in the case of Africa, southern Arabia, and regions of India, animal domestication may have preceded the domestication of plants. Wild animal species would have been slow to adapt to domestication, and humans would have needed to select for desirable traits such as optimal size, rate of growth, age at death, reduced aggression, horn size, ability to manage, etc. In addition, initial attempts to domesticate animals may have been futile or extremely difficult, and because the difference in nutritional value between domesticated and wild species was minimal, humans may have initially preferred hunting over domestication.

Although the transition from hunting and gathering to the domestication of plants and animals is referred to as a revolution, the evidence now seems clear that the transition did not occur rapidly and did not resulted in immediate changes in economy, politics, increased population size, etc. In addition, these ultimate changes did not occur at the same rate or time in different locations and among different cultures.

THE BEGINNINGS OF AGRICULTURE

Most researchers agree that the agricultural revolution developed independently in at least 11 different places around the world, with two major periods of importance: (1) the transition to the Holocene (the time since the last ice age) 12–9 kya and (2) the middle Holocene, between 7 and 4 kya. Around 12 kya, the Paleolithic Ice Age ended and ushered in the Holocene period (the most recent period that continues today).[12] During the early Ice Age, hunter-gatherers may have attempted to grow plants, but their efforts failed because of extreme and rapid climate fluctuations.[2] As the Holocene period began, temperatures increased, glaciers melted, sea levels rose, and ecosystems were rearranged.[4] The climate became warmer, wetter, and more reliable. The Younger Dryas (an abrupt cooling of the Northern Hemisphere between 12.9 and 11.7 kya) interrupted this transition period between the Ice Age and the Holocene.[13] Many scientists hypothesize that the end of the Younger Dryas and the onset of the Holocene sparked the development of agrarianism.[14] This theory asserts that the Holocene's gradually warmer temperatures and carbon dioxide–rich environment allowed farmers to successfully grow desirable crop plants.

Several other ideas have also been proposed to explain the development of agricultural practices. One widespread hypothesis explaining the development of agriculture is that increased population densities and limited food resources caused communities to explore other means of food production.[14] An alternative theory suggests that hunter-gatherer societies aimed to protect specific plants from overharvesting, and as a result, began to domesticate them.[15]

Climate change was most likely the major catalyst for agriculture's inception. Moderate temperatures and climatic stability created conditions that facilitated the onset of farming. A reliable source of food production sustained an increase in population and the development of homesteads and cities. However, it is still debated as to whether an increase in population led to a need for agriculture or if agriculture produced larger populations. As discussed in more detail below, agriculture not only enabled an improved standard of living, the development of cities and governments but also reshaped human society eventually leading to problems with nutrition and infectious disease.

GLOBAL REVOLUTIONS

Evidence from various archeological sites and radiocarbon dating, used to estimate the age of early-domesticated plants and animals,[4] indicate that the Middle East region in particular, served as a major epicenter from which the agricultural revolution spread in all directions. In most instances, it appears that the transition from nomadic hunter-gatherers to sedentary farming societies initially resulted from the selective cultivation of plants and then turned to the selective breeding of animals.[3] Most likely, hunter-gatherers who lived close to some of the first farmers learned about the specialized food production, and as farming societies occupied foraging territories, agriculturalists imposed their new techniques onto the nomadic tribes.[7]

There is still disagreement as to what extent the practice of agriculture and domestication resulted from acculturation or migration of people carrying the new technologies of farming. Humans migrated not only with their families but also with samples of the plants and animals that allowed them to survive in new locations. The transport of crops and animals provides useful evidence for how and where humans migrated and the affinities among human groups. For example, archeological and genetic evidence from domesticated dogs, pigs, and chickens have provided data tracing the movement of humans from East Asia into the Pacific and the Americas.[16] Research into the relative contributions of cultural diffusion and human migration (gene flow) to the spread of agriculture differ from study to study as well as location of genesis and direction of spread. Today, most scientists acknowledge some degree of migration, originating in the Near East and subsequently introducing agriculture in Europe and Asia, which led to the establishment of homesteads and eventually cities (Fig. 13.1). The spread of agriculture may have been purely cultural or demographic in which agriculturists displaced the hunter-gatherer communities.

FERTILE CRESCENT

One of the epicenters of agricultural emergence was the Fertile Crescent, established during the Neolithic Revolution. The Neolithic Revolution, a term popularized in 1932 by Australian archaeologist Vere Gordon Childe, refers to a series of agricultural revolutions that occurred in the Middle East where hunter-gatherers began settling in permanent farming communities.[14] The period, encompassing the Neolithic Revolution, existed from around the beginning of the Holocene era, approximately 12 kya, to the beginning of the Chalcolithic era approximately 6 kya. The Fertile Crescent acquired its name from James Henry Breasted

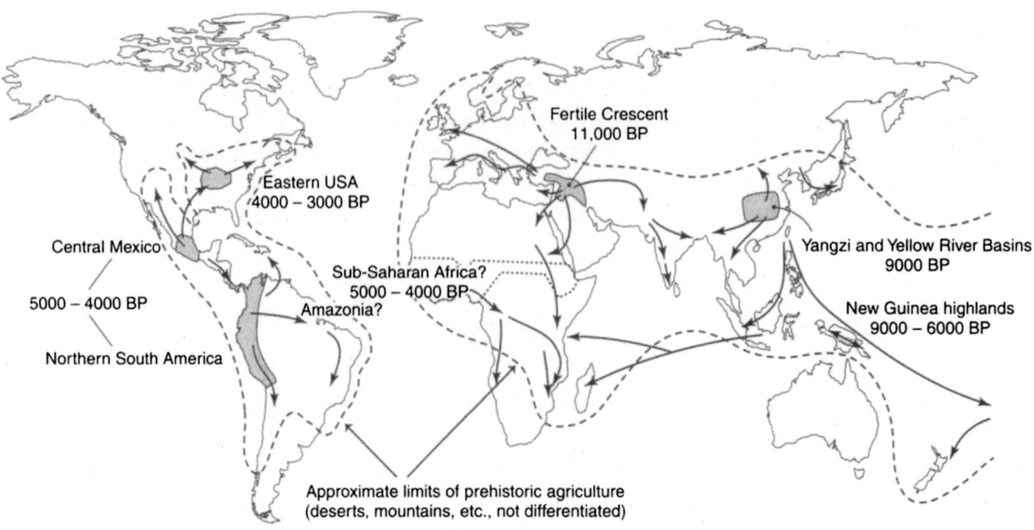

FIGURE 13.1 The global spread of agriculture. The map shows key regions of agriculture origins and the spread of agriculture. As agriculture spread many of the crops, animals and agricultural technology followed. *From Diamond J, Bellwood P. Farmers and their languages: The First Expansions. Science 2003;* ***300****:597–603. http://science.sciencemag.org/content/300/5619/597.full.*

who described the quarter moon shape it forms over the map in the Middle East (Fig. 13.2). This half-moon region is surrounded by the Mediterranean Sea to the Northwest, deserts and grasslands to the south, and forests to the north and east.[7] Modern-day descriptions of the region include Mesopotamia, the Levant, the Eastern Coast of the Mediterranean, and the land around the Tigris and Euphrates rivers. The Fertile Crescent, nicknamed the "cradle of civilization," is known as the birthplace of urbanization, writing, trade, science, glass, the wheel, organized religion, and agriculture. It was ideal for the beginning of agriculture because of its climatic stability and adequate rain.[17] Preagricultural societies in this area survived by foraging wild plants and animals, which were abundant because of the Tigris, Nile, and Euphrates rivers that infiltrate the landscape. These rivers create a flood plain resulting in a specialized alluvial soil, which allows plants to grow without complex irrigation.[12]

Investigations studying the Neolithic Revolution primarily concentrate on the emergence of farming in present-day areas of Iraq, Israel, Egypt, Jordan, Syria, and Lebanon and provide crucial evidence about the gradual evolution from foraging to farming, the significant changes affecting communities, and the transition from a nomadic to a more nonmigratory lifestyle.

Some of the earliest Neolithic cultures are traced to the southern region of the Fertile Crescent and descended from Natufian communities. The Natufians were sedentary hunter-gatherers who lived between 10.2 and 12.5 kya.[4] Notably, Natufians resided in subterranean houses composed of stone and wood (Fig. 13.3). Investigators credit the Natufians as pioneering the development of agriculture in the Fertile Crescent most likely in response to increasing population densities and climatic changes. Natufians developed some primitive agricultural techniques and tools, including the sickle for harvesting, picks for digging, elaborately decorated mortars to grind grain, leather bags, and round buildings for storage. Their cultivation

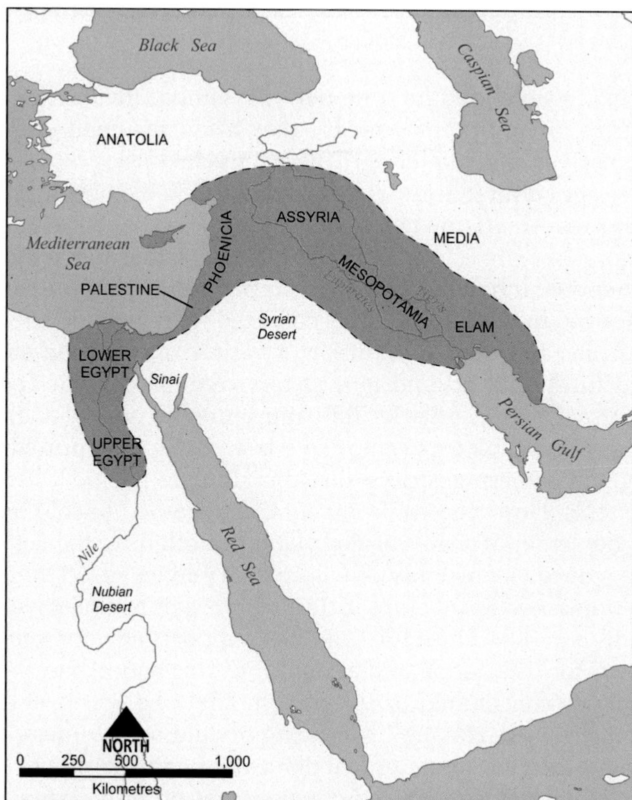

FIGURE 13.2 Map of the Fertile Crescent. The map shows the crescent shape region that is often referred to as the "cradle of civilization." Modern-day descriptions of the region include Mesopotamia, the Levant, the Eastern Coast of the Mediterranean, and the land around the Tigris and Euphrates rivers. *From https://www.ancient.eu/image/169/.*

FIGURE 13.3 Natufian houses. Natufians were sedentary hunter-gatherers who resided in subterranean houses composed of stone and wood. Anthropologists credit the Natufian's as pioneering the development of agriculture in the Fertile Crescent. *From https://s-media-cache-ak0.pinimg.com/736x/71/f1/c4/71f1c45d8e2a7a26df946ed3dc90eb84.jpg.*

of grains allowed them to store food for long periods without the threat of decay associated with other food sources. In addition to cereals, they also consumed goats, fox, rabbit, and birds, sustaining themselves on a hunter-agricultural model.[17,18]

Two of the earliest agricultural areas discovered by archeologists in the Fertile Crescent are Netiv Hagdud and Tell Abu Hureyra. Netiv Hagdud was abandoned around 9.5 kya after 300 years of occupation. This site is located in modern-day Israel and is thought to have supported only several hundred inhabitants.[19] Scientists have unearthed and identified plant and animal remains indicating that the inhabitants collected more than 50 species of wild plants including legumes, fruits, and nuts. In addition, a variety of tools and storage units (made from mud brick structures built of limestone slabs) were discovered. The storage facilities most likely stored extra grains. Similarly, Tell Abu Hureyra, an ancient settlement located in Northern Syria, is believed by experts to have been occupied approximately 6–13 kya.[20] Originally a hunting and gathering society, the Abu Hureyra people lived in pit houses and manufactured stone tools. However, environmental changes led to cold and arid conditions that resulted in the depletion of many of the plants collected by the settlers. Investigators hypothesize that the reduced availability of wild plants may have led the Hureyra people to experiment with the domestication of crops including rye, lentils, einkorn wheat, and barley.[20]

The Netiv Hagdud and Tell Abu Hureyra sites support the contention that the Fertile Crescent was a nucleus for civilization and agriculture. This region had several environmental advantages that likely contributed to cultivation. First, the Fertile Crescent was located within a zone of Mediterranean climate[21] consisting of mild, wet winters, as well as hot, dry summers.[22] This climate selected for perennial plants that had adapted to the long, dry summer and then resumed normal growth and development once the rains commenced in the winter. Many of the plants found in the Fertile Crescent, including the cereals, were annuals. (Annuals complete a life cycle in 1 year from the early onset of germination to the final production of the seed.) Annuals were extremely advantageous because the plants germinated in the wet winter months and then could be harvested in the dry season. The quick growing season reduced the expenditure of soil nutrients with the limited water sources and allowed the plants to proliferate in the region. The second advantage to the Fertile Crescent was that the wild ancestors of the domesticated crops were already abundant and thriving in the region so that humans would only initially have to select for those plants that produced large edible seeds, plants that did not shed their seeds rapidly on maturity, and seeds that could easily be harvested.[7] The endemic nature of these species facilitated the transition from wild plants to cultivated crops. Lastly, the Fertile Crescent consisted of a wide range of altitudes and landscapes within a small area.[23,24] The variety of habitats harbored a high diversity of plants and animals, which protected against potential environmental threats and supplemented a diversified and rich diet. Because of the unique and desirable nature of this region, historically it has been a region of great diversity and the object of numerous territory wars that continue to the present day.

Eastern Asia

Agriculture also evolved independently in Eastern Asia. It has been established that early farming in China can be divided into the northern practices of cultivating millets (Figs. 13.4 and 13.5) and the southern practices of cultivating rice (Fig. 13.6). It is not clear, however, as to

FIGURE 13.4 A crop of millet ready for harvest. *From https://upload.wikimedia.org/wikipedia/commons/f/f0/Grain_millet,_early_grain_fill,_Tifton,_7-3-02.jpg.*

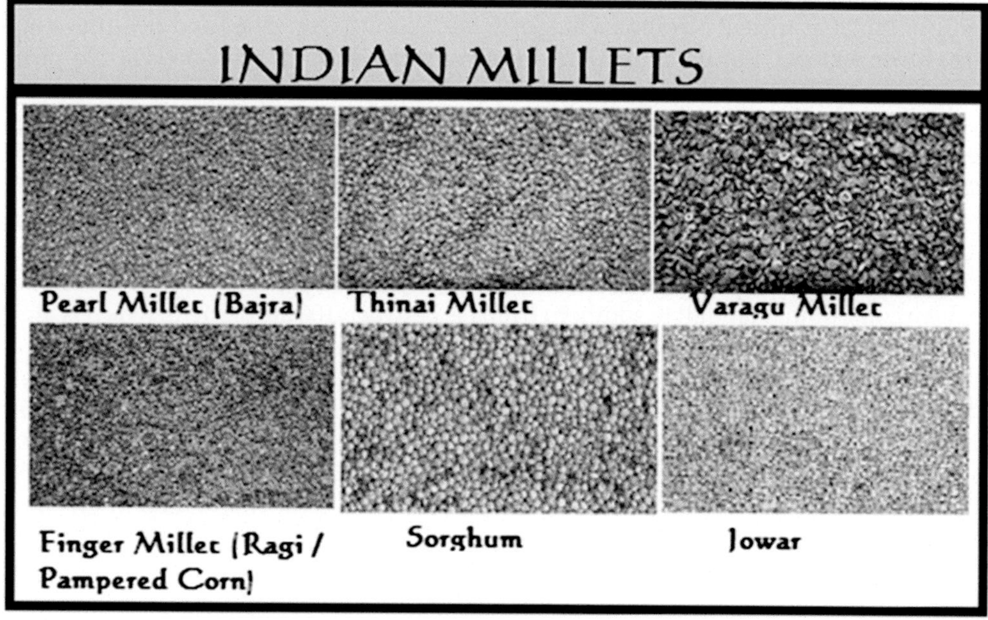

FIGURE 13.5 Various forms of Indian millet showing the variety of colors. *From https://upload.wikimedia.org/wikipedia/en/d/db/Indian_Millets.png.*

FIGURE 13.6 Rice plants ready for harvest. Rice was domesticated in Southern China. *From http://ricewisdom.org/images/rice-wisdom-004-home.jpg.*

how these practices intertwined (i.e., did the northern practices spread south or the southern practices spread north). The earliest farming areas are associated with China's main rivers: the Yellow River in the north and the Yangtze River in the south.[25] Archeologists have found evidence of millet in ancient Neolithic villages in the Yellow River valley and rice grains in the Yangtze River region.[26] Burnt rice remains were found embedded in 10,000-year-old pottery from the Yangtze River region suggesting the origins of rice farming. The roots of agriculture in eastern Asia are linked to so-called "affluent foragers" that were people who combined gathering of fruits and nuts with the cultivation of cereals.[27] This cosubsistence model was probably universal in the transition from hunting and gathering to agriculture.

The first attempt to cultivate rice may have resulted from people's efforts to take advantage of the naturally productive swampy habitats. Southeast China's heavy yearlong rainfall (concentrated in the summer months) and subtropical climate provides ideal conditions for rice cultivation. Annual rainfall in southern China is between 100 and 150 cm, which is almost double the rainfall of northern China. Early farmers built dams that would retain the water from heavy rainfall[3] allowing them to cultivate wild rice by planting seeds in the manufactured paddies at the end of the rainy season. By planting and harvesting the paddies, farmers emulated the seasonal germination and growth cycles of wild rice.[28] By changing the dry land to a continually flooded environment, the cultivators increased the rice yields. In southern China, this early practice of habitat modification may have laid the foundation for the transition from foraging to farming.

People of north China cultivated millet rather than rice. Archeologists believe that the earliest common millet cultivation was established 10 kya[29] in the western regions near the Yellow River. In this area, approximately 40 early millet settlements were discovered. These sites, which encompassed small villages complete with houses, storage pits, and burial sites, were

TABLE 13.1 Approximate Dates for the Appearance of Domesticated Species in Several Regions of the World

Region and Organism	Date of Appearance (In Thousands of Years)
1. SOUTHWEST ASIA	
1.1. Plants	11.5
1.2. Animals	10.5
2. CHINA	
2.1. Millet	10.0
2.2. Rice	>7.0
3. SOUTH ASIA	
3.1. Plants	5.0
3.2. Animals	8.0
4. AFRICA	
4.1. Plants	5.0
4.2. Animals	9.0
5. NEW GUINEA	
5.1. Plants	>7.0
6. EASTERN NORTH AMERICA	
6.1. Plants	5.0
7. MEXICO	
7.1. Corn	9.0
8. SOUTH AMERICA	
8.1. Plants	10.0
8.2. Animals	6.0

Modified from Price TD, Bar-Yosef O. The origins of agriculture: new data, new ideas. Wenner-Gren symposium. Curr Anthropol *2011;***52***(Suppl. 4):S161–S510.*

once inhabited by people, known as the P'ei-li-kang. These people lived in small round houses (2–3 m in diameter), in villages that were 1–2 ha in size. Based on archeological evidence from plant remains, the P'ei-li-kang communities cultivated broomcorn and foxtail, two different millet varieties, as well as rapeseed and cabbage.[30] The P'ei-li-kang also domesticated dogs and pigs. Wild plants in the region included walnuts, hazelnuts, and various fruits (Table 13.1). Based on the population size of the P'ei-li-kang communities, they could not have survived based solely on hunting and gathering.[31] However, some investigators believe the transition from foraging to farming was not initiated by the P'ei-li-kang people, but by an earlier community around 9 kya. This theory comes from archeological evidence demonstrating that the area was already well developed before the P'ei-li-kang people occupied the region.[30] To increase the yield and reliability of their food source, these previous inhabitants may have supplemented their hunting and gathering diets with domesticated millet (Table 13.2).

Rice and millet were pioneer crops in Asia and are believed to have served pivotal roles in the global spread of agriculture. Today, rice comprises about half of the food eaten by 1.7 billion people worldwide. In 2014–15, more than 484 metric tons of rice were consumed worldwide, comprising 21% of the total calories that humans consume. The transition from harvesting rice and millet in Asia to the widespread dominance of these two food sources

TABLE 13.2 Approximate Dates of Earliest Signs of Domestication in Different Regions of the World

Region and Organism	Date of Earliest Signs of Domestication (In Thousands of Years)
1. SOUTHWEST ASIA	
1.1. Plants	
1.1.1. Wheat	11.0
1.1.2. Barley	10.5
1.1.3. Pea	10.0
1.2. Animals	
1.2.1. Sheep	9.8
1.2.2. Goat	9.8
1.2.3. Pig	9.7
1.2.4. Cattle (taurine)	10.3
1.2.5. Cat	4.0
2. SOUTH ASIA	
2.1. Plants	
2.2.1. Rice (indica)	4.0
2.2. Animals	
2.2.1. Cattle (zebu)	8.0
2.2.2. Water buffalo	4.5
3. EAST ASIA	
3.1. Plants	
3.1.1. Rice (japonica)	7.6
3.1.2. Soybean	5.5
3.1.3. Melon	4.0
3.2. Animals	
3.2.1. Pig	8.5

TABLE 13.2 Approximate Dates of Earliest Signs of Domestication in Different Regions of the World —cont'd

Region and Organism	Date of Earliest Signs of Domestication (In Thousands of Years)
3.2.2. Silkworm	5.4
3.2.3. Horse	5.5
3.2.4. Bactrian camel	4.5
3.2.5. Duck	1.0
3.2.6. Chicken	4.0
4. NEW GUINEA	
4.1. Plants	
4.1.1. Banana	4.0
5. AFRICA AND SOUTH ARABIA	
5.1. Plants	
5.1.1. Sorghum	4.0
5.1.2. Rice (African)	2.0
5.2. Animals	
5.2.1. Cattle (taurine)	7.7
5.2.2. Donkey	5.5
5.2.3. Dromedary camel	3.0
6. NORTH AMERICA	
6.1. Plants	
6.1.1. Squash	5.0
7. MESOAMERICA	
7.1. Plants	
7.1.1. Squash (pepo)	10.0
7.1.2. Maize	9.0
7.1.3. Common bean	3.0
7.2. Animals	
7.2.1. Turkey	2.0
8. SOUTH AMERICA	
8.1. Plants	
8.1.1. Peanut	5.0
8.1.2. Cotton	6.0

(Continued)

TABLE 13.2 Approximate Dates of Earliest Signs of Domestication in Different Regions of the World —cont'd

Region and Organism	Date of Earliest Signs of Domestication (In Thousands of Years)
8.1.3. Coca	8.0
8.1.4. Manioc	7.0
8.1.5. Quinoa	3.5
8.1.6. Yam	5.5
8.2. Animals	
8.2.1. Llama	6.0
8.2.2. Alpaca	5.0
8.2.3. Guinea pig	5.0

Reproduced from Larson G, et al. The modern view of domestication. Special feature. Proc Natl Acad Sci USA *2014;**111**:6139–97.*

was most likely facilitated by trade. Hunter-gatherers in parts of southern Europe quickly adopted Asian cereal crops and livestock around 6 kya, and this practice eventually spread to Central Europe around 5 kya.[6] About this time, hunting and gathering had become less productive in Southern and Central Europe, which most likely helped accelerate the adoption of farming.

Xinjiang, a northwest region in China, for example, was a trading hub during the rise of the Egyptian, Mesopotamian, and Indus civilizations.[32] Trade took place by sea routes via India and the Mideast and overland on the Silk Road through Persia, resulting in the transportation of peach, mulberry, and citrus to the west. The cultural exchange accompanying these trade routes was mutually advantageous. Whereas other civilizations gained the knowledge of rice and millet cultivation and other unusual fruits, Chinese communities benefitted from receiving new technologies, which, in turn, bolstered the developing rice and millet communities. Xinjiang enjoyed an ideal location, where people passed through the region when traveling east and west. Genetic studies have linked European millet to Asian millet, providing evidence that this crop was spread through trade rather than by isolated origins of cultivation.[32,33]

In addition to the development and domestication of crops, ornamental horticulture became part of the Chinese culture and spread throughout Asia with the development of gardens for various emperors. Flower cultivation became an art, and China continues to remain the world's largest consumer of horticultural products.

Europe

Approximately 8000 years ago,[3] agriculture spread from the Fertile Crescent to southern Europe (via traversal of the Balkans to the shores of the Mediterranean) and southward to North Africa. Europe and Africa both exhibited climatic extremes that limited the success of earlier attempts at farming. Northern Europe was too cold, whereas much of the Sahara desert in North Africa was too dry.

Southern Europe most likely established agriculture by trade, not through migration. Archaeological evidence for the transition from hunting and gathering to agriculture comes from the Franchthi Cave in southern Greece. Investigators have determined that the Franchthi Cave was occupied from the Upper Paleolithic through the Neolithic periods. It is one of the few sites that show almost continuous occupation over these periods. Studies of the cave indicated that the cave dwellers were hunter-gatherers during the Paleolithic Period. In the Neolithic Period, the Franchthi people began to herd sheep and goats and planted emmer wheat and barley.[34] Archaeologists discovered grinding and sickle tools dating to 8 kya, which were likely used for processing grain and harvesting plants.[35] Similar sites displaying transitions from foraging to farming were discovered in Sicily, Italy and Chateauneuf, France. Archaeologists speculate that the Mediterranean people did not evolve to a full sedentary agricultural lifestyle until 7 kya.[3]

Africa

Agriculture arose in the Nile Valley and eastern Africa about 1000 years after farming settlements developed in southern Europe. Some investigators believe that the emergence of agriculture in the Nile Valley was because of trade and migration from southern Europe. Evidence for this includes the adoption of Mediterranean style farming and European crops in North Africa.[4]

The Nile Valley receives very little moisture, and the Nile River provided an oasis in the middle of the desert and an essential component to the emergence of farming in North Africa. The Nile, with its predictable seasonal flooding in September and August, and the subsequent creation of fertile soil, created ideal conditions for the production of crops that were planted in October and harvested in March and April. The development of agriculture in this region was an essential component for the development of wealth and power in ancient Egypt.[36]

From the Nile Valley, agriculture diffused to communities of southern Africa. Areas of the Sahara had previously existed as grounds for foragers because of moderate rainfall and the grasslands, which allowed hunter-gatherers to be successful in the arid terrain by congregating around the limited water resources.[36] Around 4 kya, agricultural economies began to prosper in South Africa where there is evidence of slash and burn agricultural practices brought by the Bantu Expansion (see Chapter 11).

Before the advent of pastoralists and farmers, hunting and gathering communities were small and spatially separated. The development of agriculture transformed African societies into large and more complex groups because of their ability to feed large groups of people, control land, and store food. This transformation changed the fate of hunter and gatherer groups that began to occupy the lowest rung on the hierarchical scale. During the transition to agriculture in Africa, animal domestication appears to be the earliest event, with the production of grains, and tropical plants following. The recovery of ancient cattle bones associated with pottery 10 kya in Egypt suggests that wild cattle were hunted or pastoralists domesticated them. When cattle remains were discovered, various theories about the origin of cattle in Africa were proposed. One idea was that the cattle were indigenous to the region, whereas alternative theories proposed that the cattle were derived from various Levantine stock that were brought down to Africa by pastoralists from the Fertile Crescent. In 2014, scientists completed a genetic study of 134 cattle breeds from around the world. They discovered that the African domesticated cattle originated in the Fertile Crescent, providing evidence that cattle were introduced to Africa as farmers migrated south and discrediting theories of the independent evolution of cattle husbandry in Africa.[37]

Americas

During the Pleistocene epoch, the Bering Strait land bridge connected Siberia and Alaska. Around 13–14 kya, humans migrated to the Americas over the land bridge. New World civilizations, including the Mayans, Aztecs, and Incas, have a long agricultural history; however, all of these agrarian civilizations are relatively young in comparison with those of Asia, Africa, and Europe. Of the three, Mayans are the oldest that show a civilization based on farming and date back to 3.8 kya. The Aztecs were a nomadic tribe who moved from North Mexico to Mesoamerica and did not start farming until around the 13th century. The Incas did not begin to build their empire until the 12th century.[38] Most of the early agriculturists in the western world focused on two plants, squash and maize.

Although the earliest evidence of agriculture used by the Mayan civilization dates to around 3.8 kya there is earlier evidence for domestication of several species including maize and squash (Table 13.2). The evidence comes from the excavation of the Guilá Naquitz Cave in Oaxaca, Mexico, where archeologists uncovered evidence of plant domestication, which included plant remains of maize, squash, beans, avocados, and chilies.[39,40] A charcoal sample of the squash from this territory was radiocarbon dated to 9800 years ago. Squash is believed to be the earliest agricultural crop of the Americas followed by maize and beans. The evidence for the early domestication of squash comes from the analysis of plant material from the cave in Oaxaca, and the fact that the plant material found included, increased seed length and peduncle diameter, as well as changes in fruit shape and color, in comparison to wild squash of the region. Maize, squash and beans made up the Three Sisters agricultural system (a system of companion planting where winter squash, maize, and tepary beans are planted close together).[41] In this system, maize is planted first, and after several weeks of growth, beans and squash are planted with the maize. The maize provides a structure for the beans to climb, whereas the beans also provide a natural source of nitrogen. The squash grows prostrate along the ground helping to shade any weeds that might develop and help to retain soil moisture (Fig. 13.7). The combination of plants also provides a nutritious diet with a variety of carbohydrates, proteins, and fatty acids. The Three Sisters planting system is thought to have been developed by Native Americans 5–6.5 kya and was used throughout North America. Modifications of this system, which included slightly different species of plants, were used in different climates. One example is the Four Sisters system, which used the Rocky Mountain bee plant *(Cleome serrulata)* to attract bees for pollination.[41]

As stated previously the domestication of squash likely occurred 10 kya.[42] The main species of squash cultivated in the Americas is *Cucurbita pepo*, which exists in three varieties (sometimes referred to as subspecies) easily identified by their colors (green, yellow, and orange). The precursor to the cultivated *C. pepo* is still disputed. Although all squash is thought to have been indigenous to Mexico, there were two different origins of domestication. The green and yellow varieties were domesticated in the United States (moving from Texas to the Mississippi River Valley), whereas the pumpkin (orange) was first cultivated in Oaxaca, Mexico. All squash is thought to have originated in what is today Mexico and then later domesticated in various parts of the United States.[41]

Genetic analysis suggests that maize was domesticated in Mexico from its precursor, teosinte, *Zea mays parviglumis* (Figs. 13.8 and 13.9). Evidence for this belief comes from numerous studies that have collected teosinte samples from the western hemisphere and compared their DNA

FIGURE 13.7 Three Sisters planting system. Corn, squash, and rice are grown together. The system not only provides structure, natural source of nitrogen and shading to prevent weeds, but it is also a nutritious combination. *From http://bloominthyme.com/wp-content/uploads/2012/04/school-squash-and-corn.jpg.*

FIGURE 13.8 Teosinte plant showing male flowers similar to those seen in modern-day corn; however, teosinte is much smaller than the modern corn plant. *From http://www.ars.usda.gov/sp2UserFiles/ad_hoc/36222000DiverseMaize Research/images/teoF1earzone.JPG.*

FIGURE 13.9 Teosinte seeds and fruiting structure. The fruiting structure is synonymous with the cornhusk of modern-day corn; however, it is much smaller and contains only a few seeds. The seeds are very hard and cannot be eaten without mechanical crushing. *From http://oregonstate.edu/instruct/css/330/six/images/LTeosinteDoebS.jpg.*

sequences with different types of maize. The results suggest that maize is genetically similar to a teosinte variety from the Central Balsa River Valley located in the southern regions of Mexico.[43] Moreover, the genetic distance between modern maize and Balsa teosinte corroborates archaeological evidence, indicating that maize domestication occurred around 9 kya.[44] Stone milling tools with maize residue dating to 8.7 kya discovered during an excavation of the Xihuatoxtla Cave in Guerrero, Mexico provided further evidence for this time frame. Another study that examined the process of domestication at the genetic level determined that teosinte and modern maize exhibited differential expression of 600 genes including a subset of loci that were likely the targets of human artificial selection.[45] Most likely, the early maize cultivators were members of small, seasonally nomadic groups that transformed teosinte into maize by careful selection for the most desirable characteristics resulting in a plant that was easily harvested and produced high yields.[46]

European expansion, during the 15th and 16th centuries, into the Americas brought together the Old and New Worlds through the transport of goods and cultural practices. Europeans transported tobacco, potato, peanuts, maize, and chili peppers into both Europe and Asia, whereas Europeans brought sheep, cattle, and horses to the Americas. In addition, cotton and sugarcane from the Americas became valuable trade crops, along with gold and silver mined in the Americas.[47]

OTHER DOMESTICATED PLANTS OF INTEREST

Banana ranks close to rice, wheat, and maize in terms of their importance as a food plant; however, the domestication of the banana remains complex and indirect. Most investigators have focused on the Kuk Valley of New Guinea as a region where humans may have first domesticated banana around 8 kya; however, domestication may have also occurred in

Southeast Asia (Table 13.2). The next appearance of the banana occurred in the Philippines and then dispersed into India, Indonesia, Australia, and Malaysia. The difficulties in tracing the domestication are the appearance, disappearance then reappearance of banana in different regions, and the asexual and hybrid forms of banana that were developed during its domestication.[48]

Recent studies have determined that domestication occurred as a series of crossings and selections leading to seed suppression and parthenocarpy (artificially induced fruit without fertilization).[49] Most edible bananas are diploid or triploid hybrids (containing two or three sets of chromosomes, respectively) from *Musa acuminata* alone or from hybridization with *Musa balbisiana*. The current global production of more than 100 million tons is based on large-scale vegetative propagation of a small number of genotypes, which derive from only a few ancient events. Understanding the origins of the modern banana is important to help identify primary species that could be used to introduce new genetic material into modern cultivars and avoid the danger of new diseases, and pests, because of the limited genetic variability of modern crops.

Grape, another crop, is the most valuable horticultural crop in the world and various species of grape are native to North America, Asia, Europe, and North Africa. Grapes grow in clusters and come in a variety of colors, including black, blue, yellow, green, orange, red, white, and pink. Grapes are consumed as fresh and dried fruit or used to make wine. The main product, wine has played a significant role in the development of human culture. Historically, wine has been referred to as the "drink of the gods," used in religious practices, and has always been a significant part of Mediterranean culture. Archeological data suggest that cultivation of the domesticated grape, *Vitis vinifera vinifera*, began around 6–8 kya in the Near East with the earliest archeological evidence for winemaking 8 kya, in Shulaveri in Georgia in the former Soviet Republic.[50,51] *V. venifera* is the only species used in the wine industry and is the only species of the genus that is indigenous to Eurasia (thought to have originated over 65 million years ago). Thousands of *V. vinifera* cultivars exist (used for wine grapes, table grapes, and raisins), but only a few are used by the wine industry. A study of 1000 samples from the US Department of Agriculture[52] detected limited genetic diversity in wine grapes due to thousands of years of widespread vegetative propagation. Although some genetic diversity has been maintained through clonal somatic mutations, and the occasional development of different varieties, the crop faces severe pathogen pressures.

ANIMAL DOMESTICATION

Animal domestication has followed a similar trajectory to plant domestication. Humans intervened in the natural life cycle of selected animals by managing their living conditions, food supply, and reproduction.[53] Evidence indicates that most animal domestication originated in the Holocene period.[46] Scientists believe that animal domestication, similar to plant cultivation, was made possible by warmer, more hospitable climates exhibiting more reliable seasonal patterns. Although domestication may have spread due to contact with major domestication sites, it is most likely that the animal husbandry efforts were independent. Animal domestication appears to have originated in geographically distinct areas and spread outward to different communities.[53]

The ideal candidate for domestication was an animal that provided an important food source, did not have rapid flight abilities, and was social and able to tolerate breeding and feeding restrictions.[46] Three types of domesticates can be identified: (1) commensals, adapted for companionship (dogs, cats, guinea pigs); (2) prey for food (cows, sheep, pigs, goats, ducks, chickens); and (3) those reared for work and nonfood resources such as wool, leather, and transportation (horses, camels, donkeys, oxen, llamas).[54]

The earliest evidence of animal domestication is that of dogs. The actual means of domestication and the initial site and time of domestication remain a mystery. Initial investigations placed dog domestication in the Middle East. Evidence for this came from a jawbone found in a cave in Iraq dated to about 12 kya. Later, genome studies of wolves and domesticated dogs led by Wayne and Novembre determined that dogs were domesticated from now extinct wolves approximately 11–16 kya, predating agriculture.[55,56] More recent mitochondrial DNA evidence suggests domestication in Europe,[57,58] whereas genomic DNA evidence from Chinese investigators suggests that dogs were domesticated 33 kya in Southeast Asia and then migrated into the Middle East 15 kya, reaching Europe around 10 kya.[59] The high levels of genetic diversity found in dogs from Southeast Asia supports the idea that they may be the most primitive forms. In addition, archeological support for the Chinese genetic studies comes from two dog skulls dated to 33 kya that were discovered in the Altai Mountains of Siberia and in Belgium. There are several possibilities explaining the different data and interpretation of where and when dogs were domesticated. One is that humans traveled great distances 33 kya and dogs followed seeking food, and they were eventually domesticated, or that domestication occurred repeatedly in different locations at different times.

All genomic data support the notion that dogs are descended from relatives of the gray wolf (*Canis lupus*). Genetic studies suggest that modern domesticated dogs are the result of admixture between *C. lupus* and extinct species of the genus *Canis*. Genetic evidence further suggests that none of the modern gray wolves are those that were first domesticated.[56,58,59] Interestingly, dogs are the only large carnivores ever to have been domesticated. Dogs may have been domesticated by hunter-gatherers for assistance during hunting.[2] The selective breeding that shows an amazing variety of shapes, sizes, and abilities (herding, swimming, hunting, aggression, etc.) occurred within the last several hundred years.[56]

Aside from the domestication of dogs, other major animal husbandry occurred in the Middle East. Domestication of the goat, pig, and sheep first transpired in the Fertile Crescent approximately 11 kya possibly because of unpredictable availability of wild game from overhunting and/or changing ecological conditions in the region. Goats, one of the first animal domesticates of this area, originated from the wild *Capra aegagrus* (Fig. 13.10). Pigs were initially domesticated from the wild boar (*Sus scrofa*) and then domesticated in Europe and Asia around 9 kya (Fig. 13.11). As with many animal species the exact wild ancestors and the methods of domestication of sheep is uncertain. Most investigators believe sheep were initially domesticated from the wild mouflon (*Ovis orientalis*) in Mesopotamia between 11 and 9 kya (Fig. 13.12). The early domestics were raised for milk, meat, and skins for clothing, with wooly sheep developed around 6 kya in Iran.[53,60]

Cattle domestication occurred multiple times and in various places but first occurred approximately 500 years after the Fertile Crescent witnessed its first surge of major animal domestication.[61] Cattle were domesticated at least twice with domestication first seen in Turkey around 10.5 kya from wild aurochs (*Bos primigenius*), then again approximately 7 kya,

FIGURE 13.10 Wild goat *Capra aegagrus*, thought to be the ancestor of domesticated goats. *From https://upload.wikimedia.org/wikipedia/commons/c/cf/Carmel_Hai-Bar_-_Capra_aegagrus_creticus_(3).JPG.*

FIGURE 13.11 The wild boar or Eurasian wild pig (*Sus scrofa*) originated in Southeast Asia (Indonesia and the Philippines) during the Pleistocene. From there it spread into Eurasia and Africa. It is thought to be the ancestor of most modern-day pig breeds. *From https://upload.wikimedia.org/wikipedia/commons/d/d1/20160208054949%21Wildschein%2C_Nähe_Pulverstampftor_%28cropped%29.jpg.*

FIGURE 13.12 Male Cyprus mouflon (*Ovis orientalis*) with ewes. *From https://www.telegraph.co.uk/finance/financialcrisis/9312154/Italian-austerity-forces-region-to-sell-its-rare-mouflon-sheep.html.*

in the Indus Valley. Cattle from Turkey gave rise to the taurine line (European cattle) and those from the Indus Valley gave rise to the indicine line (humped cattle); both are descendants of wild aurochs. Scientists have concluded that trade likely spread the domesticated of cattle globally.[61–63] Cattle are useful domesticates for food, milk, and their secondary products including clothing and tools from hides and bones (Fig. 13.13).

Domestication of horses and camels occurred 4–6 kya.[6] Horses provide food, facilitate transportation, and enhance warfare capabilities. Genomic studies suggest that the process of domestication involved closely related male animals mating with a variety of females. This evidence comes from genome studies showing there is limited sequence diversity in horse Y chromosomes (passed on from males), which contrasts with a high diversity of mitochondrial DNA (passed on by females).[64] Camels provided transportation, milk, meat, and hair for clothing. Camels are classified as Old and New World camelids. The Old World camelids are those most familiar to people as the one-humped camel (dromedary) that is native to the Middle East and the two-humped camel (Bactrian) that is native to Central Asia. New World camelids are the llama, alpaca, and guanaco of South America. Dromedaries may have been the first to be domesticated in Arabia around 5 kya.[65]

Chickens are one of the most common domesticated animals and a worldwide food source making up a variety of cuisines, where they are primarily used as a source of meat and eggs. Although it is a major food source today, many archeologists think that is was originally used for cock fighting and not as a food source.[66] The use for cock fighting was most likely a major reason for its spread and only later was it recognized as a food source. The domestication of the chicken dates back to 4–8 kya in Southeast Asia and most likely resulted from the wild red jungle fowl *Gallus gallus*. A lack of data makes it difficult to

FIGURE 13.13 *Bos primigenius taurus* originally from Turkey gave rise to the taurine line of cattle referred to as modern-day European cattle. *From http://www.naturephoto-cz.com/photos/andera/highland-cattle-xxx2010.jpg.*

pinpoint the exact timing, but it is clear that from Southeast Asia the bird spread into India where a number of varieties were developed and exported. The export of chickens from India is thought to have given rise to the modern-day chickens of Europe, the Middle East, and the Americas.[67]

Animal domestication also helped humans through the use of manure as fertilizer and fuel for fires.[6] Until the Industrial Revolution and the invention of the railroads, large mammals served as the main mode of transportation. Before animal domestication, humans transported goods and people on their own backs. Animals such as donkeys, horses, and llamas made it possible to carry heavy goods over longer distances with minimal resources.[6]

DIET

Changes in diet and food availability, due to environmental conditions or migration patterns, represented selective pressures that acted on both biological processes and anatomical features. Genetic studies have shown that change in diet is an important epigenetic force (see chapter 1) that can have effects on gene regulation and expression. This type of environmental change (animal domestication, agriculture, disappearance of longstanding food sources, appearance of new food sources) would have led to a rapid response in the biological mechanisms underlying nutritional processes and metabolism.

The transition from foraging to farming affected the human diet. Hunter-gatherers had a diverse diet because of the variety of game and plants, which were high in protein, fiber, and vegetable-based carbohydrates and low in saturated fats.[68] When these foragers transitioned to agriculture, they lost nutritional variety and quality[2] by focusing on cultivating certain staple crops that were grown in surpluses but were high in calories and low in vitamins and minerals.[69] The inadequate amounts of iron, calcium, zinc, B12, or thiamine, and other nutrients led to malnutrition, poor growth, and immunosuppression.[70] One example of immunosuppressive effects can be seen in pregnant women (maternal immunosuppression). The immunosuppressive effects are the result of diet suppression due to nausea susceptibility, olfaction responses, as well as the development of cravings which accompany many pregnancies. Nausea and olfaction responses are thought to be an evolutionary mechanism for time-limited gestational avoidance of meat (which could have carried heavy loads of pathogens). Gestational cravings on the other hand target substances that may have higher nutritional value and influence immune functioning. Other examples include the immune response associated with Celiac disease and the consumption of wheat and various vitamin deficiencies that can reduce the immune response and lead to increased susceptibility to pathogens.[71–73]

Furthermore, many of the staple crops were high in carbohydrates leading to a positive correlation between the development of agriculture and the presence of cavities. Starch sticks to teeth and causes plaque formation that traps bacteria on the enamel, and the bacteria excrete acids that dissolve teeth. Cavities that penetrate in the gum layer can affect the nerves and cause fatal infections.[74] Cavities were rare among foragers but are widely seen among ancient farmers and people today. In addition, the dietary changes associated with modern agriculture have increased obesity and chronic diseases, including diabetes, cancer, and cardiovascular disease.

Farming, and hunting and gathering both have advantages and disadvantages. Farming can lead to the loss of nutritional variety, and farming communities can become dependent on agricultural products, and that dependence cultivates individuals who are not capable of hunting and gathering their own food (an example would be the number of individuals who could sustain themselves today if there suddenly was no agriculture). In addition, crops are more susceptible to long-term shortages from drought, flood, war, and destructive weather, leading to famines.[2,69] On average, farming is more reliable and consistent; however, when farming fails, the effects can be detrimental compared with hunting and gathering. Hunter-gatherers have less extreme shortages and famines because they are not confined to certain staple crops, and a particular area of land, allowing them to shift to alternative food sources when needed. Being nomadic allowed foragers to be more flexible in times of low resources.

Multiple years of crop failure usually results in famines. A modern example is the Irish Potato Famine. In the 17th century, potatoes were imported from South America to Ireland.[75] The potato quickly became the staple crop because it was easily cultivated in the Irish terrains. Potatoes also provided a large amount of calories and helped produce a rapid increase in population. In 1945, a fungus (*Phytophthora infestans*) spread throughout Ireland killing approximately 75% of the harvest. The famine lasted for 6 years resulting in millions of deaths due to starvation and the emigration of millions of individuals from Ireland.[75]

Although agriculture resulted in the loss of nutritional variety, the loss of quality and diversity, and sometimes led to long-term depletion of resources, animal and plant domestication allowed communities to have food surpluses. These surpluses were fundamental to changes in demographics and the progression of civilizations.

EFFECTS OF POPULATION CHANGES

Agriculture has permanently altered the human population and demographics. At the beginning of plant domestication, approximately 6 million people inhabited Earth compared with more than 7 billion today.[76] This means that in the 11,000 years since agriculture began, the population has multiplied 12,000 times. Before agriculture, limited food sources restricted the number of children that traditional hunter-gatherers could raise. A hunter-gatherer's mother in a nomadic community could only carry one child and her possessions. Foraging families often limited the number of children by means of sexual abstinence and/or the practice of infanticide. In contrast, the birth interval for farming families was about 2 years.[6] Farmers benefited from many offspring because children were a useful labor force that could work on the farm and within the household.

Scientists have identified the demographic changes associated with the emergence of agriculture as the Neolithic Demographic Transition (NDT).[77] The NDT represents a shift during the Mesolithic–Neolithic periods toward higher fertility values and higher mortality rates. The increased mortality rate of agricultural societies can be attributed to several factors. These include the rise of new pathogens, increased incidence of contamination, disease transmission associated with higher population densities, and occasional famine. The data for the NDT are based on the proportion of immature skeletons in cemeteries from Europe and Africa, the rise in the economy, and the subsequent rise in births. The cause of the NDT is attributed to the shift to agriculture leading to increased maternal intake of high calories relative to foragers, the sedentary lifestyle leading to decrease female mobility, and a decrease in lactation period. The time difference between the increased fertility and mortality rates resulted in a rapid growth in population.[77]

As population densities increased, disease transmission across the communities intensified. Unlike hunter-gatherers, farmers were sedentary and as a result, they built villages that allowed them to easily trade resources without having to travel great distances. However, the close proximity of farmers to animals and to their farming communities was ideal for transmissible diseases to spread between humans and between humans and their livestock. Clear examples of this can even be seen today in regions where people are closely associated with farm animals. The origin of the bird flu, and swine flu are the result of viruses that are transmitted from humans to animals, where the virus resides in the animal and mutates (by mixing with other viruses already present in the animal). In many cases these new combination viruses do not cause symptoms in the animals but if transmitted back to humans through close proximity, they can become deadly because of their acquired variation (mix of human and animal viral characteristics). These instances are usually rare and are usually contained, but if the virus attains the ability to become human to human transmissible (instead of just animal to human transmissable) then this can lead to severe epidemics or pandemics. Survival of many diseases depends on the number of hosts that are infected and the ability of the disease to spread from one host to another.[78] Villages and farms epitomized conditions for the spread of diseases. Furthermore, early villages were not equipped to handle sanitation. Poor sanitation led to the increased presence of flea-infested rodents, many of which harboring pathogens that caused diseases such as the plague and typhus.[2,78] Other zoonotic diseases including tuberculosis, measles, diphtheria, leprosy, and influenza[2] (and more recently swine flu, SARS, and bird flu) arose because of the increased contact with various domesticated and nondomesticated animals. Agricultural communities were also affected by nutritional deficiencies resulting in dietary disorders such as scurvy, pellagra, and anemia.[2] Furthermore, long-term food storage was a breeding ground for aflatoxins. These harmful compounds produced by fungi that lived on cereals and nuts caused liver damage, cancer, and neurological diseases.[79]

Early agricultural villages were two-to-six times larger than prefarming communities.[3] These increased population densities advanced civilization; however, sedentary communities faced challenges regarding the organization and control of a greater number of people with a finite amount of resources. Food surpluses allowed for the origin of social elite.[6,76,77] Hunter-gatherer societies were likely egalitarian with little political organization because all individuals needed to devote similar time to finding food. In farming societies, as food surpluses were collected, some individuals accumulated wealth and land, whereas others became workers to manage the farms. The social elite (those with resources) began to dominate their communities by focusing on politics and commerce rather than agriculture. The beginnings of social elite gave rise to social stratification, which in many cases eventually led to slavery, oppression, war, and famine.[6,76] Slavery began as larger farms and cities began to develop. Early on, primitive farmers had no use for extra mouths to feed and sustained only family members. As farms became larger and food surpluses grew, cities emerged and under those conditions, cheap, reliable labor became desirable. Cheap laborers only need minimal food and lodging, and eventually they become slaves to those who supply their essential needs.

More food sources allowed for a greater number of mouths to feed resulting in a population that needed to be governed, and this in turn led to disputes between different communities. Land became a desired commodity and war over land and resources became a main source of slaves. During wars, useful captives were often taken as slaves, whereas others, deemed as unfit workers, were killed. Criminals, the impoverished, or groups of individuals deemed incompetent by the social elite were also candidates for slavery.[80]

Famine among the masses could also result from social hierarchy as elites continued to consume food during hard times, whereas those deemed socially unfit suffered the consequences of droughts, floods, and other environmental disasters that reduced crop yields or reduced livestock numbers.

Overall, domestication created a rapid increase in population size. More people allowed for the advent of complex societies with sophisticated culture, but an increased population also led to spread of diseases and the start of social stratification. Today's modern societies are a result of early agricultural efforts and the development of technology. Before agriculture, the environment limited the growth of populations. Once plants and animals were domesticated, humans had more time to develop technologies to overcome environmental barriers. These technologies allowed humans to spread into regions that would not normally sustain agriculture or human habitation. Today because of technology, we change our environment instead of adapting and as a result have large populations living under a variety of environmental conditions that would not have sustained our ancestors (one obvious example are the large populations living in the deserts of the Southwestern United States).

AGRICULTURAL STRESSES

An agricultural economy led to a sedentary lifestyle, which facilitated social complexity, expanded trade, introduced new diseases, and indirectly affected the genetics of populations. These technical, sociopolitical, medical, and evolutionary changes were often accompanied by increased levels of mental, physical, and emotional stress.

For example, scientists use height as an indication of the health of prehistoric people. Height measurements are indicators of overall health, including nutrition and disease status; moreover, this physical characteristic may also reveal societal patterns. Height is used because growth plates elongate during normal childhood and adolescent development, resulting in increased height. If a child is malnourished, or needs to use his or her energy to fight infections, he or she will have less energy devoted to growth. Height measurements of skeletal remains associated with early farming settlements in the Middle East suggest that the adoption of agriculture was beneficial to people's health. Around 11000 years ago, average height increased by approximately 1.5 inches in males and females. However, the observed increase in height has not been continuous over time and remained the same or diminished until the middle of the 19th century.[81] Examinations of European skeletons show no significant differences in height from the dawn of agriculture through the early 1800s. Why average height began to increase during the 18th and 19th centuries is not completely clear but may be linked to climate change and improvements in agriculture. During this time the temperature increased by 2–3°C allowing an extension of the growing season by 3–4 weeks and allowing farming at higher elevations.

Until the 19th century, on average, people diminished in size in some agricultural economies. In the Neolithic era, the populace of Asia decreased in height on average by 3.1 in., and in Mesoamerica, height decreased by approximately 2.5 in.[2] Despite agriculture's increased production in food, the initial overall health of humans seems to have decreased most likely as a result of children spending more energy fighting infections, shortages of food, and overall stress, which have been shown in many studies to affect growth and development.

For example, during World Wars I and II, when hunger was frequent in the German civilian population, the heights of children declined, but then average height returned in individuals born after the war. Today in poor countries, or among the poor in developing nations, people who are biologically stressed, or nutritionally deprived, show stunted growth. Today, height is still used in some countries as an indicator of socioeconomic division and can reveal discrimination within social, ethnic, economic, occupational, and geographic populations.[82]

Today even with advanced medical care and nutrition, height has leveled off, suggesting that an upper limit to height is dictated by our genes regardless of environmental improvements.

Osteology studies have shown that foragers experienced more daily biomechanical stresses than farmers. Members of agricultural communities experienced a fundamental reduction in physical workload. This is supported by a decreased level of osteoarthritis,[83] which is the inflammation of bones, particularly at joints, due to repetitive wear and tear. Scientists discovered a higher incidence of joint swellings in hunter-gatherers versus farmers, which suggests that osteoarthritis rates decreased with the adoption of agriculture.[84]

Skeletal remains show changes in craniofacial growth and development that lead to decreased head and jaw size as farmers transitioned to diets with less meat. It is thought that jaw size was reduced because the new food staples produce by agriculture were easier to chew. This reduction in jaw size increased tooth crowding because the number of teeth did not reduce at the same rate. The need for braces among younger populations today and the removal of wisdom teeth arose from the transition to farming as cranial skeletal sizes have reduced, yet teeth have not changed in size.[85]

As discussed above, an agricultural lifestyle may have resulted in health problems in the past and present. An example that is of interest today is the age of first menstruation, which is influenced by nutrition. Over the last century the age of puberty in human females has shown a decrease with an overall increase in nutrition. Some scientists believe that the increase in teenage pregnancies may be the unanticipated consequence of improved nutrition.

GENE–CULTURE COEVOLUTION

Culture influences genetics and reciprocally, genetics influences cultural practices. Scholars have coined the term for the mutual evolution of genes and culture as gene–culture coevolution. The emergence and advancement of agriculture increased the size of the population and exposed people to new environments (diets, diseases, climates, etc.). Environmental modification created new selection pressures that people adapted to through new mutations and epigenetic modifications, some of which became fixed in some populations.

The association of genes and culture works together so that along with the increase in population density come genetic consequences affecting evolution. Population genetics predicts that within a large population, the number of random mutations will increase, whereas small populations have relatively fewer new random mutations. For example, if a mutation has a probability of one in a billion occurring per gene per generation (10^{-9}/gene/generation) then with 7 billion people in the world today the likelihood of a mutation at this gene is 14 occurrences per generation (two copies of each gene—one from each parent, one on each chromosome). The steady increase in population size predicts that soon every single-step mutation will occur at least once in every generation. Most mutations will be neutral (have no effect) or deleterious, though a few may be beneficial. However, as the population increases, so will the number of deleterious mutations. Assuming the deleterious genes are under selection, natural selection should remove deleterious mutations, thus maintaining and driving advantageous mutations to higher frequencies. This could lead to a substantial increase in the frequency of an advantageous allele. For example, if a new mutation conferred a selective advantage that increased the number of offspring by 5% in 423 generations (approximately 10,000 years), 99.5% of the population would carry the allele.

An example of genetic mutation that was selected by an agriculturally modified diet is the ability of adult humans to digest milk. Before agriculture, humans did not consume milk after they finished nursing.[83] After nursing, children stopped producing lactase, the enzyme that breaks down the sugar, lactose, in milk.[84] By age five or six, children could no longer drink milk without experiencing symptoms of lactose intolerance including vomiting and diarrhea. The domestication of cattle in Europe, which followed the origins of agriculture, led to an increased prevalence of a mutated copy of the lactase gene. This mutation allows people to produce the lactase enzyme past their nursing years, so that they can break down lactose in adulthood.[2] Milk provided the benefits of calcium, vitamins, and the carbohydrate lactose to those early agrarian populations that inherited the mutation. Because of these benefits, natural selection favored this mutation; however, DNA extracted from the bones of ancient Europeans have identified that these people remained intolerant to lactose for 5000 years after they adopted agricultural practices. Today, approximately 90% of Europeans carry the mutated lactase gene and can drink cow's milk because of early domestication events.[53]

Another example of gene–culture coevolution relating to agriculture is a disease called sickle cell anemia. The disease is caused by a genetic mutation that prevents the formation of normal hemoglobin (the protein that carries oxygen in the blood and gives red blood cells their red color). The mutation in the hemoglobin of red blood cells causes them to be sickled or flattened in shape and also causes the red blood cells to be destroyed in the liver earlier than normal, causing anemia (thus the name of the disease—sickle cell anemia). Although sickle cell anemia can be a deadly disease, the sickled cells can prevent another more deadly disease, malaria, by inhibiting the malaria parasite from invading and replicating in the sickled red cells. Malaria is caused by several species of the *Plasmodium* parasite that is transmitted via mosquitoes. Once transmitted to humans, the *Plasmodium* parasite infects normal red blood cells. The symptoms of malaria can include constant chills, fever, blood loss, seizures, organ failure, and even death. Approximately 1 million people each year die of malaria with a heavy concentration of the deaths occurring in Africa.[84] Some people have acquired a genetic resistance to malaria by acquiring one copy the allele for sickle cell anemia.

Sickle cell protection against malaria has been traced back to early cultivation of yams in Africa. To grow yams, farmers had to cut down trees.[85] The removal of trees increased the amount of standing water when it rained, which produced an ideal condition for mosquito breeding. Owing to the rapid proliferation of mosquitos, there was an increase in malaria. More cases of malaria led to a condition where the sickle cell gene became beneficial.[58,85] The allele for sickle cell is recessive and people with two copies of the recessive gene do not normally live past their midtwenties but are resistant to malaria. Individuals who have one copy of the recessive gene show resistance to malaria.[60] Therefore, in areas of the world that are greatly affected by malaria, natural selection has increased the frequency of sickled cell genes in the population.

A third example of gene–culture coevolution is obesity. The emergence of social elite with access to food surpluses likely triggered the initial onset of obesity, but obesity can also arise from a sudden change in diet. As previously stated, the typical diet of a hunter-gatherer would today be considered healthy because it consisted of high levels of proteins and fiber but low levels of simple carbohydrates and fats.[61] The major problem associated with preagrarian and preindustrial diets was their susceptibility to food shortages. However, some evidence links communities that experienced food shortages, and then subsequent dietary changes to high levels of obesity. For example, the Pima Native American tribe has a high prevalence of obesity and type 2 diabetes today.[86] Historically, this tribe had frequent food shortages and a diet that consisted of high levels of fiber. Approximately 800 years ago, this tribe experienced droughts that led to major food stresses at least once every 25 years. Natural selection favored certain genes that allowed them to store calories in times of surpluses. For the majority of the Pima people, the storage of calories was needed, and females with greater fat reserves enjoyed an advantage during food shortages for themselves and for the children they carried during pregnancy.[86] When the Pima no longer experienced food shortages and started consuming the high fat and high carbohydrate diet of most Americans the selective advantage for carrying more fat predisposed the population to obesity leading to high blood pressure and type 2 diabetes.[63]

Most modern societies have not experience frequent food shortages that previously stressed their metabolism; however, obesity continues to be a concern especially in the United States. This phenomenon is rapidly permeating other Western societies and even developing countries. One important factor in human evolution was the increased consumption of animal tissues.[87] Advantages to eating a diet rich in animal protein include higher density caloric content and concentrated micronutrients; however, increased animal fat ingestion can also lead to medical problems especially when accompanied with a sedentary lifestyle. The combination of reduced activity and increased caloric consumption has resulted is a higher prevalence of obesity and chronic metabolic disorders (adult type 2 diabetes, hypertension, and elevated serum triglycerides and cholesterol) in industrialized societies. According to the Centers for Disease Control and Prevention, in 2015, the prevalence in the United States of age-adjusted rates of obesity was 47.8% among non-Hispanic African-Americans, 42.5% among Hispanics, 32.6% in non-Hispanic Caucasians, and 10.8% in East Asians. Obesity is higher among middle age adults, 40–59 years, and less frequent in young adults, 20–39 years, or adults more than 60 years. With the low nutritional quality of diets and the genetic predisposition to being overweight, global obesity is an epidemic.[88]

Agriculture continues to influence human societies. The ability to consume milk, the evolution of a defense mechanism against malaria, and the global epidemic of obesity, among others, are related to early agricultural efforts. Plant and animal cultivation has advanced society in many ways, as seen with the development of complex civilizations, but has also been detrimental, as shown with the decrease in a nutritional diet and the increase of diseases. Although early agricultural practices were practiced by small groups and in a variety of environment, today there are several centers for agriculture that supply food around the world (Fig. 13.14). Many of the regions used for modern-day food production are controlled

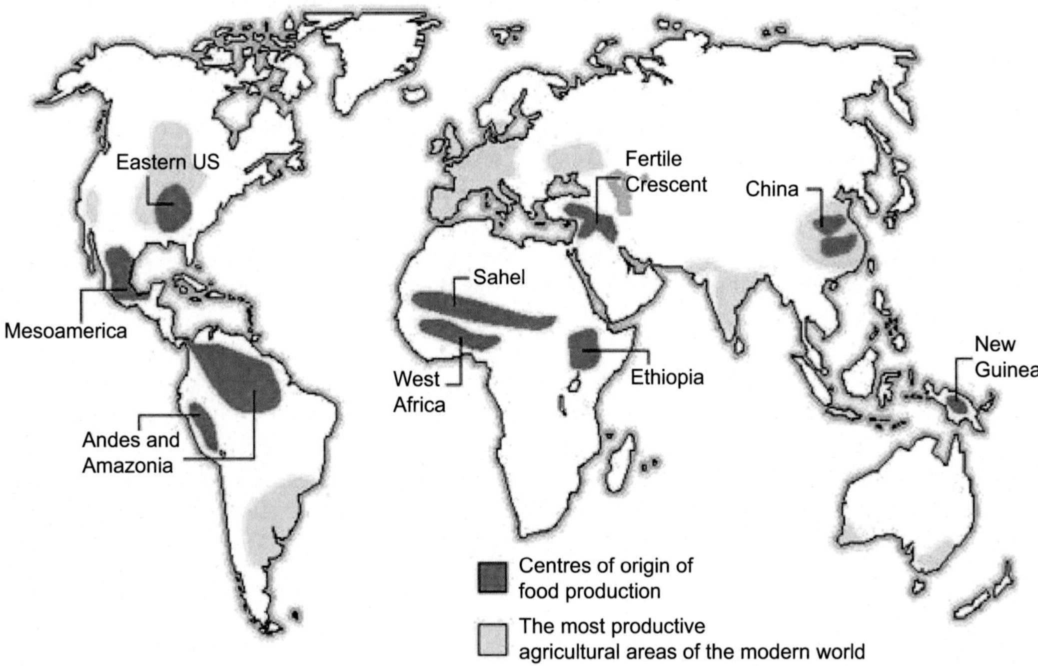

FIGURE 13.14 Centers for the origin of food production and modern-day centers for food production. Most of the farming today is done by large international companies (agribusinesses). *From http://www.nature.com/nature/journal/v418/n6898/images/nature01019-f2.2.jpg.*

by large international companies (agribusinesses) that have displaced many small farmers and small farming communities.

CONTEMPORARY REVOLUTIONS

In addition to the Neolithic Revolution, other agricultural developments have occurred. The Industrial Revolution marked a turning point in agriculture in which technological innovations and governmental policies shaped the mechanization and commercialization of farming.[89] The Green Revolution, which refers to research and development from 1930 to 1970, attempted to improve crop performance and eliminate world hunger through the development of high yielding cereal crops, synthetic fertilizers, pesticides, development of hybrid seeds, and new methods of cultivation and mechanization. The Green Revolution was not unique to North America but occurred in various regions around the world, including Europe, Asia, South America, and parts of Oceania.[90]

During the 1970s the field of agricultural biotechnology began to flourish with the advent of DNA manipulation and interest in the genetic modification of plants and animals. Although hundreds of species of plants have been domesticated and used, modern societies only rely on 20 species of plants for food.[4] These crops became the focus for biotechnology in an attempt to reduce herbicide and pesticide use, improve nutritional quality, and incorporate disease resistance to crops. Since then scientists have genetically modified several essential crops to make more efficient harvesting systems, increased plant resistance to herbicides, resistance to viruses, resistance to high salt concentrations, and improved nutritional quality, just to name a few. Maize, cotton, papaya, squash, alfalfa, sugar beets, canola, and soybeans have all been genetically modified and are currently grown in the United States.

The current advancements in biotechnology especially the development of CRISPR technology, also applies to animals. Genetic engineering through CRISPR is able to introduce desirable traits, or remove undesirable traits from animals' and plant genomes.[91,92] In the past molecular technology was able to increase milk production in cattle, increase growth rates in salmon, and produce human pharmaceuticals in plants and farm animals. The advent of CRISPR and the fact that it has become more and more efficient in genetically manipulating organisms has raised ethical, environmental, and evolutionary concerns about the long-term consequences of genetic manipulation. In addition, evidence that diet, and obesity cause epigenetic modifications that can be passed down to subsequent generations, continues to concern scientists and medical professionals about the immediate and long-term consequences of agriculture and cultural practices on human evolution.[93] The independent origins of agriculture worldwide, and at different times, cemented the mutualistic relationship between humans, domesticated animals, and plants. Future advances in agriculture, genetic manipulation of plants, animals, and humans and their epigenetic consequences will bring more gene–cultural adaptations to human populations.

CONCLUSION

Archeological and biological evidence illustrates that, starting in the Holocene epoch approximately 12 kya, the domestication of plants and animals developed in separate global locations most likely triggered by climate change and local population increases. The development of agriculture marks a major turning point in human history and evolution. In several independent domestication centers, cultivation of plants and animals flourished according to the particular environmental conditions of the region, whereas human migration and trade propelled the global spread of agriculture. This change in subsistence provided surplus food that accumulated during the summer and fall for storage, and winter consumption as well as domesticated animals for meat and dairy products. Because these new survival strategies no longer required the need for relocation and migration in search of food, humans were able to establish homesteads, towns, and communities, which, in turn, caused rapid increases in population densities and lead to the emergence of civilizations. The Middle East region in particular, served as a major epicenter from which the agricultural revolution spread in all directions initially resulting in the selective cultivation of plants and then turned to the selective breeding of animals. The agricultural economies that developed in Europe and Africa differed in their pioneer crops. In Europe, wheat and barley dominated, introduced to Europeans through Asian societies, whereas in Africa, millet, sorghum, and African rice were the premier cultigens. These three indigenous crops remain important food sources for African and Asian people today. In the Americas, early agriculturists cultivated the indigenous crops, squash and maize, and later beans, chillies, and avocados. The development of agriculture has led to both good and bad. Dependence on plant and animal domestication entailed a number of environmental adaptations including deforestation, irrigation, and the allocation of land for specific crop cultivation. It also triggered various other innovations including new tool technologies, commerce, architecture, intensified division of labor, defined socioeconomic roles, property ownership, and tiered political systems. This shift in subsistence mode provided a relatively safer existence and in general more leisure time for analytical and creative pursuits resulting in complex language development, and the accelerated evolution of art, religion, and science. It also changed dietary habits leading to obesity and in difficult environments starvation. Recently genetic evidence suggests that diet and other environmental factors can have epigenetic consequences. In the 1970s, crops became the focus for biotechnology in an attempt to reduce herbicide and pesticide use, improve nutritional quality, and incorporate disease resistance to crops. Genetic engineering has introduced certain desirable traits into animals' genomes, including increased milk production, higher growth rates, and the production of human pharmaceuticals in farm animals. These recent advances have raised ethical, environmental, and evolutionary concerns about the long-term consequences of genetic manipulation.

References

1. Lee R, Daly R. *The Cambridge encyclopedia of hunters and gatherers*. Cambridge, UK: Cambridge University Press; 1999. p. 1–22.
2. Lieberman D. *The story of the human body: evolution, health, and disease*. Pantheon Books; 2013. p. 67–298.
3. Smith B. *The emergence of agriculture*. New York: Scientific American Library; 1995. p. 1–208.
4. Rindos D. *The origins of agriculture: an evolutionary perspective*. Orlando: Academic Press; 1984. p. 1–285.
5. Barnard A. *Anthropology and the Bushman*. Oxford: Berg; 2007. p. 4–7.
6. Campbell BG, et al. *Humankind emerging*. 9th ed. Boston, MA: Pearson; 2006. p. 443–4.
7. Diamond J. *Guns, germs, and steel: the fates of human societies*. W. W. Norton & Company; 2005. p. 83–292.
8. Conner SE, Kvavadza EV. Modelling late quaternary changes in plant distribution, vegetation and using pollen data from Georgia, Caucasus. *J Biogeogr* 2009;**36**:529–45.
9. Byrd BF. Reassessing the emergence of village life in the Near East. *J Arch Res* 2005;**13**:231–90.
10. Metcalf J, et al. Synergistic roles of climate warming and human occupation in Patagonian megafaunal extinctions during the Last Deglaciation. *Sci Adv* 2016;**6**:e1501682.

11. Fuller DQ. Emerging paradigm shift in the origins of agriculture. *Gen Anthropol* 2010;**17**:1–6.
12. Walker M, et al. Formal definition and dating of the GSSP (Global Stratotype Section and Point) for the base of the Holocene using the Greenland NGRIP ice core, and selected auxiliary records. *J Quat Sci* 2009;**24**:3–17.
13. Richardson P, et al. Was agriculture impossible during the Pleistocene but mandatory during the Holocene? A climate change hypothesis. *Soc Am Archaeol* 2001;**66**:387–411.
14. Munro N. *Small game, the younger Dryas, and the transition to agriculture in the southern levant*. 2003.
15. Anderson DG. Multiple lines of evidence for the possible human population decline/settlement reorganization during the Early Younger Dryas. *Quat Int* 2011;**242**:570–83.
16. Herrera RJ, et al. *Genomes, evolution and culture: past, present, and future of humankind*. West Sussex, UK: Wiley Blackwell; 2016. p. 198–200.
17. Brown T. The complex origins of domesticated crops in the fertile crescent. *Trends Ecol Evol* 2009;**24**. https://doi.org/10.1016/j.tree.2008.09.008. Cell Press.
18. Banning E. The neolithic period: triumphs of architecture, agriculture, and art. *Near E Archaeol* 1998;**61**:188–237.
19. Bar-Yosef O, et al. Netiv: Hagdud: an Early Neolithic village site in the Jordan Valley. *Maney Publishing: J Field Archaeol* 1991;**18**:405–24.
20. Moore G, et al. The excavation of Tell Abu Hureyra in Syria: a preliminary report. *Proc Prehist Soc* 1975;**41**: 50–77.
21. Diamond J. Location, location, location: the first farmers. *Science* 1997;**278**:1243–4.
22. Lewin R. A revolution of ideas in agricultural origins. *Science* 1988;**240**:4855.
23. Zohary D. Monophyletic vs. polyphyletic origin of the crops on which agriculture was founded in the Near East. *Genet Resour Crop Evol* 1999;**46**:133–42.
24. Scheffler T. Fertile crescent, orient, Middle East: the changing mental maps of southwest Asia. *Eur Rev Hist* 2003;**10**:253–72.
25. Bellwood P. *First farmers: the origins of agricultural societies*. Blackwell Publishing; 2006. p. 89–97.
26. Jones M, Liu X. Origins of agriculture in east Asia. *Science* 2009;**324**:730–1.
27. Fuller DQ. Contrasting patterns in crop domestication and domestication rates: recent archaeobotanical insights from the Old World. *Ann Bot* 2007;**100**:903–24.
28. Silva F, et al. Modelling the geographical origin of rice cultivation in Asia using the rice archaeological database. *PLoS One* 2015. https://doi.org/10.1371/journal.pone.0137024.
29. Houyuan L, et al. Earliest domestication of common millet (*Pancium miliaceum*) in east Asia extended to 10000 years ago. *Proc Natl Acad Sci USA* 2009. https://doi.org/10.1073/pnas.0900158106.
30. Nai H. *New archaeological finds in China*. IstitutoItaliano per l'Africa e l'Oreinete; 1979. p. 7–13.
31. Anderson E. *The food of China*. Yale University Press; 1988. p. 1–253.
32. Lawler A. Bridging east and west. American association for the advancement of science. *Science* 2009;**325**:940–3.
33. Lawler A. Millet on the move. *Science* 2009;**325**:942–3.
34. Hansen J, Renfrew J. Paleolithic-Neolithic seed remains at Franchthi cave, Greece. *Nature* 1978;**271**:349–52.
35. Jacobsen T. *Franchthi cave and the beginning of settled village life in Greece*. American School of Classical Studies; 1981. p. 303–19.
36. Janick J. Ancient Egyptian agriculture and the origins of horticulture. p. 23–39. *Proc inter sympos Medit Hort. Acta hort*, vol. 582. 2002. p. 55–9.
37. Decker JE, et al. Worldwide patterns of ancestry, divergence, and admixture in domesticated cattle. *PLoS Genet* 2014;**10**:e1004254. https://doi.org/10.1371/journal.pgen.1004254.
38. Curry A. Ancient migration: coming to America. *Nature* 2012;**485**:30–2.
39. Smith BD. The initial domestication of *Cucurbita pepo* in the Americas 10,000 years ago. *Science* 1997;**276**:932–4.
40. Benz BF. Archaeological evidence of teosinte domestication from Guilá Naquitz, Oaxaca. *Proc Natl Acad Sci USA* 2005;**98**:2104–6.
41. Bushnell GHS. The beginning and growth of agriculture in Mexico. *Phil Trans Roy Soc Lond* 1976;**275**:117–20.
42. Sanjur OI. Phylogenetic relationships among domesticated and wild species of Cucurbita (Cucurbitaceae) inferred from a mitochondrial gene: implications for crop plant evolution and areas of origin. *Proc Natl Acad Sci USA* 2002;**99**:535–40.
43. Weller K. *Maize (corn) may have been domesticated in Mexico as early as 10,000 years ago*. 2008. www.sciencedaily.com/releases/2008/06/080627163156.htm.
44. Price D. Ancient farming in eastern north America. *Proc Natl Acad Sci USA* 2009;**106**:6427–8.
45. Swanson-Wagner R, et al. Reshaping of the maize transcriptome by domestication. *Proc Natl Acad Sci USA* 2012;**109**:11878–83.

46. Piperno DR, et al. Starch Grain and phytolith evidence for early ninth millennium B.P. maize from the Central Balsas River Valley, Mexico. *Proc Natl Acad Sci USA* 2009;**106**:5019–24.
47. Mayne RJ. *History of Europe*. [Britannica Online Encyclopedia]. 2010.
48. Kennedy J. Pacific Bananas: complex origins, multiple dispersals. *Asian Perspect* 2008;**47**:75–94.
49. Perrier X, et al. Multidisciplinary perspectives on banana (*Musa* spp.) domestication. *Proc Natl Acad Sci USA* 2011;**108**:11311–8.
50. McGovern P. *Ancient wine: the search for the origins of viniculture*. Princeton University Press; 2003.
51. Lacombe P, et al. Historical origins and genetic diversity of wine grapes. *Trends Genet* 2006;**22**:511–9.
52. Myles S, et al. Genetic structure and domestication history of the grape. *Proc Natl Acad Sci USA* 2011;**108**:3530–5.
53. Gamba C, et al. Genome flux and stasis in five millennium transect of European prehistory. *Nat Commun* 2014;**5**:5257. https://doi.org/10.1038/ncomms6257.
54. Herrera RJ, et al. *Genomes, evolution and culture: past, present, and future of humankind*. West Sussex, UK: Wiley Blackwell; 2016. p. 148–51.
55. Freedmen A. Dogs domesticated before farming. *Nature* 2014;**505**:589–90.
56. Freedman AH, et al. Demographically-based evaluation of genomic regions under selection in domestic dogs. *PLoS Genet* 2016;**12**:1–23.
57. Thalmann O. Complete mitochondrial genomes of ancient Canids suggest a European origin of domestic dogs. *Science* 2013;**342**:871–4.
58. Larson G, Bradley DG. How much is that in dog Years? The advent of canine population genomics. *PLoS Genet* 2014;**10**:e1004093.
59. Skoglund P. Ancient wolf genome reveals an early divergence of domestic dog ancestors and admixture into high-latitude breeds. *Curr Biol* 2015;**25**:1515–9.
60. Hiendleder SK, et al. Molecular analysis of wild and domestic sheep question current nomenclature and provides evidence for domestication from two different subspecies. *Proc Biol Sci* 2002;**269**:893–904.
61. Friend J. *Cattle of the world*. Blanford Press; 1978. p. 1–147.
62. Vigne J. The origins of animal domestication and husbandry: a major change in the history of humanity and the biosphere. *C R Biol* 2011;**334**:171–81.
63. McTavish EJ, et al. New World cattle show ancestry from multiple independent domestication events. *Proc Natl Acad Sci USA* 2013;**110**:1398–406.
64. Achilli A, et al. Mitochondrial genomes from modern horses reveal the major haplogroups that underwent domestication. *Proc Natl Acad Sci USA* 2012;**109**:2449–54.
65. Scarre C. In: Kindersley D, editor. *Smithsonian timelines of the ancient world*. 1993. p. 176.
66. Al-Nasser A, et al. Overview of chicken taxonomy and domestication. *World Poultry Sci J* 2007;**63**:285–300.
67. Eriksson J, et al. Identification of the yellow skin gene reveals a hybrid origin of the domestic chicken. *PLoS Genet* 2008. https://doi.org/10.1371/journal.pgen.1000010.
68. Brown P, Konner M. An anthropological perspective on obesity. *Ann N Y Acad Sci* 2006;**499**:29–46.
69. Cochran G, Harpending H. *The 10,000 year explosion*. NY: Basic Books; 2009. p. 65–84.
70. Jew S, et al. Evolution of the modern human diet: linking our ancestral diet to modern functional foods as a means of chronic disease prevention. *J Med Food* 2009;**12**:925–34.
71. Fessler DMT. Reproductive immunosuppression and diet: an evolutionary perspective on pregnancy sickness and meat consumption. *Curr Anthropol* 2002;**43**:19–61.
72. Nisheeth K, et al. Cardiomyopathy associated with celiac disease. *Mayo Clin Proc* 2005;**80**:674–6.
73. Gingrich RE. Effects of diet and immunosuppression of *Mus Musculus* on infestation, survival, and growth of *Hypoderma* Lineatum Diptera: Oestridae. *J Med Entomol* 1973;**10**:482–7.
74. Tayles N, et al. Agriculture and dental caries? The case of rice in prehistoric Southeast Asia. *World Arch* 2000;**32**:68–83.
75. O'Neill J. *The Irish potato famine*. ABDO Publishing Company; 2010. p. 6–23.
76. Bocquet-Appel J. When the World's population took off: the springboard of the neolithic demographic transition. *Science* 2011;**333**:560–1.
77. Fracchia H. The neolithic demographic transition and its consequences. *Can Stud Popul* 2010;**37**:613–5.
78. Cochran G, Harpending H. *The 10,000 year explosion*. NY: Basic Books; 2009. p. 85–128.
79. Goldbatt L. *Aflatoxin*. Academic Press; 1969. p. 1–40.
80. Larsen C. The agricultural revolution as environmental catastrophe: implications for health and lifestyle in the Holocene. *Quat Int* 2006;**150**:12–20.

81. Ezzati M, et al. A century of trends in adult human height. *eLife* 2016;**5**:e13410. https://doi.org/10.7554/eLife.13410.
82. Steckel RH. Health and nutrition in the PreIndustrial era: insights from a millennium of average heights in northern Europe. In: Allen RC, Bengstsson T, Dribe M, editors. *Living standards in the past: new perspectives on wellbeing in Asia and Europe*. Oxford: Oxford University Press; 2005. p. 227–53.
83. Goldman J. *How human culture influences our genetics*. British Broadcasting Corporation; 2014.
84. Robayo-Torres C. *Genetics home reference: LCT*. United States National Library of Medicine; 2008.
85. Mayo Clinic Staff. *Diseases and conditions: malaria*. Mayo Clinic; 2015.
86. Gilbert R, Mielke J. *The analysis of prehistoric diets*. New York: Academic Press; 1985.
87. Minnis P. *Social adaptation to food stress: a prehistoric Southwestern example*. University of Chicago Press; 1985.
88. Hawks J, et al. Recent acceleration of human adaptive evolution. *PNAS* 2007;**104**:20753–8.
89. Truswell A, Hansen J. Diet and nutrition of hunter-gathers. In: *Health and disease in tribal societies*. Ciba Foundation; 1977. p. 213–26.
90. Floud R, McCloskey D. *The economic history of Britain since 1700: agriculture during the industrial revolution*. Cambridge University Press; 1994. p. 96–132.
91. Evenson R, Gollin D. Assessing the impact of the Green revolution, 1960 to 2000. *Science* 2003;**300**:758–62.
92. Ma H, et al. Correction of a pathogenic gene mutation in human embryos. *Nature* 2017;**548**:413. https://doi.org/10.1038/nature23305.
93. Huypens P. et al. Epigenetic germline inheritance of diet-induced obesity and insulin resistance. *Nature Genetics* 2016;**48**:497–500.

Further Reading

Wollstonecroft M. Investigating the role of food processing in human evolution: a niche construction approach. *Archeol Anthropol Sci* 2011;**3**:141–50.

Ormandy E, et al. Genetic engineering of animals: ethical issues, including welfare concerns. *Can Vet J* 2011;**52**: 544–50. United States National Library of Medicine National Institutes of Health.

CHAPTER

14

The Silk Roads

I can see clothes of silk, if materials that do not hide the body, nor even one's decency, can be called clothes... Wretched flocks of maids labour so that the adulteress may be visible through her thin dress, so that her husband has no more acquaintance than any outsider or foreigner with his wife's body. ***Lucius Annaeus Seneca***[1]

SUMMARY

Although the Silk Roads are currently associated with images of caravans traveling through the deserted landscapes of Central Asia carrying silk and other precious cargo, these routes have even more ancient origins. In fact, the Silk Roads can be traced back to the paths used by pastoralist nomads, such as the Sredny Stog, Kurgan, and Scythians, who were involved in the domestication and trading of horses. The Steppe Roads used by these horse cultures date back to approximately 10 kya at the end of the Upper Paleolithic, approximately 5k years preceding the silk trade. The territory of these early merchants extended longitudinally from the Far East to the Danube River in Eastern Europe, a distance of approximately 10,000 km. Although these wandering groups are generally associated with a culture that existed somewhere between primitive hunter-gatherers and sedentary agrarian civilizations, a stunted stage in cultural evolution, in recent years, it has become increasingly clear that horse cultures, such as the Scythians, were much more than simple pastoralists moving around, always looking for finer pastures in Central Asia. They were appreciative of fine arts that they traded for their fine horses. With the domestication of the silk moth *Bombyx mandarina* and the artificial selection of *Bombyx mori*, a new commodity emerged. Silk became a luxury item and a symbol of wealth and beauty. This new commodity was a key factor in the conversion of the established conduits of the Steppe Roads into the Silk Roads. The Silk Roads encompassed a network of passages covering some 7000 km extending from the Yellow River Valley in East China to the Mediterranean Sea. The Silk Roads were used not only to transport silk but also a number of other items, such as ceramics, glass, precious metals, gems, livestock, textiles, spices, grains, vegetables, fruits, animal hides, tools, woodwork, metalwork, religious objects, and artworks. In addition, the Silk Roads were channels for the dissemination of ideas, beliefs, and technologies. Over time, the cities and oases connected by the Silk Roads became centers of intellectual

Ancestral DNA, Human Origins, and Migrations
https://doi.org/10.1016/B978-0-12-804124-6.00014-8

and cultural exchange that eventually developed into hubs of learning. In general, science, technology, arts, and literature moved across Eurasia along the Roads at a speed not seen before in human history, ushering in a new age of information transfer. The outcome of this technological/information revolution was a cultural unity among different populations practicing diverse survival modes. These cultural connections enabled nomads, woodland foragers, pastoralists, and agrarian communities to interact. These interactions became even greater during the dynastic period in China and the empires of the West.

Yet, not all the impacts of the Silk Roads were positive for the cultures that relied on it. On the negative side, it is known that pandemics, such as the bubonic plague, anthrax, and leprosy, traveled and spread through the Roads. Another negative outcome of the Roads was its appeal to hordes of nomads, particularly the Mongols who totally devastated cities, many of which were never able to recover and were eventually abandoned.

The Chinese were very diligent in keeping the process of silk making a secret. It is quite significant that even after three centuries subsequent to the discovery of silk, Romans had no clue of what silk was or the process of growing the silk moth. It was not until the Byzantine period that the silk secret was procured out of China and taken to Constantinople by two priests commissioned by Justinian I. Since then, silk production and distribution was no longer a Chinese monopoly.

The decline of the Silk Roads came about from a series of events in tandem. The downfall started with the An Shi Rebellion of AD 755–762 during the Tang Dynasty. After the uprising, the Silk Roads started to deteriorate with large expanses in Central Asia falling into the control of Tibet and neighboring empires. Further deterioration occurred as the Mongolian Empire disintegrated after the Mongol emperor Möngke Khan (the fourth Khan) died in AD 1259 in battle. Without a declared successor, this event initiated a cascade of internal feuds within the line of descent to the title of Great Khan, which culminated in the Toluid Civil War. The Age of Discovering or Exploration, which established new, faster maritime routes by a number of European powers, provided the *coup de grâce* or last blow to the Silk Roads.

ON THE NATURE OF THE ROADS

How Did It Work?

Since the term *Die Seidenstrassen* (the Silk Road) was coined by the German geographer Ferdinand von Richthofen in the late 1800s,[2] the name is usually employed in the singular. Generally speaking the Silk Roads connected the Yellow River Valley in what is currently northeastern China to the Mediterranean Sea. It included a number of regions that today are known as Korea, China, India, Pakistan, Tibet, Uzbekistan, Tajikistan, Kyrgyzstan, Turkmenistan, Afghanistan, Kazakhstan, Persia, Iraq, Turkey, Greece, and Italy. It also extended through latitudinal routes into India, the Near East, Africa, the Mediterranean basin, and northern Europe. In addition to the terrestrial set of roads, sea routes were also established.

In recent times, it has become increasingly clear that it was not a single road but many paths, constantly changing and branching out in different directions as needed as a result of climatic changes, wars, or geopolitical issues. The different trajectories were modified with time as circumstances and challenges shifted. During its existence, sections of these arteries grew at times only to be truncated later. Looking back in time, these properties seem to reflect a system very responsive to the existing environment. For example, traders from the Roman Empire tended to bypass Parthian (Persian Empire in power from 247 BC to AD 224) territory, enemies of Rome in wartime, by traveling routes to the north, across the Caucasus and over the Caspian Sea. In addition, as the networks of rivers dried out or overflowed in Central Asia, the caravans adjusted their paths accordingly. In retrospect, it seemed as if this ever-changing network of communication and trade routes had a life of its own.

Thus, today many scholars refer to this dynamic ancient transportation system in the plural. Geographically, these roads encompassed numerous reticulations that at times branched out and at other times converged latitudinally across the northern steppes of Central Eurasia 40° north of the equator, spreading out approximately 10° to the north and south. These conduits generally tried to avoid the harsh conditions of the Tibetan Plateau and the Taklamakan Desert. Although goods were transported routinely along the entire east–west axis of the Silk Roads, individual caravans generally only covered and specialized in much shorter sections of the entire length of the Roads. Usually, any given caravan transported items between key adjacent towns or oases. Once the convoy reached a certain oasis, goods were unloaded and exchanged, and the merchandise was moved to the next destination by a different band of merchants. Traders normally paused and rested once they arrived and conducted business at specific towns.

Caravansaries (inns with large courtyards for the overnight stay of animals) were large guesthouses available and suited to accommodate traveling merchants and their cargo during their stopover in the trading posts. Caravansaries were strategically located within a day's trip from each other, about 30–40 km apart. These relatively short distances were manageable, guarded, and maintained by locals providing safety to the convoys. These establishments played multiple roles. Foremost, it facilitated the movement of people and goods along the routes. In addition, these temporary quarters not only allowed the consummation of deals among traders but also provided informal meeting places for merchants and travelers to exchange information, for example, the best routes, and supply and demand of specific items in different locations along the Roads. Possibly more impacting, in the long run, caravansaries allowed people from different places to meet and exchange ideas, cultures, and languages. These informal meeting places facilitated learning foreign languages and costumes needed to carry out business efficiently and to prevent misunderstandings and insulting traders of different backgrounds.

It is likely that after resting for a few days, traders replenished their exhausted supplies, picking up fresh merchandise, and then started traveling back to their point of departure or to a new destination. It is also likely that specific items were exchanged with rural local populations for fresh supplies as the convoy moved in between oases. Imagine a continuous back and forth movement of goods along routes that eventually generated a net flow of items that reached the extremes of the Silk Roads in Europe and East Asia. Although most of the items traded were considered luxurious and only a small sector of the populace were able to acquire and enjoy them, the manufacturing, transportation, and trading of the goods benefited a much wider spectrum of the population, and thus, the impact of the trading had more profound consequences on the common people.

Today it is known that the Silk Roads were not the original transcontinental transportation system. Indeed, at least 5000 years before the Silk Roads emerged, the precursor known as the

Steppe Roads were initiated and developed by Eurasian pastoralists, approximately 10 kya at the end of the Upper Paleolithic. They linked Eastern Europe, Central Asia, China, South Asia, and the Near East economically, politically, and culturally.[3] The Steppe Routes spread approximately 6000 miles from the mouth of the Danube River to East Asia. Even from these early years the Steppe Roads were a source of unity linking Eurasian populations.

The Silk Roads involved many different types of goods and exchanges. Although silk was the single and most unique commodity transported, due to its beauty, light weight, high value, and multiple uses, other items were carried as well, such as ceramics, glass, precious metals, gems, livestock, textiles, spices, grains, vegetables, fruits, animal hides, tools, woodwork, metalwork, religious objects, artworks, as well as genes.[4] Fig. 14.1 presents a general visual representation of the valuables traded in the main stations along the land routes. These precious items were transported over very long distances in caravans. When rivers were encountered, the merchandise was loaded on rafts and moved to the opposite shore as humans and animals crossed the watercourses and rivers.

Complementing the land trajectories, a number of maritime routes existed, which specialized in the transportation of spices (the Spice Roads), such as cinnamon, pepper, ginger, cloves, and nutmeg, from Indonesia as well as fabrics, woodwork, precious stones, metalwork, incense, timber, and saffron from the west coast of Japan to the east coast of China, Southeast Asia, and India, eventually reaching the Near East and the Mediterranean.

Ideas, Beliefs, and Technologies

The traffic of ideas between groups was multidirectional, and the extent of the generated transcontinental unity among populations is specifically illustrated in the ancient Oldowan and Acheulean lithic traditions shared by Paleolithic people, such as the Venus figurines made by early *Homo sapiens* throughout Eurasia. Later in the Roads of the middle Neolithic, numerous ideas, beliefs, and inventions were transmitted. In China, paper and, later, the printer were invented.[5] Paper was invented during the Han Dynasty (206 BC–AD 220), when the Silk Road trade was beginning to flourish. In Asia, these discoveries allowed Buddhist priests to duplicate their religious books. In the West, the two creations

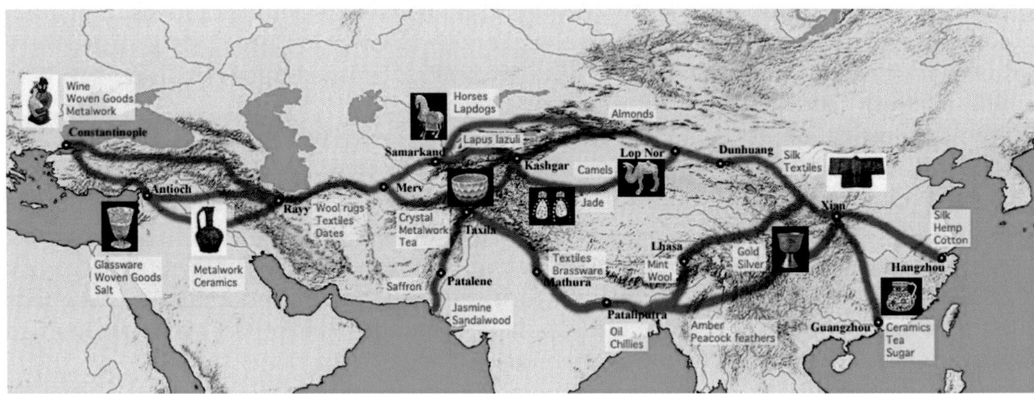

FIGURE 14.1 Goods traded along the Silk Roads. *From http://depts.washington.edu/silkroad/exhibit/trade/trade.html.*

were very effectively combined to duplicate documents of different types. Islamic traders transmitted medical innovations such as inoculations via the Roads. Military technologies, including certain siege weapons and gunpowder, were invented in China and transported to the West, courtesy of the Mongols. The noria (a series of buckets on a wheel for raising water) or irrigation waterwheel was created in Roman Syria and subsequently spread across Eurasia from West Europe to China. Edible plant species also traveled the Roads. For example, apples traveled from the steppes in both directions, oranges were dispersed from China to the Mediterranean basin by marine routes, and in the opposite direction grapes reached China.

Religions were also spread eastward into China from their place of origin in the Near East or India. All major religions disseminated throughout Eurasia via the Silk Roads. Judaism, for example, penetrated West Asia and Europe utilizing Roman routes, part of the trade Roads. Buddhism traveled the Silk Roads spreading Buddhist art and building shrines along its paths in distant locations such as Bamiyan in Afghanistan, Mount Wutai in China, and Borobudur in Indonesia. In the case of Buddhism, the religion experienced a rebirth when it dispersed into the Far East as it was dwindling in India, its birthplace. Chinese Buddhist priests routinely traveled the Route in pilgrimages to India bringing back with them religious manuscripts, artifacts, and recollections of their trips that today constitute a highly valuable source of information. Christianity, Islam, Hinduism, Zoroastrianism, and Manichaeism used the Roads to carry their messages as the faithful encountered travelers and merchants who became devotees and passed the religions to their homelands. A well-known example of religious transmission along the sea Roads is the case of how Hinduism and then Islam were transported into Indonesia and Malaysia via the maritime trade routes from India and Arabia to Southeast Asia and Oceania. As a result, today the islands of Indonesia represent a collage of various proportions (depending on the island in question) of these two major religions.

It is also important to underscore that a good number of the cities connected by the Silk Roads became centers of intellectual and cultural exchange that eventually developed into hubs of learning. In general, science, technology, the arts, and literature were disseminated across Eurasia along the routes at a speed not seen before in human history, promoting a quantum leap in information transfer comparable with that of the technological/information revolution that impacts our societies today.

Negative Impacts

On the negative side, it is known that pandemics, such as the bubonic plague, anthrax, and leprosy traveled and spread through the Roads. In the case of the bubonic plague, its transmission along the Silk Roads was facilitated because the infection is endemic to local rodents. It has been theorized that fur containing plague flea eggs were transported to Near Eastern ports. There they hatched into fleas that then infected local rats that were taken on boats to Italy and then moved all over Europe. In London, two-thirds of the populace died as a result of the plague.

Another negative outcome of the Roads was its appeal to hordes of nomads, particularly the Mongols. Mongols sacked and massacred the populace of towns along the Silk Roads. Sadly, some of those besieged cites were never able to recover from the damage inflicted and have remained abandoned ever since.

Beyond the Trade of Goods

In addition to the mere economic importance of the Roads providing a system for the exchange of goods, which led to progressive and accelerated prosperity of the people, it was a system of communication among a diversity of people living in different ecosystems and separated by large distances. Thus, it could be argued that the connections among different groups of people, facilitated by the Roads over the millennia of traffic, have fostered some degree of ideological tolerance. Maybe not necessarily out of choice, this flexibility with other fellow humans stems from the necessity to do efficient business. Thus, clearly these migrational networks served a multitude of functions. Yet, perhaps the most enduring byproduct of the Roads was their contribution to the cultural cohesiveness among populations within Eurasia and around the world.

Furthermore, it may be argued that the Roads promoted some degree of cultural unity among different populations practicing diverse survival modes. Initially, during the end of the Paleolithic and into the Bronze Age, the transecological exchanges linked all the regions within the networks. These cultural connections provided interaction among nomads, woodland foragers, pastoralists, and agrarian communities. In addition to military conflicts and genocide, it could be theorized that these networks provided for cross-fertilizations of ideas and likely ushered the Agricultural Revolutions in West and East Eurasia. This unity generated by the Roads also explains the common threads that are seen today and throughout history in the continent in the form of common technologies, styles, cultures, religions, and pandemics. It is known that many travelers ventured onto the Silk Roads not as traders but for adventure and to learn through intellectual and cultural exchange.

HOW OLD ARE THE SILK ROADS?

To estimate the age of the Silk Roads, we must first address the question of the Steppe Roads' age. To do this, it is necessary to first define what the Steppe Roads were. In their simplest form the Steppe Roads may be characterized as trading routes among communities. The genesis of these communities resulted from the surplus of food and tools resulting from the development of pastoralism, agriculture, and domestication at the end of the Paleolithic. These developments were linked to changes in climatic conditions subsequent to the last glacial maximum (LGM). Thus, in its broadest sense, the Roads were initiated, as some nomadic groups became semisedentary, signaling the beginning of community life. Irrigation farming and human-driven artificial selection of animal and plant stocks by some of these early communities generated excess food and supplies, which in turn motivated independent communities to trade. These events had profound impacts on the development of a novel economic system. It led to the exchange of experiences and resource materials. Initially, during the Late Paleolithic, this new way of life involved small-scale trade in precious stones, such as jade, shells, and other stones or weapons, among communities separated by small distances. Furthermore, it has been postulated that in addition to farming and animal husbandry, the development of craftsmanship and specialized labor in the form of highly skilled artisans contributed to more stable settlements and more defined routes of trade connecting specific communities who were most adept at providing specific goods and skills with each other.

FIGURE 14.2 Steppe at the East Ayagoz region of Kazakhstan. *From en.wikipedia.org/wiki/Steppe_Route.*

Subsequent to the LGM and deglaciation about 14.5 kya, as Central Asia became warmer and more humid, the terrain traversed by the Steppe Roads was transformed into grasslands (Fig. 14.2). This increase in temperature and wetness peaked about 7000 ya. These extensive prairies provided more reliable food resources in the form of wild cereals and game. Several of the wild floras eventually became agriculturally domesticated, the most notable of these being wheat and barley. In addition to the agricultural uses of the plains, these open landscapes served as pastures for a number of ancestral animal species that subsequently became stock breeds, such as horses, sheep, goats, and donkeys. Specifically, domestication of certain animals, such as the Mongolian or Przewalski's horse (*Equus przewalskii*) and the two-humped Bactrian camel (*Camelus bactrianus*) about 6 kya, certainly helped create a new era of communication among settlements along the Central Asian steppes by increasing the speed of movement and facilitating the transportation of heavy loads over long distances. Without the domestication of these species the Silk Roads would have never come to fruition.

HORSE PASTORALISM

Nomadic Cultures of the Steppes

During the Neolithic, as Eurasian settlements grew in number and size, a number of nomadic groups (Greek term for roaming about for pastures) in the Steppes of Central Asia began to engage in horse pastoralism and trade, which involved the periodic movement of herds to different locations in search of pasture.[6] This chapter posits that these activities initiated trade that eventually culminated in the Silk Roads thousands of years later. Many of these nomadic communities were not primitive barbarians, as is sometimes suggested in the literature. Together with agrarian populations, nomads were the next sociocultural group that stemmed from the Agricultural Revolution. The two types of existence evolved side by side and were interdependent. The farmers produced the goods, whereas the nomads were the marketers and distributors of the goods as well as the providers of quality horses.

Importantly, the growth of these nomadic horse cultures was one of the earliest developments leading to the creation of the Steppe Roads. In fact, it can be argued that these nomadic populations were at the roots of both the Steppe and Silk Roads. In other words, these trading routes had their genesis not in what is today China but in the grasslands of Central Asia, and the commerce related to the domesticated horse. From there, as trade grew in geographical scope and diversity of products, these migrant communities became richer and more influential as intermediates and distributors.

Sredny Stog People

The Sredny Stog culture was a pre-Kurgan group of people that lived approximately 7–6 kya in what is today Ukraine. By approximately 6000 ya, traveling Sredny Stog communities from Eastern Ukraine and south Russia, a region considered by many to be the homeland of the Indo-European languages, began to domesticate the small wild Przewalski's horse.[7] The Sredny Stog people not only initiated horse domestication, but they were also involved in trade with a number of agrarian societies, notably the Cucuteni–Trypillian culture to the west in what is currently western Ukraine and northeastern Romania. The Cucuteni–Trypillians prospered during the Neolithic–Eneolithic (7–5.5 kya) period and built the largest homesteads of the Neolithic in Europe.[8] Some of their towns were impressive, made up of several thousands of homes and populated by as many as 40,000 people. Some of the homes were two-story structures, the lower level made up of mud and the upper one of clay to keep the weight of the structure to a minimum. A reconstruction of a Cucuteni–Trypillian dwelling is illustrated in Fig. 14.3. The size of these Cucuteni–Trypillian settlements suggests considerable growth and prosperity in these sedentary communities as well as the potential of surplus goods to trade with outsiders, such as the Sredny Stog groups of the steppes. The level of sophistication and ingenuity of these agricultural cultures, as demonstrated in their beautiful art (Fig. 14.4), likely attracted the interest of refined nomad horse traders from the steppes to the east, such as the Sredny Stog. Scenarios may be envisioned in which the Sredny Stog nomads provided domesticated horses while the Cucuteni–Trypillians gave comestibles, tools, and objects of beauty in return. Curiously, the Cucuteni–Trypillians periodically burned their cities. Typically, their settlements lasted less than 100 years and the new structures were simply erected on top of the old ones.[9] Remarkably, the town of Poduri in Romania was burned and rebuilt a total of 13 times. There must have been some strong compelling motivation to justify this type of destructive behavior considering the time and effort it takes to rebuild an entire city. It is possible that the burning was done to eradicate epidemics or to put an end to phenomena the inhabitants perceived as malevolent.

Kurgan Horse Culture

The Kurgan Horse Culture is thought by some to derive from Sredny Stog groups. Kurgan origins can be traced back to the early part of the fourth millennium before the current era. They were likely the first speakers of the Proto-Indo-European mother language that they

FIGURE 14.3 Reconstruction of a Cucuteni–Trypillian home. *From http://ukrainaincognita.com/sites/default/files/u3/trypillia_p_muzei6.jpg.*

subsequently spread throughout extensive areas of Eurasia starting about 5000 ya.[10] Although the Kurgan Culture is considered by some to encompass *all* horse cultures from the Neolithic characterized by burial mounds, most scholars recognize distinct nomad groups from the steppes such as the Bug-Dniester (8 kya), Samara (7 kya), Khvalynsk (7 kya), Dnieper-Donets (6 kya), Maikop-Dereivka (5 kya), Yamna (4 kya), and Usatovo (6 kya). Originally from a region just north of the Black Sea, the Kurgans like the Sredny Stogs were also involved in extensive horse trading, and by about 5 kya, Kurgans were doing business throughout the Steppe Roads and Eastern Europe.

Typically, Kurgan populations inhabited flat steppe grasslands near wooded areas and water, which provided pasture for their herds and protection. Their utensils were made from elk antlers, cattle and sheep bones, and boar tusks. A common tool was a modified multi-purpose elk antler presumably used for striking and digging. The findings of wooden bows, arrows, bone harpoons, and fishhooks among their possession indicate that in addition to trading, they were involved in hunting and fishing. The discovery of grindstones and pestles as well as millet grains and melon seeds in their camps may indicate that although Kurgans were not farmers, they may have processed grains into flour. The grains were obtained from agrarian communities in exchange for horses. Although the Kurgans of the Caucasus region in their western range were capable of collecting gold and silver in pure form from panning the rivers, and Kurgan artisans could have created some of the beautiful art in the form of vases, beads, rings, and animal figurines, it is likely that many of the artistic pieces were procured through trade with sedentary communities.

FIGURE 14.4 Pottery and art of the Cucuteni–Trypillian culture. *From https://svasticross.blogspot.com/2015/07/swastika-cucuteni-trypillian.html; http://www.pysanka.com/en/images/T-Fox%20Hunt%20$35.JPG.*

Their movements along the steppes were also facilitated by their use of primitive chariots, two- and four-wheeled wagons (Fig. 14.5 illustrates a toy model) with big wheels of solid wood without spokes. Their range of operation was huge. To the west, they extended into what is today Azerbaijan, Georgia, Armenia, and the Near East where they traded with ancient farming groups. To the East, the Kurgans entered the steppes of the Sayano–Altai Mountains in Central Asia and Kazakhstan. Southward in the Turkmenistan and Aral steppes, they interacted with agricultural communities in what is today northern Iran and Afghanistan. Considering this extensive area of operation within Eurasia, it is not surprising to find a longitudinal range of different proportions of European and Mongoloid morphological characteristics among Kurgan skulls.

FIGURE 14.5 Kurgan toy representing a four-wheeled chariot, clay model art. *From https://www.pinterest.se/pin/557109416389471447/; http://mek.oszk.hu/02100/02185/html/img/5_165a.jpg.*

The Kurgan culture is known for their pit burials or tumulus, which at times were treasure graves containing items such as gold, silver, and precious stones procured from their agrarian trading partners. The name Kurgan derives from a Türkic word that means burial mound or castle. The tradition of burial mounds was transmitted to a number of subsequent cultures including the Scythians, which will be discussed in the next section. Although the common people were placed in simple graves with a few ordinary, everyday personal possessions, such as arrow tips, beads, and amulets, leaders of the Kurgan communities were buried in pits made to resemble their homes. These tombs were loaded with items for survival and comfort in the afterlife. The bodies were placed in an enclosure made out of wood. The corpse was nicely dressed with gold ornaments, including images of animals and the sun. It seems that the amount of precious metal accompanying the dead was directly proportional to the social status of the individual while alive. Skeletons of women and children have been found in these tumuli, suggesting sacrifice of wife(s) and children to accompany the deceased.

For the Kurgans, their ponies were a multipurpose commodity, used as a source of food (i.e., meat and milk) and a currency for trade with distant agrarian settlements in exchange for edibles and precious objects produced by sedentary groups. Because cattle, sheep, pig, and goat bones have been recovered from Kurgan sites, it is likely that the goods obtained from farmers included livestock. The Kurgans went beyond the Sredny Stogs in developing into highly successful groups of nomads that combined horse husbandry, migrant existence, and long-distance trading. Their way of life provided for long-distance travel, contacts, and influences that extended east to west over large expanses of Eurasia. In addition, they practiced a primitive metal industry that was used to manufacture weapons and utensils. In retrospect, considering all the characteristics exhibited by the Kurgan culture, it can be argued that these nomads built an early precursor to the future Silk Roads.

In addition to utilizing horses as a mode of transportation, Kurgans also employed domesticated Bactrian double-humped camels. Camels are well adapted to the arid environment of Central Asia and are capable of traveling up to 160km per day at a speed of 16km per hour. Caravans carrying a heavy burden could cover a distance of 30–40km at a speed of about 3.5km per hour. It is thought that the Bactrian camel was domesticated approximately 5kya in the region east of the Zagros Mountains in what is today central Iran. From there the domesticated *C. bactrianus* was adopted by the civilizations of the steppes. In the later stages of the Silk Roads, this animal became an iconic symbol of the caravans.

Scythians—Sophisticated Nomads

The Scythians or Saka were another group of mobile pastoralists that flourished and prospered in the steppes of Central Asia. As a culture, they originated in what is today Iran about 5kya and exerted their influence on the Steppe Roads as traders until as recently as the Middle Ages.[11] Some of their characteristics as a culture, such as a keen appreciation for precious metals as well as their use of burial mounds, argue for an ancestral connection to the Kurgan culture that thrived centuries before. They were a heterogeneous group of people and spoke a number of Iranian languages, and with time, they expanded and occupied large expanses of the central Eurasian steppes all the way to East Asia. Initially, their realm of control only involved Persian routes, but later in their tenure, they reached China, actively trading with Chinese merchants. At the height of their culture, they extended west to the Near East at the border with current-day Egypt. In addition to trading, these nomads were well versed in mounted warfare. In this mode of cavalry attacks, Scythians used blitzkrieg tactics to surprise the enemy with bows and arrows on horseback with maneuvers similar to the ones employed by the Mongols a millennium later.

Scythians represent a more recent stage in the evolution and growth of nomadic trading communities that preceded and eventually evolved into the Silk Roads. As such, they improved on the speed, distance, and efficiency of transactions. Scythian culture is characterized by more sophisticated weapons, riding gear, and exquisite animal art, some of which are periodically unearthed from their distinctive burial mounds. Yet, considering the large expanse of their territory of operations, it is not clear the degree of sociocultural homogeneity that existed among Scythian populations and their *lingua franca* (if any), and thus, it is possible that their vast dominion was made up of discrete and unique populations. At the pinnacle of their culture, Scythians controlled the entire steppes from Central Europe in their western fringe to Central China to the east. As traders, they operated a complex network of routes, some of them likely inherited from previous migrant groups, such as the Kurgans. Some scholars attribute the commerce with the Scythians to the growth and successes of a number of powerful dominions, such as the Persian, Greek, Indian, and Chinese emperies.

Like the other previously mentioned nomadic horse cultures of the steppes, Scythians have been traditionally envisioned as primitive barbarians who moved from site to site in search for fertile land for their herds. To the ancient Greeks, for instance, the name nomad meant the people who roam about for pasture. This notion was not only popular during ancient times but in some respects still persists today, as the term nomad is generally associated with wandering groups socially somewhere in between the primitive hunter-gatherers and sedentary agrarian civilizations, a stunted stage in cultural evolution, cultures that have not learned yet to settle down and work the land. Yet, in recent years, it has become increasingly clear that horse cultures such as the Scythians were much more than simple pastoralists moving around, looking for fine pasture in Central Asia.

Instead of simple wandering nomads, Scythians were sophisticated merchant traders with a fancy for precious metals and an appreciation for beautiful work of art. They fueled their prosperity by maintaining exchange networks along the steppes 3 kya when they facilitated trade involving Persians and Chinese. Evidence of their wealth and tastefulness for beautiful things is seen in their tombs.

These burial mounds are time capsules illustrating the life of the Scythian elite. Similar to the tumulus of the Kurgan culture, the Scythian interments were designed to provide the upper class of the society with the material and spiritual world that they were accustomed to while alive. These tumuli represent microcosms of a privileged life and an attempt to preserve it even in the afterlife. Although the pits varied in size, typically they were approximately 100 feet in diameter, 10–15 feet deep, and reached a height of up to 10 feet from the ground level with trunks delineating the periphery (Fig. 14.6). Inside these tombs items such as gold ornaments, carved wood, horns, leather, saddles with embroidered cloth, gold plaques with figures of real and mythical animals, textiles, ceramics, bronzes, and horse sacrifices were placed to accompany the deceased. Some of the artifacts show stylistic influences from the different trading partners along the extensive steppe trading networks that they controlled. For example, their art often portrays images of griffins that originated in the Persian Empire to the west and effigies of dragons from China to the east. In some cases, it is not clear if the items were acquired through trading or created by Scythian artisans using inspiration and

FIGURE 14.6 Scythian grave. Tumulus or burial mound. *From www.nytimes.com/2012/03/13/science/from-their-graves-ancient-nomads-speak.html.*

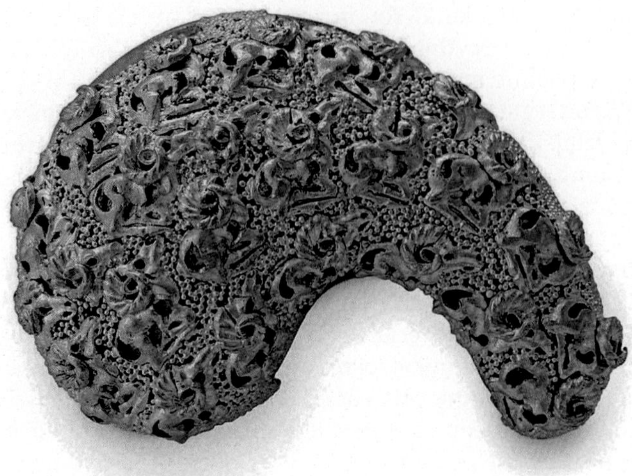

FIGURE 14.7 A teardrop-shaped gold plaque is one of the objects that show the strong social differentiation of nomad society. *From www.nytimes.com/2012/03/13/science/from-their-graves-ancient-nomads-speak.html.*

artistic elements from various trading partners. Scythian artisans also modified the art from different regions to fit their tastes and beliefs. For instance, they created fantastic versions of creatures such as boars curled in teardrop shapes (Fig. 14.7) and two-headed beasts. In addition, cultural influences streamed in the opposite direction, eastward, as Chinese artists began to imitate the autochthonous Scythian style depicting animals interlocked in combat using the traditional Chinese jade as a medium.

The abundance of precious objects found within these interments and the simplicity of common tombs suggest highly socially stratified communities. The items recovered from these people are testament to a nomadic culture who would be the logical ancestors to the future traders of the Silk Roads: highly specialized in long-distance trading with a remarkable fancy for beautiful precious objects.

IMPACT OF EARLY STEPPE ROAD NOMADS ON THE GENETIC MAKEUP OF EURASIA

Just as linguistic affiliations, cultural affinities and anatomical traits are routinely used to assess ancestral relationships among populations; genetic characteristics are also useful. Human genetic markers are analyzed by probing two categories of DNA. One type is the uniparental DNA that is inherited from only one parent; for example, mitochondrial DNA (mtDNA) is passed strictly from mother to child while the Y chromosome is paternally derived. Thus, uniparental markers are excellent for following maternal and paternal lineages. The other type of DNA is referred to as whole genome, which reflects the cellular DNA inherited from both parents (see Chapter 5, section on *DNA analysis* and Chapter 10 section on *Three main types of DNA* for description of DNA types).

Before the incursions of Central Asians into Europe, two major groups of humans penetrated and settled West Eurasia. First, the original Paleolithic anatomically modern humans (AMHs) dispersed into the continent about 45 kya (see Chapter 12), which led to confrontations with our sister species, the Neanderthals, possibly contributing to their demise. The original AMHs were the direct descendants of the out of Africa migrants of around 80 kya that eventually made their way to the Near East and then Europe. This dispersion is still apparent in the genetic constitution of present-day European populations. Y chromosome types I, C, F, and E as well as mtDNA groups U5, U4, HV, and I provide evidence of this original settlement.[12]

Later, about 10 kya, Europe experienced another major wave of migrants, this time from the Near East made up of agriculturists from the Fertile Crescent (see Chapter 12). This event marks the introduction of agriculture, a new technology, and social norms into Europe. These Neolithic populations moved into territories already occupied by the Paleolithic hunter-gatherers resulting in amalgamation with each other. The genetic signature of this second wave is also palpable today.[13] Y chromosome types, such as E1b1b1, G2a, and J as well as mtDNA J, N1a, T1, and U3, signal this invasion. It is noteworthy that about 80% of the present-day European Y chromosome types are from Neolithic agrarian descent, yet only about 11% of the mtDNA is linked to agriculturists.[14] This dichotomy involving paternal and maternal heritage from famers may be the result of reproductive advantage of farming males relative to indigenous hunter-gatherer males subsequent to the invasion and/or a disproportionally larger number of male than female Neolithic migrants.

Genetic studies employing uniparental and whole genome markers also provide evidence in support of extensive migrations of the nomadic populations along the Steppe Roads to the west into Europe and to the east into East Asia. Along these lines, a number of studies of Kurgan and Scythian ancient DNA suggest a major impact on the genetic constitution of Europe dating to the times of the Steppe Roads. Specifically, a genome-wide DNA component shared by the Bronze Age (about 5.8 kya) Central Europeans and nomadic populations, but lacking in older European farmers and hunter-gatherers, was also found in South Asian populations. Remarkably, these data suggest east–west migratory movements possibly driven by Kurgan and Scythian traders moving longitudinally in Eurasia.[13]

It is likely that these longitudinal genetic signals represent trading, yet the massive contribution of nomad DNA also suggests migration of people settling and interbreeding with the native hunter-gatherer and agrarian populations. In other words, it is likely that the Central Asian nomads not only migrated back and forth along the steppes, but some of them converted to a sedentary existence when they confronted the agrarian European populations.

Genome-Wide Markers

In a recent study of 69 ancient Europeans who lived between 8 and 3 kya and using about 400,000 whole genome DNA markers, investigators discovered the signature of a sudden massive migration from the steppes of Central Asia into Europe.[15] The authors attributed the spread to Indo-European-speaking populations travelling into Western and Northern Europe from the Steppes in Central Asia north of the Black Sea. These substantial dispersions of nomads impacted a number of European populations during the Bronze Age. Their genetic impact is particularly seen in the Corded Ware (after the cordlike imprints on their pottery) pastoral populations of 5 to 4 kya in Northern Europe, Central Europe, and Eastern Europe in an area delineated by the Rhine to the Volga rivers. In the Corded Ware populations, the contributions by the three main population components, Central Asians, western hunter-gatherers, and early Neolithic farmers, were 79%, 4%, and 17%, respectively. This nomadic genetic input was so significant that today Europeans overall are a mixture of three main ancient groups: the original hunter-gatherers derived from the Paleolithic Out of Africa dispersal, the Neolithic farmers from the Near East, and the pastoralist nomads from the Central Asian steppes. It is significant that by 3 kya, the Eastern Europe populace was largely replaced by steppe nomad groups. It has been suggested that this nomadic genetic flow from Central Asia is partially the result of the longitudinal movement of horse cultures involved in trading with Europe along the Steppe Road corridor.

In addition, these nomadic migrations extended eastward as well, into Asia and the Altai region carrying with them the Indo-European languages. The genetic signals from these incursions by nomadic cultures into present-day China were gradually diluted by admixture or replaced by East Asian populations that dispersed into East Asia later.[16] This substantial genetic contribution that accounted for about 75% of the DNA of the Corded Ware groups was traced to the nomadic Yamnaya people of the steppes and resulted from rapid migration rather than a long protracted gene flow process. The Yamnaya are considered the ancestors of the Scythians (Fig. 14.8).[17] Both the Yamnayas and Scythians exhibited genetic characteristics from the western steppes (the Caucasus) and South Asia. These genetic contributions corroborate long-distance contacts involving these nomads and territories at the extreme ends

FIGURE 14.8 Scythian horseman. *From http://rolfgross.dreamhosters.com/IndianArtArchitecture/GreekEmpires/Greek%20 Empires.html.*

of their longitudinal ranges and provide support for the idea that Indo-European languages originated with populations such as the Sredny Stog, Yamnayas, Kurgan, and Scythians from the northern fringes of the Black Sea, which subsequently migrated and traded with Europe. Furthermore, it adds credence to the hypothesis that the Indo-European languages were introduced into Europe not by acculturation (cultural transmission without migration of people) but by actual movement of people from Central Asia.[16] The introduction of Indo-European languages to the West and East throughout the steppes indicates that in addition to trading goods and horses, these nomads disseminated their vernacular by way of their travel networks.

In terms of specific genes that may have been passed to European populations, the Steppe nomads may have contributed to the higher stature of the current Northern Europeans. The Steppe nomads, such as the Yamnaya, are responsible for about half of the ancestry of Northern Europeans today.[18] This possibility is congruent with the fact that the migrants from Central Asia were taller than other populations of the Bronze Age.

Inexplicably, the nomad genetic component in European populations (especially in Southern Europe) experienced a decline during the Bronze Age (about 5 kya). Presently, the genetic imprint from the Central Asian nomads is still evident and distributed throughout Europe.

Uniparental Genetic Markers—The Y Chromosome

Early seminal investigations on the distribution of the R1a-M420[19] and R1b-M343[20] Y chromosome types in Eurasia were based primarily on contemporary populations. These studies indicate that present-day Europeans possess high frequencies of R1a and R1b Y chromosome types. In recent years, it has become increasingly evident that the genetic profiles of populations change in time and space, and thus extrapolating genetic relationships based only on existing groups is risky and could lead to errors.

In spite of this limitation, early studies primarily based on present-day populations indicated that R1a-M420 and R1b-M343 originated along the steppes. And although the groups are closely related to each other and largely responsible for the introduction of the Indo-European languages in Europe, these two Y chromosome types migrated from unique locations in Central Asia. It is thought that R1a-M420 populations inhabited the forest and tundra just north of the steppes while R1b-M343 tribes lived to the south in the open steppes. It has been proposed that these two Y chromosome types started to genetically diversify, giving rise to a number of subtypes within the steppes of Western Central Asia in what is today Iran or in Southern Siberia about 5.8 kya and subsequently moved into Europe, partially displacing the autochthonous Paleolithic and Neolithic Y chromosomes. The indigenous Y chromosomes present in Europe before the migration of Central Asian nomads were mainly of the type I and its derivatives. The origin and migration of these people from the steppes carrying the R1a-M420 marker and its derivatives is clearly seen in a map of Eurasia, illustrating the longitudinal distribution of this Y chromosome type (Fig. 14.9). Dark colors in Fig. 14.9 indicate higher abundance of R1a-M420 and may be indicative of geographical origins. As the distances from these darker centers increase, the frequencies of R1a-M420 and its derivatives gradually decrease suggesting dilution of the markers as migrants interbreed with the indigenous hunter-gatherer and agricultural populations. Notice that these R1a-M420 Y chromosome distributions extended

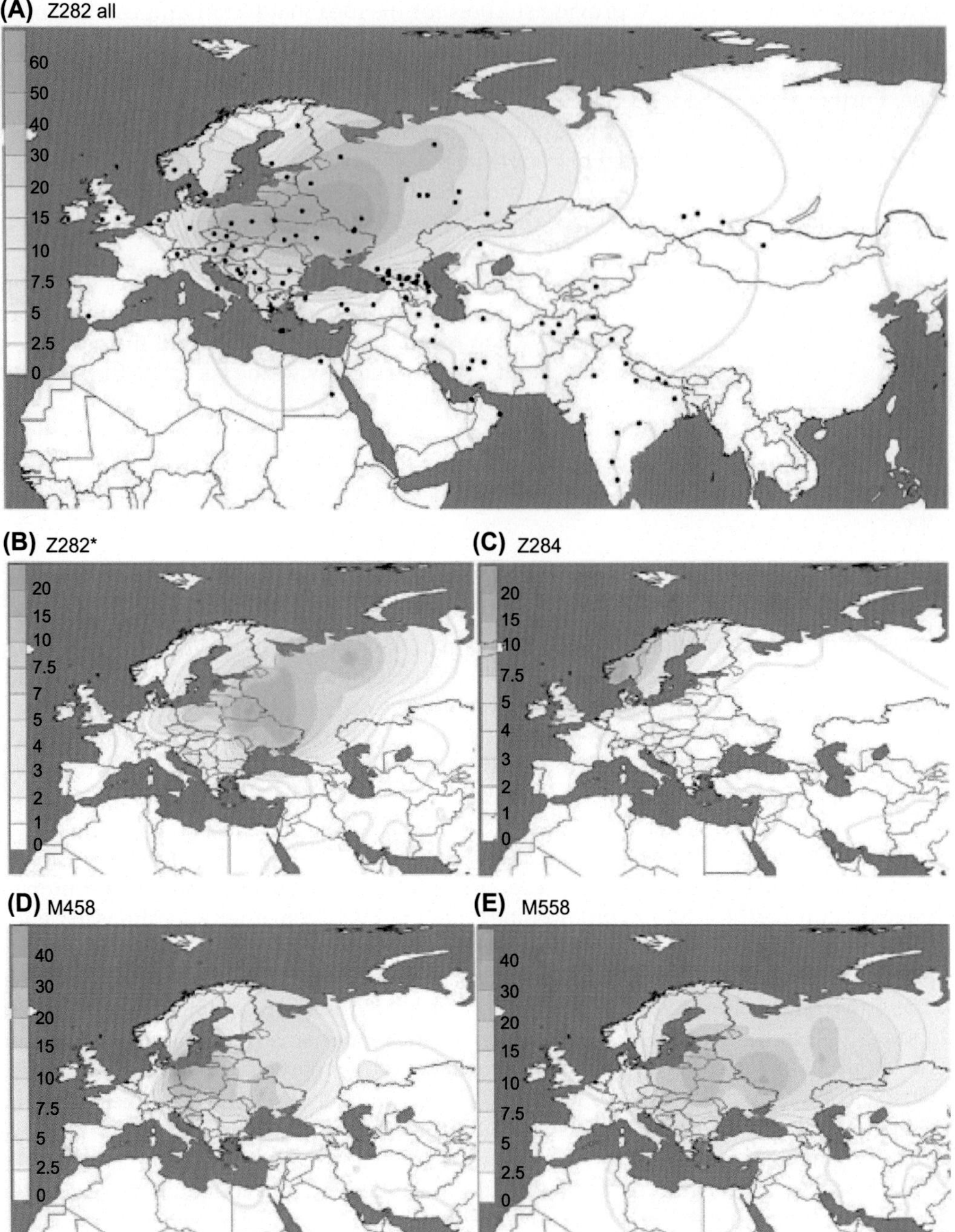

FIGURE 14.9 East–West frequency gradients of R1a Y chromosomes. Increasing color intensity indicates abundance, and dots indicate locations of populations sampled. *From http://www.nature.com/ejhg/journal/v23/n1/fig_tab/ejhg201450f2.html#figure-title.*

horizontally along the East–West axis not only in the direction of Europe but also toward East Asia (Fig. 14.9). Similarly, the R1b-M343 Y chromosome type emerged in Central West Asia and rapidly spread with its derivatives into Europe and East Asia.

The impact of the steppe migrants on the genetic constitution of contemporary Europe is evident in the relative proportions of their DNA before and after their arrival. Before the existence of the Steppe Roads, ancient DNA analyses indicate minimal presence (approximately 1.4%) of the R1b-M343 Y chromosome type in Europe as compared to 60% subsequent to the arrival of the nomads in Europe, excluding Russia. Today, R1a-M420 and R1b-M343 are the most abundant Y chromosome types in many European populations. Recent studies indicate that this replacement of hunter-gatherer Paleolithic and agrarian Neolithic DNA occurred after 5000 years, resulting from migrations via the Steppe Roads.[15]

Uniparental Genetic Markers—mtDNA

The Neolithic farmers, when they arrived with their agrarian mode of subsistence, brought with them a number of uniparental markers, such as mtDNA type N1a and Y chromosome G2a. These genetic signals practically disappeared during the Bronze Age when the Steppe nomads migrated into Europe. The N1a mtDNA type was almost totally replaced by mtDNA haplogroups I, T1, U2, U4, U5a, W and derivatives of H, such as H4 and H6. Also, C and D mtDNA haplogroups have been implicated with migrations from eastern Asia and southern Siberia to eastern and northeastern Europe during the middle Holocene (7000 to 5000 ya) when the nomads were active. A typical ancient group recently studied was a Hungarian Medieval population from a contact zone in Central Europe.[21] The comparisons of the maternally and paternally derived inheritance before and after the nomadic incursions indicate that both sexes participated in the migrations.

FROM THE STEPPE ROUTES TO THE SILK ROADS

The Steppe Roads started with the trading of horses not silk. The domesticated horse was the main commodity used for transactions throughout the networks. It was a system that benefited all parties involved. To the Chinese, the domesticated horse represented a quantum leap in advancing their military campaigns against rival kingdoms and barbarians outside their borders, mainly to the north. The horse provided the Chinese with a cavalry and a new dynamic mode of mounted warfare that gave them a substantial military advantage, especially during attacks against infantry on foot. Although the Chinese did breed horses, but the nomads' horses were larger and more powerful than Chinese ones. To the nomads, Chinese products were precious magical items and symbols of power. Thus, this mutually beneficial commerce expanded as traders and consumers flourished. As discussed in the previous sections, nomads played a key role in the genesis and growth of these extensive trading networks. It could be argued that these early exchanges along the early Steppe Road networks fueled the cultural and economic growth of the societies that eventually evolved into the dynasties of Iran, China, and Korea. In addition, this scenario, together with the domestication of the wild silk moth or *B. mandarina*, set the stage for the transition from the Steppe Roads to the Silk Roads. By the time silk was mass produced by the domesticated *B. mori*, the Steppe Road networks utilized by the nomads for several millennia were well developed and poised to be used in the commerce of different types of commodities, such as silk, spices, precious metals, and art. By the time the silk trade started, cultural, linguistic, and political nexuses among populations were in place, and the Silk Roads were able to flourish and expand even further.

DOMESTICATION OF *BOMBYX MANDARINA*

The transformation of the wild silk moth *B. mandarina* to the domesticated form, *B. mori*, and the continuous artificial selection by humans to maximize the quantity, quality, and diversity of silk, as well as the creation of infection-resistant strains while keeping them genetically healthy, figure among one of the most successful breeding programs in the history of humanity. Undoubtedly, it represents one of the greatest achievements in the field of genetic manipulation by selective breeding or artificial selection.

It is likely that the domestication process started when an early Chinese agrarian community made keen observations on the properties of silk produced by the wild species, *B. mandarina*, and realized the potential of the product, especially if it could be improved as it was done previously with domesticated species, such as rice, millet, pig, and sheep. Although *B. mandarina* made only limited amount of silk with some unfavorable characteristics, these early breeders probably envisioned the utility of the fiber for a multitude of uses. It is plausible that they realized that silk was a promising medium for art or as a smooth and delicate fabric for clothing and initiated the systematic selection of specific desirable traits in their breeds. Like any artificial selection process, it took place in a premeditated fashion, where every generation of organisms producing the desirable traits were allowed to breed and the ones that were undesirable were not allowed to reproduce. Instead of the nature's survival of the fittest, artificial selection is the survival of those organisms with the most desirable traits. Over the course of centuries, strains with the appropriate combination of characteristics were generated. In this conscious selection process, a number of different strains with various combinations of attributes were generated to please the desires and esthetic inclination of the consumers. Fig. 14.10 illustrates a number of silk cocoon types. Notice how the size, shape, color, and the texture of the cocoons vary reflecting various strains of *B. mori*. This process of artificial selection designed to produce larger amounts, better quality, and/or different types of silk still continues in a number of breeding centers around the world as novel strains of *B. mori* are created or modified.

The current geographical range of *B. mandarina* includes inland China, Korea, Japan, and the far eastern regions of Russia. Also throughout East Asia, there are different species of wild silk moths exhibiting considerable anatomical differences (Fig. 14.11). Even to the untrained eye, the various forms would seem as very different species. These morphological distinctions are also reflected on their DNA. Even within the *B. mandarina* species, the northeastern populations at the extreme of its distribution range are unique from the ones in inland China, based on their chromosome morphologies. The various regional varieties also differ in their mtDNA sequences. The domesticated moth, *B. mori*, shares greater chromosomal and mtDNA sequence similarities with the wild *mandarina* of inland China than groups from other areas. This suggests that *mori* derives from inshore Chinese populations.[22] The genetic data also point to a single domestication event during a short period of time involving a relatively large number of *B. mandarina* organisms as

FIGURE 14.10 Various silk types produced by different strains of *Bombyx mori*. *Personal photo (RJH).*

the initial inbreeding stock. Yet, it is not certain whether domestication took place in a single or several nearby locations and where exactly in China this process took place.

In addition to creating a productive and docile organism, the artificial selection process created a species incapable of reproducing and surviving in the wild. The adult moths cannot fly due to their vestigial wings, and they are so clumsy that they cannot find mates by themselves. Without human assistance the species would cease to exist. In addition, it seems that in the process of selecting for highly productive organisms capable of making good quality silk, a number of other traits were chosen or incorporated as well, such as blindness, no fear of predators or humans, compromised sensitivity to environmental odors, no color camouflage (Fig. 14.12), and sluggish and clumsy behavior. Some of these traits are certainly beneficial to the moth farmers because they allow for easy manipulation of the imago (adult moth), selective mating, and prevention of the moths from flying away and escaping. They are so complacent to manipulation that it is rather simple to engage specific pairs, that exhibit certain desired traits, in copulation, thus facilitating artificial selective mating. Thus, it is likely that the early Chinese who practiced sericulture selected a number of traits that allowed them to easily manipulate the moths, while other traits that served no apparent purpose to breeders were likely introduced by chance.

Genetic studies suggest that the domestication of the wild species *B. mandarina* started around 7.5 kya and the active process of selection to generate *B. mori* continued until about 4 kya[23] according to historical accounts from the Shang Dynasty. Yet, older findings of silk also exist from the banks of the Yellow River in Shanxi Province (4.6–4.3 kya); silk ribbons, fibers, and fabrics from the providence of Zhejiang (5 kya); and silk fabrics were found in Henan Province (5–6 kya). Additional archeological findings date silk production to even earlier

FIGURE 14.11 Different species of wild silk moths. *From https://www.pinterest.com/pin/522487994250992701/.*

FIGURE 14.12 *Bombyx mori* (left) and *Bombyx mandarina* (right) females. Notice size difference and lost of camouflage (color) in the domesticated species. *From https://www.mpg.de/7623868/silkmoths-captivity.*

times. For example, an ivory mug with images of *B. mori* larva was made about 6–7 kya and spinning tools, silk fibers, and silk fabrics from locations along the lower Yangzi River suggest even more ancient dates for the start of sericulture. It is not clear whether these ancient reports represent wild *B. mandarina* silk production or silk from fully domesticated *B. mori* or intermediate stages during the selection process of *B. mori*. But generally, the genetic, archeological, and historical data corroborate the genesis of silk from domesticated moths during the Chinese Agricultural Revolution about 5 kya.

The initiation of *B. mori* domestication also coincides with the husbandry of other livestock by different agrarian societies around the world during the Neolithic Agricultural Revolutions, 5–10 kya. As a result of intense artificial selection pressure, domestication of the moth triggered a number of bottleneck episodes that periodically reduced the genetic variability of the inbred stocks. The domestication process also involved interbreeding events with the wild species *B. mandarina* to keep domesticates genetically healthy. *B. mori* and *B. mandarina* are capable of hybridization and production of fertile offspring when copulation is mediated by humans (juxtaposition of male and female genitalia) (Fig. 14.13). This suggests the two populations are a single species and not two separate species.

It has been postulated that periodic gene flow or introgression with the ancestral wild species, *B. mandarina*, was crucial for the creation of *B. mori* because it provided for genetic variability. This process of systematic out breeding to keep varieties genetically strong is routinely done in husbandry to keep the stock genetically vigorous. The practice of introgression with the wild species during the artificial selection period, from the start to the end of domestication, explains why *B. mori* only lost approximately 17% of the genetic diversity present in the

FIGURE 14.13 Copulating male (left) and female (right) silk moths of the species *Bombyx mori*. *Personal photo (RJH).*

wild ancestor *B. mandarina*. Because *B. mori* cannot fly due to its vestigial wings and is incapable of surviving in nature and often requires human assistance to reproduce, it is likely that interbreeding with *B. mandarina* was intentional and mediated by humans.

The processes of human-mediated selection for the moth did not stop 4000 ya. Artificial selection still continues until the present in different parts of the world, and today region-specific strains with unique characteristics are being generated and maintained in a number of repositories. *B. mori* farms and breeding centers exist in Japan, Korea, China, Southeast Asia, India, Australia, and Europe, among others. Considering the economic value of the stocks, the *B. mori* repositories around the world are carefully managed to keep the integrity of their various precious strains and prevent accidental escapes and admixture among the pure strains. Just as the Chinese did in antiquity, the facilities guard zealously the strains to prevent the illegal transport of organisms out of the complexes and accidental crossbreeding between strains. In most countries, it is a serious federal offense to transport silk moths across international borders.

Although *mori* caterpillars can feed on an artificial diet, they prefer their natural food, the mulberry leaf. So, most farms grow their larva on mulberry leaves because it is more cost productive and the larvae like the plant better. They feed on different species of mulberry, but their favorite leaf is from *Morus alba*, the white mulberry. As a result, *mori* farms and breeding centers are integrally coupled with extensive mulberry fields not far away from the moth colonies.

THE SILK

Silk is made up of insoluble proteins by a number of insect species, including the larvae of the moth genera *Bombyx*, *Antheraea*, *Cricula*, *Samia*, and *Gonometa* as well as spider species.

The life cycle of *B. mori* involves a number of steps that includes an adult form or imago (Fig. 14.13). In the adult stage, sexual reproduction occurs. After internal fertilization, over 100 eggs are usually laid on a leaf or a piece of paper (Fig. 14.14). About 10 days after the eggs are deposited, the young larvae hatch, and approximately 2 weeks later, they reach full size (Fig. 14.15). The larvae are sensitive organisms, and their environment is carefully monitored preventing loud sounds, air currents, and scents. Soon after they attain full size, the caterpillars start the process of spinning the cocoons (Fig. 14.16), which takes about 6 days. The anatomy and physiology of making silk is complex requiring a number of tissues, as well as a combination of mechanical and biochemical processes. The final product is an oval case or capsule made up of silk designed to protect the larva as it undergoes metamorphosis during the next 12–14 days, a process in which tissues, organs, and organ systems are destroyed, modified, or created, and the larva is transformed into a pupa (Fig. 14.16).

The silk is routinely harvested at this stage by cutting open the cocoons, extracting and disposing of the pupae, and boiling the cocoons to partially clean and dissipate the silk. Depending on the strain, about 3000 *B. mori* caterpillars consume roughly 104 kg of mulberry leaves to produce about 1 kg of silk, enough to make a blouse and a skirt.

In addition to providing the raw material for the manufacturing of luxurious items of great beauty, such as garments and works of art, silk exhibits certain very valuable properties. For example, although silk can be degraded by a number of bacterial species from the genera *Amycolatopsis* and *Saccharotrix*, silk is pretty much immune to chemical attacks from most bacteria. Likewise, early sericulturists also soon observed that silk was highly resistant to fungal

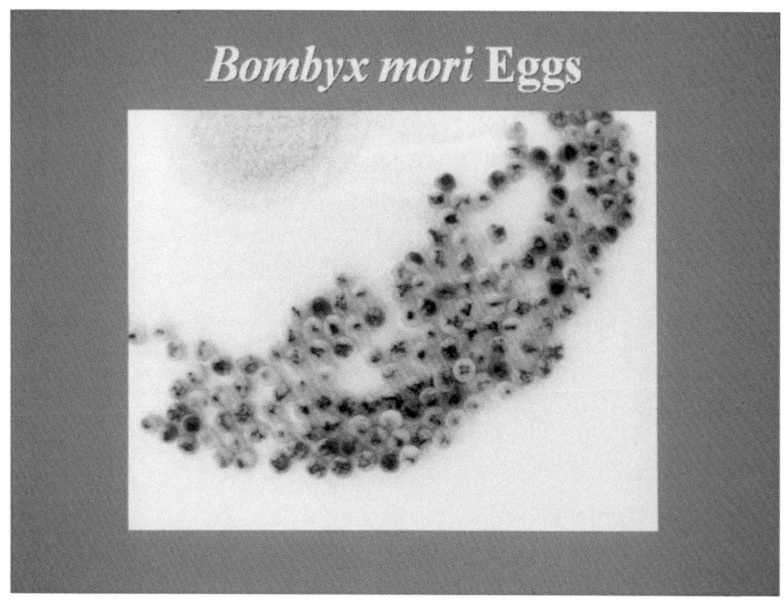

FIGURE 14.14 *Bombyx mori* eggs. *Personal photo (RJH).*

FIGURE 14.15 Full-grown larva of *Bombyx mori*. Newly hatched caterpillar about 1 mm in length can be seen on the top-front of the full-grown larva. *Personal photo (RJH).*

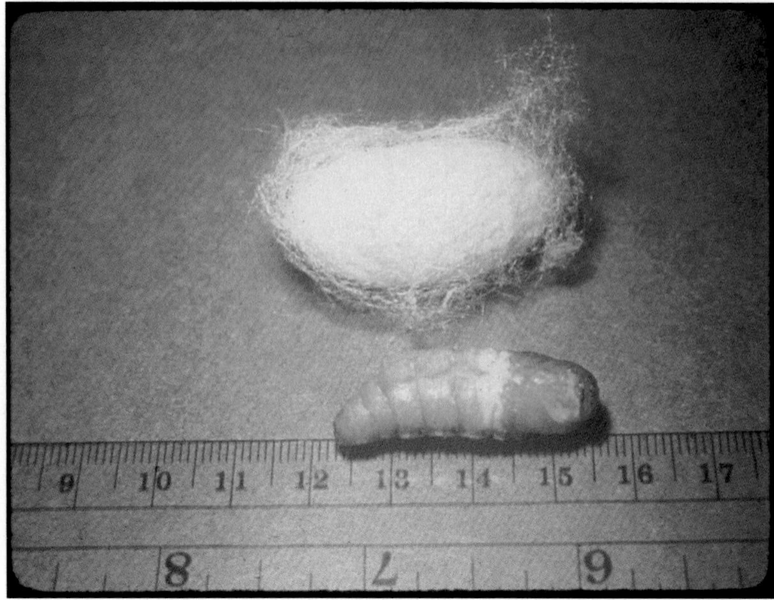

FIGURE 14.16 Cocoon and pupa. *Personal photo (RJH).*

growth and naturally repels dust mites, mold, and mildew. These characteristics provide for more durable and cleaner clothing. In addition, when worn, silk fabrics facilitate moisture and heat regulation. Some of these properties offer advantages over other fabrics used in wound dressings, which makes silk clinically useful for the treatment of lesions and promotion of healing. In addition, silk does not burn easily, and when it is actually set on fire, it does not fuse with skin, as polyester does. In more recent times, silk has been credited with being hypoallergenic, providing relief to sensitive or affected skin suffering from conditions such as eczema, atopic dermatitis, asthma, shingles, psoriasis, and other skin allergies. It is unclear if ancient societies were aware of some of these beneficial properties of silk, but it is clear that silk has vast uses and advantages.

Perhaps, a less well-known attribute of silk is its tensile strength (stress that a material can withstand while being stretched or pulled before breaking) that can be stronger than the tensile strength of steel. Although a number of organisms produce their own type of silk with specific properties, they all are light in weight and possess high tensile strength and extensibility (capacity to extend). For example, the strength of the dragline silk of the spider *Nephila clavipes* (the banana or American golden orb-web spider) is comparable to high-grade alloy steel (450–2000 megapascals or MPa) and about half as strong as Kevlar (3000 MPa). Thus, it is no surprise that leaves can be seen suspended in midair by some seemingly invisible string, and only upon further observation, that a single silk strand is seen holding up the leaf in spite of the wind force. This combination of high tensile strength and extensibility allows silk to absorb large amount of energy before fracturing, making it particularly useful in the manufacturing of clothing as well as protective gear, such as explosive- and bullet-proof vests and clothing. Although spider silk is tougher than *B. mori's*, spiders have not been domesticated, and they spin most of their fibers in the form of webs, not cocoons, thus their fibers cannot be harvested for commercial uses. Efforts in genetic engineering designed to introduce the superior spider silk genes into *B. mori* moths to produce the arachnid fibers as part of a cocoon have not been commercially successful.

The strength of silk derives from its basic chemical structure. In *B. mori* the fiber is made up of two main proteins, fibroin and sericin. The fibroin molecule is a simple polymer (a molecule made up of repeated subunits) composed of cassettes (blocks) of six amino acids, glycine-serine-glycine-alanine-glycine-alanine, repeated in tandem head to tail (Fig. 14.17). Glycine, serine, and alanine are abbreviated as Gly, Ser, and Ala, respectively. The high content of glycine and alanine provides silk with its rigidity and extraordinary tensile strength. Molecularly, two filaments of fibroin in parallel are covered with an adhesive layer of sericin that keeps them together (Fig. 14.18). Thousands of these linear sequences of amino acids (building blocks of proteins) are stacked together to form a filament. A cocoon is made of a single filament 600–900 m in length. Several of these silk filaments are twisted together by humans to generate a thread, and threads are woven into fabrics.

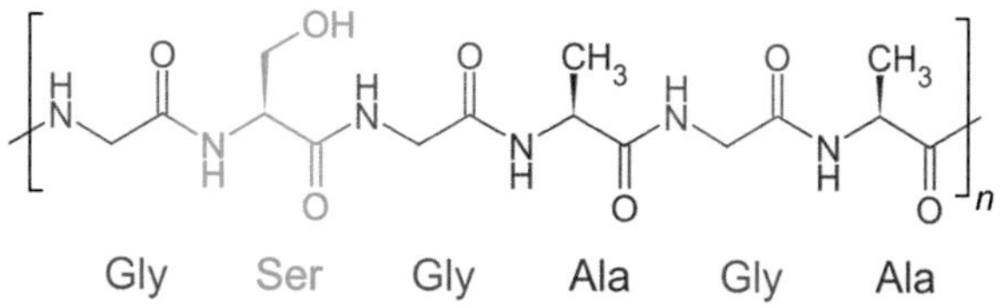

Gly Ser Gly Ala Gly Ala

FIGURE 14.17 Primary structure of *Bombyx mori* silk or fibroin. The (Gly-Ser-Gly-Ala-Gly-Ala)^n repeating unit of amino acids forms the basic fiber.

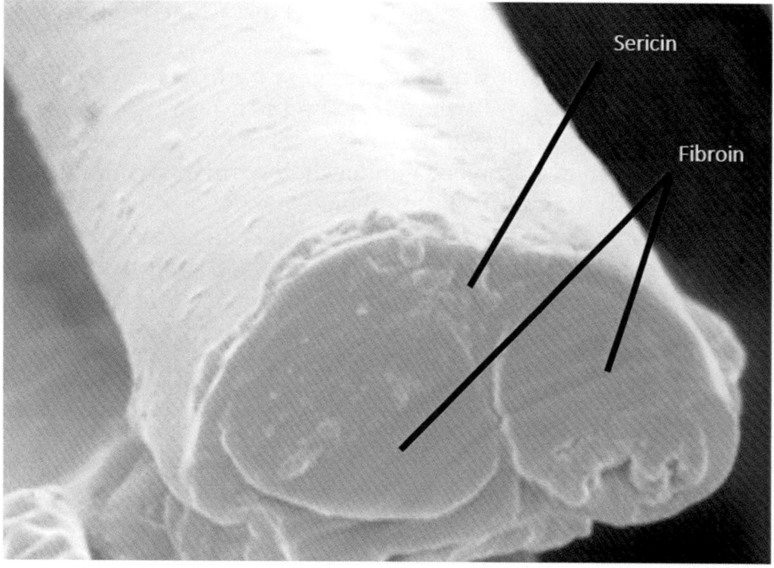

FIGURE 14.18 Scanning electron microscopic image of *Bombyx mori* silk. Silk produced by the caterpillar consists of two main proteins, fibroin and sericin. Two stack sheets of fibroin at the structural center of the silk, and sericin being the sticky substance (smaller protein) surrounding and keeping them together. *From uwimona.edu.jm/courses/CHEM2402/Textiles/Animal_Fibres.html.*

SECRET, ESPIONAGE, AND INTRIGUE

Silk production was a well-guarded process. In fact, it was one of the most zealously kept secrets in history. The *B. mori* organism, at all of its developmental stages, was kept under strict vigilance. The rearing of this domesticated species, the cultivation of mulberry trees, and the manufacturing of silk fibers and fabrics were strictly controlled by imperial decree, and any dissemination of information pertaining to silk was punishable by death. As a result of the efforts to keep silk a secret, the fiber was confined to China for thousands of years. Foreigners were clueless about the makeup of silk and how it was made into such an exquisite delicate fabric. Romans, for example, thought silk was related to soft fine down feathers that were collected from leaves with the help of water. This notion is mentioned by the Roman author Pliny the Elder in his book *Naturalis Historia* (Natural History).[24]

> The Seres (Chinese) are famous for the woolen substance obtained from their forests; after a soaking in water they comb off the white down of the leaves... So manifold is the labor employed, and so distant is the region of the globe drawn upon, to enable the Roman maiden to flaunt transparent clothing in public.

It is quite significant that even after three centuries after the discovery of silk, Romans had no idea of the process, which illustrates the extensive effort by the Chinese to guard the secret.

Initially, possessing silk was a privilege only for the emperor and his immediate family. They dressed in white silk robes inside the palaces, whereas outside the royal compounds they wore yellow silk. Then, as production increased, the nobility was allowed to wear silk, but it was still forbidden by the commoners to own it. Silk clearly represented a unique status symbol. In these early days, the foreign aristocracy had to pay exorbitant prices to acquire the fiber. And it was not until the Qing dynasty (1644–1911) that Chinese farmers, the producers of the silk, were allowed to wear silk.

Silk production in ancient China was, for the most part, women's work. Even today women mostly staff silk factories. Whether at home or in factories, women did reeling, spinning, braiding, dyeing, and embroiling. To keep the *B. mori* eggs warm, women went to the extent of keeping them in cotton bags under their garments as they went about their daily chores. It was a family tradition, and mothers were devoted to training their daughters in all aspects of growing the moths, harvesting the silk, and making it into fabrics. Even more recently, during the time of the Byzantine Empire and after silk production spread to Constantinople, many factories only employed women and forced them to stay within the workshop complexes. There are stories of female weavers sequestered from the city of Thebes in Byzantium and transported to Sicilian silk workshops when the Normans sacked the city in 1147.

It is thought that it was women's genius that made possible the improvement of the spindle wheel, the treadle-operated loom, and the silk-reeling frame in households of China.[25] Every spring, it was customary for the empress to open the silk-making season as a symbol of the female-oriented nature of the work. After silk production was no longer a secret, woman in Central and Western Asia began to grow their private moths and harvest their own silk, selling their best fabrics in the market. Women's roles in the ancient silk trade are testament to their impact on human entrepreneurship and progress.

Soon after its invention, silk became the raw material for a number of products that jump-started and drove the Chinese economy. Around 2600 ya, six Chinese provinces were already engaged in silk making. Although in the West the use of silk was basically restricted to the manufacturing of fabrics for clothing and art, in China, silk was almost immediately employed for a number of utilitarian applications, including paper, bowstrings, fishing lines, and strings for musical instruments, among others.

Yet, silk was in such a high demand, and there was so much money to be made in producing and trading it that the secret was destined to become public knowledge sooner or later. China eventually lost its monopoly on the fiber. The secret likely escaped China in a number of ways. Yet, some of the known escape scenarios are more credible than others. For example, it got to Korea when Chinese laborers began to settle the peninsula about 2220 ya. Subsequently, knowledge of the silk also got to India around the 3rd century of the current era and then to Persia.

One of the best-documented accounts indicates that around 550 of the current era, two Nestorian (doctrine born in Constantinople that underscores the duality of Jesus, its divine, and human nature) monks from the Byzantine Justinian court managed to smuggle *B. mori* eggs out of China within their hollow bamboo walking sticks.[26] According to reports, these two priests were preaching in India when they decided to travel to China in the year 551 CE. It is not clear what motivated them to undertake such a trip or whether in fact they were sent by Justinian I to steal the secret. Nevertheless, this trip brought about a landmark discovery that benefited Byzantium tremendously.

A historical account of the events derives from the writings of the Roman historian Procopius Palaestina Prima (provided below) who lived in the mid-6th century AD.[27] At the time of this document, Justinian I had been on the throne for about 25 years in the city of Constantinople, which is in present-day Istanbul, Turkey.

> About the same time there came from India certain monks; and when they had satisfied Justinian Augustus that the Romans no longer should buy silk from the Persians, they promised the emperor in an interview that they would provide the materials for making silk so that never should the Romans seek business of this kind from their enemy the Persians, or from any other people whatsoever. They said that they were formerly in Serinda, which they call the region frequented by the people of the Indies, and there they learned perfectly the art of making silk. Moreover, to the emperor who plied them with many questions as to whether he might have the secret, the monks replied that certain worms were manufacturers of silk, nature itself forcing them to keep always at work; the worms could certainly not be brought here alive, but they could be grown easily and without difficulty; the eggs of single hatchings are innumerable; as soon as they are laid men cover them with dung and keep them warm for as long as it is necessary so that they produce insects. When they had announced these tidings, led on by liberal promises to the emperor to prove the fact, they returned to India. When they had brought the eggs to Byzantium, the method having been learned, as I have said, they changed them by metamorphosis into worms, which feed on the leaves of mulberry. Thus began the art of making silk from that time on in the Roman Empire.

At the time of Justinian I, the Byzantines thought that silk was made in India. Thus, at this point in time, the Romans were still uninformed about silk. Clearly the Chinese were doing a good job in keeping silk a secret. But, while in China, the two monks finally not only discovered that silk was made there, as opposed to India, but also observed how *B. mori* was raised and the procedures for making the silk fabric. When the priests returned to Constantinople in the year 552 CE, they informed the emperor of their amassing findings, and it seems that

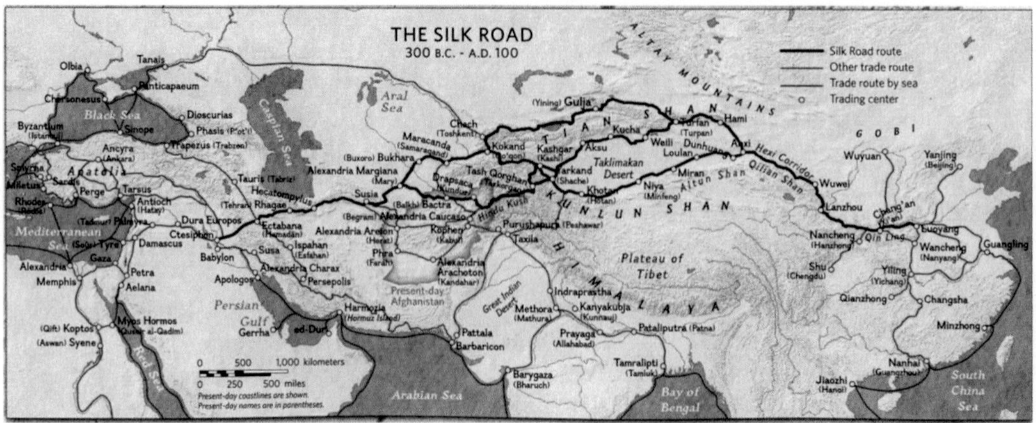

FIGURE 14.19 Silk Roads during the Han Dynasty. *Reproduced from Silkroadcatalyst.*

Justin I instructed them to return to China and smuggle the organism to Byzantium. It is likely that for the return trip from China they used the northern Silk Route (Fig. 14.19), taking advantage of the Black and Caspian Seas as waterways and sources of food, as well as the trans-Caucasus path to end up in Constantinople.[26] In addition to transporting eggs and/or larvae to Byzantium, it was necessary to import mulberry leaves, their only diet. Without mulberry plants it was not possible to establish a successful colony back home. Yet, it is not clear how the mulberry plant got to Byzantium.

It is thought that the round-trip of this covert mission took about 2 years. Considering the delicate and sensitive nature of silk moth eggs or larvae, it is difficult to envision how they would have survived such an arduous and lengthy trip through the Central Asian deserts without making the trip in stages. So, it is likely that the monks stopped periodically in oases to hatch and grow the larvae and complete the moth's life cycle, producing eggs in their dormant stage before resuming the trek. In the dormant stage, *B. mori* eggs are less sensitive to dehydration and death. Alternatively, it is possible that dormant eggs survived the 2-year journey if they were kept warm and moist to prevent hatching. What is clear is that by 527–565 CE, Byzantium broke the silk enigma and got a start in the production and trade of silk, keeping the secret away from other Europeans. Soon after the return of the two monks with their precious cargo, silk workshops began to open throughout the Byzantine Empire finally breaking the Chinese and Persian silk monopoly in Europe. Silk production and trade became paramount for the Byzantine economy. This monopoly persisted until 1204, when Muslim Constantinople was sieged and sacked by crusaders (Fourth Crusade) under Baldwin of Flanders.[28] Byzantium, the new player in the silk industry, began to undercut the higher-grade Chinese silk, especially in the Near East. Yet, the demand for the higher-quality Chinese silk in the West continued until the Mongol period in the 13th century because the centers of production in the West did not supply enough to meet the local demands. In addition, there were always issues of price and style of the European product that probably kept a demand for the imported Chinese silk.

Another curious story of intrigue and deception involving silk relates to an event about 100 years earlier when a prince from Khota, an oasis and Silk Road stop town in southwestern Xinjiang Province, Turkestan, or Western China, obtained *B. mori* eggs from a Chinese princess and future wife as part of her dowry. According to the tale, the princess smuggled out the moth eggs in 140 BC by concealing them within her voluminous hairpiece.

What is clear is that by 1400 ya, the Persians also got into the silk industry. And starting during the second Christian Crusades in the 13th century, sericulturists began to grow *B. mori* in Europe. Today, silk is harvested worldwide, and the finest silk fabrics are produced in European countries, such as Italy and France.

THE AGE OF DYNASTIES AND EMPIRES

Before the age of dynasties in China, a number of culturally similar Bronze Age societies or chiefdoms occupied the middle and lower basins of the Yellow and Yangtze Rivers. At these centers, agriculture and domestication were initiated. Initially, agrarian communities flourished along the floodplains of these waterways. Later, the use of flood control and irrigation allowed cities to grow and political power to set in. It was in these agrarian settings that Chinese dynasties had their beginnings. During the second dynasty or Shang Dynasty (1600–1046 BC), written records first appeared and the use of bronze became common. This was also the time that the Steppe Roads transitioned to the Silk Roads. This major transformation coincided with a period of rapid progress and the domestication of the wild moth *B. mandarina* and the creation of the highly productive, captive species *B. mori*.

It is difficult to assess if there was a cause and effect relationship between these two events. In other words, did the genesis of *B. mori* and silk production about 5 kya during the Chinese Agricultural Revolution trigger the development of central governments and dynasties in China? Was it a contributing factor? What is clear is that the juxtaposition of these two developments initiated a dramatic expansion in the number of paths, and merchandise trafficked within the Roads.

Trajectories

The Silk Roads extended approximately 7000 km, of which 4000 km are found within China. The Roads connected the Yellow River Valley in East China to the Mediterranean Sea (Fig. 14.19). It took about 6 months for caravans to go from Korea to Constantinople (Istanbul) in Turkey at the junction of the Mediterranean and the Black Sea. Within China, the Roads traversed the provinces of Gansu, Ningxia, Qinghai, and Xinjiang. Most depictions of the Silk Roads do not indicate a single simple straight line running longitudinally East–West. Scholars have learned that the Silk Roads were more like a mesh of roads made up of a multitude of major paths as well as minor passages branching off and looping out in different directions (Fig. 14.19). It was more an overland network of routes that enabled traders to travel from Xi'an (the ancient city of Chang'an) in East China to the Near East. It is likely that the myriad of tracks thus far identified represent only a small fraction of the passages that once existed because extensions of the Roads were constantly being created, modified, or truncated.

A major destiny city in the Near East was Constantinople. From there a number of sea routes connected the Silk Roads to a number of main European ports, such as Rome and Venice. From these key European cities, the merchandise was routinely distributed to other European trading partners, such as Spain and France. In addition to the East–West land trajectories, a number of other secondary land and sea branches developed over the years, including the roads to India and Persia as well as North and East Africa, reaching as far as Zanzibar and Mombasa.

Maritime routes also transported merchandise between China and Southeast Asia, across the Indian Ocean to Africa, and between India and the Near East. The sea routes have a long history and derive from ancestral contacts that can be traced back for thousands of years of commerce between the Arabian, Mesopotamian, and the Indus Valley Civilizations. These sea trades intensified during the Muslim Empire as Islam experienced a surge in the sciences and the arts. Specifically, the sea routes increased in efficiency and importance with a number of navigational innovations and improvements including the astrolabe (instrument used to measure the altitude of the sun or stars) that the Greeks invented, and Arabs perfected in the 8th century. The Omani Empire, for example, played a key role in this market connecting East Africa and West India. In the Middle Ages, this traffic developed further with the expansion of Islam in Southeast Asia, North and East Africa, and Southwest Europe. In this sea-driven commerce, the ports of Goa in India, Zanzibar in East Africa, Alexandria in North Africa, and Muscat in the Arabian Peninsula were key maritime trading cites. These cities involved in maritime trade flourished rapidly as financial centers and hubs of transfer of ideas, philosophies, languages, and religions. The flexible nature of sea routes, providing numerous destinations, contributed to interactions among merchants and sailors as well as different populations with diverse backgrounds.

As part of the mainland Roads, two cities, Hangzhou and Guangzhou, connected the Chinese east coast to the interior city of Xi'an, the traditional start of the Silk Roads (Fig. 14.19). Additional routes joining Korea and Southeast Asia to Xi'an were also in place. The caravans routinely met in Xi'an to travel in groups for protection. From Xi'an, traders traveled along the Hexi Corridor following a path north of the Qilian Shan Range to Anxi, a trip of about 2000 km lasting approximately 2 months. Beyond Anxi, the traders were faced by the Tarim Basin (Fig. 14.20), which is mostly occupied by the Taklamakan Desert. This depression is surrounded by high mountains and is one of the most inhospitable places in the world. The conditions are so unfavorable that no permanent settlements exist. Archeological data indicate that due to the extreme environment of the Tarim Basin, it was one of the last locations reached and colonized by humans in Asia. Although a number of streams flow into the desert from mountain glaciers, providing water to the oases within the basin, which are used by the locals, none of these waterways reach the sea as the water is quickly evaporated and absorbed by the arid soil or deposited into small salt ponds and sloughs. One of the peaks is K2, also known as Mount Godwin-Austen or Chhogori, on the China–Pakistan border, the second highest peak in the world after Everest.

The Taklamakan Desert was clearly a formidable natural barrier to the caravans. Poised at the oasis city of Anxi, the convoys were confronted with three choices. One consisted in frontally penetrating the arid Tarim basin (Fig. 14.20) and risking death. Alternatively, most convoys used routes that circumvented the desert by traveling north or south of it, the Northern and Southern Routes, respectively. The third alternative was to use the Middle Tarim bypass connecting the North and the South Routes. This bypass provided a more direct

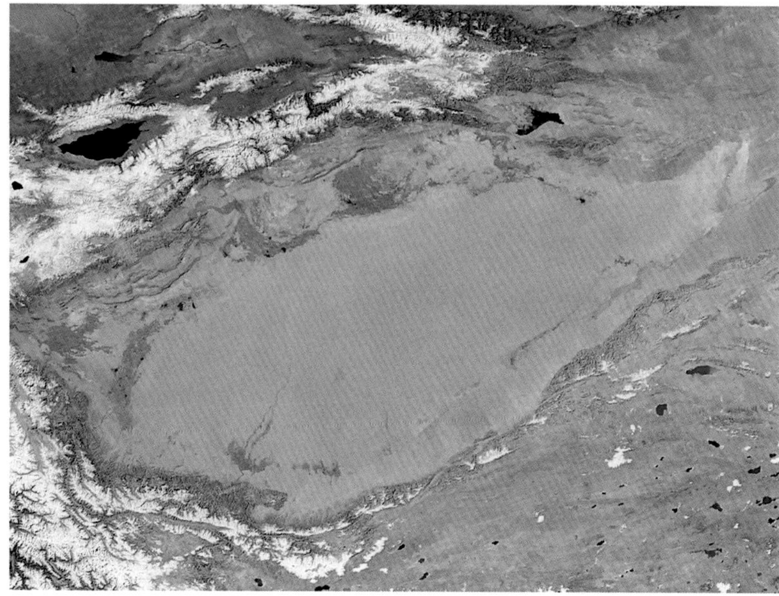

FIGURE 14.20 Satellite image of the Tarim Basin. *From https://en.wikipedia.org/wiki/Tarim_Basin.*

approach, although it was more precarious. Specifically, going from west to east, this middle route reduced the traveling distance and the transit time by connecting the oasis city of Korla (region east from the city of Kucha) on the Northern Route to the oasis city of Loulan (also known as Krorän) on the Southern Route and over to Dunhuang (Fig. 14.19). The trip from Loulan to Dunhuang, both on the Southern Route, went across the Lop Nor region, currently a dry salt lake and uninhabited since the 4th century, but once home to the thriving Tocharian culture (approximately 4–1 kya). After Dunhuang, caravans going east reached Xi'an where the Northern and Southern Routes connected. The Middle Route was not a separate path, but just a detour that allowed the journeyers to skip a region on the Northern Route that included the oasis cities of Weill, Turfan, and Hami. The Middle Route has not been in use since the 6th century AD because of increased aridity in the region.

The Northern Route ran longitudinally along a path delineated by the Taklamakan Desert/Tarim Basin to the south and the Tian Shan Mountain Range to the north (Fig. 14.19). Starting in the oasis of Kashgar and traveling eastward, the North Route went over the cities and supply/trading stations of Yarkand, Aksu, Kucha, Korla through the Iron Gate Pass to Karasahr, and over Weill, Turfan, Hami to Anxi (Fig. 14.19). As part of the Silk Roads, the 7-km-long Iron Gate Pass or Tiemen Pass was susceptible to attacks from the north, and thus it was heavily protected. During the Tang Dynasty, for example, a permanent military presence was established at the pass. The North Route along the Taklamakan Desert was preferred over the South Route because of the greater number of oases in it that provided much-needed rest and supplies.

The South Route running east from Kashgar went over Yarkant, Karghalik, Pishan, Khotan, Keriya, Niya, Qarqan, Qarkilik, Miran, and Dunhuang to Anxi (Fig. 14.19). Thus, in a west to east direction, both North and South Routes met at Anxi before reaching Xi'an while at the west end of the Taklamakan Desert/Tarim Basin the two Routes met at Kashgar

(Fig. 14.19). From Kashgar, convoys continued west through the Hindu Kush pass (western Himalayan Range). After crossing the Himalayas, caravans followed a trans-Caucasus path dotted by major oasis cities in Central Asia including Maracanda, Bukhara, Alexandria Margiana (founded by Alexander the Great in AD 327), and Rhagas. Further westward in the Mediterranean basin, the traders reached major urban centers such as Constantinople, Palmyra, Antioch, Tarsus, and Smyrna (present-day coastal city of Izmir Eastern Turkey). In coastal cities, such as Constantinople, Tarsus, and Smyrna, the goods were routinely embarked into boats and distributed to trading centers and destinations throughout the Near East and Europe. Once in Europe, items were transported to a number of continental locations including Northern Europe.

Throughout the existence of the Silk Roads, the trajectories through the Tarim Basin have shifted periodically. For example, during the early days of the Silk Roads the convoys traveled through the Southern Route, whereas during the Han Dynasty, it changed to the Middle bypass (i.e., Jade Gate–Loulan–Korla) (Fig. 14.19). Yet when the Tarim River changed its course around AD 330, most caravans moved through the Northern Route via Hami. Furthermore, when there was war on the Gansu Corridor, trade entered the basin near Charkilik from the Qaidam Basin. The routes to India originally commenced in Yarkand and Kargilik, but it was later modified to start in Kashgar south through the Karakoram (cordillera extending the boundaries of Pakistan, India, and China) pass.

Also to the west of Kashgar via the Irkeshtam crossing, at the border between Kyrgyzstan and Xinjiang, China, the Road connected to the Alay Valley and then Persia.

The Shang Dynasty

The trade between the nomads of the steppes and China resulted in the progress of the Chinese civilization during the Shang dynasty, also known as the Yin dynasty. The Shang dynasty controlled a sizeable region of current-day China (Fig. 14.21). Three major innovations that arrived from the west with the traders contributed to the growth of the Shang dynasty: (1) wheeled transport; (2) stronger, larger horses; and (3) metallurgy. Central Asian horses or Turkic horses had thick bands of muscles on both side of the spine that made them easier to ride, and the Chinese preferred the imported animals because local horses were smaller and not as strong. The Shang Dynasty is unique because during its tenure the transformation from trading domesticated horses and chariots for goods and consumables to transactions involving silk, spices, precious stones, metals, and art occurred (around 4 kya). This date closely corresponded with the end of the domestication period of the silk moth and the beginning of mass production of the fiber. This temporal coincidence allowed the Shang Dynasty to participate early in the silk trade and benefit from it, which contributed to the rapid growth of the empire.

Archeological findings and historical records indicate that the Shang period was the second oldest of the Chinese dynasties dating from about 1600 to 1046 BC and it occupied the Yellow River valley of the northern Chinese plains. It succeeded the Xia Dynasty that was in power from 2700 to 1600 BC, and it was followed by the Zhou Dynasty that reigned from 1046 to 256 BC. Early in the Shang Dynasty, silk began to be mass produced with the help of looms and noncomplicated weaving methods. From the very beginning, silk production was female driven and conducted from individuals at private homes (Fig. 14.22).

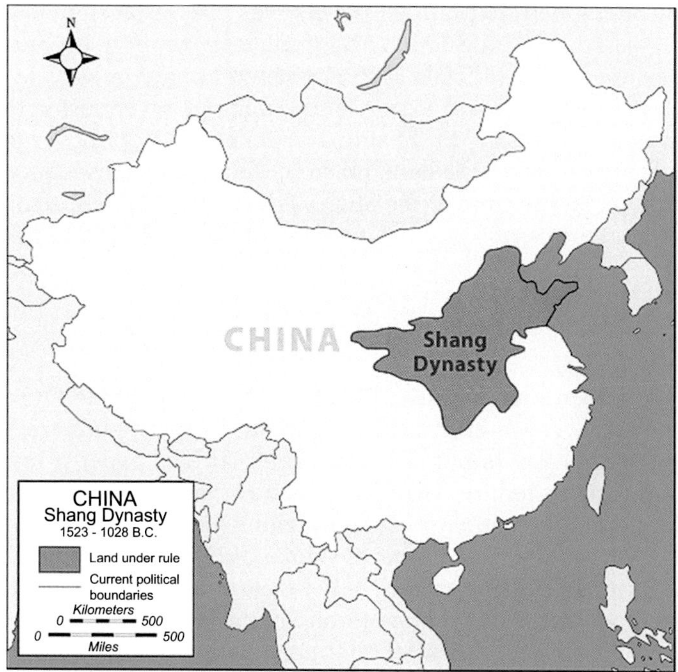

FIGURE 14.21 Territory occupied by Shang Dynasty. *From https://learnodo-newtonic.com/shang-dynasty-facts.*

FIGURE 14.22 Silk production by women at home during the Shang Dynasty. *From http://www.silk-road.com/artl/silkhistory.shtml.*

Beginning with the Shang period, plain white silk was used as currency and as a mode to pay taxes. As time passed, silk production and trade became more organized and centralized, and the government and officials started to increasingly supervise and control the process. Soon, silk became the main export and source of revenue for the Chinese who considered it a dynastic patrimony exercising a complete monopoly. As the word got out that this wonderful fabric existed, the demand from the west increased as well as the funds coming into the empire. At the time of the Shang Dynasty, it took approximately 6 months for goods to travel from the capital city of Gyeongju of the Silla Kingdom in Korea to reach Constantinople.

The Han Dynasty

Encompassing more than four centuries (206 BC–AD 220), the Han Dynasty is known as one of China's golden ages. It was an epoch of economic prosperity unparalleled by previous dynasties, and a number of key important discoveries, such as paper printing and the seismometer, occurred during its tenure. The Han period commenced when the insurgent commander Liu Bang deposed the emperor of the previous dynasty (Qin Dynasty 221–207 BC) after 4 years of civil war and became Emperor Gaozu of Han. During the Han Dynasty, silk ceased to be just a fabric and became an item with innate value. Silk started to be employed as currency to pay for goods, services, and favors, and prices were given in terms of quality and amount of silk. Even in transactions across international borders involving various empires, silk was used as a method of payment.

In this prosperous economic environment, the Silk Roads rapidly flourished starting about 100 BCE and peaked soon after. This sudden increase in traffic along the Roads closely coincided with the emergence of a number of adjoining powers with a key interest in trade, such as the nomadic tribes of the Xiongnu alliance, the Roman Empire, the Parthian Empire, and the Kushan Empire. This prosperity, which was fueled by the trading along the Silk Roads, occurred in the midst of a number of significant scientific and technological discoveries, such as paper, the printing press, ship rudders, negative numbers in mathematics, the seismometer, the spherical astrolabe (model of objects in the sky with Earth or the Sun at the center), and relief maps (three-dimensional image of terrains).

In addition, during the Han Dynasty, taxes were levied to afford expanding the empire westward including extensions and protection of the Silk Roads. Fortifications, including watchtowers, were built to secure merchants and goods along the Silk Roads (Fig. 14.23). These structures were made of available resources in the desert, such as compressed gravel. Sentinels and guards were stationed permanently at strategic places to guard the traffic. These new Silk Road branches established by the Hans permeated into the Tarim Basin in Central Asia. The Hans also invaded and extended the Silk Roads to Nanyue and Dian territories to the south and the Korean Peninsula to the north. The economic success of the Silk Road trade also fueled conflicts with nomads to the north. For example, conflicts with the Xiongnu people of the Asian steppes in the 1st century AD led to the escalation of hostilities. At its peak, the Xiongnu Empire controlled Manchuria, Mongolia, and the Tarim Basin, an area extremely critical for the flow of goods along the Silk Roads. A cessation of hostilities was reached in 198 BCE when the Hans agreed to pay tribute to the Xiongnu in the form of large amounts of silk fabric, food, and wine as part of the Heqin agreement.[29] Yet, the lures

FIGURE 14.23 Han watchtower made up of compressed soil at Dunhuang, Gansu province, the eastern portion of the Silk Roads. *From https://en.wikipedia.org/wiki/Han_dynasty#/media/File:Summer_Vacation_2007,_263,_Watchtower_In_The_Morning_Light,_Dunhuang,_Gansu_Province.jpg.*

from the riches transported along the Northern Silk Roads were too much to resist, and the Xiongnus resumed raiding the Han posts. It was not until 119 BCE that the Han Dynasty managed to expel the Xiongnus from the Tarim Basin and not until 121 BCE that they gained control of the critical Hexi Corridor and the Lop Nur region along the Southern Route by military action and reinstated order on the Central Asian area of the Silk Roads. The Hexi Corridor runs diagonally from the Yellow River to the Tarim Basin via the Northern Silk Road. A string of oases peppered this main path for caravans to travel from Northeast China to the Silk Roads.

During this period of Han expansion into Central Asia, the Dynasty started establishing contacts with Rome, Dayuan (Fergana) in eastern Uzbekistan in southern Central Asia, Sogdiana (Iran), Bactria in northern Afghanistan near the critical Hindu Kush Himalayan pass, Shendu in the Indus River valley of North India, and Anxi in the Parthian Empire (Iran and Iraq). Official ties were initiated with Rome by Marcus Aurelius and Emperor Huan in AD 166 when an embassy was established in the Han court.[30] All of these regions enacted formal relationships with the Han Dynasty that facilitated trading through the Silk Roads.

The Roman Empire

The strong connections established by the Han Dynasty with the West eventually allowed the convoys to reach the Near East, Rome, and its trade partners within the empire to bring silk from China and transport glass items, jewelry, and coins back to the East.[31] Within the Roman Empire, Chinese silk from the Han period has been recovered in Northern Africa in ancient Egypt.

Sometime during 1st century BCE, the Roman populace was introduced to silk, and in a short period of time, Roman citizens became mesmerized with the novel fabric. The Romans' encounter with this new material can best be described as unsuppressed fascination. This obsession for silk extended to the Roman provinces as well. Women were particularly impressed with this new and unique exotic luxury, and the Empire was forced to regulate prices by decree to control the import of silk. The craze for silk within the Empire prompted some alarmed Roman citizens (especially males) to voice their frustration. A particularly colorful account was expressed by moralist philosopher and writer Lucius Annaeus Seneca the Younger in AD 65.

> I can see clothes of silk, if materials that do not hide the body, nor even one's decency, can be called clothes... Wretched flocks of maids labour so that the adulteress may be visible through her thin dress, so that her husband has no more acquaintance than any outsider or foreigner with his wife's body.[1]

Seneca went on to argue that silk was a conduit for exoticism and eroticism because women were naked when wearing it. The Roman state also tried to outlaw men from using silk tunics. A good portion of society felt that it was in poor taste to wear expensive imported fabrics from the East.[32]

It is possible that these negative feelings toward the foreign silk were related to the value many privileged Romans placed on trade in general. Real state possessions and agriculture were more valued to Romans than trade. This perception changed somewhat during the Imperial Period (27 BCE to AD 284) at the height of the Silk Roads when Rome became more active in commerce with the Han Dynasty and its provinces.[33]

From archeological remains, coins, and shipwrecks throughout the Mediterranean, scholars have been able to assess what the Romans exchanged, the amounts, and who was involved. The goods traded included all kinds of consumables, such as edible plants, utensils, building materials, slaves, and luxury items, including silk. Within the Empire, specific regions specialized in particular items. For instance, Spain produced olive oil and gold, whereas France was heavily vested in wine making. The surplus derived from each region was transported to other provinces at relative cost ratios of approximately 1:5:28 by sea, rivers, and land routes, respectively. The vessels used within the Mediterranean usually had a capacity of 75–300 tons of cargo. Although Rome had its own merchant fleet, some of the trade was performed by private entities. Yet, all items were carefully stamped by the government, providing information on the authenticity of goods and the collection of taxes. Romans were particularly meticulous in documenting details on the merchandise. The labels included the weight of the empty craft, the weight of the goods, location of origin of goods, the name of the trader(s), and the names and signatures of the Roman officials who performed the inspection. This meticulous Roman bookkeeping practice allows scholars today to track trade routes and content primarily from shipwrecks. The duties collected were used to subsidize the army and its expanding campaigns. It is estimated that the largest allocation of funds was the army, which consumed about 70% of the state's revenues.

In addition to trading within the Roman Empire, luxury items were exchanged with foreign powers, primarily the Han Dynasty, the Persian Empire, Arabia, India, and East Africa. The magnitude of the internal (within the Empire) and international trade was unparalleled within the ancient world, and it was only matched in recent times during the Industrial Revolution.

Parthians and Romans: A Love–Hate Relationship

The link to the Roman Empire from the main branch of the Silk Roads started in the Parthian city of Rhagas (present-day Tehran, Iran), just south of the Caspian Sea (Fig. 14.19). From there the Roads connected to a number of cosmopolitan centers within the Parthian realm and the Roman Empire, such as Constantinople, Palmyra, Antioch, Tarsus, and Smyrna. This portion of the trek is described in detail in a treatise entitled "Parthian Stations" written by the Greco-Roman geographer Isodorus of Charax dated to about AD 1.[34] At the time, the Persian (Parthian) and Roman Empires were the two most dominant players along the Silk Roads, after the Hans, and as such their behavior toward each other had profound repercussions on the efficiency of transportation of goods.

Isodorus of Charax describes that Antioch caravans traveled through the Syrian desert to Palmyra and from there to Ctesiphon, the Parthian ancient capital city located on the eastern bank of the Tigris River about 32 km southeast of present-day Baghdad. Thereupon, the merchandise went to city ports, such as Spasinu Charax at the head of the Persian Gulf, a major center of maritime commerce, where it was transported throughout the Parthian Empire. Many of the overland routes within the Near East ended up in eastern Mediterranean ports, such as Smyrna and Ephesus, and were then transported throughout the Roman Empire by sea. Also, with ports in the Red Sea, Romans transported goods over land to the Nile River and then to Alexandria on the North African coast. Other caravan paths moved through the Arabian peninsula and reached major metropolitan centers, such as Petra, Gaza, and Damascus. Alternatively, from the same port city of Spasinu Charax mentioned above, sea routes took merchandise to South Arabia, Somalia, and South India. All of these cities in the Near East prospered primarily as centers of trade, networking and providing traveling convoys with supplies, rest, and information. They also guarded the Roads against muggers.

It is interesting that during much of this golden era of the Silk Roads, Romans and Parthians, two of the main empires playing major roles in the transport and protection of the Roads, were at war with each other. During the period spanning 66 BC to AD 217, the Roman and Parthian Empires were bitter enemies, experiencing alternating periods of war and tense diplomacy. They participated in a number of altercations known as the Roman–Parthian Wars and coexisted in an uneasy relationship as dominant powers in the region throughout this time period. It is significant that although Rome controlled the entire Mediterranean basin and most of the Near East, Rome was never able to conquer the Parthians. These conditions of tension between these two superpowers persisted due to mutual benefit to both parties. On one hand, Rome was infatuated with silk and the others luxuries from the Han Dynasty and reluctantly was willing to pay the Parthians good money for the passage of goods to their dominion. Because the Parthians controlled the routes to the Far East, they acted as middlemen and demanded high prices for goods to pass through their territory and reach the Roman Empire. Although the Parthians disliked the Romans, they were willing to put up with them because the Roman's insatiable demands for Chinese items meant more revenues for them. Because many Parthian cities acted as relay stations for caravans, the entire Parthian Empire also benefited from the traffic of goods to Rome in the form of revenues from supplies and accommodations provided to the traders. This mutually beneficial relationship created a deterrent from continued and constant aggression in spite of the mutual hatred between the two powers. The Parthians and Kushans (Empires of Western and South Asia from AD

30 to AD 375 included present-day Afghanistan, Pakistan, and Northern India) were happy to act and remain as intermediates in this very lucrative trade along the Silk Roads, and they did their best to inhibit and discourage any direct contact between the Han Dynasty and the Roman Empire.

Both Romans and Parthians were polytheistic and practiced the cult of Mithra (Zoroastrian angelic divinity and protector of truth), as most of the ancient world. Both empires were tolerant of different cultures and religious beliefs; however, the Romans and Parthians were very different types of people. They exhibited strong cultural and ideological differences. The Parthian Empire extended from Mesopotamia far into the east with routes reaching deep into China and as a result they enjoyed a highly productive economy resulting from the trade with Central Asia including a number of unstructured nomad tribes. Thus, their attitudes toward life were influenced by those nonsedentary populations. In addition, their technologies varied considerably. To illustrate, Persians employed quanats (sloping underground ducts that move water from a source a lake or well to the surface), whereas Romans built impressive aboveground aqueducts to transport water from the mountains. Romans possessed marine and land trade routes, whereas Persians only moved merchandise by land. But, fundamentally, the most profound difference relates to how the Persians and Romans dealt with diverse groups of people. Through their conquests, both of these empires encompassed a number of very unique populations, originally separated by cultural, geographical, and ecological divides, but they differed dramatically in the degree of inclusion of these heterogeneous groups into their governing systems. The Persians were exclusionary in nature, whereas the Romans were inclusionary. The Romans believed in inclusion and significantly in promotion by merit. Philosophically, Romans were convinced that the merit-based system promotes organizational effectiveness by resiliency and flexibility across time.[35]

Although the Parthian Empire was dissolved early in the 3rd century AD when Sasanian forces from southwest Iran conquered their dominion, the stalemate with the Romans persisted, and it was not until the Muslim defeat of the Sasanians and massive territorial losses to the Byzantine Eastern Roman Empire that the two superpowers stopped fighting. Altogether, the tensions between Romans and the Parthians/Sasanians superpowers persisted for seven centuries.

Tang Dynasty

If the Han Dynasty was the Chinese golden age of economic prosperity and key scientific discoveries, the Tang Dynasty (AD 618–907) was the golden age of poetry, painting, and the arts in general. Exquisite tricolored glazed pottery and woodblock printing characterized the Tang period. The Tang artisans began to experiment with various media, including fabrics, metal, and ceramics utilizing new techniques, patterns, and forms. Foreign motifs were adopted into the arts, and in the assimilation process, traditions were changed, generating a more cosmopolitan culture.[35]

In addition, during this period, the Silk Roads reached their pinnacle in both reach and amounts of goods exchanged. The silk trade developed further relative to the Han Dynasty, attaining greater vigor. A greater importance was given to the Northern Route to bypass the massive Pamirs mountain range. In addition, additional maritime venues were developed between the Chinese seaport of Guangzhou, just northwest of the current city of Hong Kong, and South India, the Persian Gulf, and East Africa. This increase in the silk trade by the Tangs is attributed to a number of factors, including an increment in production, a more open mind to foreign ideas and people, and a key interest in trade as well as a desire to expand westward into Central Asia by building new Silk Road branches. Overall, compared with the Hans, Tang monarchs were more sophisticated and savvy. During this period, the capital city of Xi'an (Fig. 14.19), already the gate to the Silk Roads, became a hub to foreign dignitaries, religious envoys, and marketers.

In this prosperous environment in AD 646, the dynasty started expanding into the northern territories of the Mongolian Plateau, and by an arranged marriage in AD 641 between Princess Wencheng and Sontzen Gampo, the ruler of Tibet, the Tangs augmented their jurisdiction into the Tibetan Plateau. Also about the same time, the empires to the West, such as the Byzantines, Persians, and Muslims, became relatively stable and interested in establishing ties with China. During this epoch, unfashioned silk and paper printing technology managed to reach all the way West to the Roman provinces of Portugal and Spain.

It was not until the An Shi Rebellion of AD 755–762 that the Tang Dynasty began to experience a decline. After the uprising, the Silk Roads started to deteriorate, with large expanses in Central Asia falling into the control of Tibet and neighboring empires.

The Decline of the Silk Roads

No single event caused the demise of the Silk Roads. Yet a number of episodes in tandem, subsequent to the collapse of the Western Roman Empire, contributed synergistically to its decline. The lack of law and order, corruption, inflation, and overexpansion of the Roman Empire; the overall stagnation of sociocultural development during the Dark Ages; the decadence of the Ming Dynasty; competition from European producers; and the discovery of less expensive and more efficient sea routes discovered during the Age of Discovery and Exploration contributed to the slowdown in demand for silk.

On September 4, 476, Odoacer, leader of the Germanic invading forces in Rome, ousted Romulus Augustulus, the last Roman emperor. Odoacer, in fact, was a general for Rome before his insurrection. Generals taking over the Empire were not new in Rome. This landmark occasion put an end to the last vestiges of the already collapsing Western Roman Empire. Although this was an important event in the downfall of the Empire, the signs of Rome's decline were visible long before. Many scholars, for example, consider the ascension of Commodus as emperor in AD 180 as the beginning of the end for the Empire. The unclear succession system was always a problem for the Empire, but this issue reached new heights during the period of AD 68–69 when four emperors were in power in succession. This lack of definition delineating the parameters of succession was symptomatic of governmental instability and respect for the system. Situations like this generated anguish and consternation throughout the Empire. The large Roman Empire began to exhibit signs of exhaustion when it started to fail enforcing its rules and when its expanse commenced partitioning into distinct regions with some degree of sovereignty. Other indications of deterioration and general weakening included currency debasement, inflation, constant harassment by barbarians, corruption among the elite Praetorian troops, insufficient funds to support troops throughout a large Empire, and large economic and social inequality among classes. Although the Roman Empire continued existing as the Eastern Roman Empire or the Byzantine Empire with its capital in Constantinople, the collapse of the western branch had a profound impact on the demand for goods, including silk, in the West. The imperial downfall abruptly affected the 1000 years of insatiable Roman appetite for luxurious items from the Far East. The final debacle of the Western Roman Empire negatively affected the Silk Roads by dramatically reducing the demands for goods from the East.

In spite of the downfall of the Western Roman Empire and the start of the Dark Ages in Europe, leading to a significant decrease in the demand for silk, silk production reached an all-time high in China. At the beginning of the Middle Ages in Europe during the Tang Dynasty (AD 618–907), both the quantity and quality of the silk produced improved. In the subsequent Song (AD 960–1279) and Yuan Dynasties, with further advances in silk manufacturing technologies and newly created varieties of *B. mori*, new colors and types of silk emerged. Fresh artistic styles, such as the brocade, ornamental gold fabrics, and tapestries (from Persia), became additional luxury items for the privileged class. Yet, a period of decadence followed during the last part of the Middle Ages during the Ming Dynasty (AD 1368–1644) as high taxes and general disinterest for the Chinese silk permeated Europe. In addition, silk was no longer the primary item transported along the Roads. At that point in time, Europeans were interested in a number of other items, such as precious stones, spices, medical products, ceramics, and carpets. The deteriorating conditions of the silk market emanating from the Far East continued throughout the Qing Dynasty (1644–1911).

Another factor that contributed to the diminishing importance of the Roads was the downturn in silk imports from China that occurred with the introduction of silk technology by Muslims to Southern Europe when they invaded Iberia in the 7th century. This means that the Europeans were finally free from their dependence on silk from the East. In addition to silk production, the Islamic influence in Iberia had an enormous impact on Medieval Europe. The huge body of knowledge forgotten by the West during the depressive Dark Ages was revived as Muslims translated and transported the classic Greek and Roman texts back to Europe via Spain. It was as if Europe woke up from an intellectual, cultural, and scientific amnesia. Although rarely recognized in today's textbooks and treatises, Muslim scholars nourished knowledge-deprived Europe with resurrected wisdom from the classics and new discoveries from India and China in medicine, mathematics, astronomy, and their own advances in the arts and sciences. This infusion of old, lost, and new knowledge was the spark that ignited the Renaissance, which is generally attributed entirely to Florence. The synergistic multicultural society of Iberia at the time, involving Muslim/Eastern technology, Jewish merchants/money lenders, and Christian labor, increasingly helped Europe become self sufficient in silk at the expense of China and Persia.

By the end of the Middle Ages European silk centers relocated from Sicily and the Near East, and from there to a number of cities in Northern Italy, such as Venice, Florence, and Genoa, and became part of the artistic expressions of the times. From Northern Italy the silk was distributed all over Europe. At the time, Italy and other European territories were partitioned into city-state systems, and these competed furiously and even fought for the control of the market. This competition was based primarily on the economic strength of the local industry, silk. This rivalry in the *Arte della Seta* (the Art of the Silk) took the form of intimidation, imitation of techniques and designs as the various city-states did their best to prevent diffusion of ideas and technologies to other regions. Just as the Chinese did centuries earlier, the city-states wanted to keep their silk technologies and traditions a secret, and they threatened silk workers with death if caught taking their trade to other centers. Initially, the city of Lucca reigned supreme as a silk weaving center, but after it was attacked and sacked by rivals the industry relocated to more powerful city-states, particularly Venice, Genoa, and Florence. In the various northern Italian city-states the industry developed differently. Florence, for example, became notorious for their exquisite velvets bearing gold and silver. Genoa, on the other hand, specialized in polychrome silk velvets made by employing the *alto e basso* (pile on pile velvet) technique that created three-dimensional figures. And in Venice, the weavers tended to imitate Eastern designs, possibly because a substantial amount of the silk produced was exported to the East. In fact, the patterns originating in Venice were at times undistinguishable from Persian designs. Ironically, what started as an ancient Chinese secret, during this period of European control of the silk market, was now traveling in the opposite direction, from West to East.

Pax Mongolica

Embedded within the Late Middle Ages, the Mongol Empire (AD 1206–1368) was a bittersweet chapter in human history impacting civilization and commerce, including the silk trade. Although relatively short lived, the Mongol Empire occupied the largest dominion in history extending longitudinally from East Asia to Eastern Europe. Just like the Silk Roads, the main axis of their territory ran West–East. So, the management and control of the silk trade came rather naturally to them (geographically speaking). Also, the huge size of the land occupied by the Mongol Empire provided political uniformity that translated into minimal restrictions in moving goods across the land. These conditions improved the flow of goods along the Silk Roads. Furthermore, the Mongols not only valued and enjoyed silk but also profited from it. As they occupied new territories, they took over existing factories, built new ones, and collected silk as tribute and taxation.

Mongols were extremely brutal. Populations went into panic and collective hysteria when the news of their impending arrival reached them.[36] The Mongols' success in conquering people depended not so much on their real troop numbers or strength but on the perception that the Mongols' were a superior fighting force for which the odds of repelling the Mongol army were insurmountable. These feelings were justified. Cities such as Herat, Kiev, Baghdad, Nishapur, Vladimir, and Samarkand were almost obliterated by the Mongols. The city of Merv (major oasis city in Central Asia on the Silk Roads), for example, was so devastated that it never recovered, and it has remained since then a deserted devastated ruin. The sacking of Baghdad (from January 29 until February 10, 1258) imparted a psychological shock from which the Muslim Empire never recovered.

During the Mongols' tenure, it is estimated that about 5% of the Eurasian populace died as a result of both their direct and indirect actions. Specifically, it is estimated that the population of Persia was reduced from 2.5 million to 250 thousands, China's 120 million decreased to 60 million, and 50% of Hungary's citizens were exterminated as a result of Mongol attacks. Notably, Mongols were the first army that employed biological warfare to conquer cities by catapulting infectious dead bodies over the city walls. The use of infected corpses during the siege of the city of Feodosia or Caffa (seaport in Crimea on the northern shores of the Black Sea) is credited for the spread of the bubonic pandemic in Eurasia.

The destruction went beyond directly killing people. Some cities were burned to the ground, and countless works of art and historical texts were obliterated.[37] In the battle of Baghdad, for example, homes, libraries, hospitals, and temples were burned to the ground, and so many books were tossed into the Euphrates River that the ink turned the water black for days. The sacking of Baghdad was pivotal as it signaled the end of the Islamic Golden Age, a dominion that extended from Iberia to Central Asia. In Persia, the destruction of aqueducts led to a collapse of agriculture, and famine ensued. The destruction was so massive that the Persian irrigation system did not recover until the end of the Middle Ages.

Yet, despite the devastation the Mongols inflicted, during most of the 162 years of Mongol control, the Silk Roads flourished. With a mixture of control, power, and cruelty, Mongols exerted leadership in different sociocultural spheres, including commerce. They established a sort of Pax Mongolica that prevailed throughout their domain and allowed extensive and long-range commercialism. And in spite of their brutal tendencies, the Mongols were pragmatic people, tolerant to diverse religions and artistic expressions.

During the terminal stages of the Mongol Empire, as it was disintegrating, it experienced instability resulting from discrepancies within the Empire and competition with other powers throughout Eurasia. Internal struggles within the Mongols started in 1259, 1 year after the capture of Baghdad, when the Mongol emperor Möngke Khan (the fourth Khan) died in battle trying to capture Diaoyu castle in Hechuan, China. This defeat was the first major blow to the Mongols' world conquest ambitions and the initiating event that started their downfall. Without a declared successor, this event initiated a cascade of internal feuds within the line of descent to the title of Great Khan that culminated in the Toluid Civil War. These internal conflicts weakened the entire Mongol dominion considerably, to the point of fracturing it into independent khanates (regions controlled by a Khan). Their separate interests and goals drove each khanate to bitter fights. These tensions also led to wars between the Mongols and other neighboring powers. Due to the large expanse of the Empire, its partitioning brought about deterioration of the trade along the Silk Roads. At this point in time, Christians, Muslims, Buddhists, and Hindus were competing for state support within the Mongol khanates. This competition involved conflicts among religions and commercial systems and affected the integrity and functioning of the Silk Roads.

Mongolian DNA

Mongols not only heavily impacted the course of civilization by conquering and controlling vast territories as well as providing and allowing the Silk Roads to run smoothly but also dramatically influenced the genetic makeup of populations in Eurasia. Remarkably, it is known that about 10% of males living today within the land, previously controlled by the Mongol Empire, and approximately 0.5% (16 million or 1/200) worldwide carry Genghis Khan's Y chromosome.[38] These numbers underscore how prolific Genghis and his male heirs were after conquering. Remember that the Y chromosome is inherited paternally from father to sons and paternal uncles also share the same Y chromosomes, so the dispersion of Genghis' Y chromosomes may not have come directly from him.

Along these lines, other genetic studies based on mtDNA, Y chromosome, and whole-genome markers indicate that certain populations from Eastern Europe possess considerable amounts of East Asian DNA. Dependent on the specific European population in question, it has been assessed that 9%–76% of the Y chromosomes and over 30% of mtDNA lineages are of East Asian and Siberian origins.[39,40] More recent work also corroborates previous accessions indicating that the penetrations by Turkic-speaking populations were of Mongolian descent and occurred between the 9th and 17th centuries, dates that overlap with known historical events, including the infiltration of Central Asians along the Silk Roads and the expansion of the Mongolian Empire into Europe.[41] Signatures of these longitudinal dispersals from the East are seen in the contemporary Lipka Tatar (Slavic-speaking Sunni Muslim minorities residing in modern Belarus, Lithuania, and Poland) populations that arrived in Central Europe during Medieval times with their uniparental and biparental DNA and became sedentary after arrival.[42] When examining these genetic findings, it is important to realize that not all of these dispersal waves from the East were driven by the Silk Roads. It is highly likely that the motivation behind some of these migration waves resulted from simple demographic expansions driven by searches for cities and homesteads to pillage.

THE AGE OF DISCOVERY

The Age of Discovery or Age of Exploration encompassed a period of time from the end of the 15th century to the 18th century. It started with the Portuguese voyages into West Africa and India as well as the rediscovery of America by Columbus. It was also an age characterized by surge in nationalistic tendencies and identity in Europe. European states started identifying themselves not so much as independent city-states but as the nations we know today.

With the disintegration of the unifying Mongol Empire and the increasing conflicts involving the new independent European and Central Asian states, land routes, and specifically the Silk Roads, were unsafe and not cost productive. Each nation with its own set of rules and regulations added to the difficulty of the trek. This grave state of affairs sparked a keen interest in establishing novel connections with the Near and Far East as well as Oceania. It was clear that new safe ways with minimal harassment and cost needed to be created and maintained. This situation set the stage for a number of European powers to set out to sea. Portugal and Spain were among the first nations to address this problem. Portugal with its well-established maritime traditions and maritime know-how was the first poised to seek a route to the East by following a coastal path along the African west and east coast. Portuguese sailors were particularly knowledgeable of prevailing directional currents and winds. These attempts prove to be very fruitful for the Portuguese who started transporting goods, especially spices and condiments from the East. Portugal's pioneering efforts increased the amount and diversity of goods transported from the Near East and India, giving them a commercial advantage over other European countries and a lucrative monopoly of the trade. Other European nations took notice, especially Spain. It was this desire to reestablish commerce with the East that prompted nations such as Spain to support Columbus to find a shorter passage to *Las Indias* (Asia). As a result of this interest in discoveries and control over trade, sea routes expanded during the Age of Discovery beyond the levels seen during the Silk Roads' golden age. Some experts believe that these events during the Age of Exploration constituted the *coup de grâce* or last blow to the Silk Roads.

CONCLUSION

Today, humanity thinks of the Silk Roads with a mixture of amazement and romanticism. Images of lonely caravans crossing a desert come to mind when the Silk Roads are mentioned. In search of this remarkable history, present-day tourists visit portions of the Roads and vestiges of oasis cities in hope to experience some residue of the past. And today, in an age of international communication and commerce, we are bewildered at the endurance of the traders and travelers who made the Silk Roads possible for thousands of years. The Roads not only ushered in a dramatic period of prosperity that changed the cultural and biological evolution of our species but also created an environment that was conducive to understanding and tolerance among human populations.

References

1. Stockwell F. Seneca the Younger c. 3 BCE-65CE, Declamations vol. 1. Definition of Silk Road, Chinese silk in the Roman Empire. http://www.wordiq.com/definition/Silk_Road.
2. Drège J-P, Buhrer EM. *The Silk Road saga*. New York: Facts on File; 1989.
3. Sariadini V. History of civilization of central Asia. *Food-producing and other neolithic communities in Khorasan and Transoxonia: eastern Iran Soviet central Asia and Afghanistan*, vol. I. UNESCO Publishing; 1992.
4. Kuzmina EE. *The prehistory of the Silk Road*. Philadelphia: University of Pennsylvania Press; 2008.
5. Tsuen-Hsuin T-H, Needham J. Science and Civilisation in China. part 1 Paper and printing. *Chemistry and chemical technology*, vol. 5. United Kingdom: Cambridge University Press; 1985.
6. Davis-Kimball J, Bashilov VA, ÎAblonskiĭ LT. *Nomads of the Eurasian steppes in the early Iron age*. Berkley, USA: Zinat Press; 1995.
7. Mallory JP. In the search of Indo-Europeans. Distribution of the Sredny Stog and Novodanilovka sites. *J Indo Eur Stud* 1989;**18**:198.
8. Ellis L. The Cucuteni-Tripolye culture: study in technology and the origins of complex society. *BAR international series*, vol. 217. Oxford, UK: British Archaeological Reports (B.A.R); 1984.
9. Aleksandr D, Menotti F. The gravity model: monitoring the formation and development of the Tripolye culture giant-settlements in Ukraine. *J Archaeol Sci* 2012;**39**:2810–7.
10. David AW. *The horse, the wheel, and language: how bronze-age riders from the Eurasian steppes shaped the modern world*. Princeton: Princeton University Press; 2007.
11. David AW. *The horse, the wheel, and language: how bronze-age riders from the Eurasian steppes shaped the modern world*. Princeton: University Press; 2010.
12. Regueiro M, Rivera L, Damnjanovic T, Lukovic L, Milasin J, Herrera RJ. High levels of Paleolithic Y-chromosome lineages characterize Serbia. *Gene* 2012;**498**:59–67.
13. Underhill PA, Kivisild T, Balaresque P, Bowden GR, Adams R, et al. Use of y chromosome and mitochondrial DNA population structure in tracing human migrations. *Annu Rev Genet* 2007;**41**:539–64.
14. David P. A predominantly Neolithic origin for European paternal lineages. *PLoS Biol* 2010;**8**(1):e1000285.
15. Haak W, Lazaridis I, Patterson N, et al. Massive migration from the steppe is a source for Indo-European languages in Europe. *Nature* 2015;**522**:207–11.
16. Allentoft ME, Sikora M, Willerslev E. Population genomics of bronze age Eurasia. *Nature* 2015;**522**:167–72.
17. Unterlander M, Palstra F, Lazaridis I, et al. Ancestry and demography and descendants of Iron age nomads of the Eurasian steppe. *Nat Commun* 2017;**8**:1–10.
18. Mathieson I, Lazaridis I, Rohland N. Eight thousand years of natural selection in Europe. *bioRxiv* 2015. https://doi.org/10.1101/016477.

19. Underhill PA, Poznik GP, Rootsi S, et al. The phylogenetic and geographic structure of Y-chromosome haplogroup R1a. *Eur J Hum Genet* 2014;**23**:124–31.
20. Myres NM, Rootsi S, Lin AA. A major Y-chromosome haplogroup R1b Holocene era founder effect in Central and Western Europe. *Eur J Hum Genet* 2011;**19**:95–101.
21. Csákyová V, Szécsényi-Nagy A, Csősz A. Maternal genetic composition of a medieval population from a Hungarian-Slavic contact zone in central Europe. *PLoS One* 2016;**11**(3):e0151206.
22. Arunkumar KP, Metta M, Nagaraju J. Molecular phylogeny of silkmoths reveals the origin of domesticated silkmoth, *Bombyx mori* from Chinese *Bombyx mandarina* and paternal inheritance of *Antheraea proylei* mitochondrial DNA. *Mol Phylogenet Evol* 2006;**40**:419–27.
23. Yang SY, Han MJ, Kang LF, et al. Demographic history and gene flow during silkworm domestication. *BMC Evol Biol* 2014;**14**:185.
24. Healy JF. *Pliny the elder on science and technology*. UK: Oxford University Press; 2004.
25. Kuhn D. In: Needham J, editor. *Textile technology: spinning and reeling in science and civilization in China*. UK: Cambridge University Press; 1988.
26. Hunt P. *Late Roman silk: smuggling and espionage in the 6th century CE*. USA: Stanford University; 2013.
27. Procopius, on the wars, internet medieval sourcebook, procopius: the Roman silk industry c. 550. Available at: http://www.fordham.edu/halsall/source/550byzsilk.html.
28. Muthesius A. Silk in the medieval world. In: Jenkins D, editor. *The Cambridge history of western textiles*. UK: Cambridge University Press; 2003.
29. Ying-shih Y. *Trade and expansion in Han China: a study in the structure of Sino-barbarian economic relations*. Berkeley, USA: University of California Press; 1967.
30. Hill JE. *Through the Jade Gate to Rome: a study of the silk routes during the Later Han Dynasty, 1st to 2nd centuries AD*. Charleston, South Carolina, USA: BookSurge; 2009.
31. An J. When glass was treasured in China. In: Juliano AL, Lerner JA, editors. *Silk Road Studies VII: nomads, traders, and holy men along China's Silk Road*. Belgium: Brepols Publishers; 2002.
32. Frankopan P. *The Silk Roads: a new history of the world*. New York: Penguin Random House; 2015.
33. Cartwright M. *Trade in the Roman world. Ancient history encyclopedia*. 2013. Retrieved from http://www.ancient.eu/article/638/.
34. Department of Ancient Near Eastern Art. Trade between the Romans and the empires of Asia. In: *Heilbrunn timeline of art history*. New York, USA: The Metropolitan Museum of Art; 2000. http://www.metmuseum.org/toah/hd/silk/hd_silk.htm.
35. Svyantek DJ, Mahoney KT, Brown LL. Diversity and effectiveness in the Roman and Persian empires. *Int J Organ Anal* 2002;**10**:260–83.
36. Diana L. *Chinese migrations: the movement of people, goods, and ideas over four millennia*. Lanham, Maryland, USA: Rowman & Littlefield; 2012.
37. Saunders JJ. *The history of the Mongol conquests*. Philadelphia: University of Pennsylvania Press; 2001.
38. Zerjal T. The genetic legacy of the Mongols. *Am J Hum Genet* 2003;**72**:717–21.
39. Comas D, Plaza S, Wells RS, et al. Admixture, migrations, and dispersals in Central Asia: evidence from maternal DNA lineages. *Eur J Hum Genet* 2004;**12**:495–504.
40. Quintana-Murci L, Chaix R, Wells S, et al. Where West meets East: the complex mtDNA landscape of the Southwest and Central Asian corridor. *Am J Hum Genet* 2004;**74**:827–45.
41. Yunusbayev B, Metspalu M, Metspalu E, et al. The genetic legacy of the expansion of Turkic-speaking nomads across Eurasia. *PLoS Genet* April 2015;**11**:e1005068.
42. Pankratov V, Litvinov S, Kassian A. East Eurasian ancestry in the middle of Europe: genetic footprints of Steppe nomads in the genomes of Belarusian Lipka Tatars. *Sci Rep* 2016;**6**:30197. https://doi.org/10.1038/srep30197.

Further Reading

Juliano AL, Lerner JA. *Monks and merchants*. Asia Society Museum; 2001. asiasocietymuseum.org.

Index

'*Note*: Page numbers followed by "f" indicate figures, "t" indicate tables.'

F

G

M

Q

R

T

Printed in the United States
By Bookmasters